APPLICATIONS AND EXPERIENCES OF QUALITY CONTROL

Edited by **Ognyan Ivanov**

Applications and Experiences of Quality Control
Edited by Ognyan Ivanov

Published by InTech
Janeza Trdine 9, 51000 Rijeka, Croatia

Publishing Process Manager Katarina Lovrecic

Technical Editor Teodora Smiljanic

Cover Designer InTech Design Team

Additional hard copies can be obtained from orders@intechopen.com

Applications and Experiences of Quality Control, Edited by Ognyan Ivanov
p. cm.
ISBN 978-953-307-236-4

For EU product safety concerns:
IN TECH d.o.o., Prolaz Marije Krucifikse Kozulić 3, 51000 Rijeka, Croatia,
info@intechopen.com or visit our website at intechopen.com.

Meet the editor

Ognyan Ivanov graduated from the Physics Department of the Sofia University "St. Kliment Ohridski". He received his PhD degree from the Georgi Nadjakov Institute of Solid State Physics (ISSP), Bulgarian Academy of Sciences in Sofia, Bulgaria, where he is currently employed. His research activity is in the area of electromagnetic interactions with solids and is concentrated on the creation of various methods for control of materials and processes, mainly in the field of quality control. The methods can be used for the control of solids, liquids and gases. Dr Ivanov is an author and co-author of 54 publications in scientific journals and books, including four invited papers, and he has participated in 26 international conferences. He has worked as a member or leader in 14 projects of ISSP with organizations outside of the Academy. Prof. Ivanov also teaches physics at the University of Architecture, Civil Engineering and Geodesy in Sofia. He is a member of national and international scientific organizations, editorial boards and has also received awards at exhibitions for inventors and innovations.

Contents

Preface

Quality control is a principle which has become a part of our way of living. One can even say that it is present everywhere in the world around us. Quality control is a process that every one of us does voluntarily and involuntarily. We carefully choose our food, clothes and the environment that we live and work in. Even our bodies have quality procedures and if they don't like something there is always some kind of symptom. Our cells check every macromolecule and protein they produce, thus ensuring our wellbeing.

But what is quality control? It is a process selected to guarantee a certain level of quality in a product, service or process. It may include whatever actions a business considers as essential to provide for the control and verification of certain characteristics of its activity. The basic objective of quality control is to ensure that the products, services, or processes provided meet particular requirements and are secure, sufficient, and fiscally sound.

Essentially, quality control requires the examination of an entity for certain minimum levels of quality. An essential aim of quality control is to recognize products or services that do not meet a company's specified standards of quality. If a problem concerning quality control is detected, temporal interruption of production may be required. Depending on the particular service or product, as well as the type of problem identified, production or implementation may not cease entirely.

Competition, technological progress, new customer requirements and other factors have forced many companies to start applying total quality management. The aim of this approach is the increase of the efficiency and competitiveness of each and every product through a customer satisfaction quality criterion. By "quality", not only the technical characteristics of a product or service are meant, but also the capability to meet customer requirements, a reasonably low cost, continuously maintained high quality, etc. The quality of the product or service is a sum of two components: technical characteristics and customer service. Particularly from a customer's point of view quality means: perfect customer service, high technical parameters, low price and more. Meeting such high demands is achieved through application of technical and scientific methods which ensure on one hand, quality, and on the other hand - low production costs and optimal usage of resources, time, energy, etc.

The main goal in total quality control is to prevent mistakes, weaknesses or deviations, especially in the initial planning of an activity applying to the functioning of a system, organization, product or service. In this context, quality control can be defined as the sum of functional procedures, confirming the quality of a product or service, based on particular specifications. An important component of quality control is the control during the production of a device. It is essentially a comparison of a product to certain specifications, which has gradually evolved into a collection of methods and techniques known as quality control. The classic approach to quality control is based on this control during production (tests, measurements, etc.) of the device after its manufacture and removal of defects. Quality control recognizes defects in the planning and construction of a product, whereby the defective units can be found out and repaired.

The development of production is an ever changing process to some extent and there are variations in the manufacturing specifications. Because of these variations there needs to be a probability model and statistical methods for product quality analysis. If all major sources of variation are under control during the manufacturing process, i.e. if there is good quality control, then no serious problems arise.

It can generally be said that there are three types of control: preliminary - focuses on preventing inconsistencies; real-time control - monitors any operations taking place; control through feedback - a corrective procedure aimed at improving either the process of regulating the production resources, or the actual operations during production.

Quality control can deal with not just products, services and processes, but also people. Employees are a crucial factor in any company. If a company has employees that don't have sufficient competence and techniques or training, have difficulty comprehending directions, or are inexperienced, quality may dangerously decline. When quality control is regarded in terms of human beings, it relates to correctable issues.

A leading aspect of quality control is to determine well-defined controls. These controls help standardize both production and responses to quality concerns. Confining possibilities for error by specifying which production activities are to be completed by which personnel decreases the chance that employees will undertake or be assigned tasks for which they do not have adequate training.

Quality control can be achieved by the following common steps: establish what parameter is to be controlled; ascertain its importance and whether it needs to be controlled before, during or after results are produced; determine a specification for the parameter to be controlled which provides limits of acceptability and units of measure; develop strategies for control which specify the methods by which the characteristics will be achieved and variation detected and removed; coordinate resources to accomplish the strategies for quality control; install a sensor at an applicable point in the process to detect variance from specification; gather and transmit data to a location for analysis; verify the results and identify the cause of variance; suggest solutions and select the procedures needed to restore the usual parameters; perform the agreed procedures and confirm that the variance has been corrected.

The quality control can be divided into three major parts. One part consists of the procedures to determine an optimum level of performance analysis by professionals. The proper supervision over field work, assessment of internal control, and exercising commonly accepted auditing standards are also considered. The monitoring of a company's system of quality control by a peer reviewer incorporates examination of the effectiveness and relevance of the company's procedures, practices and conformity thereto, potential for professional development, and quality of the company's practice aids. Another part is the practices and methods used to affirm that some level of performance has been attained. Controls in design and inspection are included. Variances from established standards are discovered and corrected. The last part comprises procedures in manufacturing to deliver a chosen level of satisfaction of the service or product that is being developed. A certain amount of experiments and evaluations may be required to certify that a product, service or process meets the desired requirements.

Most manufacturers would always disregard the possibility of excluding the quality control as a part of their production processes. Without quality control, the number of defective products that must be modified, demolished or sent back would significantly increase. Almost all service organizations observe closely the quality of the services they deliver to maintain their reputations, guarantee the satisfaction of their customers. After all, nobody would keep a product that disintegrates the day after it is purchased or choose a service that hasn't provided some guarantees for its quality. For instance, lowering the quality of any given procedure at any health center may lead to a fatal end for a great number of patients.

New communication technologies provide new opportunities for quality control. Systems for electronic tracing of the source and whereabouts of a product are developed. Certain categories of products including food, cosmetics and household appliances cannot be sold without identification and tracking codes. Customers can obtain information about each product by entering its code in a particular Internet address or by phone. Such a system is currently implemented in China. This is an important step to ensure product quality.

Quality control is an important component of management and related policy. The purpose of this policy is to comply with the requirements of customers through the stages of the manufacturing cycle of a product without wasting material, financial or personal resources of the enterprise. Product quality is the result of careful planning, impeccable execution and consistent control. The goal in planning the quality is in all stages from the creation to the application of the product to address as accurately as possible all quality related problems, issues and expenses.

In this context, quality control is establishing whether a product satisfies the quality requirements. In the case given, the parameters affecting the quality of the finished product are being tested.

Quality control is the basis for regulating the quality and the quality assurance. Regulation of quality includes the activities to prevent errors, control and adjustment of processes in the production order, so that the quality requirements can be met. Typical for the regulation of the quality is that it lies in the planning stage and uses the test results of the quality control. The quality assurance activities include establishing a trust for the fulfillment of quality requirements through the systematic presentation of the ongoing events in an enterprise quality management in various stages of the manufacturing process.

The system of quality management must have a leading role with respect to all other systems. It plays an important role for the quality control department. This department is independent from other departments. Its task is to assist and coordinate the activity of quality management. It conducts the inspection of the processes, the quality improvements and improves the quality-related costs. This is done through periodic sampling, checking product quality, writing reports for the plant management and proposals for improving production quality. During certain periods, it inspects the activities of the quality system and establishes whether the system complies with the requirements. This department does not assume full responsibility for production quality. It must ensure the optimal functioning of the quality system by direct assignment of management.

As mentioned at the beginning, quality control is a very broad topic, which is present in areas very distant from each other. In this book some of the quality control aspects in the public health care, biology, mining, processing and manufacturing industry, the development of sensors for quality control and some of the computer and mathematical approaches are reviewed. The chapters in the book are divided into three sections: Quality control in Public health and Biology, Computer and Mathematical approaches for Quality control and Quality control in the Industry.

In public health, the quality of the procedures performed is of highest importance. There are different errors than can be made, human error being the more common. But human errors can be avoided with proper training and strict quality control. The quality of the lab results can be of a crucial significance for the wellbeing of the patient. With the proper procedures, the accuracy of the laboratory equipment can be brought to high levels, but if the samples are not properly acquired the apparatuses will give false results that can be fatal to the patient. So, the pre-analytical quality control is of great importance. In section Quality control in Public health and Biology, the topics that are referred to below are introduced.

In the book, the necessity to choose the reliable indicators of the quality of new techniques for colonoscopy procedures is examined. The goal is to choose a limited number of quality criteria that now appear unavoidable and must be systematically evaluated (indication of the colonoscopy withdrawal times, patient consent, comorbidities, quality of preparation, cecal intubation, rate of adenoma by colonoscopy, immediate complications).

The physical aspects of quality control in brachytherapy are discussed, and how essential it is that the medical physicists possess the knowledge and skills necessary to make the sometimes delicate decisions concerning this type of approach.

Presented are the challenges brought by the application of quality control methods to surgical procedures in cardiac surgery, general surgery and orthopaedic surgery and interventional procedures such as endoscopy, anaesthesiology, and reproductive medicine.

Safety considerations for the patient related to endoscopy are also introduced. They include the mandatory use of a protocol for all parts of endoscopy procedures, control in the endoscope maintenance and the importance of auditing in the endoscopy unit.

Reviewed are a number of problems related to Dual energy X-ray absorptiometry applications resulting from body composition terminology and data acquisition validity.

Dual energy X-ray absorptiometry is principally designed to provide information for the accurate diagnosis of osteoporosis, but also determines the whole body composition.

Described are two strategies in two stages that are used to develop urodynamic standards for quality control. Those strategies involve the establishment of a tool for quantitative plausibility check and quality control, and another tool for qualitative plausibility check and quality control. The two stages include quality control during urodynamic investigation and that of retrospective analysis.

Guidelines for developing automated quality control procedures for brain magnetic resonance images acquired in multi-center clinical trials are provided. The emphasis is placed on demonstrating the need for appropriate quality control procedures, defining quality, determining the objectives, developing a framework to facilitate the creation of quality control procedures for magnetic resonance images, and methods to design, develop, test, and validate quality control procedures.

Studied are the physiological and pathological roles of an endoplasmic reticulum molecular chaperone and folding enzyme, binding immunoglobulin protein, based upon several in vivo studies in mammals, which may reveal diverse functions of endoplasmic reticulum quality control in vivo.

Analyzed is the significance of quality assurance in the pre-analytical phase, which includes all processes from the physician's request for laboratory test to the moment the sample is ready for testing, since any errors that occur at this stage often become apparent later in the analytical and post-analytical phases.

Alexandria University clinical laboratories share their experience with the critical factors, involving the imprecision, inaccuracy, and instability of a measurement procedure, as well as the error detection and false rejection characteristics of a statistical quality control procedure that affect the quality of laboratory test results.

Explained is the demand for external quality assessments as the only way to homogenize the level of performance of the molecular diagnosis of congenital toxoplasmosis among laboratories in France.

Presented is a summary of two processes that can improve the quality and safety of the pediatric and neonatal intensive care units: the detection of nosocomial infections and early use of end respiratory pressure support and noninvasive ventilation in newborn infants of very low birth weight.

Represented is an overview of the requirements in quality control and quality management in certified clinical laboratories as well as a description of the different stages in clinical testing procedures and a highlight on the critical steps that need to be monitored via quality assurance indicator tools.

Conducted is a study to design suitable M13-tagged primers to amplify the coding exons of any human gene following standard buffer and cycling conditions and adopting the Applied Biosystems software package, Variant Reporter®, to enable high throughput quality-controlled sequence analysis.

In order to increase the quality control, some computer and mathematical approaches should be applied. They are cost-effective and easily maintained once properly implemented. These approaches greatly lower the probability of human error and increase the precision of the tests. In section Computer and Mathematical approaches for quality control, the topics that are referred to below are introduced.

A mathematical model is required for unbiased quality control. In practice, in the absence of databases (which happens during design and development of new products), it is convenient for mathematically imprecise quality control to refer to probability assignment functions, the basis of which is some or another differential law of distribution, and in the first place – the Gaussian function. Such reference enables to note that probability assignment functions synergistically integrate the approach based on probability theory and mathematically imprecise approach to quality control.

The properties of nonlinear differential equations for image processing purposes are studied, and more precisely - contrast enhancement, noise filtering or image segmentation, which constitute the most common pre-processing tasks performed to enhance pattern recognition and classification applications.

Nowadays the laboratory results of the clinical laboratories are determined by automatic analyzers. In terms of quality control, it is obvious that checking the reliability of each laboratory result is unavoidable. Quality control can be performed only for the equipment and the analytical methods. The only way to do this is to check the reliability of certain samples which are running on the analyzers with the analytical methods of the laboratory. The whole procedure is called "statistical quality control".

Water quality control from the perspective of water supply system users' safety contributed to create drinking water production and quality control procedures. The system of hazard analysis concern critical control points and its main principles have been described.

An approach which combines a statistical process control system with potential failure mode and effect analysis repository can help determine the operators' decision-making process when a quality problem has been detected. The prototype named "intelligent quality control system" has been developed, and its application shows that the efficiency of quality control can be improved. This system provides an effective way for manufacturing process continuing improvement.

A complete picture of model-driven adaptive quality control in service-oriented architectures is provided, starting with the concept of quality of service and quality of experience management. The primary goal of this research is quality of service and quality of experience control in complex software systems under changing business processes execution requirements.

New predictive-maintenance methodology that is optimal for developing a quality-control analysis of attributes is researched. Results show that Visual Basic for Applications (VBA) for marine engineering applications is the most adequate language that can reduce the programming time and is compatible within the available limited resources.

Illustrated is a VBA-based ship technical information management system that has been developed for application during the ship's working periods, and tested in real ship conditions by experienced marine engineers. Results reveal real improvement in corrective maintenance like failure detection and location of replacement parts in the storage room.

Methodologies to perform quality assessments of seafloor relief mapping with multibeam echo sounder are explained. The final product in seafloor mapping is often a digital elevation model. From this model, different cartographic products can be derived, such as contour line maps, perspective views, shaded relief maps or maps with coloured depth intervals. Since the construction of a digital elevation model can be based on different interpolation methods, an assessment of common methods is also presented.

In industry, the quality of a product is as important to the customer as it is to the manufacturer. Increasing the quality of the manufacturing process increases the quality of the product and lowers the waste, thus lowering the price. The control should be exerted on all stages of the production process: acquiring the raw materials, their processing and the manufacturing of the finished product. In section Quality control in the Industry, the topics that are referred to below are introduced.

The growing use of biofuels has triggered an increasing demand for analytical methods that allow the monitoring of their quality. The electrochemical analytical methods (potentiometric, conductometric, coloumetric, voltammetric and amperometric) generally have a high detectability, high portability and a comparatively lower cost compared to other methods, and that is why they are presented in this book. There is also an increasing need to find the right system to appropriately distribute the power generated from all renewable energy sources. What are the components of one such particular micro-grid power system and what are its main advantages is also described.

Addressed is the implementation of quality control in manufacturing companies and how the application of quality control helps the companies to ensure high quality and reliable products according to the needs of the customer. The need to increase the quality of their products forces manufacturers to increase the level of control on finished and semi-finished parts, both qualitatively and quantitatively. The adoption of optical systems based on digital image processing can be a cheap but versatile way to conduct quality control by on-contact inspection of injection-molded parts.

Introduced are the applications of near-infrared spectroscopy to monitoring the polymers' polymerization, processing, curing and production of polymer prepregs. For each application, the principle, the operation method and the results are described in detail.

Field-matter interactions are very attractive for practical applications in contactless and rapid analysis. Two types of effects, induced by irradiation with electromagnetic waves, are considered: acoustoelectric effect and surface photo-charge effect, together with the possibilities for their application in a large area of quality control activities. Experimental approaches and set-ups for detecting the studied effects are described as well as the main experimental results, highlighting the advantages and disadvantages of the quality control methods based on them.

The best quality control practices are compared with some current practices observed on international exploration and mining projects, and their direct implications on resource estimation and classification are discussed. Modern mining projects use hundreds of thousands of samples for obtaining quantitative characteristics of the geological attributes. Sampling errors can appear at any stage - planning sampling program, samples extraction, preparation and their eventual analytical assaying. A special quality control technique commonly used in exploration and mining geology for verification of mineralization grades, and its basic principles are demonstrated.

The design and implementation of a smart calibration system based on a manifold absolute pressure sensor as an example for automotive application is demonstrated, although the proposed approach can be extended to other types of sensors. A concept for a novel, closed-loop calibration system, which enables analysis of sensor properties and optimization of the calibration procedure, is presented.

There is a demonstration of hyperspectral imaging spectroscopy which integrates conventional imaging and moderate resolution spectroscopy, and of the feasibility of spectral imaging for the high speed, in-line inspection and enhanced quality control essential in commercial and industrial applications; it is accompanied by a discussion on the difficulties that may arise from real scenarios such as deviation in the illumination spectrum or instabilities in the background radiation.

An optical setup for automatic quality control of microlens arrays is shown together with the results of its computer simulations and experimental creation, and the setup's main assets and drawbacks are summarized.

The rich palette of topics set out in this book provides a sufficiently broad overview of the developments in the field of quality control. I hope this issue will be beneficial to all professionals with an interest in this field, regardless of their specific expertise. By providing detailed information on various aspects of quality control, this book can serve as a basis for starting interdisciplinary cooperation, which has increasingly become an integral part of scientific and applied research.

Ognyan Ivanov
Georgi Nadjakov Institute of Solid State Physics (ISSP)
Bulgarian Academy of Sciences
Sofia, Bulgaria

Part 1

Quality Control in Public Health and Biology

1

Quality Indicators for Colonoscopy Procedures

Romain Coriat[1,2], MD, MSc, Elise Pommaret[1,2], MD, Sarah Leblanc[1,2], MD, and Stanislas Chaussade[1,2], MD, PhD
[1]*Service de Gastroentérologie, Hôpital Cochin, GHU Ouest, Paris,*
[2]*Université Paris Descartes, Paris,*
France

1. Introduction

Colonoscopy remains the gold standard for morphologic colon. Despite the development of new methods of morphologic bowel, colonoscopy is still considered the « gold standard » because of its ability at detecting small neoplasic lesions as well as adenomas. Unlike other methods, colonoscopy has the great advantage of carrying out the same time the removal of polyps.

Colonoscopy also has a number of limitations. Studies have confirmed that the colonoscopy examination was an improvement over the performance review that fluctuates depending on the quality of it. Thus Pickardt et al showed that colonoscopy could miss up to 10% of polyps greater than 10mm (1). Also, it should be noted that interval cancers after colonoscopy is not uncommon (2). These results underpin the idea that colonoscopy is an examination of improvement and it is necessary to define quality criteria.

The most famous of all is the detection rate of adenoma. This simple criterion was used to compare the performance of endoscopists (3). To reduce variation between endoscopists and to generalize the practice of colonoscopy quality, we must have reliable and easily measurable criteria for assessing the quality of examinations. These criteria should ensure that consideration is medically justified. It is carried out by using standard validated, that lesions are diagnosed correctly and appropriate treatments are made. All of it should be done with minimal risk to patients. Moreover, these criteria must evaluate the entire examination and not just the technical act. Those criteria must also take into account: the information provided to the patient, risk assessment, and conditions of the act.

Indications for colonoscopy and appropriate intervals have been established by the taskforce in 2006 between the American College of Gastroenterology and the American Society of Digestive Endoscopy (4).

2. Quality criteria before colonoscopy

Indication

Indications for colonoscopy vary by country, particularly in terms of policies in place for colorectal cancer mass screening. In France, in contrast to the USA, screening colonoscopy is not recommended for mass screening, that is to say persons without familial or individual risk factors. But it is recommended for persons in high or very high risk of colorectal cancer.

Colonoscopy surveillance is warranted consideration by the patient's personal history, such as a history of polypectomy for adenoma, or a history of colorectal cancer. In France, diagnosis colonoscopy is justified when there are digestive symptoms or if the screening test (Hemoccult ® test) is positive.

Indications list for colonoscopy must be validated by an expert committee and must be clearly indicated in the report of examination. An audit carried out in France in 2006-2007 has shown that the colonoscopy indication was consistent with the recommendations of the ANAES in 94% of cases (5).

Risk factors for complications

Colonoscopy is an examination potentially at risk. This risk must be assessed by endoscopist. Quality criteria should take into account the ground, comorbidity, current treatments include anticoagulant and antiplatelet agents.

Informed consent, including information on risks of the examination must be obtained in all cases. The gastroenterologist, possibly with the assistance of the anesthesiologist must identify possible risk factors related to land and salaries made by the patient, including anticoagulants or antiplatelet agents that need to be managed with the help of cardiologists. (6, 7). It is the same for antibiotic prophylaxis. American Society of Anesthesiologists (ASA) score risk assessment anesthetic could be a simple criterion of evaluation of gesture. ASA score or "Physical status score" was developed in 1941 by American Society of Anesthesiologists. This score assesses both risk of anesthesia and predict mortality and perioperative morbidity. Ideally this score should be briefed on the report of colonoscopy, as well as taking antiplatelet or anticoagulant treatment, and implementation of prophylactic antibiotics.

Block and the staff of endoscopy

There were many recommendations on traceability of the material over the past 10 years. This aspect is now under control and regular monthly monitoring. To justify the validity of washing, the date and time of washing equipment must be indicated on the record. Similarly, any use of disposables or not must be indicated on the record with reference material used.

3. Quality criteria related to the procedure

The quality of bowel preparation

The quality of the preparation has been a recent development (8). If the quality of preparation for colonoscopy is arguably dependent patients, it does not mean totally independent gastroenterologist. It is the responsibility of preparing gastroenterologist adjusted according to the patient to be considered good bowel preparation in review. (Picture 1a,b,c) The gastroenterologist must explain why the patient issues of preparation is mandated and how to get a good preparation of the colon. The impact of colon cleanliness assessment on endoscopists' recommendations for follow-up colonoscopy has been evaluated by Ben-Horin et al (9). They showed that clinical évaluations of the colon cleanliness vary considerably among endoscopists. Also, poor preparation exhibited at risk of missing lesions (10), to extend the duration of the examination and have an incomplete review. This might explain in part the observed differences in performance between endoscopists. The type of bowel preparation used, and any difficulties encountered by the patient to prepare (nausea, vomiting, failure to take prescribed amount in full) should be included in the record review. It would take them into account when the next review, and avoid the failures of preparation at the 2nd colonoscopy (11).

The difficulty is that there is no standardized system for evaluating the quality of preparation, to define what an inadequate preparation, and at what point should repeat the test. It was shown that while 23.4% were deemed unprepared colons by endoscopy, colonoscopy was considered to redo that 6% of cases (11). In other cases insufficient preparation of the colon, are known to gastroenterologists tend to shorten the interval between examinations (9), without, however, this attitude has not been validated (12). Another difficulty is the subjective nature of interpretation. While it is well established that the same preparation can be evaluated differently by endoscopists (9), it is interesting to note that endoscopists with the best performance are generally the most demanding quality the bowel preparation (13).

The rate of complete colonoscopy

A colonoscopy is called complete when the endoscope has reached the cecum. We can be certain of having reached the lowest depths when cecal ileocecal valve and appendiceal orifice was clearly visualized. In case of doubt, the valve must be crossed. Reaching the bottom caecum should be stipulated in the record review.

The average rate of complete colonoscopy must be calculated annually by the gastroenterologist and / or the endoscopy unit to which he belongs. According to U.S. guidelines, the rate of complete colonoscopy should be above 90%, and 95% for colonoscopy screening (4). The reasons for the failure of cecal intubation should be included in the report. It may be the poor quality of bowel preparation, technical difficulties related to the anatomy of the colon, the existence of a marked diverticulosis, sedation insufficient, or because of stenosis.

The detection rate of adenomas

This is the best criterion for quality of colonoscopy, because it is the purpose of this examination to diagnose and to resect colorectal neoplasic lesions. According to U.S. guidelines, the detection rate of adenomas should be greater than or equal to 25% in men and 15% of women submitting to a first screening colonoscopy after 50 years (4).

Recently, the detection rate of adenoma has been recognized as the main criterion of quality of colonoscopy. Similarly, the authors acknowledge that this criterion of quality allowed to decrease the risk of interval cancer (14).

The detection rate of adenomas is an indicator still difficult to be applied by all gastroenterologists or all endoscopy centers, for histological data are not available at the time of writing minutes of colonoscopy. Circumvent this difficulty involves the availability of suitable software to return to reporting and enrich it with pathological results. A standardization of this test is possible with a possible justification for the quality of the endoscopist by this single criterion like the "pay-forperformance" in force in the U.S. (15).

The time of withdrawal of the endoscope

It is the study of Barclay et al. (16) which attracted particular attention on the relationship between detection rate of adenomas and time of withdrawal of the endoscope. The authors reported detection rates of adenomas among endoscopists significantly different according to whether they had a withdrawal time greater or less than 6 minutes. The withdrawal time was used as a quality criteria and is now consider as a criteria in colonoscopies without injury.

This study confirmed the work from the Mayo Clinic showed that 50% of polyps were diagnosed an average withdrawal time of 6.7 minutes, and 90% of polyps to a withdrawal time of 12 minutes (17). A recent observational study conducted among 315 gastroenterologists practicing in 17 U.S. states has confirmed the results of Simmons et al. study, showing that those with an average withdrawal time equal to 6 min detected 1.8 times more polyps than other (18).

A different question is whether the application of rule 6 min. is likely to improve the performance of endoscopists. A work of Barclay et al. (19) has responded positively to this question. He was asked to 12 endoscopists to have a withdrawal time of at least 8 minutes, that is to say, look for at least 2 minutes each of four segments: right colon, transverse, left, and rectosigmoid. The performance obtained after introduction of the recommendations were compared to those recorded during a previous period. In this study it was observed significant increase in performance of endoscopists in terms of rate of colonoscopy with adenomas, number of adenomas or advanced adenomas by colonoscopy (19). Conversely, a study group in Boston has shown that the establishment of an institutional policy requiring a withdrawal time equal to 7 minutes did not alter the performance of a group of 42 endoscopists performing more than 23,000 colonoscopies (20). Another study of 43 gastroenterologists in two cities in Minnesota concluded the same way the lack of improvement in performance over time despite awareness programs (21).

Moreover, the interpretation of an average time is difficult when performing colonoscopy. It is therefore useful to estimate its average withdrawal time as recommended by the American College of Gastroenterology (4) but do not consider the withdrawal time of less than 6 minutes in a patient as a cry of poor quality examination. The threshold is an average of 6 minutes, and not a criterion required for each examination. If the withdrawal time is not longer a guarantee of performance, it is nevertheless witnessed a conscientious and thorough examination, and is likely to improve its performance. In total, if a withdrawal time of less than 6 minutes should not be considered at the individual level as a criterion of poor quality, a time longer than 6 minutes may be a factor in favor of a careful examination and quality. For this, the withdrawal time of colonoscopy and the total time of the review should be indicated on the record.

The Record Review

It must contain certain information relating to prior colonoscopy criteria (indication, sedation, quality of preparation, ASA score, total time of examination, removal time and comorbidities), information relating to the review (cecal intubation , number of polyps sized location, treatment biopsies), but also information on the after colonoscopy (what to do, operative risk). Lesions should be described precisely (number, location, shape, size).

Complications

Serious complications of colonoscopy such as perforation, or those of endoscopy in general, should be regularly recorded by endoscopy center, and be discussed. Immediate complications are easily identified in the report of the review.

The latest series of the literature showed that the rate of perforation secondary to colonoscopy is currently the order of 1 case of perforation from 1000 to 1400 examinations (22). Lower rates (1 per 4900 examinations) have been reported recently in Germany in a series of 269,000 colonoscopies (23). Three quarters of the perforations are diagnosed immediately or early (<24 hours) (22). Achieving a gesture of polypectomy increases the risk of perforation by a factor more than the polyp is located on the right colon and the size of the polyp is greater than 1 cm (22, 24). In cases of perforation during endoscopy, endoscopic treatment should be considered with endoscopic clipping (25, 26).

The quality of postmarketing surveillance review

Recommendations for future monitoring of the colon usually are not included in the minutes of the colonoscopy, because of lack of histological findings at its completion. When the patient is discharged, an appointment with the doctor should always be given if biopsies were performed in order to communicate the results to the patient. Following this consultation, a report must be sent to the doctor recalled the reason for the colonoscopy, the findings of the examination and histological results. This letter must be concluded with recommendations concerning monitoring (the next review date) and the necessary treatment.

Items dependent on the colonoscopy procedure *(included systematically in colonoscopy or pathology reports)*	
	Quality of the colonic preparation
	Completeness of the procedure
	Number of adenomas or adenocarcinoma found per procedure
	Colonoscopy difficulty
	Sedation
Items independent of the colonoscopy procedure *(noted prospectively on colonoscopy checklists)*	
	Patient characteristics (specific information about colonoscopy risk determined by the gastroenterologist)
	Information consent about Creutzfeldt-Jakob disease
	Comorbid condition (valvulopathy)
	Treatment with drugs with a bleeding risk:
	Antiplatelets
	Aspirin
	Vitamin K antagonist
	Appropriateness of the colonoscopy procedure (6 items)
	Digestive haemorrhage
	Functional bowel disorder
	Screening colonoscopy
	Digestive symptoms refractory to symptomatic treatment
	Personal history of colon cancer or adenoma or inflammatory bowel disease
	Familial history of adenoma or colon cancer
Quality criteria to analyse post colonoscopy	
	Adenoma detection rate
	Time Withdrawal (Mean upper than 6 minutes)
	Early complications rate
	Late complications rate
	Good preparation rate
	Cecal intubation rate

Table 1. Quality criteria for colonoscopy

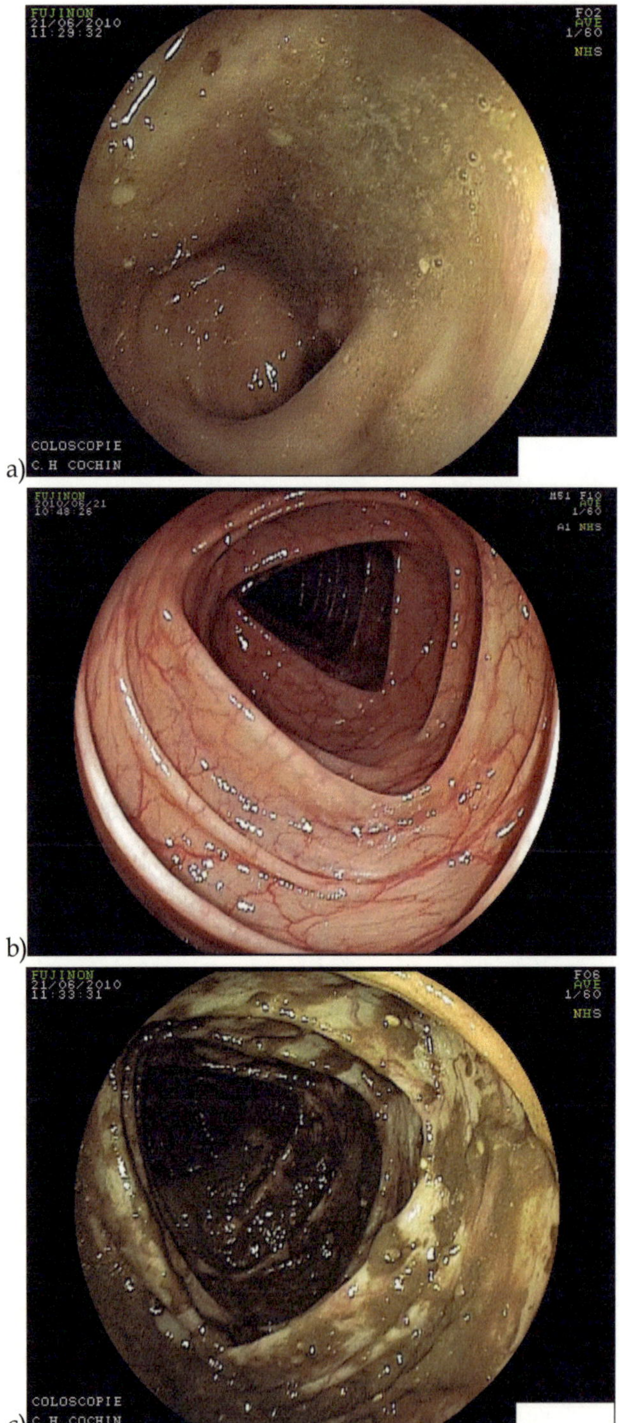

Fig. 1. Photographs depict representative luminal views of colon cleanliness of the large bowel. The cleanliness was not sufficient (a), intermediate (b), and good (c)

4. Conclusion

Colonoscopy is still the gold standard exploration for the colon. But at the présent time were new explorations techniques of the colon are developed, it seems important to justify the quality of the review criteria with simple reliable repeatable and standardized. To do this, you have to use reliable indicators that must be systematically integrated into the reporting review. The goal is to choose within the center in a limited number of quality criteria relevant to the population management. Certain criteria now appear unavoidable and must be systematically evaluated (indication of the colonoscopy withdrawal times, patient consent, comorbidities, quality of preparation, cecal intubation, rate of adenoma by colonoscopy, immediate complication).

5. References

[1] Pickhardt PJ, Nugent PA, Mysliwiec PA, Choi JR, Schindler WR. Location of adenomas missed by optical colonoscopy. *Annals of internal medicine*. 2004 Sep 7;141(5):352-9.

[2] Bressler B, Paszat LF, Chen Z, Rothwell DM, Vinden C, Rabeneck L. Rates of new or missed colorectal cancers after colonoscopy and their risk factors: a population-based analysis. *Gastroenterology*. 2007 Jan;132(1):96-102.

[3] Chen SC, Rex DK. Endoscopist can be more powerful than age and male gender in predicting adenoma detection at colonoscopy. *The American journal of gastroenterology*. 2007 Apr;102(4):856-61.

[4] Rex DK, Petrini JL, Baron TH, Chak A, Cohen J, Deal SE, et al. Quality indicators for colonoscopy. *The American journal of gastroenterology*. 2006 Apr;101(4):873-85.

[5] Coriat R, Pommaret E, Chryssostalis A, Viennot S, Gaudric M, Brezault C, et al. Quality control of colonoscopy procedures: a prospective validated method for the evaluation of professional practices applicable to all endoscopic units. *Gastroenterologie clinique et biologique*. 2009 Feb;33(2):103-8.

[6] Napoleon B, Boneu B, Maillard L, Samama CM, Schved JF, Gay G, et al. Guidelines of the French Society for Digestive Endoscopy (SFED). *Endoscopy*. 2006 Jun;38(6):632-8.

[7] Veitch AM, Baglin TP, Gershlick AH, Harnden SM, Tighe R, Cairns S. Guidelines for the management of anticoagulant and antiplatelet therapy in patients undergoing endoscopic procedures. *Gut*. 2008 Sep;57(9):1322-9.

[8] Parente F, Marino B, Crosta C. Bowel preparation before colonoscopy in the era of mass screening for colo-rectal cancer: a practical approach. Dig Liver Dis. *2009* Feb; 41(2):87-95.

[9] Ben-Horin S, Bar-Meir S, Avidan B. The impact of colon cleanliness assessment on endoscopists' recommendations for follow-up colonoscopy. *The American journal of gastroenterology*. 2007 Dec;102(12):2680-5.

[10] Lai EJ, Calderwood AH, Doros G, Fix OK, Jacobson BC. The Boston bowel preparation scale: a valid and reliable instrument for colonoscopy-oriented research. *Gastrointestinal endoscopy*. 2009 Mar;69(3 Pt 2):620-5.

[11] Ben-Horin S, Bar-Meir S, Avidan B. The outcome of a second preparation for colonoscopy after preparation failure in the first procedure. *Gastrointestinal endoscopy*. 2009 Mar;69(3 Pt 2):626-30.

[12] Bond JH. Should the quality of preparation impact postcolonoscopy follow-up recommendations? *The American journal of gastroenterology*. 2007 Dec;102(12):2686-7.

[13] Thomas-Gibson S, Rogers P, Cooper S, Man R, Rutter MD, Suzuki N, et al. Judgement of the quality of bowel preparation at screening flexible sigmoidoscopy is associated with variability in adenoma detection rates. *Endoscopy*. 2006 May;38(5):456-60.

[14] Kaminski MF, Regula J, Kraszewska E, Polkowski M, Wojciechowska U, Didkowska J, et al. Quality indicators for colonoscopy and the risk of interval cancer. *The New England journal of medicine*. May 13;362(19):1795-803.

[15] Johnson DA. Pay for performance: ACG guide for physicians. *The American journal of gastroenterology*. 2007 Oct;102(10):2119-22.

[16] Barclay RL, Vicari JJ, Doughty AS, Johanson JF, Greenlaw RL. Colonoscopic withdrawal times and adenoma detection during screening colonoscopy. *The New England journal of medicine*. 2006 Dec 14;355(24):2533-41.

[17] Simmons DT, Harewood GC, Baron TH, Petersen BT, Wang KK, Boyd-Enders F, et al. Impact of endoscopist withdrawal speed on polyp yield: implications for optimal colonoscopy withdrawal time. *Alimentary pharmacology & therapeutics*. 2006 Sep 15;24(6):965-71.

[18] Overholt BF, Brooks-Belli L, Grace M, Rankin K, Harrell R, Turyk M, et al. Withdrawal times and associated factors in colonoscopy: a quality assurance multicenter assessment. *Journal of clinical gastroenterology*. Apr;44(4):e80-6.

[19] Barclay RL, Vicari JJ, Greenlaw RL. Effect of a time-dependent colonoscopic withdrawal protocol on adenoma detection during screening colonoscopy. *Clin Gastroenterol Hepatol*. 2008 Oct;6(10):1091-8.

[20] Sawhney MS, Cury MS, Neeman N, Ngo LH, Lewis JM, Chuttani R, et al. Effect of institution-wide policy of colonoscopy withdrawal time > or = 7 minutes on polyp detection. *Gastroenterology*. 2008 Dec;135(6):1892-8.

[21] Shaukat A, Oancea C, Bond JH, Church TR, Allen JI. Variation in detection of adenomas and polyps by colonoscopy and change over time with a performance improvement program. *Clin Gastroenterol Hepatol*. 2009 Dec;7(12):1335-40.

[22] Panteris V, Haringsma J, Kuipers EJ. Colonoscopy perforation rate, mechanisms and outcome: from diagnostic to therapeutic colonoscopy. *Endoscopy*. 2009 Nov;41(11):941-51.

[23] Bokemeyer B, Bock H, Huppe D, Duffelmeyer M, Rambow A, Tacke W, et al. Screening colonoscopy for colorectal cancer prevention: results from a German online registry on 269000 cases. *European journal of gastroenterology & hepatology*. 2009 Jun;21(6):650-5.

[24] Crispin A, Birkner B, Munte A, Nusko G, Mansmann U. Process quality and incidence of acute complications in a series of more than 230,000 outpatient colonoscopies. *Endoscopy*. 2009 Dec;41(12):1018-25.

[25] Coriat R, Cacheux W, Chaussade S. Iatrogenic colonoscopic perforations: clipping or calling for a surgeon? *Digestion*. 2008;78(4):214-5.

[26] Lohsiriwat V. Colonoscopic perforation: incidence, risk factors, management and outcome. *World J Gastroenterol*. Jan 28;16(4):425-30.

2

The Need for Quality Control in High Dose Rate Brachytherapy

Carlos E. deAlmeida Ph.D., ABMP, FAAPM
Full Professor in Medical Physics,
State University of Rio de Janeiro
Brazil

1. Introduction

Brachytherapy refers to the delivery of radiation directly into or onto the surface of the area to be treated. Radiation sources can be used in body cavities (e.g., the uterus, vagina, bronchus, esophagus, and rectum), can be placed on the surface of tumors in the skin or may be placed directly into a tissue by interstitial techniques such as those used in the head and neck region, prostate, and breast.

One of the main objectives of brachytherapy is to ensure an accurate and safe dose delivery to a target volume while avoiding unnecessary delivery to surrounding healthy tissue. [1]

In order to ensure the optimal treatment of patients, much effort is required during the commissioning phase of new brachytherapy equipment and during its clinical lifetime. The institution must therefore develop a suitable Quality Control (QC) program not only for brachytherapy sources and equipment but also for the physical and clinical procedures.

In 2000, the IAEA published its Report No. 17 entitled "Lessons learned from accidental exposures in radiotherapy". [2] Although brachytherapy is applied only in about 5% of all radiotherapy cases, 32 of the 92 accidents reported in this booklet were related to the use of brachytherapy sources. Errors in the specification of source activity, dose calculation or the quantities and units involved resulted in doses that were up to twice the prescribed dose. Some of the accidents were clearly related to human error. The same document demonstrates the need for well-designed QC programs for brachytherapy. For the conception of such programs, one must consider the consistency in the management of each individual treatment, the realization of the clinical prescription, and the control of safe execution of the treatment with regard to the patient and to others who may be involved with, or exposed to, the sources during treatment. [3, 4]

As a result of several accidents recently reported involving the use of advanced technology in radiation oncology, QC programs involving independent quality audits are also seen as a preventive action. [5,6,7,8]

The consequences of errors that are bound to occur in a radiation oncology clinical environment may be caused by the radiation oncologist, physicist, dosimetrist or radiation therapist. Of these errors, the most grievous to the patient are the systematic errors made by the physicists without his or her perception.

Therefore, the type of mistakes made by each staff member is different, as is the magnitude of the impact to the patient, as specified below:

- if a physician makes a mistake, it usually affects one patient;
- if a dosimetrist makes a mistake, it affects one patient or one tumor location;
- if a technologist makes a mistake, it normally affects one fraction of the treatment;
- if a physicist makes a mistake, it may affect all patients in the clinic during a given period of time.

A sound QC program must be in place to reveal and prevent these mistakes and may include as an important component independent Quality Audits.

2. The brachytherapy technology

During the past 15 years, high dose rate ^{192}Ir sources have become available as an efficient substitute for ^{137}Cs sources. Several advantages have drawn attention to this new technique, among which are the possibility to treat several new clinical sites due to the small dimension of the source, the viability of out-patients treatment, the lesser number of treatment sessions required, the complete remote-controlled source management, which increases the staff safety, and a computerized treatment planning system that offers the 3D-volume dose calculation.

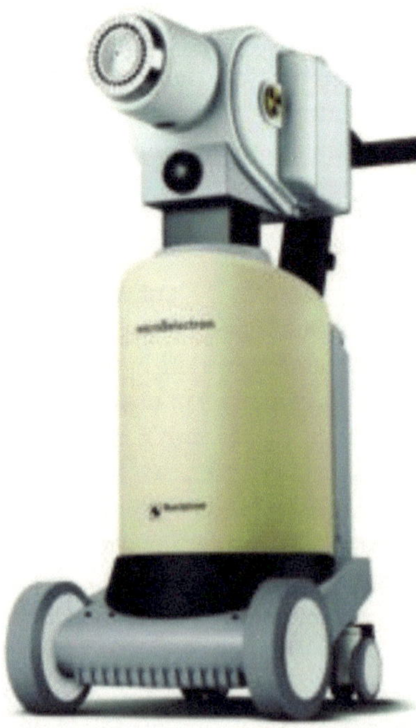

Fig. 1. Typical configuration of commercially available HDR treatment unit

Since there are several types of HDR sources available in the market, Table 1 presents some of the main physical characteristics that are useful for entry-independent calculation analytical and numerical calculation methods.

Model:	Active length	Active diameter	Total diameter	Distance from active edge to tip of the source	Encapsulation
MicroSelectron Nucletron (new design)	3.6	0.65	0.9	0.2	Stainless steel
VariSource	10	0.35	0.61	1	Ni-Ti
Varian	5	0.34	0.59	1	Ni-Ti
Buchler	1.3	1	1.6	1	Stainless steel
Gamma Med 12i	3.5	0.6	1.1	0.86	Stainless steel
Gamma Med Plus	3.5	0.6	0.9	0.62	Stainless steel
BEBIG	3.5	0.6	1.0	0.9	Stainless steel

Table 1. Specific characteristics of the ^{192}Ir high dose rate sources. Dimensions are in mm. Adapted from [3]

3. Main treatment sites

High dose rate machines may have clinical indications in the treatment of a variety of different organs either as a primary treatment or as a complementary therapy. As a result, there are a large number of applicators and accessories designed specifically to fit the geometrical needs of each treatment site.

Among the numerous applications [1] for HD brachytherapy, the three main areas are:

Prostate: Two techniques are in use: 1. Temporary implants using a stepping ^{192}Ir HDR source to deliver large single-dose fractions has gained acceptance with current radiobiological models that predict a low [alpha]/[beta] ratio for prostate cancer, and 2. Permanent implantation, mainly with small radioactive seeds of ^{125}I source with a half-life of 59 days is a second option. In some centers, the ^{121}Pd is also used. The latter is not in the scope of this chapter.

Breast: Two techniques for partial breast implant irradiation are in use: 1. multicatheter brachytherapy is used as a conventional brachytherapy to cover the tumor bed with a 2- or 3-plane interstitial implant, and 2. the single-catheter technique that uses a new applicator marketed as the Mammosite is essentially a single-line flexible HDR afterloading catheter with an inflatable balloon at the end. The latter is not in the scope of this chapter.

Gynecological: This is focused mainly on cancer of the endometrium and cervix using specific types of applicators, for instance a ring system that allows multiple source positions including in the upper vagina.

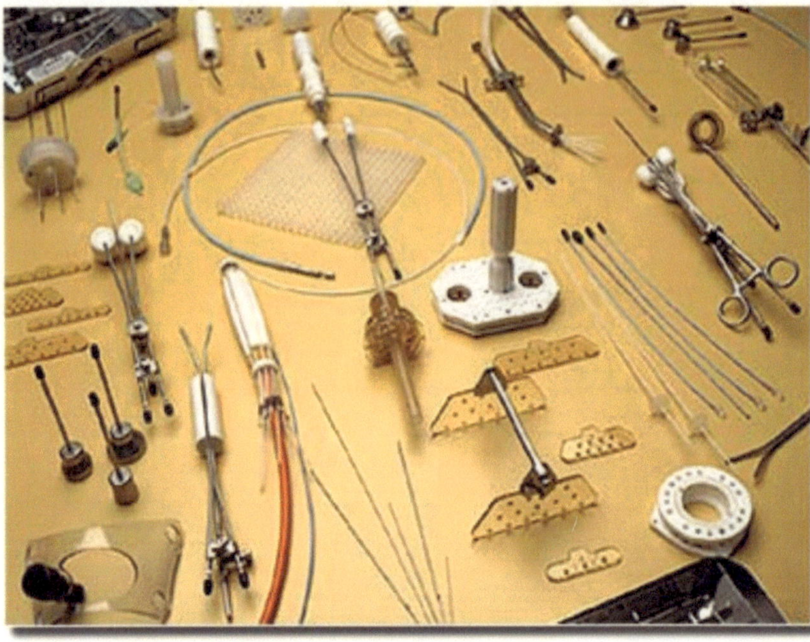

Fig. 2. Typical set of applicators, accessories, catheters and sleeves used for HDR treatments

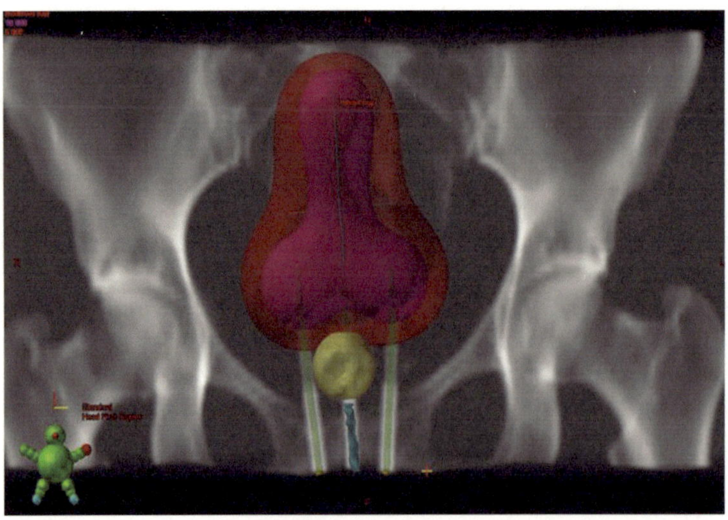

Fig. 3. Typical examples of a gynecological intracavitary treatment and a 3D-dose distribution around it

4. The scope of the quality control process

The emphasis in this chapter is to cover the physics aspects of a quality control program for HDR brachytherapy and to promote awareness of the tolerances and frequencies of the basic tests required. Those tests are based on the likelihood of a malfunction and the seriousness or potential consequences of an unnoticed malfunction that would affect the patients and/or to the personnel involved in the clinical procedure if a malfunction occurs and is not identified during normal treatment applications. The purpose of the QC program is to guarantee that all operational procedures are being performed properly and in accordance with the licensing specifications.

The QC must consider the compliance aspects of planning through a comprehensive management plan, documentation of the actual operations, and the monitoring and reporting requirements. [8]

Tolerances and action levels

Performance within the tolerance level provides an acceptable degree of accuracy in any situation. If performance is outside the action level, it may demand an action to fix the situation.

A quality control test should use appropriate measuring equipment, which must itself have been subjected to maintenance and quality control. Irradiation conditions and measuring procedures should be designed to be suitable to each task, and they are expected to give the best estimate of the particular measured parameter. Because each measurement has an associated uncertainty that depends on several factors, the tolerances established must take that into account.

Action levels are related to tolerances but it may afford sufficient flexibility to monitoring and adjustment; for example, if a measurement indicates a result between the tolerance and action levels, it may be acceptable to allow clinical use to continue until this is confirmed by measurement the next day before taking any further action.

As a normal procedure, the action levels must be activated depending on each particular situation, such as;

- if a daily measurement is within tolerance, no action is required;
- if the measurement falls between tolerance and action levels, this may be considered acceptable until the next daily measurement;
- if the measurement exceeds the tolerance level, immediate action is necessary and the machine may not be used clinically until the problem is corrected and verified by measurement;
- if repeated measurements remain consistently between tolerance and action levels, adjustment is required;
- any measurement that is outside the action level requires immediate investigation and, if confirmed, rectification.

As in general, action levels are set at about twice the tolerance level, while some critical parameters may require tolerance and action levels to be set much closer to each other or even at the same value.

Test frequencies must be considered in the context of the acceptable variation during a treatment course and also the period of time over which a parameter varies or deteriorates. Frequencies may be modified in the light of experience of the performance and stability of a given piece of equipment, initially setting a nominal frequency that may be subsequently reviewed in the light of observation. As machines get older, this may need further review.

The staff resources available to undertake the tests may limit what can be checked, which may have an effect on the structure of the quality control program. Tests should be designed to provide the required information as rapidly as possible with minimal time and equipment.

Whenever available, quality control protocols developed by national organizations should be applied. The following sections provide some examples of parameters, test frequencies, and tolerances currently used for different items of radiotherapy equipment.
For consistency, the values listed are based on the AAPM TG 40 [9] and ESTRO [3], and in some cases, on recently published additional reports on quality control in radiotherapy [10]. In any case, those protocols should be referred to for more details and adapted for local circumstances.

5. General concepts

It is essential that the performance of treatment equipment remain consistent within accepted tolerances throughout the clinical life of the equipment, as patient treatments will be planned and delivered on the basis of the performance measurements made during the equipment acceptance and commissioning tests.
An ongoing quality control program shall include regular checks beginning as soon as the equipment is commissioned. This program should be able to identify departures from expected performance and to activate corrective actions, and it shall consider the following points:

- the parameters to be tested and the tests to be performed;
- the specific equipment to be used to perform the tests;
- the geometry of the tests;
- the frequency of the tests;
- the staff group or individual performing the tests, as well as the
- individual supervising and responsible for the standards of the tests and for actions that he or she may deem necessary if problems are identified;
- the expected results;
- the tolerance and action levels;
- the actions required when the tolerance levels are exceeded

6. The quality control program

The organization of a program requires as a foundation that the institution is willing to be prepared for organizational rearrangements if needed in order to lay down very clearly the duties and responsibilities of each staff member.

The institutional organization

It is important to clearly describe not only the general aspects of the institutional position in the organization chart, but also the interaction with similar departments, the legal aspects and services provided, and the formal delegation of individual tasks to the staff. It is also crucial to detail the information flow, scheduling, list of staff members involved in the treatment, information given to the patient, and the radiation protection equipment and procedures in the treatment area.
A set of recommendations is proposed to be an integral part the patient file in order to promote appropriate treatment, to help the reviewing process. [3].
Patient ID, all documents, films, prints, plots provided with treatment, source strength matched with its decayed value, identification of the customizing file and source library, correct use of magnification factors and source-film distances, the position of sources or

applicators on the plots with radiographs compared with treatment (volume) prescription, correct use of units for all quantities, correct use of shielding or other correction factors, correct use of treatment parameters, such as step size, catheter length, begin- and end points along the catheters, consistent use of prescription criterion and optimization routines, according to the physician's intent and the possibilities of the implant geometry, statement about the dose uniformity and the dose differentials, dose and dose per fraction clearly stated according to prescription, dose to normal structures, high dose areas, constraints clearly fulfilled, identification of position of reference points, patient points, applicator points on the plots, match with those measured on the film, step size, catheter length, dwell times on treatment unit according to plan and for subsequent treatments program card, stored standards, or equivalent settings matching the first treatment.

Clinical procedures

It is of fundamental importance to clearly specify the control points (as described in Table 2) related to each of the procedures currently in use, including the procedures for CT and MRI imaging acquisition protocols, imaging reconstruction techniques, data transfer to the treatment planning system (TPS), dose validation methods involving the patient simulation and treatment, the frequency of each test, training, chart rounds, peer review, and incident reports.

Item	Material and Methods	Frequency
Test of outcome of HDR calculation	Apply the methods developed at the institution for specific clinical cases	each patient
Treatment protocols	All types of applications should be described in detail	each patient
Standard forms	To be developed for each application	each patient
Independent check	Ensure that a second person checks the work of the first planner	each patient

Table 2. Simple guidelines to be considered as part of the design of the treatment operational flux for each patient. Adapted from [3]

Physical procedures

The control points defined for the QC program shall also include the physical data either taken from tables, publications, guidelines, measurements involving the machine acceptance and commissioning, source calibration, patient data (protocols for imaging acquisition and data entry to the treatment planning system), dose calculation for the patient, dose validations or patient chart checks.

The early detection of errors in the operational process following the steps proposed in Table 3 is very important in order to allow the elimination of errors and to promote the necessary modification of the routine procedures.

Item	Material and Methods
Patient identification	All documents, films, prints, plots provided for a treatment
Dose prescription	Delivered dose vs. prescribed dose, evaluation of uniformity of the dose, location of prescribed dose, dose distribution/differentials in dose, begin and end positions correctly along the catheter
Dose to normal structures	Identification of the location of high-dose areas, location of normal structures and constraints to be fulfilled
Program identification	Identify the algorithms used, version number, shielding and correction factors
Program verification	Identify the source strength, step size and tip length
Transfer of data	Identify the correct position of each dwell position, dwell time, total time and correct channels

Table 3. The main sources of detectable errors and preventive actions recommended in orderto avoid unnecessary sources of errors. Adapted from [3]

Safety aspects

Protection of the patient and the staff is the most important objective to be considered in the proper treatment to the patient. One shall consider the physical conditions and the calibration certificates of the physics equipment and review the current procedures performed, as well as the emergency procedures, drills, source storage disposal issues, posting, surveys including the results of the wipe tests and its appropriate register.

In addition, one must consider items such as electrical interlocks, source exposure indicator lights on the after loader, control console, viewing and intercom systems, emergency response equipment, radiation monitors, timer accuracy, clock (date and time) in unit's computer and decayed source activity in unit's computer.[8]

A set of minimum requirements for specific actions are presented in Table 4, including the frequency of verification of the most important parameters related to the safe operation of the clinical procedures.

It is the physicist's task to inspect the performance history of the system meticulously, using the data in the logbook noted during the clinical lifetime of the equipment.

An important component of safety is the proper training of the staff to deal with unexpected situations. The knowledge of the equipment design and its components, including access to the emergency knobs, buttons and tools as shown in Fig.4, may help to speed the resolution of a particular incident.

Description	Minimum requirements	
	Test frequency	Action level
Safety systems		
Warning lights	Daily/3M*	-
Room monitor	Daily/3M*	-
Communication equipment	Daily/3M*	-
Emergency stop	3M	-
Treatment interrupt	3M	-
Door interlock	W	-
Power loss	3M	-
Applicator and catheter attachment	W	-
Obstructed catheter	W	-
Integrity of transfer tubes and applicators	3M	-
Timer termination	Daily	-
Contamination test	A	-
Leakage Radiation	A	-
Emergency equipment (forceps, survey meter)	Daily/3M*	-
Practicing emergency procedures	6M	-
Hand crank functioning ***	A	-
Hand-held monitor	3M/A**	-
Physics parameters		
Source calibration	SE/3M	>5%
Source position	Daily/3M*	>2 mm
Length of treatment tubes	6M	>1 mm
Irradiation timer	6M	>1 %
Date, time and source strength in treatment unit	Daily	-
Transit time effect	3M	-

(3M–quarterly; 6M–biannual; A–annual; SE–source exchange; W=weekly Adapted from [3] *Daily checks are assumed to be an implicit part of normal operation and registered in a logbook. ** Verify the proper function of the hand-held monitor, e.g., with a known source of radiation. "Action level" reflects the upper limit in clinical conditions

Table 4. Recommended types, frequencies and tolerances of the quality control tests. Adapted from [3]

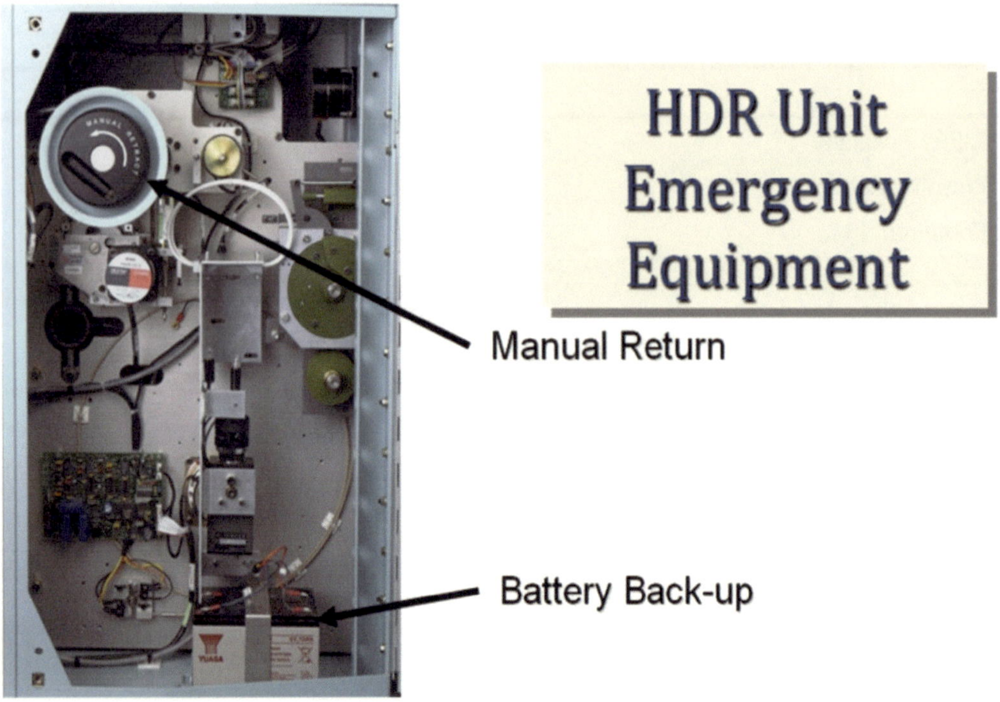

Fig. 4. An inside view of one of the machines available on the market illustrating a manual system to retrieve the source and the battery back-up that keeps the equipment operational in case of a power failure

7. Dose calculation

The dose calculation algorithm of the treatment planning systems uses the TG 43 data [11,12], which is the best approach at the present time. This step should include a review of the patient identification data, the imaging acquisition protocols, the data transfer to the TPS and the results of the independent checks.

Two important steps must be considered to validate and periodically control this parameter; first, the basic dose calculation steps as described in Table 5, and second, the dose volume histograms, including optimization, as shown in Table 6. Both tables contain parameters of significant relevance to allow a correct clinical decision for treatment and long term follow-up.

In brachytherapy, it is not redundant to reinforce that the concept of recommended tolerance levels should be interpreted such that when the tolerance level is reached, it is essential that appropriate actions be taken.

By all means, the treatment unit should not be used clinically when the tolerance level is exceeded unless after careful consideration by the medical physicist and with the agreement of the radiation oncologists and radiation technologists, a joint decision is made to carry on.

For medical physicists in this position, it is essential that they possess the awareness and skills necessary to make a sound judgment based on all available information and following an appropriate assessment of the risks involved. This type of approach may also help with a better understanding the concept of tolerance levels.

Item	Material	Frequency
Point dose calculation	Identify relevant dose points around the source for which a dose rate table is available, compare results, tolerance level is at ±2%, analyze in detail if deviations are > 5%	Initially and with software updates
Source selection	Check that the system performs the source selection from the library properly	Initially and with software updates
Check dose distribution calculated by TPS	Clinical benchmark cases shall be used	Initially and with software updates
Check dose distribution calculated by TPS of multiple source geometries	Compare with previously selected clinical benchmark cases	Initially and with software updates
Shielding	Check dose distribution of sources near the applicator shielding	Initially and with software updates

Table 5. The main steps involved in control of the basic dose calculations

Item	Material	Frequency
Volume calculation from 3-D imaging data	Calculate the volume of well-defined "organs"	Initially and with software updates
DVH of an isotropic point source	Cumulative differential and natural dose volume histograms	Initially and with software updates
Anisotropic sources	Calculate the DVHs with anisotropy correction	Initially and with software updates
Clinical examples	Calculate DVHs for benchmark clinical cases	Initially and with software updates

Table 6. The critical parameters related to dose volume histograms and their optimization. Adapted from [3]

- Treatment-planning system

In addition to the computer itself, the major hardware devices associated with a planning system are the digitizer and the plotter. Their accuracy should simply be routinely checked with a benchmark case that checks the overall accuracy of source reconstruction, including translation from film to Cartesian coordinates, rotations, and corrections for magnification. [13,14].

The frequencies should follow those described in Table 9.7

- The consistency between quantities and units

A main source of error in dose distribution calculations is the incorrect use of quantities and units required by the dose calculation software. It is essential to verify the correct labeling of the input and output quantities and units. The strength of the sources (activity) may be specified in one of several alternative units, and the user should pay particular attention to this important parameter. Inconsistent use of units could lead to severe errors in the treatment and possibly compromise the expected treatment outcome.

- Computer versus manual dose calculation for a single source

The computer-calculated dose distribution around one linear source must be compared with published dose rate values for a similar source or with manually calculated values using the Sievert integral. Additional tests may include summing of the dose for multiple sources and the check of decay corrections with a proper choice of units.

- Source Tests

An important topic to be considered is the check of a new incoming source that is usually changed 3-4 times a year, its comparison with the institutional data and the data provided by the manufacturer.

- Wipe tests

All sources that are maintained in a permanent inventory are required to be wipe-tested for any leakage of radiation.

Radiation levels should be measured and recorded both at the surface and at a 1-m distance. Individual encapsulated sources should be wipe-tested for possible leakage or contamination. A source is considered to be leaking if ~200 Bq (~5 nCi) of removable contamination using a cotton swab is measured in a GM counter or a scintillation well counter.

- Calibration process and its metrological traceability

Due to still some unresolved issues regarding the absolute standardization of the calibration procedures of ^{192}Ir HDR sources [15,16], several methods (using the interpolative approach) are still being used, though traceable to the calibration laboratories. [17].

Fig. 5. A typical well-type ionization chamber, including a tool to performs tests of source positioning within 0.3 mm, timer accuracy, and consistency of source activity

It is recommended that brachytherapy sources have their source strength calibrations traceable to a national standards laboratory by using a calibrated well-type ionization chamber such as the one shown in Fig. [18]

The activity of all sources should be measured on receipt with a local well-type chamber and compared with the manufacturer's certificate of source strength. In case it is above the tolerance level, another measuring system should be used until the reason for the deviation is fully explained.

- Constancy check of a calibrated dosimeter

The consistency of the response of the dosimetric system (electrometer, cable and chamber) may be checked by periodic measurement using a source with long half-life, such as ^{137}Cs placed inside the chamber in a reproducible position. Alternatively, one may use a radiation beam from a Linac or a Cobalt teletherapy unit fixing the irradiation parameters such as field size, geometry, distance source to surface of chamber and number of MU. [16].

- Regular check of the source positioning.

The source positioning the catheter is a very important control point before each treatment is conducted. A very simple device may be used with the dummy source as shown in Fig.

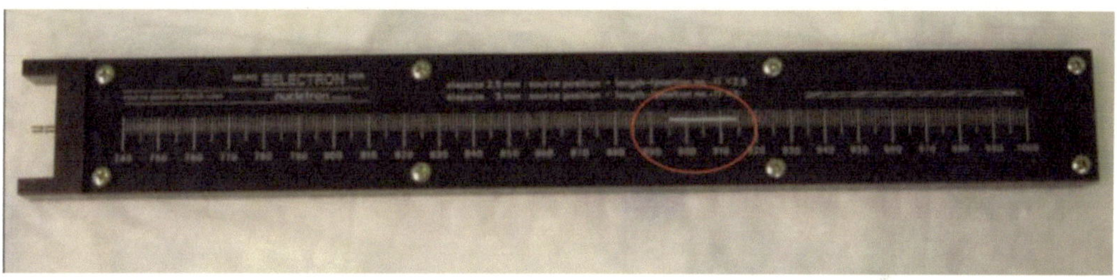

Fig. 6. The source position ruler showing the source in a red circle

- Regular checks of applicators

The applicators are used with very high frequency and go through severe handling, cleaning and sterilization processes.

Fig. 7. Typical Gyn intracavitary applicator ring type with an intrauterine tandem

Periodic inspection and radiographic evaluation of all applicators should be performed at some pre-established frequency as proposed by and described in Table 6.2

Gynecological appliances	Element to be tested	Frequency
Tandems	Flange screws function	Each use
	Curvature	Each use
	Closure caps function	Each use
	Plastic sleeve and rod fit and slide	Each use
Fletcher-type ovoids	Source carrier function	Each use
	Integrity of welds	Each use
	Position of shields	Semi-annually or after repair
	Identification markers	Each use
	Bridge integrity/thumb screws	Each use
Tandem-based cylinders and tandem checks	Flanges function	Each use
	Identification markers	Each use
	Cylinders fit snugly	Each use
Solid cylinders	Source carriers function	Each use
	Closure caps function	Each use
Intra luminal catheters	Integrity	Each use (after sterilization)
	Strength of tip	Each use (after sterilization)

Table 8. Quality control procedures for brachytherapy appliances

8. Independent auditing

Quality audits are now considered an essential component of a QC program, and there is a strong trend for these audits to become an operational requirement by the licensing authorities. The quality audit is a way of independently verify the correctness and effectiveness of an ongoing program. This may be performed by individuals with qualified proficiency or by an accredited organization or institution. The first program available was offered by the Radiological Physics Centre (RPC) [6,1], which has been funded by the National Cancer Institute (NCI) continuously since 1968 to provide quality auditing of dosimetry practices at institutions participating in NCI cooperative clinical trials. These services may or not include site visits when the discrepancies need to be resolved. Similar services are provided in Europe by the EQUAL, which is funded by ESTRO [7]. In Brazil, this activity is conducted by the LCR [8], and recently the IAEA promoted this conception worldwide as part of the QUATRO program [10].

Other institutions providing quality audits are the American College of Radiology and American College of Radiation Oncology. These two programs are more involved and include site visits to the radiation oncology department being reviewed [19,20,21,22].

9. Final remarks

Brachytherapy is indeed a very important treatment modality of several of the malignant diseases that allows conformal treatment without complex technological involvement.
Although the basic principles of brachytherapy have not changed much during the past 100 years of radiotherapy, the advent of remote after-loading made brachytherapy more efficient for the patient and much safer for staff from the radiation protection point of view.
In terms of human resources, brachytherapy treatment requires considerably more involvement than an average external beam patient.
Quality control programs must be flexible to permit additional testing whenever it seems necessary following repair, observed equipment behavior or indications of problems during the regular quality control tests. To diminish treatment interruption due to non-regular interventions or additional quality control measurements, it is essential to maintain the test and measurement equipment in good order and subject this equipment to its own quality control program, as well as to have alternate equipment readily available.

10. References

[1] P. J. Hoskin and Brownes, P. Innovative Technologies in Radiation Therapy: Brachytherapy Seminars in Radiat. Oncol. (2006)16:209-217

[2] INTERNATIONAL ATOMIC ENERGY AGENCY, Lessons Learned from Accidental, Exposures in Radiotherapy, *Safety Report Series* No. 17, IAEA, Vienna (2000).

[3] EUROPEAN SOCIETY OF THERAPEUTICAL RADIOLOGY AND ONCOLOGY, A practical guide to quality control of brachytherapy equipment, *ESTRO Booklet* 8, Brussels (2004)

[4] EUROPEAN SOCIETY OF THERAPEUTICAL RADIOLOGY AND ONCOLOGY, Practical guidelines for the implementation of a quality system in radiotherapy, ESTRO, *Physics for Clinical Radiotherapy Booklet* No. 4, ESTRO, Brussels (1998).

[5] INTERNATIONAL ATOMIC ENERGY AGENCY, Standardized Quality Audit Procedures for On-site Dosimetry Visits to Radiotherapy Hospitals, *SSDL Newsletter* No. 46, IAEA, Vienna (2002).

[6] *Radiological Physics Center History*. Available from: http://rpc.mdanderson.org/RPC/home.htm.

[7] A. Roué, I. Ferreira, J. Dam, H. Svensson, J. Venselaar, The EQUAL-ESTRO audit on geometric reconstruction techniques in brachytherapy', *Radiotherapy and Oncology*, 2006) Volume 78 Issue 1, Pages 78 - 83 (http://cat.inist.fr/ and http://linkinghub.elsevier.com)

[8] C.E. deAlmeida. The Brazilian National Quality Program in Radiation Oncology. National Institute of Cancer. (1998).

[9] G. J. Kutcher, L. Coia, M. Gillin, W. F. Hanson, S. Liebel, R. J. Morton, J. R. Palta, J. A. Purdy, L. E. Reinstein, G. K. Svensson, M. Weller, and L. Wingfield, Comprehensive QA for radiation oncology: Report of the AAPM Radiation Therapy Committee Task Group 40, *Med. Phys.* .(1994) 21, 581-618.

[10] INTERNATIONAL ATOMIC ENERGY AGENCY, Comprehensive Audits of Radiotherapy Practices: A Tool for Quality Improvement, Quality Assurance Team for Radiation Oncology (QUATRO). *IAEA*, March 2007.

[11] P. H. Halvorsen, I.J. Das, M. Fraser, et al., AAPM Task Group 103 report on peer review in clinical radiation oncology physics. *J Appl Clin Med Phys*, (2005). 6(4): p. 50-64.

[12] R. Nath, L. L. Anderson, G. Luxton, K. A. Weaver, J. F. Williamson and A. S. Meigooni, Dosimetry of interstitial brachytherapy sources: recommendations of the AAPM Radiation Therapy Committee Task Group No. 43. American Association of Physicists in Medicine, *Med. Phys.* (1995).22, 209-234.

[13] M. J. Rivard, B. M. Coursey, L. A. DeWerd, W. F. Hanson, M. S. Huq, G. S. Ibbott, M. G.Mitch, R. Nath and J. F. Williamson, Update of AAPM Task Group No. 43 Report: A revised AAPM protocol for brachytherapy dose calculations, *Med Phys* (2004)31, 633-674.

[14] AMERICAN ASSOCIATION OF PHYSICISTS IN MEDICINE, Report of AAPM TG 53, Quality assurance for clinical radiation treatment planning, *Med. Phys* 25 (1998) 1773–1829.

[15] INTERNATIONAL ATOMIC ENERGY AGENCY, Commissioning and Quality Assurance of Computerized Treatment Planning Systems for Radiation Treatment of Cancer, *Technical Reports Series* No. 430, IAEA, Vienna (2004).

[16] C. E. deAlmeida, A. J. Pereira, M. H. Marechal, J. C. Cruz, J. C. Ferraz, A. J. Giordani, C.M. Khalil, E. H. Martins, G. Menegussi, D. Moreira, J. R. Rocha and M. A. Pinto, Intercomparison of calibration procedures for 192Ir HDR sources in Brazil, (1999). *Phys.Med. Biol.* 44, N31-N38

[17] R. Di Prinzio and C.E. deAlmeida Air kerma standard for calibration of well type chambers in Brazil using ^{192}Ir HDR sources and its traceability. *Med. Phys.* (2009) 36 (3), 953-60.

[18] IAEA, Calibration of brachytherapy sources: Guidelines of standardized procedures for the calibration of brachytherapy sources at secondary standard dosimetry laboratories (SSDL) and hospital, *Report N° TECDOC 1079, IAEA*, Vienna, 1999.

[19] N. A. Ellerbroek, M. Brenner, P. Hulick, et al., Practice accreditation for radiation oncology: quality is reality. *J Am Coll Radiol*, (2006). 3(10): p. 787-92

[20] R. Ochoa, F. Gome, I. H. Ferreira, F. Gutt and C. E. deAlmeida; CE. Design of a phantom for the quality control of high dose rate ^{192}Ir source in brachytherapy. *Radiotherapy and Oncology*, (2007). (82), p. 222-228.

[21] N. A. Ellerbroek, M. Brenner, P. Hulick, et al., Practice accreditation for radiation oncology: quality is reality. *J Am Coll Radiol*, (2006). 3(10): p. 787-92.

[22] G. W. Cotter and R.R. Dobelbower, Jr., Radiation oncology practice accreditation: the American College of Radiation Oncology, Practice Accreditation Program, guidelines and standards. *Crit Rev Oncol Hematol*, (2005). 55(2): p. 93-102.

[23] P. A. Johnstone, D.C. Rohde, B.C. May, et al., Peer review and performance improvement in a radiation oncology clinic. *Qual Manag Health Care*, (1999). 8(1): p. 22-8.

3

Applications and Experiences of Quality Control to Surgical and Interventional Procedures

David Biau[1,3], Lionel Dessolle[2] and Raphaël Porcher[1,3]
[1,3]*Département de Biostatistique et Informatique Médicale, Hôpital Saint-Louis, Assistance Publique Hôpitaux de Paris, Paris*
[3]*UMR-S 717, Inserm; Université Paris Diderot – Paris 7, Paris*
[2]*Service de gynécologie obstétrique, Hôpital Tenon, Assistance Publique Hôpitaux de Paris, Paris*
France

1. Introduction

Statistical process control have emerged in the medical literature after a wide expansion in the industry. In the nineteen-twenties, Walter A. Shewart, with a group of experts, established the scientific ground for quality control at Bell Telephone Laboratories Shewhart (1931). It is only in the nineteen fifties that quality control procedures were employed to ensure the precision of hospital laboratory machines used in biology, nuclear medicine, drug industry and other medical domains Anderson (1982); Batson et al. (1951); Brookeman (1978); Hollinger & Lansing (1956); Loynd (1962); Pribr & Shuster (1967); Waid & Hoffmann (1955). Later, in the nineteen seventies, the use of these methods shifted to the monitoring of the effect of treatments on patients Kinsey et al. (1989); Morgan et al. (1987); Robinson & Williamson (1974); Walters & Griffin (1986); Wohl (1977), and then to other more complex levels such as the performance of departements Chamberlin et al. (1993); The Inquiry into the management of care of children receiving complex heart surgery at the Bristol Royal Infirmary (2001), hospitals Sellick (1993), regions Tillett & Spencer (1982) or nation-wide processes Hand et al. (1994). It is only recently that these methods have been used to monitor the performance of physicians themselves in various disciplines. In this case, the success of the procedure is directly imputed to the the competency of the care provider. Statistical process control methods have been used in surgery de Leval et al. (1994); Novick & Stitt (1999); Van Rij et al. (1995), obstetrics Lane et al. (2007), endoscopy Williams et al. (1992), anaesthesiology Kestin (1995); Lagasse et al. (1995), etc.

Recently the Institute of Medecine, Washington DC, USA, in its report "Crossing the Quality Chasm: A New Health System for the 21st Century" stated that *"reducing waits and sometimes harmful delays for both those who receive and those who give care"* was one of the six aims proposed to adress key issues of today's health care system of Medicine (2001). The implementation of quality control procedures are meant for that purpose: detecting inadequate performance, should it occur, as soon as possible so that corrective actions may be undertaken and patient care improved.

The use of quality control methods in medicine presents some specificity and challenges when compared to that of the industrial context. When one monitors the performance of a care provider, the manufactured good is biologic: the patient. First, the raw product onto which the procedure will be performed, the patient, presents characteristics whose only a very limited subset are usually known. For instance when a physician is interested in monitoring endoscopic cholecystectomy, the patient undergoes a CAT scan and a few blood tests which provide very limited information on the product. Second, since some significant characteristics are retrieved before an intervention, we know that based on this limited information only, patients are nonetheless very different from one another. These differences may lead to variation in the outcome which should not always be attributed to a change in the performance of a surgeon for instance. Third, the process under scrutiny is poorly contained and under the effect of numerous unknown or unanticipated variables, either seen as chance or assignable causes. A surgeon who desires to monitor the implantation of total knee replacements to ensure that his or her performance is correct, may see failures because of poor bone quality, because of a traumatic fracture, or because of an infection, all being unrelated to his or her surgical skills. Last, the measure of quality is equivocal, difficult and subject to variation. For instance, measuring the success of an interventional procedure such as fiberoptic intubation may include the actual intubation per se, the time required to complete the procedure, the occurrence of a complication, or a composite of these criteria.

Another very specificity with using quality control methods in medicine is that often physicians are interested in determining when one can say, with some confidence, that a trainee has reached a certain level of performance. In that case, we are monitoring a process that we know is out of control at the begining of the observation period and we want to determine when this process reaches an in control state. In standard control chart usage this corresponds to the phase I of control chart application where engineers gather data, define control limits, identify and correct assignable causes and repeat this cycle until the process is brought to a stable, in control performance state Montgomery (2005). In medicine, this situation corresponds to the learning curve of a trainee and a new control chart has been recently developed for that purpose Biau & Porcher (2010).

In the sequel, we will present the main statistical process control methods used to monitor surgical and interventional procedures with their specificity such as the possibility for case-mix adustment. We will also present a new method specifically designed to monitor the learning curve Biau & Porcher (2010). Some historical papers will be presented and an example will be detailed.

2. Statistical process control methods

2.1 Monitoring a process from an in control to an out of control state

2.1.1 Control charts

Walter A. Shewart developped a control chart in the nineteen-twenties while working for Bell Telephone Laboratories Shewhart (1931). A control chart is a graphical representation of the measure of some characteristics of a product, or of a sample of products, against the sequence of these measures. Usually a control chart has a central line which corresponds to the expected value of the characteristics when the process monitored is in control and one or two control limits which define in and out of control zones. Different control charts exist: variables control charts when we are interested in a continuous variable or attributes control charts when we are interested in a dichotomous variable; control charts without memory such as the Shewart chart or with memory such as the exponentially weigthed moving average (EWMA) that averages the data in a way that gives less and less weight to data as they are further removed in time, etc Page (1954); Roberts (1959); Shewhart (1931); Wald (1945). But for all, the objective is to display the variation over time of the process under scrutiny.

The choice of control limits will determine the capacity of the chart to detect a change in the process, and therefore the user will require fine adjustement of these limits before the start of monitoring. Typically, given a period of monitoring the user will have to set the risks he or she is willing to take not to detect a change in the performance of the process when it occurs or, on the contrary, to emit a false alarm when no significant change in the performance of the process has occured. The upper and lower control limits, UCL and LCL have a general definition of the form:

$$\begin{cases} UCL = \mu_w + L\sigma_w \\ LCL = \mu_w - L\sigma_w \end{cases} \quad (1)$$

where μ_w and σ_w are the mean and standard deviation of the statistic used to monitor the process, and L the distance of the limit to the mean. A well known limit used in quality control is the 3 sigma limit where $L = 3$; in that case, when the process is in control and if the statistic used follows a normal distribution, 99.7% of the measures will be found in between the upper and lower control limits. Numerous control charts have been developed over time such as the cumulative summation (CUSUM) chart, the EWMA, etc.

2.1.2 the CUSUM test

2.1.2.1 Definition

The CUSUM test was developed in the nineteen-fifties by Page Page (1954) after Wald's work on sequential tests Wald (1945). It was originally developed to monitor the quality of manufactured products and later was found to be attractive to the medical context. Let us note $X_1, X_2, X_3, \ldots,$ the sequence of observations of the process under surveillance and assume no serial correlation. We denote by μ the mean of the process. These could be the times to complete appendectomies, the outcomes of neonatal arterial switch operations, or any other quantity that one would wish to monitor. The CUSUM sequentially tests after each observation $X_t (t > 0)$ the following hypothesis H_0: $\mu = \mu_0$, ie the process is in control, versus H_1: $\mu \neq \mu_0$, ie the process is out of control. The value μ_0 is often referred as the target. In its one sided formulation, where H_0: $\mu = \mu_0$ is tested against H_1: $\mu = \mu_1$ ($\mu_1 > \mu_0$) the test is based on the statistic S_t computed after each observation X_t as:

$$S_t = \max(0, S_{t-1} + W_t), \qquad S_0 = 0 \quad (2)$$

where the sample weight W_t depends on the observation X_t, μ_0, and μ_1. Moustakides Moustakides (1986) showed that optimal choices for W_t are proportional to the log-likelihood ratio. The test statistic S_t is compared to a predefined limit h. If S_t equals or exceeds h, the null hypothesis is rejected. In quality control wording the CUSUM test is said to emit an alarm indicating that the process is out of control. As long as S_t remains below h, the null hypothesis cannot be rejected and monitoring continues under the assumption that the process is in control.

2.1.2.2 Normally distributed data

The monitoring of normally distributed data is uncommon in the medical literature. Examples are the monitoring of the alignement of knee prosthetic replacements Nizard et al. (2004) and the laxity of anterior cruciate ligament reconstructions Biau et al. (2010). For normally distributed data of known standard deviation σ, because deviation may occur above or below the target value, two CUSUM tests are used in parallel to test H_0 against H_1: $\mu = \mu_1 = \mu_0 + g\sigma$ and H_{-1}: $\mu_{-1} = \mu_0 - g\sigma$ (with $g > 0$). The tests statistics S_t^+ and S_t^- used are defined as:

$$\begin{cases} S_t^+ = \max(0, S_{t-1}^+ + W_t^+) \\ S_t^- = \min(0, S_{t-1}^- + W_t^-) \end{cases} \tag{3}$$

with the following sample weights,

$$\begin{cases} W_t^+ = (X_t - \mu_0)/\sigma - g/2 \\ W_t^- = (X_t - \mu_0)/\sigma + g/2 \end{cases} \tag{4}$$

With such scores, S_t^+ and S_t^- only build up if X_t deviates from μ_0 by $g\sigma/2$ or more. The null hypothesis is rejected if $S_t^+ \geq h$ or $S_t^- \leq -h$. In practice σ is not known but as monitoring is prospective and parameters defined a priori, σ represents the expected standard deviation of the series and not the estimated standard deviation.

2.1.2.3 Binomial data

Medical data monitored are most often expressed in a success/failure dichotomy and therefore, the CUSUM test for binary data is the most frequently used Biau et al. (2007). Examples include monitoring the quality of implantation of hip prostheses Biau et al. (2009), or monitoring the 30-day mortality after complex cardiac surgery Steiner et al. (1999). In this case the mean of the process μ is equal to the probability of the event monitored (usually the failure), which will be noted p. The CUSUM tests after each procedure the following hypothesis $H_0 : p = p_0$ vs $H_1 : p = p_1$, where p_0 is the probability of failure when performance is adequate, and p_1 the probability of failure when performance is inadequate. Inadequate performance failure rates are usually 2 to 5 times higher than the target rates Bolsin & Colson (2000); de Oliveira Filho (2002); Kestin (1995). In this setting, the log-likelihood ratio score in equation 2 is obtained from the Bernoulli distribution as:

$$W_t = \log[p_1^{X_t}(1-p_1)^{1-X_t}/p_0^{X_t}(1-p_0)^{1-X_t}] \tag{5}$$

where X_t, the outcome, is equal to 0 for success and 1 for failure.

2.1.2.4 Time to event data

CUSUM test have been developed for monitoring time to event data such as the failure of a prosthesis, or that of kidney and liver transplants, over time Biswas & Kalbfleisch (2008); Hardoon et al. (2007); Sego et al. (2009). The reason for using time to event data is that compared to a CUSUM test based on the Bernoulli distribution, one does not have to wait the end of a period of time to assess failures and successes. Therefore, the outcome of each procedure contributes to the test statistic instantaneously. Depending on the situation, the test is based on the deviation of the process from a constant hazard or from one that varies over time such as that from a Weibull model Hardoon et al. (2007); Sego et al. (2009). The method may be adjusted and based on a discrete time Hardoon et al. (2007) or continuous time detection Biswas & Kalbfleisch (2008); Sego et al. (2009). However, in practice it should be noted that eventually the responsiveness of the CUSUM test developped will be limited by the technical possibilities of collecting, analysing, and reporting the data on a regular and frequent basis. For instance the CUSUM test based on the likelihood ratio statistic will have the following sample weight Hardoon et al. (2007):

$$W_j = O_j \log(\mathrm{HR}) - (\mathrm{HR} - 1)E_j \tag{6}$$

where O_j and E_j represent the number of failures observed and expected on interval j under the null hypothesis, and HR represents the hazard ratio of departure from target that one wants to detect. For instance, the number of failures expected on interval j based on the Weibull distribution where the hazard function is $h(t) = \lambda\gamma t^{\gamma-1}$ Collett (2005), with $\lambda > 0$, $\gamma > 0$, and $0 \leq t < \infty$, is:

$$E_j = \lambda_0 \sum_{i=1}^{n} \{(t_{2i} - t_{0i})^{\gamma} - (t_{1i} - t_{0i})^{\gamma}\} \tag{7}$$

where λ_0 is the scale parameter of the Weibull distribution under the null hypothesis, γ is the shape parameter under both the null and alternative hypothesis, t_{0i} is the time of realisation of the procedure on a time scale, t_{1i} is the maximum between t_{0i} and the lower boundary of interval j, and t_{2i} is the minimum between the time of failure of the procedure and the upper boundary of interval j.

2.1.2.5 Adjusted CUSUM

To avoid unduly penalizing a care provider or a center owing to disadvantageous case-mix selection, adjusted CUSUM tests have been proposed Sego et al. (2009); Steiner et al. (2000). In that case, each procedure is attributed a risk of failure based on the values of a set of covariates, and the statistic used is weighted on these covariates. Therefore, say a surgeon after some years in practice sees his or her practice shifts towards operating more and more complex cases. Although this surgeon has gained experience over the years, his or her results may tend to be pulled downwards and a standard monitoring method may detect inadequate performance when it is in fact the opposite. Adjusting the CUSUM on the difficulty of the cases which the surgeon operates on will avoid unecessary alarms and audits. As an illustration, let us consider the monitoring of a binary (success or failure) outcome. Assume that for patient t the pre-operative surgical risk may be determined from a set of covariates, and is noted p_t. The null and alternative hypotheses are now defined in terms of an odds ratio (OR), since each patient has a different baseline risk. The CUSUM test repeatedly tests H_0: OR = OR_0 versus H_1: OR = OR_1, and one may use the following sample weights Steiner et al. (2000):

$$W_t = \begin{cases} \log\left[\frac{(1-p_t+OR_0 p_t)}{(1-p_t+OR_1 p_t)}\right] & \text{if } X_t = 0 \quad \text{(success)} \\ \\ \log\left[\frac{(1-p_t+OR_0 p_t)OR_1}{(1-p_t+OR_1 p_t)OR_0}\right] & \text{if } X_t = 1 \quad \text{(failure)} \end{cases} \tag{8}$$

There are caveats in using adjusted CUSUM tests though. First, the set of covariates used for the adjustment must be unambiguous and consensual, and cannot be determined from the sample being monitored. Second, adjusted CUSUM tests explicitly infer that the results expected are different from on patient to another, and at some point, this goes against the philosophy of quality control which is to get rid of variability; indeed, as Douglas Montgomery puts it *"Quality is inversely proportional to variability"* Montgomery (2005). The variability seen should not necessarily be accounted for in the statistical method, and one may have to deal with it in the care of the patients.

2.1.2.6 Performance of the test

The CUSUM test has the particularity to display a holding barrier at 0 and to never accept the null hypothesis. Eventually this latter hypothesis is rejected regardless of the true performance of the process. Therefore, when the number of observations tends to infinity the CUSUM test has type I error rate and power of 100% in terms of traditional hypothesis tests. Performances of CUSUM tests are thus expressed differently. Average run lengths (ARL) are commonly used to report the performance of a CUSUM test: the ARL_0 and ARL_1 are the mean number of observations before an alarm occurs if the process runs under the null and alternative hypotheses respectively. Ideally, one wants to choose a limit so as to yield an important ARL_0 and a low ARL_1; therefore when the process is in control, false alarms are rare, and when the performance shifts, the test is responsive. However, both ARLs vary in the same direction as the limit h is moved upwards or downwards and there is a necessary trade-off between the probability of false alarms under the null hypothesis and the speed at which alarms are emitted under the alternative. Concerns have been raised in the literature regarding the use of average run lengths. First, run lengths have a skewed distribution that renders the average an inadequate measure of the underlying performance of CUSUM tests Woodall (1983). Second, run lengths are sometimes misleading and lead to the choice of inadequate limits Biau et al. (2009) and physicians usually prefer dealing with type I and type II errors Frisen (1992). In that case the performances of CUSUM tests are expressed in terms of true and false discovery rates (TDR and FDR), namely the probability of an alarm to be emitted under the alternative and null hypotheses within a defined number of observations Marshall et al. (2004). We now prefer this latter approach.

In some medical applications no decision boundaries are used and the sample weight W_t is simply plotted against the sequence of observations; this is called a CUSUM graph. In that case performance is informally judged by looking at the slope of the line plotted and compared to the theoretical slope expected under H_0.

2.1.3 Other methods commonly used

Numerous statistical methods have been developed to monitor the performance of a process and, consecutively, have been used in surgery and interventional procedures. The sequential probability ratio test (SPRT) was developed by Wald and is sometimes used in the medical domain Wald (1945). The SPRT is based on the following test statistic:

$$Y_t = Y_{t-1} + W_t, \qquad Y_0 = 0,\ t = 1, 2, 3, ... \tag{9}$$

with

$$W_t = \log\left\{\frac{l_{1t}}{l_{0t}}\right\}, \tag{10}$$

where l_{1t} et l_{0t} correspond to the likelihood of observation t under H_1 and H_0, respectively. W_t is thus defined similarly as for the CUSUM tests. However, one major difference with the CUSUM test is that the SPRT has no holding barrier and presents two decision limits, one to accept the null hypothesis, and the other one to reject the alternative hypothesis. The ability of the SPRT to accept the null hypothesis, i.e. to reject the alternative, has led physicians to use it to monitor the learning curve Biau et al. (2007). However, there are caveats in using such a procedure. We have found no practical interest so far in detecting, at the same time, a shift towards inadequate or adequate performance. It is one or the other that is of interest. If the performance monitored is assumed to be in control, then we are interested in detecting when it becomes inadequate and there is no need in detecting adequate performance only to continue the monitoring. If the performance is assumed to be out of control, such as for a trainee for instance (see below), there is no need for a test that indicates that the performance is indeed inadequate and we are only interested in knowing when it reaches a certain level of adequacy. If we don't know the status of the process being monitored, then we should assume it is no adequate and use a LC-CUSUM test (see below) and once an in control status has been reached, a conventional CUSUM test can be used.

The Exponentially Weighted Moving Average (EWMA) was developed by Roberts in the late fifties and, later, developed further by other authors Hunter (1986); Roberts (1959). In this method, the older the observation the less weight it conveys. The test is based on the following statistic:

$$Z_t = \lambda X_t + (1-\lambda)Z_{t-1}, \qquad Z_0 = \mu_0,\ 0 < \lambda \leq 1,\ t = 1,2,3,\ldots \tag{11}$$

With the following limits:

$$\begin{cases} h^+ = \mu_0 + L\sigma\sqrt{\frac{\lambda}{2-\lambda}[1-(1-\lambda)^{2i}]} \\ h^- = \mu_0 - L\sigma\sqrt{\frac{\lambda}{2-\lambda}[1-(1-\lambda)^{2i}]} \end{cases} \tag{12}$$

where L is the difference between the limits. The parameters L and λ, usually chosen as $0.05 \leq \lambda \leq 0.25$ and $L = 3$, allow to define the test performance. Overall, EWMA and CUSUM charts have been reported to have similar efficiency Lucas & Saccucci (1990), but choosing the appropriate exponential function remains however difficult in the absence of previous information. An example of the use of this method in medicine may be found in an intensive care unit Pilcher et al. (2010).

2.2 Monitoring a process from an out of control to an in control state: the LC-CUSUM test

Often physicians are interested in determining the time at which point we can assume a trainee has reached a certain level of performance so that supervision, or even training, may stop and resources be redirected to another trainee Biau et al. (2007). The LC-CUSUM test has been developped for that purpose Biau & Porcher (2010); Biau et al. (2008). The LC-CUSUM test was designed in a similar manner to the two one-sided tests procedure for equivalence clinical trials Chow & Shao (2002). This time, the null hypothesis H_0 is that the process is not in control, or in the present case that performance is not adequate. In terms of measurements, the null hypothesis corresponds to a process that is not centered on a target μ_0, but deviates on average at least by δ. This can be rewritten as H_0: $|\mu - \mu_0| \geq \delta$. The alternative hypothesis is that the process does not deviate from μ_0 by more than δ, namely $H_1 : |\mu - \mu_0| < \delta$. The null hypothesis is the union of the two one-sided hypotheses $H_{01} : \mu - \mu_0 \geq \delta$ and $H_{02} : \mu - \mu_0 \leq \delta$, and H_0 has to be rejected when both H_{01} and H_{02} are rejected. In practice, simple hypotheses are set to parameter the test: H_{01}: $\mu = \mu_0 + \delta$ and H_{02}: $\mu = \mu_0 - \delta$ under the null hypotheses and H_1: $\mu = \mu_0$ under the alternative hypothesis. A CUSUM test statistic is then constructed for each hypothesis in a traditional way (see eq. 2). When the process can only deviate on one side, the corresponding hypothesis H_{01} or H_{02} only is tested. For instance, when monitoring binary data (success/failure) one is generally interested in testing only whether the failure rate of a trainee does not deviate by more than an acceptable amount above adequate performance. In that situation, the hypothesis H_{02} is ignored. This situation can be regarded as akin to non-inferiority trials. The purpose of the CUSUM test is to reject an inadequate performance level; the purpose of the LC-CUSUM test is to signal when performance is sufficiently far away from an inadequate performance level to consider the performance as adequate.
For normally distributed data, with $\delta = g\sigma$ ($g > 0$), the following sample weights are used:

$$\begin{cases} W_t^1 = (X_t - \mu_0)/\sigma - g/2 \\ W_t^2 = (X_t - \mu_0)/\sigma + g/2 \end{cases} \tag{13}$$

and for dichotomous data:

$$W_t^1 = \log\left[\frac{p_0^{X_t}(1-p_0)^{1-X_t}}{(p_0+\delta)^{X_t}(1-p_0-\delta)^{1-X_t}}\right] \tag{14}$$

Given the fact that one may start monitoring with an LC-CUSUM test and then, once adequate performance has be shown, continue monitoring with a CUSUM test, confusion may arise in the definition of the adequate and inadequate performance levels under the null and alternative hypotheses for the CUSUM and LC-CUSUM tests. Therefore, it is preferable to determine the adequate and inadequate performance levels and the acceptable deviation based on clinical information such as previous data, literature, and opinions of expert and then decide what levels will be chosen under the null and alternative hypothesis for both tests. For instance, in endoscopic retrograde cholangiopancreatographies one will decide that the adequate performance is 10% failure, that 25% failure is inadequate and that 5% is an acceptable deviation from the adequate performance level. Therefore, for the LC-CUSUM test, the performance considered under the null hypothesis is H_0: $\mu = \mu_0 + \delta = 0.15$ and under the alternative is H_1: $\mu = \mu_0 = 0.1$; and for the CUSUM test, the performance considered under the null hypothesis is H_0: $\mu = \mu_0 = 0.1$ and under the alternative is H_1: $\mu = \mu_1 = 0.25$.
Obviously the acceptable deviation from target, δ, considered for the LC-CUSUM may vary between 0 and $|\mu_1 - \mu_0|$. As proposed by Jones et al Jones et al. (1996) for equivalence trials, choosing this acceptable deviation to be $\delta \leq (\mu_1 - \mu_0)/2$, will reasonnably allow a buffer between what is deemed as acceptable for the LC-CUSUM test and the performance considered as unacceptable for the CUSUM test. Performances of LC-CUSUM tests may also be expressed in terms of true and false discovery rates (TDR and FDR).

3. Applications

3.1 Applications to surgical procedures

Marc R. de Leval, a cardiothoracic surgeon, was the first to report the monitoring of his own performance in the nineteen-nineties de Leval et al. (1994). In his landmark paper, he used a sequential probability ratio test to monitor the 30-day mortality rate after 104 neonatal arterial switch operations. During this period, he experienced a cluster of surgical failures with 7 deaths from the 53th to the 68th observation. The probability of observing 7 deaths or more in 16 patients when the expected mortality rate is 10% is very small (0.0005, by the binomial distribution), and this cluster was attributed to inadequate performance. He concluded that control charts were useful to detect unfavorable trends in performance and correct surgical technique if needed. He also introduced the concept of near-miss, a surrogate for a failed procedure, which was used in civil aviation to monitor the skills of pilots, and that could play a role in monitoring the performance of surgeons.
The Bristol Royal Infirmary case was another major determinant in the interest that quality control procedures would raise at the begining of the 21st century The Inquiry into the management of care of children receiving complex heart surgery at the Bristol Royal Infirmary (2001). A public inquiry was conducted between 1998 and 2001 on the management of the care of children receiving complex cardiac surgery at the Bristol Royal Infirmary between 1984 and 1995. The mortality rate at that center proved to be twice the rate in England and did not demonstrate the decrease seen over time in the rest of the country. Despite having enough information from the late 1980s onwards to raise significant questions about the mortality rate, there was no systematic mechanism for monitoring the clinical performance of care providers or of the center. Therefore, it took many years before clinical activity was stopped and an audit conducted. The report therefore recommended that *"for the future there must be effective*

systems within hospitals to ensure that clinical performance is monitored. There must also be a system of independent external surveillance to review patterns of performance over time and to identify good and failing performance". An application of quality control methods to the data collected during the inquiry showed that inadequate performance could have been detected as early as 1991 Spiegelhalter et al. (2003). Numerous applications have then been published around the world with various control charts, adjustement on patient risk factors, and use of near-misses Albert et al. (2004); Novick et al. (2001); Poloniecki et al. (2004); Rogers et al. (2005); Tsang et al. (2009). The use of quality control procedures has spread to other surgical disciplines. Although the enthousiasm is not as strong as that seen in cardiac surgery, there are initiatives using these methods and improving patient care in other areas. In general surgery these methods have been used to monitor the quality of, among other procedures, laparoscopic cholecystectomy Bartlett & Parry (2001), parotidectomies Sharp et al. (2003), ileal pouch-anal anastomosis Tekkis, Fazio, Lavery, Remzi, Senagore, Wu, Strong, Poloneicki, Hull & Church (2005), laparoscopic colorectal resections Tekkis, Senagore, Delaney & Fazio (2005), or thyroidectomy Connolly & Watters (2010). In orthopaedics, these methods have been used to monitor the quality of implantation of knee and hip replacement Biau et al. (2009); Nizard et al. (2004), the failure of hip implants Hardoon et al. (2006) and reconstruction of the anterior cruciate ligament Biau et al. (2010).

3.2 Applications to interventional procedures

Numerous applications of quality control methods have been reported in interventional disciplines. Parry Parry & Williams (1991) was one of the first to report the monitoring of his experience of 305 unassisted colonoscopies using a CUSUM chart (no limits were used in this report). Success was defined as completion of the procedure and a target performance of 10% failure rate was aimed at. Overall, 250 procedures (82%) were completed. The analysis showed that there was a steep improvement in performance during the first 100 procedures and that the target performance was reached around the 200th procedure; however, as noted earlier, the CUSUM chart does not allow a precise determination of when the performance can be considered as adequate as opposed to the LC-CUSUM test.

Anaesthesiologists who perform many critical and highly technical interventional procedures for their practice have also shown a particular interest in using quality control procedures to monitor their performance. Kestin Kestin (1995) reported in the mid-nineties the use of the SPRT to monitor residents performing obstetrics extradurals, spinal anaesthesia, central venous and arterial cannulations. The levels of performance varied from one procedure to another from 5% to 20% and 20% to 40% for the adequate and inadequate levels respectively. The same year Lagasse Lagasse et al. (1995) reported the use of a Shewart chart to monitor 13 clinical indicators of inadequate care, such as the unplanned admission to intensive care unit of a patient or the occurrence of a postural headaches following spinal or epidural analgesia. 13,389 procedures were monitored over a 12 months period, and 116 complications were reported. Only one indicator revealed inadequate performance. Afterwards, other authors have monitored the outcome of various interventionnal procedures such as labour epidurals Naik et al. (2003), nerve blocks Schuepfer & Johr (2005), or orotracheal intubation in a mannequin Correa et al. (2009).

3.3 Applications to the learning curve

From the beginning, statistical process control methods have been used to monitor the performance of trainees because physicians were not satisfied with the usual methods available for evaluation. Traditionally, either a predetermined number of cases was prescribed or direct observation by the tutor was necessary. In the former case, the method was not felt to be individualized to the trainee with some trainees having completed the required number of procedures but who were not proficient yet and others showing competence early and thus rendering supervision unecessary Dagash et al. (2003). In case a tutor was used to evaluate the trainee, subjectivity was a problem Sloan et al. (1995). Therefore, most studies using statistical process control mehods to monitor the performance of surgical or interventional procedures have been directed at monitoring the learning curve Biau et al. (2007). The idea with using statistical process control methods is that one would be able to determine when a trainee has reached proficiency in an objective and individual manner at the same time. Numerous methods have been used: some, such as Parry Parry & Williams (1991) for instance, have used a CUSUM chart and based on the slope of the curve they decide approximately when an adequate level of performance has been reached. This method provides a visual help but it is not accurate. Some, such as Kestin Kestin (1995), have used a sequential probability ratio test and use one of the boundaries to detect adequate performance whilst the other boundary detects inadequate performance. Although this test is attractive, it is inadequate to monitor the learning curve because of the inconsistency in the emission of alarms when processes with similar performance are monitored. It is for that reason that we have designed the LC-CUSUM test as described above. To date it has been used to monitor the competence of an endoscopist performing endoscopic retrograde pancreatography Biau et al. (2008), that of a gyneacologist performing vitrification of embryo Dessolle et al. (2009) and embryo transfers Dessolle et al. (2010), and that of radiologist in the diagnosis of endometriomas by transvaginal ultrasound Bazot et al. (2010).

4. Worked example

4.1 Monitoring the ongoing performance of a surgical procedure

Say a surgeon wants to monitor the quality of implantation of total hip replacements. She knows that the longevity of the implant is associated with the technical precision with which she will perform the procedure. She defines ten criteria (see table **??**) to assess the overall quality of the procedure. Some, such as the cup abduction angle, are a direct measure of the technical aspect of the procedure, whilst others, such as a revision during hospital stay, are surrogates for the whole process; however, she feels that all criteria are of importance. In the absence of previous information on case-mix she decides not to use an adjusted method and to weight each criterion identically. Therefore, the procedure will be considered as failed if any of the ten criteria is failed. Based on previous information Biau et al. (2009) and on consensus among other surgeons she decides that 18% failure represents an adequate level of performance (target performance) and that 36% failure rate would indeed indicate inadequate performance. She decides to use a CUSUM test and based on computer simulation she sets the decision limit $h = 5.6$, yielding a TDR of 99.8% and FDR of 3% for 200 observations. Figure 1 shows the CUSUM test for the monitoring period. The test is based on equation 5, with $p_0 = 0.18$ and $p_1 = 0.36$. Five meetings were held during that year (indicated by the arrows). Overall 106 of the 500 total hip replacements (21.2%) were considered as failed. The score increases with each failure and decreases otherwise. The holding barrier at 0 prevents the score from drifting away from the decicion limit with the accumulation of successes; therefore, the test remains responsive to the occurrence of inadequate performance at all times. The CUSUM test emmited an alarm after 87 procedures. After an audit, she identifies the main reason for failures and corrects her surgical technique accordingly. Monitoring is restarted and no alarm is emitted afterwards.

4.2 Monitoring the learning curve of a trainee

Say a program director in obstetrics and gynecology desires to assess the competence of trainees wishing to learn the use of fetoscopic-directed laser therapy for the treatment of twin-twin transfusion syndrome. Given the syndrome is rare only few centers around the world perform this procedure and the resources required to train a gynecologist are limited. Also, the procedure is technically complex and the life of fetuses at risk. Therefore, he would like to make sure that each trainee is only allowed to perform the procedure with no supervision once he or she is competent and that, given the limited number of cases available for teaching, as soon as a trainee is competent the resources may be directed to another trainee. After a literature search and consensus among senior practitioners in the country they decide to set a nation-wide program that will allow fellows to be trained and during which their performance will be monitored with a LC-CUSUM test. An adequate performance level of 10% failure rate, inadequate performance of 25% and an acceptable deviation from adequate performance of 5% will be considered. Because the outcome of a poorly performed procedure threatens the life of the fetus, they decide to allow no more than 5% FDR. Given the limited number of cases they cannot provide more than 100 procedures per trainee during training. Therefore, they choose a limit $h = 1.25$ so that the TDR is 85% over these 100 procedures. Three fellows decide to enter the program. All procedures they perform are under supervision and at the end of each procedure a debrief is conducted. The procedure is considered as failed if none of the fetus survives at birth or if the trainee was not able to complete the procedure without any assistance. Therefore, in case the tutor considers the procedure is not performed adequately by the trainee, and that this puts the fetus life at risk, he or she takes on the case and completes the procedure in an adequate manner so that the level of care provided to patients remains adequate. The LC-CUSUM test for the three trainees is depicted figure 2. One trainee demonstrated competency at the 52nd procedure, another one at the 83rd procedure, and the last one did not demonstrate competency during the alloted number. The case of the last trainee should be discussed among seniors and with the trainee before he or she can resume monitoring. One of the reason being that statistically, if the trainee is given more cases with the same limit, the risk of a false detection increases, and eventually, regardless the true performance of the trainee, an alarm will always occur.

5. Conclusion and perspectives

To date, numerous reports using quality control methods have been published in the surgical and interventional literature. These reports provide good evidence that these methods are useful in ensuring that adequate performance level is maintained over time. Also, and more importantly, they provide some assurance that should inadequate performance occur, it will not go undetected for an undue period of time. Ensuring that the quality of care provided is adequate at all time is paramount from the patient, physician and health authorities perspectives. However, there are non statistical issues that deserve more attention before we see these methods unveil their full potential.

First, these methods are underused because of the lack of knowledge of most physicians that such methods exist. These methods were transferred only recently to the medical domain and are quite separated from the natural understanding of physicians. They are not taught in curriculums and it is most of the time by an incidental discovery that a surgeon or an endoscopist will encounter them. As quality improvement initiatives make their way into the medical curriculum hopefully the awareness of physicians will increase accordingly Boonyasai et al. (2007).

Second, there is a genuine fear that these methods, which can measure performance at an individual level, will be misused. Physicians may feel these methods will be used to rank and compare them and ultimately be used in a coercitive way to impose changes they do not agree with. This is not what these methods should be used for. These methods are not meant for deterrence but for improvement. Quality control methods should be part of a quality improvement program and, as such, it is paramount that they remain in the hands of those who are under observation to improve the quality of care they deliver. They should be aimed at detecting under-performing units in order to improve them, not in order to blame them, and at all time they should be centered on the patient's need. It is only by stepping in and implementing these methods first that physicians will limit their use to what they want so that these methods are not imposed blindly by healthcare authorities or management. Also, the use of such methods to detect inadequate performance does not mean that only a small subset of units will benefit from quality improvement methodology; ideally the modification of the process under observation should benefit every care provider, or every center, not just the lower end of them.

Third, there is a lack of consensus among centers and practitionners regarding numerous aspects of monitoring which restrain the spread of these methods. Processes are not always clearly defined when, for instance, it may involve the technical act *per se* or include the perioperative period. For a similar process, the outcome measures may vary from one center to another or from one practitioner to another. The levels of performance may also be different. If we agree that discrepancies are expected with surgeons or interventionists willing to see different aspects of a process being monitored or different levels being targeted, there is an urge for those who spearhead the implementation of quality control methods to meet and discuss their fundamentals.

Last, although the statistical methods of quality control are constantly developed, the major difficulty physicians will encounter is the practical aspect to implement these methods. It is time and resource consuming to collect, analyse and report data. Ideally, these methods should be functioning in an almost automated manner, smoothly, in the background so as not to interfer with clinical practice. And it is only at the time of regular reports that physicians should come aware of them and, if necessary, change the process that was under observation. Only a close relation between information technology and clinical practice will allow that.

6. Figures and tables

Subset		Inadequate if	Method of measurement
Technique	1	fracture of the acetabulum requiring fixation or change in the technique	as per operative note
	2	fracture of the femur requiring fixation or change in the technique	as per operative note
	3	-1 cm > leg length difference > 2 cm	X-ray AP pelvis
Stem	4	-4°> stem inclinaison angle > +4°	X-ray AP pelvis
	5	stem cement mantle modified Barrack C or D class	X-ray AP pelvis
Cup	6	+30°> cup abduction angle > +55°	X-ray AP pelvis
	7	0°> cup version angle > +30°	X-ray AP pelvis and hip
	8	lucent line $\geq$ 1*mm* in one zone or more for cemented cups or a gap $\geq$ 3*mm* for uncemented cups	X-ray AP pelvis
Clinical	9	Dislocation or reoperation for any reason during hospital stay	as per hospital record
	10	Death regardless the reason during hospital stay	as per hospital record

Table 1. Performance criteria considered for the success of total hip replacement

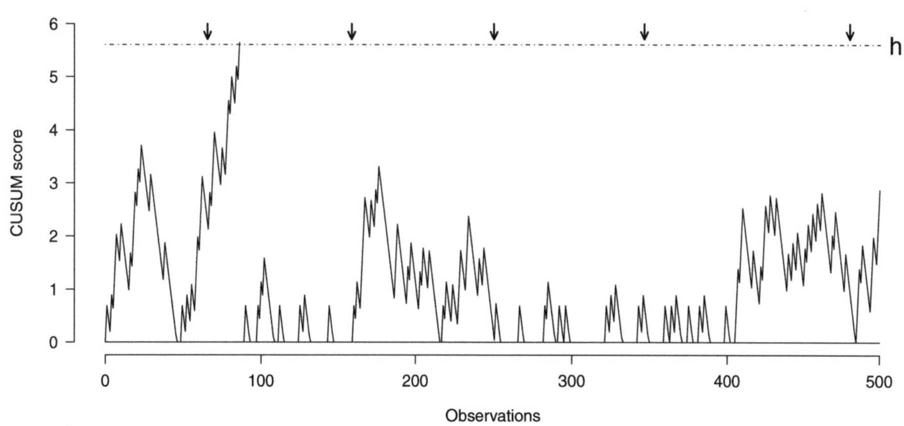

Fig. 1. CUSUM test for a surgeon performing 500 total hip replacements with $p_0 = 0.18$, $p_1 = 0.36$, and $h = 5.6$. Arrows indicate meetings where the results were presented. The CUSUM test emmited an alarm after the 87 procedures.

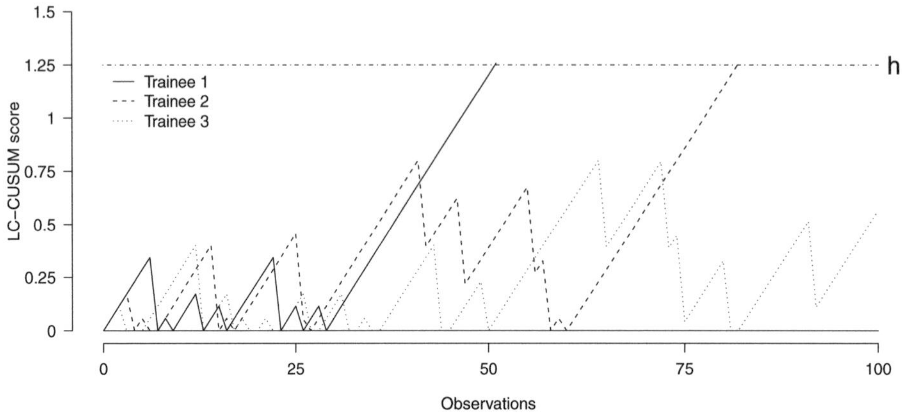

Fig. 2. LC-CUSUM test for 3 trainees performing fetoscopic-directed laser therapy for the treatment of twin-twin transfusion syndrome with $p_0 = 0.1$, $\delta = 0.05$, and $h = 1.25$. The LC-CUSUM test emmited an alarm after 52 and 83 procedures for trainee 1 and 2 respectively. No alarm was raised for trainee 3.

7. References

Albert, A. A., Walter, J. A., Arnrich, B., Hassanein, W., Rosendahl, U. P., Bauer, S. & Ennker, J. (2004). On-line variable live-adjusted displays with internal and external risk-adjusted mortalities. A valuable method for benchmarking and early detection of unfavourable trends in cardiac surgery, *Eur J Cardiothorac Surg* 25: 312–319.

Anderson, D. J. (1982). The quality control of medical gases, *Hosp Eng* 36: 18–19.

Bartlett, A. & Parry, B. (2001). Cusum analysis of trends in operative selection and conversion rates for laparoscopic cholecystectomy, *ANZ J Surg* 71: 453–456.

Batson, H. C., Brown, M. & Oberstein, M. (1951). An adaptation of quality control chart methods to bacterial vaccine potency testing, *J Bacteriol* 61: 407–419.

Bazot, M., Darai, E., Biau, D. J., Ballester, M. & Dessolle, L. (2010). Learning curve of transvaginal ultrasound for the diagnosis of endometriomas assessed by the cumulative summation test (LC-CUSUM), *Fertil Steril* .

Biau, D. J., Landreau, P., Graveleau, N. & Gravelau, N. (2010). Monitoring surgical performance: an application of industrial quality process control to anterior cruciate ligament reconstruction, *Knee Surg Sports Traumatol Arthrosc* 18: 1263–1268.

Biau, D. J., Milet, A., Thevenin, F., Anract, P. & Porcher, R. (2009). Monitoring surgical performance: an application to total hip replacement, *J Eval Clin Pract* 15: 420–424.

Biau, D. J. & Porcher, R. (2010). A method for monitoring a process from an out of control to an in control state: Application to the learning curve, *Stat Med* 29: 1900–1909.

Biau, D. J., Resche-Rigon, M., Godiris-Petit, G., Nizard, R. S. & Porcher, R. (2007). Quality control of surgical and interventional procedures: a review of the CUSUM, *Qual Saf Health Care* 16: 203–207.

Biau, D. J., Williams, S. M., Schlup, M. M., Nizard, R. S. & Porcher, R. (2008). Quantitative and individualized assessment of the learning curve using LC-CUSUM, *Br J Surg* 95: 925–929.

Biswas, P. & Kalbfleisch, J. D. (2008). A risk-adjusted CUSUM in continuous time based on the Cox model, *Stat Med* 27: 3382–3406.

Bolsin, S. & Colson, M. (2000). The use of the Cusum technique in the assessment of trainee competence in new procedures, *Int J Qual Health Care* 12: 433–438.

Boonyasai, R. T., M., W. D., Chakraborti, C., S., F. L., Rubin, H. R. & Bass, E. B. (2007). Effectiveness of teaching quality improvement to clinicians: a systematic review., *JAMA* 298(9): 1023–1037.

Brookeman, V. A. (1978). Computers and quality control in nuclear medicine, *Semin Nucl Med* 8: 113–124.

Chamberlin, W. H., Lane, K. A., Kennedy, J. N., Bradley, S. D. & Rice, C. L. (1993). Monitoring intensive care unit performance using statistical quality control charts, *Int J Clin Monit Comput* 10: 155–161.

Chow, S. C. & Shao, J. (2002). A note on statistical methods for assessing therapeutic equivalence, *Control Clin Trials* 23(5): 515–520.

Collett, D. (2005). *Modelling Survival Data in Medical Research*, New York: Chapman & Hall/CRC.

Connolly, T. M. & Watters, D. A. (2010). Monitoring performance in thyroidectomy: cumulative sum analysis of outcomes, *Thyroid* 20: 407–412.

Correa, J. B., Dellazzana, J. E., Sturm, A., Leite, D. M., de Oliveira Filho, G. R. & Xavier, R. G. (2009). Using the Cusum curve to evaluate the training of orotracheal intubation with the Truview EVO2 laryngoscope, *Rev Bras Anestesiol* 59: 321–331.

Dagash, H., Chowdhury, M. & Pierro, A. (2003). When can I be proficient in laparoscopic surgery? A systematic review of the evidence, *J. Pediatr. Surg.* 38: 720–724.

de Leval, M. R., Fran¸cois, K., Bull, C., Brawn, W. & Spiegelhalter, D. (1994). Analysis of a cluster of surgical failures. Application to a series of neonatal arterial switch operations, *J Thorac Cardiovasc Surg* 107: 914–923.

de Oliveira Filho, G. R. (2002). The construction of learning curves for basic skills in anesthetic procedures: an application for the cumulative sum method, *Anesth. Analg.* 95: 411–416.

Dessolle, L., Biau, D. J., de Larouziere, V., Ravel, C., Antoine, J. M., Darai, E. & Mandelbaum, J. (2009). Learning curve of vitrification assessed by cumulative summation test for learning curve (LC-CUSUM), *Fertil. Steril.* 92: 943–945.

Dessolle, L., Freour, T., Barriere, P., Jean, M., Ravel, C., Darai, E. & Biau, D. J. (2010). How soon can I be proficient in embryo transfer? Lessons from the cumulative summation test for learning curve (LC-CUSUM), *Hum. Reprod.* 25: 380–386.

Frisen, M. (1992). Evaluations of methods for statistical surveillance, *Stat Med* 11(11): 1489–1502.

Hand, R., Piontek, F., Klemka-Walden, L. & Inczauskis, D. (1994). Use of statistical control charts to assess outcomes of medical care: pneumonia in Medicare patients, *Am J Med Sci* 307: 329–334.

Hardoon, S. L., Lewsey, J. D., Gregg, P. J., Reeves, B. C. & van der Meulen, J. H. (2006). Continuous monitoring of the performance of hip prostheses, *J Bone Joint Surg Br* 88: 716–720.

Hardoon, S. L., Lewsey, J. D. & van der Meulen, J. H. (2007). Continuous monitoring of long-term outcomes with application to hip prostheses, *Stat Med* 26: 5081–5099.

Hollinger, N. F. & Lansing, R. K. (1956). Standards, control charts, and reliability in the determination of serum protein, *Am J Med Technol* 22: 365–375.

Hunter, S. J. (1986). The exponentially weighted moving average, *J Qual Technol* 18: 203–210.

Jones, B., Jarvis, P., Lewis, J. A. & Ebbutt, A. F. (1996). Trials to assess equivalence: the importance of rigorous methods, *BMJ* 313: 36–39.

Kestin, I. G. (1995). A statistical approach to measuring the competence of anaesthetic trainees at practical procedures, *British Journal of Anaesthesia* 75(6): 805–809.

Kinsey, S. E., Giles, F. J. & Holton, J. (1989). Cusum plotting of temperature charts for assessing antimicrobial treatment in neutropenic patients, *BMJ* 299: 775–776.

Lagasse, R. S., Steinberg, E. S., Katz, R. I. & Saubermann, A. J. (1995). Defining quality of perioperative care by statistical process control of adverse outcomes, *Anesthesiology* 82: 1181–1188.

Lane, S., Weeks, A., Scholefield, H. & Alfirevic, Z. (2007). Monitoring obstetricians' performance with statistical process control charts, *BJOG* 114: 614–618.

Loynd, H. J. (1962). Quality control in the drug industry, *J Indiana State Med Assoc* 55: 236–238.

Lucas, J. & Saccucci, M. (1990). Exponentially weighted moving average control schemes: properties and enhancements, *Technometrics* 32(1): 1–12.

Marshall, C., Best, N., Bottle, A. & Aylin, P. (2004). Statistical issues in the prospective monitoring of health outcomes across multiple units, *J Roy Stat Soc A* 167: 541–559.

Montgomery, D. C. (2005). *Introduction to Statistical Quality Control, Fifth edition*, New York: John Wiley and Sons.

Morgan, G. J., Thomas, S., Cavill, I. & Bentley, D. P. (1987). Detection of relapse in acute lymphoblastic leukaemia by cusum analysis of peripheral blood-count, *Lancet* 2: 1274–1275.

Moustakides, G. V. (1986). Optimal stopping times for detecting changes in distributions, *Ann Statist* 14: 1379–1387.

Naik, V. N., Devito, I. & Halpern, S. H. (2003). Cusum analysis is a useful tool to assess resident proficiency at insertion of labour epidurals, *Can J Anaesth* 50: 694–698.

Nizard, R. S., Porcher, R., Ravaud, P., Vangaver, E., Hannouche, D., Bizot, P. & Sedel, L. (2004). Use of the Cusum technique for evaluation of a CT-based navigation system for total knee replacement, *Clin. Orthop. Relat. Res.* pp. 180–188.

Novick, R. J., Fox, S. A., Stitt, L. W., Swinamer, S. A., Lehnhardt, K. R., Rayman, R. & Boyd, W. D. (2001). Cumulative sum failure analysis of a policy change from on-pump to off-pump coronary artery bypass grafting, *Ann. Thorac. Surg.* 72: S1016–1021.

Novick, R. J. & Stitt, L. W. (1999). The learning curve of an academic cardiac surgeon: use of the CUSUM method, *J Card Surg* 14: 312–320.

of Medicine, I. (2001). *Crossing the quality chasm: a new health care system for the 21st century*, National academy press, Washington, DC.

Page, E. S. (1954). Continuous inspection schemes, *Biometrika* 41: 100–115.

Parry, B. R. & Williams, S. M. (1991). Competency and the colonoscopist: a learning curve, *Aust N Z J Surg* 61: 419–422.

Pilcher, D. V., Hoffman, T., Thomas, C., Ernest, D. & Hart, G. K. (2010). Risk-adjusted continuous outcome monitoring with an EWMA chart: could it have detected excess mortality among intensive care patients at Bundaberg Base Hospital?, *Crit Care Resusc* 12: 36–41.

Poloniecki, J., Sismanidis, C., Bland, M. & Jones, P. (2004). Retrospective cohort study of false alarm rates associated with a series of heart operations: the case for hospital mortality monitoring groups, *BMJ* 328: 375.

Pribr, H. C. & Shuster, M. (1967). Improving quality control in the hospital laboratory, *J Med Soc N J* 64: 213–217.

Roberts, S. W. (1959). Control chart tests based on geometric moving averages, *Technometrics* 42: 97–102.

Robinson, D. & Williamson, J. D. (1974). Letter: Cusum charts, *Lancet* 1: 317.

Rogers, C. A., Ganesh, J. S., Banner, N. R. & Bonser, R. S. (2005). Cumulative risk adjusted monitoring of 30-day mortality after cardiothoracic transplantation: UK experience, *Eur J Cardiothorac Surg* 27: 1022–1029.

Schuepfer, G. & Johr, M. (2005). Psoas compartment block (PCB) in children: Part II–generation of an institutional learning curve with a new technique, *Paediatr Anaesth* 15: 465–469.

Sego, L. H., Reynolds, M. R. & Woodall, W. H. (2009). Risk-adjusted monitoring of survival times, *Stat Med* 28: 1386–1401.

Sellick, J. A. (1993). The use of statistical process control charts in hospital epidemiology, *Infect Control Hosp Epidemiol* 14: 649–656.

Sharp, J. F., Cozens, N. & Robinson, I. (2003). Assessment of surgical competence in parotid surgery using a CUSUM assessment tool, *Clin Otolaryngol Allied Sci* 28: 248–251.

Shewhart, W. A. (1931). *Economic Control of Quality of Manufactured Products*, New York: Van Nostrand.

Sloan, D. A., Donnelly, M. B., Schwartz, R. W. & Strodel, W. E. (1995). The Objective Structured Clinical Examination. The new gold standard for evaluating postgraduate clinical performance, *Ann. Surg.* 222: 735–742.

Spiegelhalter, D., Grigg, O., Kinsman, R. & Treasure, T. (2003). Risk-adjusted sequential probability ratio tests: applications to Bristol, Shipman and adult cardiac surgery, *Int J Qual Health Care* 15: 7–13.

Steiner, S. H., Cook, R. J. & Farewell, V. T. (1999). Monitoring paired binary surgical outcomes using cumulative sum charts, *Stat Med* 18: 69–86.

Steiner, S. H., Cook, R. J., Farewell, V. T. & Treasure, T. (2000). Monitoring surgical performance using risk-adjusted cumulative sum charts, *Biostatistics* 1: 441–452.

Tekkis, P. P., Fazio, V. W., Lavery, I. C., Remzi, F. H., Senagore, A. J., Wu, J. S., Strong, S. A., Poloneicki, J. D., Hull, T. L. & Church, J. M. (2005). Evaluation of the learning curve in ileal pouch-anal anastomosis surgery, *Ann. Surg.* 241: 262–268.

Tekkis, P. P., Senagore, A. J., Delaney, C. P. & Fazio, V. W. (2005). Evaluation of the learning curve in laparoscopic colorectal surgery: comparison of right-sided and left-sided resections, *Ann. Surg.* 242: 83–91.

The Inquiry into the management of care of children receiving complex heart surgery at the Bristol Royal Infirmary (2001). The Report of the Public Inquiry into children's heart surgery at the Bristol Royal Infirmary 1984–1995, *Technical report*. URL: *http://www.bristol inquiry.org.uk/*

Tillett, H. E. & Spencer, I. L. (1982). Influenza surveillance in England and Wales using routine statistics. Development of 'cusum' graphs to compare 12 previous winters and to monitor the 1980/81 winter, *J Hyg (Lond)* 88: 83–94.

Tsang, V. T., Brown, K. L., Synnergren, M. J., Kang, N., de Leval, M. R., Gallivan, S. & Utley, M. (2009). Monitoring risk-adjusted outcomes in congenital heart surgery: does the appropriateness of a risk model change with time?, *Ann. Thorac. Surg.* 87: 584–587.

Van Rij, A. M., McDonald, J. R., Pettigrew, R. A., Putterill, M. J., Reddy, C. K. & Wright, J. J. (1995). Cusum as an aid to early assessment of the surgical trainee, *Br J Surg* 82: 1500–1503.

Waid, M. E. & Hoffmann, R. G. (1955). The quality control of laboratory precision, *Am J Clin Pathol* 25: 585–594.

Wald, A. (1945). Sequential Tests of Statistical Hypotheses, *Ann Math Stat* 16: 117–186.
Walters, S. & Griffin, G. E. (1986). Resolution of fever in Staphylococcus aureus septicaemia–retrospective analysis by means of Cusum plot, *J Infect* 12: 57–63.
Williams, S. M., Parry, B. R. & Schlup, M. M. (1992). Quality control: an application of the cusum, *BMJ* 304: 1359–1361.
Wohl, H. (1977). The cusum plot: its utility in the analysis of clinical data, *N Engl J Med* 296: 1044–1045.
Woodall, W. H. (1983). The distribution of the run length of one sided cusum procedures for continuous random variables, *Technometrics* 25: 295–301.

4

Whole Body Composition by Hologic QDR 4500/A DXA: System Reliability versus User Accuracy and Precision

Aldo Scafoglieri, Steven Provyn, Joanne Wallace, Olivia Louis, Jonathan Tresignie, Ivan Bautmans, Johan De Mey and Jan Pieter Clarys
Vrije Universiteit Brussel (VUB)
Belgium

1. Introduction

Accurate and precise measurement of human tissue composition is both important and imperative in individual health assessment. Although body composition (BC) data acquisition and analysis are both popular and important, selecting an appropriate method or technique for accurate and/or precise assessment of individuals and/or groups remains a challenging task within various sectors of public health. Since 1950s and 1960s, with the pioneer work of Keys & Brozek (1953), Forbes et al. (1956), Siri (1956), Brozek et al. (1963), Behnke (1963), Durnin & Rahaman (1967), BC almost became a scientific discipline profiling itself with the development of many methods, techniques, and equipments. Popular approaches have been criticized over the years because they are subject to measurement errors and/or violation of basic assumptions underlying their use such as hydrodensitometry (Clarys et al., 2010c; Clasey et al., 1999; Elowsson et al., 1998; Heyward, 1996; Johansson et al., 1993; Prior et al., 1997) or anthropometry, e.g., skinfolds (Beddoe, 1998; Clarys et al., 1987, 2005, 2010a; Martin et al., 1985, 1992; Scafoglieri et al., 2010a) and the universally accepted new method of choice, the dual energy X-ray absorptiometry (DXA) (Bolotin, 1998, 2007; Bolotin & Sievanen, 2001; Bolotin et al., 2001; Clarys et al., 2010b; Provyn et al., 2008; Scafoglieri et al., 2010c).

1.1 New gold standard for BC data acquisition?

After reviewing the literature of DXA application, one obtains a controversial impression of this new method. On the one hand, we find an important number of validation and application studies that support the DXA technique as convenient, as the criterion for fat percentage, lean body mass (LBM), and bone mineral content (BMC) (Clasey et al., 1999; Haarbo et al., 1991; Johansson et al., 1993; Prior et al., 1997; Pritchard et al., 1993). A number of authors as mentioned in Provyn *et al.* (2008) suggest DXA as the gold standard for validation of other techniques essential for the measurement of BC (Eston et al., 2005; Poortmans et al., 2005; Salamone et al., 2000). On the other hand it needs reminded that DXA, likewise hydrodensitometry, anthropometry, bioelectrical impedance, air, gas, and water displacement methods, is an indirect *in vivo* technique for measuring BC. Validation or even cross-validation in between indirect methods cannot guarantee both accuracy and reality precision. Perfect correlations and low coefficients of variation allow for good predictions and assumptions only (Bolotin & Sievanen, 2001; Clarys et al., 2010b; Provyn et al., 2008).

1.2 Data acquisition quality issues

Possibly the greatest problems with accuracy/precision in DXA are found with fat and lean tissue estimates (Prentice, 1995), with its projected areal bone density (Bolotin, 2007; Bolotin et al., 2001; Clarys et al., 2008, 2010b; Provyn et al., 2008; Scafoglieri et al., 2010c) and with the basic confusion between overall BC terminology, e.g., fat, adipose tissue (AT), fat-free mass (FFM), LBM, lean and AT free mass (ATFM), bone mineral density (BMD), surface and volume density, BMC, ash weight, actual mineral content, and BMC, with or without soft tissue (ST) covering (Clarys et al., 2010b; Martin et al., 1985; Provyn et al., 2008; Scafoglieri et al., 2010b; Wadden & Didie, 2003). These issues give rise to concern, but the accuracy of absorptiometry can be affected by the choice of calibrating materials. As a consequence, both absolute and relative values can differ substantially between manufacturers, between instruments and the ad hoc software used (Clasey et al., 1999; Prentice, 1995). Despite the multitude of DXA validation studies and despite the related controversy of its measuring quality, it is being reaffirmed that there have been comparatively few validation experiments of accuracy, and precision of either bone or BC measurements by cadaver and/or carcass analysis. More of these validations against direct values are necessary before we can be confident about the accuracy of absorptiometry (Prentice, 1995; Clarys et al., 2010b).

1.3 State of the art

A review of the state of the art of carcass studies related to DXA (Clarys et al., 2008, 2010b; Scafoglieri et al., 2010c) reveals validation attempts with rhesus monkeys (Black et al., 2001), mice (Brommage, 2003; Nagy & Clair, 2000), piglets (Chauhan et al., 2003; Elowsson et al., 1998; Koo et al., 2002, 2004; Picaud et al., 1996; Pintauro et al., 1996), pigs (Lukaski et al., 1999; Mitchell et al., 1996, 1998), pig hindlegs (Provyn et al., 2008), chickens (Mitchell et al., 1997; Swennen et al., 2004), and with dogs and cats (Speakman et al., 2001). The majority of these validation studies were based on chemical analysis and only a few on direct dissection comparison. Although almost all studies indicated perfect correlations for all variables with DXA, approximately half of the results of the various variables were found to be significantly different. In approximately a third of these studies, DXA was suggested to be valid and accurate for all its variables, whereas two studies indicated significant differences and/or erroneous data at all levels and for all variables. However, two important statements resulting from these studies are retained: (i) dissection and direct comparison combined with bone ashing are considered the most accurate and direct validation technique (Elowsson et al., 1998) and (ii) further research with direct dissection and ashing is needed (Prentice, 1995), in particular, with focus on the influence of abdominal and thoracic organs associated with dispersed gas/air pockets and internal panniculus adiposus (Provyn et al., 2008; Clarys et al., 2010b).

1.4 Aim

Although BC measurements by DXA are increasingly used in clinical practice, no study has been giving clarity yet about the content and meaning of "lean" as produced by DXA. Because direct dissection is the best possible direct measure, different soft tissue combinations, e.g. skin, muscle, and viscera will be related to the DXA-lean variable. The exact knowledge of what is the content of the meaning of "lean" as measured by DXA is mandatory. In this chapter the reliability and the validity of Hologic QDR 4500/A DXA system will be determined based on direct *in vitro* measurements. A number of problems related to DXA applications resulting from BC terminology and data acquisition will be discussed.

2. Hologic QDR 4500/A DXA system

The DXA system is principally designed to provide information for the accurate diagnosis of osteoporosis, but also determines whole BC. A QDR 4500/A upgraded to Discovery Hologic DXA device (Hologic, Waltham, MA) (Figure1) utilizes a constant X-ray source producing fan beam dual energy radiation with effective dose equivalents (EDE) of 5μSv (e.g., to situate this low radiation in terms of example: a one-way transatlantic flight produces ±80μSv EDE and a spinal radiograph ~700μSv EDE) (Prentice, 1995). The estimations of fat and lean mass are based on extrapolation of the ratio of ST attenuation of two X-ray energies in nonbone containing pixels. The two X-ray energies are produced by a tungsten stationary anode X-ray tube pulsed alternately as 70 kilovolts (peak) (kVp) and 140 kVp. The software (for Windows XP version 12.4.3) performs calculations of the differential attenuations of the two photon energies and presents data of percentage of fat, fat mass (g), lean mass (g), BMC (g), BMD (g/cm^2), and total mass (g). According to the manufacturer, a coefficient of variation for human BMD of 0.5% can be expected during repeated measurements. DXA equipment was calibrated daily with a spine phantom (supplied by the manufacturers) to assess stability of the measurements, but also calibrated weekly using a step phantom to allow for correction of sources of error related to, e.g., skin thickness.

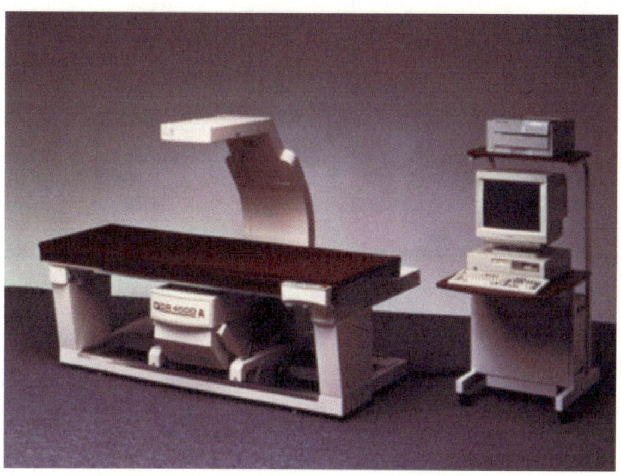

Fig. 1. Hologic QDR 4500/A DXA system

3. System reliability of DXA

Today it is generally accepted that using different models of DXA scanners (e.g. QDR 4500 Series against QDR 1000 Series), different software versions and different scanning speeds may reduce precision of body composition measurement (Barthe et al., 1997; Guo et al., 2004; Litaker et al., 2003)

Consequently the use of different DXA devices in longitudinal or multicenter studies may lead to measurement error and bias. On the other hand quality control studies proved that the intrainstrument, intra- and interoperator reliability for soft and rigid tissue mass estimates is high (Bailey et al., 2001; Burkhart et al., 2009; Glickman et al., 2004; Godang et al., 2010; Guo et al., 2004; Koo et al., 2002).

In order determine the reliability of the QDR 4500/A DXA system, 22 human cadavers (seventeen males and five females, mean age ± sd, 79.6 ± 8.8 y; mean weight ± sd, 69.589 ± 13.194 kg) were scanned three times consecutively. The first two scans were taken by a single experienced operator without repositioning the cadaver. From these data, the intramachine (test-retest) reliability for all DXA variables was calculated (Table 1). A third scan was taken and analyzed by a second experienced operator after repositioning the cadaver. This allowed for the calculation of the interrater reliability for whole BC assessment by DXA (Table 1).

The measurement results revealed no significant differences for the test-retest and between the operators, except for total mass. With coefficients of variation ranging from 0.2% to 3.5%, and with intraclass correlation coefficients ranging from 0.99 to 1.00 the Hologic QDR 4500/A DXA system showed to be highly reliable for BC assessment.

Variable	Reliability type	d ± sd	*P*	95% LOA	CV(%)	ICC
Total mass (g)	Intramachine	-49.70 ± 168.66	>0.05	± 330.6	0.2	1.000
	Interrater	-71.94 ± 162.38	=0.05	± 318.2	0.2	1.000
Fat (g)	Intramachine	-37.30 ± 338.67	>0.05	± 572.2	0.7	0.999
	Interrater	-132.71 ± 483.45	>0.05	± 912.4	1.0	0.999
Fat (%)	Intramachine	-0.02 ± 0.41	>0.05	± 0.81	1.9	0.999
	Interrater	0.11 ± 0.67	>0.05	± 1.31	3.5	0.998
Lean (g)	Intramachine	-10.77 ± 291.91	>0.05	± 663.8	1.9	0.999
	Interrater	58.34 ± 465.48	>0.05	± 947.6	3.5	0.999
BMC (g)	Intramachine	-1.63 ± 37.03	>0.05	± 72.5	1.6	0.997
	Interrater	2.43 ± 50.73	>0.05	± 99.4	2.4	0.995
BMD (g/cm^2)	Intramachine	0.002 ± 0.019	>0.05	± 0.037	1.6	0.992
	Interrater	-0.014 ± 0.021	>0.05	± 0.040	1.8	0.990

Table 1. Intramachine and interrater reliability for whole body DXA variables (d = mean difference, sd = standard deviation, *P* = t-test significance level, LOA = limits of agreement, CV = coefficient of variation, ICC = intraclass correlation)

4. Validation of DXA for BC measurement

Twelve 6- to 18-month-old "Belgian Native" pigs were prepared for human consumption and were acquired within 2-day intervals, immediately after electroshock slaughter (six females and six castrated males, mean weight ± sd, 39.509 ± 4.335 kg). Special permission was obtained from the Belgian Directorate General of Public Health, Safety of the Food Chain and Environment, for the transport of the carcasses and for the nonremoval of abdominal and thoracic content, which is a normal procedure in consumption matters. The carcasses were exsanguinated and decapitated between the atlas and the occipital bone. To minimize further dissection error, front and hindlegs were disarticulated distal from humeri and femora, e.g., on elbow and knee levels, respectively. The mean weight ± s.d. of the remaining carcass plus viscera was 33.051 ± 3.324 kg (whole carcass weights being taken with a digital hang scale (KERN-HUS-150K50) accurate to 50 g). The composition of the carcasses was studied in the following order.

4.1 Dissection procedure

After the DXA measurements, the carcasses were dissected into their various components as expressed on the tissue-system level: skin, muscle, AT, viscera, and bones (Wang et al., 1992). Muscle included tendon, blood vessels, and nerves belonging to the actual muscle. The subcutaneous, intramuscular (mostly intratendon), and intravisceral ATs were combined as one tissue. Again, blood vessels and nerves within AT were attributed to AT. Bones were carefully scraped, ligaments were added with muscle tendons to muscle tissue, and cartilage remained part of the bone tissue. Seven expert prosectors and anatomists worked simultaneously and each dissected particle was collected under cling film and kept in color-labeled, continuously covered plastic containers (12 × 10 × 10 cm) of known weight in order to minimize or eliminate evaporation (Clarys et al., 1999, 2010b; Provyn et al., 2008; Scafoglieri et al., 2010a). Full container mass was measured during the dissection by two researchers using Mettler Toledo digital scales (Excellence XS precision balance model 40025) accurate to 0.01 g. Once a bone was fully prepared, the same procedure was followed but completed with its hydrostatic weight while placed in a wire cradle suspended to the same scale allowing for the volume-based bone density (g/cm^3) calculation.

4.2 Chemical fat and hydration analysis

After the dissection and multiple weighing procedures, samples of all tissues of ~100–150 g (min–max) were deep-frozen. Small parts were cut off and weighed in recipients of known weight before lyophilization overnight. With dried samples, the water content was measured after storing into metal cells, and fat (lipids) extracted with technical hexane using a Dionex accelerated solvent extractor. After the hexane evaporation of the extraction, total (final) lipid content was determined (weighed). Part of the dissection protocol of the 12 porcine carcasses was the total defleshing of the skeleton, including the removal of extraosseous soft tendon and ligament tissue by scraping. Cartilage and intraosseous tissue (e.g., intervertebral discs) remained intact.

4.3 Ashing

The whole skeleton was diamond-cut into pieces in order to fit in the ashing furnace (type Nabertherm; Nabertherm, Lilienthal, Germany). After incineration, each sample was heated using a ramped temperature protocol of 2 hours to 800 °C and ashed for 8 hours, as determined by prior pilot work. Before weighing on the Mettler Toledo precision scale (accurate to 0.01 g), the ash was cooled undercover and collected in a main container. The ashing of one full porcine skeleton took between 50 and 60 hours.

5. Validity of DXA data acquisition

Although the interpretation of DXA results is generally straightforward, it is important to be aware of common pitfalls and to maintain rigorous quality assurance. The purpose of this part is to compare directly and indirectly obtained data of masses and densities (e.g., of whole body bone, adipose, and nonadipose tissue) using two different techniques and to provide information on the terminology as used in the respective methodologies.

5.1 Assumptions regarding data acquisition outcome

Table 2 shows an overview of terminology used per technique as applied and that are assumed to measure the same values. Although the basic assumption of equality of outcome and despite the different terminology used, knowledge of the ad hoc mass and density names will create a better understanding of the respective data acquisitions (e.g., Table 3).

Dissection	DXA	Biological background
Total mass (g)	Total mass (g)	-
Total tissue mass (g)	Total mass (g)	The Σ of all dissected tissue masses
Adipose tissue (AT)(g)	Fat (g)	AT is an anatomical issue FAT is a chemical issue (e.g. lipids)
Adipose tissue free mass (ATFM)(g)	Lean or lean body mass (LBM)(g)	ATFM is an anatomical concept LBM = Fat Free Mass plus essential lipids
Skeleton mass (g)	Bone mineral content (BMC)(g)	Skeleton or bone mass are morphological issues; BMC suggests the Σ of all mineral constituents of the skeleton
Ash weight (g)	BMC (g)	Ash weight = bone mass minus total bone hydration
Skeleton density (g/cm^3)	Bone mineral density (g/cm^2)	Volume (g/cm^3) based versus surface (g/cm^2) based density

Table 2. Different terminologies assumed to measure a similar outcome (DXA = dual energy X-ray absorptiometry)

Given the basic reasoning that the measurement of whole-body adiposity (in g or %), or nonadipose tissue (in g) and density (in g/cm^3) with different techniques using different equipment should produce similar, if not identical results on the same individuals, cannot be supported because of underlying assumptions, models, or approaches of the techniques and/or equipment are different (Beddoe, 1998; Heyward, 1996; Martin et al., 2003). This can be perfectly illustrated by the experiment on one single healthy male subject of which its % adiposity or fatness was measured with four different techniques on the same day, e.g., with an anthropometric formula, with dual energy X-ray absorptiometry, with bioelectrical impedance and with hydrostatic weighing including the calculation via the Siri (1956) formula. According to the original basic reasoning, the results of these four measures of adiposity should be in agreement. On the contrary, one notices with an anthropometric formula 12.5%, with DXA 17.5%, with bioelectrical impedance 21.5%, and with Siri 26.8% of adiposity was found (Table 3).

Method	Predicted whole body %fat
Anthropometry (Jackson and Pollock, 1978)	12.5
Dual energy X-ray absorptiometry	17.5
Bio-electrical Impedance Analysis	21.5
Hydrodensitometry (Siri formula)	26.8

Table 3. Predicted %fat by 4 different methods on one single male subject on the same day

5.2 Erroneous interchangeable use of BC terminology

The reality is that one is measuring different adiposity approaches with the same confusing terminology, e.g., % whole-body fat. Body fat (BF) is defined as the etherextractable constituent of body tissues, and must be considered as a chemical component of the body. This is already known since Keys and Brozek (1953). The interchangeable use of the terms BF and AT has led and is leading still to ambiguities and serious error. Among all DXA validation studies, only a few (Elowsson et al., 1998; Nagy & Clair, 2000) have defined the meaning of its adiposity variables mentioning or precising as DXA fat and lean against chemical (CHEM fat and CHEM lean).

In the Anglo-Saxon literature, in particular, the term "fat" is commonly used quantitatively when referring to the degree of obesity of a body and with "fatness" qualitatively referring to the appearance of the body that results from the deposition of AT (Wadden & Didie, 2003). Technically, "fat" may be defined biochemically as extractable lipid that consists of depot lipids such as the triglycerides and free fatty acids from AT and also so-called "essential" lipids such as structural phospholipids of cell membranes and nervous tissues, lipids of bone marrow, and a small moiety of other lipid-based compounds. Because of the confusion surrounding terms that are used both colloquially and technically, the terms "fat" and "fatness" should not be used. The term "fat" used in a biochemical sense should be replaced by the term "lipid," and the term "fatness" will be replaced by "adiposity" when referring to the quantity of AT in the body. ATs are masses separable by gross dissection and includes not just lipid but also the nonlipid constituents of cells, such as water and protein, and of course, the bulk of the subcutaneous AT and tissue surrounding organs, viscera and variable amounts between muscles, e.g., the intramuscular AT. These phenomena are known since 1950s and 1960s (Brozek et al., 1963) and were reinforced in 1980s by extensive direct data acquisition (Clarys & Marfell-Jones, 1986; Clarys et al., 1999). Table 2 indicates other discrepancies, e.g., for the nonadipose terminology. ATFM is an anatomical concept and lays in the continuation of the AT vs. FM. DXA pretends to measure lean or LBM as opposed to FFM, which could be expected because manufacturers claim to measure chemical components. In attempts to identify the physiologically relevant tissues, the concept of LBM was introduced more than half a century ago (Behnke et al., 1942). This consists of the FFM plus the essential fat specification that has varied from 2 to 10% for the FFM. Because of its imprecise definition, this term also has led to much confusion in the literature and is often erroneously used as a synonym for FFM. In addition to FFM and LBM, the anatomical concept of ATFM was proposed as a normalizing approach for interpopulation comparisons (Clarys & Martin, 1985; Clarys et al., 2010b).

5.3 Accuracy and precision of DXA data acquisition outcome

Table 4 combines the data acquisition of all directly obtained measures and the complete set of indirect estimates made by DXA. The purpose of this Table 4 is to evaluate the predictive quality of DXA, but also to evaluate precision and accuracy between direct and indirect values. For a good understanding and despite the significance of a correlation found, this study considers $r \geq 0.90$ as a good, $r \geq 0.80$ as a medium, and $r \geq 70$ as an average (mediocre) indicator of prediction. The t-statistics are considered as an indicator of precision or accuracy. Significant differences are set at $P < 0.05$. If not significantly different with the dissection reference, one can assume an acceptable level of measurement precision. Table 4 confirms that for almost all soft tissue (ST) comparisons, including total masses, a majority of good correlations ($r \geq 0.90$) and one medium correlation ($r \geq 0.80$) was found. Despite this

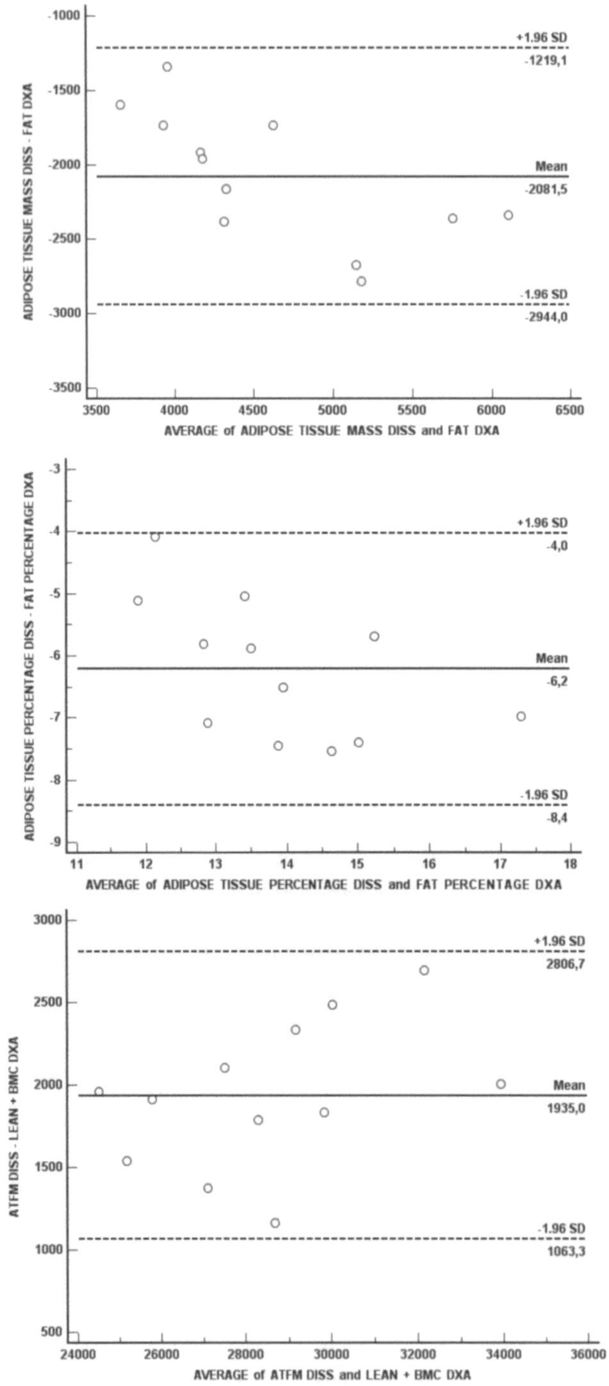

Fig. 2. Bland–Altman plots comparing adipose tissue (AT) and adipose tissue free mass (ATFM) by dissection to dual energy X-ray absorptiometry (DXA) measures with assumed similar outcome (BMC, bone mineral content; DISS, dissection)

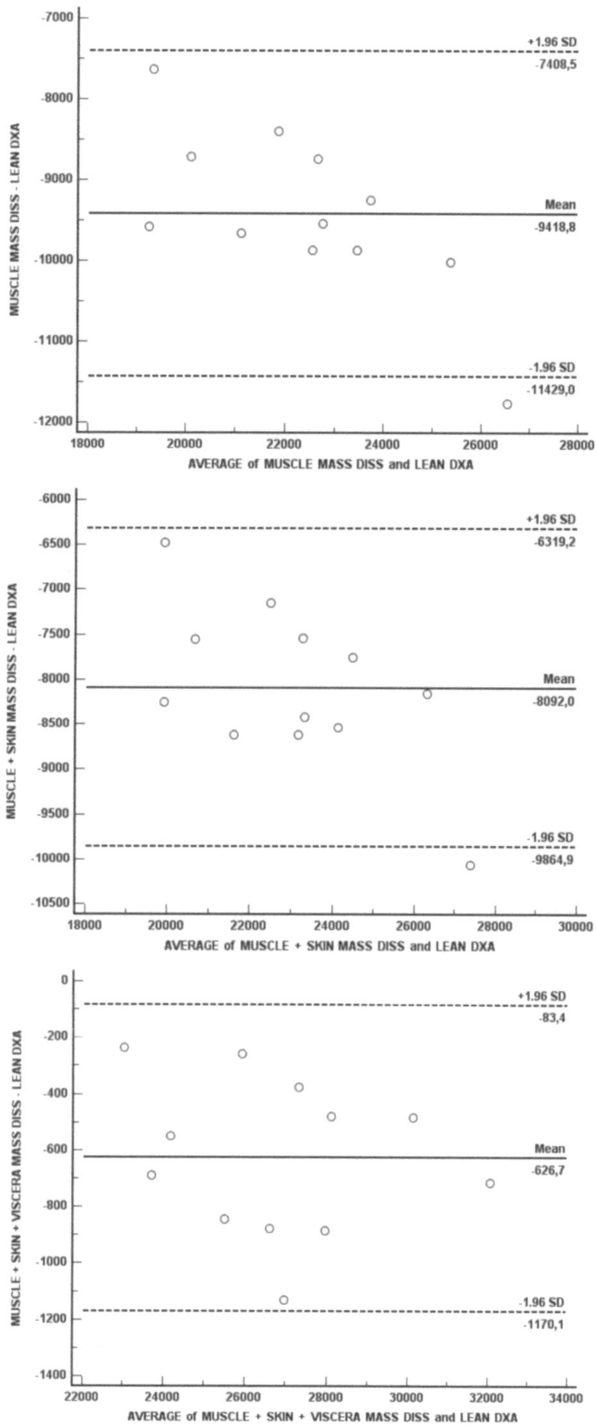

Fig. 3. Bland–Altman plots comparing "LEAN" by dual energy X-ray absorptiometry (DXA) to dissection measures with assumed similar outcome (DISS, dissection)

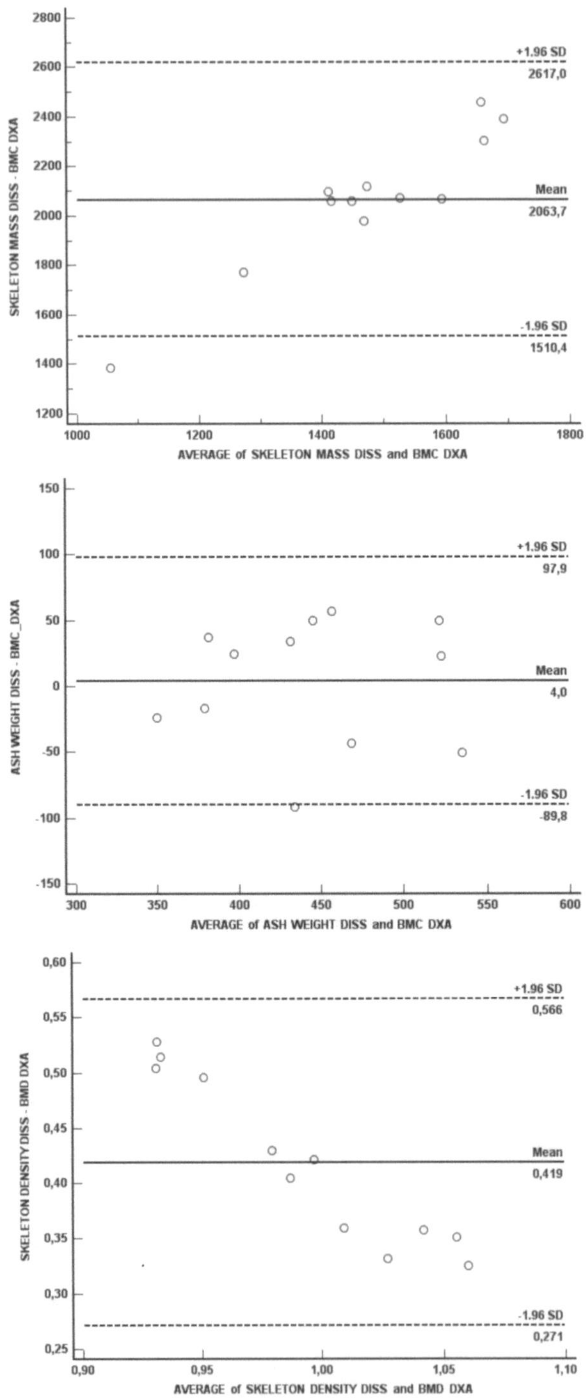

Fig. 4. Bland-Altman plots comparing different skeleton measures by dissection to dual-energy X-ray absorptiometry (DXA) measures with assumed similar outcome (BMC, bone mineral content; BMD, bone mineral density; DISS, dissection)

majority of good prognoses for prediction related to the dissection reference, we do find significant differences in accuracy for total masses, adiposity (g and %) for all nonadipose ST combinations, and for all bony comparisons except for the ashing which indicates an acceptable precision and comparability with DXA-BMC. Agreement and disagreement of DXA data acquisition with the dissection reference of the compared tissues and tissue combinations are shown in Figures 2–4 with 9 Bland–Altman (1986) plots. The solid lines indicate the mean difference deviation from zero and the dashed lines the individual variation between ±1.96 s.d.

Variable	Dissection x ± sd	DXA x ± sd	r	*P*
Total mass (g)	33051.3 ± 3323.8	33192.3 ± 3336.6	1.00	<0.01
Total tissue mass (g)	32723.4 ± 3427.0	33192.3 ± 3336.6	1.00	<0.001
Adipose tissue/Fat (g)	3571.6 ± 632.8	5653.1 ± 934.1	0.91	<0.001
Adipose tissue/Fat (%)	10.8 ± 1.27	17.0 ± 1.87	0.81	<0.001
ATFM/Lean + BMC (g)	29479.7 ± 2874.7	27544.7 ± 2681.5	0.99	<0.001
Muscle/Lean (g)	17684.3 ± 1908.8	27103.1 ± 2647.3	0.95	<0.001
Muscle + skin/Lean (g)	19011.1 ± 2092.3	27103.1 ± 2647.3	0.95	<0.001
Muscle + skin + viscera/Lean (g)	26476.4 ± 2593.8	27103.1 ± 2647.3	0.99	<0.001
Skeleton mass/BMC (g)	2505.3 ± 317.5	441.6 ± 64.6	0.62	<0.001
Ash weight/BMC (g)	445.6 ± 66.2	441.6 ± 64.6	0.73	NS
Skeleton density(g/cm^3)/BMD(g/cm^2)	1.201 ± 0.02	0.782 ± 0.09	0.68	<0.001

Table 4. Comparison between direct dissection data values with the corresponding DXA values (DXA = dual energy X-ray absorptiometry, x = mean, sd = standard deviation, r = Pearson correlation coefficient, *P* = t-test significance level, ATFM = adipose tissue free mass, BMC = bone mineral content, NS = not significantly different)

The Bland–Altman plots confirm the findings as shown in Table 4. If we look at the mean value level of the respective variables, there cannot be any doubt that DXA produces anatomical-morphological quantities, evidently at all adipose and nonadipose combinations. (Clarys et al., 2010b)

5.4 Assumptions regarding chemical tissue composition constancies

The chemical tissue composition of the dissection masses was determined according to anatomic segmentation into upper limb, lower limb, and trunk (e.g., for skin, muscle, and bone). For AT, additional differentiation was made for subcutaneous (e.g., external) and visceral (e.g., internal) trunk AT. For each segment, the water content and the fat (e.g., lipid) content was determined for the respective tissues and presented as % of the studied mass per tissue in Table 5.

With the confounding effect of the high variability of AT removed, the composition of the ATFM shows smaller deviations of its components and smaller differences between males and females than when body mass is used as a reference (Martin & Drinkwater, 1991). Conversion from FFM to LBM and/or to ATFM is susceptible to significant error because we are dealing with two totally different models. In addition, DXA does not take into account the water content and lipid content variations (Table 4) of both its adipose and nonadipose constituents. Small variation of tissue hydration may explain important differences of ad hoc estimates (Prior et al., 1997; Wang et al., 1999, 1995). In DXA, and other newer technologies (Muller et al., 2003), body fat is calculated on the constancy assumption that ≈73% of LBM (e.g., lean or lean + BMC) is water (31). This assumed constancy of hydration, e.g., the observed ratio of total body water to FFM was confirmed in humans by Wang *et al.* (1999). However, this assumption is subject to some questions that highlight the need for more research on the matter. Viewing tissue water content obtained by lyophilization in several human tissue studies, one can make two observations: (i) assuming a constant % of water in FFM may be jeopardized by the variable tissue water content within and between the tissues that compose FFM; and (ii) water content in AT is highly variable, e.g., ranging from ±17% to ±50% in humans (Table 6) (Provyn et al., 2008; Clarys et al., 2010b).

Tissue	Segment	Water content(%) x±sd	Lipid content(%) x±sd	r
Skin	Up limb	61.0 ± 8.6	4.6 ± 6.0	- 0.73
	Lo limb	60.7 ± 4.9	4.3 ± 1.4	- 0.55
	Trunk	50.1 ± 9.3	10.2 ± 7.4	- 0.20
Adipose	Subcut Up limb	47.2 ± 7.0	15.0 ± 7.0	- 0.72
	Subcut Lo limb	47.2 ± 6.6	15.6 ± 6.9	- 0.84*
	SubcutTrunk	21.0 ± 5.3	29.0 ± 7.3	- 0.16
	Visceral Trunk	50.1 ± 10.6	19.0 ± 6.7	- 0.70
Muscle	Up limb	75.4 ± 1.4	1.4 ± 1.0	- 0.86*
	Lo limb	74.5 ± 2.7	3.1 ± 3.2	0.16
	Trunk	73.8 ± 3.9	3.7 ± 2.3	- 0.70
Bone	Up limb	39.0 ± 8.2	10.9 ± 2.7	- 0.84*
	Lolimb	39.5 ± 8.1	9.8 ± 1.9	- 0.71
	Trunk	49.4 ± 2.4	7.7 ± 3.3	- 0.20

Table 5. Water (lyophilization) and lipid (ether extraction) content of different tissues and relationship (x = mean, sd = standard deviation, r = Pearson correlation coefficient, Up = upper, Lo = lower, Subcut = subcutaneous, *P<0.01) (Clarys et al., 2010b)

	Muscle	Skin	Viscera	Bone	AT
Forbes & Lewis (1956) (n=2)	1) 67.5	53.7	73.4	26.8	26.2
	2) 68.2	51.8	72.0	31.6	18.3
Mitchell et al. (1945) (n=1)	79.5	64.7	76.6	31.8	50.1
Cooper et al. (1956) (n=2)	1) 68.9	53.5	73.7	30.2	16.8
	2) 77.3	72.5	77.8	39.5	83.9
Forbes et al. (1953) (n=1)	70.1	57.7	73.3	28.2	23.0
Clarys et al. (1999) (n=6)	70.8	63.2	79.1	---	21.6

Table 6. Water content (%) of lean and adipose tissue masses in humans (lyophilisation) (AT = adipose tissue)

This is confirmed in this study on animal corpses with % whole-body water content ranging from ±20 to ±50% (Table 5) repeating that the constancies claimed by DXA cannot be maintained (e.g., with fluid ranging between ±50 and ±61% for skin, between ±39 and ±49% for bone, but little variability for muscle). Because no total tissue lipid extraction was carried out because technical circumstances allowed sample fractionation only, lipid content is expressed as % of the measured sample mass. From sample masses being identical for hydration and lipid fractionation (Table 5), we learn that lipid content of tissues is variably related to its ad hoc fluid content, but if the extremities are considered separately, one notices an apparent constancy both in hydration and lipid fractionation. The fact that all trunk tissue data (e.g., in skin AT, muscle, and bone) deviate both, but nonsystematically in hydration and lipid content from the upper and lower extremities indicate the importance of the trunk as discriminating segment and the associated abdominal/metabolic syndrome theories. As Elowsson *et al.* (1998) and Provyn *et al.* (2008) were previously evaluating the accuracy of DXA with dissection in animals, both studies motivated the choice of using plain carcasses (decapitated pigs without abdominal and thoracic organs) or just hindlegs to minimize various errors. According to Elowsson *et al.* (1998), with DXA this would marginally increase DXA's underestimation. This can no longer be supported; on the contrary, not measuring the internal trunk will just increase the error because of an assumption of segment constancy of hydration and ad hoc lipid fractionation. Wang *et al.* (1999) examined *in vitro* and *in vivo* studies allowing a review and critical appraisal of the importance of hydration of FFM and confirming the findings of Provyn *et al.* (2008). They conclude that, even though methodological limitations preclude a highly accurate analysis, adult mammals, including humans, share in common, a relatively constant hydration of FFM. The segmental data presented in Table 5 within a four-component dissection model dismisses the idea of constant hydration of FFM. In addition, the assumed ad hoc constancy of 0.73 cannot be retained. The question whether the hydration status of FFM, LBM, or ATFM reflects physiologic regulatory mechanisms (Going et al., 1993; Wang et al., 2005) cannot be answered, but it seems that trunk nonadipose tissues may affect hydration differently than the lean tissues of the extremities or vice versa (Table 5). (Clarys et al, 2010b)

5.6 What does DXA measure?

Regardless of the existing mechanisms and regardless of the hydration and lipid (fat) content of nonadipose tissue, this study has not been able to detect what the content is of the DXA nonadipose variables, e.g., "lean" and/or "lean + BMC." We still do not know what DXA is exactly measuring under these ad hoc headings. "Lean" compared with muscle tissue, with muscle plus skin tissue, and with muscle plus skin plus viscera (dissection) resulted in equally high correlations (r values between 0.95 and 0.99) assuming a good prediction estimate but with systematic significant difference confirming its imprecision, "lean plus BMC" is certainly not measuring ATFM (e.g., skin + muscle + viscera + bone), although its high $r = 0.99$, but again with a significant difference ($P < 0.001$) indicating a lack of precision and accuracy. Contrary to Bloebaum *et al.* (2006), but in agreement with Louis *et al.* (1992), BMC seems a good estimate ($r = 0.73$) with no significant difference of its ash weight. The impression is given, however, that DXA nonadipose values are expressed as anatomical-morphological values combined with chemical elements. This study cannot confirm what the nonadipose component of DXA is measuring, but it does confirm that all the DXA components are subject not only to measurement error but also to terminology error and violation of basic assumptions. It is known since many decennia that density in its weight/volume quantification (g/cm^3) can be considered as an additional and separate dimension of BC. The DXA-derived BMD, however, is a weight/surface quantification (g/cm^2) and therefore not a true density, nor the density based on which indication of osteoporosis classifications were studied in the past (Bolotin, 1998, 2007; Bolotin & Sievanen, 2001; Bolotin et al., 2001; Lochmuller et al., 2000). In a pilot (dissection) study using porcine hindlegs in which DXA BMD was compared with bone covered with muscle, AT, and skin tissue, and compared with scraped bones only (Clarys et al., 2008; Provyn et al., 2008), it was found that DXA BMD underestimates true density with more than 40%. In this study (Table

4), under whole-body conditions, one notices a similar level of high underestimation of DXA but with a better correlation, e.g., $r = 0.68$ for the whole-body value against $r = 0.39$ for the hindleg study. The extensive work done by Bolotin (2007) shows DXA-measured BMD methodology (*in vivo*) to be an intrinsically flawed and misleading indicator of bone mineral status and an erroneous gauge of relative fracture risk. The transfer of their findings to the *in situ* carcass situation of this study confirms that the DXA methodology cannot provide accurate, quantitatively precise, meaningful determinations of true bone densities and proper bone mass because of the contamination of independent ST, e.g., fluid and lipid content contributions. (Clarys et al., 2010b)

6. Conclusions

The majority of present consensual acceptance and understanding of the DXA estimate quality rests solely upon a number of well-established, multiconfirmed, *in vivo* and *in situ* significant high correlations. In terms of true "reality precision" measures, DXA produces inaccurate and misleading values at all levels of its output. Both the adipose and nonadipose components of DXA ignore the ad hoc lipid content, and the nonadipose variables do not take into account the true composing tissues. Lean and lean + BMC of DXA do not correspond to anatomical-morphological tissue combinations, nor to chemical values. It cannot be determined what DXA really measures. Bone mineral content versus ash weight is the single variable with a close reality and nonsignificant difference output. DXA is based on a series of constancies within tissues, regardless of segments, hydration, and lipid content variability. These results suggest that clinical precision essential in individual diagnosis of health status is at risk. Erroneous terminology, constancy assumptions related to hydration and lipid distribution, and two-dimensional data acquisition principles are the major causes of these problems. The hypothesis that DXA methodology provides accurate, precise, and relevant BC determinations is proven to be unwarranted and misplaced.

7. References

Bailey, B.W.; Tucker, L.A.; Peterson, T.R. & LeCheminant, J.D. (2001). Test-retest reliability of body fat percentage results using dual energy X-ray absorptiometry and the Bod Pod. *Med Sci Sports Exerc*, 33, 174.

Barthe, N.; Braillon, P.; Ducassou, D. & Basse-Cathalinat, B. (1997). Comparison of two Hologic DXA systems (QDR 1000 and QDR 4500/A). *Brit J Radiol*, 70, 728-739.

Beddoe, A.H. (1998). Body fat: estimation or guesstimation? *Appl Radiat Isot*, 49(5-6), 461-463.

Behnke, A.J.; Feen, B. & Welham, W. (1942). The specific gravity of healthy men. Body weight divided by volume as an index of obesity. *J Am Med Assoc*, 118, 495-498.

Behnke, A.R. (1963). Anthropometric evaluation of body composition throughout life. *Ann N Y Acad Sci*, 110, 450-464.

Black, A.; Tilmont, E.M.; Baer, D.J.; Rumpler, W.V.; Ingram, D.K.; Roth, G.S. & Lane, M.A. (2001). Accuracy and precision of dual-energy X-ray absorptiometry for body composition measurements in rhesus monkeys. *J Med Primatol*, 30(2), 94-99.

Bland, J.M. & Altman, D.G. (1986). Statistical methods for assessing agreement between two methods of clinical measurement. *Lancet*, 1, 307–310.

Bloebaum, R.D.; Liau, D.W.; Lester, D.K. & Rosenbaum, T.G. (2006). Dual-energy x-ray absorptiometry measurement and accuracy of bone mineral after unilateral total hip arthroplasty. *J Arthroplasty*, 21(4), 612-622.

Bolotin, H.H. & Sievanen, H. (2001). Inaccuracies inherent in dual-energy X-ray absorptiometry in vivo bone mineral density can seriously mislead diagnostic/prognostic interpretations of patient-specific bone fragility. *J Bone Miner Res*, 16(5), 799-805.

Bolotin, H.H. (1998). A new perspective on the causal influence of soft tissue composition on DXA-measured in vivo bone mineral density. *J Bone Miner Res*, 13(11), 1739-1746.

Bolotin, H.H. (2007). DXA in vivo BMD methodology: an erroneous and misleading research and clinical gauge of bone mineral status, bone fragility, and bone remodelling. *Bone*, 41(1), 138-154.

Bolotin, H.H.; Sievanen, H.; Grashuis, J.L.; Kuiper, J.W. & Jarvinen, T.L. (2001). Inaccuracies inherent in patient-specific dual-energy X-ray absorptiometry bone mineral density measurements: comprehensive phantom-based evaluation. *J Bone Miner Res*, 16(2), 417-426.

Brommage, R. (2003). Validation and calibration of DEXA body composition in mice. *Am J Physiol Endocrinol Metab*, 285(3), E454-459.

Brozek, J.; Grande, F.; Anderson, J.T. & Keys, A. (1963). Densitometric analysis of body composition: Revision of some quantitative assumptions. *Ann N Y Acad Sci*, 110, 113-140.

Burkhart, T.A.; Arthurs, K.L. & Andrews (2009). Manual segmentation of DXA scan images results in reliable upper and lower extremity soft and rigid tissue mass estimates. *J Biomech*, 42, 1138-1142.

Chauhan, S.; Koo, W.W.; Hammami, M. & Hockman, E.M. (2003). Fan beam dual energy X-ray absorptiometry body composition measurements in piglets. *J Am Coll Nutr*, 22(5), 408-414.

Clarys, J.P. & Martin, A. (1985). The concept of the adipose tissue-free mass. In Norgan, N. (Ed.), *Human body composition and fat distribution.*, pp. 49-61, Wageningen: Wageningen Agricultural University.

Clarys, J.P. & Marfell-Jones, M.J. (1986). Anthropometric prediction of component tissue masses in the minor limb segments of the human body. *Hum Biol*, 58, 761-769.

Clarys, J.P.; Martin, A.D.; Drinkwater, D.T. & Marfell-Jones, M.J. (1987). The skinfold: myth and reality. *J Sports Sci*, 5(1), 3-33.

Clarys, J.P.; Martin, A.D.; Marfell-Jones, M.J.; Janssens, V.; Caboor, D. & Drinkwater, D.T. (1999). Human body composition: A review of adult dissection data. *Am J Hum Biol*, 11(2), 167-174.

Clarys, J.P.; Martin, A.D.; Marfell-Jones, M.J.; Janssens, V.; Caboor, D. & Drinkwater, D.T. (1999). Human body composition: A review of adult dissection data. *Am J Hum Biol*, 11(2), 167-174.

Clarys, J.P.; Provyn, S. & Marfell-Jones, M.J. (2005). Cadaver studies and their impact on the understanding of human adiposity. *Ergonomics*, 48(11-14), 1445-1461.

Clarys, J.P.; Provyn, S.; Wallace, J.; Scafoglieri, A. & Reilly, T. (2008). *Quality control of fan beam scanning data processing with in vitro material.*, In: Transactions of 2008 IEEE International Conference on Industrial Engineering and Engineering Management (IEEM), Singapore.

Clarys, J.P.; Scafoglieri, A.; Provyn, S. & Bautmans, I. (2010a). Body Mass Index as a measure of bone mass. *J Sports Med Phys Fitness*, 50:202-206.

Clarys, J.P.; Scafoglieri, A.; Provyn, S.; Louis, O.; Wallace J.A. & De Mey, J. (2010b). A macro-quality evaluation of DXA variables using whole dissection, ashing, and computer tomography in pigs. *Obesity*, 18, 1477-1485.

Clarys, J.P.; Scafoglieri, A.; Provyn, S.; Sesboüé, B. & Van Roy, P. (2010c). The hazards of hydrodensitometry. *J Sports Med Phys Fitness*, 0(0), 000-000.

Clasey, J.L.; Kanaley, J.A.; Wideman, L.; Heymsfield, S.B.; Teates, C.D.; Gutgesell, M.E.; Thorner, M.O.; Hartman, M.L. & Weltman, A. (1999). Validity of methods of body composition assessment in young and older men and women. *J Appl Physiol*, 86(5), 1728-1738.

Cooper, A.R.; Forbes, R.M. & Mitchell, H.H. (1956). Further studies on the gross composition and mineral elements of the adult human body. *J Biol Chem*, 223(2), 969-975.

Durnin, J.V. & Rahaman, M.M. (1967). The assessment of the amount of fat in the human body from measurements of skinfold thickness. *Br J Nutr*, 21(3), 681-689.

Elowsson, P.; Forslund, A.H.; Mallmin, H.; Feuk, U.; Hansson, I. & Carlsten, J. (1998). An evaluation of dual-energy X-Ray absorptiometry and underwater weighing to estimate body composition by means of carcass analysis in piglets. *J Nutr*, 128(9), 1543-1549.

Eston, R.G.; Rowlands, A.V.; Charlesworth, S.; Davies, A. & Hoppitt, T. (2005). Prediction of DXA-determined whole body fat from skinfolds: importance of including skinfolds from the thigh and calf in young, healthy men and women. *Eur J Clin Nutr*, 59(5), 695-702.

Forbes, G.B. & Lewis, A.M. (1956). Total sodium, potassium and chloride in adult man. *J Clin Invest*, 35(6), 596-600.

Forbes, R.M.; Cooper, A.R. & Mitchell, H.H. (1953). The composition of the adult human body as determined by chemical analysis. *J Biol Chem*, 203(1), 359-366.

Forbes, R.M.; Mitchell, H.H. & Cooper, A.R. (1956). Further studies on the gross composition and mineral elements of the adult human body. *J Biol Chem*, 223(2), 969-975.

Glickman, S.G.; Marn, C.S.; Supiano, M.A. & Dengel, D.R. (2004). Validity and reliability of dual-energy X-ray absorptiometry for the assessment of abdominal adiposity. *J Appl Physiol*, 97, 509-514.

Godang, K.; Qvigstad, E.; Voldner N.; Isaksen, G.A.; Frøslie, K.F.; Nøtthellen, J.; Henriksen, T. & Bollerslev, J. (2010). Assessing body composition in healthy newborn infants: reliability of dual-energy x-ray absorptiometry. *J Clin Densitom*, 13, 151-160.

Going, S.B.; Massett, M.P.; Hall, M.C.; Bare, L.A.; Root, P.A.; Williams, D.P. & Lohman, T.G. (1993). Detection of small changes in body composition by dual-energy x-ray absorptiometry. *Am J Clin Nutr*, 57(6), 845-850.

Guo, Y.; Franks, P.W.; Bookshire, T. & Tataranni P.A. (2004). The intra-and inter-instrument reliability of DXA based on ex vivo soft tissue measurements. *Obes Res*, 12, 1925-1929.

Haarbo, J.; Gotfredsen, A.; Hassager, C. & Christiansen, C. (1991). Validation of body composition by dual energy X-ray absorptiometry (DEXA). *Clin Physiol*, 11(4), 331-341.

Heyward, V.H. (1996). Evaluation of body composition. Current issues. *Sports Med*, 22(3), 146-156.

Johansson, A.G.; Forslund, A.; Sjodin, A.; Mallmin, H.; Hambraeus, L. & Ljunghall, S. (1993). Determination of body composition--a comparison of dual-energy x-ray absorptiometry and hydrodensitometry. *Am J Clin Nutr*, 57(3), 323-326.

Keys, A. & Brozek, J. (1953). Body fat in adult man. *Physiol Rev*, 33(3), 245-325.

Koo, W.W.; Hammami, M. & Hockman, E.M. (2002). Use of fan beam dual energy x-ray absorptiometry to measure body composition of piglets. *J Nutr*, 132(6), 1380-1383.

Koo, W.W.; Hammami, M. & Hockman, E.M. (2004). Validation of bone mass and body composition measurements in small subjects with pencil beam dual energy X-ray absorptiometry. *J Am Coll Nutr,* 23(1), 79-84.

Litaker, M.S.; Barbeau P.; Humphries, M.C. & Gutin, B. (2003). Comparison of Hologic QDR-1000/W and 4500W DXA scanners in 13-to18-year olds. *Obes Res,* 11, 1545-1552.

Lochmuller, E.M.; Miller, P.; Burklein, D.; Wehr, U.; Rambeck, W. & Eckstein, F. (2000). In situ femoral dual-energy X-ray absorptiometry related to ash weight, bone size and density, and its relationship with mechanical failure loads of the proximal femur. *Osteoporos Int,* 11(4), 361-367.

Louis, O.; Van den Winkel, P.; Covens, P.; Schoutens, A. & Osteaux, M. (1992). Dual-energy X-ray absorptiometry of lumbar vertebrae: relative contribution of body and posterior elements and accuracy in relation with neutron activation analysis. *Bone,* 13(4), 317-320.

Lukaski, H.C.; Marchello, M.J.; Hall, C.B.; Schafer, D.M. & Siders, W.A. (1999). Soft tissue composition of pigs measured with dual x-ray absorptiometry: comparison with chemical analyses and effects of carcass thicknesses. *Nutrition,* 15(9), 697-703.

Martin, A.D. & Drinkwater, D.T. (1991). Variability in the measures of body fat. Assumptions or technique? *Sports Med,* 11(5), 277-288.

Martin, A.D.; Drinkwater, D.T.; Clarys, J.P.; Daniel, M. & Ross, W.D. (1992). Effects of skin thickness and skinfold compressibility on skinfold thickness measurement. *Am J Hum Biol,* 4, 453-460.

Martin, A.D.; Janssens, V.; Caboor, D.; Clarys, J.P. & Marfell-Jones, M.J. (2003b). Relationships between visceral, trunk and whole-body adipose tissue weights by cadaver dissection. *Ann Hum Biol,* 30(6), 668-677.

Martin, A.D.; Ross, W.D.; Drinkwater, D.T. & Clarys, J.P. (1985). Prediction of body fat by skinfold caliper: assumptions and cadaver evidence. *Int J Obes,* 9 Suppl 1, 31-39.

Mitchell, A.D.; Rosebrough, R.W. & Conway, J.M. (1997). Body composition analysis of chickens by dual energy x-ray absorptiometry. *Poult Sci,* 76(12), 1746-1752.

Mitchell, A.D.; Scholz, A.M.; Pursel, V.G. & Evock-Clover, C.M. (1998). Composition analysis of pork carcasses by dual-energy x-ray absorptiometry. *J Anim Sci,* 76(8), 2104-2114.

Mitchell, H.H.; Hamilton, T.S.; Steggerda, F.R. & Bean, H.W. (1945). The chemical composition of the adult body and its bearing on the biochemistry of growth *Journal of Biology and Chemistry,* 158, 625-637.

Müller, M.J.; Bosy-Westphal, A.; Kutzner, D. & Heller, M. (2003). Metabolically active components of fat free mass (FFM) and resting energy expenditure (REE) in humans. *Forum Nutr,* 56, 301-303.

Nagy, T.R. & Clair, A.L. (2000). Precision and accuracy of dual-energy X-ray absorptiometry for determining in vivo body composition of mice. *Obes Res,* 8(5), 392-398.

Picaud, J.C.; Rigo, J.; Nyamugabo, K.; Milet, J. & Senterre, J. (1996). Evaluation of dual-energy X-ray absorptiometry for body-composition assessment in piglets and term human neonates. *Am J Clin Nutr,* 63(2), 157-163.

Pintauro, S.J.; Nagy, T.R.; Duthie, C.M. & Goran, M.I. (1996). Cross-calibration of fat and lean measurements by dual-energy X-ray absorptiometry to pig carcass analysis in the pediatric body weight range. *Am J Clin Nutr,* 63(3), 293-298.

Poortmans, J.R.; Boisseau, N.; Moraine, J.J.; Moreno-Reyes, R. & Goldman, S. (2005). Estimation of total-body skeletal muscle mass in children and adolescents. *Med Sci Sports Exerc,* 37(2), 316-322.

Prentice, A. (1995). Application of dual-energy X-ray absorptiometry and related techniques to the assessment of bone and body composition. In Davis, P.S.W. & Cole, T.J. (Eds.), *Body composition techniques in health and disease.*, pp. 1-13, New York: Cambridge University Press.

Prior, B.M.; Cureton, K.J.; Modlesky, C.M.; Evans, E.M.; Sloniger, M.A.; Saunders, M. & Lewis, R.D. (1997). In vivo validation of whole body composition estimates from dual-energy X-ray absorptiometry. *J Appl Physiol,* 83(2), 623-630.

Pritchard, J.E.; Nowson, C.A.; Strauss, B.J.; Carlson, J.S.; Kaymakci, B. & Wark, J.D. (1993). Evaluation of dual energy X-ray absorptiometry as a method of measurement of body fat. *Eur J Clin Nutr,* 47(3), 216-228.

Provyn, S.; Clarys, J.P.; Wallace, J.; Scafoglieri, A. & Reilly, T. (2008). Quality control, accuracy, and prediction capacity of dual energy X-ray absorptiometry variables and data acquisition. *J Physiol Anthropol,* 27(6), 317-323.

Salamone, L.M.; Fuerst, T.; Visser, M.; Kern, M.; Lang, T.; Dockrell, M.; Cauley, J.A.; Nevitt, M.; Tylavsky, F. & Lohman, T.G. (2000). Measurement of fat mass using DEXA: a validation study in elderly adults. *J Appl Physiol,* 89(1), 345-352.

Scafoglieri, A.; Provyn, S.; Bautmans, I.; Van Roy, P. & Clarys, J.P. (2010a). Direct relationship of body mass index and waist circumference with tissue distribution in elderly persons. *JNHA,* 14:000-000.

Scafoglieri, A.; Provyn, S.; Bautmans, I.; Wallace, J.; Sutton, L.; Tresignie, J.; Louis, O.; De Mey, J. & Clarys, J.P. (2010b). Critical appraisal of data acquisition in body composition: evaluation of methods, techniques and technologies on the anatomical tissue-level system. In: *Data Acquisition,* pp. 000-000.

Scafoglieri, A.; Provyn, S.; Louis, O.; Wallace, J.A.; De Mey, J. & Clarys, J.P. (2010c). A macro-quality field control of dual energy X-ray absorptiometry with anatomical, chemical and computed tomographical variables. *IFMBE Proceedings,* 1, Volume 29, Part 3, pp. 549-553.

Siri, W.E. (1956). The gross composition of the body. *Adv Biol Med Phys,* 4, 239-280.

Speakman, J.R.; Booles, D. & Butterwick, R. (2001). Validation of dual energy X-ray absorptiometry (DXA) by comparison with chemical analysis of dogs and cats. *Int J Obes Relat Metab Disord,* 25(3), 439-447.

Swennen, Q.; Janssens, G.P.; Geers, R.; Decuypere, E. & Buyse, J. (2004). Validation of dual-energy x-ray absorptiometry for determining in vivo body composition of chickens. *Poult Sci,* 83(8), 1348-1357.

Wadden, T.A. & Didie, E. (2003). What's in a name? Patients' preferred terms for describing obesity. *Obes Res,* 11(9), 1140-1146.

Wang, M.C.; Bachrach, L.K.; Van Loan, M.; Hudes, M.; Flegal, K.M. & Crawford, P.B. (2005). The relative contributions of lean tissue mass and fat mass to bone density in young women. *Bone,* 37(4), 474-481.

Wang, Z.; Deurenberg, P.; Wang, W.; Pietrobelli, A.; Baumgartner, R.N. & Heymsfield, S.B. (1999). Hydration of fat-free body mass: review and critique of a classic body-composition constant. *Am J Clin Nutr,* 69(5), 833-841.

Wang, Z.M.; Heshka, S.; Pierson, R.N., Jr. & Heymsfield, S.B. (1995). Systematic organization of body-composition methodology: an overview with emphasis on component-based methods. *Am J Clin Nutr,* 61(3), 457-465.

Wang, Z.M.; Pierson, R.N., Jr. & Heymsfield, S.B. (1992). The five-level model: a new approach to organizing body-composition research. *Am J Clin Nutr,* 56(1), 19-28.

5

Quality Control in Endoscopy Unit: Safety Considerations for the Patient

Jean-Francois Rey
Institut Arnault Tzanck
Hepato-Gastroenterology Department
06721 St Laurent du Var Cedex
France

1. Introduction

Safe endoscopic procedures have been a daily concern in every endoscopy unit since the early eighties. At first range, prevention of crossed contamination was the primary goal for endoscopy units' director. But, step after step, all parts of endoscopy procedures have to be monitored and the use of protocols is now mandatory. In this chapter, we will review: patient information and protocol of all endoscopy related procedures, check list in endoscopic theatre, control in endoscope maintenance and the importance of audit in endoscopy unit.

2. Patient information

In the last decade, patient information has been a very hot topic in case of complications issues. In some countries, it is the most important point in case of malpractice. The relationship between practitioners and patients has changed and the paternalist time is behind us. Patients are well informed, perhaps too much in some cases, through multiple media and mainly Internet.

Patient information is linked very closely with oral and written information. As proved by (Felley et al., 2008) in a randomized trial between combined written and oral information prior to a gastrointestinal endoscopy versus oral information alone. Written and oral is better than oral only. Written information should develop: preparation of the procedure, expectations and risks. It is important to rely on this issue on national endoscopy and gastroenterology scientific societies. The leaflets are written with accuracy and caution but is will be a mistake to handle it to the patient without oral comments, sometimes asking him to sign by a staff nurse. Oral information has to be adapted to the patient level of knowledge, language and culture; it is sometimes difficult. A letter to the patients' general practitioner is a wise mean to complete patient information and it is a proof in the practitioner file.

The content of information should deal with the following points:

- Pre-procedural considerations:
    - Comorbidity, body habitus (mobility, airway)
    - Anticoagulation and antiplatelet therapy
- Specific considerations for ERCP

- Radiation with events relying to pregnancy
- Complications
- For colonoscopy, considerations have to be underlined on:
  - Bowel preparation
  - Cardiac and renal functions with issue related to the preparation in case of colonic lavage
- Post procedure informations are mandatory:
  - Drowsiness: driving, operating machine, complete stases,
  - Side effects and complications: chest pain, bleeding, vomiting, breathless, abdominal pain or fever. In case of any discomfort or complications, how the patient could contact the endoscopy unit or emergency department and bring the suitable endoscopy report.
  - Follow-up appointment
  - Recommencing if necessary, anticoagulation and antiplatelet therapy.

For more risky procedures as ERCP and therapeutic biliary pancreatic endoscopy, ASGE makes a detailed recommendation (Table 1). If variances occur outside of these recommendations, careful medical justification should be recorded.

A. Jaundice thought to be the result of biliary obstruction
B. Clinical and biochemical or imaging data suggestive of pancreatic or biliary tract disease
C. Signs or symptoms suggesting pancreatic malignancy when direct imaging results are equivocal or normal
D. Pancreatitis of unknown etiology
E. Preoperative evaluation of chronic pancreatitis or pancreatic pseudocyst
F. Sphincter of Oddi manometry
G. Endoscopic sphincterotomy
1. Choledocholithiasis
2. Papillary stenosis or sphincter of Oddi dysfunction causing disability
3. Facilitate biliary stent placement or balloon dilatation
4. Sump syndrome
5. Choledochocele
6. Ampullary carcinoma in poor surgical candidates
7. Access to pancreatic duct
O. Stent placement across benign or malignant strictures, fistulae, postoperative bile leak, or large common bile duct stones
P. Balloon dilatation of ductal strictures
Q. Nasobiliary drain placement
R. Pseudocyst drainage in appropriate cases
S. Tissue sampling from pancreatic or bile ducts
T. Pancreatic therapeutics
VARIANCE REQUIRES JUSTIFICATION

Table 1. ASGE recommendation (Indications for ERCP)

Finally all informations could be related in a consent form which deals with multiple issues. (Bottrell et al., 2000) underlining the main questions:

- What are the local legal requirements? This topic is highly variable to one country to another.
- Ethical requirements
- Physicians-Patients relationships

All elements related to the procedure has to be recorded (Elfant et al.,1995):

- Nature of the procedure and its handling.
- Procedure benefits: treat bleeding, identify source of bleeding, pain, discomfort, diagnosis, remove polyps, inserting stents.
- Procedure related risks: bleeding, infection, perforation, cardiopulmonary, allergy.
- Alternatives with disadvantages versus advantages. Do we have an alternative treatment?

Ideally, this information and consent form request have to be carried on days before procedure. The patient should have time to read, understand and make its mind. (Song et al., 2010) has recently published an inquiry on the benefit of the informed consent (Table 2).

Did you think that the informed consent was a useful and necessary procedure?	74 (65.49)
What did you think was the main objective of the informed consent procedure?	
. Improving a patients' understanding	62 (54.87)
. Evidence of understanding and mutual agreement	54 (47.79)
. Protection of the legal rights of the doctors	26 (23.01)
. Senseless formality	4 (3.54)

Table 2. Informed Consent (Song et al., 2010)

3. Protocols

Written protocols have to be implemented in each endoscopy unit. It is the last way to keep high standard in a daily routine whatever the procedure is carried out by a senior or a junior endoscopist (Naylor et al., 2003; Faigel et al., 2006). In most of the cases, written protocols are issued from national or international guidelines but adapted to the local situation. Protocol means program for the systematic monitoring and evaluation of the various aspects of a project, service or facility to ensure that standards of quality are being met. The program for quality assurance has been described by (O'Mahony et al., 2000) with the following topics:

1. Improve the overall quality of care
2. Limit inappropriate procedures
3. Limit morbidity and mortality
4. Improve training in endoscopy
5. Limit patient complaints and litigation
6. Contain costs
7. All of the above

In building your protocol, you should identify quality, indicators, members of your team and how to build the process. This lead to raise multiple questions (O'Mahony et al., 2000):

1. How to set it up so that it works in everyday practice?
2. What information do we need to collect and how should we analyse it?
3. How much will it cost, who will pay for it?
4. How do we set quality standards?
5. How do we deal with underperforming doctors?

Some are obvious as how to use properly and set adequately an electrosurgical unit in relation with the device (Rey et al., 2010). For therapeutic endoscopy a closed monitoring of patient coagulation is also mandatory as old patients are often treated with anticoagulation drugs or even more dangerous some very effective antiplatelet drug such as Clopidogrel. Condition risk for thromboembolic has been described by a national society (ASGE, 2009), (Table 3) as the embolism risk after a brief interruption of anticoagulation (Veitch et al., 2008), (Table 4). But we have to accept various recommendations for the same topic if guidelines have been designed by cardiology or gastroenterology scientific societies. Antibio-prophylaxy protocol has to be designed in requirements of national recommendations and adapted with a better knowledge of patient condition. Level of competency for endoscopists practitioners is also variable from one country to another as reminded by (Faigel et al., 2009), (Table 5).

Higher-risk condition Low-risk condition	**Low-risk condition**
- Atrial fibrillation associated with valvular heart disease, prosthetic valves, active congestive heart failure, left ventricular ejection fraction <35%, a history of a thromboembolic event, hypertension, diabetes mellitus, or age >75 y	- Uncomplicated or paroxysmal nonvalvular atrial fibrillation
- Mechanical valve in the mitral position	- Bioprosthetic valve
- Mechanical valve in any position and previous thromboembolic event	- Mechanical valve in the aortic position
- Recently (<1 y) placed coronary stent	
- Acute coronary syndrome	- Deep vein thrombosis
- Non stented percutaneous coronary intervention after myocardial infarction	

Table 3. Condition risk for thromboembolic event (ASGE Standards of Practice Committee, 2009)

High risk	Low risk
- VTE, embolic stroke ≤12 wk ago	- DVT/PE ≥12 wk ago
- AF with valvular HD	- NV AF, no CCF/embolism
- Mechanical mitral/multiple valves, previous TEE	- Bio-prosthetic valve
- Thrombophilia	- Mechanical aortic valve
- IHD with bare stent ≤12 wk ago . DE stent 6-12 mo ago	- IHD, no coronary stents
	- Cerebrovascular disease (CVD)
	- PVD
- Embolic risk if anticoagulation interrupted in low risk condition 1-2/1000 - Avoid elective procedure for 12 wks after acute DVT/PE	

Table 4. Embolism risk after brief interruption of anticoagulation (Veitch et al., 2008)

	USA *	Australia =	Canada	Poland	India	Europe ‡
Procedure						
Sigmoidoscopy	30		30			50
Colonoscopy	140	100 to cecum	150	500	120	150
EGD	130	200	150	500	190	200
ERCP	200	200	200	200	140	150
EUS	150	200				150

EGD, esophagogastroduodenoscopy; ERCP, endoscopic retrograde cholangiopancreatography; EUS, endoscopic ultrasonography

* American Society for Gastrointestinal Endoscopy (ASGE) Guideline

= Colonoscopy: cecal intubation in >90% of the last 50 logged procedures. ERCP: unassisted, with intact papilla, to include 80 sphincterotomies and 60 stent placements.

‡ European Board of Gastroenterology: colonoscopy numbers include polypectomy, and assume competency in EGD first (10).

Table 5. Threshold numbers of endoscopic procedures before competency can be assessed by direct observation or other objective measures, as required in different countries or regions. (Faigel et al., 2009).

4. Time out

Endoscopy procedures are very close to surgical procedures, it is why endoscopy suites have to be carried on as a surgical theatre taking in account endoscopy specificities (Table 6). Recently a new process has been issued as time out or also called "checklist" (Haynes et al., 2009). This process is issue from airplane pilot cockpit where, before each flight commandant and first officer revised orally and click on a sheet, all airplane equipment. Of course, it is much more easily with machine elements than with a human body. The main goal is to avoid errors in patient labelling, type of the procedures, where procedures should be applied, patient treatment and more specifically, anticoagulation and antiplatelet therapy. In endoscopy, the clinical benefit of the time out or checklist has to be proved but it becomes more important in a very busy unit.

Sign in
Before induction of anesthesia, members of the team (at least the nurse and an anesthesia professional) orally confirm that: The patient has verified his or her identity, the surgical site and procedure, and consent. The surgical site is marked or site marking is not applicable The pulse oximeter is on the patient and functioning All members of the team are aware of wheter the patient has a known allergy The patient's airway and risk of aspiration have been evaluated and appropriate equipment and assistance are available If there is a risk of blood loss of at least 500ml (or 7ml/kg of body weight, in children), appropriate access and fluids are available
Time out
Before skin incision, the entire team (nurses, surgeons, anesthesia professionals, and any others participating in the care of the patient) orally: Confirms that all team members have been introduced by name and role Confirms the patient's identity, surgical site, and procedure Reviews the anticipated critical events Surgeon reviews critical and unexpected steps, operative duration, and anticipated blood loss Anesthesia staff reviews concerns specific to the patient Nursing staff reviews confirmation of sterility, equipment availability, and other concerns Confirms that prophylactic antibiotics have been administrated ≤60 min before incision is made or that antibiotics are not indicated Confirms that all essential imaging results for the correct patient are displayed in the operating room
Sign out
Before the patient leaves the operating room: Nurse reviews items aloud with the team Name of the procedure as recorded That the needle, sponge, and instrument counts are complete (or not applicable) That the specimen (if any), is correctly labelled, including with the patient's name Wether there are any issues with equipment to be addressed The surgeon, nurse and anesthesia professional review aloud the key concerns for the recovery and care of the patient

* the checklist is based on the first edition of the WHO Guidelines for Safe Surgery (Geneva, World Health Organization, 2008).

Table 6. Elements of the Surgical Safety Checklist* (Haynes et al., 2009)

5. Control in endoscope maintenance

Quality control in servicing and repairing endoscopes is an important business for each endoscopy unit. It is why we have been required to write a guideline on behalf of ESGE (Rey et al., 2004). This work has been carried on with the help of the personal from three major endoscopes makers, Fujinon, Olympus and Pentax. But it is clear that repairing services could be provided both by endoscopes manufacturers and so called third party as independent repaired company.

In issuing guidelines on repairing endoscopes, the aim of the European Society of Gastrointestinal Endoscopy (ESGE) Guidelines Committee is to draw the attention of those who manage endoscopy units to the quality control requirements involved in having endoscopes repaired. Information on endoscope repair is important not only for medical doctors, nurses and assistants, but also for medical engineers and - even more so - for hospital administrators. The aim of these ESGE guidelines is to draw the attention of endoscope users to both the technical and the legal implications relating to endoscope repair. The sources of spare parts and the exact repair procedures used should be clearly stated in the repair contract. If second-hand spare parts are used, traceability and hygiene issues should be taken into consideration. For the ESGE, it is clear that repair services can be provided both by endoscope manufacturers and by other repair service providers (known as "third parties"), as long as they follow the same quality control processes and guarantees as those followed in endoscopy units.

The intention in these guidelines is to provide users of medical endoscopes with information about the appropriate selection of service providers, including information on how to reduce user liability relative to compliance with the Medical Device Directive (MDD) during the lifespan of endoscopic equipment and devices.

The Medical Device Directive (MDD) regulates the requirements for the design, manufacture and sale of new medical devices and equipment very precisely. A clear sign for the user that the equipment or device complies with MDD regulations is the CE mark. A reference to the original manufacturer is also provided by the identification label on each product. The basic aim of these measures is to ensure the safety of users, patients, and others, as well as to ensure compliance with technological standards. Before the first use of any new endoscope, the MDD requires proof that these requirements have been met, based on both clinical and nonclinical testing procedures. The original manufacturer maintains a master configuration file, which documents the design configuration, manufacturing configuration, and test results. This ensures seamless configuration control.

After the product has been sold, the liability and responsibility for ensuring safe usage of the product is transferred to the user. The degree of user liability is likely to depend on the contract between the user's organization and the selected service provider. The transfer of liability, and all associated risk, depends on the person or organization undertaking the service and on the way in which maintenance and repair work is carried out. If a failure in use occurs after maintenance or repair and this failure leads to serious injury to a patient or user, there is a greater likelihood that the user and service provider may be held legally responsible for any injuries caused, if the endoscope was not repaired in accordance with the manufacturer's instructions and operating standards.

Users need to ensure that they follow an appropriate strategy for maintaining and repairing the equipment. Options for maintenance and repair services include choosing either:

- The original manufacturer of the endoscope

- An in-house biomedical engineering department
- An independent service provider
- A managed service provider.

Usually, the original manufacturer is the primary source for maintenance and repair services. However, not all manufacturers carry out repair and maintenance of their devices. Whoever is carrying out the maintenance and repair activity, the written service documentation provided by the manufacturer should always be used. If any organization other than the original manufacturer undertakes repair or maintenance work, the organization concerned should ensure that the work and inspection is carried out in accordance with the latest manufacturer's instructions and specifications, and according to defined standards.

The most important issues to be considered as selection criteria for any service provider are in the four core areas:

- General competence
- Workmanship
- Parts and materials
- Verification, inspection, and risk management.

However, in addition, the following criteria are important elements that should also be taken into account when choosing a service provider:

- A clear analysis and understanding of the relationship between apparent financial savings and increased liability risks, and calculation of total ownership costs relative to repair frequency
- The frequency with which maintenance and repairs are likely to be needed and the time required to respond to endoscope malfunction:
- Uptime requirements
- Requirements for loan equipment
- Requirements for a mobile service.

It is advisable for the owner of the endoscopic equipment to always make a clear contractual agreement with the chosen service provider. This agreement should clearly describe the level of service to be provided to the owner organization, and it is recommended that it should include:

- Reference to the original manufacturer's written instructions
- Information on availability, source, and traceability of spare parts
- Notification of any changes in parts, procedures, tests performed, etc., including the use of parts and methods other than those specified by and originating from the original manufacturer
- Details of the training and qualifications of the technicians and inspectors
- Details on the way in which requirements for adequate record-keeping regarding traceability are ensured
- Regulations on the use of subcontractors employed by the ser- vice provider.

A repair organization that complies with both the comprehensive legal and quality aspects as well as with specific customer requirements will be able to respond positively to the following questions.

a) Competence

- What is the experience and customer reputation that the ser- vice provider has in repairing and maintaining the specific type of endoscopic or electronic equipment? (List of customers)

- Is a quality assurance system in place? (ISO 9001 - 2000 as a minimum; preferably ISO EN13488 to show compliance with MDD)
- How are the requirements for adequate record keeping and traceability ensured?
- Is the availability of loan equipment adequate to meet customer demand, and is the loan equipment chargeable or not?
- What is the response time and what are the regular service hours?
- Is a mobile service offered, and where is it located?
- Is the service organization sufficiently flexible to adapt to customers' uptime requirements?
- Is a risk management strategy in place?
- Does the service provider have reliable access to the original manufacturer's information (manuals, safety information, and device specifications) and spare parts?

b) Workmanship

- Are the original manufacturer's repair procedures in place and regularly updated, (preferably by the original manufacturer) and what is the source for the relevant documentation?
- How does the update procedure work, and how is it documented?
- Is a written agreement with the original manufacturer available?
- Is the level of documentation appropriate to the level of repair?
- Are the technicians regularly trained on the basis of the original manufacturer's manufacturing and service standards (preferably by the original manufacturer's approved trainers)?
- What certificates and records are available for repair and inspection training with regard to the specific equipment type and the extent of repair training that has been performed?
- What are the training intervals for technicians?
- What regular reviewing of the training level is done, and is it done at appropriate intervals?
- Are the tools that are being used in accordance with the original manufacturer's jigs and special tools specifications, or are they originally provided by the manufacturer?
- Are specific repair and inspection procedures in place, regularly updated, and followed (preferably by the origin manufacturer)?
- If repair and inspection procedures other than the original manufacturer's are being used:
  - Are the methods and procedures documented and regularly updated and appropriately detailed on the basis of the required level of maintenance and repair?
- Is a system in place to inform the user organization of any alternative methods used?
- What certificates and records are available for inspection?
- Have the risks been identified and documented?
- Is a risk management strategy for this documentation in place?

c) Parts and Materials

- Are the criteria for the definition of part defects documented and verified as being in accordance with the original manufacturer's criteria and with MDD requirements?
- Are all the parts and materials used for repairs the certified original manufacturer's parts, and are they purchased from the original manufacturer or any authorized agency?

- If no original parts and materials (e.g., adhesives) are used, is the service provider willing to certify the origin of the parts and materials used for repairs?
- Is a risk management strategy for these parts and materials in place and the appropriate documentation available?
- Have the risks been identified and documented?
- Are certificates of biocompatibility available?
- Do the parts and materials comply with the original manufacturer's specifications of compatibility with reprocessing agents?
- Is the sourcing of these parts covered by a contract (to ensure continuity of supply?

d) Verification, Inspection and Risk Management

- Have criteria for functionality and safety testing been defined, and are they available and verified as being in accordance with MDD requirements and the original manufacturing standards?
- Have the methods and equipment used for quality assurance inspection been defined and are they appropriate for verifying the quality of the repaired endoscope according to the origin manufacturer's standards? Is a quality assurance tracking report available?
- Will the repair contractor provide a statement of conformity with the original manufacturer's functional, safety, and quality standards after an endoscope has been repaired?
- Is the repair process documented and traceable?
- Process steps
- Technicians involved
- Parts used
- How long will the necessary documentation be kept available?
- What notification and approval procedures are in place to ensure that the user organization is aware of any changes in spare parts and methods other than those covered by the contract?
- With regard to the whole repair process, is a risk management system in place and who is the person responsible for it?
- Is a procedure in place for reporting conditions to the user organization that have a potential to cause a device failure or otherwise compromise the intended use of the endoscope?

e) Technical consequences of any repairs services

To carry out any repairs, it is necessary to disassemble an endoscope either partly or wholly. During this procedure, the functionality and condition of subcomponents or parts inside the endoscope can be inspected in addition to verifying the originally detected problem. Reassembling the parts, including internal adjustments of subcomponents, also requires specialized understanding and expertise in the manufacturer's procedures. Performing the repairs in accordance with the manufacturer's standards ensures the continuing functionality and safety of the endoscope.

If any changes are made during a repair procedure with regard to the materials, spare parts used, or characteristics of the medical endoscope, this may invalidate the related CE marking. In case of doubt, a new CE marking procedure carried out by a registered medical device manufacturer for the endoscope concerned will be required.

Responsibility for the safety of the device lies primarily with the owner or his or her delegated person, regardless of who carries out any repairs.
Effectively, this means that in the event of endoscopic equipment being subjected to any work, processing, etc., that is not in accordance with the original manufacturer's specifications and/or quality standards, the equipment may have performance standards that deviate from (i.e., are inferior to) those for which the original CE approval by a notified body and its registration was issued.
This would therefore completely invalidate any continuing liability on the part of the original manufacturer even for equipment that might still normally be within the manufacturer's warranty period. Any liability for claims arising from subsequent malfunctioning or harm to patients and/or users would therefore pass to the owners of the equipment.
In summary, endoscopy maintenance is an important point in order to keep high standard endoscopy unit procedure. The organisation of repairing service may vary from one country to another depending of the local organisation in national legal requirement. For accessories, it is even more important with legal and technical concerns. With single use equipment, the safety is linked to one time use accessory, for reusable device, we should follow standard for cleaning and disinfection (Beilenhoff et al., 2008). Of course in many countries, single use accessories are reused, it becomes a legal issue if forbidden by law or a civil concern in case of inadequate reprocessing.

5. Endoscopy unit audit

This point has been particularly developed for monitoring of cleaning and disinfection procedures, but also, more generally in the British National Health Service (NHS) for all endoscopy units accredited for colorectal cancer screening (the JAG).
Cleaning and disinfection of endoscopes is a complex and multi-tasks process. Mistake in handling could occur at any steps. Manual pre-cleaning has to be developed on immediately after the end of the endoscopy procedure. Cleaning and disinfection could be achieved with manual cleaning or water disinfection machine. But in all cases, defects may occurred and lead to an infective process. For example, water supply or washer-disinfector could be infected with pseudomonas. It is why regular monitoring, every 3 months with specimen analysis at each step, is recommended. In case of reprocessing accessories, cleaning and disinfection have to be monitored very closely and it is why, in many endoscopy units, single use devices are a primary choice.
Due to the poor reputation of NHS, huge financial investments have been implemented five years ago in order to improve efficacy, safety and reduce waiting lists especially with the development of colorectal cancer screening policy. Each endoscopy unit had to develop accreditation process for all procedures, physicians have been also audited by senior endoscopists. In a couple of years, marked improvements have been obvious. This policy should be strongly recommended for each country (Faigel et al., 2009). Avoid side effects should be a daily goal for every endoscopy unit director but also for all practitioners and staff.

6. Conclusion

Quality control is becoming a very important point in each endoscopy unit. It is a multifactor process where medical and non-medical staffs have to work altogether very closely. It is also a primary goal and question issue for our patients and patients' associations: Doctor, is it safe to be scoped in your unit?

7. References

ASGE Standards of Practice Committee (2009). Management of antithrombotic agents for endoscopic procedures. *Gastrointest Endosc.* 70, 6, pp.1060-1070

Beilenhoff, U. et al. (2008). ESGE-ESGENA guideline: Cleaning and disinfection in gastrointestinal endoscopy-Update 2008. *Endoscopy*, 40, pp. 939-957

Bottrell, M. et al. (2000). Hospital informed consent for procedure forms. *Arch Surg*, 135, 1, pp. 26-33

Elfant, A. et al. (1995). Recall of informed consent after endoscopic procedures. *Dis Col Rect*, 38, 1, pp. 1-3

Faigel, D.O. et al. (2006). Quality indicators for gastrointestinal endoscopic procedures: an introduction. *Gastrointest. Endosc.* 63, (4 Suppl), pp. S3-9

Faigel D.O. et al. (2009). London OMED Guidelines for credentialing and quality assurance in digestive endoscopy. *Endoscopy*, 41, pp. 1069-1074

Felley, C. et al. (2008). Combined written and oral information prior to GI endoscopy compared with oral information alone: a randomized trial. *BMC Gastroenterology*, 8, pp. 22

Haynes A.B. et al. (2009). A surgical safety checklist to reduce morbidity and mortality in a global population. N. Engl J Med, 360, 5, pp.491-499

Joint Advisory Group on Gastrointestinal Endoscopy (the JAG). http://www.thejag.org.uk

Naylor, G. et al. (2003). Setting up a quality assurance program in endoscopy. *Endoscopy*, 35, 8, pp.701-707

O'Mahony, S.; Naylor, G. & Axon A. (2000). Quality assurance in gastrointestinal endoscopy. *Endoscopy*, 32, 6, pp. 483-488

Rey, J.F. et al. (2004). ESGE Guidelines for Quality Control in Servicing and Repairing Endoscopes. *Endoscopy*, 36, pp. 921-923

Rey, J.F. et al. (2010). European Society of Gastrointestinal Endoscopy (ESGE) guideline: the use of electrosurgical units. *Endoscopy*, 42, pp. 764-771

Song, J.H. et al. (2010). Acceptance and understanding of the informed consent procedure prior to gastrointestinal endoscopy by patients: a single-center experience in Korea. *Korean J Intern Med*, 25, 1, pp. 36-43

Veitch, A.M. et al. (2008). Guidelines for the management of anticoagulant and antiplatelet therapy in patients undergoing endoscopic procedures. Gut, 57, pp.1322-1329

6

Development of Urodynamic Standards for Quality Control

Limin Liao[1] and Werner Schaefer[2]
[1]*Department of Urology, China Rehabilitation Research Center, Rehabilitation School of Capital Medical University, Beijing 100068,*
[2]*Continence Research Unit, University of Pittsburgh, Pittsburgh PA 15213*
[1]*China*
[2]*USA*

1. Introduction

1.1 Anatomy of the lower urinary tract

Initially, a brief summary of the anatomy of the lower urinary tract will be presented.

The bladder is a hollow muscular organ in pelvis lined by a mucous membrane and covered on its outer aspect partly by peritoneal serosa and partly by fascia (Lich et al. 1978; Dixon and Gosling 1987). The muscularis of the bladder is formed of smooth muscle cells which comprise the detrusor muscle. The detrusor muscle is often described as consisting of three layers, the muscle fibres of the outer and inner layers tending to be orientated longitudinally, while those of the middle layer are circularly disposed. Woodburne (1960) has clearly shown that the musculature of the bladder is a "muscular meshwork". The constituent muscle bundles frequently branch and reunite with one another to form this interlacing meshwork. The bladder is profusely supplied with autonomic nerve fibres which form a dense plexus among the smooth muscle cells (Dixon and Gosling 1987).

In the male bladder neck, the smooth muscle cells form a complete circular collar which forms the internal urethral sphincter and extends distally to surround the preprostatic portion of the urethra. In the female bladder neck, muscle bundles extend obliquely or longitudinally into the urethral wall. There is the rich sympathetic innervation in the male bladder neck, while the female bladder neck is well supplied with the parasympathetic innervation (Gosling et al. 1977; Dixon and Gosling 1994).

The male urethra is considered in four regional parts: preprostatic, prostatic, membranous and spongiose. In the membranous urethra, muscle coat consists of a relatively thin layer of smooth muscle bundles and a layer of circularly orientated striated muscle fibres forming the external urethral sphincter. In the female urethra, the muscle coat consists of an outer sleeve of striated muscle, which forms the external sphincter together with an inner coat of smooth muscle fibres (Dixon and Gosling 1994).

1.2 Physiology of the lower urinary tract

Briefly the physiologic functions of the lower urinary tract will be described.

The lower urinary tract functions as a group of interrelated structures whose joint function is to bring about efficient bladder filling and urine storage and its voluntary expulsion. During bladder filling at physiologic rates, intravesical pressure initially rises slowly despite large increase in volume. Physical distensibility or higher compliance, slowly physiologic filling rate and inhibition of detrusor contraction are the major characters of physiologic function of the bladder during filling phase. There is a gradual increase in proximal urethral resistance, and urethral pressure is always greater than intravesical pressure during the bladder filling. Many factors have been thought to contribute to proximal and distal urethral closure mechanism (Arsdalen and Wein 1991; Torrens 1987).

Normal voiding is a voluntary act which results in sustained contraction of the detrusor muscle and relaxation of the urethra until the bladder is empty. Voiding can be divided into the initiation of voiding, the continuation of voiding and the termination of micturition. At the initiation of voiding, urethral closure pressure progressively drops, and this pressure drop occurs slightly before a corresponding increase in detrusor pressure with urinary flow when the detrusor pressure exceeded the urethral pressure. Then, the bladder pressure begins to increase. Descent of the bladder base is initiated by relaxation of the pelvic floor muscles and continued by relaxation of the urethra, and the bladder neck opens. This process is accompanied by the bladder outlet assuming a funneled shape (Tanagho and Miller 1970; Torrens 1987). At some specific intravesical pressure, the sensation of bladder distention is perceived, and micturition is initiated voluntarily at an appropriate time, proceeding to completion and involving a decrease in bladder outlet resistance and contraction of the bladder smooth musculature. Whilst the initiation of voiding is a voluntary act, its continuation depends on a more automatic activity. Micturition can be initiated by voiding reflexes almost regardless of the capacity of the bladder.

Normally, the bladder is very accurate in its ability to void exactly the capacity within itself with a voiding pressure which is maintained at a very constant level. Occasionally, the bladder seems to over- or under-estimate the capacity it has to deal with. Under-estimation results in the bladder continuing to contract after it is empty, resulting in an isometric increase in contraction pressure known as "after contraction"; this seems to have no identified pathologic significance (Torrens 1987).

The physiologic functions of lower urinary tract can be summarized as two aspects (Arsdalen and Wein 1991).

Bladder filling and urine storage

- increasing volumes of urine at a low intravesical pressure and with appropriate sensation must be accommodated.
- the bladder outlet must be closed at rest and remain so during increases in intraabdominal pressure.
- no involuntary bladder contraction can occur.

Bladder emptying

- a coordinated contraction of the bladder smooth musculature of adequate magnitude must occur.
- a concomitant lowering of resistance must occur at the level of the smooth sphincter and the striated sphincter.
- no anatomic obstruction can occur.

1.3 Urodynamics

Urodynamics is the medical science concerned with the physiology and pathophysiology of urine transport from the kidneys to the bladder as well as its storage and evacuation (Susset

1985). There are two basic aims of urodynamics: to reproduce the patient's symptomatic complaints and to provide a pathophysiological explanation for the patient's problems (Abrams 1997).

1.3.1 The history of urodynamics

Urodynamics can trace its roots to the 1800s, when instrumentation was first developed and described for the measurement of bladder pressure and urine flow rate; however, the term *urodynamics* was only recently coined by Davis (Davis 1954; Perez and Webster 1992).

1.3.1.1 The history of cystometry

Even prior to the invention of the cystometer, intravesical pressure had been measured by several European investigators, with Dubois in 1876 perhaps being the first (Smith 1968). In 1882, Mosso and Pellacani had shown that bladder pressure rose due to the contraction of the detrusor muscle (Perez and Webster 1992). In 1897, Rehfisch described an apparatus used for the simultaneous measurement of vesical pressure and urinary volume (Derezic 1988). In 1927, Rose coined the term cystometer and described its development and clinical usefulness (Rose 1927; Perez and Webster 1992); and he found cystometry to be a much more accurate way to determine the neurologically abnormal bladder than cystourethroscopy. Up to 1933, Denny-Brown and Robertson used a specially designed double catheter and a photographic recording method to measure pressure in the bladder, urethra and rectum (Denny-Brown and Robertson 1933). They showed in humans that bladder pressure is independent of intraabdominal pressure, and they were also the first to note and name the after contraction. In 1948, Talbot used the terms stable and unstable bladder detrusor in his study on spinal cord injury patients (Talbot 1948). In the modern era, technologic advances undoubtedly wrote the history of cystometry, in particular emerging of computerized urodynamic system. The application of the computer technology and electronic pressure transducers (external transducer and microtransducer) made the measurement and analysis in cystometry much more accurate.

1.3.1.2 The history of uroflowmetry

Before the invention of the uroflowmeter, Rehfisch in 1897 recorded the timing of onset and finish of micturition (Derezic 1988). In 1922, Schwartz and Brenner measured indirectly urethral exit pressure, then calculated the velocity of exiting urinary stream (Smith 1968). In 1925, Gronwall made the first recordings of unimpeded urinary flow (Smith 1968). None of these investigators made accurate calculation of flow rate. In 1948, Drake reported the development of the *uroflowmeter*; he developed an instrument that measured the increasing weight of urine against a factor of time on a kymograph and called the graphs produced uroflowgrams (Drake 1948; Perez and Webster 1992). In 1956, von Garrelts reported the use of electronics to record voiding rates; he used a transducer to convert changes in pressure of urine collected onto a photokymograph (von Garrelts 1956).

1.3.1.3 The history of pressure-flow study

With von Garrelts' classic article and Davis' book entitled *Mechanisms of Urologic Diseases*, the 1950s represented the infancy of modern urodynamics. Davis' book inspired much interest in the simultaneous measurements of uroflowmetry and cystometry (Perez and Webster 1992). In 1956, von Garrelts reported the normal micturition pressures electronically in men; in 1963, Zinner and Paquin did that in women (Zinner and Paquin 1963). In 1960, Murphy and Schoenberg reintroduced the measurement of micturition pressures by using suprapubic cystometry (Murphy and Schoenberg 1960). In 1962, Gleason and Lattimer reported the use of cystometry and uroflowmetry in combination to determinate bladder outlet strictures indirectly that was called as the *pressure-flow study* (Gleason and Lattimer 1962), and drew back the prologue of modern urodynamic studies on bladder outflow obstruction (BOO). In 1971, Griffiths introduced the concept of *fluid mechanics of collapsible tubes*, in relation to the conduit function of the lower urinary tract (Griffiths 1971). In the early 1980s, Schäfer introduced the concept of *passive urethral resistance relation, PURR* (Schäfer 1981; 1983), making the understanding to physiology of micturition much more profound. These were mathematically and biophysically elegant, computerized statements of the then-current understanding of bladder, bladder neck and urethral physiology and pathophysiology. They hearkened back to the call the 1968 meeting on "*The Hydrodynamics of Micturition*" for models that were based on the best urology and bioengineering (Boyarsky 1998). Up to now, these concepts have still being used in clinical and basic research widely, and have a role of guiding. The achievement of basic research must be applied to clinical practice, and must serve it. In 1979, Abrams and Griffiths reported a pressure-flow rate plot for classifying bladder outflow as obstructed, equivocal and unobstructed conditions (Abrams, Griffiths 1979). This Abrams-Griffiths (A/G) nomogram has been used for classification of obstruction in clinical practice. Afterwards, Schäfer developed and reported a nomogram for grading BOO using PURR principle (Schäfer et al. 1989; 1990). He further simplified PURR and introduced the *Linear PURR (L-PURR)* concept so that the clinical use of his nomogram became easier (Schäfer 1990). Schäfer nomogram divided bladder outflow obstruction into 7 grades from 0 to VI. As a semi-quantitative method for assessment of BOO, it has been known well by clinician, and has been applied to clinical practice widely. Basing on these nomograms mainly, the International Continence Society (ICS) recommended a nomogram as a standard for assessment of BOO (Griffiths et al. 1997).

1.3.1.4 The history of urethral pressure measurements

In 1923, Bonney reported crude measurements of bladder and urethral pressures; then, Kennedy described an innovative method to measure urethral resistance in 1937 (Perez and Webster 1992). In 1953, Karlson reported the simultaneous measurements of pressure in the bladder and internal and external urinary sphincters (karlson 1953). In 1969, Brown and Wickham reported a simple method, *the urethral pressure profile (UPP)*, to measure the pressure exerted by the urethral wall at every point of its length (Brown and Wickham 1969). For some years, UPP as a tool for the evaluation of patients with incontinence and BOO; however, in the current, it enjoys only limited application in its simple form (Perez and Webster 1992).

1.3.1.5 The history of Videourodynamics

In 1930, Thomsen reported the first series of lateral voiding cystourethrograms in female patients; and Muellner and Fleischner used the fluoroscope to study normal and abnormal micturition extensively (Muellner and Fleischner 1949; Perez and Webster 1992). In 1967, Miller described that truly popularized the use of cinefluoroscopy in conjunction with the other lower urinary tract urodynamic studies (Miller 1967). This was the birth of what has become known as videourodynamics, which includes the simultaneous recording and projection of fluoroscopic and functional data (Perez and Webster 1992). In 1970, Bates et al. reported the simultaneous cinefluoroscopy and pressure-flow studies, and discovered the combination of studies necessary for the evaluation of a variety of micturition dysfunction (Bates et al. 1970).

1.3.2 Urodynamic laboratory techniques

1.3.2.1 Uroflowmetry

Uroflowmetry is the simplest and least invasive test. With the developments in urine flowmeters, the clinical use of uroflowmetry has become widespread. The common flowmeter involves urine falling onto a disc rotating at a constant speed. The mass of the urine tends to slow the rotation, but a servomotor keeps the speed constant. The power necessary to do this is proportional to the urine flow rate. The urine volume can be derived by integration of the flow rate (Torrens 1987). Urine flow may be described in terms of rate and pattern and may be continuous or intermittent. Flow rate is defined as the volume of fluid expelled via the urethra per unit time. It is expressed in ml/s. The maximum flow rate (Q_{max}) is the only value so far submitted to an extensive quantitative investigation. Voided volume, patient environment and position, way of filling (by diuresis or by catheter) as well as type of fluid can influence the results of uroflowmetry (Abrams et al. 1988). von Garrelts reported that there was a correlation between the Q_{max} and the square root of the volume voided (von Garrelts 1957). It is now quite clear that uroflowmetry give evidence of urinary dysfunction, and is a good screening test. However, urine flow rate is a combined produce of detrusor contractility and outlet resistance. Then, uroflowmetry cannot offer a precise diagnosis as to the cause of abnormal flow; for example it has poor diagnostic specificity for BOO, and cannot be used alone except in clearly defined situations (Abrams et al. 1997).

1.3.2.2 Cystometry

Cystometry is used to study both the storage and the voiding phases of micturition in order to make a diagnosis which enables effective treatment to be given (Abrams 1997).

1.3.2.2.1 Cystometry during filling

Cystometry during filling is the method by which the pressure/volume relationship of the bladder is measured. All systems are zeroed at atmospheric pressure. For external transducers the reference point is the level of the superior edge of the symphsis pubis. For catheter mounted transducers the reference point is the transducer itself. Cystometry is used to assess detrusor activity, sensation, capacity and compliance. Present techniques allow the continuous recording of pressure within the bladder during artificial or natural filling. Before starting to fill the bladder the residual urine may be measured. Intravesical pressure is the pressure measured within the bladder. Abdominal pressure is the pressure surrounding the bladder and is usually measured as rectal pressure. Detrusor pressure (P_{det}) is calculated by electronically subtracting the abdominal pressure (P_{abd}) from the intravesical pressure (P_{ves}). The simultaneous measurement of abdominal pressure is essential for the interpretation of the intravesical pressure trace. During cystometry, any variations should be specified. Access for pressure measurement is most commonly by transurethral catheterization; occasionally a percutaneous suprapubic catheter is used. In current practice, the fluid medium is usually liquid (saline). The temperature of fluid is usually the same with that of room. The positions of patient may be in supine, sitting or standing; the different positions of can result in the different abdominal pressures, but detrusor pressure is constant. Certain cystometric parameters may be significantly altered by the speed of bladder filling. For the general discussion, the following terms for the range of filling rate may be used: up to 10 ml/min is slow filling; 10 to 100 ml/min is medium filling; over 100 ml/min is rapid filling (Abrams et al. 1988). Bladder sensation is difficult to evaluate; it usually assessed by questioning the patient in relation to the fullness of the bladder during cystometry. In patients with normal sensation, maximum cystometric capacity is the volume at which the patient feels he or she can no longer delay micturition. Compliance indicates the change in volume for a change in pressure. Compliance is calculated by dividing the volume change (ΔV) by the change in detrusor pressure (ΔP_{det}) during that change in bladder volume ($C=\Delta V/\Delta P_{det}$). Compliance is expressed as mls per cm water.

1.3.2.2.2 Cystometry during voiding

Pressure-flow study of micturition is a method that the abdominal, intravesical, and detrusor pressures and flow rate are simultaneously recorded during the voiding phase of cystometry. With regard to the methodology, ICS has published an updated report on pressure-flow standardisation (Griffiths et al. 1997). The currently available urodynamic equipment is adequate for the accurate recording of pressures and flow. During investigation, the patients must be in the position as the usual voiding; and catheters should be as thin as possible, for example 6F double lumen. As standard transurethral double lumen catheter has been suggested to use; only in children and patients with severe constrictive obstruction a suprapubic pressure recording may have advantages. A rectal balloon catheter is recommended to use for abdominal pressure recording (Schäfer 1998). Many investigators have focused on the analysis of pressure-flow study. Because the ideas about the mechanical behavior of the urethra and the bladder were new, it was not immediately evident how to apply them in practice and so a number of competing approaches to the analysis of pressure-flow studies were developed (Abrams and Griffiths 1979; Schäfer W 1983; 1990; Griffiths et al. 1989; Spangber et al. 1989; Höfner et al. 1995). All of them share a similar fundamental basis, but there are differences in detail and in objectives. The results of pressure-flow studies may be used for various purposes, for example, for objective diagnosis of BOO or for statistical testing of differences in urethral resistance between groups of patients. For these purposes, methods have been developed to quantify pressure-flow plots in terms of one or more numerical parameters. The parameters are based on aspects such as the position, slope, and curvature of PURR of the plot. A/G nomogram was based on data from 117 patients. The upper boundary line, separating clearly obstructed patients from others, was arrived at by a combination of theoretical insight and clinical judgment. A lower boundary separates clearly unobstructed patients from others. The equivocal region between these two boundaries contained a mixture of obstructed and unobstructed patients (Abrams and Griffiths 1979). Working independently, Schäfer developed the L-PURR method of grading BOO and modified to a nomogram: Schäfer nomogram (Schäfer 1990). Based on studies of the urodynamic changes following TURP, it provided 7 grades ranging from not obstructed to severely obstructed, 4 grades for the strength of the contraction ranging from very weak to strong. Grade II fulfills a similar function to the equivocal zone of A/G nomogram: the upper between obstructed and equivocal or slightly obstructed are identical in A/G nomogram and Schäfer nomogram. The position of the lower boundary of grade II suggests that the equivocal zone in A/G nomogram is too large at the lower flow levels. Combining these aspects, ICS recommended a provisional nomogram; it is important for comparing results from different centers. Therefore, it is recommended that the upper boundary line of ICS nomogram should be used to distinguish between clearly obstructed patients and others; Schäfer nomogram may be used to grade the severity of obstruction; detrusor pressure at Q_{max} or AG number may be used to represent urethral resistance (Abrams et al. 1997).

1.3.2.3 Urethral pressure measurement

The urethral pressure and the urethral closure pressure are idealized concepts which represent the ability of the urethra to prevent leakage. The urethral pressure may be measured by a number of different techniques which do not always yield consistent values. Not only do the values differ with the method of measurement but there is often lack of consistency for a single method. For example, the effect of catheter rotation when urethral pressure is measured (Abrams et al. 1988). On the other hand, there still are some problems resulting in that distinction of physiological facts and urodynamic artifacts are difficult. One of the problems is caused by bring a probe into a closed system; another one is related to the dynamics (Schäfer 1998). These problems limited its application and interpretation for urethral closure function. Intraluminal urethral pressure may be measured: at rest, with the bladder at any given volume; during coughing or straining; during voiding. Measurement may be made at one point in the urethra over a period of time, or at several points along the urethra consecutively forming a urethral pressure profile (UPP).

1.3.2.4 Videourodynamics

Videourodynamics is the simultaneous radiological visualization and urodynamic measurement of lower urinary tract. Some centers use it as a first line investigation, this is unnecessary, hazardous and expensive. Videourodynamics is indicated when structural information is required as well as functional information. In neuropathic patients there is an increased prevalence of bladder shape abnormalities, vesicoureteric reflux, and urethral sphincter abnormalities; therefore, videourodynamics is the investigation of choice in suspected neuropathic vesicourethral dysfunction. It also is indicated in failed surgery for stress incontinence and in men who develop incontinence after prostatic surgery. It allows the clinician to differentiate between incontinence, secondary to sphincter damage and that due to detrusor instability (Abrams 1998). Videourodynamics, by combing the simultaneous measurement of detrusor pressure, flow rate and radiological visualization provides the most comprehensive urodynamic evaluation; however, there is no evidence that video adds clinical benefit, beyond that given by pressure-flow studies, in elderly men with suspected BOO (Abrams et al. 1997).

1.3.3 Quality control of urodynamic data

Undergoing the development, urodynamics has been applied to clinical practice widely, and has more and more important role at present. The aim of clinical urodynamics is to reproduce patient's symptoms under the condition of precise measurement in order to identify the underlying causes for the symptoms and to provide a pathophysiological explanation for them (Schäfer 1998). In clinical urodynamic practice, one of the most important problems is whether or not a reliable diagnosis can be made. A reliable diagnosis relies on a good urodynamic practice, which has a precise measurement with data quality control and accurate analysis of results. However, considerable data quality problems were found when traces from a study of multi-centers were examined. In analysis on the data from the ICS "benign prostatic hyperplasia (BPH)" study, Schäfer et al found that up to 60% of traces had significant technical errors and obvious artifacts. One of these could be easily corrected, and were due to common problems, such as, a difference in pressure transmission to the P_{ves} and P_{abd} tracings, incorrect position of the zero reference line, and spikes and other irregularities in the P_{ves} tracing. One third of artifacts were less easy to correct, such as periodic loss of a signal, pressure rising above full scale deflection, slow drift in a pressure tracing, and loss of the urethral catheter during voiding. A small percentage (10%) of traces

could not be analyzed due to a lack of scaling or indicated zero position, or complete loss of a pressure or flow signal (Schäfer et al. 1994). Although ICS published a series of reports on standardisation of urodynamics, some investigators did not perform urodynamic tests according to them; therefore, considerable technical errors and artifacts were produced. These indicate that data quality control has not received enough attention or has lacked consensus, and the urodynamic practice is badly in need of the standards for quality control. Data quality control has substantial contents, it mainly involves the following aspects: equipment set-up and configuration before measurement; signal testing, plausibility check, pattern recognition and artifacts correction during investigation; and retrospective analysis and artifacts correction for results after study. Quality control and plausibility check during investigation are the best way to avoid and to correct artifacts at an early stage. Quality control relies on pattern recognition and knowledge of typical values (Schäfer 1998).
Quality control during urodynamic investigation can avoid and eliminate various artifacts and technical errors. However, it is difficult to acquire a perfect measurement in clinical urodynamic practice. Therefore, there are either more or less artifacts and errors existed in urodynamic data. For these artifacts existed in data, retrospective quality control and correction are necessary. Especially, computer's application to clinical urodynamic practice makes the retrospective quality control more important. The computer-based urodynamic systems have gradually replaced traditional ones, have been playing a role in many aspects of urodynamics. However, computer's application has also brought some problems into urodynamics. Up to now, a true urodynamic expert system has not yet been developed. Many computer printouts are inferior to traditional paper-chart records. Computer is not able to pick up technical artifacts and human errors. Some investigators accept the automated result of computer without question (Lewis et al. 1997). The studies of manual correction in uroflowmetry have been performed by some investigators. Rowan et al found that up to 20% uroflow traces showed artifacts (Rowan et al. 1987). Grino et al compared manual and automated values, and found consistently lower values of Q_{max} in manual readings (Grino et al. 1993). Madsen et al compared manual and computerized values of Q_{max} and detrusor pressure at Q_{max} ($P_{det.Qmax}$) in a small group of patients, and found some different pressure-flow results between manual and computerized groups (Madsen et al. 1995). From these views, quality control in retrospective data analysis is necessary.

1.3.4 Standard for quality control of urodynamic data

The study on quality control of urodynamic data is lacking in the published literature. In order to carry out quality control, the urodynamic standards for it are crucial. Schäfer et al drafted the ICS standards of "*Good Urodynamic Practice*" which have been presented and discussed at two ICS meetings in 1997 and 1998, and was published in 2002 (Schaefer et al. 2002). This report has provided us with the standards and guideline for quality control of urodynamic data. Certainly, the project of this dissertation has been carried out along this line.

1.4 Objectives

The aim for this dissertation is to develop the urodynamic standards for quality control. This aim will be achieved by two strategies in two stages:

- Quality control during investigation: there are two strategies:
    - To establish the typical value ranges (TVR) as the tool for quantitative plausibility check and quality control;

  - To describe the typical signal patterns (TSP) as the tool for qualitative plausibility check and quality control.
- Quality control in retrospective analysis:
  - To recognize and correct the technical errors and artifacts in computerized urodynamic data using the above-mentioned strategies;
  - To evaluate the impact of the technical errors and artifacts on the outcome by comparing the computerized results with manual one, and to indicate the significance of retrospective quality control.

2. Materials and methods

A total of 181 elderly males with lower urinary tract symptoms (LUTS) was recruited in the study. The mean age of the males was 65.3 years, with a range of 43 ~ 86 years.
All cystometric measurements were done in standing or sitting position with 30 ml/min infusion rate using Dantec Menuet urodynamic system. During cystometry, the patient was asked to cough before and at beginning of filling, at regular intervals during filling phase, and before and after voiding. For each patient, a free uroflowmetry was recorded before cystometry. Methods, definitions and units comforted to the standards proposed by the ICS except where specifically noted (Abrams et al. 1988).
The study was retrospective. A total of 606 cystometric traces from the males was reviewed. The traces that were non-interpretable and non-correctional because of various artifacts and technical errors were excluded in the study; then, a total of 582 cystometric traces was included for further analysis. All traces were manually read, and various technical errors and artifacts were recognized and corrected.

2.1 Establishing the typical value ranges

For each trace, P_{ves} and P_{abd} estimated from rectal pressure were recorded simultaneously; and P_{det} was calculated by the electronic subtraction of P_{abd} from P_{ves}. The overactive detrusor factors were ruled out when the filling end point was defined. Maximum cystometric capacity (MCC), Q_{max}, $P_{det.Qmax}$ and volume voided (V_{void}) were recorded and read. Compliance of bladder was calculated by dividing the volume change by the change in detrusor pressure during that change in bladder volume. The values of P_{ves}, P_{abd} and P_{det} were read before, at beginning and at end of filling, and at end of voiding respectively. For each moment above-mentioned during cystometry, the mean value, standard deviation, median, 95% confidence interval (CI), and 50%, 80% and 95% TVR as well as upper limit of 99% range were calculated using computer in each parameter. The technical errors and atypical changes that were relevant to TVR were classified and given some examples to indicate the role and significance of quantitative plausibility check using TVR in cystometry.

2.2 Describing the typical signal patterns

For describing TSP, the signals' scales recorded were 40 cm H_2O per cm for P_{ves}, P_{abd} and P_{det}, and 4 ml/s per cm for uroflow rate; the scales of time axis were 1 minute per cm during filling and 15 seconds per cm during voiding. The changes corresponding to cough tests were observed and compared among P_{ves}, P_{abd} and P_{det} tracings. In data analysis, we classified the signal patterns of pressures into four types: I. fine structure (noise); II. minimal dynamic changes caused by breathing, talking and moving (minor changes); III. major changes due to regular cough tests; IV: typical major changes related to straining, detrusor overactivity, rectal contractions and detrusor contraction. The fine structure of pressure signal shows a "live" signal tracing with some minimal amplitude signal variations (noise). Straining is characterized by pressure increases on the P_{ves} and P_{abd} tracings but not on the P_{det} tracing; detrusor overactivity has the unstable waves recorded on P_{ves} and P_{det} tracings but not on P_{abd} tracing; rectal contraction shows a positive wave on the P_{abd} tracing and a negative artifact or dip on the P_{det} tracing but not on P_{ves} tracing. Typical pattern of detrusor contraction is that P_{ves} and P_{det} tracings increase and decrease with uroflow tracing smooth and steady. According to it, we can identify several special patterns of detrusor contraction: after-contraction, fluctuation contraction and unstable detrusor voiding. These four types of signal patterns were compared among P_{ves}, P_{abd} and P_{det} tracings at beginning of filling, during filling, before, during and after voiding respectively. Then, typical signal patterns of pressures were described; the relevant errors and artifacts were given some examples and analyzed to indicate the role and significance of qualitative plausibility check using TSP recognition in cystometry.

2.3 Quality control in retrospective analysis

In the study on retrospective quality control, all traces were printed out, and were manually read. The readers were blinded to the computer results. For each trace, artifacts during filling and voiding were examined according to typical value ranges and typical signal patterns. During filling cystometry, artifacts involved mainly the wrong initial resting pressures, spikes related to test-coughs and periodic signal loss or stepwise changes. During voiding cystometry, artifacts in uroflow and pressures were recognized and were corrected to indicate the effect of quality control. The recognition and correction of Q_{max} artifacts contain two aspects of value and location. Firstly, Q_{max} must be located at the highest plateau on a main uroflow curve. The additional modifications in flowrate and the spike artifacts on a main uroflow are smoothed and corrected to get Q_{max} value. Secondly, the spike artifact located at beginning or end of uroflow tracing, which is recognized as Q_{max} by computer, must be corrected. Reader has used the following two specifications to read manually Q_{max}: 1. Q_{max} must be measured at the highest plateau or peak of the flow curve that lasted for 2 seconds or more; 2. Q_{max} value must be read to the nearest 0.5 to 1.0 ml. per second. Various artifacts and errors during voiding cystometry may occur on P_{ves} and P_{abd} tracings, then influence P_{det} tracing. The pressure artifacts have been classified into technical and physiologic ones. The technical artifacts may be caused by phasic signal loss, signal stepwise changes and catheters' dislocations and others. The physiologic artifacts may be the spikes and dips on P_{det} tracing resulted from the different causes. We have described three common causes. The first one is the spikes due to the different transmissions between P_{ves} and P_{abd} tracings corresponding to straining. The second one is the dips caused by rectal contractions. The last one is the spikes or the dips due to urethral sphincter overactivity during voiding, which is sphincter contraction or relaxation. In the study, any rapid rising and dropping of P_{det} tracing were recognized as spike and dip artifacts, and were smoothed and corrected manually.

In analysis of pressure-flow data, various parameters and different methods were employed. As a continuous quantitative parameter, obstruction coefficient (OCO) developed by Schäfer et al was used to detect the difference in urethral resistance between manual and computerized results, and was calculated according to the following formula: OCO = $P_{det.Qmax}/40+2Q_{max}$ (Schäfer and Sterling 1995). Schäfer nomogram was used to grade the degree of obstruction and to evaluate the changes of obstructed grade after correction

(Schäfer 1990). International Continence Society (ICS) nomogram were used to classify and diagnose obstruction and to find the shifts of classifications due to correction (Griffiths et al. 1997).

The different statistical analyses were performed using computer. The correlation analyses between manual and computerized results were done in the following variables: Q_{max}, $P_{det.Qmax}$ and OCO. In above mentioned variables, the variations between manual and computerized values were evaluated by the matched-pairs z test for a big sample. The percentages in various grades of Schäfer nomogram and classification of ICS were calculated. The variations in classification of ICS nomogram and in grades of Schäfer nomogram between manual and computerized readings were examined by chi-square test and relative to an identified distribution (Ridit) analysis respectively. In above mentioned statistical analyses, $p<0.05$ was considered significant.

3. Results

The results were shown from three aspects: establishing TVR, describing TSP and retrospective quality control.

3.1 Establishing the typical value ranges

At the different moments during cystometry, TVR for various pressures and other parameters were shown in the following tables. Taking 50% as a usual range, we can find that before and at beginning of filling TVR for P_{ves} and P_{abd} were 31~42 cmH_2O and 28~39 cmH_2O in standing or sitting position respectively; and that of P_{det} was 0~4 cmH_2O with a mean of 2.3 cmH_2O, which was very close to zero (Table 1). These ranges were TVR for initial resting pressures. The upper limits of 95% and 99% ranges for P_{det} were 9 and 13 cmH_2O respectively (Table 1); then, we took 10 cmH_2O as the upper limit of a maximum possible resting value for P_{det}. With these TVR for initial resting pressures, we can check the technical errors and artifacts occurred in zero setting and a pressure reference level establishing retrospectively.

	Mean±SD	Median	95% CI	50%	80%	95%	99% upper limit
P_{ves} (cmH_2O)	35.4 ±10.7	37	0.87	31~42	24~46	7~51	63
P_{abd} cmH_2O)	33.1±10.9	35	0.88	28~39	20~44	5~49	59
P_{det} (cmH_2O)	2.3±3.5	2	0.29	0~4	0~6	0~9	13

Table 1. The mean value, standard deviation, median, 95% confidence interval (CI), and 50%, 80% and 95% typical value ranges as well as upper limit of 99% range for initial resting pressures in cystometry

According to a definite value: P_{det} cannot be negative and a relatively definite value: initial resting P_{det} is rarely over 10 cmH_2O, we can divide these errors relating to initial resting pressures into three types. Type I error has a normal initial resting P_{det}, but both P_{ves} and P_{abd} are wrong; type II has a negative initial resting P_{det}; and type III error has a too high initial resting P_{det} (over 10 cmH_2O). Concerning type I error, there are two sub-types: a. both initial resting P_{ves} and P_{abd} are too low comparing with their TVR (Fig. 1a); b. both P_{ves} and P_{abd} are too high comparing with their TVR (Fig. 1b).

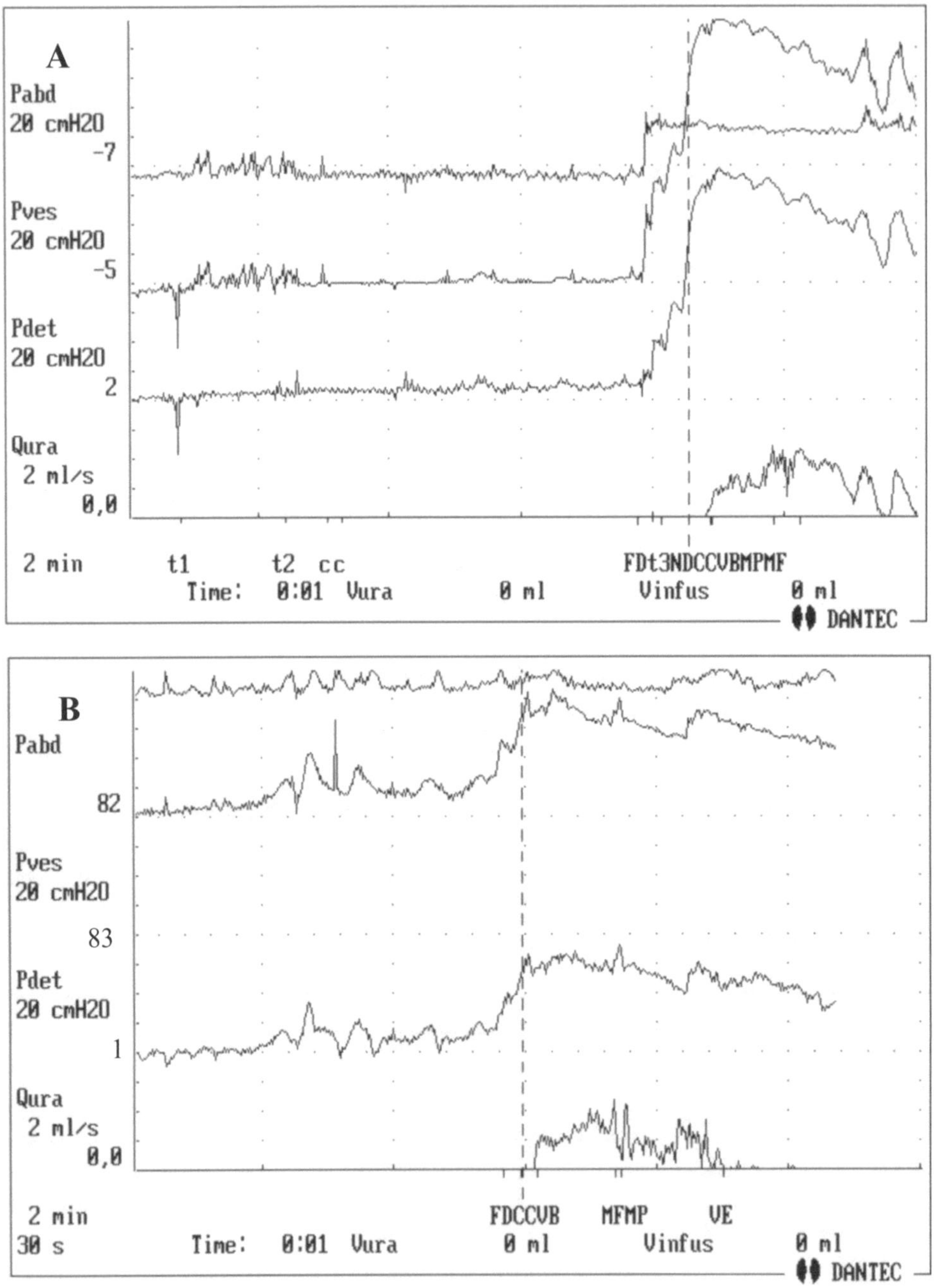

Fig. 1. Type I error related to typical value ranges for initial resting pressures. **A**: type Ia error has a normal initial resting P_{det}, but both P_{ves} and P_{abd} are too low comparing with their TVR; in this case, P_{det} is 2 cmH_2O, but P_{ves} and P_{abd} are -5 cmH_2O and -7 cmH_2O respectively. **B**: type Ib error has a normal initial resting P_{det}, but both P_{ves} and P_{abd} are too high comparing with their TVR; in this case, P_{det} is 1 cmH_2O, but P_{ves} and P_{abd} are 83 cmH_2O and 82 cmH_2O respectively

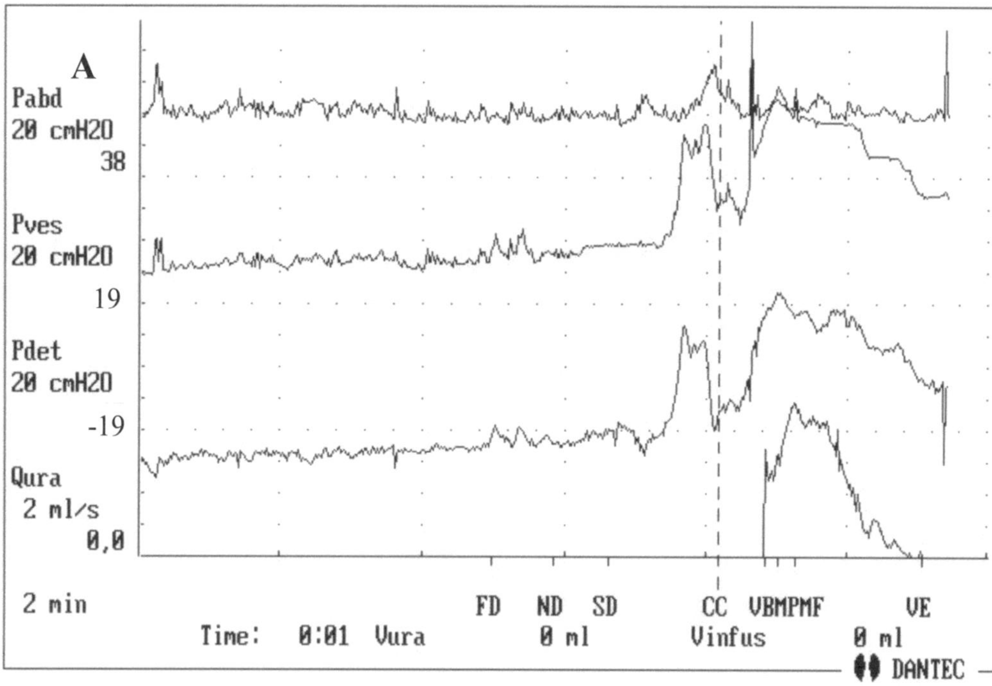
A
Pabd
20 cmH2O
38
Pves
20 cmH2O
19
Pdet
20 cmH2O
-19
Qura
2 ml/s
0,0
2 min
FD ND SD
CC VBMPMF
VE
Time: 0:01 Vura 0 ml Vinfus 0 ml
DANTEC

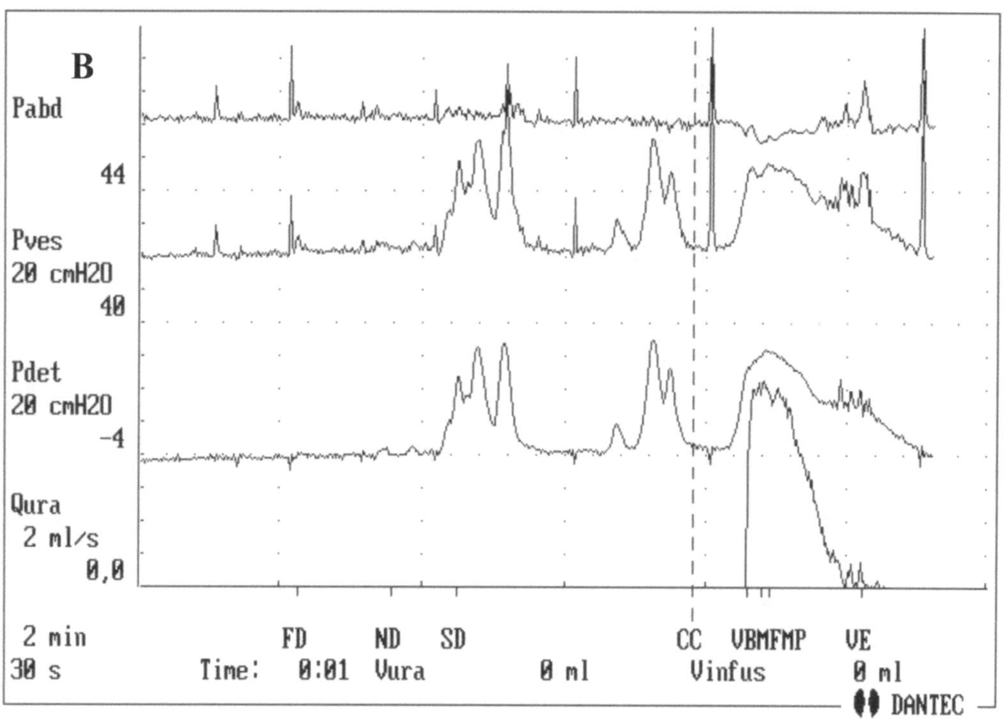
B
Pabd
44
Pves
20 cmH2O
40
Pdet
20 cmH2O
-4
Qura
2 ml/s
0,0
2 min
30 s
FD ND SD
CC VBMFMP VE
Time: 0:01 Vura 0 ml Vinfus 0 ml
DANTEC

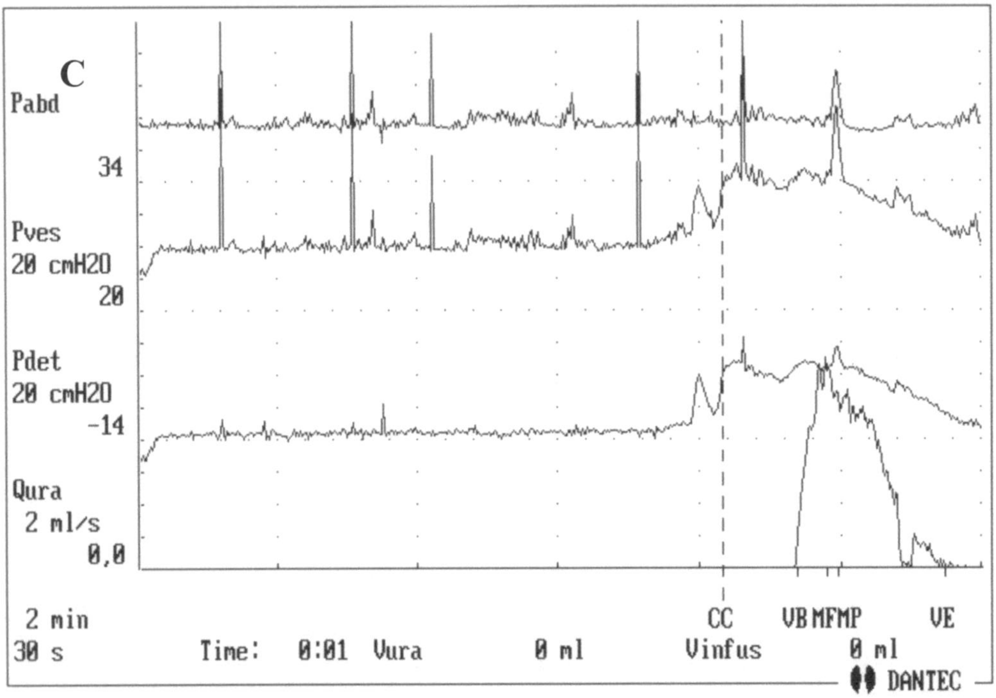

Fig. 2. Type II error related to typical value ranges for initial resting pressures. **A**: type IIa error has a negative initial resting P_{det} caused by a too low P_{ves} comparing with its TVR; in this case, P_{det} is -19 cmH_2O, P_{ves} and P_{abd} are 19 cmH_2O and 38 cmH_2O respectively. **B**: type IIb error has a negative initial resting P_{det} caused by a too high P_{abd} comparing with its TVR; in this case, P_{det} is -4 cmH_2O, P_{ves} and P_{abd} are 40 cmH_2O and 44 cmH_2O respectively. **C**: in this case, initial P_{det} is negative (type IIa error), then becomes positive during initial 30 seconds of filling. At beginning of filling, P_{det} is -14 cmH_2O, P_{ves} and P_{abd} are 20 cmH_2O and 34 cmH_2O respectively

There are two reasons leading to type II error: a. initial resting P_{ves} is too low comparing with TVR while initial resting P_{abd} is in TVR (Fig. 2a); b. initial resting P_{abd} is too high while P_{abd} is in TVR (Fig. 2b). In type II error, P_{det} value may become positive during initial 30 seconds of filling in some traces (Fig. 2c).

Also, there are two reasons resulting in type III: a. initial resting P_{ves} is too high comparing with TVR while initial resting P_{abd} is in TVR (Fig. 3a); b. initial resting P_{abd} is too low while P_{abd} is in TVR (Fig. 3b).

In the traces analyzed, the incidences of I, II and III errors were 9.8%, 4.5% and 1.4% respectively. P_{det} signal of 11.5% traces with type II error returned to TVR during initial 30 seconds of filling.

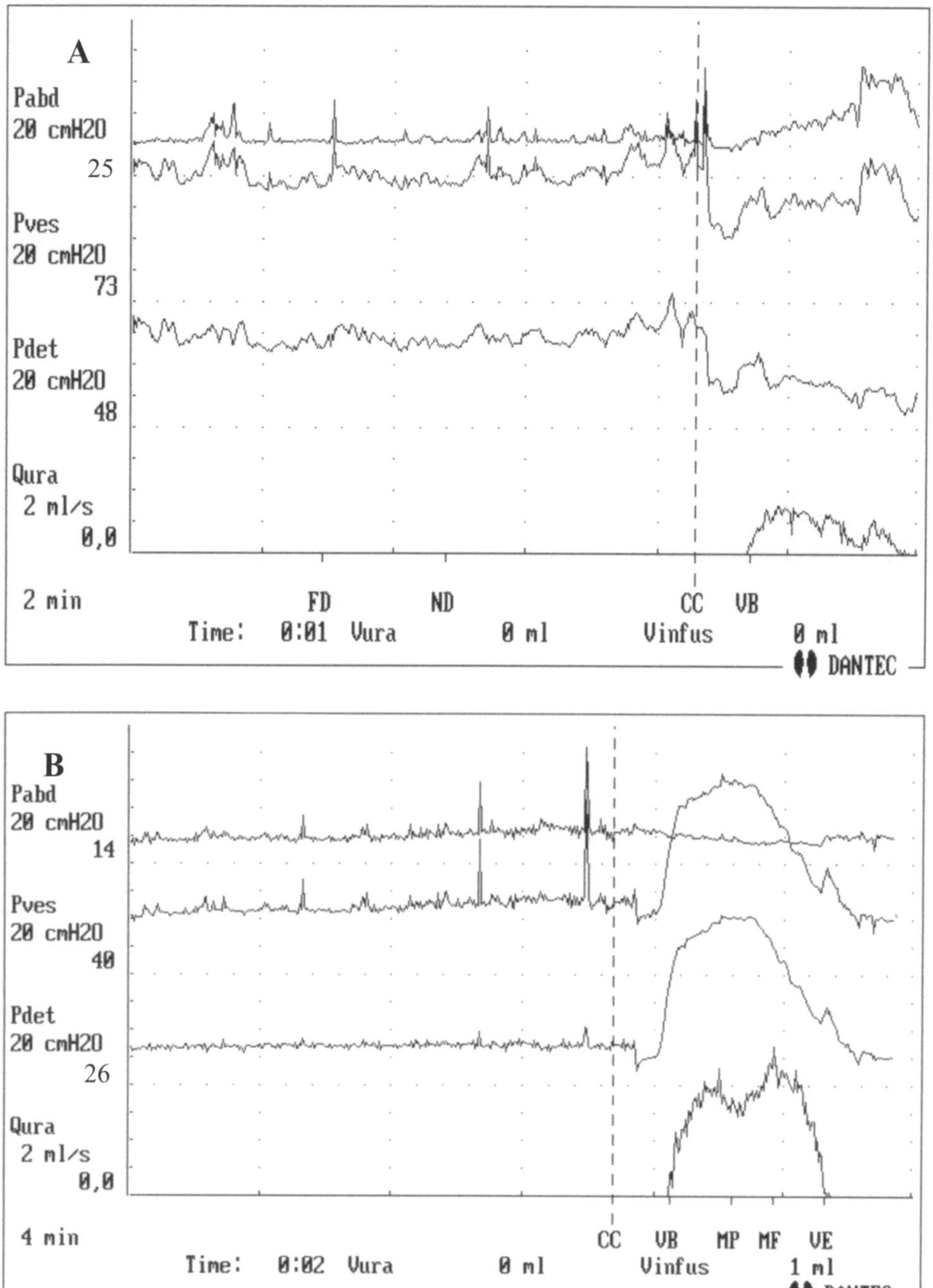

Fig. 3. Type III error related to typical value ranges for initial resting pressures. A: type IIIa error has a high initial resting P_{det} caused by a too high P_{ves} comparing with its TVR; in this case, P_{det} is 48 cmH_2O, P_{ves} and P_{abd} are 73 cmH_2O and cmH_2O respectively. B: type IIIb error has a high initial resting P_{det} caused by a too low P_{abd} comparing with its TVR; in this case, P_{det} is 26 cmH_2O, P_{ves} and P_{abd} are 40 cmH_2O and 14 cmH_2O respectively

At end of filling, TVR of P_{ves}, P_{abd} and P_{det} were 38~50 cmH_2O, 30~41 cmH_2O and 5~10 cmH_2O respectively. TVR of MCC and compliance were 157~345 ml and 26.7~70.8 ml per cmH_2O (Table 2). It means that P_{det} increases 5~6 cmH_2O during filling phase.

	Mean±SD	Median	95% CI	50%	80%	95%	99%
P_{ves} (cmH_2O)	42.4±12.5	43	1.02	38~50	29~56	10~62	80
P_{abd} (cmH_2O)	34.2±12.3	36	1.0	30~41	20~47	2~51	63
P_{det} (cmH_2O)	8.2±4.9	7	0.40	5~10	4~13	2~18	28
MCC (ml)	261.6±136.9	244	11.13	157~345	105~441	49~587	651
C (ml/ cmH_2O)	58.5±61.1	41.5	4.99	26.6~70.8	17.8~122.8	7~220	345

Table 2. The mean value, standard deviation, median, 95% confidence interval (CI), and 50%, 80% and 95% typical value ranges as well as upper limit of 99% range for pressures, maximum cystometric capacity (MCC) and compliance (C) at end of filling in cystometry

During voiding, TVR of P_{abd} at relaxation was 25~38 cmH_2O (Table 3). A typical relevant error, called type IV error, was that P_{abd} became negative at relaxation during voiding; and this error lead to a meaningless P_{det} value that was higher than P_{ves} (Fig. 4). In the traces analyzed, the incidence of this type of error was 0.7%. TVR of Q_{max}, $P_{det.Qmax}$ and V_{void} were 5.5~9 ml/s, 57~92 cmH_2O and 167~315 ml respectively (Table 3). On the other hand, TVR of Q_{max} and V_{void} in free uroflowmetry were 8~9.2 ml/s and 167~301 ml (Table 3). It means that Q_{max} and V_{void} during cystometry are comparable with those of free uroflowmetry.

	Mean±SD	Median	95% CI	50%	80%	95%	99%
Voiding cystrometry							
$P_{det.Qmax}$ (cmH_2O)	76.5±31.7	70	2.57	57~92	42~118	37~159	216
$P_{det.min.void}$ (cmH_2O)	44.6±18.9	40	1.62	30~53	23~70	20~92	107
$P_{abd.relax}$ (cmH_2O)	31.5±10.9	32	0.88	25~38	18~44	8~50	69
Q_{max} (ml/s)	7.3±2.6	7	0.21	5.5~9	4~10.7	2.9~13	15
V_{void} (ml)	250.8±119.9	234	10.09	167~315	114~406	63~560	628
Free uroflowmetry							
Q_{max} (ml/s)	7.9±2.8	8	0.36	8~9.2	4.2~11.4	3.1~15	16.8
V_{void} (ml)	242.4±109.9	233	13.99	167~301	120~374	87~493	689

Table 3. The mean value, standard deviation, median, 95% confidence interval (CI), and 50%, 80% and 95% typical value ranges as well as upper limit of 99% range for pressures, Q_{max} and voided volume (V_{void}) during voiding cystometry and free uroflowmetry

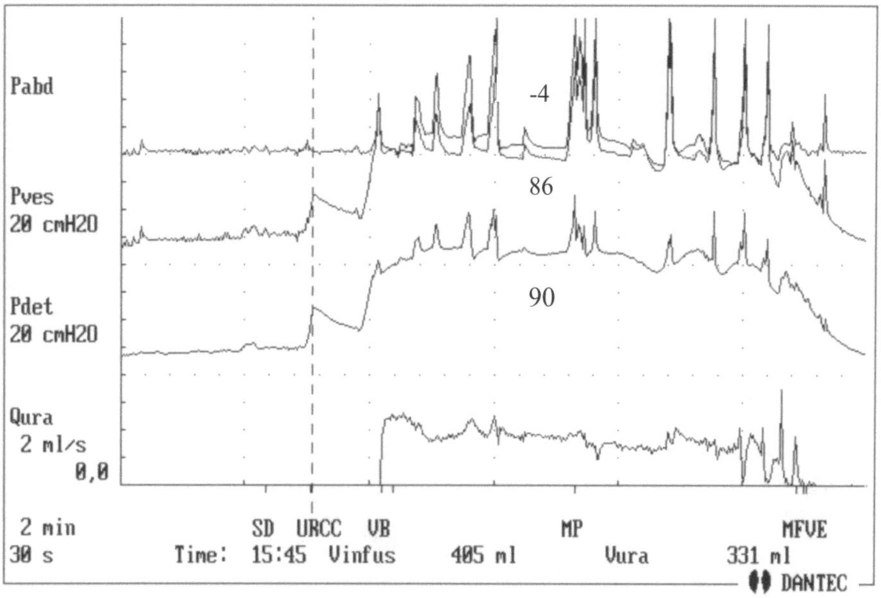

Fig. 4. Type IV error related to typical value ranges for P_{abd} during voiding. This type of error shows that P_{abd} becomes negative during voiding due to over relaxation of pelvic floor, and it leads to a meaningless P_{det} value that is higher than P_{ves}. The error usually has a low initial resting P_{abd}. In this case, during voiding P_{abd} around Q_{max} is -4 cmH_2O, P_{ves} and P_{det} are 86 cmH_2O and 90 cmH_2O respectively; and initial resting P_{abd} is 5 cmH_2O, initial P_{ves} and P_{det} are 9 cmH_2O and 4 cmH_2O respectively (type Ia error)

At end of voiding, TVR of P_{ves}, P_{abd} and P_{det} were 40~55 cmH_2O, 30~41 cmH_2O and 10~14 cmH_2O respectively (Table 4).

	Mean±SD	Median	95% CI	50%	80%	95%	99%
P_{ves} (cmH_2O)	48.5 ±13.4	47	1.09	40~55	35~64	26~79	108
P_{abd} (cmH_2O)	34.3±10.0	35	0.81	30~41	22~45	10~50	61
P_{det} (cmH_2O)	14.2±11.3	11	0.92	10~14	13~19	16~29	47

Table 4. The mean value, standard deviation, median, 95% confidence interval (CI), and 50%, 80% and 95% typical value ranges as well as upper limit of 99% range for post-void pressures in cystometry

Comparing these TVR with those before voiding, we found P_{abd} had little change, P_{ves} and P_{det} were close to the levels before voiding. Also, there were two types of relevant errors. One (type V error) was that P_{ves} and P_{det} after voiding still kept high levels beyond their TVR while a high post-void residual volume was ruled out. There may be two reasons leading to this error: a. post-void P_{ves} is too high but P_{abd} normal, and P_{ves} tracing becomes a high level "dead" line during voiding due to the signal loss or urethral catheter dislocation (Fig. 5a); b. post-void P_{abd} is too low due to the rectal catheter dislocation or the signal loss of P_{abd} tracing during voiding (Fig. 5b).

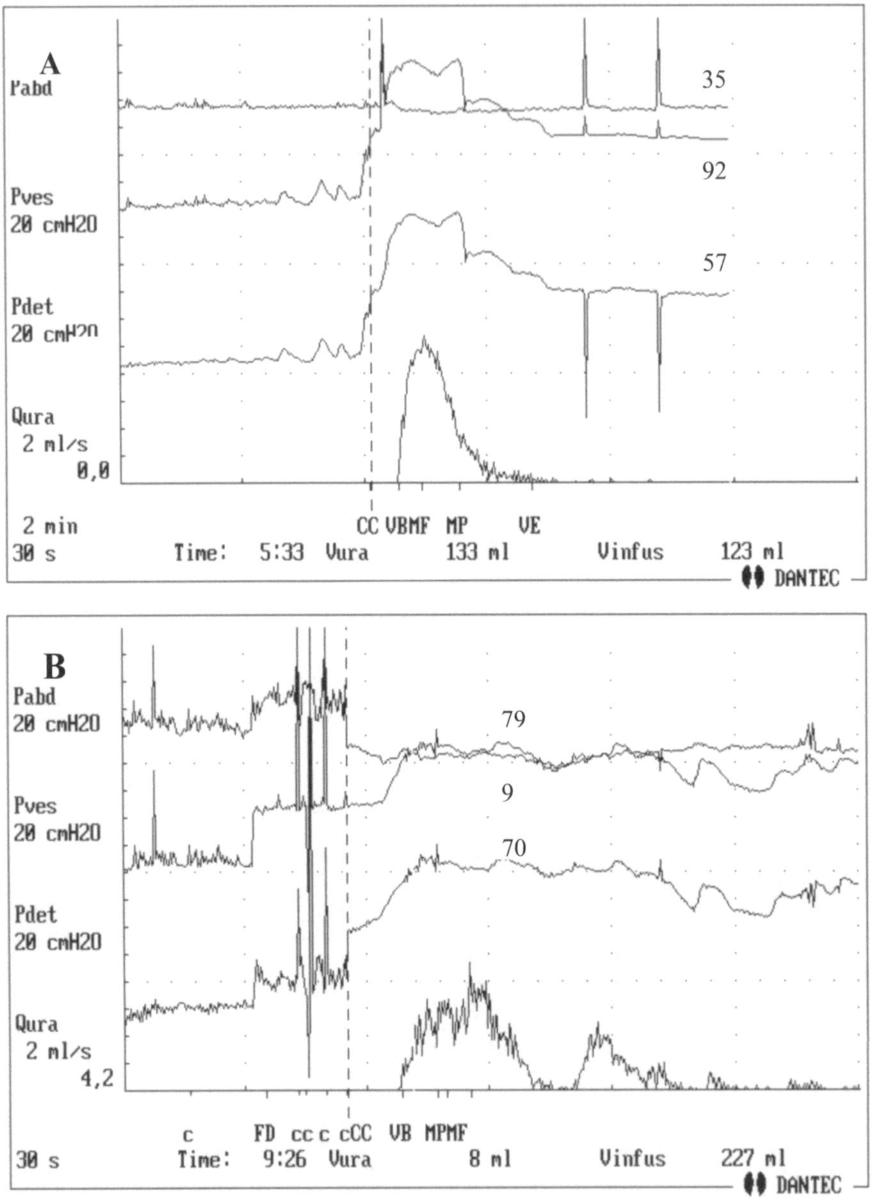

Fig. 5. Type V error related to typical value ranges for post-void pressures. **A**: type Va error shows that P_{det} after voiding still keeps high levels beyond their TVR while a high post-void residual volume is ruled out. It has a too high P_{ves} but a normal P_{abd}; the reason is that P_{ves} curve became a high level "dead" line due to the signal loss or the urethral catheter dislocation during voiding. In this case, post-void P_{det} is 57 cmH_2O, P_{ves} and P_{abd} are 92 cmH_2O and 35 cmH_2O respectively. **B**: type Vb error shows that P_{det} after voiding still keeps high levels. It has a too low P_{abd} because of the rectal catheter dislocation or the signal loss of P_{abd} tracing during voiding. In this case, post-void P_{det} is 70 cmH_2O, P_{ves} and P_{abd} are 79 cmH_2O and 9 cmH_2O respectively

Another one (type VI error) was that P_{ves} and P_{det} became negative because the urethral catheter was voided with the urine stream during voiding (Fig. 6).

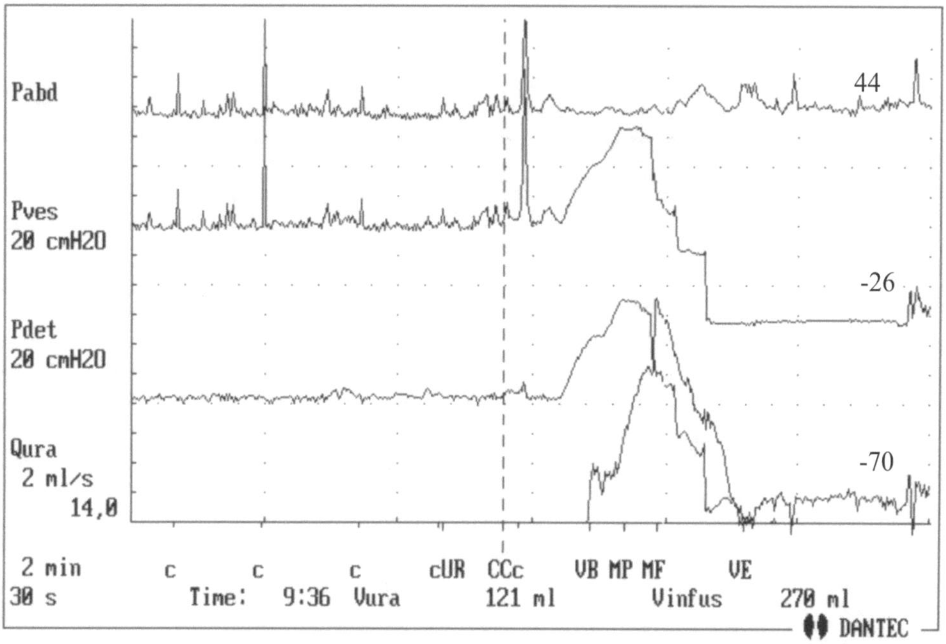

Fig. 6. Type VI error related to typical value ranges for post-void pressures. This type of error shows that P_{ves} and P_{det} become negative after voiding because the urethral catheter has been voided with the urine stream during voiding. In this case, post-void P_{det} is -70 cmH_2O, P_{ves} and P_{abd} are -26 cmH_2O and 44 cmH_2O respectively

3.2 Describing the typical signal patterns

Comparing the various types of signal patterns among P_{ves}, P_{abd} and P_{det} tracings in different cystometric phases, the results were shown in the following tables. From the statistical analyses, we were able to describe typical signal patterns during different cystometric phases. At beginning of filling, 91.8% of traces showed the identical fine structure and microscopic changes between P_{ves} and P_{abd} tracings while P_{det} tracing did not have this fine structure and the microscopic changes. 7.7% of traces showed the different patterns between P_{ves} and P_{abd} due to the problems of signal transmission (Table 5). In the cough tests of initial filling, 74.8% of traces had the equal pressure changes corresponding to the test-coughs between P_{ves} and P_{abd} while P_{det} did not show any change and deflection. 17.5% showed the similar changes between P_{ves} and P_{abd}, which lead to some small biphasic deflections on P_{det} tracing; these biphasic spikes were acceptable. 7.7% had the different changes between P_{ves} and P_{det}, which resulted in the obvious up- or down-deflections; these obvious spikes suggested the problems of pressure transmission (Table 5). In initial phase of filling, only 3.1% of traces showed the macroscopic changes. 0.9% had straining, which was characterized by the identical changes on P_{ves} and P_{abd} tracings but not on P_{det} tracing. 2.2% had rectal contractions, which showed the different changes among P_{ves}, P_{abd} and P_{det}

tracings: a positive wave on the P_{abd} tracing and a negative artifact on the P_{det} tracing but not on P_{ves} tracing (Table 5).

	P_{ves} and	P_{abd}		P_{det}			P_{ves} and P_{det}	
	Identical	Similar	Different	Without	Little	Obvious	Identical	Different
	no (%)	no (%)	no (%)	no (%)	no (%)	no (%)	no (%)	no (%)
Pattern I: fine structure	534 (91.8)	3 (0.5)	45 (7.7)	534 (91.8)	3 (0.5)	45 (7.7)	-	-
Pattern II: microscopic changes	534 (91.8)	3 (0.5)	45 (7.7)	534 (91.8)	3 (0.5)	45 (7.7)	-	-
Pattern III: cough tests changes	435 (74.8)	102 (17.5)	45 (7.7)	435 (74.8)	102 (17.5)	45 (7.7)	-	-
Pattern IV: macroscopic changes								
a: straining (no=5)	4 (80)	1 (20)	0 (0)	4 (80)	1 (20)	0 (0)	-	-
b: detrusor overactivity (no=0)	0	0	0	0	0	0	0	0
c: rectal contractions (no=13)	0 (0)	0 (0)	13 (100)	0 (0)	0 (0)	13 (100)	0 (0)	13 (100)

Table 5. Comparing the various of signal patterns among P_{ves}, P_{abd} and P_{det} tracings at beginning of filling in 582 cystometries

During filling phase, 98.3% of traces showed the identical fine structure and microscopic changes between P_{ves} and P_{abd} while P_{det} trace did not have this fine structure and the microscopic changes. 1.5% of traces showed the different patterns between P_{ves} and P_{abd} due to the problems of signal transmission (Table 6). In the cough tests at regular intervals during filling, 67.0% of traces had the equal pressure changes corresponding to the test-coughs between P_{ves} and P_{abd} while P_{det} did not show any changes and deflections. 31.5% showed the similar changes between P_{ves} and P_{abd}, which lead to some biphasic deflections on P_{det} tracing; these biphasic spikes were acceptable. 1.5% had the different changes between P_{ves} and P_{det}, which resulted in the obvious up- or down- deflections; these obvious spikes suggested the unequal pressure transmission (Table 6). There were various types of typical macroscopic changes occurred during filling. 8.3% of 582 traces showed straining. 91.7% of 48 straining traces were characterized by the identical changes on P_{ves} and P_{abd} tracings but not on P_{det} tracing; but 8.3% showed the changes on P_{det} tracing due to the different transmission to P_{ves} and P_{abd} corresponding to straining. 33.7% of 582 traces had detrusor overactivity, which showed that single or multiple unstable waves recorded on P_{ves} and P_{det} tracings but not on P_{abd} tracing. 17.4% of 582 had rectal activity, which showed single or multiple rectal contractions recorded on P_{abd} tracing; and the changes among P_{ves}, P_{abd} and P_{det} tracings were different: a positive wave on the P_{abd} tracing and a negative artifact on the P_{det} tracing but not on P_{ves} tracing (Table 6).

	P_{ves} and P_{abd}			P_{det}			P_{ves} and P_{det}	
	Identical	Similar	Different	Without	Little	Obvious	Identical	Different
	no (%)	no (%)	no (%)	no (%)	no (%)	no (%)	no (%)	no (%)
Pattern I: fine structure	572 (98.3)	1 (0.2)	9 (1.5)	572 (98.3)	1 (0.2)	9 (1.5)	-	-
Pattern II: microscopic changes	572 (98.3)	1 (0.2)	9 (1.5)	572 (98.3)	1 (0.2)	9 (1.5)	-	-
Pattern III: cough tests changes	390 (67.0)	183 (31.5)	9 (1.5)	390 (67.0)	183 (31.5)	9 (1.5)	-	-
Pattern IV: macroscopic changes								
a: straining (no=48)	44 (91.7)	4 (8.3)	0 (0)	44 (91.7)	4 (8.3)	0 (0)	-	-
b: detrusor overactivity (no=196)	0 (0)	0 (0)	196 (100)	0 (0)	0 (0)	196 (100)	196 (100)	0 (0)
c: rectal contractions (no=101)	0 (0)	0 (0)	101 (100)	0 (0)	0 (0)	101 (100)	0 (0)	101 (100)

Table 6. Comparing the various of signal patterns among P_{ves}, P_{abd} and P_{det} tracings during filling in 582 cystometries

Before voiding, 94.0% of traces had the equal rises in P_{ves} and P_{abd} corresponding to cough tests; 1.9% showed the small biphasic spikes; and 4.1% had the significant spikes, which suggested the different pressure transmission (Table 7).

	P_{ves} and P_{abd}			P_{det}		
	Identical	Similar	Different	Without	Little	Obvious
	no (%)	no (%)	no (%)	no (%)	no (%)	no (%)
Pattern I: fine structure (after voiding)	531 (91.2)	5 (0.9)	46 (7.9)	531 (91.2)	5 (0.9)	46 (7.9)
Pattern III: cough tests changes						
a: before voiding	547 (94.0)	11 (1.9)	24 (4.1)	547 (94.0)	11 (1.9)	24 (4.1)
b: after voiding	509 (87.5)	24 (4.1)	49 (8.4)	509 (87.5)	24 (4.1)	49 (8.4)

Table 7. Comparing the various of signal patterns among P_{ves}, P_{abd} and P_{det} tracings before and after voiding in 582 cystometries

During voiding, 91.2% of traces still kept the "live" signal on P_{ves} and P_{abd} tracings, which had the same fine structure and microscopic changes; but 7.9% showed the "dead" signals on

P_{ves} or P_{abd} tracing due to various causes, which showed the different fine structures and microscopic changes (Table 8). Also, there were several types of typical macroscopic changes occurred during voiding. 95.2% of traces showed the identical and simultaneous increase and decrease on P_{ves} and P_{det} when detrusor contracted to void; 4.8% had the atypical patterns due to urethral catheter dislocation or signal loss of P_{ves} during voiding. However, there were three special patterns of detrusor contraction in our data: 0.9% after-contraction, 3.4% fluctuation contraction and 3.1% unstable detrusor voiding. The traces with after-contraction showed that P_{ves} and P_{det} increased after uroflow tracing; ones with fluctuation contraction showed that P_{ves} and P_{det} tracings fluctuated corresponding to the simultaneous changes of uroflow tracing; ones with unstable detrusor voiding showed that P_{ves} and P_{det} increased rapidly before uroflow tracing, then fell suddenly as soon as flow started. 53.3% of 582 traces showed the different degrees of straining during voiding. 71.6% of 310 straining traces were characterized by the identical changes on P_{ves} and P_{abd} tracings but not on P_{det} tracing. 28.4% showed various changes on P_{det} tracing, which were spikes and dips on P_{det} curve due to the different transmission to P_{ves} and P_{abd} corresponding to straining. 2.1% of 582 traces had rectal contractions during voiding, which lead to different patterns between P_{ves} and P_{det}, and were characterized by some dips on P_{det} curve. 15.3% of 582 traces showed the different degrees of relaxation of pelvic floor during voiding, which were characterized by the different degrees of decreases on P_{abd} curve, and resulted in the similar or identical changes on P_{ves} and P_{det} curves (Table 8).

	P_{ves} and P_{abd}			P_{det}			P_{ves} and P_{det}	
	Identical	Similar	Different	Without	Little	Obvious	Identical	Different
	no (%)	no (%)	no (%)	no (%)	no (%)	no (%)	no (%)	no (%)
Pattern I: fine structure	531 (91.2)	5 (0.9)	46 (7.9)	531 (91.2)	5 (0.9)	46 (7.9)	-	-
Pattern II: microscopic changes	531 (91.2)	5 (0.9)	46 (7.9)	531 (91.2)	5 (0.9)	46 (7.9)	-	-
Pattern IV: macroscopic changes								
a: detrusor contraction (no=582)	0 (0)	0 (0)	582 (100)	0 (0)	0 (0)	582 (100)	554 (95.2)	28 (4.8)
b: straining (no=310)	220 (71.0)	2 (0.6)	88 (28.4)	220 (71.0)	2 (0.6)	88 (28.4)	-	-
c: rectal contractions (no=12)	0 (0)	0 (0)	12 (100)	0 (0)	0 (0)	12 (100)	0 (0)	12 (100)
d: relaxation (no=89)	0 (0)	0 (0)	89 (100)	0 (0)	89 (100)	0 (0)	89 (100)	0 (0)

Table 8. Comparing the various of signal patterns among P_{ves}, P_{abd} and P_{det} tracings during voiding in 582 cystometries

After voiding, 91.2% of traces had the same fine structure between P_{ves} and P_{abd} tracings, which were "live" signal; but 7.9% showed the "dead" signals on P_{ves} or P_{abd} tracings due to various artifacts occurred during voiding. These traces had the different fine structures and microscopic changes on P_{ves} and P_{abd} tracings during and after voiding. After voiding, 87.5% of traces had the equal pressure rises in P_{ves} and P_{abd} corresponding to cough tests; 4.1% showed some acceptable biphasic spikes on P_{det} tracing; and 8.4% did obvious spikes suggesting the problems of signal quality during voiding (Table 7).

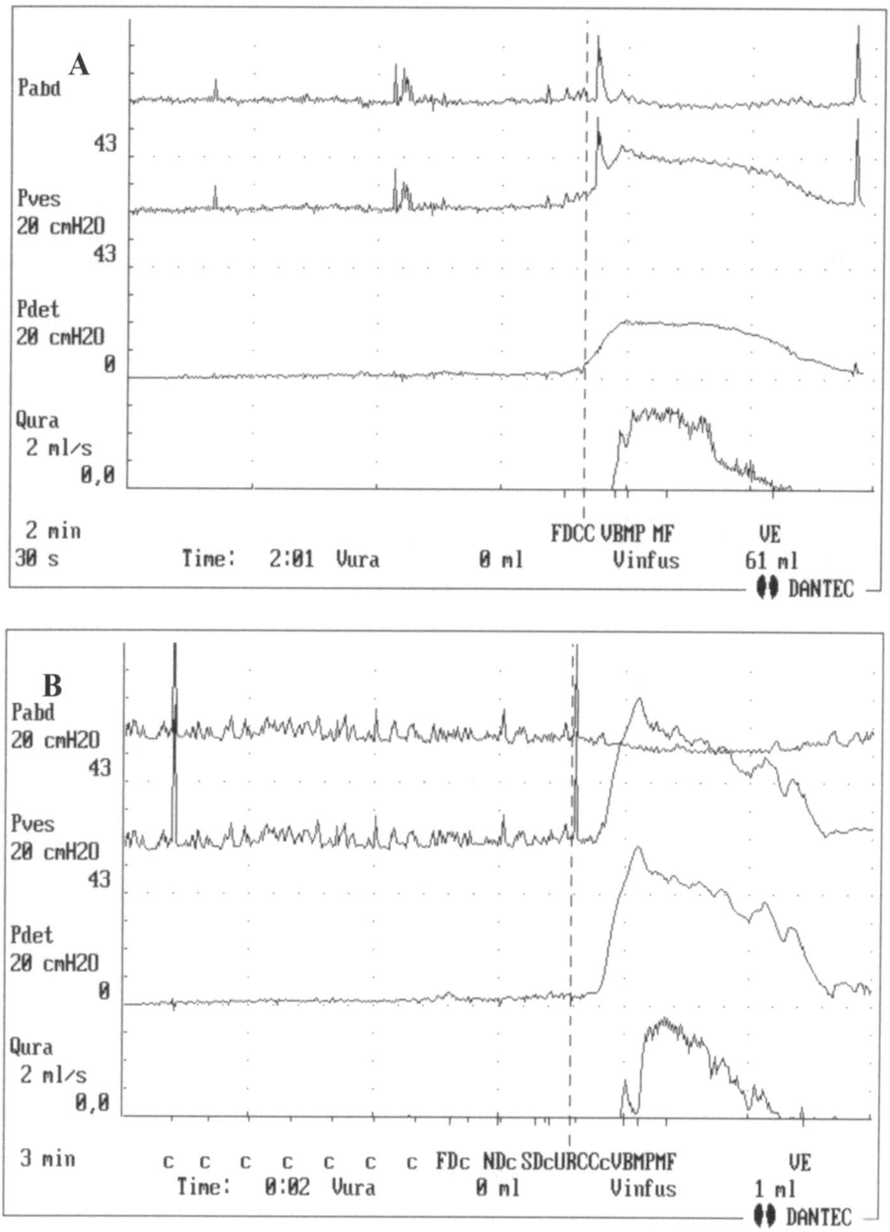

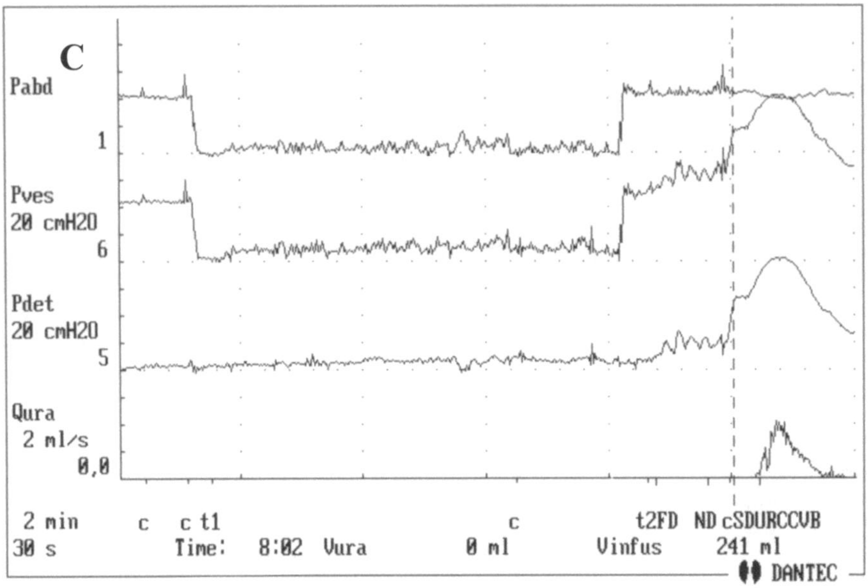

Fig. 7. The fine structure (pattern I) and microscopic changes (pattern II) of signals. **A**: there are the same fine structures on P_{ves} and P_{abd} tracings at all stages of investigation, which show the "live" signals with some minimal variations (noise); but P_{det} tracing does not have fine structure. **B**: fine structure becomes the stronger signal activity called microscopic changes with patient's breathing and talking; P_{ves} and P_{abd} tracings show the identical microscopic changes, but P_{det} tracing has no change. **C**: P_{ves} and P_{abd} tracings keep identical changes with patient's moving or position change, P_{det} tracing does not show obvious changes

On the base of above-mentioned results, we can describe TSP of pressures during cystometry from four gradations. At beginning of and during filling, P_{ves} and P_{abd} tracings have the identical fine structure, which showed the "live" signals with some minimal variations (noise). The fine structure become the stronger signal activity corresponding to the patient's breathing, talking or moving, which means that there are the same microscopic changes on P_{det} and P_{abd} tracings. P_{det} tracing dose not show any fine structure and microscopic changes (Fig. 7a, b and c). When the cough tests are undergone at beginning of filling or at regular intervals during filling, the equal major pressure changes on P_{ves} and P_{abd} tracings are produced corresponding to the test-coughs; P_{det} tracing dose not show any changes, or at most, has some small biphasic spikes; these points suggest a high signal quality (Fig. 8a, 8b). There may be several types of typical macroscopic signal changes. Straining is characterized by the identical pressure changes on P_{ves} and P_{abd} tracings in response to strains but not on P_{det} tracing (Fig. 9a). Detrusor overactivity shows that single or multiple unstable waves due to detrusor contractions are recorded on P_{ves} and P_{det} tracings but not on P_{abd} tracing (Fig. 9b). Rectal activity, which results from single or multiple rectal contractions, is characterized by the different changes among P_{ves}, P_{abd} and P_{det} tracings: a positive wave on the P_{abd} tracing and a negative artifact on the P_{det} tracing but not on P_{ves} tracing (Fig. 9c). Before voiding, P_{ves} and P_{abd} have the equal response to the cough tests; and there are may be some biphasic spikes on P_{det} tracing but without obvious spikes (Fig. 8b). During voiding, P_{ves} and P_{abd} tracings still keep the "live" signals, and show the same fine structure and microscopic changes; but P_{det} tracing does not have (Fig. 10).

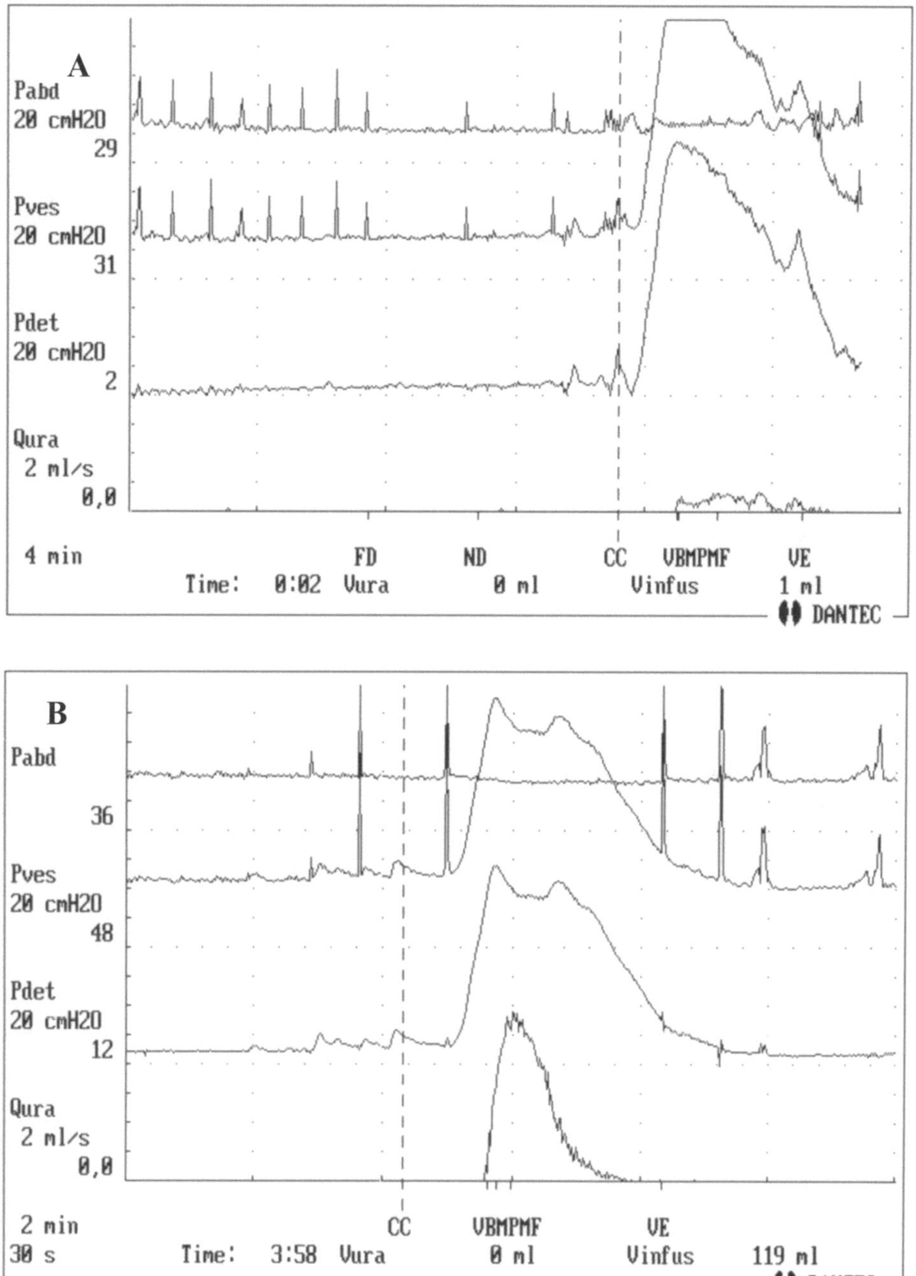

Fig. 8. The major changes of signals corresponding to cough tests (pattern III). **A**: P_{ves} and P_{abd} tracings show equal pressure changes corresponding to cough tests at beginning of filling and during filling at regular intervals; there is no obvious changes on P_{det} tracing. **B**: P_{ves} and P_{abd} tracings show equal pressure changes corresponding to cough tests before and after voiding; there is no obvious changes on P_{det} tracing

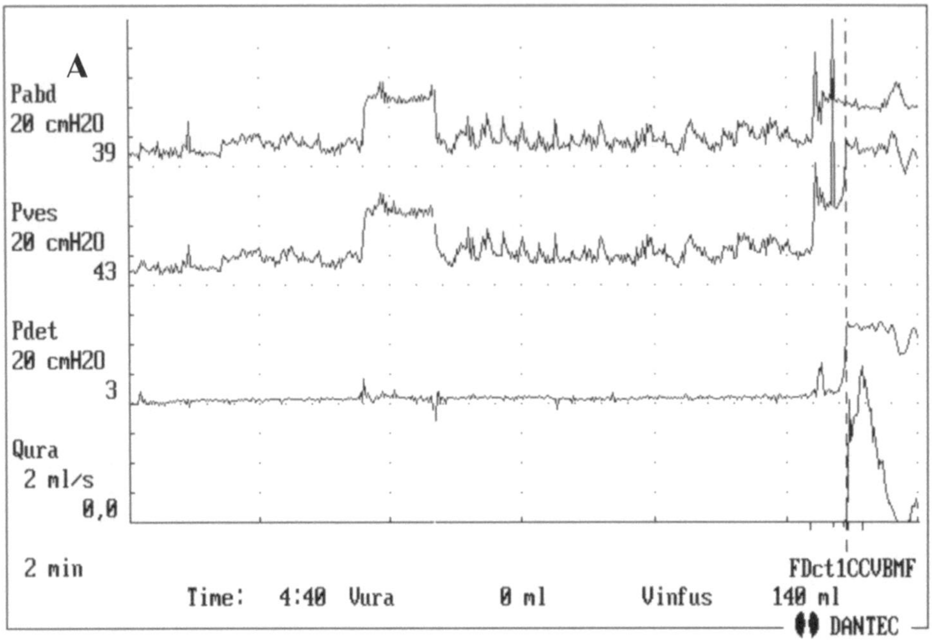
A
Pabd
20 cmH20
39
Pves
20 cmH20
43
Pdet
20 cmH20
3
Qura
2 ml/s
0,0
2 min
FDct1CCVBMF
Time: 4:40 Vura 0 ml Vinfus 140 ml
DANTEC

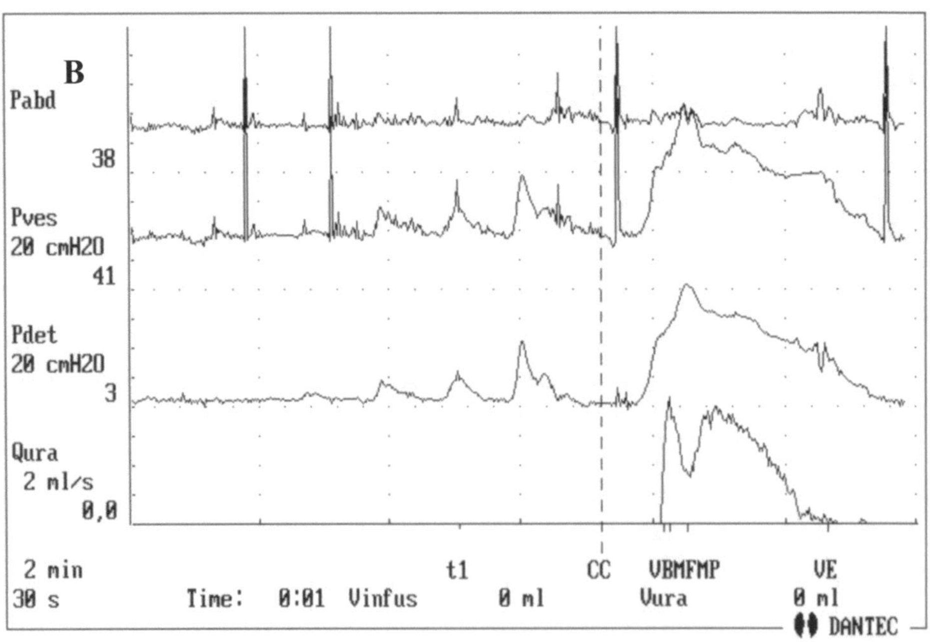
B
Pabd
38
Pves
20 cmH20
41
Pdet
20 cmH20
3
Qura
2 ml/s
0,0
2 min
30 s
t1
CC
VBMFMP
VE
Time: 0:01 Vinfus 0 ml Vura 0 ml
DANTEC

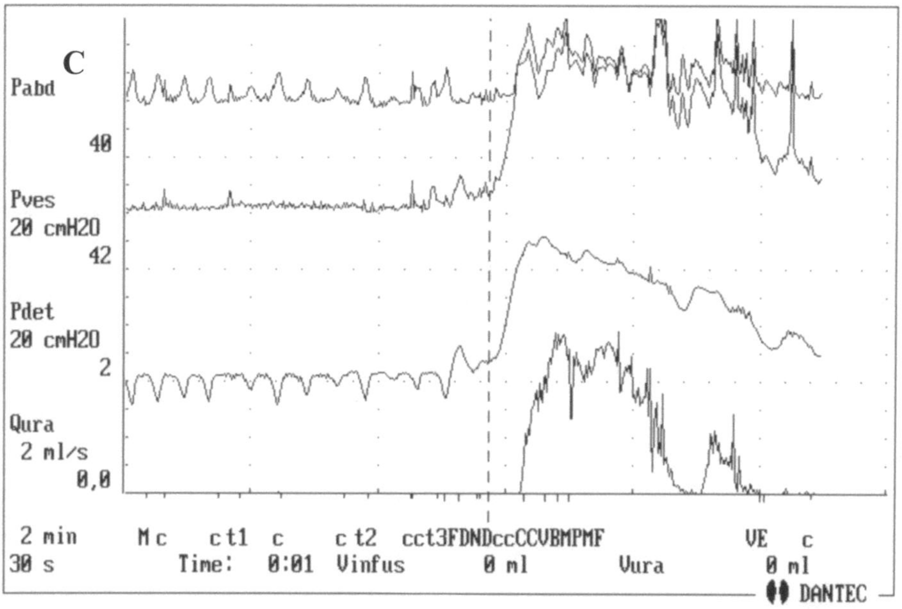

Fig. 9. The macroscopic changes of signals during filling: straining, detrusor overactivity and rectal contractions (pattern IV). **A**: straining is characterized by the identical signal changes on P_{ves} and P_{abd} tracings corresponding to strains but not on P_{det} tracing. **B**: detrusor overactivity shows that the unstable waves on P_{ves} and P_{det} tracings but not on P_{abd} tracing. **C**: rectal contractions are characterized by the different changes among P_{ves}, P_{abd} and P_{det} tracings: the positive waves on P_{abd} tracing and the negative dips on P_{det} tracing but not on P_{ves} tracing

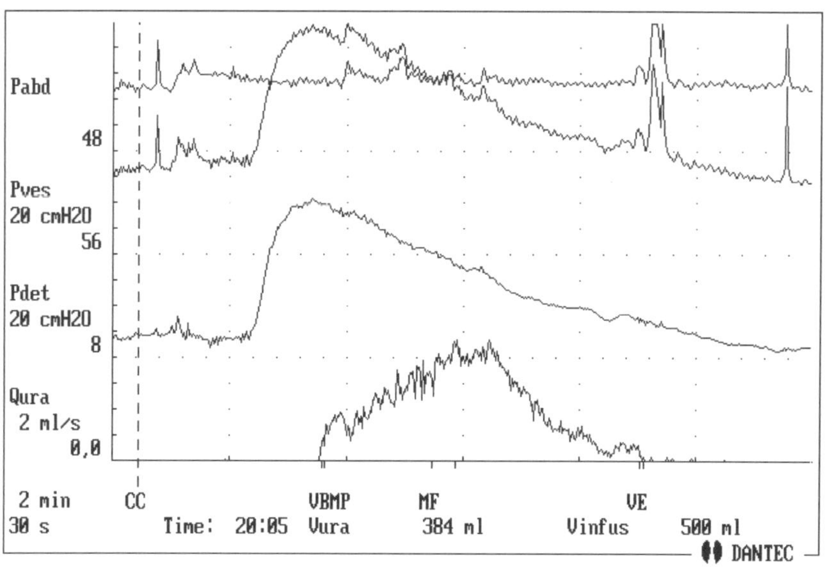

Fig. 10. The fine structures and microscopic changes during and after voiding (pattern I and II). During and after voiding, P_{ves} and P_{abd} tracings still keep the "live" signals, and show the same fine structure and microscopic changes; but P_{det} tracing does not have

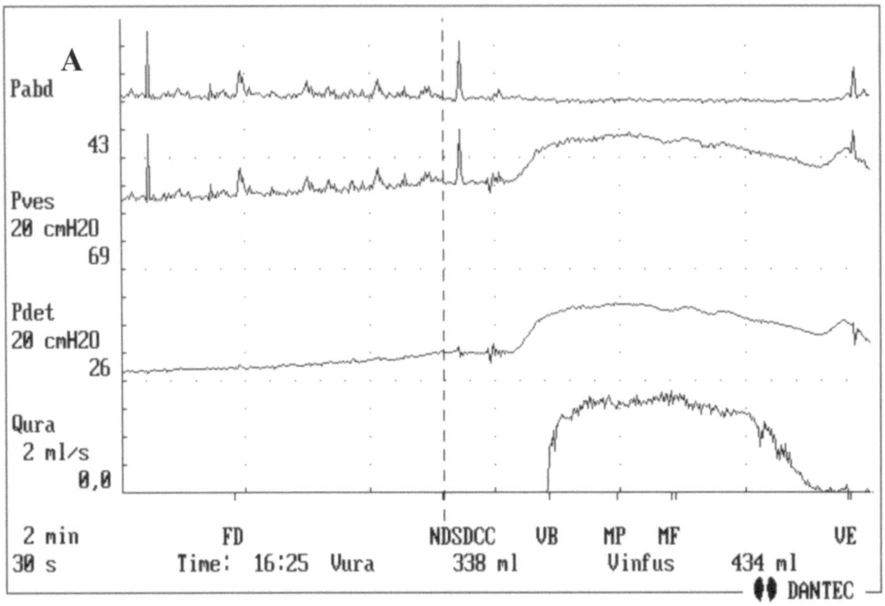
A
Pabd
43
Pves
20 cmH2O
69
Pdet
20 cmH2O
26
Qura
2 ml/s
0,0
2 min
30 s
FD
NDSDCC
VB
MP
MF
VE
Time: 16:25 Vura 338 ml
Vinfus 434 ml
DANTEC

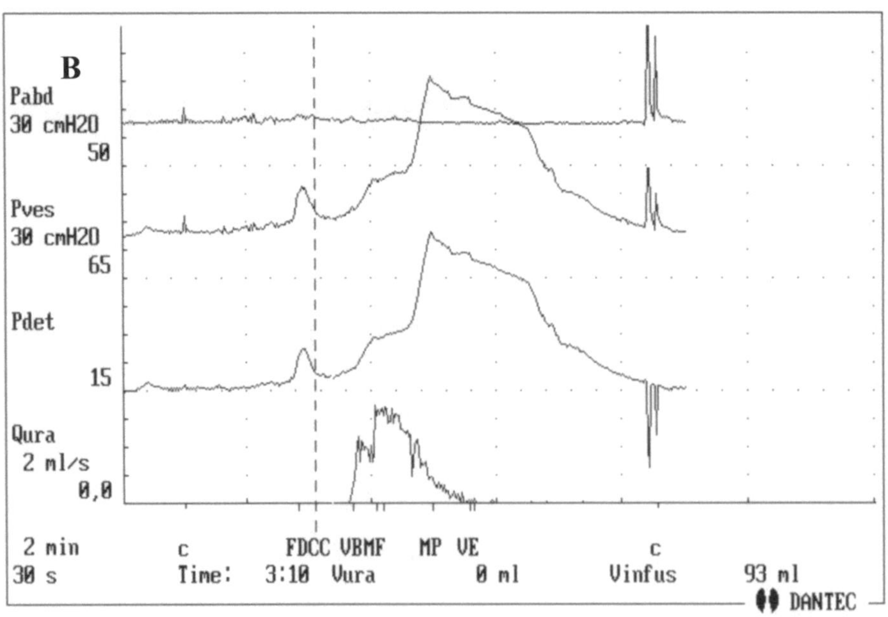
B
Pabd
30 cmH2O
50
Pves
30 cmH2O
65
Pdet
15
Qura
2 ml/s
0,0
2 min
30 s
c
FDCC VBMF
MP VE
c
Time: 3:10 Vura 0 ml
Vinfus 93 ml
DANTEC

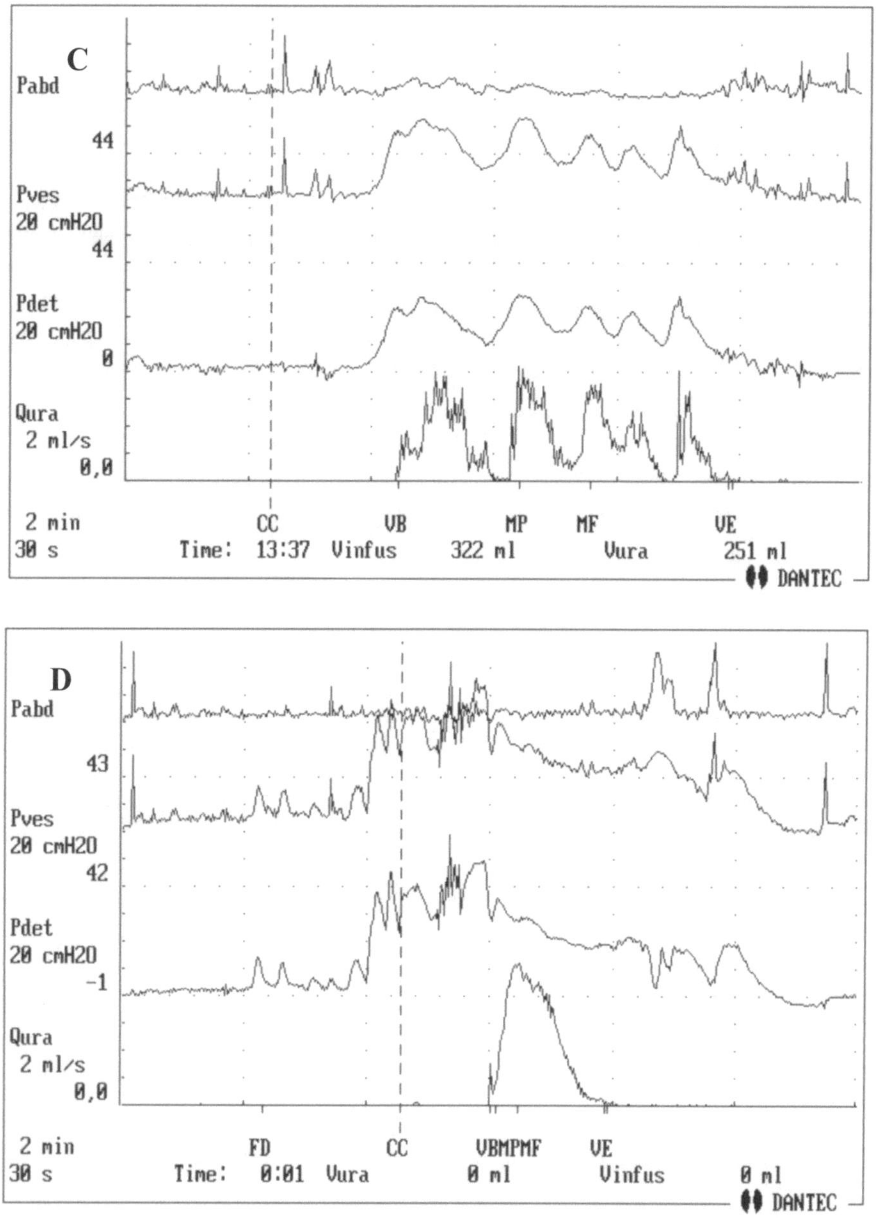

Fig. 11. The macroscopic changes of signals during voiding (pattern IV): the different patterns of detrusor contraction. **A**: typical pattern: typical detrusor contraction shows the smooth and identical pressure increase and decrease on P_{ves} and P_{det} tracings corresponding to the simultaneous changes on uroflow tracing. **B**: special pattern: after-contraction is characterized by that P_{ves} and P_{det} increase after uroflow tracing. **C**: special pattern: fluctuation contraction shows that P_{ves} and P_{det} tracings fluctuate corresponding to the simultaneous changes of uroflow tracing. **D**: special pattern: unstable detrusor voiding shows that P_{ves} and P_{det} increase rapidly before uroflow tracing, then fall suddenly when flow starts

There may be several types of typical macroscopic changes occurred during voiding. Typical detrusor contraction shows identical and simultaneous pressure increase and decrease on P_{ves} and P_{det} tracings (Fig. 11a); but there are some special patterns of detrusor contraction, for example, after-contraction, fluctuation contraction and unstable detrusor voiding. After-contraction is characterized by that Pves and Pdet increase after uroflow tracing (Fig.11b); fluctuation contraction shows that Pves and Pdet tracings fluctuate corresponding to the simultaneous changes of uroflow tracing (Fig. 11c); unstable detrusor voiding shows that Pves and Pdet increase acutely before uroflow tracing, then fall suddenly as soon as flow starts (Fig. 11d). Straining is characterized by the identical pressure changes on Pves and Pabd tracings but not on Pdet tracing (Fig. 12a). Rectal contractions lead to different patterns between Pves and Pdet, and are characterized by some dips on Pdet curve (Fig. 12b). Relaxation of pelvic floor is characterized by the pressure decreases of different degrees on Pabd tracing, and the changes on Pves and Pdet tracings are similar or identical (Fig. 12c). After voiding, Pves and Pabd tracings still have the same fine structure and microscopic changes with "live" signal, and keep the equal response to the cough tests; they indicate that there is a good signal quality during voiding (Fig. 8b, Fig. 10).

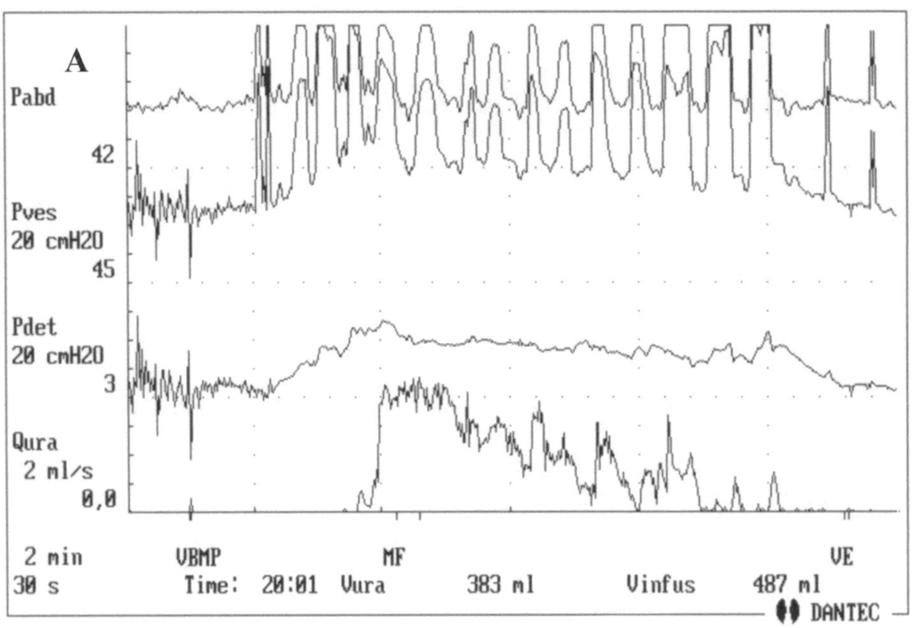

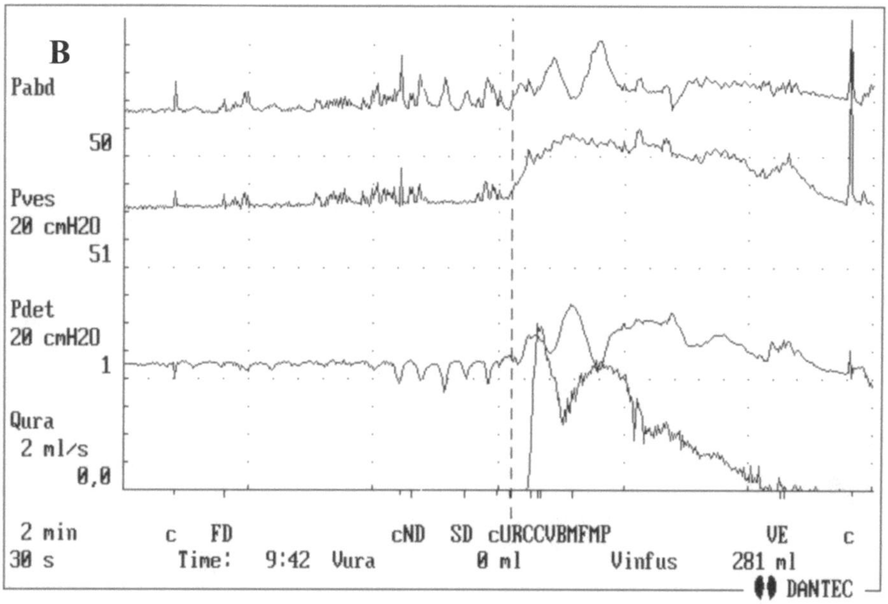

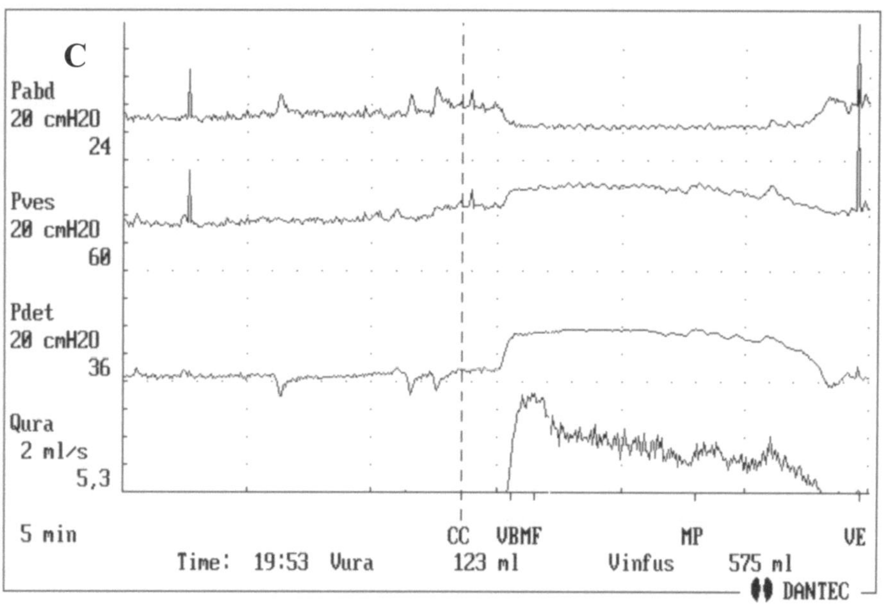

Fig. 12. The macroscopic changes of signals during voiding (pattern IV): straining, rectal contractions and relaxation of pelvic floor. **A**: straining is characterized by the identical pressure changes on P_{ves} and P_{abd} tracings but not on P_{det} tracing. **B**: rectal contractions lead to different patterns between P_{ves} and P_{det}, and are characterized by two dips on P_{det} curve. **C**: relaxation of pelvic floor shows the pressure decreases of different degrees on P_{abd} tracing, and the changes on P_{ves} and P_{det} tracings are similar or identical

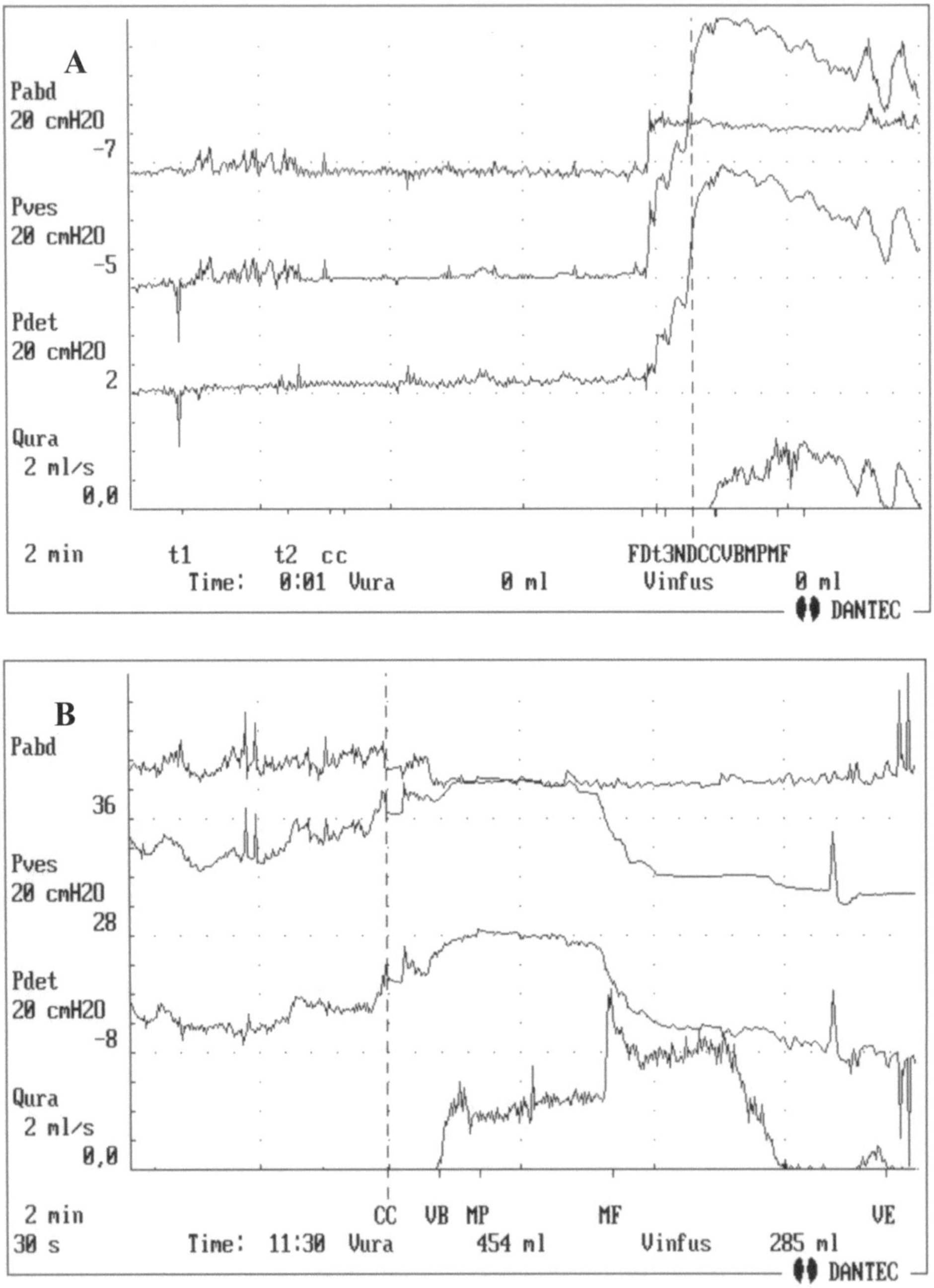

Fig. 13. The errors and artifacts related to fine structure and minor changes. **A**: fine structure and minor changes on P_{ves} tracing disappeared during filling phase, suggesting the problems of signal quality. **B**: fine structure and minor changes on P_{ves} tracing disappeared during voiding phase, suggesting the problems of signal quality or other artifacts occurred during voiding

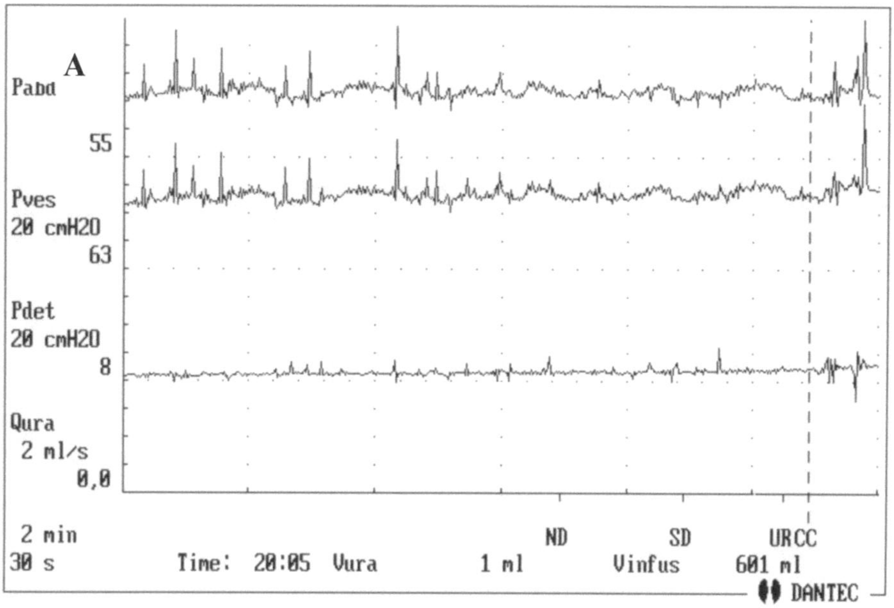
A
Pabd
55
Pves
20 cmH2O
63
Pdet
20 cmH2O
8
Qura
2 ml/s
0,0
2 min
30 s
ND
SD
UR CC
Time: 20:05 Vura 1 ml Vinfus 601 ml
DANTEC

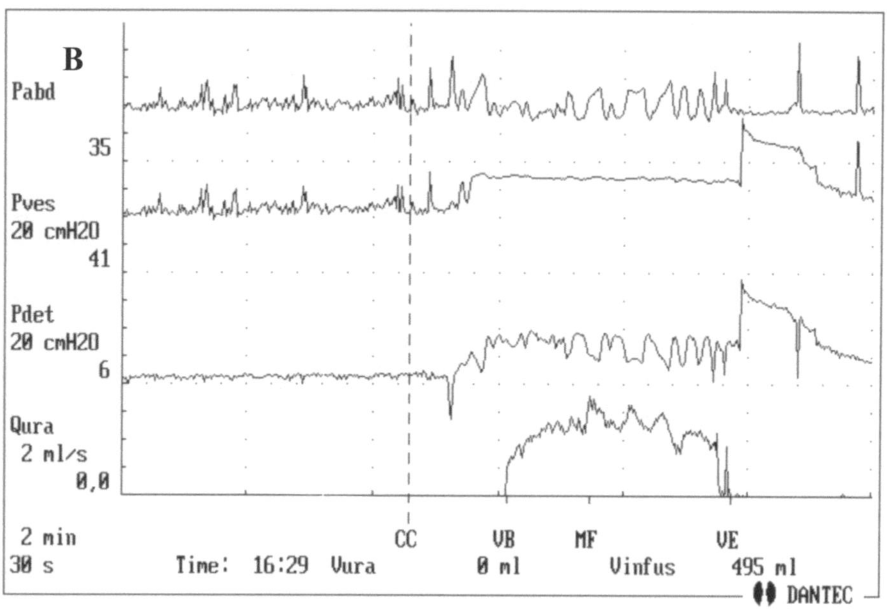
B
Pabd
35
Pves
20 cmH2O
41
Pdet
20 cmH2O
6
Qura
2 ml/s
0,0
2 min
30 s
CC
VB
MF
VE
Time: 16:29 Vura 0 ml Vinfus 495 ml
DANTEC

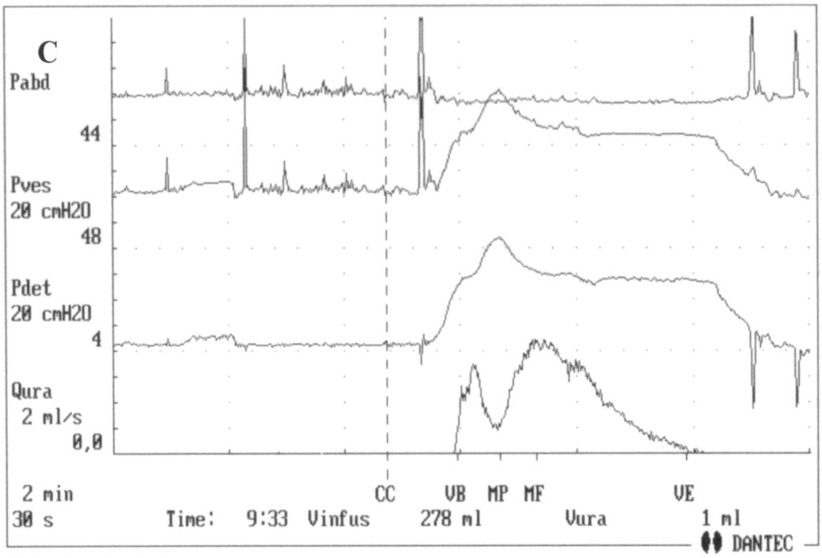

Fig. 14. The errors and artifacts caused by the different responses to cough tests between P_{ves} and P_{abd} tracings during different phases. **A**: there are some up- or down deflections and biphasic spikes on P_{det} tracing due to the different pressure transmissions between P_{ves} and P_{abd} tracings during filling. **B**: P_{ves} tracing does not respond to cough-test before voiding, then a P_{ves} tracing with "dead" signals appear during voiding. **C**: P_{ves} tracing does not respond to cough-tests after voiding, this suggests the signal loss on P_{ves} tracing during voiding

According to above-mentioned TSP, we gave some examples for atypical signal patterns or artifacts to indicate the role and significance of TSP in quality control during cystometry. As the indicator of "live" signals, fine structure and minor changes should exist at all stages of investigation; their disappearance indicated the certain problems of signal quality. Fig. 13a and 13b showed that fine structure and minor changes on P_{ves} tracing disappeared during filling and voiding phases respectively. Cough test is a powerful tool to examine the signal quality during different cystometric phases; the different response to test-coughs between P_{ves} and P_{abd} tracings suggested the problems of pressure transmission and other artifacts. Fig. 14a showed some deflections and biphasic spikes on P_{det} tracing due to the different pressure transmissions during filling; Fig.14b showed that P_{ves} did not respond to cough-tests before voiding, then a P_{ves} tracing with "dead" signal appeared during voiding; Fig.14c showed that P_{ves} did not respond to cough-tests after voiding, this suggested the signal loss on P_{ves} tracing during voiding. Therefore, the cough tests before and after voiding are especially important for signal quality control during voiding. The different response to straining between P_{ves} and P_{abd} can lead to some artifacts on P_{det} tracing; Fig.15a showed this type of artifact occurred during filling; Fig.15b showed some spikes on P_{det} tracing due to the different transmissions to straining during voiding. The sudden pressure drops on both P_{ves} and P_{det} tracings often suggest urethral catheter loss during voiding (Fig.16). The scale of signal is the base of TSP recognition; a changed scale often leads a wrong impression. Fig.17 suggested that an enhanced scale on P_{det} lead to an obvious fine structure on P_{det} tracing, which was not equal to the subtraction between P_{ves} and P_{abd} tracings.

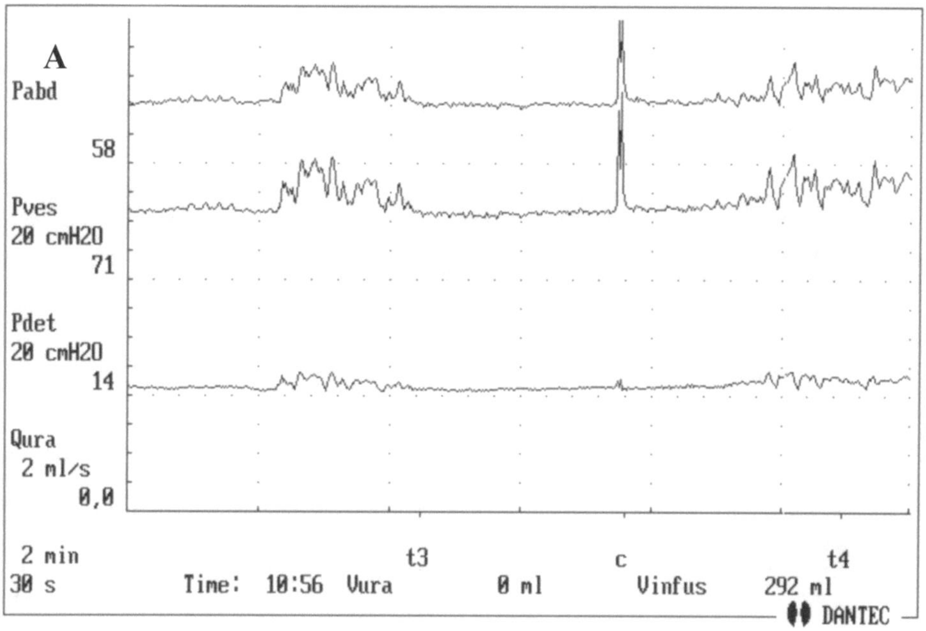

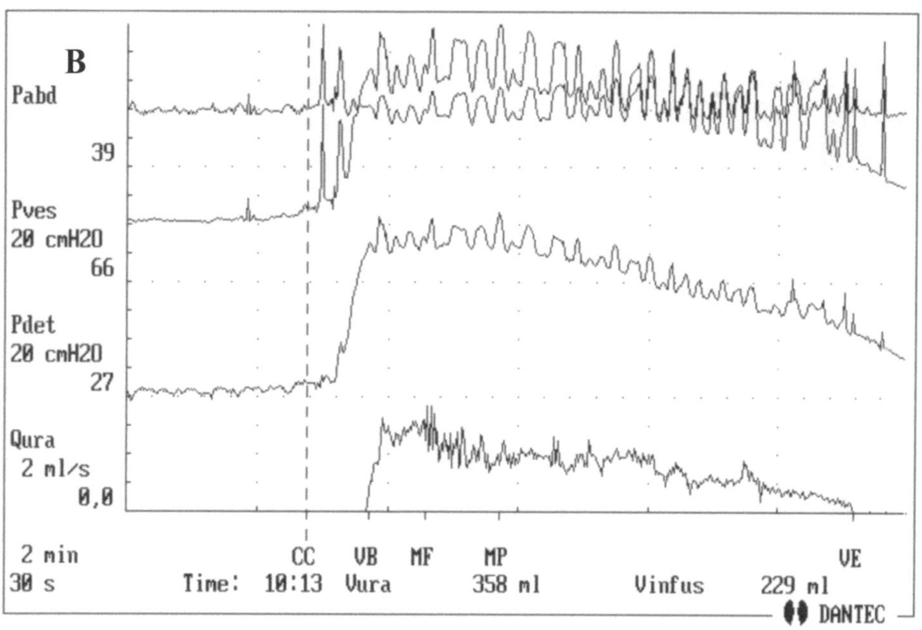

Fig. 15. The errors and artifacts resulted from the different responses to straining between P_{ves} and P_{abd} tracings. **A**: there are some artifacts on P_{det} tracing due to the different transmission to straining during filling. **B**: some spikes on P_{det} tracing due to the different transmission to straining during voiding

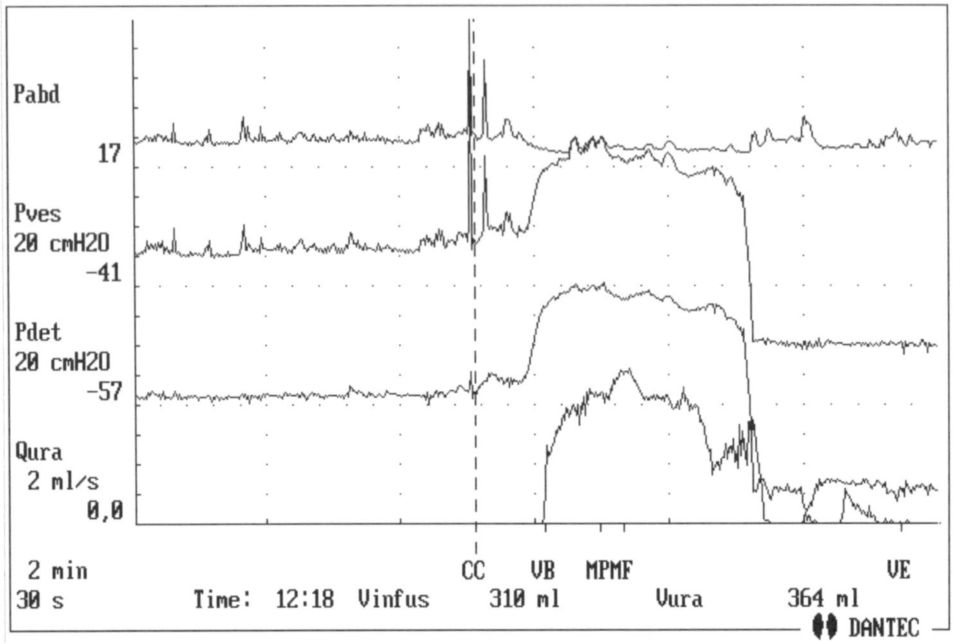

Fig. 16. The sudden drops on both P_{ves} and P_{det} tracings suggest urethral catheter loss during voiding

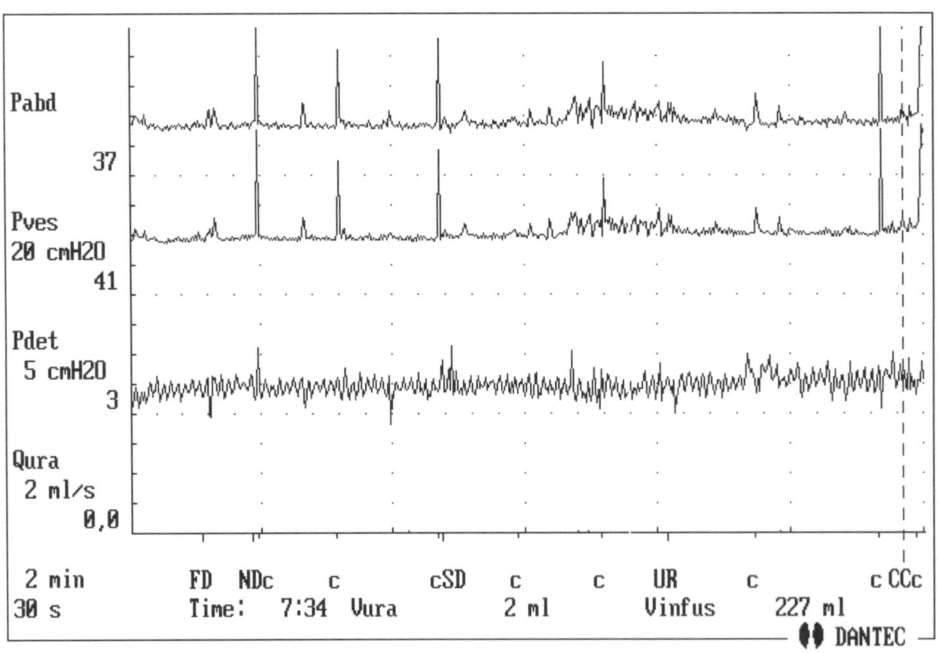

Fig. 17. The artifacts related to the signal scale: a changed scale often leads a wrong impression. This case shows that an enhanced scale on P_{det} leads to an obvious fine structure on P_{det} tracing, which is not equal to the subtraction between P_{ves} and P_{abd} tracings

3.3 Retrospective quality control of urodynamic data

582 (96.0%) of 606 pressure-flow traces were included and analyzed; 24 (4.0%) traces had to be discarded due to non-interpretable and non-correctional artifacts.

3.3.1 Artifacts during filling phase

4.5% of 582 traces had the negative initial resting P_{det} (Fig. 2); and 1.4% had over high initial resting P_{det} (Fig. 3). 31.5% of 582 traces showed the biphasic spikes, and 1.5% showed obvious deflections (14a). 1.6% of 582 traces showed the phasic loss of pressure signals; 1.0% showed stepwise changes on P_{ves} or P_{abd} tracings.

3.3.2 Artifacts during voiding phase

81.8% (476/582) traces showed obvious artifacts of Q_{max} (Fig. 18a, b); 23.1% of 476 traces were corrected for the location of Q_{max} on uroflow tracing (Fig. 19a, b).

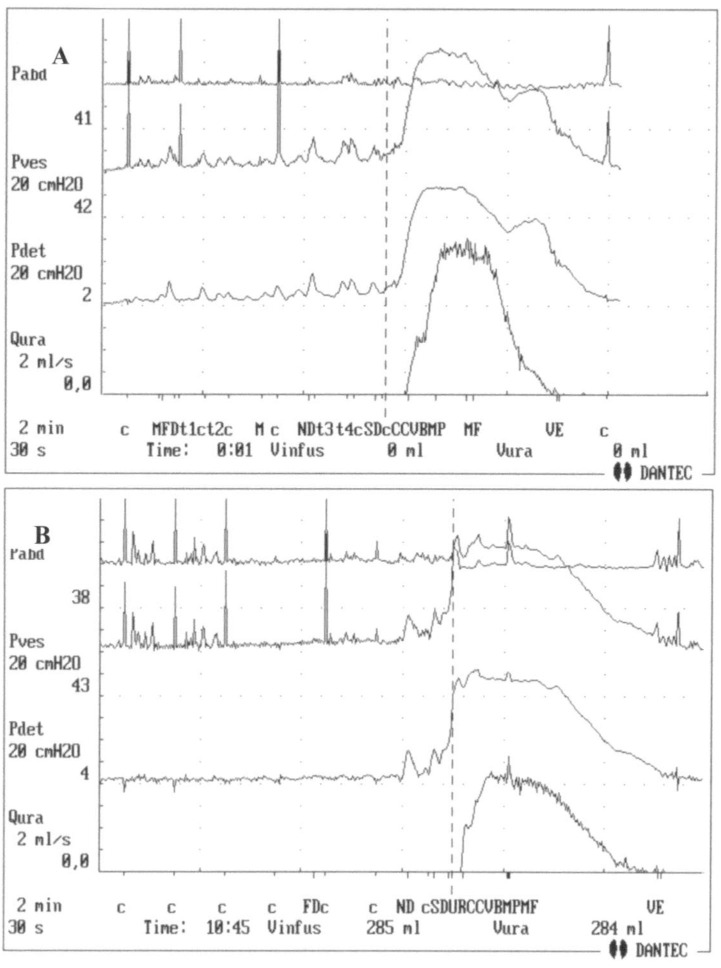

Fig. 18. Artifacts of Q_{max} on a main uroflow curve. **A**: the additional modifications in flowrate showed many small spikes on uroflow curve due to uroflowmeter. **B**: a spike artifact located at the highest plateau on a main uroflow curve due to straining

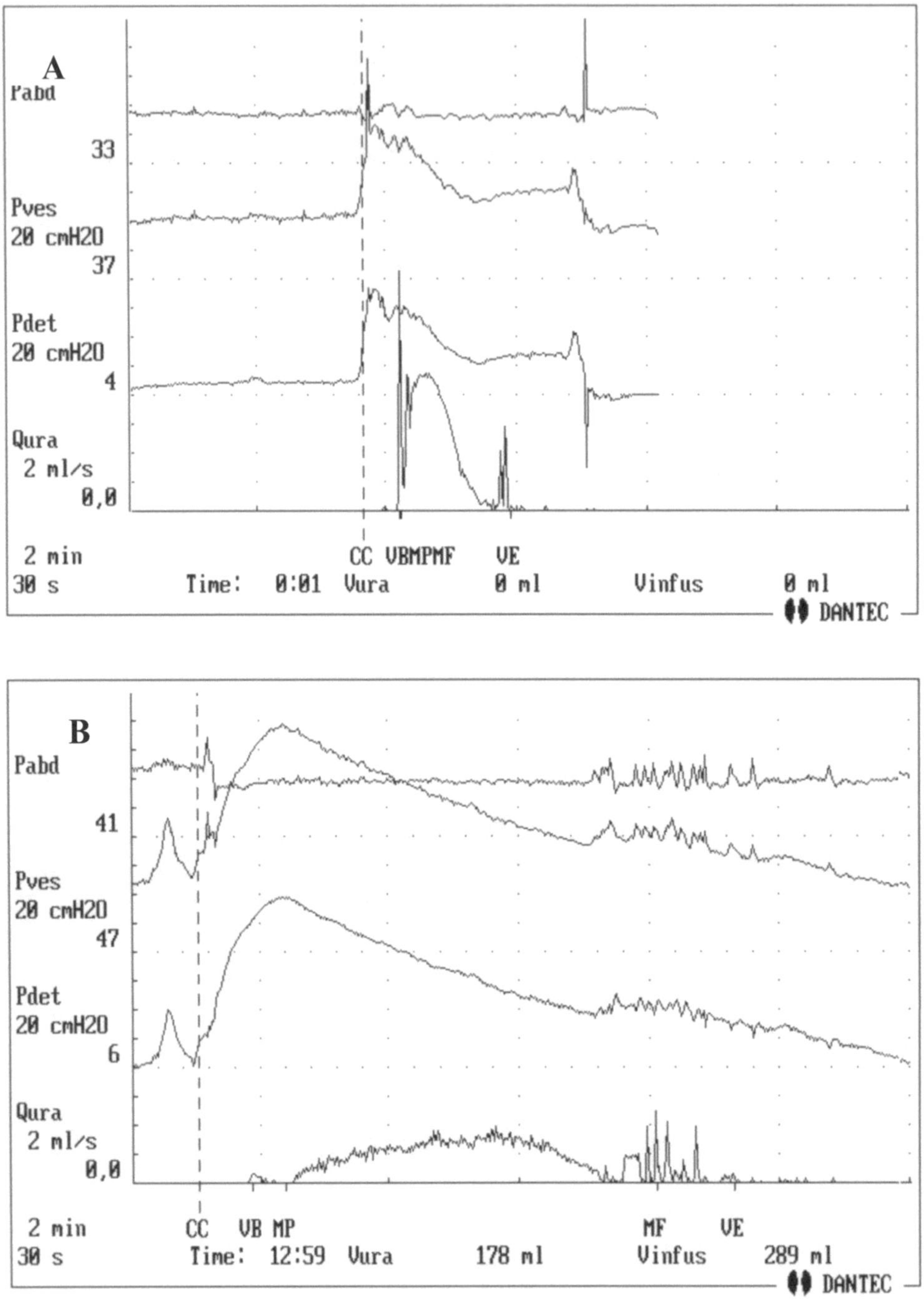

Fig. 19. Artifacts of Q_{max} at beginning and end of voiding. **A**: a spike artifact occurred at beginning of uroflow changed the location of Q_{max} in computerized data analysis. **B**: the spike artifacts appeared at end of uroflow tracing due to straining changed the location of Q_{max} in computerized data analysis

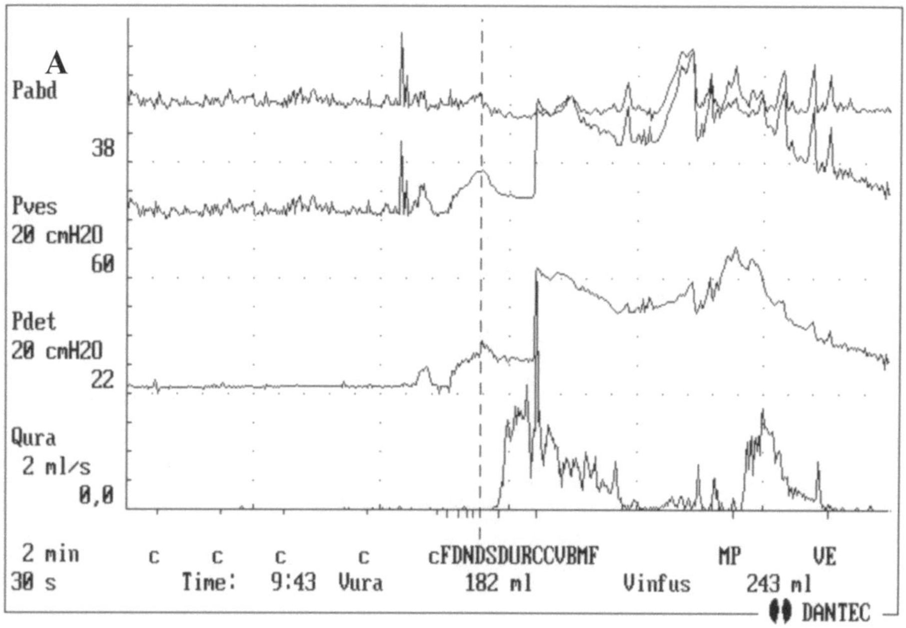
A
Pabd
38
Pves
20 cmH2O
60
Pdet
20 cmH2O
22
Qura
2 ml/s
0,0
2 min
30 s
c c c c cFDNDSDURCCVBMF MP VE
Time: 9:43 Vura 182 ml Vinfus 243 ml
DANTEC

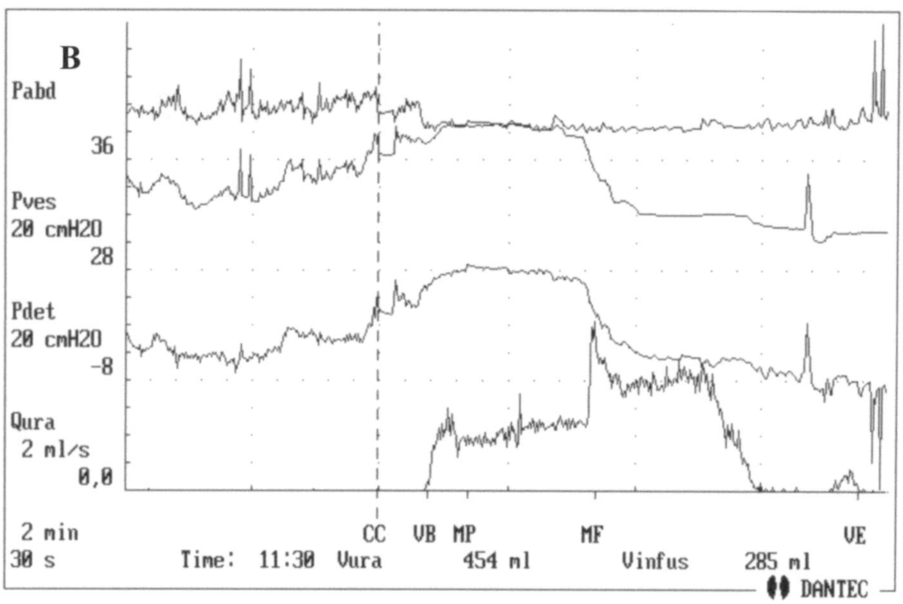
B
Pabd
36
Pves
20 cmH2O
28
Pdet
20 cmH2O
-8
Qura
2 ml/s
0,0
2 min
30 s
CC VB MP MF VE
Time: 11:30 Vura 454 ml Vinfus 285 ml
DANTEC

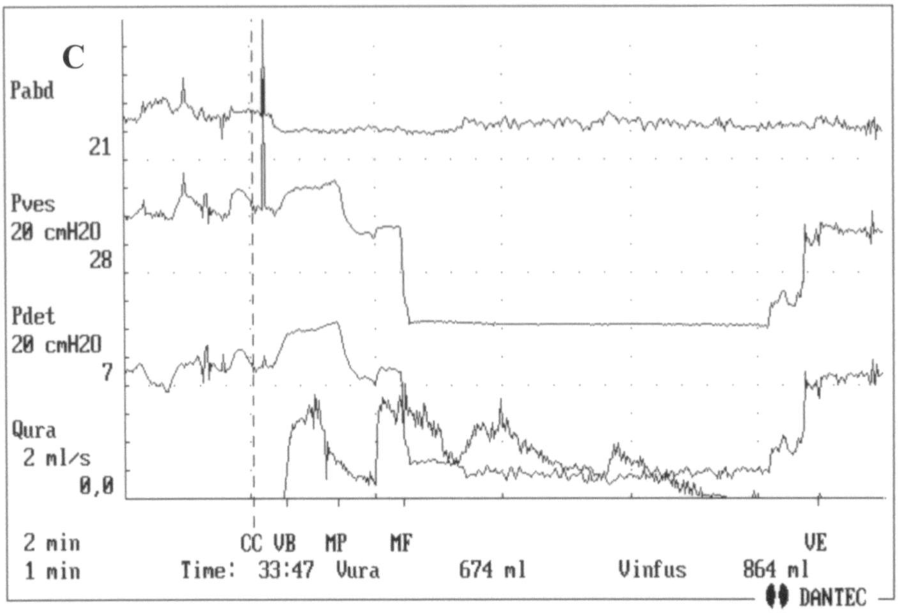

Fig. 20. The technical artifacts on pressure tracings. **A**: a periodic signal loss on P_{ves} tracing occurred at earlier stage of voiding. **B**: the signal stepwise changes on P_{ves} tracing appeared during voiding. **C**: urethral catheter was fallen out during voiding

Concerning technical artifacts during voiding, 1.4% of 582 traces showed the phasic loss of pressure signals (Fig. 20a); 3.4% had stepwise changes on P_{ves} tracing (Fig. 20b); 0.9% traces were found that urethral catheter was fallen out during voiding (Fig. 20c); and 2.2% traces were with the dislocation of rectal catheter.

With respect to the physiologic artifacts, 15.1% of 582 traces showed the spikes on P_{det} tracing due to straining (Fig. 21a); 2.1% had the dips on P_{det} tracing due to rectal contractions (Fig. 21b); 10.1% showed the spikes or dips on P_{det} tracing caused by urethral sphincter overactivity (Fig. 22a, b).

3.3.3 Effects of quality control

Comparing the manual values of pressure-flow data with computerized ones, we found the changes in parameters and urethral resistance, and in grading, classifying and diagnosing of obstruction after correction. Q_{max} had a consistently significant decrease ($p<0.001$), with a mean of 1.17 ml/sec, and had a changed range of -0.5~10.4 ml/sec. $P_{det.Qmax}$ had inconsistent changes with a slight systematic increase, with a mean of 0.75 cmH_2O, but no significant variation was demonstrated ($P>0.05$). Concerning the changes of $P_{det.Qmax}$ after manual correction, 321 (55.2%) of 582 traces had a significant increase ($p<0.01$), with a mean of 4.90 cmH_2O; 184 (31.6%) had a no significant decrease ($p>0.05$), with a mean of 6.16 cmH_2O; 77 (13.2%) did not change; and 505 (86.8%) underwent intra-individual changes with a range of -70 ~ 56 cmH_2O. OCO underwent a systematically significant increase by 0.067 on average ($p<0.05$); intra-individual changes were inconsistent, with a range of -1.379 ~ 0.995 (Table 9,10 fig. 23, 24, 25).

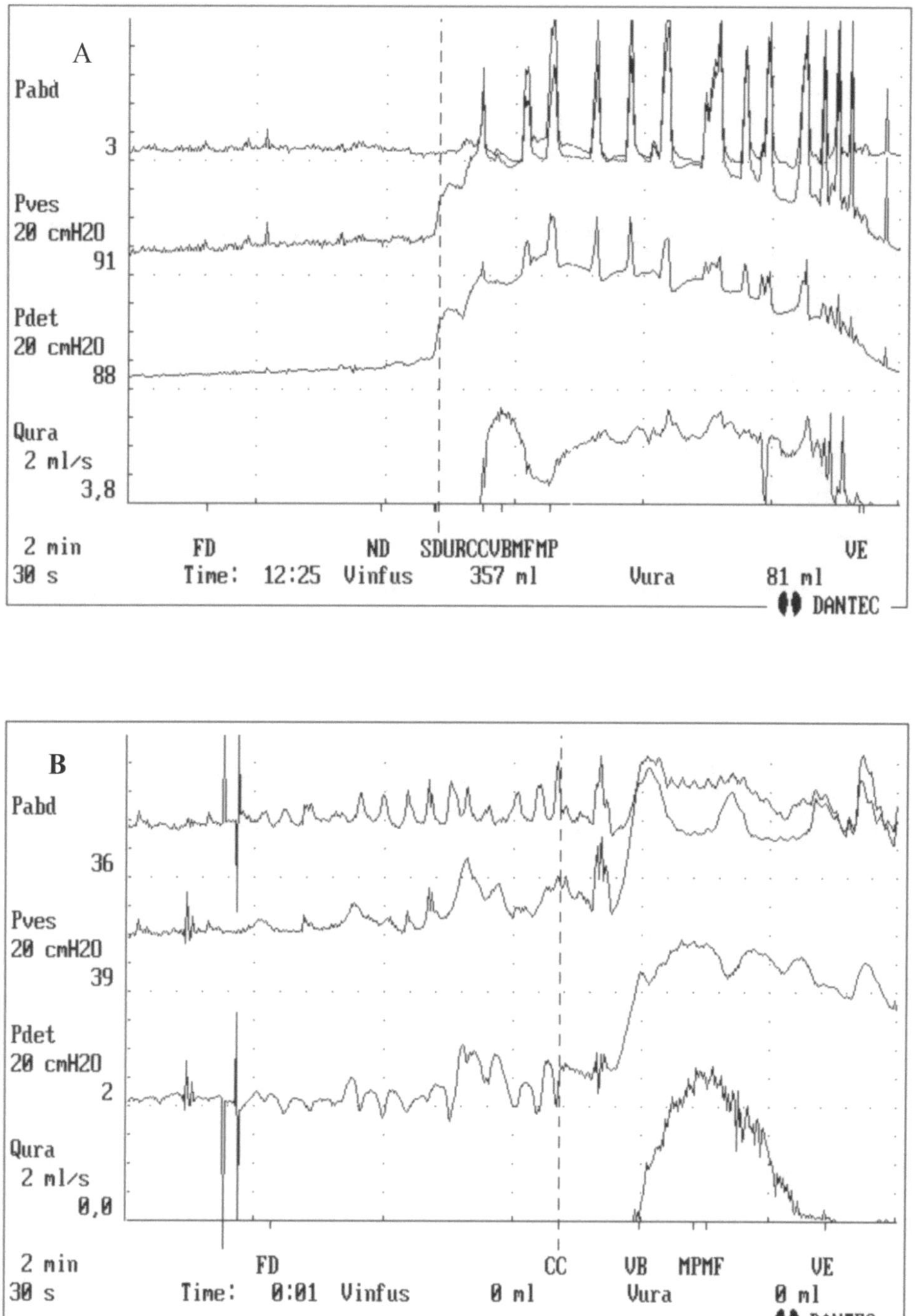

Fig. 21. The physiologic artifacts on pressure tracings. **A**: the spikes on P_{det} tracing due to a different transmission to P_{ves} and P_{abd} corresponding to straining during voiding. **B**: the dips on P_{det} tracing caused by rectal contractions

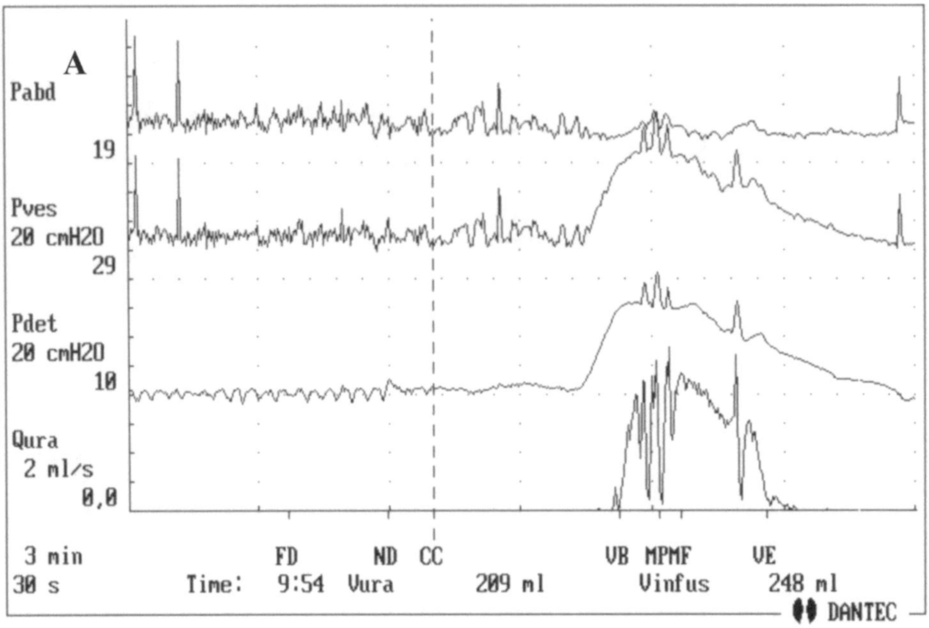

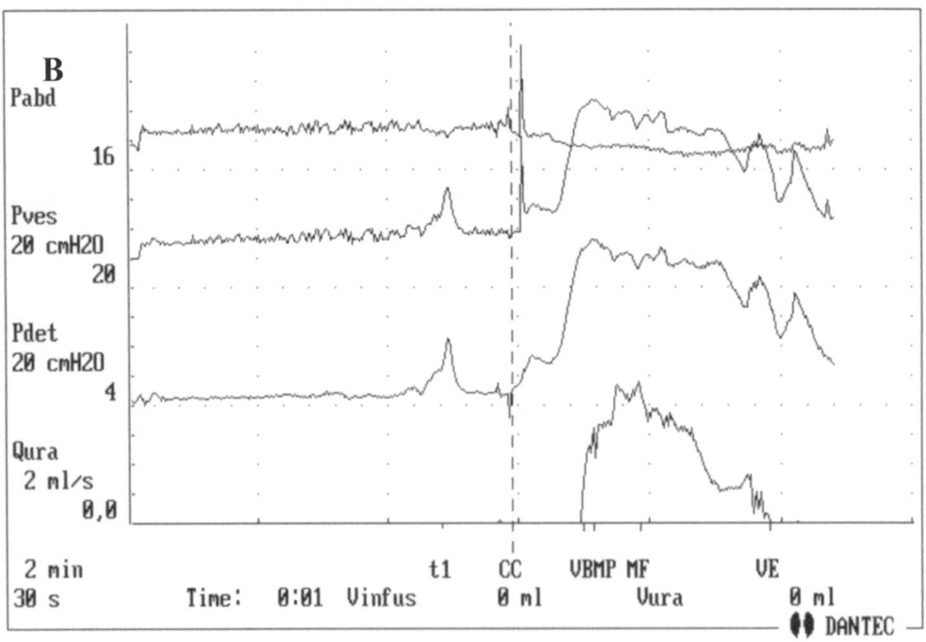

Fig. 22. The physiologic artifacts on pressure tracings due to urethral sphincter overactivity during voiding. **A**: the spikes on P_{det} tracing caused by urethral sphincter contractions during voiding. **B**: the dips on P_{det} tracing resulted from urethral sphincter relaxation during voiding

Correlation coefficients (r) of Q_{max}, $P_{det.Qmax}$ and OCO between computerized and manual readings were 0.909, 0.969 and 0.970 respectively (Table 9).

	Computerized	Manual	Change	p Value	r
Q_{max} (ml/sec)	8.46 ± 2.87	7.29 ± 2.62	1.17 ± 1.20	<0.001	0.909
$P_{det.Qmax}$ (cmH_2O)	75.75 ± 33.34	76.50 ± 31.67	-0.75 ± 8.31	0.346	0.969
Increased $P_{det.Qmax}$ (cmH_2O)	68.93 ± 25.44	73.83 ± 25.38	-4.90 ± 6.71	0.007	0.965
Unchanged $P_{det.Qmax}$ (cmH_2O)	74.22 ± 35.32	74.22 ± 35.32	0		1
Decreased $P_{det.Qmax}$ (cmH_2O)	88.28 ± 40.48	82.13 ± 38.65	6.15 ± 7.86	0.068	0.981
OCO	1.359 ± 0.664	1.426 ± 0.652	-0.067 ± 0.162	0.040	0.970

Table 9. Changes of parameters of pressure-flow studies after manual correction (mean± SD)

	Computerized values		Manual values		Changes	
	Median	Range	Median	Range	Median	Range
Q_{max} (ml/sec)	8.2	1.6 ~ 18.6	7	1.2 ~ 16.7	0.9	-0.5 ~ 10.4
$P_{det.Qmax}$ (cmH_2O)	69	2 ~ 264	70	20 ~ 246	-1	-70 ~ 56
Increased $P_{det.Qmax}$ (cmH_2O)	67	2 ~ 159	70	28 ~ 161	-3	-70 ~ -1
Unchanged $P_{det.Qmax}$ (cmH_2O)	62	29 ~ 246	62	29 ~ 246		
Decreased $P_{det.Qmax}$ (cmH_2O)	75	33 ~ 264	70	20 ~ 240	3	1 ~ 56
OCO	1.223	0.03 ~ 4.87	1.30	0.35 ~ 4.9	-0.06	-1.38 ~ 1.0

Table 10. Changes of parameters of pressure-flow studies after manual correction

With respect to the decreased degree of Q_{max} after correction, the percentages of decrease of ≤ 0, 0.1~0.9, 1~1.9, 2~2.9, 3~3.9 and ≥4 ml/sec were 2.1%, 54.1%, 29.0%, 8.4%, 3.4% and 3.0% respectively (Fig. 23). Concerning the changed degree of $P_{det.Qmax}$, the percentages of increase of 1~9, 10~19 and ≥20 cmH2O were 49.3%, 3.4% and 2.4% respectively; the percentages of decrease of 1~9, 10~19 and ≥20 cmH_2O were 25.8%, 3.6% and 2.2% respectively (Fig. 24). Concerning the changed degree of OCO, the percentages of increase of 0.001~0.04, 0.05~0.14, 0.15~0.24, 0.25~0.49 and ≥0.5 were 22.5%, 44.0%, 7.7%, 5.0% and 2.2% respectively; the percentages of decrease of 0.001~0.24 and ≥0.25 were 5.5% and 13.1% respectively (Fig. 25).

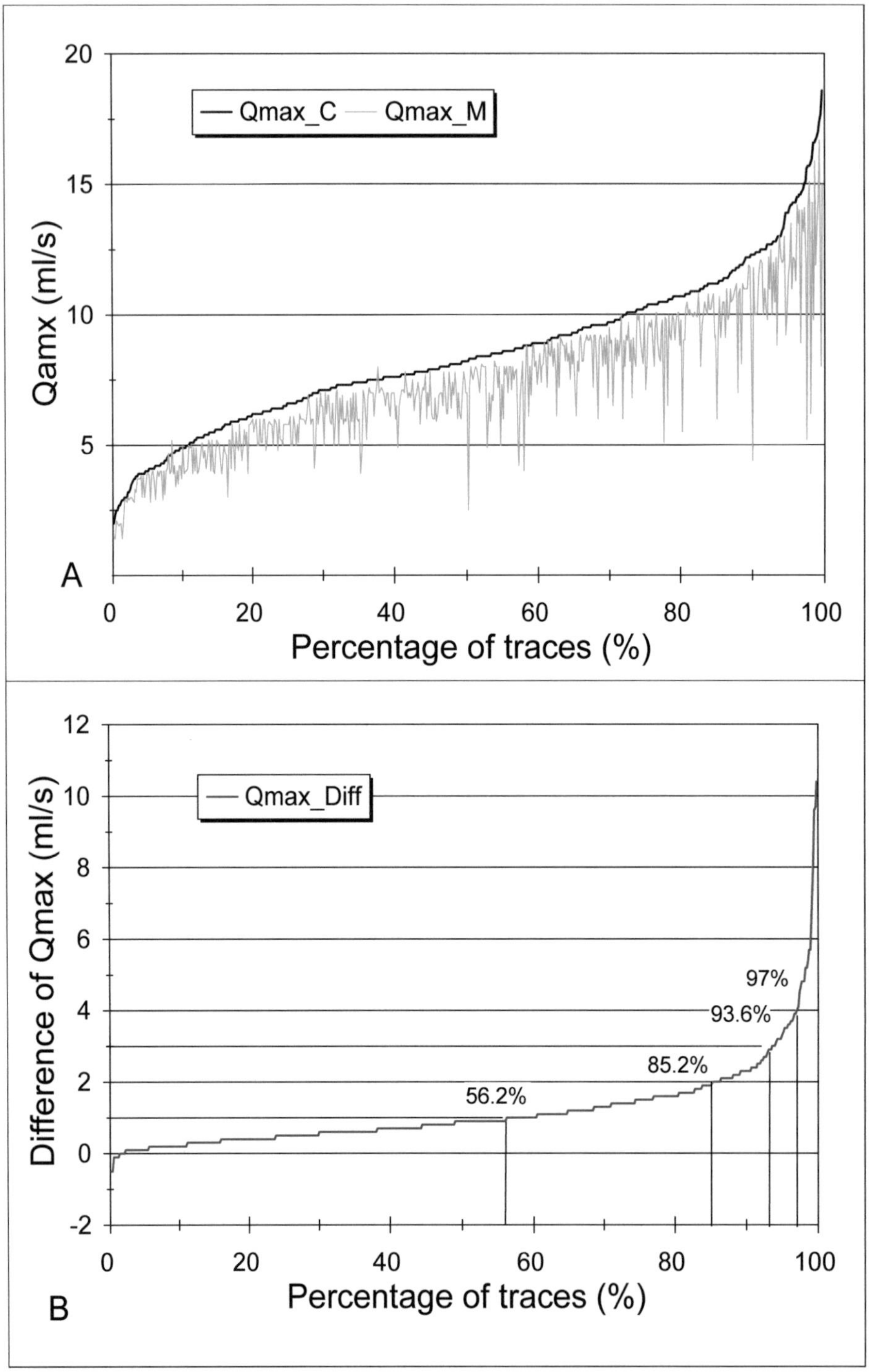

Fig. 23. **A**: Q_{max} of computerized and manual readings; **B**: the difference of Q_{max} between computerized and manual readings and the percentages of change in 582 pressure-flow traces

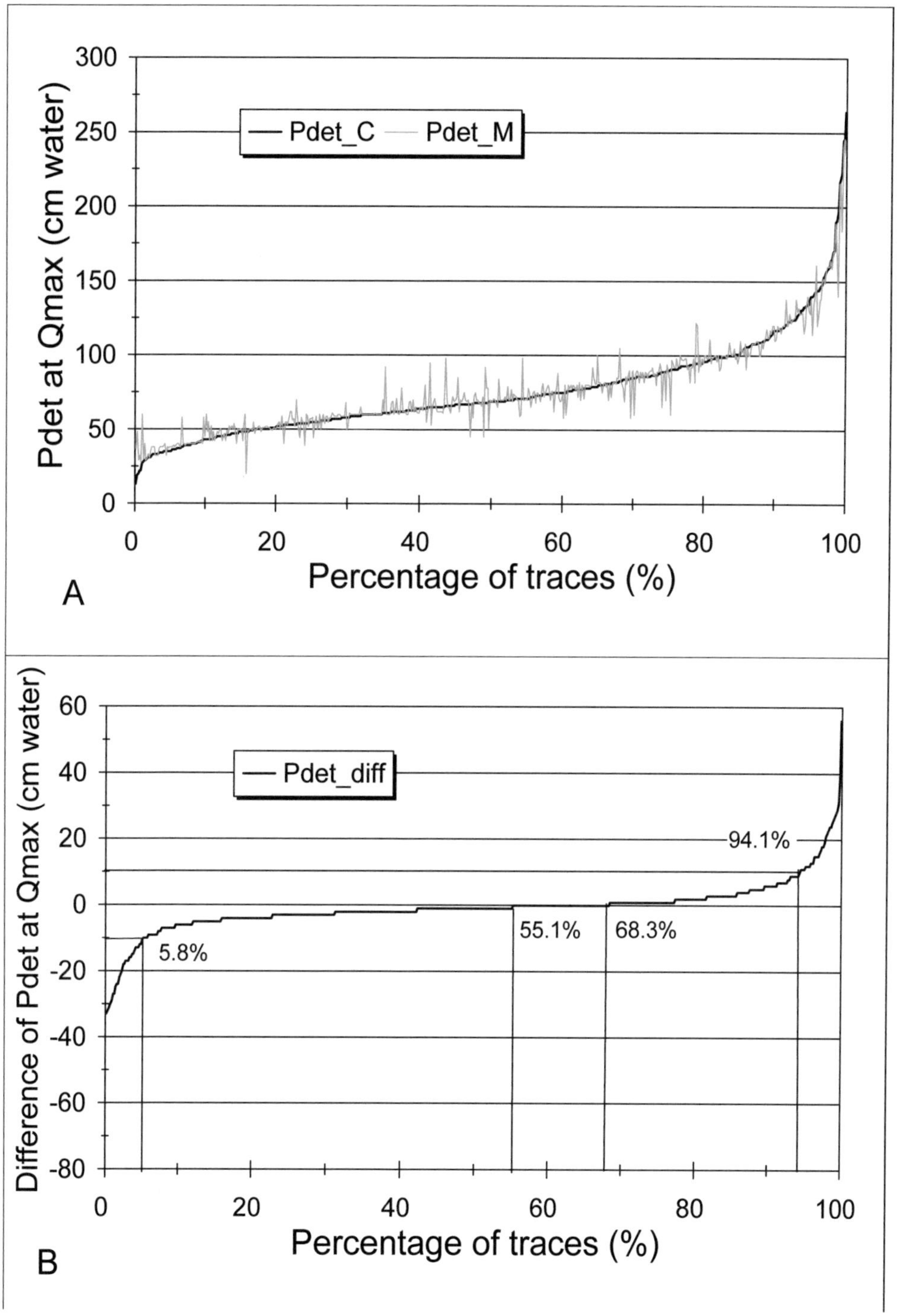

Fig. 24. **A**: $P_{det.Qmax}$ of computerized and manual readings; **B**: the difference of $P_{det.Qmax}$ between computerized and manual readings and the percentages of change in 582 pressure-flow traces

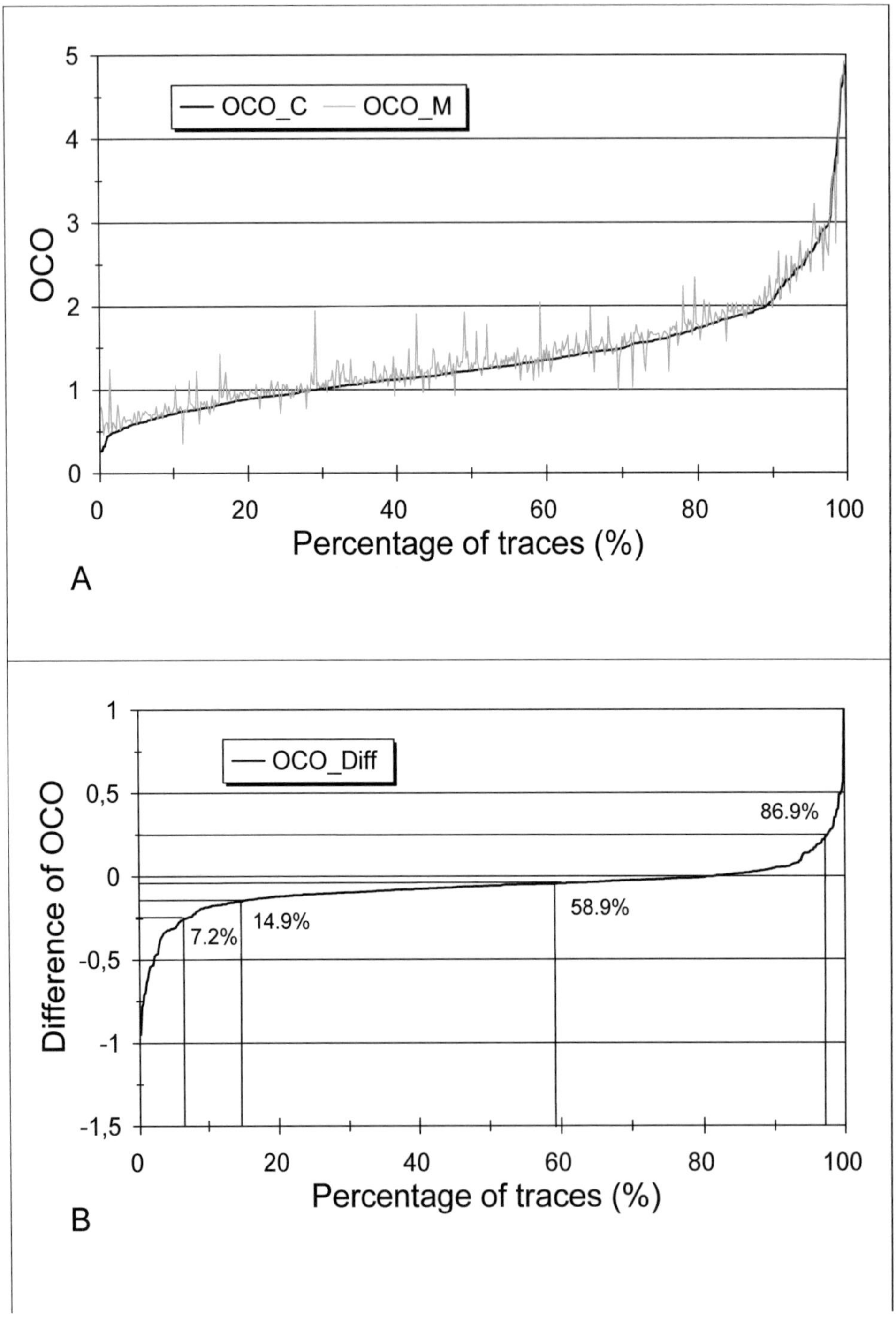

Fig. 25. **A**: OCO calculated according to computerized and manual readings; **B**: the difference of OCO between computerized and manual readings and the percentages of change in 582 pressure-flow traces

The percentages in classification using ICS nomogram and in grades using Schäfer nomogram were shown in Table 11. Comparing these percentages of manual results with computerized ones, a significant systematic difference was found. Using ICS nomogram, the obstructed percentage increased from 69.8% to 73.9% ($p<0.05$), and unobstructed one decreased from 8.8% to 5.3% ($p<0.05$). Using Schäfer nomogram, the obstructed percentage (III-VI) increased from 72.5% to 77.3% ($p<0.01$), and unobstructed one (0-I) decreased from 9.1% to 5.5% ($p<0.01$). Systematically, the distribution and degree of obstruction had a significant increase after correction. However, the intra-individual changes of classification and grade were different. After manual correction, 64 (11.0%) of 582 traces changed the classification in ICS nomogram. 53 (82.8%) of 64 traces increased obstructed degree, and 11 (17.2%) decreased one using ICS nomogram (Table 12). Using Schäfer nomogram, 168 (28.9%) of 582 traces changed the grade after correction. 143 (85.1%) of 168 traces increased obstructed degree, and 25 (14.9%) decreased one. A trace with a great change moved from 0 to IV grade, and the most of traces (94.6%) changed one grade after correction (Table 13).

	Computerized reading		Manual reading		
	No	%	No	%	p Value
ICS nomogram					< 0.05
Obstructed	406	69.8	430	73.9	
Equivocal	125	21.5	121	20.8	
Unobstructed	51	8.8	31	5.3	
Schäfer nomogram					< 0.01
0	6	1.0	1	0.2	
I	47	8.1	31	5.3	
II	107	18.4	100	17.2	
III	179	30.8	164	28.2	
IV	166	28.5	182	31.3	
V	40	6.9	64	11.0	
VI	37	6.4	40	6.9	

Table 11. Difference of percentages between computerized and manual reading in classifications of ICS nomogram and in grades of Schäfer nomogram

	No	%
Unobstructed → Equivocal	21	32.8
Equivocal → Obstructed	30	46.9
Unobstructed → Obstructed	2	3.1
Equivocal → Unobstructed	3	4.7
Obstructed → Equivocal	8	12.5
Total	64	100
Shifting into obstructed zone	32	80
Shifting out of obstructed zone	8	20
Total	40	100

Table 12. Shifts of measurements in classification of ICS nomogram after manual correction

After manual correction, 40 (6.9%) of 582 traces changed the diagnosis of obstruction using ICS nomogram; 32 (80%) of 40 traces shifted into obstructed zone, and 8 (20%) shifted out of one (Table 4, fig. 5, 6). Using Schäfer nomogram, 42 (7.2%) of 582 traces changed the diagnosis of obstruction. 35 (83.3%) of 42 traces moved from <III to ≥III grade, and 7 (16.7%) moved from ≥III to <III grade (Table 13).

	No	%
Increased 1 grade	135	80.3
Increased 2 grades	6	3.6
Increased 3 grades	1	0.6
Increased 4 grades	1	0.6
Decreased 1 grade	24	14.3
Decreased 2 grades	1	0.6
Total	168	100
< III → ≥ III grade	35	83.3
≥ III → < III grade	7	16.7
Total	42	100

Table 13. Changes of measurements in grading of Schäfer nomogram after manual correction

4. Discussion

Quality control involves in urodynamic investigation and retrospective analysis. Quality control and plausibility check during investigation are the best way to avoid and to correct artifacts at an early stage; quality control in retrospective analysis is also necessary. Quality control relies on knowledge of typical values and signal patterns recognition (Schäfer 1998). The urodynamic standards for quality control and their application in clinical urodynamic practice will be discussed from these aspects.

4.1 Quality control relies on knowledge of typical values

Quality control and plausibility check during investigation are necessary for collecting of reliable and accurate urodynamic data, which are free of the technical errors and artifacts. The typical value ranges, especially TVR for initial resting pressures, are the tool for checking the plausibility of measurement quantitatively. The correct initial resting pressures are the important factors to undergo a good cystometry; and a wrong resting pressure usually is beyond its typical value range. The proper initial resting pressures indicate two means: one is correct zero setting, another one is correct pressure reference level establishing according to the standard of ICS. The ICS has defined that all systems are zeroed at atmospheric pressure, and the pressure reference point is the level of the superior edge of the symphysis pubis (Abrams et al. 1988). However, there are still some investigators who do not obey the standard of ICS, and continue to use initial resting P_{ves} and P_{abd} as zero by balance. This is a common mistake in clinical urodynamic practice, and often leads to above-mentioned type I and type IV errors. Our study found that the relaxation of pelvic floor during voiding was a typical physiologic behavior. It appears that type I error has a normal P_{det} value. However, type Ia error has either a lower P_{abd} or a lower P_{ves} due to intra-rectal and intravesical zero setting; this lower P_{abd} often becomes negative during voiding because of over-relaxation of pelvic floor. With a negative P_{abd}, a P_{det} value higher than P_{ves} can be calculated, but this P_{det} is meaningless; this is called type IV error. Type Ib error often is caused by a wrong pressure reference level establishing. It seems that this error does not lead to any serious consequences for P_{det} is normal; but it should be corrected for the purpose of quality control and standardisation. Type I and IV errors can be avoided and solved by setting zero and establishing a pressure reference level correctly.

Type II error shows a negative initial resting Pdet caused by the atypical initial resting P_{ves} and P_{abd} values. Type IIa error has a lower P_{ves} than typical value, which often results from the problems of pressure transmission in urethral catheter and tubing, for example, air bubbles, catheter blocked and urethral catheter dislocation. It can be corrected by flushing gently through the P_{ves} transducer and tubing or adjusting the position of urethral catheter. Type IIb error has a higher P_{abd} than typical value, which often is related to an over-filling of rectal catheter balloon. Reducing the balloon filling to a proper volume and adjusting the P_{abd} to its typical range can solve this type of error.

Type III error shows a too high initial resting P_{det}. In our study, 10 cmH_2O pressure has been suggested as the upper limit of a maximum possible resting value for P_{det}. We have found that P_{det} increases 5~6 cmH_2O during filling phase according to the typical value range; then, an over 10 cmH_2O initial resting P_{det} seems to be impossible if the detrusor overactivity has been ruled out. However, this value is relatively definite. Type IIIa error has a relatively higher P_{ves} than typical range, which is often due to the problems of pressure transmission of P_{ves}, for example, urethral catheter blacked, urethral catheter dislocation. The solution is the same with that of type IIa error. Type IIIb error has a relatively lower P_{abd} than typical range, which is often related to a less filling volume of rectal catheter balloon or rectal catheter dislocation. Re-filling the balloon to a proper volume or adjusting the position of rectal catheter can correct this type of error.

From the above-mentioned analyses for the technical errors, it can be demonstrated that the suggested typical value ranges for initial resting pressures are sensitive and reliable indicator for a plausible and correct measurement. These ranges developed from our retrospective data analysis and calculation are in accord with ones recommended by

Schäfer in the ICS Standardization Report " Good Urodynamic Practice "; for example, our calculated range for initial resting P_{det} is 0~4 cmH_2O, and recommended range is 0~5 cmH_2O (Schaefer et al. 2002). It means that the initial resting detrusor pressure should be close to zero. However, our suggested TVR for P_{det} is not simply equal to a direct mathematical subtraction between two TVR of P_{ves} and P_{abd}. The reason may be that our study has been performed on the base of conventional cystometric data, and these data contain considerable cases with negative initial resting P_{det}. This leads to the difference between the ranges from statistical analysis for data and ones from the direct mathematical subtraction. However, we think the former is much more suitable to the present situation of clinical urodynamic practice. On the other hand, these suggested typical value ranges for initial resting pressures can only be used in the sitting or standing position. For supine position, the ranges for P_{ves} and P_{abd} are lower than those of standing and sitting positions; but the typical range for P_{det} is the same among all positions. Moreover, our study has found that 11.5% of traces with negative initial resting P_{det} recover the typical positive P_{det} value during initial 30 seconds of filling. This type of short-time negative initial P_{det} is often related to two causes. One is the rectal contractions during this period of time. Another one is that the contact of catheter tip with bladder wall interferes with the pressure transmission in initial filling phase. Once the tip leaves bladder mucosa with the increase of media around it, the pressure transmission will become normal. In these cases, the measurements can go on well. If a negative initial pressure continues for a longer period of time, the pressure signals will deteriorate further; and the measurement will have to be stopped to check and correct the errors. In this situation, a repeated measurement is often necessary to get a reliable result. Generally, using the typical value ranges for initial resting pressures is an effective for the plausibility check and quality control of cystometry, and is a reliable indicator of good investigation.

Our results suggest that detrusor pressure usually increases 5~6 cmH_2O during filling in 50% TVR, and TVR of detrusor compliance is 26.6~70.8 ml/ cmH_2O. Then, it is necessary to determine the dependency on the filling speed by interrupting filling for some period of time when a significant steady pressure increase beyond above-mentioned change range and TVR of compliance is observed. On the other hand, MCC recorded at end of filling should be comparable with the typical voided volume recorded in bladder diary of patient.

One of the significant parameters during voiding is P_{abd} at relaxation. In our study, TVR of this pressure value is 25~38 cmH_2O. It means that during voiding P_{abd} has a decrease with a mean of 2.7 cmH_2O due to the relaxation of pelvic floor; therefore, relaxation during voiding is a typical physiological pattern. However, it may be lead to type IV error when it combines with a wrong lower initial resting P_{abd}. At this time, a repeated measurement with correct initial resting pressures will be necessary. This study has indicated that Q_{max} and V_{void} recorded in cystometry are comparable with those of free uroflowmetry; then, they are also important values for plausibility control of voiding cystometry. It means that a free uroflowmetry with an over 150 ml V_{void} before cystometry is necessary, and Q_{max} and V_{void} from cystometry should be comparable with those of free uroflowmetry.

Although it is impossible to specify typical pressure values ranges at initial voiding, the similar typical value ranges as for the initial resting pressures may be applied for the post-void pressures. It is important to record a longer post-voiding phase until the pressure values return to the plausible resting value levels. In view of this, we have also suggested TVR for post-void resting pressures for the plausibility control of voiding phase. It is interesting that 50% TVR for post-void P_{abd} is the same with that of filling end. It means that P_{abd} has not obvious difference before and after voiding except for during voiding. TVR for post-void P_{ves} and P_{det} are slightly higher than those of filling end, and P_{det} has an increase with a mean of 6 cmH_2O. This slight increase is acceptable, and is related to the post-void residual volume. In the elderly males with LUTS, the post-void residual volumes of different degrees are typical pathophysiological changes for this population. Therefore, the suggested TVR for post-void pressures are with the specificity of population. With the view of quality control, post-void P_{ves} and P_{det} should be close to ones at end of filling as much as possible except for the factor of high post-void residual volume. If unlike this, it suggests the problems of signal quality during voiding. Using a similar strategy to the initial resting pressures, we also can define two types of errors associated with voiding phase. Type V errors show a significant high post-void P_{ves} and P_{det} that are not relevant to a high post-void residual volume. Type Va error is caused by the signal loss ("dead" signal) of P_{ves} or the urethral catheter dislocation during voiding. If this error occurs before Q_{max} arriving, or a complete pressure-flow analysis is requested, a repeated measurement will be necessary. Type Vb error is related to the rectal catheter dislocation or the signal loss of P_{abd} tracing during voiding. Because P_{abd} is relatively unchangeable before and after voiding, it seems that this type of error can be corrected using TVR for P_{abd}. However, if the accurate Q_{max} and $P_{det.Qmax}$ are required in the precise pressure-flow analysis, the repetition will be unable to be avoided. Type VI error often results from the urethral catheter loss during voiding. For this type of error, the strategy for repetition is the same with that of type Va error. Generally, TVR of post-void pressures also are important for plausibility check and quality control during voiding cystometry. It seems to be more difficult to define an unvague upper limit for TVR of post-void Pdet because of the considerable variability. Usually, the solution for the errors and artifacts during voiding phase is repetition.

4.2 Quality control relies on signal patterns recognition

Quality control of urodynamic data relies on either static resting values in typical range, which is quantitative plausibility check for signal quality, or dynamic typical signal patterns, which is qualitative plausibility check for signal quality. To acquire the high quality urodynamic data, the investigators must conscientiously observe the signal patterns and changes at all stages of the investigation together with continuous signal testing. Therefore, keeping the above-described typical signal patterns in mind is necessary; the aim is at first to avoid artifacts and at second to identify and to correct all atypical signal patterns and artifacts immediately during investigation.

We can check the signal plausibility through examining the signal quality from four gradations of TSP during any phases of study. Before or at beginning of filling, a high quality signal must have the same fine structure between P_{ves} and P_{abd} tracings (pattern I); must show the identical minor changes in response to the patient's activity, for example, breathing, talking or moving (pattern II); and must show the equal major changes corresponding to cough tests (pattern III). P_{det} tracing dose not have any significant changes. If there are not any fine structure and minor changes on P_{ves} or P_{abd} tracings, or there are some obvious up- or down-spikes on P_{det} tracing corresponding to the test-coughs, the investigation will be unable to start or will be stopped to check the cause for poor signal quality. The common causes and the correction for them can be described as the following. The bubbles existed in catheters or tubings often lead to a problem for pressure transmission; then, both the bladder and rectal lines should be flushed once more to ensure that all bubbles have been removed. The connection leaks may result in a low pressure signal. When the leaks are demonstrated by flushing, all connections in the lines must be examined and tighten to correct them. An urethral chatter located in the bladder neck or urethra, or a rectal catheter closed to the anal verge may also lead to the problems of signal quality. In these cases, the catheter positions must be adjusted to ensure urethral catheter in the bladder, or to position the rectal catheter around 10 cm above the anal verge. Sometimes, faecal loading may interfere with the signal of P_{abd}; therefore, the patient must be asked to remove it before investigation. A catheter blocked or a catheter with a kink can hinder the pressure signal transmission. In these cases, it is necessary to flush catheter or to renew one. Once these errors are corrected, the typical and proper signal patterns are observed, and equal changes corresponding to test-coughs between P_{ves} and P_{abd} are demonstrated, the bladder filling can start or recommence with the high quality signals. In the initial filling phase, there often are two types of major changes: straining and rectal contractions (pattern IV). Straining also is a tool for testing signal quality automatically; and the good-quality signals must have the equal responses to straining between P_{ves} and P_{abd}. The initial rectal contractions often are related to faecal loading or a big balloon filling volume. Rectal contraction may lead to a negative initial P_{det}. With rectal contractions, a measurement may go on as they can be corrected in retrospective analysis; however, the interference from rectal contractions should be ruled out in a perfect measurement. It is very important to observe and test signals, and to correct any errors before starting investigation. The high-quality initial resting signals are the premise to finish a precise and reliable cystometry.

Usually, the perfect signals at beginning of measurement stay throughout all investigation. However, it is possible that the signal quality deteriorates at any stage of cystometry, both good and bad initial signals. Therefore, the conscientious observation of signal quality and the plausibility control of signals during filling phase are crucial too. The strategy for TSP recognition and analysis during filling is the same with the initial filling phase; it is also carried out from four gradations of TSP. At any moment of filling phase, we must observe the signal patterns from four gradations: fine structure, minor changes, test-cough changes and major changes, and compare them among P_{ves}, P_{abd} and P_{det} tracings. If the deterioration of signal quality is occurred, or atypical signal patterns are observed at any stage, the investigation must be stopped and the causes of poor signal quality must be found. The investigation can continue after errors have been corrected. For the high-quality signals during filling, P_{ves} and P_{abd} must have the same fine structure (pattern I); must show the identical minor changes in response to the patient's activity, for example, breathing, talking or moving (pattern II); and must have the equal major changes corresponding to test-coughs (pattern III). There are no fine structure and minor changes on P_{det} tracing; at most, there may be some small biphasic spikes corresponding to test-coughs on it. If fine structure and minor changes on P_{ves} or P_{abd} tracings disappear, or the obvious up- or down-spikes appear on P_{det} tracing corresponding to the test-coughs at regular intervals, these points suggest an indicator of signal quality deterioration. In these situations, the investigation has to be stopped to investigate the causes for poor signal quality. Usually, the causes can be found as the following. Firstly, the bubbles existed in catheters or tubings do not be flushed completely, or appear again because of a loose connection. Secondly, the connection leaks do not be corrected completely. Thirdly, the urethral catheter moves into the bladder neck or urethral, or rectal catheter moves into the anal verge. Fourthly, faecal loading influences the signal of P_{abd}. For the correction of these errors, the methods are the same with those of initial filling phase. Finally, the contact of catheter tip with bladder wall often interferes with the pressure signal transmission; flushing through urethral catheter can correct this error. Usually, there are three types of typical major changes during filling: straining, detrusor overactivity and rectal contractions (pattern IV). Like the test-coughs, straining can test signals automatically; the good-quality signals show the equal changes on P_{ves} and P_{abd} tracings corresponding to straining, and no changes occur on P_{det} tracing. Unlike this, the causes of poor signal transmission must be checked. Detrusor overactivity is a typical major signal pattern during filling, which appears in the forms of single or multiple unstable detrusor contractions in 33.7% traces analyzed. We must familiarize ourselves with this typical pattern in order to discriminate from the atypical patterns. The changes on P_{ves} and P_{abd} tracings are identical when the unstable detrusor contractions occur. However, the technical factors resulting in detrusor overactivity, for example, fast filling or too cold media for filling, must be checked and removed. Rectal activity is another typical major signal pattern during filling, which is characterized by single or multiple rectal contractions. Its incidence is 17.4% in the traces analyzed. We must also recognize this typical pattern. There is a positive wave on P_{abd} tracing and a negative wave on P_{det} tracing when a rectal contraction happens. Faecal loading and a big balloon filling volume often lead to the rectal contractions. With rectal contractions, the measurement may continue because they can be recognized and corrected in retrospective analysis as artifacts. However, if the rectal contractions interfere with the data analysis, especially with the analysis for detrusor overactivity, a repeated measurement will usually be necessary. During filling, the cough tests at regular intervals, for example every minute, are the powerful tool for checking the plausibility of signals.

Usually, the plausibility control of signals during voiding is difficult; therefore, it is very important to ensure the high-quality signals to go into voiding phase. For this purpose, the cough tests before voiding are the effective tool for ensuring the good signal quality. If there are the obvious different changes between P_{ves} and P_{abd} tracings corresponding to the test-coughs before voiding, this will be an indicator of signal deterioration, and these signals can not be allowed into voiding (Fig. 14b). In this situation, the investigation must be interrupted to examine for the causes of poor signal quality, and to rule them out. The pre-voiding phase may be recorded for a period as long as possible in order to get sufficient information for a signal control and to have enough time to test signals by coughs.

As voiding phase goes on quickly, and more complex anatomical and physiological aspects are involved in this process, quality control during voiding is difficult. Comparing with before and during filling phases, we can do fewer things for signal quality control and plausibility check. However, we must still keep conscientious observation of signals and TSP recognition for there are various artifacts occurred during voiding. Firstly, there are the same fine structure (Pattern I) and minor changes (pattern II) on P_{ves} and P_{abd} tracings, which suggest that the signals are "live". When fine structure and minor changes on P_{ves} or P_{abd} tracing disappear during voiding, the tracing often shows the stepwise changes with a "dead" line or a sudden drop (Fig. 14b, 14c, Fig. 16). One of the common causes is the catheter dislocation or loss. Urethral catheter often moves into the bladder neck or the sphincter area, or is projected with the urine stream during voiding. Similarly, rectal catheter often slips down into the anal sphincter area, or falls out. If these errors appear before maximum uroflow rate (Q_{max}) or a complete voiding is request for the purpose of pressure-flow analysis, the measurement should be repeated. Secondly, typical major signal patterns (Pattern IV) must be recognized. During voiding, typical pattern of detrusor contraction is easy to identify (Fig. 11a); however, identification usually is difficult when some special patterns appear. After-contraction is a common finding and a normal phenomenon (Fig. 11b). Fluctuation contraction often associates with detrusor underactivity, which suggests that detrusor is unable to empty the bladder depending on once a contraction (Fig. 11c). This type of fluctuation contraction must be distinguished from that of detrusor-sphincter dyssynergia. The former shows the simultaneous changes among P_{ves}, P_{det} and uroflow tracings, but the latter has the rapid increases of P_{ves} and P_{det} between two urine spurts. Usually, a unstable detrusor may get to a high pressure level before voiding due to involuntary contraction, the pressure drops suddenly as soon as uroflow starts (Fig. 11b). This unstable detrusor voiding often leads to an illusion. Straining is one of the most common typical patterns during voiding, which is found in 71% traces analyzed. With the good signal quality and the equal signal transmission, there will not be any artifacts on P_{det} tracing. When the signal transmissions between P_{ves} and P_{det} tracings are unequal, there will be some spike artifacts on P_{det} tracing (Fig. 15b), which must be corrected in retrospective analysis. Rectal contraction also is a type of TSP during voiding, which occurs in 2.1% traces analyzed. Rectal contractions may lead to the dip artifacts on P_{det} tracing, which can be corrected in retrospective analysis. Relaxation of pelvic floor during voiding is another type of TSP, which appears in 15.3% traces analyzed. Relaxation indicates the significance of the high-quality recording of P_{abd}, which can eliminate the impact of perivesical pressures on P_{ves}. During voiding, P_{ves} and P_{det} tracings increase and decrease with detrusor contraction smooth and steady. If they drop suddenly, the common cause is the catheter's dislocation or loss.

The plausibility check for signals and signal tests after voiding are necessary for demonstrating the signal quality during voiding, and for ensuring the high-quality pressure-flow data. A longer post-voiding recording phase can provide us with sufficient information for quality control and enough time to test signals. In this phase, cough tests are the powerful tool to check the signal quality during voiding. After voiding, there still are the identical fine structure and minor change on P_{ves} and P_{abd} tracings, and the tracings show the equal changes corresponding to test-coughs; these points suggest the high-quality signals during voiding. If not, the causes of poor signal quality must be checked. The common causes also are the catheter's dislocation or loss, and signal loss on P_{ves} or P_{abd} tracing (Fig. 13b, Fig. 14b, 14c and Fig.16). If these errors appear before Q_{max} or a complete voiding is request for a precise pressure-flow analysis, the investigation must be repeated.

Finally, pattern recognition is based on the scales of signals. A changed scale often leads an illusion; therefore, the scales should be kept unchanged as much as possible either during investigation or in retrospective analysis (Fig. 17).

4.3 Quality control in retrospective analysis

The best way of quality control is to avoid artifacts and technical errors and to eliminate them at an early stage during investigation; the retrospective correction is the worst solution. However, retrospective correction is unable to be lacked for the artifacts existed, especially in computerized data. In our retrospective analysis, artifacts during filling were mainly related to the wrong initial resting pressures, spikes due to test-coughs and periodic signal loss or stepwise changes. The correction for these artifacts is easy comparing with those of voiding cystometry. Therefore, we mainly discussed the recognition and correction for artifacts during pressure-flow investigation, demonstrated the effects of correction, and indicated the role and significance of retrospective quality control.

Pressure-flow studies can provide us with a diagnostic standard for bladder outlet obstruction, and measure the urethral resistance and changes. In pressure-flow analysis, the obstructed degree and urethral resistance usually depend on two variables: Q_{max} and $P_{det.Qmax}$. The problem that we face is how to obtain the reliable values of Q_{max} and $P_{det.Qmax}$, which are free of various artifacts, and to ensure the clinical diagnosis and research an objective and reliable result. Therefore, quality control of pressure-flow data becomes increasingly important; it can be performed either during collection of data or in retrospective analysis of data. Quality control during collection of data can avoid, reduce or eliminate artifacts. On the other hand, the artifacts existed in data can only be corrected in retrospective analysis. This is the worst solution, but is necessary for computer results. Modern computerized urodynamic systems have presented new problems in analysis of data. Almost all machines are unable to pick up and correct artifacts. Many clinicians do not examine the traces for artifacts and accept the computer's values of parameters; this must influence the clinical diagnosis and the research result significantly. The main tasks of retrospective quality control of pressure-flow data are the typical pattern identification, artifacts recognition and correction as well as manually reading for Q_{max} and $P_{det.Qmax}$ coming from computer printouts. In the present study, 4% traces had to be discarded because of the non-interpretable and non-correctional artifacts and the pattern with multi strong strains during voiding. In the interpretable pressure-flow traces, Q_{max}, urethral resistance (OCO), grading and classifying of obstruction underwent significant systematic changes; $P_{det.Qmax}$ had no systematically significant changes, but with considerable intra-individual changes after manual correction. Q_{max} reduced consistently by 1.17 ml/sec on average, which was similar to the result (by 1.5 ml/sec on average) reported by Grino et al in 1645 uroflow measurements and the result (by 0.8 ml/sec on average) reported by Madsen et al in pressure-flow studies of 25 patients (Grino et al. 1993; Madsen et al. 1995). In this study, 81.8% traces showed obvious artifacts of Q_{max}. The decreased value of Q_{max} resulted from the correction of spike artifacts and extracorporeal modifications in flowrate. 23.1% of artifacts changed the location of Q_{max} (Fig.19a and 19b). The decreased degree of Q_{max} was variable, but 83.1% readings were with 0.1~1.9 ml/sec decrease of Q_{max}. There were 7 (1.2%) traces with increased values of Q_{max} after correction, the reason would be that the investigators changed computer results of Q_{max}. Smoothed and corrected value of Q_{max} underwent a significant decrease; still, there was a good correlation between computerized and manual values of Q_{max}. It means that manual correction has not changed the nature of Q_{max} data, the smoothed and corrected Q_{max} can reflect the condition of urethral resistance much more really. How are artifacts of Q_{max} identified? A normal uroflow curve is smooth without any rapid changes or spikes. Rapid changes in flowrate may have physiological and physical causes (Schaefer et al. 2002). The physiological spikes can result from changes in outflow resistance, for instance, sphincter and pelvic floor contraction or relaxation, or from changes in driving energy, for instance, abdominal straining. These intracorporeal physiological artifacts should be minimized during the investigation (Fig.18b, Fig.19). Extracorporeal additional modifications in the flowrate signal, which usually is small spikes, can be introduced by any funnel or collecting device of uroflowmeter (Fig.18). This type of non-physiological artifacts should be eliminated. As a simple rule of thumb, any rapid change in uroflow rate lasting less than two seconds should be smoothed and corrected as artifacts in retrospective analysis. In a recent standardization report, ICS recommended that an internal electronic smoothing with a sliding average over two seconds was used to make electronically reading value of Q_{max} more reliable, comparable and clinically useful. In manual graphical readings of Q_{max}, a graphical line smoothing to a continuous curvature for at least a period of two seconds was drawn to get a smoothed Q_{max} value. Generally, only smoothed Q_{max} that is lower than electronically reading Q_{max} is clinically meaningful. ICS agreed that as a standard only smoothed Q_{max} values were reported (Schaefer et al. 2002).

$P_{det.Qmax}$ showed a slight systematic increase by 0.75 cmH_2O on average after manual correction, but no significant variation was demonstrated. Similarly, Madsen et al reported a no significant slight decrease of $P_{det.Qmax}$, with a mean of 2.8 cmH_2O after correction (Madsen et al. 1995). The reason would be that the location of $P_{det.Qmax}$ responds to Q_{max}; therefore, intra-individual changes of $P_{det.Qmax}$ were inconsistent. 55.2% traces had a significant increase by 4.90 cmH_2O on average, and 31.6% traces had no significant decrease by 6.15 cmH_2O on average. Although there was no systematically significant change, $P_{det.Qmax}$ indeed underwent a intra-individual considerable change ranged from -70 to 56 cmH_2O after manual correction. The artifacts of $P_{det.Qmax}$ are various and complex, and sometimes are difficult to interpret. As a smooth muscle, detrusor contracts smoothly and steadily, and then any pressure change caused by detrusor contraction must show a smooth and steady pattern without rapid changes. A typical pattern of trace of detrusor pressure during voiding is that the pressure tracings rise and drop smoothly and steadily. Therefore, any rapid changes on the curve in short time should be considered as artifacts, and must be interpreted and corrected. Artifacts of detrusor pressure during voiding have a number of types, and can be produced in a variety of ways. We may classify them into technical and physiologic artifacts. The technical artifacts may be caused by phasic signal loss, signal stepwise change and urethral catheter loss and rectal catheter dislocations; the incidences in our study are 1.4%, 3.4%, 0.9% and 2.2% respectively (Fig. 20). The physiologic artifacts are mainly the spikes and dips on P_{det} tracing resulted from the different causes. The first cause is the spikes due to straining under the condition of a difference in pressure transmission to the P_{ves} and P_{abd} (Fig. 21a); the incidence of this type of artifact is 15.1% in our study. The second one is the dips caused by rectal contractions (Fig. 21b); the incidence of this type of artifact is 2.1%. The last one is the spikes or the dips due to urethral sphincter overactivity during voiding, which is sphincter contraction (Fig 22a) or relaxation (Fig 22b); the incidence of this type of artifact is 10.1%. In retrospective quality control, the spike and dip artifacts can manually be corrected by smoothing. However, it usually is difficult to correct the technical artifacts in retrospective analysis. If they occur after Q_{max}, we may acquire two key parameters of pressure-flow study: Q_{max} and $P_{det.Qmax}$. If they appear before Q_{max} or a complete pressure-flow plot is requested for the purpose of precise analysis, the traces with these artifacts will have to be discarded.

As a continuous quantitative parameter, OCO can precisely measure the urethral resistance and change. In this study, a systematically significant increase of OCO by 0.067 on average was shown after manual correction. Intra-individually, OCO changes were inconsistent, with a range from -1.379 to 0.995. The reason would be that the inconsistent changes of P_{det} could influence those of OCO. The changes of OCO indicated that manually reading lead to higher urethral resistance, and artifacts reduced urethral resistance. Therefore, we could say that OCO calculated by manually reading values could really indicate the condition of urethral resistance.

More serious is that various artifacts have influenced the diagnosis of obstruction and the assessment of obstructed degree. Generally, it seems that artifacts lead to a less obstructed degree. In our study, ICS and Schäfer nomograms were employed to evaluate this impact. After manual correction, more traces located in obstructed zone or grades. 11.0% traces changed the classification in ICS nomogram, and 28.9% did the grade in Schäfer nomogram. Using these two nomograms, 6.9% and 7.2% traces changed the diagnosis of obstruction; 5.5% and 6.0% traces shifted into obstructed zone or grades. It could be said that computerized results produced 5.5% or 6.0% false negative diagnoses of obstruction due to various artifacts in these 582 measurements. On the other hand, 1.2% and 1.4% traces shifted out of obstructed zone or grades. It could also be said that computerized readings produced 1.2% or 1.4% false positive diagnoses of obstruction. Therefore, we could say that retrospective quality control corrected considerable false diagnoses of obstruction.

From above analyses, It was found that a corrected Q_{max} should be determined at first in retrospective analyzing of pressure-flow data; then a location of P_{det} corresponding to Q_{max} can be found, and the urethral resistance parameters, such as OCO, can be calculated. It seems that a systematically significant difference of Q_{max} results in those of OCO, classifying and grading of obstruction after manual correction.

Summarily, quality control is involved in both on-line and off-line urodynamic investigation. Getting the most out of urodynamics depends not only on a good urodynamic practice, but also on the training and experience of the clinician charged with interpreting the result (Abrams 1998). In the interpretation of pressure-flow data, the clinician must meticulously examine the trace for artifacts before accepting the computer results. At present, retrospective quality control of computerized pressure-flow data is necessary; it can get rid of the impact of artifacts on Q_{max}, P_{det}, urethral resistance, classifying and grading of obstruction, and diagnosis of obstruction. The data through quality control become more objective, reliable and acceptable, and can be used for further analysis. These effects of retrospective quality control have been demonstrated in the present study.

5. Conclusions

Quality control and plausibility check during investigation are the best way to avoid and to correct artifacts at an early stage; quality control relies on knowledge of typical values and signal patterns recognition.

Typical value ranges for pressures and other parameters in cystometry are effective tools for quantitative plausibility check and quality control of data. Typical value ranges for initial resting P_{ves} and P_{abd} are 31~42 cmH_2O and 28~39 cmH_2O in standing or sitting position respectively; and that of P_{det} is 0~4 cmH_2O, which is very close to zero. The suggested typical value ranges for the initial resting pressures are sensitive and reliable indicator for plausible and correct measurement. The errors related to initial resting pressures must be recognized and corrected before or at beginning of filling. The typical ranges for other parameters during filling and voiding phases are useful for checking plausibility of measurement. Q_{max} and V_{void} in cystometry must be comparable with those of free uroflowmetry. Typical value ranges for post-void P_{ves}, P_{abd} and P_{det} are 40~55 cmH_2O, 30~41 cmH_2O and 10~14 cmH_2O respectively. After voiding, P_{abd} has little change, P_{ves} and P_{det} are close to the levels before voiding. The suggested typical value ranges for post-void pressures are also important for the plausibility control of voiding phase.

Typical signal patterns are powerful tool for the qualitative plausibility check and quality control. Combining with typical value ranges, they allow a definitive judgment of the quality of an urodynamic investigation. We have described the TSP from four gradations: fine structure (pattern I), minor changes (pattern II), major changes corresponding to test-coughs (pattern III) and major changes (Pattern IV). At any stages of investigation, there must be the same fine structure and minor changes due to patient's breathing, talking and moving on P_{ves} and P_{abd} tracings but not on P_{det} tracing. These two patterns suggest that the signals are "live". The cough tests before or at beginning of filling, during filling at regular intervals, and before and after voiding are effective tool for checking the signal quality and plausibility. P_{ves} and P_{abd} tracings must show the equal changes corresponding to the test-coughs. Straining, detrusor overactivity and rectal contractions are typical major changes during filling; and detrusor contraction with different patterns, straining, rectal contractions and relaxation of pelvic floor during voiding are typical major changes during voiding. These major patterns have their characters, and we must be able to recognize them. Analyzing signal quality from these four gradations, we can find some atypical patterns, technical errors and artifacts and can correct them immediately in order to acquire the urodynamic data with a high-quality during investigation.
In retrospective quality control, various considerable artifacts were found in the cystometric data. The systematically significant differences of Q_{max}, urethral resistance, and classifying and grading of obstruction between manual and computerized readings suggested the existing of artifacts and their interference with the clinical judgment. The manually corrected Q_{max} had a consistently lower value; a higher value of OCO was calculated, and a more obstructed degree was assessed according to the manual readings. Although no systematically significant change of $P_{det.Qmax}$ was demonstrated after manual correction, it indeed underwent intra-individual considerable changes. Manually reading corrected considerable false diagnoses of obstruction. The effects of manual correction were shown. Therefore, retrospective quality control of computerized urodynamic data is significant and necessary; only the data throughout quality control can be used and reported.

6. Acknowledgements

We hereby express our thanks for the supports from some funds, e.g. the National Technology R & G Program of China. We also appreciate the kind helps from Mr. M. Lutz, Dr. R. Kirschner-Hermanns and Prof. G. Jakse from Department of Urology of RWTH Aachen in Germany.

7. References

Abrams P: *Urodynamics*, 2th edition. London: Springer-Verlag, p. 1, 1997.

Abrams P, Griffiths DJ: The assessment of prostatic obstruction from urodynamic measurements and from residual urine. *Br J Urol* 51: 129, 1979.

Abrams P, Blaivas JG, Stanton SL, et al: The standardisation of terminology of lower urinary tract function. *Scand J Urol Nephrol suppl*: 114, 1988.

Abrams P, Buzlin JM, Griffiths DJ, et al: The urodynamic assessment of lower urinary tract symptoms. In: Denis L, Griffiths K, Khoury S, et al (eds). *Proceedings of 4th international consultation on benign prostatic hyperplasia* (BPH). Paris, p. 323, 1997.

Abrams P. Urinary incontinence and neuropathic bladder. In: *3th annual european course in urology programme and course book*. Rome, p. 25, 1998.

Arsdalen KV, Wein AJ: Physiology of micturition and continence. In: Krane RJ, Siroky MB (eds). *Clinical neuro-urology,* 2th edition. Boston: Little, Brown and Company, p. 25, 1991.

Bates CP,Whiteside BM, Turner-Warwick R: Synchronous cine/pressure/flow/cysto-urethrography with special reference to stress and urge incontinence. *Br J Urol* 42: 714, 1970.

Boyarsky S: Formative years of the urodynamics society: reminiscences. *Neurourol Urodynam* 17: 159, 1998.

Brown M, Wickham JEA: The urethral pressure profile. *Br J Urol* 41: 211, 1969.

Davis DM: The hydrodynamics of the upper urinary tract (urodynamics). *Ann Surg* 140: 839, 1954.

Denny-Brown D, Robertson EG: On the physiology of micturition. *Brain* 56: 149, 1933.

Derezic D: Disorders of voiding and bladder function presented in textbooks published during the nineteenth century. *Eur Urol* 14: 482, 1988.

Dixon JS, Gosling JA: Structure and innervation in the human. In: Torrens M, Morrison JFB (eds). *The physiology of the lower urinary tract.* London: Springer-Verlag, p. 3, 1987.

Dixon JS, Gosling JA: The anatomy of the bladder, urethra and pelvic floor. In: Mundy AR, Stephenson TP, Wein AJ (eds). *Urodynamics: principles, practice and application.* Edinburgh: Churchill Livingstone, p. 3, 1994.

Drake WM: The uroflowmeter: an aid to the study of the lower urinary tract. *J Urol* 59: 650, 1948.

Gleason DM, Lattimer JK: The pressure-flow study: a method for measuring bladder neck resistance. *J Urol* 87: 844, 1962.

Gosling JA, Dixon JS, Lendon RG: The automatic innervation of the human male and female bladder neck and proximal urethra. *J Urol* 118: 302, 1977.

Griffiths DJ: Hydrodynamics of male micturition. I. theory of steady flow through elastic-walled tubes. *Med Biol Eng* 9: 581, 1971.

Griffiths DJ: Hydrodynamics of male micturition. II. measurement of stream parameters and urethral elasticity. *Med Biol Eng* 9: 589, 1971.

Griffiths DJ, van Mastrigt R, Bosch R: Quantification of urethral resistance and bladder function during voiding, with special reference to the effects of prostate size reduction on urethral obstruction due to benign prostatic hyperplasia. *Neurourol Urodyn* 8: 17, 1989.

Griffiths DJ, Höfner K, van Mastrigt, et al: Standardization of terminology of lower urinary tract function: pressure-flow studies of voiding, urethral resistance, and urethral obstruction. *Neurourol Urodynam* 16: 1, 1997.

Grino PB, Bruskewitz R, Blaivas JG, et al: Maximum urinary flow rate by uroflowmetry: automatic or visual interpretation. *J Urol* 149: 339, 1993.

Höfner K, Kramer AEJL, Tan HK, et al: CHESS classification of bladder outflow obstruction: a consequence in the discussion of current concepts. *World J Urol* 13: 59, 1995.

Karlson S: Experiment studies on the functioning of the female urinary bladder and urethra. *Acta Obstet Gynecol Scond* 32: 285, 1953.

Lich R., Howerton LW, Amin M: Anatomy and surgical approach to the urogenital tract in the male. In: Harrison JH, Grittes RF, Perlmutter AD, et al (eds). *Campbell's Urology,* 4th edition. Philadelphia: WB Saunders, p. 3, 1978.

Lewis P, Howell S, Shepherd A, et al: Computerized urodynamics: help or hindrance? *Neurourol Urodynam* 16: 508, 1997.

Madsen FA, Rhodes PR, Bruskewitz RC: Reproducibility of pressure-flow variables in patients with symptomatic benign prostatic hyperplasia. *Urology* 46: 816, 1995.

Miller ER: Techniques for simultaneous display of X-ray and physiologic data. In: Boyarsky S (eds). *The neurogenic bladder*. Baltimore: Williams and Wilkins, p. 79, 1967.

Murphy JJ, Schoenberg HW: Observation on intravesical pressure changes during micturition. *J Urol* 84: 106, 1960.

Muellner SR, Fleischner FG: Normal and abnormal micturition: a study of bladder behavior by means of the fluoroscope. *J Urol* 61: 233, 1949.

Perez LM, Webster GD: The history of urodynamics. *Neurourol Urodynam* 11: 1, 1992.

Rose DK: Determination of bladder pressure with cystometer. *J Am Med Assoc* 88: 151, 1927.

Rose DK: Cystometric bladder pressure determination: their clinical importance. *J Urol* 17: 487, 1927.

Rowan D, James ED, Kramer AEJL, et al: Urodynamic equipment: technical aspects. *J Med Eng Tecn* 11: 57, 1987.

Schäfer W, Fischer B, Meyhoff HH, et al: Urethral resistance during voiding: I. the passive urethral resistance relation, PURR. II. the dynamic urethral resistance relation, DURR. In: *Proceedings of the XIth Annual Meeting of the International Continence Society*. London, 1981.

Schäfer W. The contribution of the bladder outlet to the relation between pressure and flow rate during micturition. In: Hinman Fjr (eds). *Benign prostatic hypertrophy*. New York: Springer-Verlag, p. 470, 1983.

Schäfer W, Waterbär F, Langen PH, et al: A simplified graphic procedure for detailed analysis of detrusor and outlet function during voiding. *Neurourol Urodynam* 8: 405, 1989.

Schäfer W. Basic principles and clinical application of advanced analysis of bladder voiding function. *Urol Clin North Am* 17: 553, 1990.

Schäfer W, de la Rosette JJMCH, Höfner K, et al: The ICS-BPH study: pressure-flow studies, quality control and initial analysis. *Neurourol Urodyn* 13: 491, 1994.

Schäfer, W. and Sterling, A.M.: Simple analysis of voiding function by coefficients: obstruction coefficient, OCO, and detrusor strength coefficient, DECO. In: *Proceedings of the 25th Annual Meeting of International Continence Society*. Sydney, p. 338, 1995.

Schaefer W, Abrams P, Liao LM, et al: Good Urodynamic Practice: Uroflowmetry, Filling Cystometry and Pressure-Flow Studies. *Neurourol Urodynam* 21: 261, 2002.

Smith JC: Urethral resistance to micturition. *Br J Urol* 40: 125, 1968.

Spangberg A, Terio H, Engberg A, et al: Quantification of urethral function based on Griffiths's mode of flow through elastic tubes. *Neurourol Urodyn* 8: 29, 1989.

Susset JG: Urodynamic society presidential address: urodynamics — a step toward experimental urology. *Neurourol Urodynam* 4: 157, 1985.

Talbot HS: cystometry, and the treatment of vesical dysfunction in paraplegia. *J Urol* 59: 1130, 1948.

Tanagho FA, Miller ER: The initiation of voiding. *Br J Urol* 42: 175, 1970.

Torrens M. Human physiology. In: Torrens M, Morrison JFB (eds). *The physiology of the lower urinary tract*. London: Springer-Verlag, p. 333, 1987.

Torrens M. Human physiology. In: Torrens M, Morrison JFB (eds). *Urodynamics*. London: Springer-Verlag, p. 277, 1987.

von Garrelts B: Analysis of micturition: a new method of recording the voiding of the bladder. *Acta Chir Scand* 112: 326, 1956.

von Garrelts B: Micturition in the normal male. *Acta Chir Scand* 114: 197, 1957.

Woodburne RT: Structure and function of the urinary bladder. *J Urol* 84: 79, 1960.

Zinner NR, Paquin AJ: Clinical urodynamics: I. studies of intravesical pressure in normal human female subjects. *J Urol* 90: 719, 1963.

7

Guidelines for Developing Automated Quality Control Procedures for Brain Magnetic Resonance Images Acquired in Multi-Centre Clinical Trials

Elias Gedamu
Department of Biomedical Engineering and Montreal Neurological Institute
McGill University
Canada

1. Introduction

Automated quality control (QC) procedures are critical for *efficiently* obtaining *precise* quantitative brain imaging-based metrics of in vivo brain pathology. This is especially important for multi-centre clinical trials of therapeutics for neurological diseases, in which brain imaging-based metrics may be used to quantify therapeutic efficacy. While there are many different types of brain imaging methods (e.g. computed tomography, magnetic resonance imaging, positron emission tomography, etc.) that have been used to quantify different aspects of in vivo pathology (e.g. presence of tumours, brain atrophy, hydrocephalus, abnormalities in blood vessels or the extravasation of blood, the depletion of receptors available for the binding of an injected substance, abnormal brain metabolism, etc.), this Chapter will focus on the automated QC procedures required to use magnetic resonance (MR) images (MRI) to yield imaging-based metrics of in vivo brain tissue pathology.

Magnetic resonance imaging is a powerful non-invasive technology that can provide in vivo images sensitive to normal and pathological brain tissue. Important strengths of MR imaging include its superior grey-matter (GM)/white-matter (WM) tissue contrast, sensitivity to WM pathology and clinical feasibility of relatively high-resolution whole-brain imaging. In conventional brain MRI, the signal intensities arise from the different relaxation characteristics of protons in water molecules present in different brain environments following radio-frequency (RF) excitation when the brain is in a magnetic field. MRI acquisition sequences vary the timing and duration of RF excitation pulses and magnetic field gradients, yielding different contrasts (termed MRI modalities) that can highlight different aspects of brain anatomy and pathology. This is illustrated in Fig. 1 using 4 conventional imaging modalities, T1-weighed (T1w) and T1w 5 min after intravenous injection of a gadolinium (Gd) contrast agent (T1w+Gd), T2-weighted (T2w), proton density weighted (PDw), and fluid attenuated inversion recovery (FLAIR) image, which were all acquired from a patient with multiple sclerosis (MS), a neurological disease that affects the brain and spinal cord. The T1w image most clearly differentiates brain GM, WM and cerebrospinal fluid (CSF). This high tissue contrast is the reason why T1w is often the optimal input modality, or included with other input modalities, for image-processing algorithms that classify the image voxels (volume elements) as WM, GM and CSF, which can be a critical step that precedes the quantification of biologically important brain characteristics (e.g. the volumes of the entire brain, individual brain structures, GM, WM, and abnormal WM and GM). In addition to the high tissue contrast, T1w MRI also informs on brain pathology. It has been shown that WM hypointensities on T1w MRI of MS patients are associated histopathologically with severe tissue destruction (Van Walderveen et al., 1998), and T1w MRI also reveals a population of hypointense lesions in the cerebral cortex of MS patients (Bagnato et al., 2006). By injecting a Gd contrast agent (Gd is paramagnetic in its trivalent state), the T1w modality can be further exploited to detect increased permeability of the "blood-brain barrier" (BBB), which under normal conditions restricts the transport of substances from the circulation into the brain, thus confining the Gd contrast agent to the blood vessels and resulting in a relatively bright intensity of blood vessels large enough to be resolved by T1w imaging. Under pathological conditions (e.g. stroke, trauma, tumour, inflammation), the permeability of the BBB may be transiently increased and the Gd contrast agent will enter the brain, resulting in a relatively bright intensity in the region of the pathology. For example, in Fig. 1 the T1w+Gd image from a patient with MS exhibits a ring of hyperintense signal that results from the increased BBB permeability associated with acute focal inflammation. The T2w image is more sensitive to different types of WM pathology (not only the severe tissue destruction) than the T1w, exhibiting abnormally hyperintense signal in regions with pathological abnormalities such as: tissue loss, injury, incomplete repair, inflammation and scarring. Important limitations of using T2w images to quantifying brain pathology is the lack of specificity of the hyperintensities (e.g. they may be oedema that may resolve quickly, they may be irreversible tissue destruction that may never repair and result in further degeneration), and the poor CSF/abnormal WM contrast (e.g. abnormal WM that abuts the CSF-filled ventricles cannot be reliably quantified). The latter limitation is addressed by acquiring PDw and FLAIR images, in which abnormal WM is hyperintense and CSF is hypointense. Furthermore FLAIR imaging has been shown to be more sensitive to focal WM MS pathology compared to standard T2w imaging (De Coene et al., 1992; Filippi et al., 1996; Geurts et al., 2005).

In clinical trials of therapies for neurological diseases (e.g. multiple sclerosis and Alzheimer's disease), various MRI modalities may be acquired within a single scanning session to quantify various aspects of brain pathology, and multiple scanning sessions may be performed on each patient throughout the trial to track the changes in MRI-derived brain pathology metrics from the baseline pre-treatment state (Fig. 2). This *multiple imaging modalities at multiple timepoints for many patients* paradigm to yield a snapshot of the brain pathology at a certain timepoint or to yield the dynamics of progressing/resolving pathology, relies upon the assumption that image intensity variations are biological. Within this assumption, MRI-derived brain pathology metrics may be calculated using an image-processing pipeline comprised of leading edge automated techniques including image intensity normalization, co-registration of different MRI modalities, registration to brain atlases, brain tissue classification, segmentation of brain structures and types of pathology (Fig. 2 - Image Processing Pipeline). The success of these automated image-processing techniques may be significantly affected by spatial and/or temporal variability in the MRI intensities resulting from methodological sources including scanner software/hardware upgrades, scanner hardware deterioration and human error (Fig. 2 - MRI Acquisition). Accordingly, the role of QC is to ensure that each MRI that enters an image-processing pipeline has been assessed and meets an acceptable level of quality (minimally affected by non-biological variability, consistent with trial protocols, and consistent with previously obtained data from the patient during the trial) to ensure the expected accuracy and precision of the MRI-derived brain pathology metrics (Fig. 2 - Quality Control).

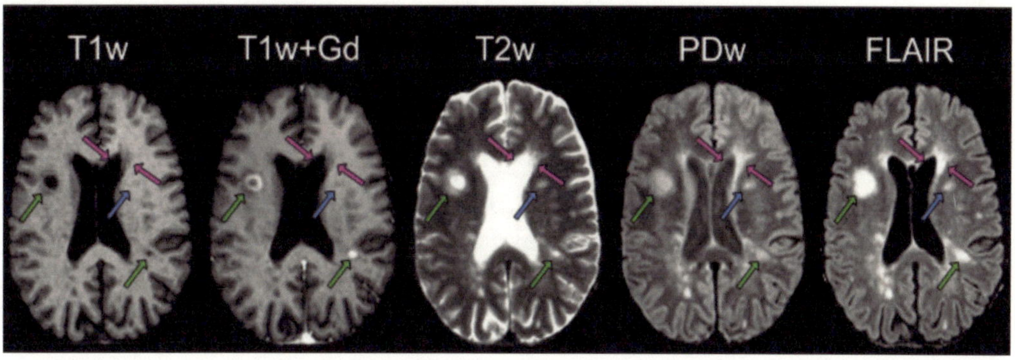

Fig. 1. Corresponding 2D slices extracted from different 3D MRI modalities acquired during a single scanning session of a patient with MS. (Left to right): T1-weighed (T1w), T1w after Gd injection (T1w+Gd), T2-weighted (T2w), proton density weighted (PDw) and fluid attenuated inversion recovery (FLAIR). Green, purple, and blue arrows highlight the advantages and disadvantages of each MR modality. Green arrows show acute increases in BBB permeability (hyperintense signal on T1w+Gd) associated with tissue destruction and inflammation (hypointense signal on T1w; hyperintense on T2w, PDw and FLAIR), and purple arrows show the weakness of T2w images in differentiating abnormal WM from adjacent CSF (better differentiation on PDw and FLAIR). Overall, the volume of abnormal WM on T2w and FLAIR modalities may be higher than on T1w, due to their high sensitivity to various pathological processes (e.g. swelling, destruction, repair, scarring)

This Chapter provides guidelines for developing an automated QC procedure for brain MRIs acquired in multi-centre clinical trials of therapeutics for neurological diseases; in particular the automated QC for multi-centre clinical trials of therapies for MS will be discussed in detail. Emphasis will be placed on: 1) demonstrating the need for appropriate QC procedures, 2) determining the objectives, 3) defining quality, 4) developing a framework to facilitate the creation of quality control procedures for MRIs, and 5) providing an example of an automated QC procedure that is used in industry. Although the focus will be on QC for clinical trials of MS therapies, the guidelines proposed in this chapter could be applied to clinical trials that use MRI-based imaging metrics to assess therapeutics for other neurological disorders such as Alzheimer's disease, epilepsy, sleep apnea, stroke, and amyotrophic lateral sclerosis.

2. Demonstrate the need for appropriate QC procedures

It may seem obvious that if an MRI scan is adequate for qualitative interpretation by a radiologist, then it should be of sufficient quality to be used to extract quantitative metrics of brain pathology, however, this is not necessarily true. Studies have been performed demonstrating the effect of specific aspects of MRI quality on specific types of MRI-based imaging metrics.

Preboske et al. (2006) compared the effect of three types of common MRI artifacts, inconsistent image contrast between serial scans, head motion, and signal-to-noise ratio (SNR), on the performance of the boundary shift integral (BSI), a method used to quantify whole brain atrophy between MRIs acquired in the same person at two different visits, by calculating the shift at the brain tissue/CSF border that may occur over the time between the visits if the brain is undergoing volume loss. They found that as image quality deteriorated due to any of the three types of artifacts, the atrophy measurement error increased. The study showed that the magnitude of error could substantially exceed the disease effect in Alzheimer's Dementia (AD) for whole brain atrophy per year (Preboske; Gunter; Ward & Jack, 2006). Blumenthal et al. (2002a) compared the effect of ringing artifacts caused by subject movement on measuring grey matter volume using ANIMAL (Kuba et al., 1999) in 180 healthy children. The authors compared the amount of ringing present (none, mild, moderate, or severe) in the MRI to the volume of brain classified as grey-matter and found that as the level of motion increased, the volume of grey matter decreased. Camara-Rey et al. (2006) examined the effect of simulated motion artifacts (ghosting, blurring, and pulsatile flow artifacts from major blood vessels like the carotid arteries) on measuring brain atrophy using SIENA (Smith et al., 2004; Smith et al., 2002). In healthy subjects they found that the presence of these artifacts could substantially affect atrophy measurements and, in some cases, have the same expected differences observed in AD patients over a 12 month period. Boyes et al. (2006) compared two methods for measuring brain atrophy measurements, the BSI and Jacobian integration (JI), using MRIs from a cohort of AD patients and healthy subjects. Three scans were acquired for each subject, a same day scan and repeat scan (re-scan) pair to determine the inherent error of each method and a scan one year later to assess the consistency of each method. Each scan was visually assessed for image quality by an experienced MRI reader based on motion artifacts and contrast differences between WM and GM, and brain and CSF. They showed that the BSI and JI techniques were susceptible to poor image quality with measurement errors exceeding three times the expected brain atrophy rate observed in normal control elderly subjects over 1 year (Scahill et al., 2003) and within the range of yearly atrophy rates observed in AD patients (Bradley et al., 2002; Fox et al., 2000).

These studies demonstrate the potential for the quality of an MRI to affect the quantification of brain metrics by adding variability that can obscure pathological changes. A QC procedure should objectively quantify the quality of an image and subsequently objectively reject images with quality metrics that do not meet software-specific a priori defined control limits.

Complexities of Developing QC procedures for Clinical Trials

While the QC studies discussed in the preceding paragraph(s) demonstrated the effect of some aspects of MRI quality on a subset of MRI-based brain metrics calculated in relatively few subjects, multi-centre clinical trials pose additional QC-related difficulties: 1) large volume of MR images acquired from multiple subjects at multiple timepoints, 2) scanner variability arising from variations in hardware performance, 3) hardware and software changes, 4) human error, 5) diversity of MRI-derived brain pathology metrics, and 6) variety of image processing methods involved in the measurement of these brain pathology metrics.

The sheer volume of MR scans that are produced by multi-centre clinical trials limits the feasibility of MRI readers to manually assess each MRI for image quality (De Stefano et al., 2010; O'Connor et al., 2009; Rovaris et al., 2008). With the human visual system and time as constraints, MRI readers cannot consistently evaluate MRI images for correct identification (Does the brain MRI actually correspond to the patient identifier?), correct MRI sequence (Do the scanning parameters match the protocol?), and acceptable image quality (Is the noise, motion, etc., within acceptable ranges of the control limits?).

Scanner variability arising from day-to-day variations in hardware performance and deliberate changes to the scanner hardware or software may result in variations in the MRI characteristics, which could introduce non-biological variability in MRI-derived pathology metrics. While a QC procedure may not be capable of detecting the most subtle variations in hardware performance (which may not significantly affect MRI-derived metrics), the QC procedure would be expected to detect failing and noted software changes.

The most common human errors in MRI acquisition in the clinical trial setting are: Mistyping of a patient identifier or using the incorrect patient identifier, acquiring MRI sequences with incorrect parameters or omitting MRI sequences, and acquiring MRI modalities in the wrong scan order (which can be critical, for example, when injection of a contrast agent is essential for a certain modality but may corrupt other modalities). An automated QC procedure can detect mistyped patient identifiers and detect incorrect patient identifiers by assessing if the present brain MRI is the same brain as other MRIs with the same identifier. Incorrect sequence parameters or missing sequences can be detected by an automated QC that compares the MRI sequences acquired in a session to the previously accepted protocol- and site-specific sequences and parameters. Incorrect acquisition order of MRI modalities can be detected by comparing the acquisition times of each scan to the previously accepted protocol- and site-specific scan order.

The diversity of the MRI-derived brain pathology metrics will also influence the QC procedure. Image quality may not be acceptable for some metrics, but may be adequate for others. For example, a localized image artifact that prevents the volume of a specific small brain structure such as the hippocampus from being measured reliably may not significantly affect the measurement of total brain white matter tissue volume. The ideal QC procedure should have the flexibility to detect and report image quality issues that prevent the reliable calculation of a specific metric, without rejecting the entire scanning session as a whole.

The variety of image processing methods to measure brain pathology metrics is another factor that can impact the QC procedure. Image quality may affect some image processing methods more than others. For example, a single-modality K-means classifier will be affected by poor SNR or ghosting more than multi-modal classifiers because they do not have complementary data. The automated QC procedure should account for the limitations of image processing tools used to calculate brain pathology metrics.

The impact that MRI quality can have on MRI-derived brain pathology metrics combined with the difficulties associated with multi-centre clinical trials demonstrates the need for appropriate QC procedures.

3. Process pipeline for multi-centre clinical trials

It is helpful to understand the intricacies of multi-centre clinical trial MRI process pipelines before proceeding to the guidelines section for developing quality control procedures for

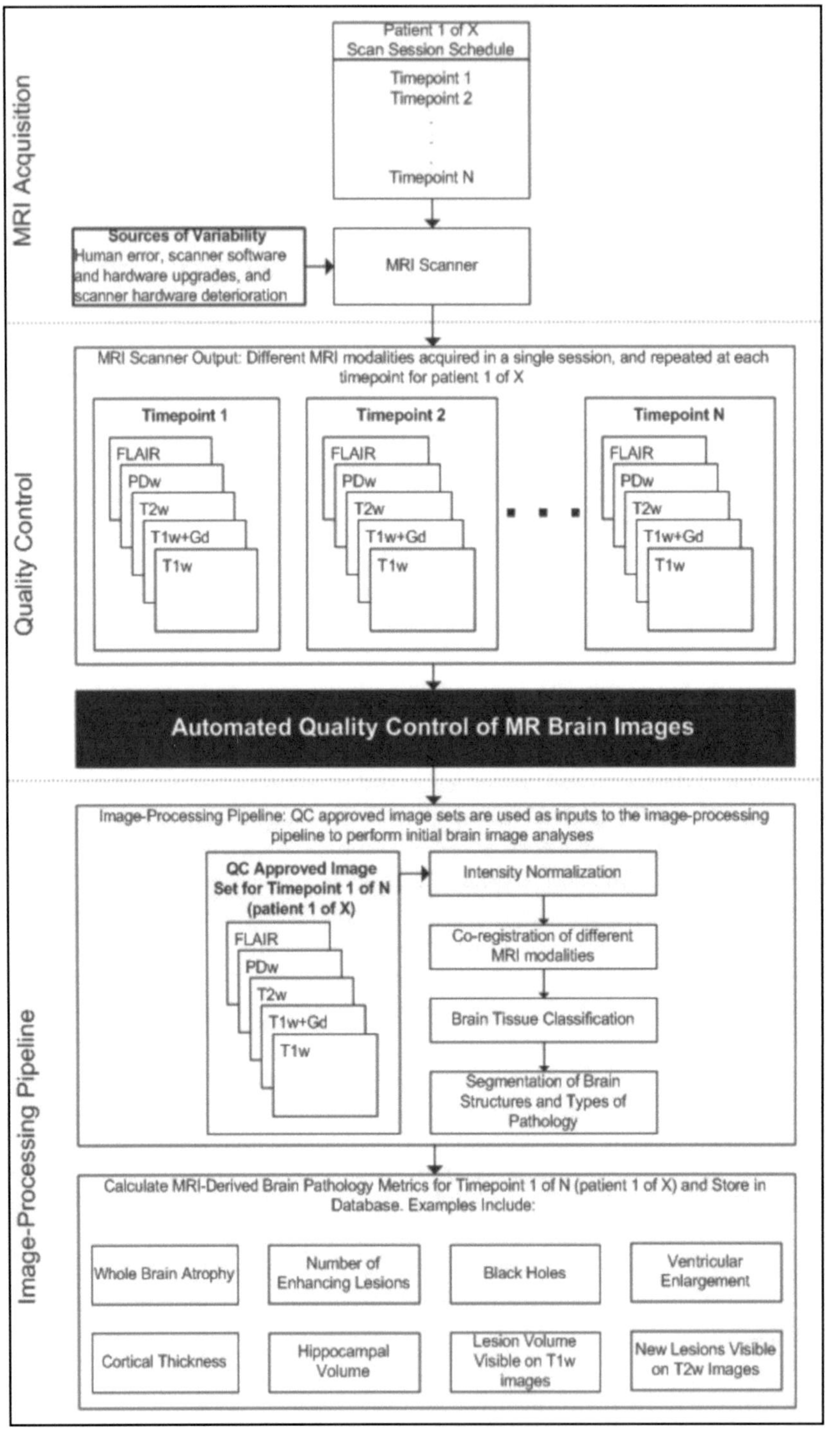

Fig. 2. An example of an MRI process pipeline for a multi-centre clinical trial. The pipeline consists of: 1) MRI Acquisition, 2) Quality Control, and 3) Image-Processing Pipeline

brain MRIs. Fig. 2 illustrates three components common to most process pipelines for multi-centre clinical trials, 1) acquisition of MRIs, 2) quality control of MRIs, 3) and image-processing of MRIs and quantification of the MRI-derived brain pathology metrics.

Acquiring multi-modal MRIs (e.g. FLAIR, PDw, T2w, T1w, T1w+Gd) from multiple subjects across multiple timepoints at various scanning sites forms the initial step of the pipeline (Fig. 2 - MRI Acquisition). Unfortunately, as discussed in Section 2, these images may be affected by human errors as well as non-biological variability introduced by the scanner. Without the quality control step (the second step in Fig. 2) these errors and non-biological variability would be propagated down the MRI process pipeline, thereby affecting the fidelity of MRI-derived brain metrics. Accordingly, quality control procedures are placed early in the MRI process pipeline, in which the image sets are submitted to a set of QC tests. Unacceptable images are flagged by comparing the QC test results to pre-determined control limits, logged in a QC database, prevented from further processing, and reviewed by imaging experts to identify the root cause of the error, while acceptable MRI sets are normalized to correct for intensity non-uniformities when appropriate, co-registered and processed using brain tissue classifiers and segmentation techniques to identify brain tissues and regions of interest (Fig. 2 - Image-Processing Pipeline). The resultant images and their corresponding maps of tissue type and locations of critical brain structures may then be used to calculate brain pathology metrics such as: total brain volume loss; increases in the CSF-filled lateral ventricles; cerebral cortical thickness; the volumes of specified brain structures; the number of white-matter lesions that are new, contrast-enhancing or associated with tissue destruction. These metrics are but a few examples of MRI-derived metrics which may be used to quantify disease evolution or therapeutic efficacy.

4. Guidelines

This section provides a set of guidelines for developing an automated QC procedure for brain MRIs acquired in multi-centre clinical trials of therapeutics for neurological diseases with a focus on multiple sclerosis. The sub-sections include defining quality and developing quality control procedures for brain MRIs.

4.1 Quality

The American Society for Quality (ASQ) states that in technical usage, quality can have two meanings: 1) the characteristics of a product or service (e.g. MRIs) that bear on its ability to satisfy stated or implied needs (e.g. accurate and reliable brain pathology measurements) and 2) a product or service free of deficiencies (ASQ, n.d.). In engineering usage, G. Taguchi provides a similar definition of quality as having two types: 1) customer quality (features what customers want, i.e. multi-center clinical trial sponsors would like MRIs that provide accurate and reliable brain pathology measurements) and 2) engineered quality (problems customer does not want) (Taguchi et al., 2000). Princeton's wordnet defines quality as the degree or grade of excellence or worth (Princeton, 2010).

In the case of multi-centre clinical trials of therapeutics for neurological diseases where therapeutic efficacy is evaluated using brain pathology metrics derived from MRIs, the definition of quality should be based on the value an MRI has towards its intented application. Accordingly, quality can be defined as the degree of worth of an MRI to measuring brain pathology metrics which is in accordance with the ASQ's and G. Taguchi's definitions of quality.

4.2 Developing QC procedures for brain MRIs

In accordance with the above definition of quality, the following sub-sections detail a framework that can be used to develop QC procedures for brain MRIs acquired in multi-centre clinical trials for neurological disease. Sub-sections include 1) factors that impact quality, 2) determining important QC tests, 3) imaging markers for QC tests, 4) determining the degree of worth of MRIs, 5) creating control limits to assess (pass and fail) MRIs, and 6) determining a course of action to take: accept, correct, or reject the MRI

Factors that Impact Quality

Non-pathological and non-physiological anomalies present on MRIs (image artifacts) and scan-to-scan variations in trial-, site-, and subject- specific acquisition protocols and sequences (longitudinal inconsistencies) are two important factors that can affect image quality. Image artifacts may result from subject movement, defective hardware, Gibb's

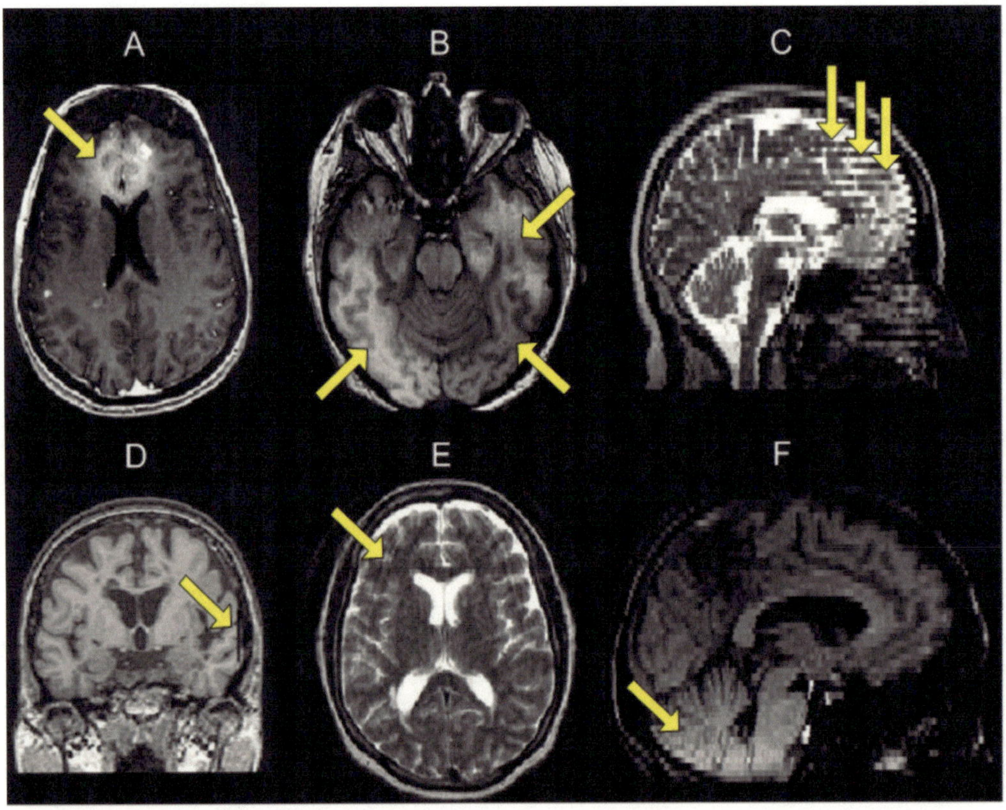

Fig. 3. Shows examples of poor quality MR images with yellow arrows highlighting artifacts: A) T1w+Gd axial slice with hyperintense artifact in the frontal lobes of the brain (caused by mucus in the nasal sinus), B) T1w axial slice with intensity non-uniformity (could not be corrected), C) T2w sagittal slice illustrating interpacket motion artifacts (sharp gradients at the edge of each slice cause by subject movement during multi-shot acquisitions), D) T1w coronal slice showing missing slices, E) T2w axial slice with motion artifacts (ringing in left/right direction), and F) T1w saggittal slice showing intensity variations in the cerebral cortex

ringing from improper sampling rate or using a small field of view (FOV), conductive objects in the FOV such as braces and tooth fillings, blood flow through major venous structures, or signal dropout from air/tissue interfaces like the nasal sinus. Fig. 3 shows examples of MRI images that are considered "poor quality" due to the presence of artifacts. These images would be expected to yield incorrect results for many image processing algorithms. While there exists algorithms to correct some of these artifacts (Ahmed et al., 2003; Blumenthal et al., 2002b; Forbes et al., 2001; Greenspan et al., 2001; Kim et al., 1999; Lötjönen et al., 2004; Malandain et al., 2004; Ourselin et al., 2000; Rousseau et al., 2005; Sled et al., 1998; Ward et al., 2000), these corrections may not be adequate to achieve the expected precision of the downstream image processing to quantify brain pathology metrics. Longitudinal inconsistencies tend to result from scanner software and hardware upgrades, scanner hardware deterioration and human errors/inconsistencies. Fig. 4 shows the effect of inconsistent patient positioning where two same-day scans were acquired, an initial scan (Scan) and repeat scan (Re-scan) after repositioning the patient in the scanner. Fig. 4 demonstrates that simply repositioning a patient and rescanning can result in non-linear changes in brain shape on the rescanned image relative to the initially acquired image that may result in loss of accuracy and precision of MRI-derived metrics. The presence of between-timepoint inconsistencies can be expected to increase the error in brain pathology metrics that compare images acquired at different timepoints (e.g. measuring change in brain volume by comparing the follow-up image to the baseline image) and decrease the power of statistical tests comparing the metrics calculated from images acquired at different timepoints (e.g. determining if there is a statistically significant difference in the volume of WM lesions at follow-up compared to the volume of WM lesions at baseline).

To develop an appropriate quality control procedure for multi-centre clinical trials of therapeutic treatments, image artifacts and longitudinal inconsistencies that affect MRI quality and the evaluation of therapeutic efficacy should be detected and controlled using appropriate tests.

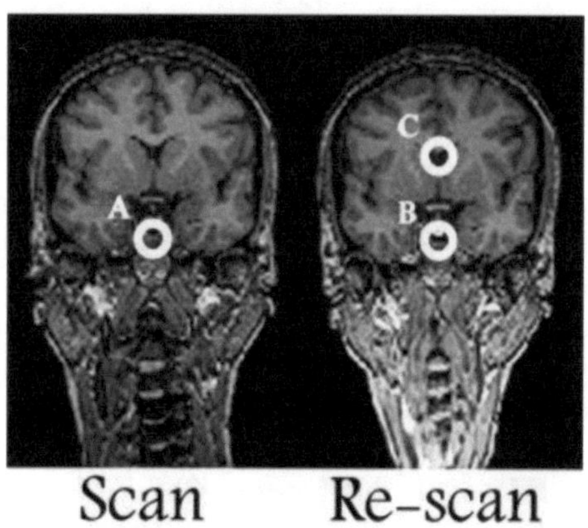

Fig. 4. Shows a scanned image with the magnet's isocenter identified with a circle and labelled A (left) and a re-scanned image with the magnet's isocenter identified with a circle and labelled B (right). The scanned image's isocenter relative to the re-scanned image's isocenter is identified on the re-scanned image with a circle and labelled C. The change in position between both images is illustrated by labels B and C. The distortion between the two images is apparent in the neck and top of the brain

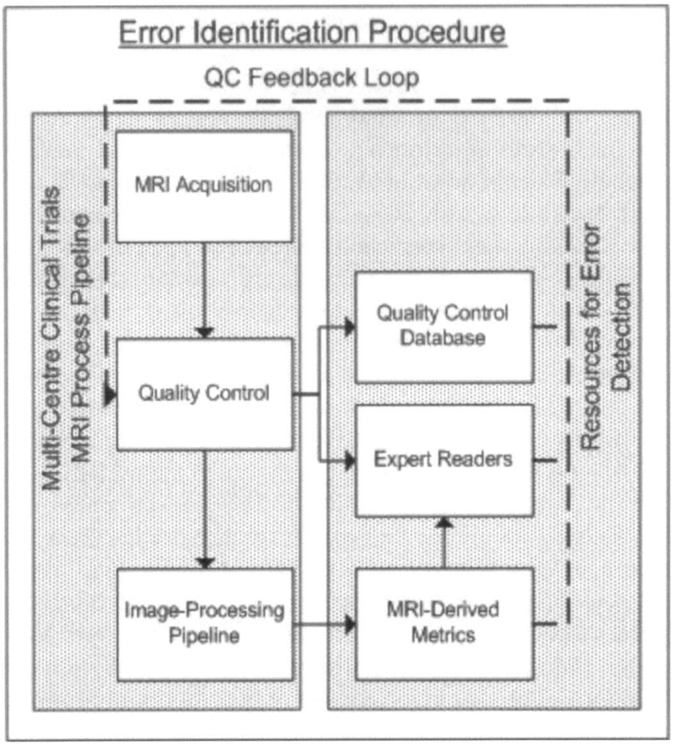

Fig. 5. An error identification procedure that is used to detect poor quality MRIs that may impact MRI-derived brain pathology metrics calculated in multi-centre clinical trials. Expert readers experienced with MRIs, information within the QC database, and MRI-derived metrics are the primary resources used to detect errors. Using a QC feedback loop, those errors are used to ensure that the tests in the QC procedure are current and effective

Determining Important QC Tests

QC tests need to be developed to detect the attributes associated with poor quality MRIs capable of affecting the accuracy and precision of brain pathology metrics. These tests can be determined using a semi-automated dynamic error identification procedure consisting of expert MRI readers, automated quality control systems and databases, and abnormal measurement variations in the MRI-derived metrics (Fig. 5). Expert MRI readers are trained professionals that have experience working on MR images that are affected by pathology and are, therefore, important to the process of identifying errors. MRI readers are an ideal resource to use for screening MRIs for image artifacts and longitudinal inconsistencies because they examine several MRIs daily, are trained to identify the pathology of the neurological disease on MRIs, and are able to distinguish between visual artifacts and expected MRI variations. Automated quality control systems and databases provide access to historical QC measurements that are especially important for identifying longitudinal inconsistencies. For example, the SNR values for serially acquired MRIs could be used to detect scanner changes when the required information in the DICOM header file is unavailable. The error identification procedures described above ensures that QC tests in the QC procedure are current and effective.

Imaging Markers for QC Tests

The QC tests use imaging markers to quantify the attributes associated with poor quality MRI that may affect the accuracy and precision of brain pathology metrics. Imaging markers are MR acquisition references that provide reliable, consistent, and representative information on the performance of the MR scanner and the fidelity of the MRI. Using image processing techniques, pertinent data in the imaging marker are identified and used to measure the level of quality in a MRI. There are three types of imaging markers that are commonly used for quality control: phantoms, external markers, and the MRI itself (normal control subjects and patient data). Consideration should be given to how the availability, feasibility, limitations, advantages, and importance of each imaging marker affects the development of a QC procedure.

An MRI phantom is brain-like in size and shape and fabricated using materials with relaxation properties conducive to MR imaging. Phantoms range from simple structures, like a sphere of agar or bottle of doped water, to more complex designs, like concentric spheres of agar where each sphere has a different concentration of agar solution. The general idea of using phantoms for quality control is that the images acquired of the phantom should be consistent with phantom images obtained at different sites involved in the trial and consistent over time at a given site. Phantoms have been developed by several groups including American College of Radiology (ACR) MRI accreditation program, European Community Concerted Action, National Electrical Manufacturers Association (NEMA), American Association of Physicists in Medicine (AAPM), and Alzheimer's Disease Neuroimaging Initiative (ADNI) (ACR, 2004, 2005; NEMA, 1988, 2001, 2003a, 2003b, 2003c; Price et al., 1990). The use of phantoms in multi-centre clinical trials requires imaging of the phantom at regular intervals using the same sequences approved for the site by the trial's MRI-analysis centre even when scanner hardware and software are stable, and also before and after every scanner-associated upgrade. The limitations of using phantoms for QC include financial and time feasibility of phantom production and repeated scanning, variability in the fabrication procedure and composition of the construction materials (affects site-to-site measurements), degradation of construction material over time (adds errors to longitudinal measurements), and the inability to represent the anatomical structures of real brain MRIs accurately (adds uncertainty to the interpretation of phantom-based measurements in the context of real brain MRIs). The advantage of using phantoms is that ground truth is known which allows for precise measurements of MR scanner performance parameters like geometric accuracy, high contrast spatial resolution, slice thickness accuracy, image intensity uniformity, percent signal ghosting, and the ability to detect low contrast objects. Additionally, phantoms can be used for correcting MRI geometric distortions caused by magnetic field inhomogeneities and gradient nonlinearities in the scanner (Jovicich et al., 2006). This is especially important for MRI-derived metrics that quantify morphological changes of anatomical structures in the brain like changes in cortical thickness, whole brain atrophy, and ventricular enlargement.

External markers for QC refer to small simple objects (e.g. cylinders, spheres) that are placed with the subject at the time of acquisition and fabricated using materials with relaxation properties that produce MR signals when scanned (e.g. tubes filled with manganese chloride or copper sulfate solution). The general idea of using external markers for QC is that they represent known quantities that are scanned under truly identical conditions (i.e. at the same time) as the brain, unlike the phantom that would be scanned in a different session when the scanner may perform slightly differently. MRI-compatible external markers such as agar are readily available and scanning with external markers is more feasible to implement in multi-centre clinical trials than phantoms since the subject and marker can be scanned simultaneously. As with the phantoms, the properties of the external markers are known, which is useful for tracking morphology and intensity changes over time resulting from MR hardware degradation and software/hardware changes, or comparing quality control parameters (e.g. SNR, contrast-to-noise ratio) for different scanners at multiple sites. External markers are susceptible to the same limitations as phantoms (i.e. variability in the fabrication procedure and composition of the construction materials, degradation of construction material over time, and the inability to represent the anatomical structures of real brain MRIs accurately). Additional limitations of using external markers for QC include the limited space they occupy that is external to the brain (i.e. cannot detect spatially varying errors within the brain) and the necessity for consistent positioning of the external markers to minimize spatial variability of measurements for QC.

MRIs of either normal control subjects (for QC only) or of the subjects enrolled in the trial can be used as imaging markers to evaluate image quality. The general idea of using human scans for QC is that, unlike phantoms, they represent the actual imaging properties of the brain under the same scanning conditions of the subjects in the trial (e.g. potential for movement, flow artifacts from the carotid arteries). The normal control subjects can be considered as "living phantoms", such that images are acquired regularly with identical sequences as prescribed by trial protocol, but not under the identical conditions as each individual patient. Unlike the man-made-phantom images, the ground truth of the normal control subject images is not known, but the biology is assumed to be stable and normal. The MRIs acquired from the subjects enrolled in the clinical trial may itself be used for QC. Despite the fact that the assumption of stable and normal biology cannot be made, QC may be performed using image characteristics that would not be changed by the presence of pathology. The advantages of using the MRIs acquired for the purposes of the trial are that 1) all scans for each modality are readily available, and 2) the measured QC parameters are indicative of the quality of the image from which the brain pathology metrics will be calculated. Since the ground truth of these images is not known, the QC strategy involves setting control limits for acceptable/unacceptable MRIs by analyzing the effect of varying QC parameters on MRI-derived metrics. For example, to define the control limit for assessing the effect of MRI motion artifacts on hippocampal volume measurements, a quantitative test can be performed by simulating MRI images with different amounts of motion artifact, calculating the hippocampal volume on these simulated images, and observing the relationship between the error in hippocampal volume and the amount of simulated motion. The control limit of MRI motion artifact for hippocampal volume measurement is thus determined as the maximum amount of motion on an MRI that can yield measurements with similar accuracy and reproducibility as the same MRI with no motion artifact.

The cost of using phantoms or normal control subjects for QC is prohibitive in many clinical trials. External markers may also be considered unfeasible, due to the additional scanning time cost associated with the placement of the markers, and analysis centre costs associated with developing image-processing tools to accommodate their presence and perform a trial-specific set of QC tests. These feasibility issues support the use of imaging markers extracted from the MRIs acquired on the subjects enrolled in the trial to measure QC parameters for quantifying, testing, and assessing image quality.

Degree of Worth of the MRIs

Determining the degree of worth of the MRIs is important because it allows us to differentiate MRIs based on levels of quality. Qualitatively, the degree of worth of the MRIs can be defined as the fidelity of an image to convey the true physiology or pathology of the subject being analyzed that is free from artifacts that could reduce the sensitivity of any MRI-derived brain pathology metric used to determine the effect of treatment on disease. MR images that have artifacts or low SNR that affect the reliability of image-processing algorithms used to quantify brain pathology metrics are considered to be poor quality. MR images that are relatively free of artifacts with acceptable SNR are considered good quality and are expected to yield brain pathology metrics that provide an accurate representation of the expected physiology and pathology.

The degree of worth of the MRIs can be determined using experienced expert MRI readers and quantitative experiments. Experienced expert readers review and analyze large volumes of MRIs, are involved in calculating MRI-derived brain pathology metrics, have knowledge on the MRIs that helped produce the derived brain pathology metrics and, accordingly, are able to assess the effect of image quality on the evaluation of metrics. These qualitative assessments can be coupled with QC test measurements to quantitatively evaluate the degree of worth of MRIs (described in the next section). Optimally, quantitative experiments that evaluate the effect of varying select QC parameters on MRI-brain pathology metrics (e.g. quantifying the effect of MRI noise levels on measuring whole brain atrophy) should be used to determine the degree of worth for each acquired MRI because these methods provide an accurate assessment of deviations in image quality on measurement error. Increases in MRI-derived brain pathology measurement errors decrease the degree of worth of the MRI and vice versa. Unfortunately, quantitative experiments are time consuming (i.e. require several steps including design, development, testing, validation, and verification), may not reflect the true image quality properties (i.e. simulated noise used to modulate SNR may be inaccurate), and difficult to incorporate in dynamic environments such as multi-centre clinical trials (i.e. time required to implement QC tests based on experimental results may be impractical since the solutions to the identified QC errors need to be incorporated promptly).

Creating Control Limits to Assess (Pass and Fail) MRIs

As previously mentioned, automated quality control for MRI brain images require control limits to define the boundary between acceptable and unacceptable MRIs based on image quality and the sensitivity to image quality of the brain pathology metric to be calculated. Control limits can be determined using receiver operating characteristic (ROC) curves, quantitative evaluations of MRI quality parameters on MRI-derived brain pathology metrics, and a set deviation from the expected value of QC parameters.

ROC curves can be used to depict the sensitivity of QC tests by comparing true positive rates (i.e. the number of MRIs that fail QC when they should actually fail) and false positive rates (i.e. the number of MRIs that fail QC when they should not fail) for a range of thresholds; truth tables may be computed if there is a gold standard. MRIs that have been evaluated either qualitatively by experienced readers (e.g. low, medium, high) or quantitatively by image-processing can be used as a gold standard. The reader evaluations can be combined with QC test measurements to generate ROCs and determine optimal image quality control limits. If reader evaluations do not exist, ROCs can also be determined using datasets where ground truth is known. For example, a MRI dataset consisting of few subjects with multiple acquisitions could be used to determine the control limits for detecting similarity of MRIs (useful to detect patient identification errors). A limitation of using ROCs based on truth tables computed using the qualitative assessments of expert readers is that the derived control limits are prone to human bias and variability.

Control limits can be determined using quantitative evaluations such as varying select image quality parameters (e.g. SNR, consistency of patient positioning) on MRI-derived brain pathology metrics (e.g. lesion volume, cortical thickness) to evaluate measurement error based on changes in MRI quality. As described earlier, the relationship between the computed value of an MRI quality parameter and the brain metric error can be used to establish a control limit reflecting tolerance of low quality only in the context that it does not result in significant brain metric error or reduced reproducibility.

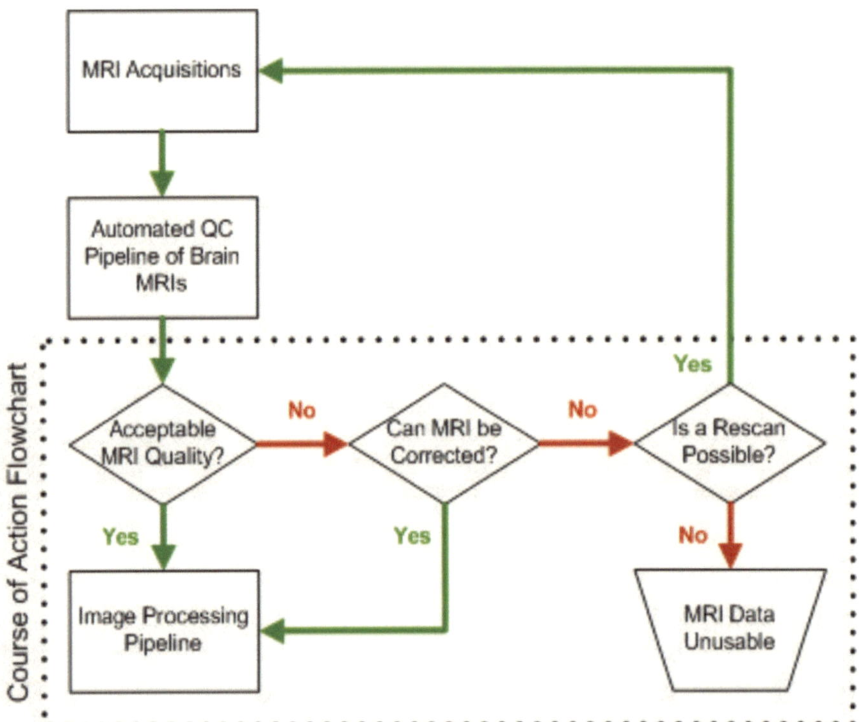

Fig. 6. Shows a QC course of action flowchart for MRIs acquired during multi-centre clinical trials for therapeutics of neurological diseases. After brain MRIs have been acquired, they are assessed for quality using an automated QC pipeline. Images that have acceptable quality progress to the image processing step where MRI-derived brain pathology metrics are calculated. MRIs that do not meet the criteria for acceptable image quality are assessed for correctability. If a correction procedure is available, MRIs are corrected and transitioned to the image processing pipeline step. If MRIs cannot be corrected, the possibility of a re-scan is investigated. In cases where the subject cannot be re-scanned the MRI data is deemed unusable. If a re-scan is possible, the newly acquired MRIs are processed using the same procedure described

Control limits can also be defined as a deviation from an expected value, which is useful for QC tests that check variables that should be constant like trial-, site-, and subject- specific acquisition protocols and sequences protocols (e.g. echo time, repetition time). Although exhaustive testing could be used to determine the effect of small deviations from acquisition parameter values like echo time (TE) or repetition time (TR) on image quality (as large deviations would be considered a breach of the approved protocol), the number of combinations required (scanner make x scanner model x software version x hardware upgrades x sequence) to perform this type of analysis makes exhaustive testing unrealistic. Instead, control limits can be defined as the expected trial-, site-, and subject- specific acquisition/sequence protocol values plus a deviation to address inherent hardware limitations (e.g. MR scanner incapable of precisely applying user selected parameter values) and differences between expected parameter values manually logged in the QC database and acquired parameter values in the DICOM header files (i.e. due to rounding errors). The deviation from the expected parameter value can be set using hardware specifications, suggestions from experienced MRI readers, analysis of QC database. As an example, the control limits of acquisition echo times could be set to TE ± 1% meaning that MRIs acquired with a measured echo time within 1% of the expected TE are considered acceptable. The deviation amount from the expected value is generally not determined using quantitative approaches and, consequently, should be set conservatively to not introduce QC errors. The aim is to ensure that the specified sequences and protocols that should have been applied were actually applied while accommodating for small variations.

Determining a Course of Action: Accept, Correct, or Reject MRIs

Once control limits for image quality are established for each test in the automated QC pipeline, they can be used to classify MRIs as either acceptable or unacceptable and an appropriate course of action can be determined (Fig. 6). MRIs with acceptable image quality progress to the image processing step where MRI-derived brain pathology metrics are calculated. MRIs that are unacceptable are reviewed by experienced readers and image-processing engineers to determine if correction procedures can be applied. MRIs that can be fixed (e.g. Fig. 3C interpacket artifacts, Fig. 3F bias field, and Fig. 4 geometric distortion) are corrected and transitioned to the image processing pipeline step where brain pathology metrics are calculated. If MRIs cannot be corrected, a MRI physicist may be consulted to determine if scanner hardware failure may be a factor, and the site will be contacted to discuss any hardware issues and the possibility of rescanning. In cases where the subject cannot be re-scanned (e.g. physical limitations, previous re-scans did not improve the MRI quality, situational circumstances) the brain metric is declared unusable. If a re-scan is possible, the newly acquired MRIs are processed using the same procedure described. Note that there are many attempts to achieve reproducible and accurate brain pathology metrics and that even if the MRI data is not adequate for the calculation of one metric it may still be of adequate quality to yield other metrics that are accurate and reproducible.

5. QC procedure for brain MRIs acquired in multi-centre MS clinical trials

The framework described in this chapter was used to create an automated quality control (aQC) procedure for brain MRIs acquired in multi-centre clinical trials for MS (Fig. 7). The aQC pipeline was composed of eight QC tests designed to increase the fidelity of MRI-derived brain pathology metrics by preventing unacceptable images from being processed.

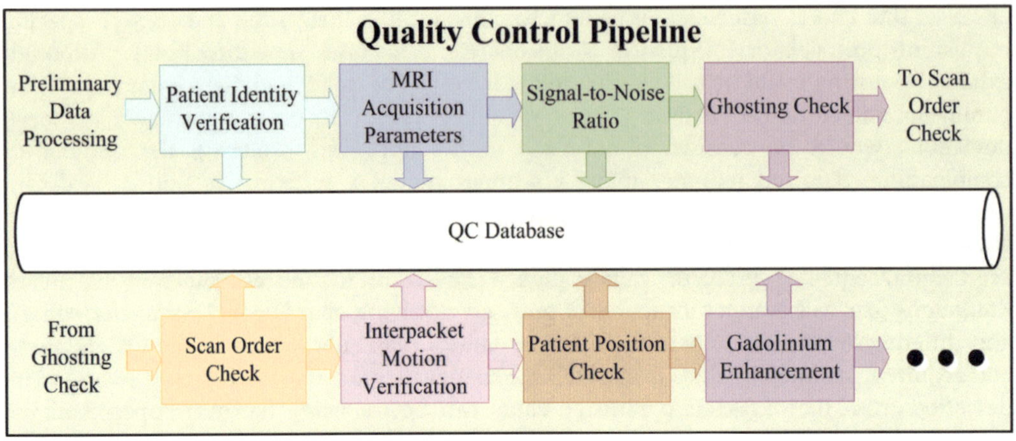

Fig. 7. Illustrates a quality control pipeline consisting of a series of eight tests designed, developed, tested, and validated for brain MRIs acquired in multi-centre clinical trials for MS. QC results and pertinent data for each test are recorded into a database and used for error checking and verifying consistency between serial acquisitions. The pipeline's efficiency can be attributed to a preliminary data processing step that optimizes operations that are shared between most QC tests. This pre-processing step minimizes the use of redundant QC test operations

Fig. 8. Shows a 2D axial slice (left) and sagittal slice (right) of the unified QC template in standard space with noise regions in orange, yellow, green, and white, WM in purple, GM in dark blue, CSF in light blue, cerebellum in dark yellow, and sagittal and straight sinus in red. The other colors are indicative of ROIs that overlap with the cerebellum

The QC test suite includes patient identity, MRI acquisition parameters, signal-to-noise ratio, ghosting, gadolinium enhancement, scan order, interpacket motion, and patient position verification; each test was identified using the error identification procedure described in section 4.3 (Fig. 5) and utilized MRI-based imaging markers and DICOM header files to measure test-specific indicators of quality (e.g. WM masks for SNR calculations, sagittal and straight sinus masks to determine if sufficient gadolinium enhancement was achieved, background noise masks to detect ringing artifacts, comparison of acquisition parameters in the DICOM header file to the requested parameters to ensure proper protocols were followed). All MRI-based imaging markers needed by the QC pipeline were consolidated into a single unified QC template (Fig. 8). Ordinarily, separate anatomical and/or background ROI would be created in a standard coordinate space (e.g. MNI-space) for each QC test and used to calculate MRI-based QC measurements (e.g. WM mask for SNR measurements). Unfortunately, this would require each ROI to be registered separately, thus, increasing the processing time of the pipeline. Using a unified QC template minimized the number of redundant operations used by each QC test and increased the overall efficiency of the pipeline. The template consisted of a superior, anterior, and lateral noise region of interest (ROI) for ghosting (Fig. 8 - white, orange, and green respectively), frontal noise ROI for SNR (Fig. 8 - yellow), sagittal and straight sinus ROI for gadolinium enhancement (Fig. 8 - red), and WM, GM, and CSF samples for SNR verification (Fig. 8 - purple, dark blue, light blue). The unified QC template was created in a standard coordinate space (MNI-space) using manually labelled ROIs (e.g. background noise and sagittal and straight sinus) and MNI-space anatomical probability maps (Mazziotta et al., 1995), tissue maps based on large sample sizes that indicate the probability of a specific tissue type being at a particular anatomical location in the image (e.g. WM, GM, and CSF). A quality control database was used to store quantitative (e.g. measured indicators of quality for each test performed) and qualitative (e.g. pass and fail flags indicating the outcome of the QC pipeline and each individual test) data as well as pertinent acquisition information found in the MRI DICOM header files (e.g. parameters used to acquire each MRI, scanner make and model, software revision). The QC database was also used for error tracking and comparing QC results from serial acquisitions for consistency. Independent sample sets populated with MRIs affected by various levels of image quality were used to train and validate each QC test, while experienced expert MRI readers and metric-based quantitative experiments were used to determine the degree of worth for each MRI. Control limits were established using ROC curves, quantitative evaluations of MRI indicators of quality on MRI-derived brain pathology metrics, and specific deviations from the expected measurement value of QC parameters. While details on the methods for each QC test has been previously described (Gedamu et al., 2008a; Gedamu et al., 2008b; Gedamu et al., 2008c), a brief description of each test procedure found in the pipeline is provided below.

Quality Control Tests

Patient Identity Verification: In clinical trials, longitudinal data often is acquired from the same subject over the course of the trial. Occasionally, such scans are incorrectly labelled, e.g., as coming from a different subject. The patient identity verification procedure verifies that serial images supposedly acquired from the same patient actually contain images of the same brain and that cross-subject MRIs within a site are unique (i.e. no two subjects have the same brain). For same-subject serial acquisitions, T1w extracted brains from two consecutive timepoints are registered together and a cross-correlation coefficient value is used to assess the similarity between both images. To ensure that cross-subject MRIs are unique within a site, the initial scans of new subjects are compared to the initial scans of all other subjects within their site using the same registration method used to verify the integrity of serial acquisitions.

MRI Acquisition Parameters Verification: In a clinical trial, it is important for data to be acquired consistently according to a pre-specified protocol in order to ensure comparability of data acquired at different sites and over time. For example, changes in echo times (TE) or repetition times (TR) can affect images contrast, and in turn, may modify the results of a tissue classification procedure. Verification of MRI parameters ensures that the acquisition values approved during site qualification, which are generally chosen to achieve consistent image characteristics for analyses, are respected. This is achieved by comparing the approved parameters that are stored in a QC database (i.e. populated during site qualification) against the received parameters recorded in the image DICOM header file.

Signal-to-Noise Ratio Verification: The processing of an image can be substantially influenced by the signal to noise ratio (SNR). The noise levels of MRIs can obscure anatomical and pathological borders between different tissue types (e.g. lesion/WM, GM/WM, lesion/CSF borders) thereby affecting the reliability of registration, classification, and segmentation procedures. The SNR verification procedure ensures that each acquired MRI is within an acceptable limit. SNR can be determined by dividing the tissue type with highest mean intensity, either WM or CSF (Fig. 8 - purple and dark blue respectively), by the standard deviation of the background noise (Fig. 8 - yellow) which has been compensated for Rayleigh distribution effects.

Ghosting Verification: Head movement during MRI examinations is a very common source of artifact, which generally appears as ringing or "ghosting" artifacts (Fig. 3E). Ring-like structures (aliasing), a characteristic trait of ghosting, produce non-uniform intensities within the brain and in the surrounding background. Consequently, confidence in anatomical borders is compromised, and the ability to discern different tissue types and pathology (e.g. lesions) decreases because the intensity coherency within each tissue type is perturbed. Ghosting artifacts can be detected by comparing the standard deviation of two independent noise regions. For 2D multi-slice acquisitions the anterior region (Fig. 8 - orange) and the left and right side of the head (Fig. 8 - green) are compared. For 3D global acquisitions the superior (Fig. 8 - white) and anterior (Fig. 8 - orange) regions are compared.

Scan Order Verification: In clinical trials, multi-modal MRIs are acquired for each subject at each timepoint (Fig. 2 - MRI Acquisition) and it is important to ensure that the order, time, and date of each modality are correct and consistent according to a pre-specified protocol. MRI modalities that should have been acquired during a single scan session but were acquired over multiple days (e.g. T2w/PDw images that were acquired days after a T1w image was acquired) could be affected by pathological/biological (e.g. appearance of a new lesion) or systemic variability (e.g. changes in patient positioning that cause geometric distortion artifacts) which could affect the reliability of brain pathology metrics. Acquisition order can be determined by comparing the approved scan order protocols that are stored in a QC database (i.e. populated during site qualification) against the actual scan times recorded in the image DICOM header file.

Interpacket Motion Verification: Inter-packet motion artifacts (Fig. 3C) are associated with subject movement during an interleaved multi-slice MR imaging sequence, a specific type of sequence where multiple 2D MRI sets, termed packets (Fig. 9 - illustrates three packets painted in green, blue, and purple), are used to construct full 3D MR volumes. Fig. 9 illustrates the effect that interpacket motion artifacts can have on MRI-derived brain metrics where three packets were acquired with the first, second, and third packets shown in green, purple, and blue respectively. Packet 1 is acquired with the brain initially rotated slightly clockwise, packet 2 is acquired with a larger rotation in the counter-clockwise direction, and packet 3 is acquired after the brain undergoes a small translation in the axial direction. The final reconstructed MRI (right) shows the effect of motion between each acquired packet as regions of missing (areas where the packets do not cover the image) and redundant data (areas where multiple packets cover the same regions). This impedes the MRI from conveying the complete anatomical and pathophysiological structures in the scanned brain and can introduce errors in subsequent MRI-derived brain metrics. This type of artifact can be determined by measuring out-of-plane motion, movement between 2 or more packets that causes missing data, and in-plane motion, movement between 2 or more packets that cause structural misalignment between 2D slices but does not result in missing data (Gedamu; Gedamu; Collins & Arnold, 2008c).

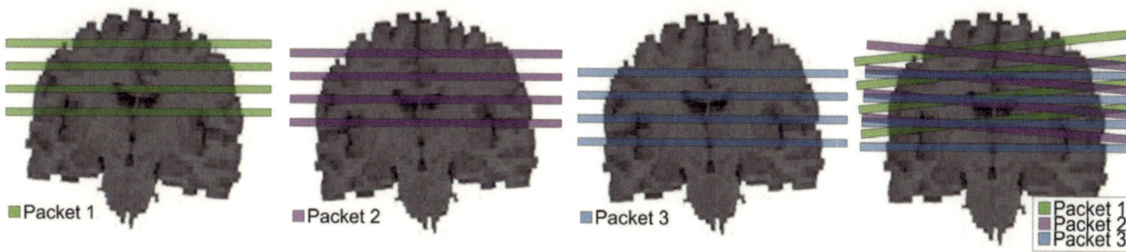

Fig. 9. Packet 1 is acquired with the brain initially rotated slightly in the clockwise direction (left), packet 2 is acquired with a larger rotation in the counter-clockwise direction (middle left), and packet 3 is acquired after the brain undergoes a small translation in the axial direction (middle right). The image on the right shows the result of registering all packets together. Notice the missing data (areas where the packets do not cover the image) resulting from motion between packets during acquisition

Patient Position Verification: Magnetic field inhomogeneities and gradient nonlinearities can alter the volume of anatomical structures in MRIs (termed geometric distortion) based on the placement of the subject in the scanner (Fig. 4). In multi-centre clinical trials, the position of the subject should be consistent for each scan and the centre of the subject's brain should be aligned with the magnet's isocenter (i.e. location least affected by geometric distortion) to minimize distortion artifacts. Subject positioning is usually approximated by aligning the center of the eye with the center of the magnet. To verify proper subject positioning during image acquisition, MRIs were registered to an average brain in standard coordinate space (MNI-space) because the center of the average brain and magnet isocenter of each MRI have the same x,y,z location, coordinates (0,0,0). Accordingly, misalignments between the center of each MRI and the magnet's isocenter were reflected in the registration transformation matrix. Deviations in the transformation matrices were also used to verify the consistency of a subject's position for serial acquisitions.

Gadolinium Enhancement Verification: In scans that require quantification of gadolinium enhancement, for example, of MS lesions, it is important to ensure the proper amount of gadolinium was injected, the scan was acquired after an appropriate delay, and the post-contrast images show appropriate enhancement of normal structures, such as blood vessels. Appropriate gadolinium enhancement was done by comparing the signal intensity of large venous structures like the sagittal and straight sinus (Fig. 8 - red) in pre and post gadolinium MRIs while the time delay between the pre-/post- contrast image acquisitions were determined using the recorded scan times in each image's DICOM header file.

QC Pipeline Optimization

Prior to running each QC test, a preliminary data processing step (Fig. 7 - Preliminary Data Processing) was done to consolidate time-consumptive operations that were redundant across most QC tests into a single operation which was shared among all tests (minimize cross-test redundancies). The preliminary data processing step comprised of using a standard registration procedure to align the unified QC template (Fig. 8) to each MRI modality that was acquired during MRI acquisition (Fig. 2 - MRI Acquisition), measuring important statistical data for each MRI modality in the regions defined by the registered unified QC template, and storing the measured statistical data and transformation matrices obtained from the registration procedure into the QC database. The standard registration procedure (Fig. 10) was performed by selecting a reference MRI among the acquired MRI modalities (e.g. T1w), calculating a transformation matrix to align a brain model in MNI-space to the selected reference image in the subject's native coordinate space, calculating a transformation matrix to align the reference MRI to the other MRI modalities (e.g. T1w-to-T1w+Gd, T2w, PDw, and FLAIR), and concatenating the transformation matrix between

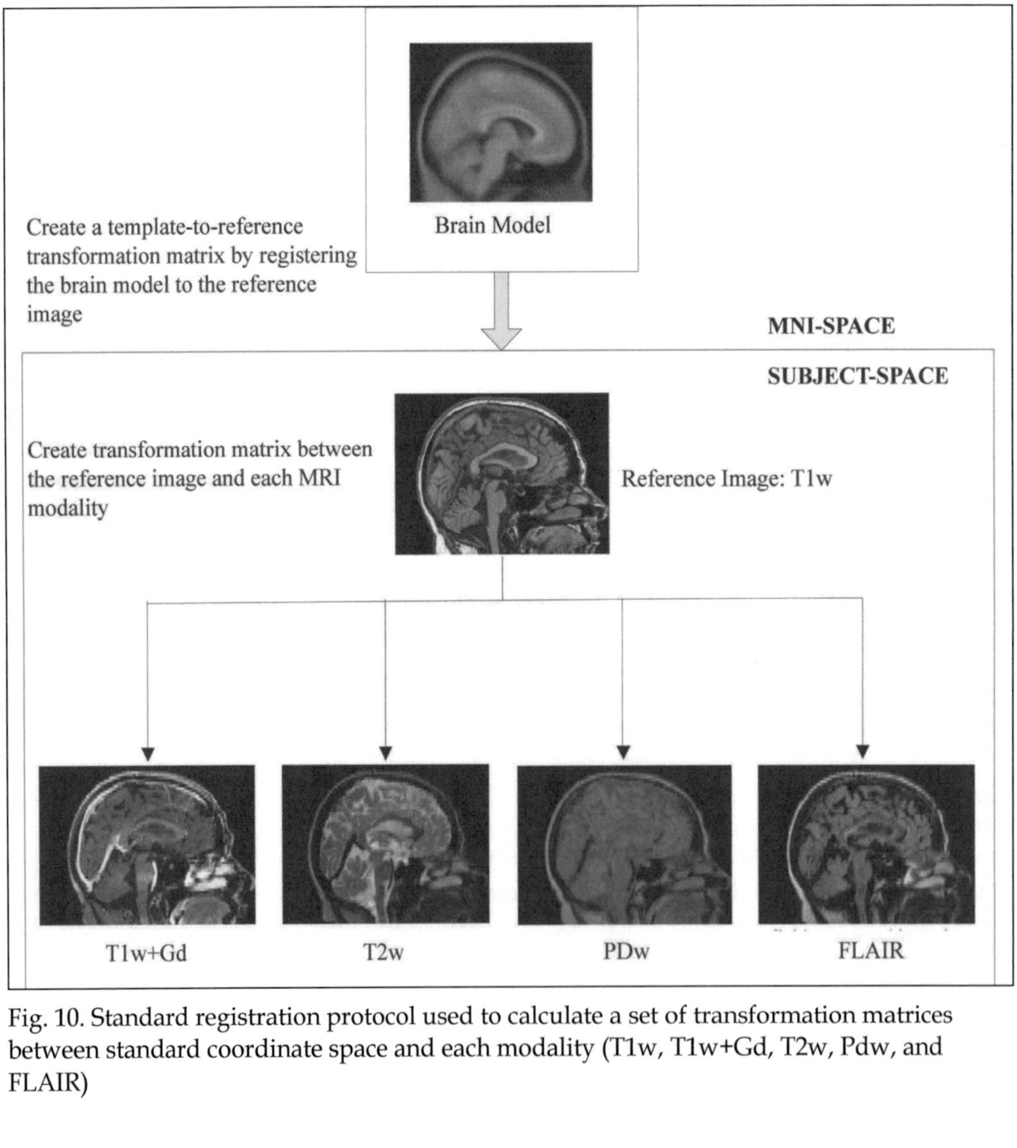

Fig. 10. Standard registration protocol used to calculate a set of transformation matrices between standard coordinate space and each modality (T1w, T1w+Gd, T2w, Pdw, and FLAIR)

the two alignments to create a set of modality-specific transformation matrices (e.g. brain model-to- T1w+Gd, T2w, PDw, and FLAIR). An MNI-space brain model, instead of the MNI-space template itself, was used to determine the transformation matrix between MNI-space and the subject's native space because the registration process requires images with similar attributes to function correctly. A T1w reference image was used because T1w images are generally acquired for clinical trial studies, making them readily available, and the brain model was T1w, which maximized the similarity between the images. The template was registered to each MRI, as oppose to aligning each MRI to the template, to ensure QC measurements were made using the original MRI data (i.e. not affected by interpolation of image data that occurs during the registration procedure). By calculating a set of transformation matrices once, in contrast to performing the same registration procedure for each test in the pipeline, reduced the number of redundant operations and increased the overall speed of the pipeline, which enabled quicker MRI quality assessments.

To accommodate for growth, the quality control procedure was designed as a series of modularized tests allowing new tests to be designed, developed, tested, and validated independently before being added to the QC pipeline. To address concerns of scalability, the pipeline was designed to minimize its load effect (i.e. number of read/write accesses) on the central MRI database. This was achieved by using computer servers to perform QC analyses locally. Essentially, the MRI database is accessed once, instead of multiple times during the course of the QC analyses, to copy all the necessary MRI data to a computer server where the QC pipeline will be run (local processing). This limits the chance of overloading the MRI database with multiple read/write accesses which could result in slow response times or even crashes (non-responsive database). Performing quality control procedures locally using server systems (i.e. copying the required MRI data locally) reduced the load on the database, increased the number of potential processing systems (i.e. 'N' possible local computer servers), and, accordingly, increased the amount of MRIs processed.

6. Conclusions

In this chapter, guidelines were provided for developing an appropriate automated QC procedure for brain MRIs acquired in multi-centre clinical trials of therapeutics for neurological diseases. In addition, these guidelines were applied to develop an aQC procedure specific to multi-centre clinical trials for MS consisting of eight QC tests (patient identity, MRI acquisition parameters, SNR, ghosting, gadolinium enhancement, scan order, interpacket motion, and patient position verification). The procedure has been applied to large scale multi-clinical trials and increased the fidelity of MRI-derived brain pathology metrics by preventing unacceptable images from being processed.

7. Acknowledgments

The Natural Sciences and Engineering Research Council (NSERC) of Canada and the Mathematics of Information Technology And Complex Systems (MITACS) funding agencies are gratefully acknowledged for their provision of financial assistance for this research.

8. References

ACR. (2004). *ACR MRI Quality Control Manual.* American College of Radiology, Reston, VA

ACR. (2005). *Phantom test guidance for the ACR MRI accreditation program.* American College of Radiology, Reston, VA

Ahmed, H. M., et al. (2003). Combined intra- and inter-slice motion artifact suppression in magnetic resonance imaging. *Proceedings of Medical Imaging 2003: Image Processing,* San Diego, CA, USA

ASQ. (n.d.). Quality, In: *Six Sigma Forum,* September 2010, Available from: http://asq.org/sixsigma/quality-information/termsq-sixsigma.html

Bagnato, F., et al. (2006). In vivo detection of cortical plaques by MR imaging in patients with multiple sclerosis. *AJNR,* Vol. 27, No. 10, pp. 2161-2167

Blumenthal, J. D., et al. (2002a). Motion artifact in magnetic resonance imaging: implications for automated analysis. *Neuroimage,* Vol. 16, No. 1, pp. 89-92

Blumenthal, J. D., et al. (2002b). Motion Artifact in Magnetic Resonance Imaging: Implications for Automated Analysis. Vol. 16, No. 1, pp. 89-92

Boyes, R. G., et al. (2006). Cerebral atrophy measurements using Jacobian integration: comparison with the boundary shift integral. *Neuroimage,* Vol. 32, No. 1, pp. 159-169

Bradley, K. M., et al. (2002). Serial brain MRI at 3-6 month intervals as a surrogate marker for Alzheimer's disease. *British Journal of Radiology,* Vol. 75, No. 894, pp. 506-513

Camara-Rey, O., et al. (2006). Simulation of acquisition artefacts in MR scans: effects on automatic measures of brain atrophy. *MICCAI,* Vol. 9, No. Pt 1, pp. 272-280

De Coene, B., et al. (1992). MR of the brain using fluid-attenuated inversion recovery (FLAIR) pulse sequences. *AJNR,* Vol. 13, No. 6, pp. 1555-1564

De Stefano, N., et al. (2010). Assessing brain atrophy rates in a large population of untreated multiple sclerosis subtypes. *Neurology,* Vol. 74, No. 23, pp. 1868-1876

Filippi, M., et al. (1996). Quantitative assessment of MRI lesion load in multiple sclerosis. A comparison of conventional spin-echo with fast fluid-attenuated inversion recovery. *Brain,* Vol. 119, No. Pt 4, pp. 1349-1355

Forbes, K. P. N., et al. (2001). PROPELLER MRI: Clinical testing of a novel technique for quantification and compensation of head motion. *Journal of Magnetic Resonance Imaging,* Vol. 14, No. 3, pp. 215-222

Fox, N. C., et al. (2000). Using serial registered brain magnetic resonance imaging to measure disease progression in Alzheimer disease: power calculations and estimates of sample size to detect treatment effects. *Archives of Neurology,* Vol. 57, No. 3, pp. 339-344

Gedamu, E. L., et al. (2008a). Automated Quality Control of MR Images in MS Lesion Analysis. *Proceedings of European Committee for Treatment and Research in Multiple Sclerosis,* Montreal, Canada, September 17-20

Gedamu, E. L., et al. (2008b). Automated quality control of brain MR images. *Journal of Magnetic Resonance Imaging,* Vol. 28, No. 2, pp. 308-319

Gedamu, E. L., et al. (2008c). MRI inter-packet movement correction for images acquired with non-complementary data. *Biomedical Imaging: From Nano to Macro, 2008. ISBI 2008. 5th IEEE International Symposium on Biomedical Imaging,* pp. 416-419

Geurts, J. J., et al. (2005). Cortical lesions in multiple sclerosis: combined postmortem MR imaging and histopathology. *AJNR,* Vol. 26, No. 3, pp. 572-577

Greenspan, H., et al. (2001). MRI Inter-slice Reconstruction Using Super-Resolution. *Proceedings of MICCAI*

Jovicich, J., et al. (2006). Reliability in multi-site structural MRI studies: effects of gradient non-linearity correction on phantom and human data. *Neuroimage*, Vol. 30, No. 2, pp. 436-443

Kim, B., et al. (1999). Motion correction in fMRI via registration of individual slices into an anatomical volume. *Magn Reson Med*, Vol. 41, No. 5, pp. 964-972

Kuba, A., et al. (1999). ANIMAL+INSECT: Improved Cortical Structure Segmentation. In: *Information Processing in Medical Imaging*, 210-223), Springer, Berlin / Heidelberg

Lötjönen, J., et al. (2004). Correction of Movement Artifacts from 4-D Cardiac Short- and Long-Axis MR Data. *Proceedings of MICCAI*

Malandain, G., et al. (2004). Fusion of autoradiographs with an MR volume using 2-D and 3-D linear transformations. *Neuroimage*, Vol. 23, No. 1, pp. 111-127

Mazziotta, J. C., et al. (1995). A Probabilistic Atlas of the Human Brain: Theory and Rationale for Its Development. *Proceedings of The International Consortium for Brain Mapping*

NEMA. (1988). EEC Concerted Research Project. Identification and characterisation of biological tissues by NMR - Part IV. Protocols and test objects for the assessment of MRI equipment. *Magnetic Resonance Imaging*, Vol. 6, pp. 195-204

NEMA. (2001). *MS1 Determination of signal-to-noise ratio (SNR) in diagnosis magnetic resonance images*. National Manufacturers Electrical Association, Rosslyn, VA

NEMA. (2003a). *MS2 Determination of two dimensional geometric distortion in diagnosis magnetic resonance images*. National Manufacturers Electrical Association, Rosslyn, VA

NEMA. (2003b). *MS3 Determination of image uniformity in diagnosis magnetic resonance images*. National Manufacturers Electrical Association, Rosslyn, VA

NEMA. (2003c). *MS5 Determination of slice thickness in diagnosis magnetic resonance images*. National Manufacturers Electrical Association, Rosslyn, VA

O'Connor, P., et al. (2009). 250 microg or 500 microg interferon beta-1b versus 20 mg glatiramer acetate in relapsing-remitting multiple sclerosis: a prospective, randomised, multicentre study. *Lancet Neurology*, Vol. 8, No. 10, pp. 889-897

Ourselin, S., et al. (2000). Block Matching: A General Framework to Improve Robustness of Rigid Registration of Medical Images. *Proceedings of MICCAI*

Preboske, G. M., et al. (2006). Common MRI acquisition non-idealities significantly impact the output of the boundary shift integral method of measuring brain atrophy on serial MRI. *Neuroimage*, Vol. 30, No. 4, pp. 1196-1202

Price, R. R., et al. (1990). Quality assurance methods and phantoms for magnetic resonance imaging: report of AAPM nuclear magnetic resonance Task Group No. 1. *Medical Physics*, Vol. 17, No. 2, pp. 287-295

Princeton. (2010). Quality, In: *About WordNet*, September 2010, Available from: http://wordnet.princeton.edu

Rousseau, F., et al. (2005). A Novel Approach to High Resolution Fetal Brain MR Imaging. *Proceedings of MICCAI*

Rovaris, M., et al. (2008). Large-scale, multicentre, quantitative MRI study of brain and cord damage in primary progressive multiple sclerosis. *Multiple Sclerosis*, Vol. 14, No. 4, pp. 455-464

Scahill, R. I., et al. (2003). A longitudinal study of brain volume changes in normal aging using serial registered magnetic resonance imaging. *Archives of Neurology*, Vol. 60, No. 7, pp. 989-994

Sled, J. G., et al. (1998). A nonparametric method for automatic correction of intensity nonuniformity in MRI data. *IEEE TMI*, Vol. 17, No. 1, pp. 87-97

Smith, S. M., et al. (2004). Advances in functional and structural MR image analysis and implementation as FSL. *Neuroimage*, Vol. 23 Suppl 1, pp. S208-219

Smith, S. M., et al. (2002). Accurate, Robust, and Automated Longitudinal and Cross-Sectional Brain Change Analysis. *NeuroImage*, Vol. 17, No. 1, pp. 479-489

Taguchi, G. i., et al. (2000). *Robust engineering*. McGraw-Hill, 0071347828, New York

Van Walderveen, M. A., et al. (1998). Histopathologic correlate of hypointense lesions on T1-weighted spin-echo MRI in multiple sclerosis. *Neurology*, Vol. 50, No. 5, pp. 1282-1288

Ward, H. A., et al. (2000). Prospective multiaxial motion correction for fMRI. *Magn Reson Med*, Vol. 43, No. 3, pp. 459-469

8

Quality Control in the Early Secretory Pathway Plays Significant Roles in vivo

Tomohiko Aoe
Chiba University Graduate School of Medicine
Japan

1. Introduction

The endoplasmic reticulum (ER) provides a folding environment for newly synthesized secretory and membrane proteins (Ellgaard and Helenius, 2003). Secretory proteins are synthesized by ribosomes and translocated cotranslationally or posttranslationally to the ER. These newly synthesized proteins interact with ER molecular chaperones, such as immunoglobulin heavy chain-binding protein (BiP), calnexin, calreticulin, and protein disulfide isomerase, to become properly folded and assembled into a mature protein complex for transport along the secretory pathway. Aberrant protein folding, due to extracellular stimuli such as ischemia and oxidative stress, or genetic mutation leads to the accumulation of misfolded proteins in the ER, which in turn evokes the unfolded protein response (UPR) (Patil & Walter, 2001). The UPR reduces the amount of misfolded proteins (Ron & Walter, 2007) by inducing the production of ER chaperones that promote protein folding, reducing general protein synthesis, and enhancing the degradation of misfolded proteins via a ubiquitin-proteasome system, termed ER-associated degradation (ERAD) (Bonifacino & Weissman, 1998). The persistent accumulation of misfolded proteins beyond the capacity for ER quality control causes cellular dysfunction and cell death, leading to diverse human disorders, such as diabetes mellitus and neurodegenerative diseases (Zhao & Ackerman, 2006).

Misfolded proteins had been believed to remain in the ER, but yeast genetic analyses have indicated that the UPR involves the whole secretory pathway (Travers et al., 2000), and that some misfolded proteins require transport between the ER and the Golgi complex for ERAD (Vashist et al., 2001). In addition, certain misfolded proteins in mammalian cells have also been reported to exit the ER and recycle between the ER and post-ER compartments, associating with ER chaperones. The KDEL receptor mediates this retrieval, suggesting that the secretion of misfolded proteins with ER chaperones from the ER and their retrieval may contribute to ER quality control (Hammond & Helenius, 1994; Yamamoto et al., 2001).

We focus on the KDEL receotor-ER chaperone retrieval system in order to explore the physiological and pathological significance of ER quality control by using culture cells and mutnat mouse models. The impairment of the retrieval system in our mutant mice caused several developmental and aging dysfunctions, which reveals that quality control in the early secretory pathway plays significant roles in vivo.

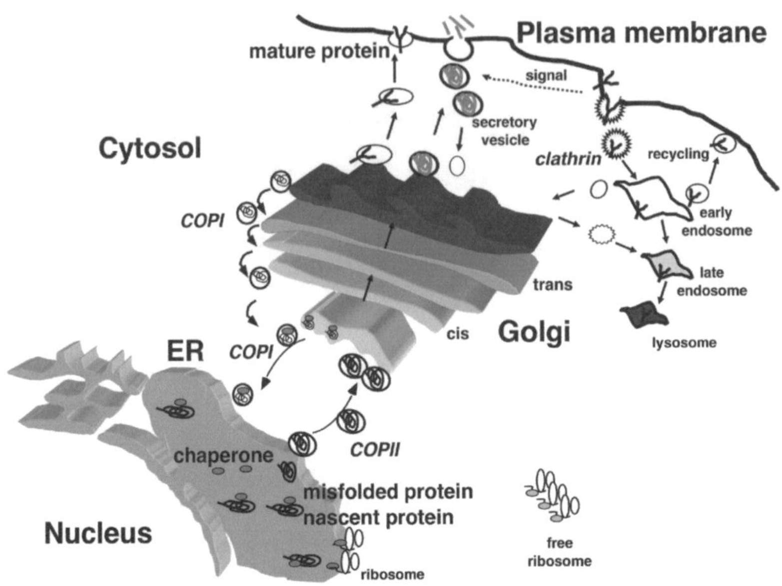

Fig. 1. Secretory pathways in the cell

1.1 Transport between the ER and the Golgi complex

Intracellular transport of proteins is mediated by membrane carriers that include coated vesicles and tubular structures. From the ER, proteins are packaged into the coat protein complex (COP) II coated vesicles to initiate their transport from the ER (Schekman and Orci, 1996). Subsequently, proteins are transferred to the intermediate compartment (IC) that then moves along microtubules to the Golgi apparatus (Bonfanti et al., 1998). Upon arrival to the Golgi complex, proteins are sorted to the peripheral compartments of the cell, such as endosomes and plasma membrane. Also, proteins can be retrieved to the ER by retrograde transport at either the IC or the Golgi complex by COPI coated vesicles (Letourneur et al., 1994).

Selective retrograde transport of proteins from post-ER compartments to the ER is achieved by multiple mechanisms. Transmembrane proteins that have a carboxyl terminal dilysine (KKXX) sequence, such as type I ER membrane proteins (Nilsson et al., 1989) and ERGIC53 (Itin et al., 1995), are transported in COPI vesicles, because the COPI coat recognizes the dilysine motif directly (Letourneur et al., 1994). However, soluble proteins that reside in the lumen of membrane compartments can also be retrieved to the ER. Ones such as BiP (or GRP78) have a carboxyl terminal Lys-Asp-Glu-Leu (KDEL) sequence (Munro & Pelham, 1987) that is recognized by the KDEL receptor in the post-ER compartments (Lewis & Pelham, 1992), and are then sorted into COPI vesicles for retrograde transport (Orci et al., 1997).

Besides its retrieval function, the KDEL receptor has been suggested to play a role in regulating COPI transport (Aoe et al., 1997; Aoe et al., 1998). ADP-Ribosylation Factor 1 (ARF1), a Ras-like small GTPase, regulates the formation of COPI vesicles (Rothman & Wieland, 1996). Like all small GTPases, the activation of ARF1 requires a guanine nucleotide exchange factor, and its deactivation requires a GTPase-activating protein (GAP). GAP acts by promoting the hydrolysis of GTP on ARF1 to GDP (Cukierman et al., 1995). This action has been demonstrated to be required for the proper formation of COPI vesicles (Hsu et al., 2009). Significantly, ligand-binding on the luminal side of the KDEL receptor also induces its interaction with ARFGAP1 on the cytoplasmic side of the receptor. As a result, ARFGAP1 is recruited from the cytosol to the membranes to activate GAP activity on ARF1 (Aoe et al., 1999). Thus, these findings suggest that the KDEL receptor does not merely function as a passive cargo protein of COPI transport, but is capable of modulating this transport pathway.

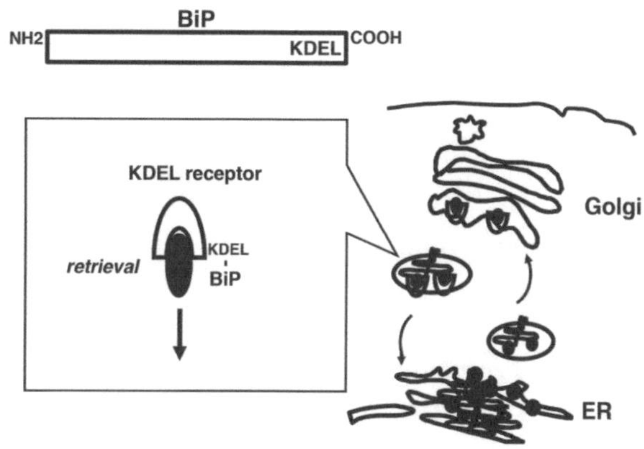

Fig. 2. ER chaperones with the KDEL sequence are retrieved by the KDEL receptor

1.2 ER chaperones and the unfolded protein response

BiP, one of the most abundant ER chaperones, plays a central role in ER function, assisting in protein translocation, folding, degradation and regulation of the UPR (Hendershot, 2004). ER chaperones are localized to the ER by two mechanisms—retention and retrieval (Sonnichsen et al., 1994). BiP is retained in the ER through interaction with other ER proteins and the ER matrix. When misfolded proteins accumulate in the ER, BiP dissociates from some ER membrane proteins, such as inositol-requiring kinase-1 (IRE1), PKR-like ER-associated kinase (PERK) and activating transcription factor 6 (ATF6). BiP dissociation activates these kinases and transcription factors and initiates the UPR (Bertolotti et al., 2000), which leads to increased expression of X-box-binding protein-1 (XBP-1) and ATF4 (Schroder & Kaufman, 2005). When BiP is secreted from the ER along with misfolded proteins (Hammond & Helenius, 1994; Yamamoto et al., 2001), the KDEL sequence of BiP is recognized by the KDEL receptor, thereby facilitating the retrieval of BiP from post-ER compartments to the ER (Lewis & Pelham, 1990; Munro and Pelham, 1987). Yeast BiP (Kar2) is essential for survival; when the retrieval sequence (in yeast: His-Asp-Glu-Leu, HDEL) is deleted, a fraction of Kar2 is secreted from the ER. However, the UPR is activated and this maintains a minimal level of Kar2 in the ER (Beh and Rose, 1995). Thus, the failure of retrieval by the KDEL receptor causes the UPR.

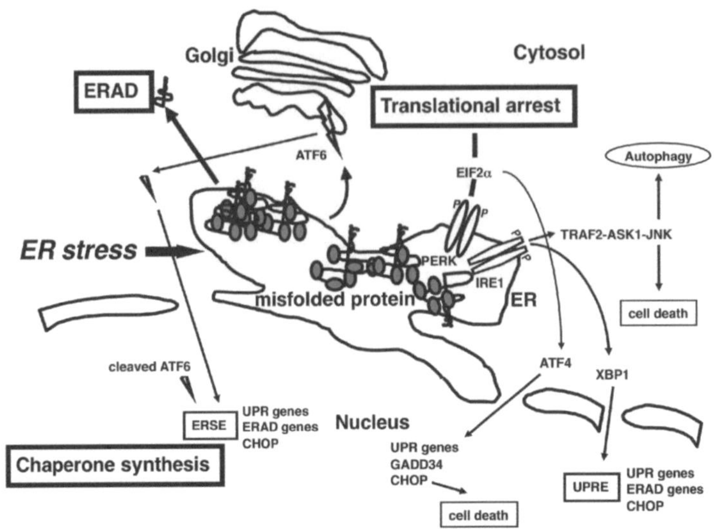

Fig. 3. ER stress causes the UPR, leading to chaperone synthesis, translational arrest and ERAD

2. The role for retrieval of misfolded proteins by the KDEL receptor in ER quality control examined by using culture cells

We examined how the KDEL receotor-ER chaperone retrieval system participates in ER quality control by studying the unassembled T cell antigen receptor (TCR) α chain as a model system (Yamamoto et al., 2001).

2.1 TCRα cycles in the early secretory system

TCR consists of at least six polypeptides (TCRα, β, CD3γ, δ, ε, and ζ). Proper assembly is required for TCR to be expressed on the cell surface (Klausner et al., 1990). A heterologous expression system has been used extensively to study the fate of unassembled TCRα chain by transfecting TCRα into COS cells that normally do not express the TCR. Unassembled TCRα has been demonstrated to be retained in the ER (Bonifacino et al., 1989) and degraded by ERAD in lymphocytes, as well as other cell types when it is expressed as a single subunit (Huppa & Ploegh, 1997).

However, because an increasing number of proteins have been appreciated to maintain their distribution in the ER by being transported out and then cycling back, we tested whether TCRα might also exhibit such a behavior. Upon treatment with bafilomycin A1, a vacuolar H^+-ATPase inhibitor that has been shown to impose a relative block on the retrograde arm of bidirectional transport between the ER and the Golgi complex (Palokangas et al., 1998), confocal microscopy revealed significantly increased colocalization of TCRα and ERGIC53. The colocalization induced by bafilomycin also involved the redistribution of ERGIC53 to a pattern that was more compact and juxtanuclear. That is, when retrograde transport was blocked, ERGIC53 could be induced to accumulate at the Golgi complex (Palokangas et al., 1998).

Because of these known effects on the itinerary of proteins cycling in the early secretory compartments, we sought to examine the distribution of TCRα in more detail by subcellular fractionation. In the control condition, when no perturbation was added, most of TCRα was in the ER fractions as a p38 form, as judged by co-fractionation with calnexin. In this control setting, we could also detect a larger form (p43) in post-ER fractions, as judged by co-fractionation with the IC marker. As previous characterization of TCRα, p38 is an immature glycoprotein that has not received Golgi-specific modification, while p43 represents a mature glycoprotein that has received Golgi-specific modification (Samelson, 1985). Consistent with the possibility that TCRα cycles out of the ER, when cells were treated with bafilomycin, TCRα accumulated markedly in post-ER fractions as the p43 form. To ascertain that the p43 form of TCRα indeed represented those outside the ER, we assessed whether it had Golgi-specific glycosylation. Immunoblotting for TCRα from the different fractions that were either treated with endoglycosidase H (endo H) or subjected to mock-treatment, we found that p38 form was endo H sensitive, while the p43 form was endo H resistant. These results, taken together with the above morphologic findings, suggest that a significant fraction of TCRα was cycling between the ER and the Golgi complex.

2.2 Retrieval of TCRα by the KDEL receptor

A dilysine (KKXX) motif on the cytoplasmic portion of transmembrane proteins mediates their retrograde transport to the ER, because COPI binds directly to this motif (Letourneur et al., 1994). While TCRα is a transmembrane protein, it contains no such motif in its cytoplasmic domain. However, another possibility is suggested by studies on the ER chaperone, BiP. BiP has been shown to leak out to the post-ER compartments, and then retrieved to the ER, because it contains a carboxy terminal KDEL motif that is recognized by the KDEL receptor (Munro & Pelham, 1987). Thus, one possibility is that the KDEL receptor can retrieve TCRα, if TCRα were bound to a KDEL ligand, such as BiP. Consistent with this possibility, co-precipitation studies revealed an association between TCRα and BiP. This association was specific, because another ER chaperone, GRP94, did not co-precipitate with TCRα.

To gain insight into the importance of the retrieval function of the KDEL receptor, we examined whether TCRα could be detected on the cell surface upon disruption of ligand-binding by the KDEL receptor. Whereas such a surface pool was undetectable when retrieval by the KDEL receptor was intact, a significant fraction was detected when retrieval was disrupted by the overexpression of lysozyme-KDEL, a chimeric KDEL ligand, that was created by appending the KDEL sequence to lysozyme.

We had shown previously that activation of the KDEL receptor through ligand-binding on its luminal side induced the KDEL receptor to interact with the ARFGAP1 on the cytoplasmic side (Aoe et al., 1998). Thus, one possibility is that the complex of TCRα with BiP represented a ligand that would also activate the KDEL receptor to interact with GAP. Upon overexpression of TCRα, we detected an increased interaction between the KDEL receptor and ARFGAP1. Thus, this finding suggested that, most likely, TCRα was being retrieved by the KDEL receptor, because it was complexed with BiP, a known ligand for the receptor.

Taken together, these observations suggested that ligand binding by the KDEL receptor would regulate the degree of COPI-mediated retrograde transport. During stress that increases the level of misfolded proteins in the ER, one possibility is that more of these misfolded proteins would leak from the ER. However, because of the mechanistic link between ligand-binding by the KDEL receptor and its interaction with ARFGAP1, an important function of the KDEL receptor appears to be in integrating increased leakage that reaches the Golgi complex to increasing COPI-mediated retrograde transport from this organelle. In this manner, any increased leakage out of the ER would not result in increased leakage out of the early secretory system. This study suggested an important role for the KDEL receptor in a post-ER retrieval mechanism that contributes to ER quality control (Yamamoto et al., 2001).

3. The role for retrieval of misfolded proteins by the KDEL receptor in ER stress response examined by using culture cells

The mammalian ER stress response is a complicated process that is coordinately induced by ER transmembrane kinases such as ATF6, IRE1 and PERK. Further complexity comes from the fact that the activation of these kinases not only enhances the expression of ER chaperones for cytoprotection but also causes apoptosis. The underlying molecular mechanism of the transition between these two opposite outcomes during the ER stress response is uncertain.

We examined how the KDEL receotor-ER chaperone retrieval system participates in ER stress response by using an expression of a mutant KDEL receptor that lacks the ability for ligand recognition. The impairment of retrieval by the KDEL receptor led to a mis-sorting of BiP, an ER chaperone that had a retrieval signal from the early secretory pathway, which induced an increase in susceptibility to ER stress in HeLa cells (Yamamoto et al., 2003).

3.1 Ligand recognition by the KDEL receptor is saturable

Transient transfection of the KDEL receptor results in various expression levels in HeLa cells. Overexpression of the KDEL receptor has been shown to enhance transport from the Golgi complex to the ER, which results in the redistribution of the whole Golgi complex to the ER (Hsu et al., 1992); however, moderate expression of the KDEL receptor does not disturb the Golgi structure and only those cells with moderate expression become stable transfectants (Aoe et al., 1998). To examine the role of the KDEL receptor in the ER stress response, we made HeLa cells that stably expressed either the wild-type or a mutant human KDEL receptor. This binding defective mutant that had a single amino acid replacement (R169N) was localized to the Golgi complex as the wild-type, but did not respond to the KDEL sequence effectively as described previously (Townsley et al., 1993). The transient expression of the lysozyme-KDEL in these stable cells induced the transport of the wild-type KDEL receptor, but not that of the mutant, to the ER. The lysozyme-KDEL was localized in the ER even in the mutant KDEL receptor cells, which indicates that minimal retrieval was maintained in these cells by the endogenous KDEL receptor.

The expression of ER chaperones is induced extensively upon ER stress, which may saturate KDEL receptor-mediated retrieval since the expression of the KDEL receptor is not inducible in mammals. We examined the effect of ER stress on the distribution of BiP, an endogenous ligand of the KDEL receptor, in the secretory pathway, using sucrose gradient analysis in cells stably expressing the wild-type or the mutant KDEL receptor. While most endogenous BiP was found in the ER in the resting state, a significant amount was detected in the post-ER fractions of these cells when the cells were treated with tunicamycin, which prevented protein glycosylation in the ER and induced the UPR. Under these circumstances, we detected more BiP being secreted to the medium in the mutant KDEL receptor cells than in the HeLa cells and the wild-type KDEL receptor cells in spite of an equivalent expression level of BiP within the cells. These results suggested that ER chaperones associating with misfolded proteins might be secreted from the ER and retrieved by the KDEL receptor. When retrieval is limited, they may escape from the early secretory pathway, especially under stressed conditions.

3.2 The loss of BiP from the early secretory pathway causes persistent UPR

The accumulation of misfolded proteins in the ER leads to the recruitment of BiP from ER transmembrane kinases such as ATF6 and IRE1, which results in the activation of these kinases and the synthesis of ER chaperones including BiP. The loss of BiP in the early secretory pathway may enhance this process. We examined the induction of BiP gene transcription with Northern blot analysis. While the BiP mRNA in all cells increased upon tunicamycin treatment, it was prominent and persistent in the mutant KDEL receptor cells. We also observed an increased phosphorylation of IRE1 upon tunicamycin treatment, even at the basal level, in the mutant KDEL receptor cells. These results indicated that retrieval by the KDEL receptor kept BiP in the early secretory pathway efficiently, and that the impairment of retrieval caused a loss of BiP there that induced intense UPR.

We assessed whether retrieval by the KDEL receptor might contribute to cell survival in the UPR. HeLa cells and the wild-type or the mutant KDEL receptor cells were treated with tunicamycin and DTT. Significantly, the mutant KDEL receptor cells were sensitive to ER stress, while cells expressing the wild-type tolerated the stress. By TUNEL assay, we detected more apoptotic features in the mutant KDEL receptor cells undergoing tunicamycin treatment. Hoechst 33258 staining revealed prominent condensed chromatins and the fragmentation of nuclei in the mutant KDEL receptor cells with tunicamycin. These features were consistent with programmed cell death by ER stress, suggesting that the retrieval of BiP and misfolded proteins from post-ER compartments by the KDEL receptor played a significant role in the ER stress response. When retrieval was limited, the susceptibility to ER stress might be increased (Yamamoto et al., 2003).

4. The effect of the impairment of the KDEL receptor-ER chaperone system in vivo

The KDEL receptor-ER chaperone system had been extensively studied in yeast and mammalian cells; however, its function and the resultant outcomes of its dysfunction in animals or humans *in vivo* were totally unknown. The impairment of the KDEL receptor was expected to perturb ER quality control, which might possibly cause diseases associated with ER stress (Hamada et al., 2004).

4.1 Generation of transgenic mice expressing the mutant KDEL receptor

We took advantage of previous studies on the KDEL receptor and made stable cell lines expressing a transport mutant human KDEL receptor (Townsley et al., 1993). The mutant KDEL receptor (D193N) recognizes the KDEL proteins; however, the receptor is not transported to the ER upon ligand recognition but rather stays in post-ER compartments (Townsley et al., 1993). We found that the mutant KDEL receptor disturbed the circulation of misfolded proteins between the ER and the Golgi complex, resulting in the accumulation of misfolded proteins in the ER. As a result, these cells became sensitive to ER stress. This finding prompted us to make transgenic (TG) mice expressing a mutant KDEL receptor possibly sensitive to ER stress. In order to investigate the function of the KDEL receptor in vivo, we created TG mice expressing the human wild-type or the transport mutant KDEL receptor, using CAG promoter. We established two lines of the wild-type and two lines of the mutant KDEL receptor TG mice. Northern blot revealed that the expression level of the mutant KDEL receptor was as high as that of the endogenous one. Western blot confirmed that the mutant KDEL receptor was expressed ubiquitously (Hamada et al., 2004).

4.2 Dilated cardiomyopathy (DCM) caused by aberrant ER quality control in the mutant KDEL receptor transgenic mice

These TG mice seemed to grow up normally until early adulthood. In the course of the study, we found that the transport mutant KDEL receptor TG mice of both lines died sporadically after the age of more than fourteen weeks. They appeared dyspneic, lethargic and motionless. These mice developed peripheral edema, ascites and cardiomegaly, and seemed to die due to heart failure. We observed marked four-chamber dilation without wall thickening. The average heart to body weight ratio of the mutant KDEL receptor TG mice was significantly higher than that of the wild-type KDEL receptor TG mice and the parental C57BL/6 mice. Histological examination revealed that the cardiomyocytes of the mutant TG mice were of varying sizes but were significantly enlarged on the average. There was also an expanded interstitial fibrosis and vacuolization. On the other hand, the histology of the wild-type TG mice appeared normal. Using TUNEL staining, we detected significantly more apoptotic cells in the hearts of the mutant KDEL receptor TG mice than in those of the control C57BL/6 mice and the wild-type TG mice. We found no significant difference in systemic arterial blood pressure between the mutant KDEL receptor TG mice and the C57BL/6 mice, suggesting that the cardiomegaly in the mutant mice was not a secondary change due to systemic hypertension. Taken together, these results suggested that the mutant KDEL receptor TG mice developed primary DCM.

4.3 Ultrastructural analyses revealed the accumulation of protein aggregates in the expanded sarcoplasmic reticulum of the mutant cardiomyocytes

In order to gain insight into the mechanism responsible for DCM in the mutant KDEL receptor TG mice, we examined the ultrastructure of the mutant cardiomyocytes. The contractile apparatus appeared intact, i.e. the arrangement of myofibril structures and the banding of myofilaments was little different in the myocardia of control C57BL/6 mice and mutant TG mice. The arrangement of mitochondria was also intact in the TG mice. On the other hand, the proliferation of sarcoplasmic reticulum (SR) was prominent around the transverse (T) tubule area, and the T tubule structure was narrowed significantly in both mutant TG mice with and without apparent cardiomegaly in comparison to the T tubules and surrounding the SR in the control. Furthermore, an aggregation of degenerated membrane structures was found among the proliferated SRs, and lamellated or further fused membrane structures were found in the T tubule area in the mutant mice. We observed electron dense materials, possibly protein aggregates in the expanded SR. These aggregates were associated with polyribosomes, suggesting that they were part of the ER network. On the other hand, the ultrastructure of the cardiomyocytes from the wild-type KDEL receptor TG mice appeared normal.

These morphological changes found in the cardiomyocytes of the mutant KDEL receptor TG mice suggested that the mutant KDEL receptor might have impaired ER quality control, which led to the accumulation of misfolded proteins in the SR.

4.4 Hearts in the mutant KDEL receptor transgenic mice might have suffered from ER stress

We took neonatal cardiomyocytes as a primary cell culture. When treated with tunicamycin, neonatal cardiomyocytes from the control mice showed a diffuse expression of BiP, apparently in the ER distribution. To the contrary, we observed cell shrinkage with uneven distribution of BiP in the cardiomyocytes of the mutant KDEL receptor TG mice, suggesting that those cardiomyocytes were sensitive to ER stress. An accumulation of misfolded proteins in the ER leads to protein degradation in a ubiquitin-proteasome system. Western blot revealed a marked accumulation of ubiquitinated protein aggregates in the hearts of the mutant KDEL receptor TG mice, not the wild-type KDEL receptor TG mice, suggesting that misfolded proteins might have saturated the ubiquitin-proteasome system during a circumstance in which the mutant KDEL receptor might have impaired the capacity for ER quality control. We found slightly increased expressions of BiP and significant accumulations of CHOP in the adult mutant hearts. Since CHOP is induced by ER stress and has been acknowledged to cause apoptosis during the UPR (Zinszner et al., 1998), this might account for the enhanced apoptosis in the mutant hearts. These results suggest that the mutant hearts might have suffered from ER stress, which might have contributed to the pathogenesis of the cardiomyopathy found in the mutant mice. Pressure load might contribute to the ER stress on the heart.

DCM is characterized by an increased ventricular chamber size and a reduced contractility of the heart typically accompanied by loss of cardiomyocytes (Narula et al., 1996). The mutant KDEL receptor TG mice demonstrated cardiac dilation with congestive heart failure, interstitial fibrosis, myocyte heterogeneity, vacuolization and apoptosis, which closely resembles the pathologic and clinical features of human DCM. Clinically and experimentally, DCM is caused by a variety of factors (Ross, 2002). ER stress brings on human disorders such as neurodegenerative diseases (Kopito & Ron, 2000); however, it was uncertain whether it also caused cardiac diseases.

Cardiomyocytes are exposed to mechanical stress throughout their lives. The mutant KDEL receptor transgene is driven by the CAG promoter and is supposed to express ubiquitously. The expression level is moderate, as high as that of the endogenous one. Although all tissues of the mutant KDEL receptor TG mice might be sensitive to ER stress, one of the most important factors that affect whether ER stress may cause a disease is the regenerative ability of the concerned tissue. Neurodegenerative diseases and diabetes mellitus are usually regarded as typical ER stress-associated diseases or conformational diseases (Kopito & Ron, 2000) because a neuron and a β cell live as long as the lifespan, and are therefore susceptible to the accumulation of misfolded proteins. In this regard, cardiomyocytes have the same lifespan, and they are exposed to harsh conditions, such as mechanical and oxidative stress. This study suggested that a cardiomyocyte might also be a possible target for ER stress (Hamada et al., 2004).

5. BiP, an ER chaperone, plays important roles in mammalian development

BiP plays a central role in ER function. Yeast BiP (Kar2) is essential for survival; when the retrieval sequence (in yeast: His-Asp-Glu-Leu, HDEL) is deleted, a fraction of Kar2 is secreted from the ER. However, the UPR is activated and this maintains a minimal level of Kar2 in the ER (Beh & Rose, 1995). Thus, the retrieval of BiP is not essential for a single cell. Complete depletion of BiP could be lethal for early embryonic cells in mice (Luo et al., 2006). We therefore produced knock-in mice expressing a mutant BiP in which the retrieval sequence was deleted by homologous recombination. These mice were used to elucidate any processes sensitive to ER stress during development and in adulthood (Mimura et al., 2007).

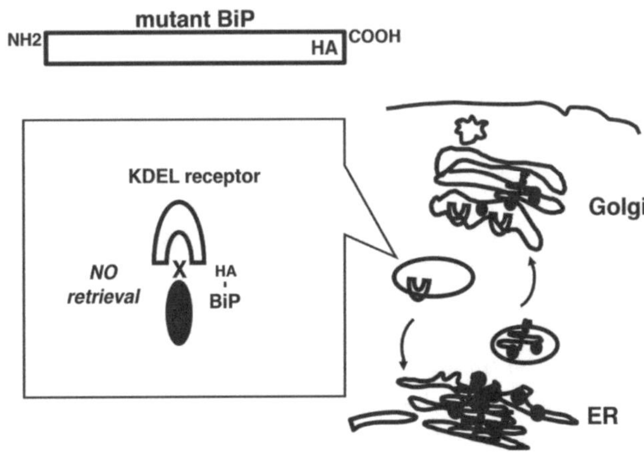

Fig. 4. The mutant BiP is defective in retrieval, leading to an escape of misfolded proteins from ER quality control

5.1 Constitutively active UPR compensates for loss of ER BiP in cultured mammalian cells

The mutant BiP contained a carboxyl-terminal HA tag in stead of a KDEL sequence. Mouse embryonic fibroblasts (MEFs) derived from homozygous BiP mutant embryos expressed the mutant BiP instead of wild-type BiP, and grew as well as wild-type MEFs. Mutant BiP localized to the ER, and its expression was enhanced by tunicamycin. These results were consistent with the expression for wild-type BiP and other ER chaperones containing a KDEL sequence. However, in metabolic labeling experiments, a significant fraction of mutant BiP was found in the medium, reflecting deletion of the KDEL sequence and impaired retrieval of mutant BiP. A fraction of wild-type BiP was also secreted into the medium during ER stress with tunicamycin treatment, indicating that retrieval by the KDEL receptor is saturable.

Tunicamycin sensitivity and expression of mutant BiP was also confirmed by Western blot. Both mutant BiP and wild-type BiP were recognized by an antibody against the amino terminus of BiP. However, an anti-KDEL antibody only recognized the wild-type BiP and GRP94, an ER chaperone with the KDEL sequence. Basal expression of XBP1, ATF4, phospho-PERK, GRP94 and another ER chaperone, calreticulin, was enhanced in homozygous mutant BiP MEFs. Basal expression of mutant BiP mRNA was also enhanced in the mutant MEFs, indicating that the UPR was constitutively active. Thus, as seen previously in yeast (Beh & Rose, 1995), constitutive UPR activation maintained a minimal level of mutant BiP in the ER of mammalian cells, thus compensating for deletion of the KDEL sequence.

5.2 Mutant BiP embryos die shortly after birth

BiP was ubiquitously expressed in both mutant and wild-type embryos. In all tissues examined, GRP94 expression was greater in homozygous BiP mutant embryos than in wild-type embryos, suggesting that homozygous BiP mutant mice might suffer from global ER stress. Homozygous mutant BiP embryos weighed less than wild-types and heterozygotes at embryonic day (E) 18.5. Homozygous BiP mutant mice were born at the expected Mendelian ratio of 1:2:1. Neonatal BiP mutants moved well and responded to painful stimuli, but they appeared pale and cyanotic. They also cried less and displayed shallow breathing. The neonatal homozygous mutants generally died within several hours of birth; thus, we suspected that the observed lethality might reflect respiratory problems. When delivered by Caesarian section at E18.5 and sacrificed prior to breathing, gross morphology of the lungs and airways from BiP mutant mice was indistinguishable from wild-type. Wild-type and homozygous BiP mutant embryonic alveoli had an equivalent distribution of alveolar type II cells expressing the surfactant protein, SP-C. However, histological examination of lungs isolated from neonatal BiP mutants several hours after birth revealed atelectasis with poor inflation of peripheral airways. Hemorrhage and cell debris were also observed in the mutant alveolar space. Alveolar epithelia in BiP mutant mice were enlarged, whereas, as expected, those in wild-type neonates were distended. These observations indicated that homozygous neonatal BiP mutants developed atelectasis and respiratory failure after birth (Mimura et al., 2007).

5.3 Respiratory distress syndrome of newborns

Respiratory distress syndrome of newborns, also called hyaline membrane disease, causes high mortality and often accompanies preterm delivery or low birth weight with reduced expression of pulmonary surfactant. Surfactant therapy combined with mechanical ventilation and other intensive care measures has reduced the mortality rate of this syndrome to below 10% (Hallman, 2004). Surfactant proteins are required for proper lung development and function.

To examine whether a deficiency of pulmonary surfactant contributes to respiratory failure in homozygous BiP mutant mice, perfluorocarbon, a substitute for pulmonary surfactant, was administered into the oropharynx. Perfluorocarbon with oxygen treatment improved the activity of neonatal BiP mutants, turned their skin color from pale to pink, and improved lung inflation. The expression of surfactant proteins in neonatal lung was examined by Western blot. Expression of SP-A and, more prominently, proSP-C, was reduced in mutant lungs compared with wild-type, but there was no significant difference in proSP-B and SP-D expression. RT-PCR analysis revealed that the marked reduction of proSP-C in neonatal mutant lung was not due to reduced transcription. Importantly, after birth, the expression of proSP-C was enhanced only in wild-type neonates, suggesting that proSP-C might be degraded post-translationally in neonatal type II cells from BiP mutants.

Mature SP-B and SP-C are transported to the lamellar body where they bind phospholipids and are then secreted into the alveolar space via regulated exocytosis, whereas SP-A and SP-

D are secreted independently of the lamellar body. The subcellular localization of SP-A and SP-C was evaluated by confocal laser microscopy. In wild-type neonatal alveolar type II cells, SP-A and SP-C (proSP-C) colocalized with BiP, and other KDEL sequence-containing ER chaperones, in the ER. SP-A accumulated in the alveolar lining area of BiP mutant mice and costained with mutant BiP. By contrast, SP-C remained in the ER, and its expression was markedly reduced in type II cells of neonatal BiP mutants. Together, these data suggested that mutant BiP impaired the secretion of pulmonary surfactant, especially secretion through the lamellar body.

Embryonic type II cells store glycogen in the cytoplasm, and this glycogen is consumed as the synthesis of pulmonary surfactant expands after birth. Type II cells in neonatal BiP mutants contained vacuole structures. Periodic acid Schiff (PAS) staining revealed cytoplasmic polysaccharides in these cells, even after birth. Ultrastructural analysis of type II cells from neonates confirmed that cytoplasmic glycogen was indeed still present in mutant, but not in wild-type, cells. More importantly, the structure of the lamellar body was abnormal in embryonic and neonatal mutant type II cells. The lamellar body in wild-type neonates had wavy, dense laminations with clefting, whereas in BiP mutant neonates the lamellar body had loosely formed lamellar structures or was almost empty. These results indicated that the biosynthesis and secretion of pulmonary surfactant was impaired in BiP mutant type II cells (Mimura et al., 2007).

5.4 Aberrant quality control in the endoplasmic reticulum causes neonatal respiratory failure in mice expressing mutant BiP

SP-C is a small, highly hydrophobic protein processed from proSP-C during its transport through the ER and the Golgi to the multivesicular body. Mature SP-C is transported further to the lamellar body where it binds phospholipids before secretion into alveolar space via regulated exocytosis. Lamellar body formation is defective in alveolar type II cells of neonatal BiP mutants. Therefore, SP-C may be degraded by endosomal/lysosomal degradation and/or the ERAD pathway. Punctate SP-C is colocalized with KDEL-containing ER chaperones in both wild-type and homozygous mutant type II cells, suggestive of ER accumulation of SP-C. The fraction of ER accumulation of SP-C in the mutant type II cells was 0.84, while that of Golgi accumulation was 0.08, evaluated by confocal colocalization images. Furthermore, expression of CHOP, a transcription factor related to cell death during ER stress, increased in homozygous mutant lungs after birth, suggesting that mutant lung tissue might be suffering from ER stress. Although endosomal/lysosomal degradation of SP-C could not be excluded, these data suggested that, in homozygous mutant-BiP type II cells, misfolded SP-C might accumulate in the ER and be degraded by the ERAD pathway.

SP-C deficits are related to acute and chronic infant lung diseases in humans (Lahti et al., 2004). Furthermore, mutations in proSP-C have been correlated with chronic interstitial pneumonia (Beers & Mulugeta, 2005). ProSP-C is an type II integral membrane protein with structural homology to the amyloidogenic BRI family of proteins, which cause neurodegenerative dementia (Kim et al., 2002). Mutant proSP-C tends to misfold and may cause protein aggregation and ER stress (Beers & Mulugeta, 2005). Thus, aberrant quality control in the mutant-BiP type II epithelial cells might have resulted in proSP-C misfolding. Misfolded proSP-C might act to generate respiratory failure by causing ER stress in mutant type II cells in concert with decreased pulmonary surfactant levels (Mimura et al., 2007).

6. BiP, an ER chaperone, plays important roles in neuronal migration and stratification

The homozygous mutant BiP neonates died after birth due to respiratory failure (Mimura et al., 2007). Besides that the mutant BiP mice displayed disordered layer formation in the cerebral cortex and cerebellum (Mimura et al., 2008).

6.1 Defective neocortical layer formation in the mutant BiP mice

The homozygous mutant BiP neonates moved, responded to painful stimuli, but were significantly smaller than the wild-type. Among the various organs, the mutant brain, including the cerebral cortex and cerebellum, was substantially smaller than that in wild-type mice, suggesting that the brain was particularly affected by the BiP mutation. In fact, the neocortical stratification at embryonic day 18 (E18), as observed with the hematoxylin-eosin staining, was defective in the mutant BiP mice. The mutant brain had a relatively high density of neurons in neocortical layer I in contrast to a low density of neuronal arrangement in the control.

Cortical neurogenesis occurs in the ventricular zone, and the new neurons migrate through other neurons to the marginal zone and then move to their final destination during embryogenesis. To further investigate the defect in layer formation during neocortical development, birth date analysis of the neocortical neurons was carried out by BrdU labeling. The results showed that in the mutant brain the earlier-born neurons reached the superficial layer and remained there, and that the later-born neurons did not reach the upper layer, remaining in the lower layer. The mutant BiP mice exhibited an outside-in pattern of neocortical layer formation in contrast to the inside-out pattern in the control (Caviness, 1982), indicating that the neocortical layer formation was impaired.

6.2 Mutant BiP mice have reduced expression of reelin

The above findings suggested that aberrant neocortical formation was due to the defects in layer formation, like a deficiency in reelin signaling in a *reeler* mutant malformation (Falconer, 1951; D'Arcangelo et al., 1995). Indeed, analysis of embryonic cerebral neocortex revealed significantly reduced reelin expression by Western blot and immunoreactivity in the superficial layer I of the mutant BiP mice. These results are consistent with the fact that reelin is a secretory protein that may interact with BiP in the ER.

Reelin is a large secreted glycoprotein (D'Arcangelo et al., 1995) produced by some cortical neurons such as Cajal-Retzius (CR) cells in the marginal zone during development. Reelin mediates cortical laminar formation through the binding to very low density lipoprotein receptor (VLDLR) and apolipoprotein E receptor type 2 (ApoER2), leading to the phosphorylation of an adaptor molecule, Dab1, on cortical neurons (D'Arcangelo et al., 1999; Tissir and Goffinet, 2003). In *reeler* mice deficient in the reelin gene (D'Arcangelo et al., 1995), the cortical neurons lack the ability to localize properly, and settle inside the earlier migrating neurons (Caviness, 1982).

Because *reeler* malformation is also well documented in the cerebellum with regard to the migration defect of Purkinje cells (Yuasa et al., 1993), the structure of the cerebellum was examined at E18. The growth of the mutant cerebellum was significantly retarded, examied by the staining with DNA dye. Though the external granular layer (EGL) was formed in the both genotypes, the development of EGL migrating tangentially from the rhombic lip was significantly retarded in the mutant BiP mice. A large number of Purkinje cells remained in the subcortical region in contrast to the cortical arrangement of Purkinje cells in the control. Hippocampal layer formation showed little defect in the mutant BiP mice.

The structure of the superficial layer of the neocortical primordium was further examined by double immunohistochemical labeling for both reelin and calretinin. Calretinin-immunopositive neurons, corresponding to CR cells in the neocortical primordium, were found in the superficial layer of the mutant BiP mice, but their number was significantly reduced and reelin immunoreactivity was barely detected, in contrast to the localization of reelin immunoreactivity in the calretinin-positive neurons in superficial layer I of the wild-type mice. Some of calretinin-immunopositive CR cells of the mutant neocortex appeared in a disorganized scattered pattern other than the marginal zone. This finding was confirmed by in situ hybridization histochemistry of the neocortical primordium by using *reelin* cRNA probe as the marker for CR cells. The cells positive for *reelin* mRNA formed the thin superficial layer in the wild-type mice. In the contrast, the cells positive for *reelin* mRNA were scattered in the upper layer of the neocortical primordium of the mutant BiP mice. These findings of in situ hybridization histochemistry corresponded well with those of calretinin-immunoreactive cells. Furthermore, these findings indicated that the transcription of *reelin* gene took place in a similar degree both in the mutant and wild-type mice, but the reelin protein was significantly reduced in the CR cells of the mutant.

While this mouse had features of a *reeler* mutant phenotype such as an outside-in pattern of neocortical layer formation and the migration defect of Purkinje cells in cerebellum, it also had other phenotypes in the brain, which were distinct from the *reeler* phenotype. These included the reduction in the size of the whole brain and the apparent scattering of reelin- and calretinin-positive neurons throughout the cortex. This is not surprising since BiP likely has a multitude of substrates significant for brain development.

6.3 Reelin secretion is impaired in the mutant BiP brain

Embryonic cerebral neocortex revealed significantly reduced reelin expression by Western blot in the homozygous mutant BiP, consistent with histological observations. Although VLDLR expression was equivalent between wild-type and mutant BiP cortex, dephosphorylated Dab1 accumulated in the mutant BiP brain, indicating that the reelin signaling pathway was inactivated there. The reelin deficiency was not a consequence of reduced transcription, because *reelin* mRNA expression did not differ in the control and mutant brains, consistent to the in situ hybridization experiment. The expression of *BiP* mRNA as well as CHOP protein was enhanced in the mutant brain, suggesting that the mutant brain suffered from ER stress.

Mutant BiP might impair the folding of reelin, leading to its degradation by the ERAD pathway or to its secretion as an immature form from the CR cells due to an escape from ER quality control. To test this possibility, we used primary neurons derived from embryonic brain and found a significant decrease in reelin secretion by the homozygous mutant BiP neurons compared with wild-type or heterozygous neurons. To investigate whether the homozygous mutant BiP neurons maintained their responsiveness to reelin stimulation, we incubated primary neurons with conditioned culture medium containing a several-fold physiological level of reelin secreted by 293T cells transiently transfected with reelin cDNA. Exogenous reelin seemed to be active on the homozygous neurons, leading to the activation of the reelin signaling pathway, as demonstrated by a reduced amount of Dab1 expression and an increased amount of phospho-Dab1 expression. On the other hand, reelin signaling pathway in the heterozygous mutant cortical neurons seemed to be constitutively active with endogenous reelin even without exogenous reelin stimulation. Thus, Dab1 expression and phosphorylation were rather unchanged in the heterozygous mutant. These results suggested that impaired secretion of reelin by the CR cells rather than defective responsiveness in the cortical neurons might be responsible for the neurological phenotype of *reeler* mutant-like malformation in the mutant BiP mice. Thus, impaired retrieval of BiP might promote the degradation of misfolded proteins by the ubiquitin/proteasome pathway. In fact, ubiquitinated proteins accumulated in the mutant cerebrum.

6.4 BiP may enhance the folding of reelin

Mutant BiP was detected in the ER (Mimura et al., 2007), but a significant fraction was also secreted from cells because of the lack of the retrieval motif (KDEL). We examined the subcellular localization of reelin to establish its relationship with mutant and wild-type BiP. Reelin co-localized with mutant BiP in the ER in primary neurons derived from heterozygous mutant BiP embryos; this was also the case in cortical neurons in the homozygous mutant postnatal brain where the expression of reelin was reduced. To obtain further insight into the interaction of BiP and reelin, we performed co-transfection experiments in HeLa cells. Co-expression of reelin and the wild-type BiP, but not the mutant BiP lacking the KDEL sequence, greatly enhanced the expression of reelin protein (reelin mRNA levels were equivalent in the two transfections). These results suggested that BiP promoted the folding of reelin.

Although the folding, intracellular transport and oligomerization of reelin have not been characterized in detail, we found that reelin protein expression was impaired in the mutant BiP mice, indicating that BiP might play a role in the maturation of reelin. Furthermore, we found that the expression of *BiP* mRNA and CHOP protein was enhanced in the mutant brain, suggesting that the mutant brain might have suffered from ER stress. We speculate that the folding of reelin protein might be vulnerable to impaired quality control in the ER and the post-ER compartments of mutant CR cells.

The mutant BiP mice revealed that a physiological increase in the production of reelin and surfactant proteins in dedicated secretory cells like CR cells and alveolar type II cells during neonatal periods may require BiP and a proper folding capacity in the ER. Neuronal migration and stratification may be sensitive to environmental insults such as viral infection, hypoxia and ischemia that perturb ER functions (Mimura et al., 2008).

7. BiP modulates the development of morphine antinociceptive tolerance

Heterozygous mutant BiP mice produce pulmonary surfactant and reelin, and grow to be apparently normal adults. However, they may be potentially sensitive to ER stress. Morphine is a potent analgesic, but the molecular mechanism for tolerance formation after repeated use is not fully understood. We tested the thermal antinociceptive effect of morphine on the heterozygous mutant BiP mice in order to elucidate physiological processes that were sensitive to BiP functions (Dobashi et al., 2010).

7.1 Morphine tolerance

Opioids are potent analgesics that are widely used to control acute and chronic pain (Somogyi et al., 2007). Although repeated administration of opioids, particularly morphine, induces tolerance that reduces the effectiveness of the analgesic, the precise molecular mechanism for the development of tolerance remains uncertain. Opioids bind to the mu opioid receptor (MOR) to activate various signaling molecules through heterotrimeric guanine nucleotide-binding proteins (G proteins), leading to a decrease in neuronal excitability by the inhibition of voltage-dependent calcium channels and the activation of inwardly rectifying potassium channels (Dickinson et al., 2000). Activation of MOR also induces the phosphorylation of MOR by G-protein-coupled receptor kinases (Johnson et al., 2005). Phosphorylated MOR is recognized by arrestins (Bohn et al., 1999), and internalized by clathrin-coated vesicles. The transient uncoupling of MOR from signaling pathways due to the phosphorylation and intracellular trafficking of MOR causes opioid desensitization. Most of the internalized MORs return to the cell surface, resulting in resensitization (Zollner et al., 2008).

Chronic morphine tolerance may be derived from adaptations in the intracellular signal transduction of post-MOR activation, as morphine does not induce effective MOR phosphorylation and internalization (Finn and Whistler, 2001). Persistent MOR activation may alter signal transduction, including changes in MOR-coupled G proteins from Giα to Gsα (Chakrabarti et al., 2005), increased activity of protein kinase C (Granados-Soto et al., 2000), and the upregulation of N-methyl-D-aspartate receptor signaling (Trujillo and Akil, 1991). These changes may contribute to the development of morphine tolerance. Chronic morphine treatment also activates the cyclin-dependent kinase 5 and glycogen synthase kinase 3β (GSK3β) signaling pathway, while the inhibition of them diminishes morphine tolerance and restores analgesia in rats (Parkitna et al., 2006). GSK3β is expressed ubiquitously and is one of the central molecules in intracellular signal transduction (Grimes and Jope, 2001). It may play an important role in diverse physiological and pathological states (Jope et al., 2007).

7.2 Morphine tolerance is attenuated in mice expressing a mutant BiP

We examined whether BiP might affect morphine analgesia using heterozygous mutant BiP mice. We evaluated morphine-induced antinociception by measuring response latencies in a hot plate test. Morphine tolerance was induced by intraperitoneal morphine injection twice a day for 5 consecutive days. We performed hot plate tests at the first and the tenth morphine treatments. The response latencies of the mutant BiP mice and their wild-type littermates before morphine treatment on day 1 were not significantly different. The time courses of response latencies of the both groups at the first morphine treatment on day 1 were almost similar, and both latencies reached the 60 s cut-off point 30 min after injection. Even just before the tenth morphine treatment on day 5, the response latencies of the both groups were not significantly different. These results indicated that the mutant BiP mice had normal sensory transmission and analgesia. After the tenth morphine treatment on day 5, the response latencies of the wild-type mice were significantly reduced, indicating that morphine tolerance had developed. However, the response latencies of the mutant BiP mice after the tenth morphine treatment were significantly longer than those of their wild-type littermates at 30, 45 and 60 min after injection. These results showed that the mutant BiP mice were impaired in the development of morphine tolerance (Dobashi et al., 2010).

7.3 Inhibition of GSK3β signaling is associated with the prevention of morphine tolerance

Since alterations in the intracellular trafficking of MOR might affect opioid analgesia, we examined the effect of the mutant BiP with the KDEL sequence deleted on the surface expression of MOR. MEFs from the wild-type and homozygous mutant embryos (Mimura et al., 2007) were transfected with a myc-tagged MOR. DAMGO, a selective peptidergic MOR ligand induces the internalization of MOR but morphine does not internalize MOR. We found the surface expression of MOR in both wild-type and mutant MEFs by confocal laser microscopy. While DAMGO induced the internalization of MOR in both types of MEFs, MOR remained on the cell surface upon morphine treatment, suggesting that the mutant BiP did not affect the transport of MOR.

Then, we speculated that the UPR signaling might attenuate the MOR signaling, which might cause the development of morphine tolerance. GSK3β is one possible candidate molecule that may play key roles in both the UPR and MOR signaling pathways. Recently the inhibition of GSK3β by the specific inhibitors SB216763 and (2'Z, 3'E)-6-bromoindirubin-3'-oxime was shown to diminish the development of morphine tolerance in rats after chronic intrathecal morphine treatment (Parkitna et al., 2006). The kinase activity of GSK3β is regulated by its phosphorylation status. Phosphorylation of residue Ser9 inactivates the activity, whereas dephosphorylation of Ser9 and phosphorylation of Tyr216 enhance the activity (Grimes and Jope, 2001).

We evaluated the phosphorylation status of GSK3β in the brain stems of wild-type and heterozygous mutant BiP mice using specific antibodies against phosphorylated Tyr216 GSK3β and phosphorylated Ser9 GSK3β. After chronic morphine injection intraperitoneally for 5 days, the wild-type mice developed morphine tolerance, whereas the mutant BiP mice remained less tolerant to morphine. Because we injected morphine intraperitoneally, both spinal and supraspinal neurons were supposed to be affected. Neurons with MOR expression in the periaqueductal gray (PAG) matter contribute to morphine tolerance (Yaksh et al., 1976; Bagley et al., 2005; Morgan et al., 2006). With repeated morphine treatment, the mutant BiP brain stems showed low levels of phosphorylation of Tyr216 in GSK3β, in contrast to the prominent phosphorylation in wild-type mice by Western blot. After chronic morphine injection intraperitoneally for 5 days in both types of mice, the brains were sectioned and double-immunostained with antibodies raised against MOR and tyrosine-phosphorylated GSK3β. MOR-immunopositive neurons in the PAG region of wild-type brains showed more enhanced expression of tyrosine-phosphorylated GSK3β significantly than those in the mutant BiP brains.

These observations suggested that chronic MOR stimulation by repetitive morphine injection might activate GSK3β, leading to the development of morphine tolerance. Mice with the mutant BiP might be defective in the activation of GSK3β. Furthermore, we examined embryonic brains of homozygous mutant BiP mice which did not express the wild-type BiP. Western blot revealed prominent phosphorylation at Ser9 of GSK3β and less phosphorylation at Tyr216 in the homozygous mutant BiP brain compared to those in the wild-type brain. These results suggested that the mutant BiP lacking the KDEL sequence might attenuate the activation of GSK3β in vivo (Dobashi et al., 2010).

7.4 Chemical chaperone attenuates the development of morphine tolerance

In order to confirm that an ER chaperone mediates the development of morphine tolerance, we examined the effect of a chemical chaperone on morphine tolerance. Tauroursodeoxycholic acid (TUDCA) is a derivative of endogenous bile acids that is thought to increase ER folding capacity and suppresses the expression of BiP (Xie et al., 2002; Ozcan et al., 2006). We administered TUDCA together with morphine twice a day for 5

days in wild-type mice, and hot plate tests were performed at the first and the tenth treatments. The response latencies of the mice receiving both TUDCA and morphine were significantly longer than those of control mice with morphine alone after the tenth treatment. Thus, TUDCA prevented the development of morphine tolerance, suggesting a mechanistic relationship between an ER chaperone and morphine tolerance.
These results suggested a novel function of BiP on the development of morphine tolerance in vivo. The modulation of morphine analgesia by TUDCA revealed a potential clinical application of chemical chaperones that could modulate ER functions for the prevention of morphine tolerance (Dobashi et al., 2010).

8. Dysfunction of the ER chaperone, BiP, may accelerates the organic injury with aging

Heterozygous mutant BiP mice express both the wild type BiP and the mutant BiP, in which the UPR is not induced in the steady state. However, it is possible that potential vulnerability to ER stress may exist in the mutant BiP mice, resulting in organic injury during chronic stress such as aging (Kimura et al., 2008).

8.1 Heterozygous mutant BiP knock-in mice developed marked tubular-interstitial lesions with aging

We found that some aged mutant BiP mice developed a severe tubular-interstitial lesion which consisted of tubular atrophy, tubular luminal dilatation and interstitial fibrosis. Mutant BiP mice over 80 weeks of age showed more severe tubular-interstitial lesions than age-matched wild type mice.
Proteinuria plays a key role in the tubular cell injury involved in human kidney disease (Remuzzi & Bertani, 1998). We used the bovine serum albumin (BSA)-overload proteinuria mouse model to clarify whether the tubular cell injury induced by proteinuria was associated with the tubular-interstitial lesions observed in aged heterozygous mutant BiP mice. BSA was injected into young mutant BiP mice (25-40 weeks of age) and control wild type mice intraperitonially 5 days per week for 6 weeks after the uninephrectomy. While the mutant BiP mice maintained renal tissue apparently as normal as the wild type mice, the BSA treatment caused severe tubular-interstitial injury only in the mutant BiP mice. The mutant BiP mice with BSA overload had a significantly higher tubular damage score than the wild-type mice. Furthermore, the serum creatinine level of the BSA-treated mutant BiP mice was also significantly higher than that of the wild-type mice, indicating that proteinuria impaired the renal function of the mutant mice. No significant difference in creatinine level was observed between the control wild type mice and the mutant BiP mice without BSA treatment.
In order to determine whether the differences in renal injury with BSA treatment between heterozygous mutant BiP mice and wild type mice could be due to different levels in urinary protein excretion, we examined the urinary protein of the two different groups. After the uninephrectomy, BSA (10 mg/body weight(g)) was injected intraperitonially into the mice once a day for 7 days and urinary protein excretion was examined on day 8. We did not find any significant differences in urinary protein excretions between the mutant BiP mice and wild type mice. Thus, these results suggested that the tubular-interstitial tissue of the mutant BiP mice was more sensitive to proteinuria than that of the wild-type mice.

8.2 Caspase-12 activation and tubular cell apoptosis occurred in the kidneys of BSA-treated heterozygous mutant BiP knock-in mice

BSA is reported to cause apoptosis in a murine proximal tubular cell line through a caspase-12 dependent apoptotic-pathway induced by ER stress (Ohse et al., 2006). In order to detect apoptosis in the BSA-treated tubular-interstitial lesions of heterozygous mutant BiP mice, we performed TUNEL staining. The number of TUNEL positive tubular cells in BSA-treated mutant BiP mice was significantly higher than that in the wild type. We also found the activation of caspase-12 in the BSA-treated mutant kidney by Western blot that showed the cleavage of caspase-12. Taken together, these results suggested that the renal tissue of the mutant BiP mice suffered from ER stress and was sensitive to proteinuria, which might cause tubular-interstitial lesions and apoptosis through ER stress pathway.

The involvement of ER stress and its downstream caspase-12 dependent apoptotic pathway on tubular cell injury induced by proteinuria have been reported (Ohse et al., 2006). Aged heterozygous mutant BiP mice showed glomelular lesion consisting of mesangial matrix increase and glomerular sclerosis. While we could not exclude the contribution of the glomerular lesion, high susceptibility to proteinuria due to potential ER dysfunction might cause the tubular-interstitial lesion of aged heterozygous mutant BiP mice. It is possible that potential vulnerability to ER stress might exist in the tubular cells of heterozygous mutant BiP mice and resulted in severe tubular injury during periods of chronic stress such as aging and proteinuria. Tubular cells are most sensitive to ER stress when mice are injected with tunicamycin that disturbs protein glycosylation in the ER and causes ER stress, which results in acute renal tubular necrosis (Zinszner et al., 1998). Consistently, we found caspase-12 activation in the kidneys of BSA-treated mutant-BiP mice, indicating that the ER stress pathway might have been involved in the BSA-mediated tubular-interstitial injury (Kimura et al., 2008).

9. Conclusion

Quality control in the early secretory pathway is a ubiquitous mechanism for adapting to ER stress, and the KDEL receptor and BiP are essential components of this system. However, at various developmental stages, some cell types may require specific quality control systems and chaperones. Mutant mouse models have revealed that the UPR plays a vital role during development by increasing protein synthesis as needed in dedicated secretory cells, such as pancreatic beta cells (Scheuner et al., 2001), plasma cells (Reimold et al., 2001) and hepatocytes (Reimold et al., 2001). Inadequate adaptation to these types of physiological demands may lead to diverse diseases.

We produced knock-in mice expressing a mutant BiP lacking the retrieval sequence to examine the effects of defects in stress response in the secretory pathway without completely eliminating BiP function (Mimura et al., 2007) - as would be the case in BiP knockout mice (Luo et al., 2006). Mutant BiP mice have a distinct phenotype, as is the case for mice lacking other ER molecular chaperones. Hsp47 is responsible for collagen biosynthesis, and Hsp47 knockout mice die on E11.5 (Nagai et al., 2000). Calreticulin and calnexin participate in glycoprotein folding in the ER. Calreticulin knockout mice are embryonic lethal and display defective cardiac development (Mesaeli et al., 1999). Calnexin knockout mice die during the early postnatal period, between birth and three months of age. These mice exhibit motor disorders due to a loss of large myelinated nerve fibers (Denzel et al., 2002). Mutant BiP predominantly affected dedicated secretory cells, such as alveolar type

II cells and Cajal-Retzius cells, in which active secretion is particularly important. Putative impairment of protein folding in these mutant cells probably caused the observed respiratory failure and neurological disorders.

Deletion of the retrieval sequence from BiP, and the consequent lack of mutant BiP recycling by the KDEL receptor, could have two possible effects. First, the folding environment in the ER and post-ER may be impaired. Mutant BiP is functional as long as it remains in the ER. Therefore, constitutive activation of the UPR could compensate for the altered folding environment by producing mutant BiP in quantities sufficient for cell survival. However, quality control in post-ER compartments may also be affected. Proper ER-to-Golgi transport and subsequent ER retrieval of proteins and lipids are thought to contribute to quality control (Hammond and Helenius, 1994; Yamamoto et al., 2001). In this regard, the folding (and therefore function) of a pulmonary surfactant protein, proSP-C, and reelin may depend on proper ER retrieval of BiP via the KDEL receptor. Second, signal transduction during UPR could be affected. In addition to retrieval, the recognition of the KDEL sequence of BiP (and other KDEL proteins) by the KDEL receptor leads to signal transduction. The activation of the KDEL receptor may trigger subsequent activation of signaling molecules such as ARF1GAP (Yamamoto et al., 2001), src (Bard et al., 2003), protein kinase A (Cabrera et al., 2003), and mitogen-activated protein kinases (Yamamoto et al., 2003). Chronic morphine administration may cause altered signal transduction through persistent MOR activation. It would be possible that the crosstalk between MOR analgesic signal transduction and BiP-KDEL receptor signal transduction may affect morphine tolerance formation in the mutant BiP mice (Dobashi et al., 2010).

The heterozygous mutant BiP mice grew up to be adults and showed apparently normal organ development. There was no significant differences in life span between the wild type mice and the heterozygous mutant BiP mice. However, potential vulnerability to ER stress may exist in the mutant BiP mice, resulting in some organic injuries with aging as described above (Kimura et al., 2008). We are interested in the relationship between ER stress and the neural degeneration with aging. In this regard, several studies have suggested the possible role of reelin in the pathogenesis of human mental disorders such as schizophrenia, autism, bipolar disorder and Alzheimer's disease (Bothwell & Giniger, 2000; Tissir & Goffinet, 2003; Fatemi, 2005). Reelin and ApoE share ApoER2 on cortical neurons (D'Arcangelo et al., 1999), and ApoE inhibits reelin signaling by competing for binding to ApoER2. Interestingly, the E4 allele of ApoE increases the risk of developing sporadic forms of Alzheimer's disease. Because reelin signaling through ApoER2 in adult brains modulates synaptic plasticity and memory formation (Beffert et al., 2005), the defective reelin signaling pathway may contribute to the pathogenesis of adult mental disorders.

In the meantime, the persistent accumulation of misfolded proteins beyond the capacity of ER quality control causes ER stress, leading to cellular dysfunction and cell death (Kaufman, 2002; Kopito & Ron, 2000). This process is thought to cause human mental disorders such as neurodegenerative diseases including Alzheimer's disease (Katayama et al., 1999) and Parkinson's disease (Imai et al., 2001), bipolar disorders (Kakiuchi et al., 2003), and ischemic neuronal injury (Tajiri et al., 2004). The involvement of impaired BiP function in neurodegenerative diseases has been reported in a mouse model where the disruption of SIL1, a co-chaperone of BiP, causes protein accumulation and neurodegeneration (Zhao et al., 2005). Thus, reelin signaling and ER quality control may be related to the pathogenesis of adult mental disorders, as seen in reeler mutant-like cerebral malformation in mutant BiP neonates (Mimura et al., 2008).

Our results suggest that ER stress would be a promising therapeutic target with which to combat chronic organ injuries. In order to treat ER stress related diseases, two kinds of strategy will be effective: the promotion of protein folding in the ER and the inhibition of an ER stress induced apoptotic pathway. Indeed, the administration of chemical chaperones that promote protein folding in the ER has been reported to be effective in treating type2 diabetes, which has been speculated to be caused by ER stress after experiments using a mouse model (Ozcan et al., 2006). We also showed that a chemical chaperone prevented the development of morphine tolerance (Dobashi et al., 2010). The heterozygous mutant BiP knock-in mice used in our experiments will be a suitable tool for investigating the relationship between ER stress and organ dysfunctions with aging.

10. Acknowledgements

I thank Drs. Victor W. Hsu and Haruhiko Koseki for helpful suggestions. I also thank Katsushi Yamamoto, Rika Fujii, Hiromichi Hamada, Naoya Mimura, Keita Kimura, Serabi Tanabe, Tamae Dobashi and Hisayo Jin as laboratory members. This work was supported by Grants-in-Aids for Science Research from the Ministry of Education, Culture, Sports, Science and Technology to T.A.

11. References

Aoe, T., Cukierman, E., Lee, A., Cassel, D., Peters, P.J., and Hsu, V.W. (1997). The KDEL receptor, ERD2, regulates intracellular traffic by recruiting a GTPase-activating protein for ARF1. *EMBO J* 16, 7305-7316.

Aoe, T., Huber, I., Vasudevan, C., Watkins, S.C., Romero, G., Cassel, D., and Hsu, V.W. (1999). The KDEL receptor regulates a GTPase-activating protein for ADP-ribosylation factor 1 by interacting with its non-catalytic domain. *J Biol Chem* 274, 20545-20549.

Aoe, T., Lee, A.J., van Donselaar, E., Peters, P.J., and Hsu, V.W. (1998). Modulation of intracellular transport by transported proteins: insight from regulation of COPI-mediated transport. *Proc Natl Acad Sci U S A* 95, 1624-1629.

Bagley, E.E., Chieng, B.C., Christie, M.J., and Connor, M. (2005). Opioid tolerance in periaqueductal gray neurons isolated from mice chronically treated with morphine. *Br J Pharmacol* 146, 68-76.

Bard, F., Mazelin, L., Pechoux-Longin, C., Malhotra, V., and Jurdic, P. (2003). Src regulates Golgi structure and KDEL receptor-dependent retrograde transport to the endoplasmic reticulum. *J Biol Chem* 278, 46601-46606.

Beers, M.F., and Mulugeta, S. (2005). Surfactant protein C biosynthesis and its emerging role in conformational lung disease. *Annu Rev Physiol* 67, 663-696.

Beffert, U., Weeber, E.J., Durudas, A., Qiu, S., Masiulis, I., Sweatt, J.D., Li, W.P., Adelmann, G., Frotscher, M., Hammer, R.E., *et al.* (2005). Modulation of synaptic plasticity and memory by Reelin involves differential splicing of the lipoprotein receptor Apoer2. *Neuron* 47, 567-579.

Beh, C.T., and Rose, M.D. (1995). Two redundant systems maintain levels of resident proteins within the yeast endoplasmic reticulum. *Proc Natl Acad Sci U S A* 92, 9820-9823.

Bertolotti, A., Zhang, Y., Hendershot, L.M., Harding, H.P., and Ron, D. (2000). Dynamic interaction of BiP and ER stress transducers in the unfolded- protein response. *Nat Cell Biol* 2, 326-332.

Bohn, L.M., Lefkowitz, R.J., Gainetdinov, R.R., Peppel, K., Caron, M.G., and Lin, F.T. (1999). Enhanced morphine analgesia in mice lacking beta-arrestin 2. *Science* 286, 2495-2498.

Bonfanti, L., Mironov, A.A., Jr., Martinez-Menarguez, J.A., Martella, O., Fusella, A., Baldassarre, M., Buccione, R., Geuze, H.J., Mironov, A.A., and Luini, A. (1998). Procollagen traverses the Golgi stack without leaving the lumen of cisternae: evidence for cisternal maturation [see comments]. *Cell* 95, 993-1003.

Bonifacino, J.S., Suzuki, C.K., Lippincott-Schwartz, J., Weissman, A.M., and Klausner, R.D. (1989). Pre-Golgi degradation of newly synthesized T-cell antigen receptor chains: intrinsic sensitivity and the role of subunit assembly. *J Cell Biol* 109, 73-83.

Bonifacino, J.S., and Weissman, A.M. (1998). Ubiquitin and the control of protein fate in the secretory and endocytic pathways. *Annu Rev Cell Dev Biol* 14, 19-57.

Bothwell, M., and Giniger, E. (2000). Alzheimer's disease: neurodevelopment converges with neurodegeneration. *Cell* 102, 271-273.

Cabrera, M., Muniz, M., Hidalgo, J., Vega, L., Martin, M.E., and Velasco, A. (2003). The retrieval function of the KDEL receptor requires PKA phosphorylation of its C-terminus. *Mol Biol Cell* 14, 4114-4125.

Caviness, V.S., Jr. (1982). Neocortical histogenesis in normal and reeler mice: a developmental study based upon [3H]thymidine autoradiography. *Brain Res* 256, 293-302.

Chakrabarti, S., Regec, A., and Gintzler, A.R. (2005). Biochemical demonstration of mu-opioid receptor association with Gsalpha: enhancement following morphine exposure. *Brain Res Mol Brain Res* 135, 217-224.

Cukierman, E., Huber, I., Rotman, M., and Cassel, D. (1995). The ARF1 GTPase-activating protein: zinc finger motif and Golgi complex localization. *Science* 270, 1999-2002.

D'Arcangelo, G., Homayouni, R., Keshvara, L., Rice, D.S., Sheldon, M., and Curran, T. (1999). Reelin is a ligand for lipoprotein receptors. *Neuron* 24, 471-479.

D'Arcangelo, G., Miao, G.G., Chen, S.C., Soares, H.D., Morgan, J.I., and Curran, T. (1995). A protein related to extracellular matrix proteins deleted in the mouse mutant reeler. *Nature* 374, 719-723.

Denzel, A., Molinari, M., Trigueros, C., Martin, J.E., Velmurgan, S., Brown, S., Stamp, G., and Owen, M.J. (2002). Early postnatal death and motor disorders in mice congenitally deficient in calnexin expression. *Mol Cell Biol* 22, 7398-7404.

Dickinson, P., Kimber, W.L., Kilanowski, F.M., Webb, S., Stevenson, B.J., Porteous, D.J., and Dorin, J.R. (2000). Enhancing the efficiency of introducing precise mutations into the mouse genome by hit and run gene targeting. *Transgenic Res* 9, 55-66.

Dobashi, T., Tanabe, S., Jin, H., Mimura, N., Yamamoto, T., Nishino, T., and Aoe, T. (2010). BiP, an endoplasmic reticulum chaperone, modulates the development of morphine antinociceptive tolerance. *J Cell Mol Med* 14, 2816-2826

Ellgaard, L., and Helenius, A. (2003). Quality control in the endoplasmic reticulum. *Nat Rev Mol Cell Biol* 4, 181-191.

Falconer, D.S. (1951). Two new mutant, 'trembler' and 'reeler,' with neurological action s in the house mouse (Mus Muscululus L.). *J Genet* 50, 192-201.

Fatemi, S.H. (2005). Reelin glycoprotein: structure, biology and roles in health and disease. *Mol Psychiatry* 10, 251-257.

Finn, A.K., and Whistler, J.L. (2001). Endocytosis of the mu opioid receptor reduces tolerance and a cellular hallmark of opiate withdrawal. *Neuron* 32, 829-839.

Granados-Soto, V., Kalcheva, I., Hua, X., Newton, A., and Yaksh, T.L. (2000). Spinal PKC activity and expression: role in tolerance produced by continuous spinal morphine infusion. *Pain* 85, 395-404.

Grimes, C.A., and Jope, R.S. (2001). The multifaceted roles of glycogen synthase kinase 3beta in cellular signaling. *Prog Neurobiol* 65, 391-426.

Hallman, M. (2004). Lung surfactant, respiratory failure, and genes. *N Engl J Med* 350, 1278-1280.

Hamada, H., Suzuki, M., Yuasa, S., Mimura, N., Shinozuka, N., Takada, Y., Nishino, T., Nakaya, H., Koseki, H., and Aoe, T. (2004). Dilated cardiomyopathy caused by aberrant endoplasmic reticulum quality control in mutant KDEL receptor transgenic mice. *Mol Cell Biol* 24, 8007-8017.

Hammond, C., and Helenius, A. (1994). Quality control in the secretory pathway: retention of a misfolded viral membrane glycoprotein involves cycling between the ER, intermediate compartment, and Golgi apparatus. *J Cell Biol* 126, 41-52.

Hendershot, L.M. (2004). The ER function BiP is a master regulator of ER function. *Mt Sinai J Med* 71, 289-297.

Hsu, V.W., Lee, S.Y., and Yang, J.S. (2009). The evolving understanding of COPI vesicle formation. *Nat Rev Mol Cell Biol* 10, 360-364.

Hsu, V.W., Shah, N., and Klausner, R.D. (1992). A brefeldin A-like phenotype is induced by the overexpression of a human ERD-2-like protein, ELP-1. Cell *69*, 625-635.

Huppa, J.B., and Ploegh, H.L. (1997). The alpha chain of the T cell antigen receptor is degraded in the cytosol. *Immunity* 7, 113-122.

Imai, Y., Soda, M., Inoue, H., Hattori, N., Mizuno, Y., and Takahashi, R. (2001). An unfolded putative transmembrane polypeptide, which can lead to endoplasmic reticulum stress, is a substrate of Parkin. *Cell* 105, 891-902.

Itin, C., Schindler, R., and Hauri, H.P. (1995). Targeting of protein ERGIC-53 to the ER/ERGIC/cis-Golgi recycling pathway. *J Cell Biol* 131, 57-67.

Johnson, E.E., Christie, M.J., and Connor, M. (2005). The role of opioid receptor phosphorylation and trafficking in adaptations to persistent opioid treatment. *Neurosignals* 14, 290-302.

Jope, R.S., Yuskaitis, C.J., and Beurel, E. (2007). Glycogen synthase kinase-3 (GSK3): inflammation, diseases, and therapeutics. *Neurochem Res* 32, 577-595.

Kakiuchi, C., Iwamoto, K., Ishiwata, M., Bundo, M., Kasahara, T., Kusumi, I., Tsujita, T., Okazaki, Y., Nanko, S., Kunugi, H., *et al.* (2003). Impaired feedback regulation of XBP1 as a genetic risk factor for bipolar disorder. *Nat Genet* 35, 171-175.

Katayama, T., Imaizumi, K., Sato, N., Miyoshi, K., Kudo, T., Hitomi, J., Morihara, T., Yoneda, T., Gomi, F., Mori, Y., *et al.* (1999). Presenilin-1 mutations downregulate the signalling pathway of the unfolded-protein response. *Nat Cell Biol* 1, 479-485.

Kaufman, R.J. (2002). Orchestrating the unfolded protein response in health and disease. *J Clin Invest* 110, 1389-1398.

Kim, S.H., Creemers, J.W., Chu, S., Thinakaran, G., and Sisodia, S.S. (2002). Proteolytic processing of familial British dementia-associated BRI variants: evidence for enhanced intracellular accumulation of amyloidogenic peptides. *J Biol Chem* 277, 1872-1877.

Kimura, K., Jin, H., Ogawa, M., and Aoe, T. (2008). Dysfunction of the ER chaperone BiP accelerates the renal tubular injury. *Biochem Biophys Res Commun* 366, 1048-1053.

Klausner, R.D., Lippincott-Schwartz, J., and Bonifacino, J.S. (1990). The T cell antigen receptor: insights into organelle biology. *Annu Rev Cell Biol* 6, 403-431.

Kopito, R.R., and Ron, D. (2000). Conformational disease. *Nat Cell Biol* 2, E207-209.

Lahti, M., Marttila, R., and Hallman, M. (2004). Surfactant protein C gene variation in the Finnish population - association with perinatal respiratory disease. *Eur J Hum Genet* 12, 312-320.

Letourneur, F., Gaynor, E.C., Hennecke, S., Demolliere, C., Duden, R., Emr, S.D., Riezman, H., and Cosson, P. (1994). Coatomer is essential for retrieval of dilysine-tagged proteins to the endoplasmic reticulum. *Cell* 79, 1199-1207.

Lewis, M.J., and Pelham, H.R. (1990). A human homologue of the yeast HDEL receptor. *Nature* 348, 162-163.

Lewis, M.J., and Pelham, H.R. (1992). Ligand-induced redistribution of a human KDEL receptor from the Golgi complex to the endoplasmic reticulum. *Cell* 68, 353-364.

Luo, S., Mao, C., Lee, B., and Lee, A.S. (2006). GRP78/BiP is required for cell proliferation and protecting the inner cell mass from apoptosis during early mouse embryonic development. *Mol Cell Biol* 26, 5688-5697.

Mesaeli, N., Nakamura, K., Zvaritch, E., Dickie, P., Dziak, E., Krause, K.H., Opas, M., MacLennan, D.H., and Michalak, M. (1999). Calreticulin is essential for cardiac development. *J Cell Biol* 144, 857-868.

Mimura, N., Hamada, H., Kashio, M., Jin, H., Toyama, Y., Kimura, K., Iida, M., Goto, S., Saisho, H., Toshimori, K., *et al.* (2007). Aberrant quality control in the endoplasmic reticulum impairs the biosynthesis of pulmonary surfactant in mice expressing mutant BiP. *Cell Death Differ* 14, 1475-1485.

Mimura, N., Yuasa, S., Soma, M., Jin, H., Kimura, K., Goto, S., Koseki, H., and Aoe, T. (2008). Altered quality control in the endoplasmic reticulum causes cortical dysplasia in knock-in mice expressing a mutant BiP. *Mol Cell Biol* 28, 293-301.

Morgan, M.M., Fossum, E.N., Levine, C.S., and Ingram, S.L. (2006). Antinociceptive tolerance revealed by cumulative intracranial microinjections of morphine into the periaqueductal gray in the rat. *Pharmacol Biochem Behav* 85, 214-219.

Munro, S., and Pelham, H.R. (1987). A C-terminal signal prevents secretion of luminal ER proteins. *Cell* 48, 899-907.

Nagai, N., Hosokawa, M., Itohara, S., Adachi, E., Matsushita, T., Hosokawa, N., and Nagata, K. (2000). Embryonic lethality of molecular chaperone hsp47 knockout mice is associated with defects in collagen biosynthesis. *J Cell Biol* 150, 1499-1506.

Narula, J., Haider, N., Virmani, R., DiSalvo, T.G., Kolodgie, F.D., Hajjar, R.J., Schmidt, U., Semigran, M.J., Dec, G.W., and Khaw, B.A. (1996). Apoptosis in myocytes in end-stage heart failure. *N Engl J Med* 335, 1182-1189.

Nilsson, T., Jackson, M., and Peterson, P.A. (1989). Short cytoplasmic sequences serve as retention signals for transmembrane proteins in the endoplasmic reticulum. *Cell* 58, 707-718.

Ohse, T., Inagi, R., Tanaka, T., Ota, T., Miyata, T., Kojima, I., Ingelfinger, J.R., Ogawa, S., Fujita, T., and Nangaku, M. (2006). Albumin induces endoplasmic reticulum stress and apoptosis in renal proximal tubular cells. *Kidney Int* 70, 1447-1455.

Orci, L., Stamnes, M., Ravazzola, M., Amherdt, M., Perrelet, A., Sollner, T.H., and Rothman, J.E. (1997). Bidirectional transport by distinct populations of COPI-coated vesicles. *Cell* 90, 335-349.

Ozcan, U., Yilmaz, E., Ozcan, L., Furuhashi, M., Vaillancourt, E., Smith, R.O., Gorgun, C.Z., and Hotamisligil, G.S. (2006). Chemical chaperones reduce ER stress and restore glucose homeostasis in a mouse model of type 2 diabetes. *Science* 313, 1137-1140.

Palokangas, H., Ying, M., Vaananen, K., and Saraste, J. (1998). Retrograde transport from the pre-Golgi intermediate compartment and the Golgi complex is affected by the vacuolar H+-ATPase inhibitor bafilomycin A1. *Mol Biol Cell* 9, 3561-3578.

Parkitna, J.R., Obara, I., Wawrzczak-Bargiela, A., Makuch, W., Przewlocka, B., and Przewlocki, R. (2006). Effects of glycogen synthase kinase 3beta and cyclin-dependent kinase 5 inhibitors on morphine-induced analgesia and tolerance in rats. *J Pharmacol Exp Ther* 319, 832-839.

Patil, C., and Walter, P. (2001). Intracellular signaling from the endoplasmic reticulum to the nucleus: the unfolded protein response in yeast and mammals. *Curr Opin Cell Biol* 13, 349-355.

Reimold, A.M., Iwakoshi, N.N., Manis, J., Vallabhajosyula, P., Szomolanyi-Tsuda, E., Gravallese, E.M., Friend, D., Grusby, M.J., Alt, F., and Glimcher, L.H. (2001). Plasma cell differentiation requires the transcription factor XBP-1. *Nature* 412, 300-307.

Remuzzi, G., and Bertani, T. (1998). Pathophysiology of progressive nephropathies. *N Engl J Med* 339, 1448-1456.

Ron, D., and Walter, P. (2007). Signal integration in the endoplasmic reticulum unfolded protein response. *Nat Rev Mol Cell Biol* 8, 519-529.

Ross, J., Jr. (2002). Dilated cardiomyopathy: concepts derived from gene deficient and transgenic animal models. *Circ J* 66, 219-224.

Rothman, J.E., and Wieland, F.T. (1996). Protein sorting by transport vesicles. *Science* 272, 227-234.

Samelson, L.E. (1985). An analysis of the structure of the antigen receptor on a pigeon cytochrome c-specific T cell hybrid. *J Immunol* 134, 2529-2535.

Schekman, R., and Orci, L. (1996). Coat proteins and vesicle budding. *Science* 271, 1526-1533.

Scheuner, D., Song, B., McEwen, E., Liu, C., Laybutt, R., Gillespie, P., Saunders, T., Bonner-Weir, S., and Kaufman, R.J. (2001). Translational control is required for the unfolded protein response and in vivo glucose homeostasis. *Mol Cell* 7, 1165-1176.

Schroder, M., and Kaufman, R.J. (2005). The Mammalian unfolded protein response. *Annu Rev Biochem* 74, 739-789.

Somogyi, A.A., Barratt, D.T., and Coller, J.K. (2007). Pharmacogenetics of opioids. *Clin Pharmacol Ther* 81, 429-444.

Sonnichsen, B., Fullekrug, J., Nguyen Van, P., Diekmann, W., Robinson, D.G., and Mieskes, G. (1994). Retention and retrieval: both mechanisms cooperate to maintain calreticulin in the endoplasmic reticulum. *J Cell Sci* 107, 2705-2717.

Tajiri, S., Oyadomari, S., Yano, S., Morioka, M., Gotoh, T., Hamada, J.I., Ushio, Y., and Mori, M. (2004). Ischemia-induced neuronal cell death is mediated by the endoplasmic reticulum stress pathway involving CHOP. *Cell Death Differ* 11, 403-415.

Tissir, F., and Goffinet, A.M. (2003). Reelin and brain development. *Nat Rev Neurosci* 4, 496-505.

Townsley, F.M., Wilson, D.W., and Pelham, H.R. (1993). Mutational analysis of the human KDEL receptor: distinct structural requirements for Golgi retention, ligand binding and retrograde transport. *EMBO J* 12, 2821-2829.

Travers, K.J., Patil, C.K., Wodicka, L., Lockhart, D.J., Weissman, J.S., and Walter, P. (2000). Functional and genomic analyses reveal an essential coordination between the unfolded protein response and ER-associated degradation. *Cell* 101, 249-258.

Trujillo, K.A., and Akil, H. (1991). Inhibition of morphine tolerance and dependence by the NMDA receptor antagonist MK-801. *Science* 251, 85-87.

Vashist, S., Kim, W., Belden, W.J., Spear, E.D., Barlowe, C., and Ng, D.T. (2001). Distinct retrieval and retention mechanisms are required for the quality control of endoplasmic reticulum protein folding. *J Cell Biol* 155, 355-368.

Xie, Q., Khaoustov, V.I., Chung, C.C., Sohn, J., Krishnan, B., Lewis, D.E., and Yoffe, B. (2002). Effect of tauroursodeoxycholic acid on endoplasmic reticulum stress-induced caspase-12 activation. *Hepatology* 36, 592-601.

Yaksh, T.L., Yeung, J.C., and Rudy, T.A. (1976). Systematic examination in the rat of brain sites sensitive to the direct application of morphine: observation of differential effects within the periaqueductal gray. *Brain Res* 114, 83-103.

Yamamoto, K., Fujii, R., Toyofuku, Y., Saito, T., Koseki, H., Hsu, V.W., and Aoe, T. (2001). The KDEL receptor mediates a retrieval mechanism that contributes to quality control at the endoplasmic reticulum. *EMBO J* 20, 3082-3091.

Yamamoto, K., Hamada, H., Shinkai, H., Kohno, Y., Koseki, H., and Aoe, T. (2003). The KDEL receptor modulates the endoplasmic reticulum stress response through mitogen-activated protein kinase signaling cascades. *J Biol Chem* 278, 34525-34532.

Yuasa, S., Kitoh, J., Oda, S., and Kawamura, K. (1993). Obstructed migration of Purkinje cells in the developing cerebellum of the reeler mutant mouse. *Anat Embryol (Berl)* 188, 317-329.

Zhao, L., and Ackerman, S.L. (2006). Endoplasmic reticulum stress in health and disease. *Curr Opin Cell Biol* 18, 444-452.

Zhao, L., Longo-Guess, C., Harris, B.S., Lee, J.W., and Ackerman, S.L. (2005). Protein accumulation and neurodegeneration in the woozy mutant mouse is caused by disruption of SIL1, a cochaperone of BiP. *Nat Genet* 37, 974-979.

Zinszner, H., Kuroda, M., Wang, X., Batchvarova, N., Lightfoot, R.T., Remotti, H., Stevens, J.L., and Ron, D. (1998). CHOP is implicated in programmed cell death in response to impaired function of the endoplasmic reticulum. *Genes Dev* 12, 982-995.

Zollner, C., Mousa, S.A., Fischer, O., Rittner, H.L., Shaqura, M., Brack, A., Shakibaei, M., Binder, W., Urban, F., Stein, C., *et al.* (2008). Chronic morphine use does not induce peripheral tolerance in a rat model of inflammatory pain. *J Clin Invest* 118, 1065-1073.

9

Quality Assurance in the Preanalytical Phase

M. Antonia LLopis, Virtudes Alvarez, Cecilia Martínez-Brú,
Rubén Gómez, Núria Barba, Mercè Ibarz, Mariano Cortés,
Montserrat Ventura and M. Jesús Alsina
The Spanish Society of Clinical Chemistry Committee for the extra-analytical quality assessment
Spain

1. Introduction

The preanalytical phase includes a set of processes that are difficult to define because they take place in different places and at different times. Classically, the preanalytical phase included all processes from the time a laboratory request is made by a physician until the sample is ready for testing. While this definition is very informative, errors that occur at this stage often become apparent later in the analytical and post-analytical phases. For example, interference effects for a given sample can be detected during the laboratory analysis or during clinical interpretation. Therefore, current recommendations call for the laboratory error to be defined as the defect occurring at any point in the cycle, from request to interpretation of results by a clinician.

The main processes that should be taken into account in the study of the preanalytical phase are: test selection; patient preparation; collection, transport, handling and preservation of the sample; and interferences. The study of the characteristics of individual patients and the biological variation for each laboratory test belong to this phase, too.

It has been documented that laboratory errors have significantly decreased in the last four decades, especially those that occur during the analytical phase; furthermore the scientific evidence reveals that most laboratory errors occur in the preanalytical phase. The large majority of laboratory errors occur in poorly standardized or manual processes. Preanalytical mistakes account for up to 70 % of total laboratory errors. The magnitude of the effect of these errors on patient care is not negligible, since information provided by clinical laboratories affects up to 60-70% of clinical decisions. The current focus of health care institutions on improving patient safety has given rise to a renewed interest in the preanalytical phase. Improvement of preanalytical processes currently constitutes a challenge to be faced by clinical laboratories.

While quality control systems designed to ensure the quality of the analytical phase are highly developed and in use at most clinical laboratories, this is not the case for the preanalytical phase. One reason may be that laboratory professionals have always considered the analytical phase (but not the preanalytical phase) to be the most important process in their profession. Outsourcing of the sampling process could be another cause. Both of these factors have resulted in a decline in quality, as evidenced by an increase in pre-analytical errors, and in turn, these errors have required that quality control systems be established for the process, such as registration and notification of the errors detected at the collection sites and sampling.

One factor that may also explain the increasing interest in quality control of the preanalytical phase is the ISO 9001: 2008 or ISO 15189: 2007 certification. These standards directly affect the need to define all laboratory processes, including preanalytical ones, and in establishing quality indicators for each process.

Study of the preanalytical phase is of emerging interest as evidenced by an increase in the number of publications in recent years. This chapter describes the most important variables involved in each of the processes of this phase, as well as those quality control mechanisms that should be established in order to minimize laboratory errors and improve patient safety.

2. Analytical test request

The most important variables requiring quality control during the test requisition process are identification of the patient and physician/clinical unit, and the tests requested. The ISO 15189 standard specifies the information that must be provided on the request form, regardless of whether the test requisition is made on paper or electronically:

a. unique identification of the patient;
b. name or other unique identifier of physician or other legally-authorized person;
c. type of primary sample;
d. tests requested;
e. relevant clinical information about the patient necessary for interpretation purposes; at a minimum, this should include gender and date of birth;
f. date and time of primary sample collection;

Failure to identify the patient, or incorrect identification of the patient could have serious consequences in the clinical decision-making process and may affect patient safety, and for this reason it should be considered a key indicator in the process.

A study published in 1993 found that up to 6.5% of preanalytical errors were due to incorrect patient identification prior to specimen collection (Renner et al., 1993). In 2002, the same authors published the results of a second study showing a large improvement—attributed mainly to participation in external quality assessment programs—in this indicator (Howanitz et al., 2002).

The most important identification errors that occur during the laboratory test request process may be due to patient identification errors (patient misidentification, errors in interpreting demographic data, or in the coincidence of having two patients in the waiting room with the same name and surname) and errors made by the physician when filling out the request form (requesting a test for a patient other than the one currently in the doctor's office). Patient identification errors can also occur during scheduling for sample extraction (Figure 1).

Failure to identify or misidentification of the physician or requesting unit may make it impossible to return the test results or the test report may be sent to the wrong physician, leading to complaints, and necessitating that a copy of the original report be sent. The implementation of computerized medical records linked to the Laboratory Information Management System (LIMS) minimizes these types of errors because the data are sent electronically to the LIMS .

Ordering inappropriate tests is another preanalytical variable that negatively impacts patient safety. Estimates of inappropriate test requisition vary considerably and range from 4.5% to 95% (Van Walraven & Naylor , 1998). Some factors that contribute to inappropriate laboratory utilization may include:

- Unnecessary repeat testing: it has been estimated that up to 30% of laboratory tests ordered each month are repeat tests (Van Walraven & Raymond, 2003)
- Overconfidence in laboratory test results
- Requests for multiple tests that are basically equivalent in their disease-detection capacity
- Quick and easy access to laboratory directory as a result of the use of new technologies.
- The addition of new tests without studies proving their clinical utility.
- Use of obsolete tests that have been replaced by tests with a greater diagnostic efficacy.
- Industry pressure.
- Little training of clinicians in understanding the diagnostic utility of laboratory tests
- Increase in patient knowledge of health issues thanks to the use of new information technologies. A survey carried out in Europe shows an important increase of internet users had used the internet at some point to look for information about health-related issues (36.3% in 2002 to 50.6% in 2005) (SIBIS Project:2002).

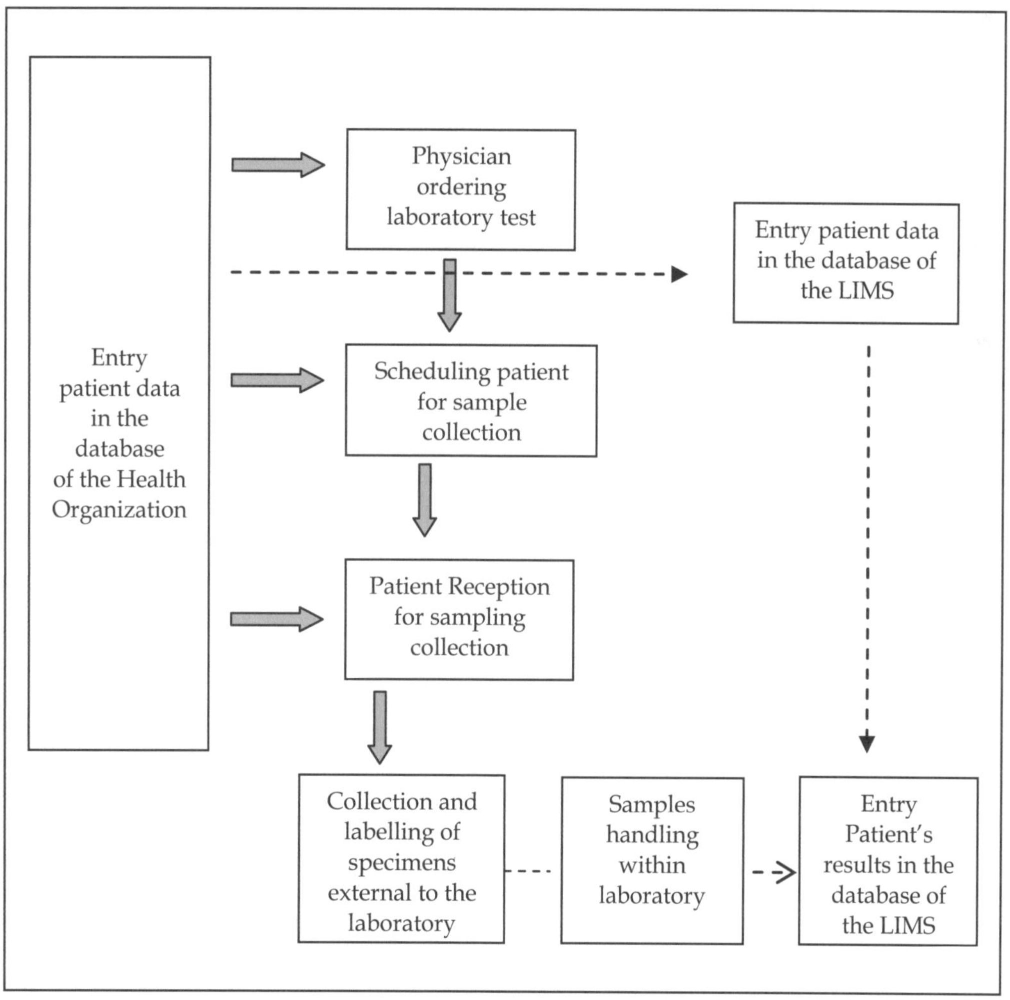

Fig. 1. Points of misidentifications patient data in laboratory diagnostic

Indiscriminate test requisition increases diagnostic uncertainty for the physician, discomfort for the patient, and laboratory expenses. Laboratory utilization can be improved through a variety of strategies:

- Development of test request protocols specific to each pathology, using recommendations from evidence-based clinical practice guidelines and based on mutual consensus between laboratories and clinicians. Despite efforts made in recent years to implement the use of these guidelines, some studies have found clinician compliance to be scant (Salvagno et al., 2007)
- Establishing the use of diagnostic algorithms (reflex testing) in laboratories depending on the disease or the patient's health status.
- The use of expert systems that serve as a connection between the clinician and the laboratory and that act as an aid in appropriate test selection and interpretation. The implementation of computerized medical records connected to the LIMS is quite useful in helping the physician to select the most relevant analytical tests. For example, attaching clinical practice guidelines to medical records improves the relevance of the requested tests.
- Information provided by laboratory specialists to clinicians about diagnostic utility of the laboratory tests, removing obsolete tests from the laboratory's menu of available tests and also those tests that have the same diagnostic value.

Another factor to consider in the test request process is the supplementary information provided by the clinician on the test request form sent to the laboratory: that is, the *individual characteristics of the patient*, such as age, gender, race, physiological conditions such as pregnancy or menopause, dietary and toxic habits, physical exercise, medications, and suspected diagnosis. These data are necessary to assign the correct reference values and avoid unnecessary test repetition in case of incongruent results that cannot be evaluated due to lack of information.

Laboratory professionals, for their part, should provide clinicians with data on the magnitude of biological variations, which may be quite useful in both preanalytical (test selection) and postanalytical (interpretation of test results) processes. It is important that clinicians understand the concept of biological variation, which includes intra-individual biological variation (fluctuation of results around a homeostatic set point of an individual) and inter-individual biological variation (variation between different individuals around a homeostatic set point). An understanding of biological variation plays an important role in selecting the most appropriate test in cases where 2 or more tests with the same diagnostic value are available, since it is always advisable to choose the test with the smallest degree of biological variation. This knowledge is also quite useful in interpreting test results, especially for follow-up; if the range of intra-individual biological variation is very small (tight range of homeostatic equilibrium), such as it happens with calcium, a minimal difference between two consecutive results might be significant; on the other hand, if the magnitude of intra-individual biological variation is large, as it occurs with amylase, the difference between two consecutive results must be much greater to reach the same level of significance.

Most laboratory test results have a high individuality, or to put it another way, a very small intra-individual biological variation compared to the inter-individual biological variation. In these cases, the use of reference intervals is of little use and therefore results should be compared with previous results in the same patient because this allows for significant changes to be detected even when the result falls within the reference interval for the population.

The advent of computerized test requesting has reduced some of the errors that occurred when requests were paper-based, as noted in a study that evaluated the evolution of quality indicators of key laboratory processes over a 5-year period; in that study, the error rate for test requests fell from 4.1% to 3.3% (Alsina et al., 2007; Llopis et al., 2010). However, due to the ease of access (i.e. electronic) to the laboratories' catalogue of services, the number of tests ordered has increased substantially, and many of these tests are not justified by the suspected diagnosis, a fact that will oblige laboratories to establish new controls over this process.

The role of the laboratory in the test requisition process should be more interactive: not only should the laboratory be involved in defining the catalogue of available tests, designing the request form, and implementing protocols based on clinical practice guidelines, but also in negotiating with clinicians to reach a consensus on these variables. Reliable indicators for test requisition quality will need to be established in order to monitor this process.

3. Patient preparation prior to sample collection

The test requisition process involves many variables, which is why it is quite complex; this complexity increases further when another variable is added, such as patient preparation to the process, as the number of variables multiplies. The difficulty resides in determining how to generate and transmit this information to the patient, and especially how to assure compliance given that the laboratory is only indirectly involved in this process. The ISO 15189 standard obliges the laboratory to have a preanalytical procedure manual that provides clear directions on the instructions given to patients prior to specimen collection. Some of the variables that patients must monitor are, among others, fasting, diet, physical exercise, stress, wakefulness, and medication. With the computerized test requisition, instructions given to the patient prior to sample collection can be individualized so that the information provided is specific to the tests ordered. Due to the lack of reliable indicators, currently clinical laboratories cannot guarantee the quality of this process; the only control that they have over the entire process is the patient's word regarding compliance. In addition, in the case of laboratories with decentralized collection centers, the laboratory does not know if all off-site specimen collection centers routinely ask patients to confirm protocol compliance. The laboratory should standardize this information in the form of a checklist and train staff to use it properly; an electronic incident reporting system should also be established. **Figure 2** shows an example of preparation instructions given to patients prior to sample extraction for prolactin measurement.

1. IMPORTANT: In order for this test to be carried out, you need to be awake 2 hours before the blood is taken, but you must not do any exercise or exert yourself during this time
2. The day before the test you should avoid food which is rich in proteins
3. The day before the test you should avoid any breast stimulation
4. You must not eat or drink anything during the 8 to 10 hours before you have the sample taken. You may drink water
5. Go to the location where you are to have a blood sample taken, at the time stated
6. If you are taking any medication, please tell the person who is taking the sample

Fig. 2. Patient preparation for serum Prolactin measurements

4. Specimen sampling and collection

It is important to differentiate between specimen extraction and collection performed by health care staff versus sample collection performed by the patient. When health care professionals are involved the process is easier to control, as these individuals are easily identifiable and accessible through the information provided to the laboratory. The ISO 15189 standard indicates that the specific instructions for extracting and handling samples should be documented and implemented by laboratory management and readily available to the sampling supervisors. The manual for primary sample collection must be part of the quality control documentation. This manual must contain instructions for:

- positive identification of the patient as well as identification, labeling and traceability between the patient, requisition, and primary sample.
- procedures for collecting primary samples, a description of the containers used, necessary additives, type and volume of the primary sample that is to be obtained, the precise time of day that sampling is to be performed, as well as patient position and time of venous occlusion when drawing a venous blood sample. The laboratory must assure that the materials used to perform extractions and collect samples are of adequate quality.
- The identity of the person who collects the primary sample and the date and hour of sample collection must be recorded.
- Safe disposal of the materials used to obtain the samples.

Primary samples that are not adequately identified should not be accepted or processed by the laboratory. When identification of the primary sample is in doubt or if instability in the sample components (cerebrospinal fluid, biopsy) is present, and the primary sample is irreplaceable or critical, the laboratory can opt to process the sample but should withhold the results until the requesting physician or the person in charge of collecting the sample takes responsibility for identifying and accepting the sample; in such cases, this person must sign the request form or the name should be included in the traceability data. If this requirement is not met, the person responsible should be identified on the laboratory report, assuming the analysis is performed.

Patient or sample identification errors could have serious consequences affecting clinical decision-making and patient safety; for this reason, these are considered key indicators in the process. Misidentification of the patient or sample can occur at the time of sample collection (the patient from whom the sample is taken does not match the patient listed on the test request form) and also when labeling the sample (a specimen collected from one patient could be mistakenly given a reference number belonging to a different patient).

The Laboratory Accreditation Program of the College of American Pathologists (CAP, 2006) and the Joint Commission on Accreditation of Healthcare Organization (JCAHO, 2007) recommend that all samples sent to laboratories be positively identified during the sample collection process, preferably by two different identifiers. When any doubt exists as to the identity of a specimen, a new specimen must be requested; if this is not possible, the laboratory result should not be disclosed.

Unsuitable specimens for testing

Most preanalytical errors occur during the sampling process: up to 60% of these errors are attributable to the sample (Lippi et al., 2006a). A retrospective analysis (2001-2005) of results obtained through the Spanish Society of Clinical Chemistry (SEQC) Quality Assessment Program for the preanalytical phase found that the most common preanalytical error was "sample not received", followed by "hemolyzed samples" (Alsina et al., 2008). However, now that electronic test requisition requires specific information to be included with each request, including the number and type of containers needed, along with the use of robots to automate this process, the number of not received samples is decreasing. The external quality control programs for preanalytical quality organized by the CAP have found that hemolyzed samples are the most commonly observed errors (Jones et al., 1997).

The main causes of hemolysis during the sample collection process are as follows (Rioja et al, 2009):

- Catheter use; this is one of the factors mentioned in the literature as a cause for the high incidence of hemolysis in hospital units.
- Needle gauge size; a decrease in the needle gauge produces an increase in flow leading to the friction that causes hemolysis. The degree of hemolysis is inversely proportional to the diameter of the needle
- Needle venipuncture and transfer to vacuum tube. Hemolysis is caused by friction in the needle due to excessive pressure in the syringe plunger during extraction or transfer and due to the presence of leaks or poor connections, which leads to turbulent flow
- Venipuncture site. The antecubital fossa is the localization with the lowest incidence of hemolysis. Venipuncture of the hand or forearm increases the incidence rate of hemolysis.
- Antiseptic. Alcohol used as a disinfectant can cause hemolysis.
- Tourniquet duration: applications lasting longer than 1 minute are associated with a greater risk of hemolysis (Saleem et al., 2007).
- Traumatic venipuncture. Venipuncture through hematomas can become contaminated by the hemoglobin released by the tissues. Venipuncture complications, including cannulation difficulties or tearing of the vein, both of which degrade the sample.
- Capillary puncture. This always involves trauma, especially if the area is massaged to produce bleeding. The use of automated capillary puncture devices reduces the degree of hemolysis.
- Type of tube: Larger volume evacuated tubes produce more hemolysis.
- Under-filled vacuum tube. Underfilled vacuum tubes are more likely to undergo hemolysis, especially during transport or centrifugation.
- Excessive or insufficient mixing of the blood and additive (anticoagulant or procoagulant).
- Experience of the personnel who perform the venipuncture.
- Workload. A recent study found that the degree of hemolysis is inversely proportional to the number of extractions performed (Hawkins, 2010)

When analyzing hemolyzed samples, the laboratory must always keep in mind that 3% of in vitro specimens will be hemolytic and this is beyond the laboratory's control (Carraro et al. , 2000).

The laboratory must control the process by rejecting all hemolytic specimens that might lead to an incorrect result. Given the high incidence of hemolyzed samples and to facilitate handling of these, in recent years the industry has developed automated systems to detect and quantify the degree of hemolysis.

Sample volume

Laboratories must document and periodically review their requirements regarding sample volume needed to guarantee that neither insufficient nor excessive quantities are extracted.

Estimates have been proposed for the minimum volume necessary for analysis, dead volume required in the analyzers, repetitions, reflex testing, and serum bank (Lippi et al., 2007).

In general, the sample volume collected from patients is insufficiently optimized. A CAP publication on external quality assessment program of sample volumes concluded that tube size has a greater impact on specimen volumes extracted by the laboratory than does the requirements of the autoanalyzer; moreover, this excessive volume, together with the ordering of unnecessary tests, could lead to iatrogenic anemia (Dale & Rudy, 2003).

5. Sample transportation

Transportation of samples from the sampling center to the clinical laboratory must meet certain requirements to assure sample stability. The 2009 ADR European norms for highway transport define the packaging requirements for transport of biological samples, which are considered infectious material (Category B). Likewise, the ISO 15189 standard states that the laboratory must monitor the transport of samples to the laboratory so that they arrive within an appropriate time frame and temperature in a manner that safety of all persons involved in transportation is ensured, and in accordance with all national and international regulations.

Neither of these two standards provides specific limits of acceptability regarding transport time or temperature. However, the guidelines published by the National Committee for Clinical Laboratory Standards (NCCLS) H5-A3 in 1994 recommend a maximum of 2 hours for transport of blood samples at a temperature range of 10-22° C. The 2009 GP16-A3 guidelines, also by the NCCLS recommend a maximum of 2 hours and a temperature range of 2-8° C for transport of urine samples.

The main variables to consider during transportation are agitation, light exposure, temperature, transport time, placement of samples within the transport container, type of packaging, and labeling.

Laboratories should document a procedure that specifies the conditions and requirements that must be met during transport in order to properly preserve the samples; in addition, rejection criteria to be used when these conditions are not met should also be defined.

At a minimum, the laboratory should implement the following control mechanisms to assure the quality of this process: control of transport temperature, packaging conditions, and time elapsed from extraction to arrival at the laboratory. Currently, quality control of sample transportation is hindered by a failure to clearly define the limits of sample stability and to register temperature fluctuation during transportation. No indicators are currently available to detect if samples have received excessive light exposure or agitation during transport.

The use of pneumatic tubes to transport samples is somewhat controversial, although most studies have found no significant changes in analytical results. Serum is the sample type most susceptible to hemolysis during transport (Gomez et al., 2009).

6. Receiving, handling, and preservation of specimens in the laboratory

The ISO 15189 standards for this preanalytical laboratory process are as follows:

- The primary samples received shall be recorded in an accession book, worksheet, computer or other comparable system.

- The date and time of receipt of samples, as well as the identity of the receiving officer, shall be recorded.
- Criteria shall be developed and documented for acceptance or rejection of primary samples at time of reception. If compromised primary samples are accepted, the final report must indicate the nature of the problem and, if applicable, that caution is required when interpreting the result.
- Always, samples received should be checked.
- If necessary, laboratories should have a documented procedure for the receipt and labeling of primary stat samples.
- Sample aliquots must be traceable to the original primary sample.
- Samples must be stored for a specified period under conditions that ensure the stability of the sample's properties in order to allow test repetition after the initial results have been issued or to carry out additional tests.

6.1 Receipt of samples

In this phase, the most important variables for quality control are verification of sample delivery by means of an automatic or manual registration system and verification of the condition of the sample on arrival, to detect and reject samples that are unsuitable for the requested analysis. One key quality indicator in this preanalytical process is the error rate for received samples (sample not sent, coagulated sample, hemolyzed sample, insufficient sample, etc).

A recent review article by G. Lippi et al. (Lippi et al., 2007) provides a series of recommendations to promote, standardize, and harmonize the detection and handling of rejected samples; the general recommendations include staff training, establishment of standardized systems to detect improper samples, and implementation of procedures to detect such samples. Of the specific recommendations given, those of greatest interest are related to the four most common situations in clinical laboratories: how to manage samples with identification errors, samples affected by interferences, coagulated samples, and samples with inadequate volume.

6.2 Sample handling

Primary samples cannot always be used directly in the analytical phase but sometimes need to undergo preparation such as centrifugation, aliquoting, or freezing prior to laboratory analysis. Control of all these processes is also fundamental, since errors affecting patient safety can occur, especially during aliquoting and manual identification of each aliquot. The ISO 15189 standard requires that each aliquot be traceable to the original sample. The Joint Commission on Accreditation of Healthcare Organization (JACHO, 2007) and the CAP (CAP, 2006) both advise against manual labeling by laboratory staff.

In recent years, many of the preanalytical processes performed in laboratories have been automated, leading to a considerable reduction in error rates (Da Rin , 2009).

6.3 Sample preservation

When samples are delivered to the laboratory, appropriate conditions must be maintained regarding preservation duration, temperature, and light exposure. The time during which a sample can be preserved will depend on its stability, defined as the capacity to maintain the initial value within specific limits for a defined time period. Multiple studies on specimen stability have been published, some of which have found conflicting results; these discrepancies may be due to the application of different study protocols as well as differing formulas used to calculate stability, and different ways of defining acceptable limits. The SEQC has published two documents to standardize stability studies performed in Spain: one of these proposed a standardized protocol for evaluation of biological stability (Alsina & Gonzalez, 2006a) and the other proposed stability limits derived from biological variation (Alsina & Gonzalez, 2006b)

The laboratory must clearly define stability time and preservation temperatures, and must also regularly monitor refrigerator and freezer temperatures where samples are stored, and also ensure that samples that have been analyzed are not stored longer than the stability limits dictate. Diligent monitoring of sample storage temperatures reduces preanalytical errors and may also prolong the stability time of some samples (Stahl & Brandslund, 2005). The SEQC performed a literature search and compiled the stability limits for several different biological specimens .The CLSI has published a series of guidelines with recommendations for collection, handling, transport, and storage of samples (**Table 1**).

H3-A6: Procedures for the Collection of Diagnostic Blood Specimens by Venipuncture; Approved Standard – Sixth Edition. October 2007
M29-A3: Protection of Laboratory Workers from Occupationally Acquired Infections; Approved Guideline - Third Edition (Vol.25, No.10)March 2005
H04-A6: Procedures and Devices for the Collection of Diagnostic Capillary Blood Specimens; Approved Standard – Sixth Edition (Vol.28, No.25) September 2008
C49-A: Analysis of Body Fluids in Clinical Chemistry; Approved Guideline (Vol.27,No.14)April 2007
H56-A: Body Fluid Analysis for Cellular Composition; Approved Guideline (Vol.26,No.26)June 2006
H18-A4: Procedures for the Handling and Processing of Blood Specimens for Common Laboratory Tests; Approved Guideline -- Fourth Edition (Vol.30, No.10) May 2010
GP16-A3: Urinalysis; Approved Guideline – Third Edition (Vol.29, No.4) February 2009
H21-A5: Collection, Transport, and Processing of Blood Specimens for Testing Plasma-Based Coagulation Assays and Molecular Hemostasis Assays; Approved Guideline - Fifth Edition (Vol.28, No.5) January 2008

Table 1. CLSI Recommendations and Guidelines related with preanalytical phase

7. Presence of interferences in the samples

Interferences in samples can be evidenced both in the preanalytical process (visual observation of hemolysis, lipemia, bilirubin), as well as in the analytical (quantification of hemolysis, lipemia, and bilirubin indices) and postanalytical phases (aberrant or unexpected results found during the validation process). However, interferences in the sample should be evaluated during the preanalytical phase since this problem is intrinsic to the sample. The ISO 15189 standard states that the laboratory shall document interferences and cross reactions for each analytical procedure. Other authors (Lippi et al., 2007) also recommend documenting: a) procedures for identifying interferences for each type of specimen, b) the analytical tests affected by the interfering substance, c) the concentration levels at which the interfering substance begins to affect the assay and, d) the laboratory's course of action when an interference is detected.

8. Quality assurance for preanalytical processes

When analyzing and comparing the quality control systems used for preanalytical processes to those used in the analytical phase, some fundamental differences become apparent:

1. Analytical quality control systems are much more developed because they were established in the 1950s whereas preanalytical controls were not established until the 1990s.
2. In the analytical phase, processes take place within the laboratory and involve few people; consequently, the variables that need to be monitored are limited and well-defined. The preanalytical phase, in contrast, involves many processes, most of them external to the laboratory; in addition, these processes are quite varied and involve many different people (patients, clinicians, nurses, shippers, administrative personnel, and laboratory staff); as a result, these multiple variables, some difficult to define, must be monitored and managed by each laboratory.
3. The material used to assess quality control in the analytical phase can be assessed much in the same way as patient specimens are evaluated, however this is not the case for preanalytical processes, in which the only option is to establish an error registration system and then calculate the quality indicators.
4. Indicators of quality for analytical processes are expressed as the percent variation for imprecision, systematic error and total error. In contrast, preanalytical quality indicators are expressed as percentages of laboratory activities (including such aspects as number of patients, test requests, quantities, samples, containers, etc), all of which make comparison between laboratories difficult.
5. While quality specifications for analytical processes have been extensively studied and defined by international consensus (Kenny et al., 1999), the same cannot be said for preanalytical processes, as published literature is still incipient and no consensus has yet been reached.

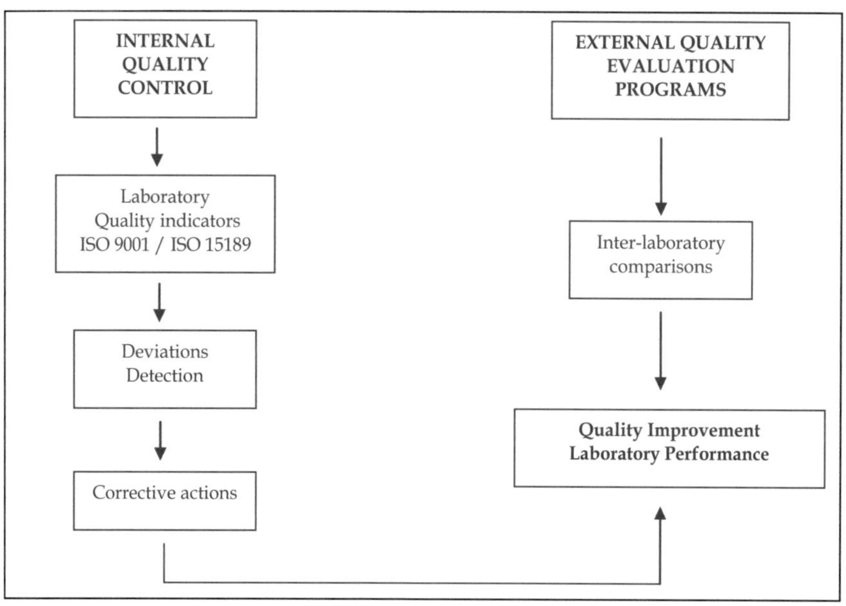

Fig. 3. Quality assurance in preanalytical phase

To attain the same level of quality control that has been developed for the analytical phase, an internal control system for preanalytical processes is needed and also laboratories must participate in external quality assessment programs for these processes **(Figure 3).**

8.1 Internal controls

Internal control is based on registering incidents and then calculating the relevant quality indicator for each preanalytical process. The UNE 66175 guide for implementation of indicator systems defines the quality indicator as the data or group of data that help in objectively measuring the evolution of a process or activity. Quality indicators should be designed in such a way as to permit the early detection of deviations from the norm, they must provide information that can be used to continuously improve these processes, and the workload required of the organization to calculate these indicators should be reasonable. The indicators should be updated regularly to adapt to changes in the processes being monitored; in this way, resources are not wasted on indicators for processes that have proven to be stable, but can be focused on processes requiring closer monitoring (CLSI Guideline GP35-P, 2009) .

The recommendations given by the Agency for Healthcare Research and Quality and the CLSI can be used for indicator classification. These organizations evaluate laboratory error incidence rates for quality management, including patient safety (Kenny et al., 1999).

Table 2 (Alsina et al., 2007) describes some of the quality indicators currently used for preanalytical processes, as well as the formulas used to calculate them. As shown in table, indicators have not been defined for all preanalytical processes, such as demand for testing, patient preparation, or specimen transport. As it has been said before, the main difficulty in defining and establishing indicators for preanalytical processes lies in the extent of their area of application, which includes aspects involving clinicians as well as patients.

All laboratories must establish quality indicators for the processes that they have the ability to control and monitor over time. It is also important to define the **quality specifications** or limits of acceptability for each indicator; when the results fall outside of these limits, corrective measures should be taken. No international consensus exists as to what the limits of acceptability for preanalytical indicators should be, but recommendations are available for some of these. According to the hierarchical model of analytical quality specifications approved by international consensus in 1999 (**Figure 4**) (Kenny et al., 1999) , acceptability limits can be defined by using the results obtained from the top 50% of laboratories in external quality assessment schemes or can be based on current state-of-the art standards. One example of this would be to use the overall rejection rate obtained by all participants in the SEQC external evaluation program for preanalytical quality over the 2001-2005 period (Alsina et al., 2006c ; Alsina et al., 2008), for the following two indicators:

- **Rejected blood samples divided by number of samples: 0.69%**
- **Rejected urine samples divided by number of samples: 2.15%**

Or alternatively, we could use those published in 2002 by the College of American Pathologists (Dale & Ruby, 2003).

- **Rejected blood samples per number of extractions: 0.30%**

The specifications based on state of the art knowledge could be obtained from these sources:

- **Published literature**
- **Results from similar laboratories**
- **Prior data from the same laboratory**

- Evaluation of the effect of analytical performance on clinical outcomes in specific clinical settings
- Evaluation of the effect of analytical performance on clinical decisions in general
- Published professional recommendations
- Performance goals set by
- Goals based on the current state of the art

Fig. 4. Analytical Quality Specifications. Consensus Statement (Stockholm 1999). Scan J Clin Lab Invest 1999; 59: 585

Indicator	Formula
Total incidences in test requests	100 * n° of requests with incidents / n° of requests
Patient data incorrect	100 * n° of requests with incorrect data / n° of requests
Patient data missing	100 * n° of request with missing data / n° of requests
Samples not received	100 * n° of each type of sample not received / n° of requests
Clotted sample	100 * n° number of clotted samples for each sample type / n° of requests
Insufficient sample	100 * n° of insufficient samples for each sample type / n° of requests
Hemolyzed serum sample	100 * n° of hemolyzed samples / n° of requests
Undetected requests with incorrect patient name	100 * n° of undetected requests with incorrect patient name / total n° of requests with incorrect name
Errors in sample management	100 * n° sample management errors / n° of samples
External control exceeds acceptance limits	100 * n° constituents exceeding 2 SD of the group using the same method / n° data evaluated
In-house laboratory reports exceed delivery time	100 * n° in-house laboratory reports exceeding delivery time / n° reports delivered
Reports from referred tests exceed delivery time	100 * n° referred test reports exceeding delivery time / n° referred tests
Copies of reports	100 * n° copies sent / n° reports

Table 2. Formulas used to calculate the indicators

One study of quality indicators for the various preanalytical processes obtained quality specifications by calculating the median result from all participating laboratories. This research was carried out by a group of laboratories whose aim was to obtain state of the art quality specifications for indicators that measure both internal and external laboratory processes for the preanalytical phase (Alsina et al., 2007).

Any laboratory that finds that its results for these indicators do not meet the acceptability limits should take measures to improve the processes. Sometimes, an individual laboratory may encounter practical difficulties applying specifications that have been published by other laboratories or groups of laboratories, a situation that can occur if there are large differences in how errors are defined or quantified or due to inter-laboratory variations in the methods used to measure the variables. In these cases, if the criteria cannot be unified, the best approach is for each laboratory to compare its own results over time. The results of this study are similar to those reported by similar studies.

- Thus, for the overall indicator of the number of erroneous samples divided by total requests, the result in this group was 5% versus 4.6% reported by Romero and colleagues (Romero et al., 2005).
- The indicator of incorrect samples per number of samples is 2.3(%), (Romero et al., 2005) similar to the 2% rate reported by Plebani (Plebani, 1997).
- The indicators for sample not received (2.9%), hemolyzed sample (0.8%), and coagulated sample (0.55%) are also similar to the results reported by Lippi et al (Lippi et al., 2006b).

Therefore, we can deduce that the quality specifications proposed by Alsina (Alsina et al., 2007) reflect current state of the art and support the recommended acceptability limits.

The same group of laboratories carried out a follow-up study (LLopis et al. , 2010) in which quality specifications for the preanalytical indicators were assessed annually over a 5-year period following their initial implementation in order to evaluate changes (Alsina et al., 2007) The authors proposed converting the specification value to the six-sigma scale in order to establish the degree of control over the process, with processes having a sigma greater than or equal to 4 considered well-controlled (Westgard, 2006). Most of the indicator values remained stable over the 5-year study period and so the mean of the inter-laboratory medians—which proved to be a robust value over the course of the study period—was recommended as a desirable specification. However, for indicators that showed a clear improvement over this period, the authors recommend that the specification be defined as the current median because changes introduced in the processes show that more demanding specifications are achievable. The research group believes that the proposed specification for processes with a sigma of 4 or less should be considered provisional because such processes need to be improved. In addition, for processes that have an important impact on patient safety (sentinel event indicator), the group recommends a specification of zero (0%). Though this value may seem unattainable, all necessary resources should be dedicated to eliminating these types of incidents, example of which include test requisitions containing incorrect patient demographic information or poorly labeled samples.

The proposed quality specifications for some of these indicators are described below:

Indicators for preanalytical processes external to the laboratory:

- *Total errors in analytical requests* (relation to total requests received): a provisional specification of 3.4% (sigma 3.4) is proposed.
- *Requests with incorrect patient demographic data*: even though the error rate in the study was only 0.11% (sigma 4.6), a specification of 0% is recommended given the importance of this indicator (sentinel event).

- *Sample not received:* the proposed specification is 0.5% of all samples. Early morning urine is the sample with the highest incidence rate, with an average median of 1.7% and a sigma of 3.7, which indicates that this process needs improvement. The other samples (serum, EDTA, citrate, 24 hours urine) have median values below 0.5% and sigma metrics ranging from 4.1 to 5.1, which reflects, in general, a process that is well under control.
- *Coagulated samples and insufficient samples*: a value of 0.1% is proposed as the specification for both indicators because the sigma ranges from 4.3 to 5.0.
- *Hemolyzed samples:* this indicator largely depends on the method used to detect hemolysis (visual vs. spectrometric), and the recommended provisional specification is 0.6% (equivalent to sigma of 4.1).

Among those indicators that indicate sample quality, the most significant are "hemolyzed sample" and "sample not submitted", although the use of computerized test requisition that includes instructions for sample pick-up will improve this indicator. The existence of various methods for detecting hemolyzed samples together with the use of varying sample rejection criteria makes it difficult to obtain robust state of the art specifications for this indicator (Gomez et al., 2009; Lippi et al., 2009).

Indicators for preanalytical processes within the laboratory:

- *Failure to detect patients with incorrect demographic data.* Even though the error rate for this indicator was quite low (0.11%; sigma 4.6), the recommended specification is 0% because this is an indicator for a potential sentinel event.
- *Errors during sample handling.* The proposed specification for this indicator is 0.03%, which is equivalent to a sigma of 5.0.

These results tell us that once the most appropriate indicators have been established for each preanalytical process and quality specifications for these defined, it will be important to calculate the indicators and evaluate their efficacy on a regular basis. In this way, improvements can be made in any process that fails to meet the defined acceptability limits.

External quality evaluation programs for preanalytical processes

External assessment programs based on inter-laboratory comparison are used to evaluate each participating laboratory in comparison with the others. Statistical analysis is used to evaluate the data provided by participating laboratories, such as the number of errors that occur in the preanalytical phase, and the results of the analysis are then sent to each participating laboratory. As a result, each laboratory can view its own results and, if necessary, take actions to improve processes found to be below standard.

These external review programs also typically provide recommendations on how processes can be improved, as well as they inform the laboratory as to the most appropriate indicators to use for each process. They can also help to establish quality specifications.

One of the greatest difficulties in statistical analysis of the data in external evaluation programs is a lack of standardization in the data collection process. Data registration in many laboratories is performed manually rather than by computer, meaning that this activity is personnel-dependent. Indicator calculation has not been standardized because there is no consensus on how the error rates should be calculated (with respect to number of patients, number of samples, number of requests, number of extractions, etc.). Other additional difficulties are that many laboratories lack experience with external review programs and, moreover, relatively few external assessment programs for the preanalytical phase are available, and even those that are available only evaluate a limited number of processes.

In 1989, the College of American Pathologists (CAP) became the first organization to carry out these types of review programs in the United States (Q Probes), and the CAP publishes its findings on a regular basis. Each study evaluates several different variables both in the preanalytical phase (identification errors in the laboratory, specimen collection volumes, accuracy of send-out-test ordering), and in the postanalytical phase (clinician and nursing satisfaction surveys, notification of critical values), and each year the variables studied are changed (Dale & Novis, 2002; Valenstein et al., 2006; Zarbo & Jones, 2002).

The SEQC (Spanish Society of Clinical Biochemistry) began a quality assurance program for the preanalytical phase in the year 2000. The salient features of the program include: anonymous participation, rigorous confidentiality, quarterly evaluation (blood and urine samples, each sent twice yearly), computerized data processing, and a duration of one calendar year. The main objective of the program is to determine the current state of quality for the preanalytical phase in clinical laboratories in Spain by quantifying the number of specimens rejected by each laboratory compared to the number of rejections in all participating laboratories.

In this program a certain number of rejections are registered (the first 100 for blood samples and the first 50 for urine samples) or the maximum number of rejections registered in one-month period. A rejection is defined as the inability of the laboratory to provide results to the clinician for one or several test requests due to causes attributable to preanalytical error.

Currently, this program only measures quality for three preanalytical processes: blood sample collection performed by health care personnel, urine sample collection/pick-up by either patient or health professionals, or transport of samples.

The assessed indicators are:

- Coagulated sample
- Sample not sent
- Insufficient sample
- Hemolyzed sample
- Unsuitable sample
- Unidentified sample
- Defective transportation
- Insufficient sample relative to anticoagulant quantity
- Contaminated sample
- Other

For each assessment, the participating laboratory is required to send the following data:

- General laboratory characteristics: hospital or primary care, public or private, etc.
- Rejection registry and characteristics of these rejections (urgent or programmed request, permanent or temporary staff)
- Laboratory activity: number of requests, number of extractions and number of samples of each type during the study period.
- Characteristics of the laboratories compared to sample collection centers: person responsible for preanalytical phase, coordinator of the off-site sample collection center, criteria for sample rejection, internal quality control of the preanalytical phase, documented extraction procedures, what information (a computerized list, sample labels or instructions on the request forms) is given to staff about the type of samples to be extracted, training of collecting personnel, and existence of off-site collection centers.

- Supplementary data, type of requests (urgent or programmed) and type of personnel (permanent staff or not).

The organizers of the program collect all the information provided by the participating laboratories, carry out statistical analysis of the data, and inform each participating laboratory of the rejection rate for the total number of samples and the type of samples obtained from each laboratory, as well as the total rejection rate by type of sample for all participating laboratories. A general report of the group data is sent to all participating laboratories. This report includes data from all participating laboratories and includes the rejection rate distribution for all participants, correlation between rejection rate and the predefined characteristics for each participant and the multiple regression analysis of the rejection rate compared to the aforementioned characteristics.

The results obtained over the 5-year period since these programs were initiated show an overall blood sample rejection rate of 0.69 %(Alsina et al., 2008), with the highest rejection rates for sodium citrate ESR and coagulation samples, where the main cause of rejection was given as "sample not received". However, for serum and plasma-heparin samples, the main cause of rejection was hemolysis. Variables related to the organization of the preanalytical phase that have a significant impact on rejection rate have also been evaluated: a positive association was found between the rejection rate and the presence of off-site collection centers, while the presence of a collection coordinator, daily registration of incidents, and regular meetings with collecting supervisors were all negatively correlated. As for the type of contract of the collection personnel, permanent staff obtained a 0.58% rejection rate versus 0.95 % for temporary staff. The heading "non-permanent staff" includes all personnel who are hired on a temporary basis or rotating basis.

In terms of recent urine samples (Alsina et al., 2006c), the overall rejection rate of 2.16 % is much higher than that for blood samples in the same period (0.69%), with the most common cause of rejection being "urine sample not sent" (81.6%), which was also much higher than in the case of blood samples (37.5 %).

The organizational management that the laboratory brings to off-site collection centers could help to minimize the risk of error that occurs in this phase. Simplification of the entire process, computerization, and robotics are all potential preventative measures for those processes involving a higher risk of compromising patient safety **(Table 3)**

RISK OF ERROR	PREVENTIVE ACTION
Incorrect identification of the patient	Process Automation
Analytical Request Errors	Computerized test requesting
Remaining Samples	Indicate exactly the number of samples per request
Misidentification samples	Automate samples labelling
Incorrect Sampling	Quality materials Information and training

Table 3. Measures to minimize the risk of error that occurs in preanalytical phase

9. References

Alsina MJ, Gonzalez Oller J. (2006a) Protocolo para el estudio de la estabilidad de las magnitudes biológicas. *Química Clínica.* Vol. 25(2):86-9.

Alsina MJ, Gonzalez Oller J. (2006b). Definición del límite de estabilidad de las magnitudes en las muestras biológicas. *Química Clínica.* Vol. 25(2):81-5.

Alsina MJ, Alvarez V, Barba N, Bullich S, Cortes M, Escoda I, Escoda MA, Martinez Brú C, Planells P, Ramon F, Ricos C, Salas A. SEQC Committee for the quality of the Extraanalytical Phase (2006c) Revisión de los resultados del Programa de Evaluación Externa de la Calidad Preanalítica. Orina Reciente (resumen 2001-2005). *Quim. Clin* Vol. 26:325-31.

Alsina MJ, Alvarez V, Biosca C, Doménech MV, Ibarz M, Minchinella J,Pastor R, Perich C, Ricós C,Trujillo G, Vilanocva C (2007). Quality indicators and specifications for key processes in clinical laboratories: a preliminary experience. *Clin Chem Lab Med.* Vol. 45 (No.5): 672-7

Alsina MJ, Alvarez V, Barba N, Bullich S, Cortés M, Escoda I, Martinez-Brú C. SEQC Committee for the quality of the Extraanalytical Phase (2008). Preanalytical quality program-an overview of results (2001-2005 summary). *Clin Chem Med Lab.* Vol. 46: 849-54.

Bonini P, Plebani M, Ceriotti F, Rubboli F. (2002). Errors in laboratory Medicine. *Clin Chem* Vol.48:691-8.

Carraro P, Servidio G, Plebani M. (2000) Hemolyzed specimens: A reason for rejections or a clinical challenge? *Clin Chem;* Vol. 46: 306-7.

Clinical and Laboratory Standards Institute (formerly NCCLS). Procedures for the Handling and Transport of Diagnostic Specimens and Etiologic Agents (1994) - Third Edition; Approved Standard (Vol.14, No.7) CLSI document H5-A3 (ISSN-0273-3099). Clinical and Laboratory Standards Institute, 940 West Valley Road, Suite 1400, Wayne, Pennsylvania 19087-1898 USA,.

Clinical and Laboratory Standards Institute (formerly NCCLS). Urinalysis; Approved Guideline (2009) -Third Edition. (Vol. 29, No. 4) CLSI document GP16-A3 (ISSN-0273-3099). Clinical and Laboratory Standards Institute, 940 West Valley Road, Suite 1400, Wayne, Pennsylvania 19087-1898 USA

Clinical and Laboratory Standards Institute (CLSI, formerly NCCLS) (2009). Development and Use of Quality Indicators for Process Improvement and Monitoring of Laboratory Quality; Proposed Guideline. CLSI document GP35-P. Clinical and Laboratory Standards Institute, West Valley Road, Wayne, Pennsylvania USA.

College of American Pathologists. Laboratory Accreditation Program. (2006). Hematology-coagulation checklist. *Available from:* http://www.cap.org/apps/docs/laboratory-accreditation/checklists/hematology_coagulation_april2006.

Comité de Garantía de la Calidad y Acreditación. Sociedad Española de Química Clínica (SEQC). Bases de datos sobre la estabilidad de las magnitudes Available from: http://www.seqc.es/es/Sociedad/7/51/8/Base_de_datos_sobre_Estabilidad_de_magnitudes_|_Bases_de_datos_y_documentos_del_Comite_de_Garantia_de_la_Calidad_y_Acreditacion_|_Comite_de_Garantia_de_la_Calidad_y_Acreditacion_/

Comité de Garantía de la Calidad y Acreditación. Comisión de Calidad Analítica. Sociedad Española de Química Clínica (SEQC) Especificaciones de calidad derivadas de los datos de variación biológica. Available from: http://www.seqc.es/bd/vb.htlm

Da Rin G(2009). Pre-anlytical workstations : a tool for reducing laboratory errors. *Clin Chim Acta;* Vol. 404:68-74.

Dale JC, Novis DA. (2002). Outpatient phlebotomy success and reasons for specimen rejection. Arch Pathol Lab Med Vol. 126:416-9

Dale JC , Ruby SG. (2003) Specimen Collection volumens for laboratory test . A College of American Pathologists Study of 140 Laboratories. *Arch Pathol Lab Med* . Vol. 127:162-8.

EN ISO 9001:2008. Quality management systems. Requirements.

EN ISO 15189:2007. Medical laboratories- Particular requirements for quality and competence.

EN ISO 66175:2003. Quality management system: Guide to implementation of indicators systems.

European Agreement concerning the International Carriage of Dangerous Goods by Road (ADR 2009) Available from http://www.unece.org/trans/danger/publi/adr/adr2009/09ContentsE.html

Gomez R, Alsina MJ, Alvarez V, Barba N, Cortes M, LLopis MA, Martinez-Bru C (2009) Hemólisis en las muestras para diagnostico. *Rev Lab Clin* Vol 2(No. 4):185-95.

Hawkins RC. (2010) Phlebotomy site haemolysis rates vary inversely with workload. *Clin Chem Lab Med;* Vol. 48 (No.7): 1049-51.

Howanitz PJ, Renner SW, Walsh MK. (2002) Continuous Wristband monitoring over 2 years decreases identification errors. A College of American pathologists Q-Tracks study. *Arch Pathol Lab Med* Vol. 126:809-15.

Joint Commission on Accreditation of Healthcare Organization. Laboratory services national patient safety goals (2007). *Available from:* http://www.jointcommission.org/PatientSafety/NationalPatientSafetyGoals/07_lab_npsgs.htm

Jones BA, Calam RR, Howanitz PJ.(1997). Chemistry specimen acceptability: a College of American Pathologists Q-Probes study of 453 laboratories. *Arch Pathol Lab Med* . Vol. 121:19-26

Kenny D, Fraser CG, Hyltoft Petersen P, Kallner A. (1999). Strategies to set global analytical quality specifications in laboratory medicine - Consensus agreement. *Scan J Clin Lab Invest* Vol.59:585.

Lippi G, Guidi GC, Mattiuzzi C, Plebani M. (2006a). Preanalytical variability: the dark side of the moon in laboratory testing. *Clin Chem Lab Med* Vol.44(4):358–65.

Lippi G, Bassi A, Brocco G, Montagnana M, Salvagno GL, Guidi GC (2006b). Preanalytic error tracking in a laboratory medicine department. Results of a 1-year experience. Clin Chem. Vol. 52:1442-3

Lippi G, Banfi G, Buttarello M, Ceriotti F, Daves M, Dolci A, Caputo M, Giavarina D, Montagnana M, Miconi V, Milanesi B, Mosca A, Morandini M, Salvagno GL (2007). Recommendations for detection and management of unsuitable samples in clinical laboratories *Clin Chem Lab Med* Vol 45 (No.6):728-36.

Lippi G, Salvagno GL, Blanckaert N, Giavarina D, Green S Kitchen S, Palicka V, Vassault AJ, Plebani M. (2009) Multicenter evaluation of the hemolysis index in automated clinical chemistry systems. Clin Chem Lab Med. Vol. 47:934-9

LLopis MA, Trujillo G, LLovet MI, Tarrés E, Ibarz M, Biosca C, Ruiz R, Alsina MJ, Alvarez V, Busquets G, Domenech MV, Figueres C, Minchinela J, Pastor R, Perich C, Ricos

C, Samsalvador M, Simon M (2010). Quality indicators and specifications for key analytical-extranalytical processes in the clinical laboratory. Five years' experience using six-sigma concept. *Clin Chem Med Lab*; in press.

Plebani M, Carraro P (1997). Mistakes in a stat laboratory: types and frequency. *Clin Chem*. Vol.43 :1348-51.

Plebani M. (2006) Errors in clinical laboratories or errors in laboratory medicine: Clin Chem Lab Med . Vol. 44(No.6):750-9

Renner SW, Howanitz PJ, Bachner P. (1993). Wristband identification errors reporting in 712 hospitals: a College of American Pathologists Q-Probe study of quality issues in transfusion practice. *Arch Pathol Lab Med*: Vol. 117:573-7.

Ricós C, Garcia-Victoria M, de la Fuente B. (2004). Quality indicators and specifications for the extra-analytical phases in clinical laboratory management. *Clin Chem Lab Med* 2004; Vol. 42 (No.6):578-82.

Romero A, Muñoz M, Ramos JR, Campos A, Ramírez G. (2005). Identification of preanalytical mistakes in the stat section of the clinical laboratory. *Clin Chem Lab Med* Vol 43 (No.9):974-5.

Saleem S, Mani V, Chadwick MA, Creanor S, Ayling RM. (2009) A prospective study of causes of haemolysis Turing venepuncture: tourniquet time should be kept to a minimum. Ann Clin Biochem 2009; 46:244-6.

Salvagno GL, Lippi G, Targher G, Montagnana M, Guidi GC. (2007) Monitoring glycaemic control : is there evidence for appropriate use of routine measurement of glycated haemoglobin?. *Clin Chem Lab Med*. Vol.45:1065-7

Stahl M, Brandslund I. (2005). Controlled storage conditions prolong stability of biochemical components in whole blood. *Clin Chem Lab Med* .Vol. 43:210-5

Statistical Indicators benchmarking the information society. eEurope2005. Source for 2002:SIBIS project. *Available from* http:// www.sibis-eu.org/files/WP4_D4-3-3_eEurope_EU15.pdf

Valenstein PN , Raab SS, Walsh MK. (2006) Identification errors involving clinical laboratory: a College of American Pathologist Q-Probes study of patient and specimen errors at 120 institutions *Arch Pathol Lab Med* Vol. 130: 1106-13

van Walraven C, Naylor CD. (1998) Do we Know what inappropriate laboratory utilization is?. A systematic review of laboratory clinical audits. *JAMA* Vol.280:550-8

van Walraven C, Raymond M. (2003). Population-based study of repeat laboratory testing. *Clin Chem* Vol.49:1997-2005

Westgard JO. Six Sigma Quality Design & Control. 2nd Edition. Westgard QC, Inc. 2006.

Zarbo RJ, Jones BA. (2002) Q-Tracks. A College of American Pathologists Program of continuous laboratory monitoring and longitudinal performance tracking. *Arch Pathol Lab Med*. Vol. 126:1036-44.

10

The Need for Quality for Better Laboratory Performance, Alexandria University Experience

Mohamed Moustafa Rizk, Abla Abou Zeid and Nermine Hossam
Alexandria university, Faculty of Medicine
Egypt

1. Introduction

Quality means meeting the pre-determined requirements of the users for a particular service. [1] The quality of laboratory services depends upon many characteristics, such as painless acquisition of a specimen, specimen analysis, correctness of the test result, proper documentation and reporting, quick turn-around time for the test, and reasonable cost. To achieve good laboratory quality; the lab has to establish the following, qualified and experienced staff, calibrated and maintained equipment, standardized methods, adequate checking, and lastly accurate recording and reporting. [2,3].

Total quality management (TQM) means that every variable that could possibly affect the quality of test results has been controlled. [2,4] The traditional framework for TQM in a laboratory emphasizes the establishment of Quality Laboratory Processes (QLP), Quality Control (QC), and Quality Assessment (QA). Also to provide a fully developed framework of a good TQM system, Quality Improvement (QI), and Quality Planning (QP) must be established. [5] The purpose of QA in laboratory practice is the right result, at the right time, at the right specimen from the right patient, with result interpretation based on right reference data, and at the right price.

2. Quality assessment activities

2.1 Pre-analytical activities

From the moment a test is ordered, quality becomes an issue critical to the outcome. The pre-analytical phase of testing is complex and is prone to the most variation and the highest proportion of mistakes, as it is not under the full control of the laboratory and involves doctors, nurses, porters and other ancillary staff. The definition of mistake in the pre-analytical phase is "any defect during the pre-analytical phase, which results in failure of further laboratory processes". This indicates that active monitoring and feedback of all defects generated by non- laboratory personnel are essential to enable the inclusion of steps outside the laboratory, in the QA program of the laboratory.[6]

Accreditation agencies require that laboratories should consider all aspects of pre-analytical variation as part of their quality assurance plans, including effective problem solving and documentation. Pre-analytic Quality assessment activities extend to the following areas ;[7] Test selection and ordering, Patient preparation, Patient identification and specimen collection procedures, Specimen transport, handling, and storage. Each laboratory should have its pre-analytical quality manual, that should be distributed to all nursing units and, therefore, be available to all medical personnel. Additionally, easy-to-understand patient handouts must be available for outpatients. [7, 8] Pre-analytical quality manual should include the following:[9]

1. Instructions to the patient in preparation for specimen collection, including fasting hours, refraining from exercises, etc.
2. Minimum sample volume needed for a laboratory test.
3. Equations to calculate sample volume for the number of tests requested for a patient.
4. Instructions for blood sample collection, order of specimen collection, and sample identification guidelines.
5. Sample processing, transportation and storage guidelines.
6. Regulations for unacceptable specimens, and how to avoid causes of rejection.
7. Guidelines for the collection of other body fluids, like 24 hours urine samples, CSF, synovial fluid, peritoneal and pleural aspirate.
8. Listing of analytes and notation on the effect of at least commonly encountered interfering factors, also drug interferences should be included.

Although the "pre-analytical quality manual" and the strict criteria for rejection of inappropriate specimens may represent useful approaches in controlling the pre-analytical variables, still laboratories need more control. Knowledge dissemination, training and education are also important key points to improve the performance. Certification of phlebotomists, including training curricula for all collection staff, preferably developed by the laboratory, is another essential part of this crucial process of standardization.

2.2 Analytical activities

Quality design in a laboratory must begin with analytical quality because it is the essential quality characteristic of any laboratory test. Although the pre-analytical and post-analytical mistakes account for 46% and 47% of the total laboratory incidents respectively; the laboratory must first be able to produce a correct test result before any other quality characteristic becomes important. For example, turnaround time is an important quality characteristic, but it doesn't matter how fast the result is reported if the result is wrong. [10]

Management of the analytical phase involves reducing inaccuracy and imprecision of test methods as much as possible. Attention to standardize test procedures and monitor method performance with a well-designed quality control system are the key elements meeting this management goal. Analytic quality assessment procedures guide and monitor all related activities, including the following: Instrument maintenance and operation, Method selection and evaluation protocol, Documentation of analytical protocols, Test calibration, Quality control, including internal and external quality control, Reagents , Supplies and Personnel.

The analytical activities that should be monitored during laboratory assessment include the following :

2.2.1 Instrument maintenance

The director must define a program that monitors the proper calibration, function check, and preventive maintenance of instruments. A schedule of daily and monthly preventive maintenance is essential; these routinely done checks should be detailed in the procedure manual. Records of routine maintenance, unscheduled maintenance, should be kept for an appropriate period of time, usually instrument maintenance records must be kept for the lifetime of the instrument and transferred with the instrument. (11)

2.2.2 Method selection and evaluation

Selection of a good laboratory method depends upon many characteristics;

Application characteristics: General factors that could affect method or instrument selection include; type of specimen required, sample volume, run size, turn around time (TAT), throughput, population tested, personnel skills, safety (chemical hazards), utilities (ionized water requirements, waste disposal), reagents storage and stability, opened versus closed system, capability of random access, instrument cost per year (calculated from the average life span of the instrument), cost per test, and space requirements. (12, 13)

Methodology and performance characteristics: These are factors that contribute to best performance, they include; calibration, sensitivity and specificity of the method, linear range of analysis, interferences, precision, and accuracy, as well as types of internal and external quality control measures applied. Both the application and performance characteristics affect method selection decisions. (14)

After selection of the method, the laboratory should validate the performance specifications for such method. College of American Pathologists (CAP) requires the validation of all the methods used by the laboratory. (11)

The following is an approach for formulating a plan for method validation. . First, establishment of the analytical quality goal should be done. Then the lab should select the appropriate experiment to reveal the expected types of analytical errors, collect the necessary experimental data, and perform statistical calculations on the data to estimate the size of analytical error. Finally, compare the observed errors with the defined allowable error, and judge the acceptability of the observed method performance. Method validation is an error detection process, the aim of all experiments performed to verify the performance specification is to detect different types of errors; either random error (RE) or systematic error (SE).(14)

2.2.3 Calibration

Calibration is a process of testing and adjusting an instrument or test system, to establish a correlation between the measurement response and the concentration of the substance that is being measured by the test procedure. The substance used for calibration is known as calibrator.

Calibration verification; refers to 2 distinct processes: confirming that the current calibration settings remain valid for a method, and validation of the analytical measurement range (AMR). (11)

All of these processes: calibration, calibration verification, and AMR validation, are required to ensure the continued accuracy of a test method.

Calibration verification is done through assaying of materials of known concentration in the same manner as patient samples to confirm that the assay system will accurately recover the activity of the analyte over the reportable range, it must be performed at least once every six months. Materials for calibration verification must have a matrix appropriate for the clinical specimens assayed by that method, and target values appropriate for the measurement system. Materials used for calibration verification may include, but are not limited to: (11)

1. Calibrators used to calibrate the analytical measurement system.

2. Materials provided by the analytical measurement system vendor for the purpose of calibration verification.
3. Previously tested unaltered patient specimens.
4. Standards or reference materials with matrix characteristics and target values appropriate for the method.

How to perform calibration verification? (15)

1. Prior to starting calibration verification, all routine quality control procedures should be performed at first to assure that the equipment and reagents are performing correctly.
2. A minimum of three standards must be tested, low or zero standard, mid-range standard, high standard, they should cover the range of clinical values reported using standard procedures for the test. Each standard should be tested at least three times, preferably five to six times.
3. The mean and SD for each standard are calculated, the mean for each standard must fall within the target range specified by the standard manufacturer.
4. If no target range is specified by the standard manufacturer, then the value for the standard must fall within ±2 SD units from the lab established mean.
5. A graph of the results is drawn by plotting the expected values on the "x" axis and the observed values on the "y" axis then plotting the upper and lower range for each observation as determined by (mean ± 2 SD). Then a straight line is drawn which passes through the upper and lower ranges for each point.

Calibration is considered verified when:

1. The mean of each standard falls within the specified range of the manufacturer or within ± 2 standard deviation units of the observed mean.
2. A straight line can be constructed between the upper and lower ranges for each standard that covers the entire reportable range of the instrument or test procedure.

Successful calibration verification certifies that the calibration is still valid; unsuccessful calibration verification requires remedial action, which usually includes recalibration and AMR revalidation.

Recalibration should be performed in the following situations;

a. A complete change of reagents that affects the range used to report patient results or QC values.
b. When QC fails to meet established criteria, after major maintenance or service.
c. When calibration verification data fail to meet laboratory acceptance criteria.
d. When recommended by manufacturer and at least every six months.

2.2.4 Quality Control (QC)

Quality control may be defined as the control of the testing process to ensure that test results meet their quality requirements. QC may be practiced prospectively and provide information about the acceptability of the most recent analytical runs, or may be practiced retrospectively and provide information about past performance. Any QA program should comprise two key components: (16)

1. Internal quality control (IQC): includes appropriate measures taken during day-to-day activities to control all possible variables that can influence the outcome of laboratory results. This is a continuous process that is operated concurrently with analysis. IQC is used in the decision to accept or reject results on patients' samples and enables the laboratory to describe and monitor the quality of its work.

2. External quality assessment scheme (EQAS): This component is necessary to ensure comparability of results among laboratories. It is carried out retrospectively and is conducted by an independent agency.

3. Internal Quality Control (IQC)

Internal Quality control means monitoring the analytical process to determine conformance to predetermined specifications, and taking any necessary corrective actions to bring the analytical process into conformance.

QC material is the daily analyzed sample with known values stated by manufacturer. This material is commercially prepared, stabilized in liquid, frozen, or lyophilized form, packaged in small bottles suitable for use on a daily basis. Ideally, control materials should have the same matrix as the specimens being tested, so that they will behave the same as the real specimen. Certain test methodologies may also influence the selection of control materials. For example, a bovine-based control material will usually give low results with a bromcresol purple albumin method, which has been optimized for human albumin. (17)

Measuring the concentration of analytes contained in QC material usually detects any deviation from the expected performance. Results of QC material must fall within "confidence limits", ie, must be at least 95% of the values stated by manufacturer, or can be graphically represented by ± 2 SD of the mean for each value in question. Deviation of an analytical measurement from expected; could be either a shift in the mean (an accuracy problem), or an increase in the SD (a precision problem). If the result measured from the QC specimen deviates significantly as defined by QC rules, routine analysis is suspended and the analytical run is investigated, and corrective action should be taken. (18)

The laboratory must establish the number, type and frequency of testing QC materials that monitor the analytical process. A minimum QC accepted is two levels per day rather than per run (8 hour's period). (30, 54)

The chance of detecting an analytical problem will depend on the size of the error occurring, the number of controls used to monitor method performance, and the sensitivity of the statistical rules being applied. In laboratories, medically important errors should be detected (Probability for error detection), and false alarms should be minimized (Probability for false rejection).

Probability for error detection (P_{ed}), describes the probability of getting a rejection signal when there is a change in the precision or accuracy of the analytical method.

Probability for false rejection (P_{fr}), describes the probability of getting a rejection signal when there is no change in method performance.

Ideally, P_{ed} should be high, near 1.00 to provide a 100% chance of detecting a medically important analytical error. And P_{fr} should be low, near 0.00 to provide a 0.0% chance of false rejections that would otherwise waste time and effort and slow the reporting of patient test results. A practical design objective is a P_{ed} of 0.90, which means there would be a 90% chance of detecting an analytical problem and P_{fr} of 0.05 or less, which means there would be only a 5% or less chance of false rejection. (19)

Analytical errors are determined by the imprecision and inaccuracy of the method evaluated, errors are classified into;

Random error (RE) or imprecision: is the lack of reproducibility or repeatability, it occurs by chance and fluctuates about the mean. It is attributed to factors affecting the reproducibility of the measurement including; instability of the instrument, variations in temperature, and variability in handling techniques such as pipetting, mixing and timing, variability in operators.

Systematic error (SE) or inaccuracy: it is a measure of the agreement between a measured quantity and the true value. It is subdivided into, constant and proportional errors. When the error stays the same over a range of concentrations, it is called constant systematic error, and when it changes as the concentration changes, it is called proportional systematic error.

Causes of systematic errors include; poorly made calibrators or reagents due to the use of bad quality distilled water or non- calibrated pipettes; failing instrumentation (lack of frequent calibration, preventive maintenance of equipment and photometer check); improper preservation and storage of regents and calibrators; and lastly poorly written procedures. (20)

Total error (TE): represents the overall error that may occur in a test result due to both the imprecision (random error) and inaccuracy (systematic error) of the measurement procedure (TE = SE + RE). The intended use of total error is to describe the maximum error that might occur in a test result obtained from a measurement procedure, if it is less than the maximum allowable error according to the CLIA, then the method over-all performance is acceptable.

4. Make it right the first time

According to Feigenbaum; Total quality control is an effective system for integrating the quality development, quality-maintenance, and quality improvement efforts of the various groups in an organization so as to enable marketing, production, and service at the most economical levels which allow for full customer satisfaction.(21) In a simple word it is the state of the art of finding defects early which will be much less costly that those found later in the product life cycle i.e "Make it right the First Time".

TQM as a philosophy states that Quality is a TOP Management responsibility with Customer Satisfaction as the primary target, continuous improvement as the credo, and the way of action being based on FACTS and not opinions, while every employee must be involved. (22)

It was our challenge in the Alexandria University Hospital (AUH) to dream for such an ideal performance; which is nowadays a daily practice this is indeed cause and effect of a well organized Support from the top level of the organization, as our TQM involves all persons in the organization starting from the dean of the faculty of Medicine to the junior staff in the lab which conform with our internal regulatory requirements. The strategy is to concentrate on simplification and improvement of processes and organizing action around the medical laboratory service offered. Use of inter-disciplinary teams for improvement and problem solving. We are continuously benchmark ourselves; train all employees in the philosophy of TQM. (23)

How do we achieve excellent performance in the AUH LABORATORY?

- Ensure quality of overall process
- Detect and reduce errors
- Improve consistency within and between laboratories
- Contain costs

This is achieved through our comittement through ISO 9001Quality Systems on the managerial level as well as the ISO 15189 Quality management in the clinical laboratory of AUH for the technical specifications and support.

The importance of achieving good organizational quality system

1. **Quality costs:** as we have a limited resources compared to the service offered which serves a large of population in our area; we had to minimize the cost. The cost of wrong or incorrect calibration test decisions can be significant. This task is well achieved by better quality practice. (24)
2. **Analytical Quality specifications or goals**: Which involves the lists of procedures for setting analytical quality specifications for laboratory methods. However, analytical goals related to biological variations have attracted considerable interest. It was suggested that the analytical SD should be less than half the intraindividual biological variation. If a subject undergoing monitoring of an analyte, the random variation from measurement to measurement consists of both analytical and biological components of variation. The total SD for the random variation during monitoring then is determined by the relation

$$A2/T = \alpha^2 \text{ within } \beta + \alpha^2 A$$

where the biological component includes the preanalytical variation.
In addition to imprecision, goals for Bias should also be considered. The allowable bias can be related to the width of the reference interval, which is determined by the combined within and between subject biological variation in addition to the analytical variation. On the basis of considerations concerning the included percentage in a interval in the presence of analytical bias, it has been suggested that

$$\text{Bias} \leq 0.25 (\alpha 2 \text{ within } \beta + \alpha^2 \text{ between B})^{0.5}$$ (20)

Analytical quality specifications for laboratory methods includes: (25)

I. Evaluation of the effect of analytical performance on clinical outcomes in specific settings.
II. Evaluation of the effect of analytical performance on clinical decisions in general.
a. Data based on components of biological variation
b. Data based on analysis of Clinician's opinions
III. Published professional recommendations:
a. From national and international expert bodies
b. from Expert local groups or individuals
IV. Performance goals set by:
a. Regulatory bodies (CLIA)
b. Organizers of EQA schemes
V. Goals based on the current state of the art:
a. Data from EQA & proficiency testing
b. Data from current publications on methodology

3. **Verification & Validation:** All quantitative procedures require a calibration of the tests using reference materials. The laboratory results are calculated via calibration curve. Semi quantitative procedures do not belong to this category (26). These results are calculated by using a so called cut off calibrator leading to an index or are obtained as titter steps.
   a. **Checking the precision and accuracy** is done by the aid of :
      I. Control material: Control materials preferably covering the three concentration ranges are used to determine or check the precision and accuracy of the

method. The concentration of the controls should be in the lower, middle and higher range of the calibration.

The following materials are used:

- Purchased control samples with known values with references methods or with known values.
- Patient pool sera may be used.

II. Calculation :

Mean and Standard deviation are calculated by using known statistical methods ; while the Coefficient of variation (CV) and the inaccuracy (INA) are calculated by using the following formulae:

$$\% \text{ CV} = \text{SD} / \text{Mean X } 100$$

$$\% \text{ INA} = \text{Measured value} - \text{target value} / \text{target value x } 100$$

b. Checking the linearity or Analytical Measurement Range (AMR)

Range of values that instrument can report directly(Less accurately called linearity).Done every six month or after major maintenance as we should verify that your test results cover the span of the analyzer claimed AMR.

c. Limit of Detection (LOD) or Verification of Analytic Sensitivity Also called lower detectable range.

Done by Running 20 blanks; if <3 exceed stated blank value, accept that blank

Run 20 low patient samples near the detection limit; if at least 17 are above the blank value, the detection limit is verified

d. Validation of Reference Ranges

The laboratory must verify that the normal values in use are appropriate for patient population served

Once or when introducing new test; if at least 18 of 20 specimen falls within recommended reference range then the run is valid. If >2 falls outside the ranges then select a new 20 samples. If greater than five specimens' falls outside the recommended ranges, the QC ranges must be corrected according to the in house mean not the manufacturer claim ones.

4. Quality Assurance / Management

Alexandria University Hospitals serve a huge community that consists not only of inhabitants of Alexandria, but the hospitals are a referral center for the northern part of Egypt and the surrounding northern coastal regions. The laboratories are located on three campuses and include the various specialties of laboratory medicine; clinical chemistry, hematology, immunology and molecular diagnostics. Most tests are performed on automated machines throughout the day. Over 80,000 routine clinical chemistry tests are performed every month and a slightly lower number of hematological tests including complete blood picture and coagulation studies are performed. Laboratory staff includes faculty staff members, interns and laboratory technologists. Being the largest reference laboratory in the north of Egypt, we felt the urge and the responsibility towards our patients to provide them with the best laboratory services possible. [27]

5. Challenges

The major challenge to introducing a quality system was, and still is, financing. The hospitals' services, both diagnostic and therapeutic, are mostly free of charge. Only a few highly specialized tests are paid for and even these are priced at a very low level that barely covers the actual cost per test. We could not increase the prices because most of our patients come from a financially disadvantaged sector of the society. Funding is mainly provided through the annual budget of Alexandria University. The tight budget means that spending areas must be prioritized and any new introductions to the existing system of laboratory management must be carefully assessed as regards the cost.

Spreading a "culture of quality" was one of the top priorities. Educating personnel and changing their attitude was one of the toughest jobs on hand. All levels of employees needed this type of education. Besides day-to-day work and collaboration, an annual workshop was organized since 2005 on laboratory management and quality assurance. These workshops focused on the introduction of the concepts of total quality management, quality assurance, laboratory safety and benchmarking among other topics. The official language of these workshops was English because medicine is taught in that language in Egypt. However, most laboratory technologists do not master the language very well, most of them are not university graduates but hold a 2-year degree from a polytechnic institute. All technologists were encouraged to attend the workshops but in addition, satellite workshops were held in Arabic for them where they attended in smaller groups and had a better chance for group discussions. Besides continuous education, annual and periodic appraisals were also effective in motivating all staff members of the laboratory to embrace the new culture.

Other challenges included a dilapidated infrastructure, lack of a laboratory information system, a very disorganized documentation system among many others.

6. Overcoming financial obstacles

Total quality management requires a considerable flow of money to ensure proper documentation, proper implementation of internal quality control and external quality assessment programs, adherence to laboratory safety procedures, not to mention the huge task of renovating the infrastructure of the laboratories.

1. Internal quality control:

Due to the high cost of quality control material, two levels of quality control were run once per day instead of with every run. This was done only in the clinical chemistry unit in the 3 hospitals. In the hematology and immunology units, no quality control material was purchased due to the very high cost. These units depend on regular inter-laboratory comparisons. They also perform the same tests on randomly chosen samples in different shifts to assess the intra-run variations. Plans are underway to introduce a proper internal quality control system in the hematology unit by the end of 2010.

2. External quality assessment:

Alexandria University laboratories started to participate in the bi-weekly external quality assessment scheme (EQAS) for clinical chemistry by BioRad since November 2006. In order to compare our lab results with the correct peer group, it was essential to provide Bio-Rad with accurate information on the methods we are currently using. Upon enrollment into the program, a Method Questionnaire was completed using the appropriate Method Classification Guide, and returned back to Bio-Rad. Every 6 months a set of 12 numbered lyophilized serum specimens were delivered to our laboratories. Every 2 weeks, the appropriate sample was reconstituted following the manufacturer's instructions and assayed for 22 different biochemical analytes. Results were emailed to Bio-Rad. Within 3 working days of each Results Receipt Date, a user-specific statistical report was sent to our lab. These reports provide statistical analyses that include specific peer group and overall comparisons. In the report, each laboratory is provided with a numerical indicator of its competence, a performance index or score, together with information on the performance of the group as a whole, enabling its proficiency relative to the group to be compared and evaluated by means of three measures. The following data were included in the report:

- Mean of Comparator: This is the mean value of the category, either 'method', 'group' (assay technology) or 'mode' (collection of methods giving similar results) against which my own result was compared for the determination of the value plotted on the Levy-Jennings chart.
- Accuracy or Z scores: The 'Accuracy Score's a measure of the percentage difference of our result from the selected 'Comparator', converted into an integer of '0'up to '10', with '0'being minimal deviation.
- Standard Deviation index (SDI) this is the number of standard deviations my result differed from the selected 'Comparator'. Group Mean–Lab Result SDI = Group SD
- A comprehensive end of cycle report was also sent to the lab.

In 2009, owing to the global financial crisis, we felt the economic strain that the bi-weekly program was exerting on our financial resources. Consequently, we shifted to a less expensive monthly EQAS program.

External quality assurance for hematological tests was done through periodic inter-laboratory comparisons.

Other measures to ensure quality assurance included appointing a quality officer in each unit whose job was mainly to evaluate accuracy and precision of automated system periodically (photometer check , pumps and pipettes i.eABS) and to check the daily performance with QC materials. Standard operating procedures (SOPs) were written for all lab procedures in an easy to understand language and kept in easily accessible places in each lab. All personnel received training in lab safety procedures. A lab information system (LIS) was installed in 2009 and is now operational. Transcriptional errors have been considerably reduced. A hospital information system is expected to operate in 2011.

7. References

[1] College of American Pathologists. Glossary. *Appendix J In: Policies and Guidelines Manual.* Northfield, IL: February 1993.

[2] Westgard JO, Burnett RW, Bowers GN. Quality management science in clinical chemistry: a dynamic framework for continuous improvement of quality. *Clin Chem* 1990; 36: 1712-1716.

[3] Green S. Good laboratory practices. [Electronic presentation]. [Updated 6 December 2000, cited 10-9-2007]. Available at: http://www.med.howard.edu/pharmacology/handouts

[4] Westgard JO, Barry PL, Tomar RH. Implementing total quality management (TQM) in healthcare laboratories. *Clin Lab Manage Rev* 1991; 353-370.

[5] Westgard JO. Basic planning for quality. Madison WI: Westgard QC Inc, 2000; 15-18.

[6] Plebani M, Chiozza ML. Audit in laboratory medicine. *Eur J Clin Chem Clin Biochem* 1996, 34:655-657.

[7] Narayanan S. The pre-analytic phase- an important component of laboratory medicine. *Am J Clin Pathol* 2000; 113: 429-452.
[8] Irjala KM and Gronross PE. Pre-analytical and analytical factors affecting laboratory results. *Ann Med* 1998; 30:267-272.
[9] Narayanan S, Guder WG. Pre-analytical variables and their influence on the quality of laboratory results. IFCC [update 3 July 2006, cited 10 September 2007]. *Electronic journal of IFCC*.vol, 13, no1; Available at: http://www.ifcc.org/ejifcc/ejifcc/vol13no1/1301200107n.htms, article ID 1301200107.
[10] Cembrowski. Thoughts on quality control system: a libertarian's perspective. *ClinicalChemistry*1997; 43: 886-892.Available at: http:/www.clinchem.org/cgi/content/full/43/5/886.
[11] Sarewitz S. College of American Pathologists. Commission on Laboratory Accreditation. Laboratory accreditation program. 2006. CAP Inspection checklists and commentaries for laboratory accreditation. [Cited 6 December 2006]. Available at:http://www.cap.org/apps/docs/laboratory_accreditation/checklists/chemistry_and_toxicology_Sep06.doc.
[12] Travers EM, Korchmal CF. A new way to determine cost per instrument. *Med Lab Obs* 1988; 59-62.
[13] Ehrmeyer S. Final CLIA Rule. Part IV: The new method validation regulations. Madison WI: Westgard QC, Inc., 2003. [cited 4 January 2007] Available at: http://www.westgard.com/cliafinalrule4.htm
[14] Westgard J. Selecting a method to validate. Madison WI: Westgard QC, Inc., 2000. [cited 4 January 2007] Available at: http://www.westgard.com/lesson20.htm.
[15] Lum G, Tholen DW, Floering DA. The usefulness of calibration verification and linearity surveys in predicting acceptable performance in graded proficiency tests. *Arch Pathol Lab* ed 1995 May; 119(5):401-8.
[16] Cembrowski GS, Carey RN. Laboratory quality management ASCP Press Chicago. 1989.
[17] Westgard JO. Method validation: The interference and recovery experiments 2000. [Cited August 2007]. Available at: http://www.westgard.com/lesson27.htm.
[18] Quam E. QC- The materials. 2000; [Cited August 2007]. Available at: http://www.westgard.com/lesson13.htm.
[19] Westgard JO. A program for assessing statistical QC procedures. *Med Lab Obs* 1994; 26(2):55, 58-60.
[20] Westgard JO. A program for assessing statistical QC procedures. *Med Lab Obs* 1994; 26(2):55, 58-60.
[21] Feigenbaum, M. J. (1978). Quantitative Universality for a Class of Non-Linear Transformations. *J. Stat. Phys.* 19: 25–52.
[22] McLaughlin CP, Kaluzny AD. Total quality management in health: making it work. *Health Care Manage Rev.* 1990 Summer;15(3):7–14.
[23] Kaluzny AD, McLaughlin CP, Jaeger BJ. TQM as a managerial innovation: research issues and implications. *Health Serv Manage Res.* 1993 May;6(2):78–88.
[24] Berwick DM. Continuous improvement as an ideal in health care. *N Engl J Med.* 1989 Jan 5;320(1):53–56.

[25] G. G. Klee. Establishment of Outcome-Related Analytic Performance Goals *Clin. Chem.*, May 1, 2010; 56(5): 714 - 722.

[26] Fraser CG, Hyltoft Petersen P, Ricos C, Haeckel R. Proposed quality specifications for the imprecision and inaccuracy of analytical systems in clinical chemistry. *Eur J Clin Chem Clin Biochem* 1992; 30:311-317.

[27] Batalden P, Smith SD, Bovender JO, Hardison CD. Quality improvement: the role and application of research methods. *J Health Adm Educ.* 1989 Summer; 7(3):577–583.

11

Quality Control for the Molecular Diagnosis of Toxoplasmosis

Emmanuelle Varlet-Marie, Yvon Sterkers and Patrick Bastien
Centre Hospitalier Universitaire (C.H.U.) of Montpellier / University of Montpellier 1, Laboratoire de Parasitologie-Mycologie, Laboratory Associated (Pôle "Biologie Moléculaire") to the National Reference Centre for Toxoplasmosis, Montpellier
France

1. Introduction

Toxoplasmosis is an endemic parasitic disease due to the protozoon *Toxoplasma gondii*. The definitive host is the cat, in which the parasite develops in the intestinal epithelium, before being eliminated as oocysts in the faeces. When oocysts, for example in contaminated soil, are ingested by humans, other mammals or birds, they pass through the stomach and excyst in the intestine; the released sporozoites invade the intestinal wall and give rise to generalized infection (Remington & Desmonts, 1995; Wong & Remington, 1993). Two tissue stages are important in pathogenesis: tachyzoites and bradyzoites. Initially, the infection consists mainly of rapidly dividing tachyzoites. Then, with developing immunity, persistent tissue cysts containing hundreds of slowly multiplying bradyzoites develop, especially in the muscle and brain. This ability to live inside cells without destroying them allows evasion of host immunity and the development of a chronic stage of infection, lasting for years. Intermediate hosts become infected either by ingesting soil, water or foodstuff contaminated with cat faeces or by ingesting raw or undercooked meat containing tissue cysts.

T. gondii is responsible for generally benign infections except when the disease occurs in pregnant women (congenital toxoplasmosis) or in immunocompromised individuals, such as human immunodeficiency virus-positive or grafted patients, in which cases the vital prognosis may be involved. In certain countries of Europe, including France, toxoplasmosis is regarded as a serious health problem. In France, the prevalence of acquired toxoplasmosis in adults is 44%, and the estimated yearly incidence of contamination in women during pregnancy and of congenital toxoplasmosis are high, respectively, 6-7/1000 and 0.1% of births (Berger et al., 2008, King et al., 2008). Prevention of congenital toxoplasmosis (CT), including prenatal diagnosis (PND), has become a national policy in France since 1978 (Thulliez, 1992). Serological screening and follow-up is established for non-immunized pregnant women, and associated with monthly ultrasound examinations. PND is proposed in case of seroconversion between 6 to 38 weeks of amenorrhea. For this, PCR-based molecular diagnostic tests using amniotic fluid, introduced in the early 90's, have rapidly eliminated the need for cordocentesis, and have competed with more classical biological methods. Wherever it has been implemented, PND of CT has clearly improved the prognosis and outcome of infected children (reviewed in (Bastien, 2002)).

Because the host immune status plays an important role in the clinically patent expression of toxoplasmosis (Luft & Remington, 1992), particularly when CD4-positive cells reach counts <200/µl (Nissapatorn & Abdullah, 2004), toxoplasmosis is also one of the most common complications in immunocompromised hosts (HIV-infected and transplanted patients). Screening for toxoplasmosis in hematology patients has therefore become the most important request for *Toxoplasma*-PCR in Parasitology departments today.
The French national prevention policy was recently reinforced by the creation of a National Reference Centre for Toxoplasmosis (CNRT) (http://www.chu-reims.fr/professionnels/cnr-toxoplasmose-1/), which includes a working group for molecular biology, of which the objectives are, among others, the improvement and standardization of the molecular diagnosis of toxoplasmosis at the national level, and of which one of us (PB) is the coordinator. In France, the PND of CT is realized essentially in University hospitals, as well as two large private biological diagnosis centers. Not all University hospital centers perform such testing, as centers and practitioners involved in this diagnosis need an official authorization for this from the national health authorities, which is granted for five years.
In total, like for many infectious diseases, PCR-based diagnostic tests have become essential in the diagnosis of toxoplasmosis. However, also like in many other fields of clinical biology, these assays suffer from a great lack of standardization, as the vast majority of centers performing this diagnosis use 'in house' or laboratory-developed PCR assays; i.e. that have been set up independently in each laboratory using different DNA targets, and customized primers and reaction conditions.. This leads to a great heterogeneity between laboratories. On top of this, 'in house' methods can largely differ at any step during the diagnostic process, such as the extraction method, the number of PCR tubes used for diagnosis, the inclusion of an internal control for detecting inhibitions of the reaction, etc.(Sterkers et al., 2009). All this may be the source of considerable interlaboratory variations in the performances of the assays, influencing the quality of the diagnosis and hampering valuable comparisons of data among different centers. Previous studies have highlighted the lack of homogeneity and performance in European countries and underlined the need for guidelines (Pelloux et al., 1998, Kaiser et al., 2007, Bastien et al., 2007, Sterkers et al., 2009, Sterkers et al., 2010). In view of this heterogeneity, standardization of PCR methods and practices has become a strong wish of both the health authorities and the community of clinical microbiologists. Such a standardization should in turn lead to improvement of the diagnosis of CT at a more global level.
However, in view of the high diversity of methods and equipments available (see below), a true standardization of methods appears impossible, rendering external quality assessments (EQAs) the only way to homogenize the level of performances of this diagnosis among laboratories. The necessity of accreditation of the clinical laboratories according to a Quality management system added weight to these requirements. Indeed, this should include "comprehensive validation of each assay using the appropriate specimen matrix, daily quality control measures, and the enrolment in an appropriate quality assurance program for each assay to ensure continued performance, the whole being associated to the recording of results to document and address any inadequacies in test performance long-term" (Beld et al., 2007). EQAs for the molecular diagnosis of CT have been the subject of a number of small-scale studies in Europe (Guy et al., 1996, Pelloux et al., 1998) before being implemented at a national (Bastien et al., 2007) or European (Kaiser et al., 2007) scale, from

2002 and 2004, respectively, onwards. Toxoplasmosis is, to our knowledge, the only parasitic infection for which EQAs for molecular diagnosis exist.
In this chapter, we shall describe the experience of eight years of implementation of the French national EQA; we shall also discuss the interest, specificities, protocols, qualitative and quantitative results, as well as the perspectives, of EQAs for the molecular diagnosis of toxoplasmosis.

2. Eight years of national EQAs

In order to implement the harmonization of prenatal diagnosis of CT in France, an early initiative for quality assurance in the molecular prenatal diagnosis of toxoplasmosis was launched by the French association of hospital practitioners and teachers in Parasitology-Mycology (ANOFEL) in 2002. Briefly, on a yearly basis, a panel of positive and negative amniotic fluid samples prepared in Montpellier was sent blinded to participating centers for PCR testing, allowing each center to assess and follow its own performances in the molecular detection of CT (Bastien et al., 2007). This EQA was then continued by the working group for molecular diagnosis of the CNRT.
An annual national survey was associated to the EQA in order to assess the diversity and evolution of methods and practices in this molecular diagnosis in France. The questionnaires included 11 queries, concerning what was considered as the most critical points of the PCR process (Sterkers et al., 2009). All answers to queries had to be given considering the routine practice of prenatal diagnosis of CT, and not what could have been realized for the EQA only.
On top of these national 'core' EQAs, more limited and specific ('scientific') EQAs aiming at refining comparative studies between PCR methods were implemented at a smaller scale. A multicentric prospective study, involving eight laboratories proficient in the molecular prenatal diagnosis of toxoplasmosis, was a first step towards the harmonization of this diagnosis among university hospitals in France (Sterkers et al., 2010). It aimed at comparing the analytical performances of different PCR protocols used for *Toxoplasma* detection, and was reproduced over two consecutive years. Each centre extracted the same concentrated *T. gondii* suspension and tested serial dilutions of the extracted DNA using its own assays. Since all reactions are not positive when the concentration of pathogens gets to the sensitivity limit of the method (Chabbert et al., 2004; Sterkers et al., 2010), original 'performance scores' were defined, taking into account the proportion of positive reactions over the number of reactions performed, which appeared as a good indicator of sensitivity. Differences in analytical sensitivities were observed between assays, particularly at low parasite concentrations ($\leq$ 2 *T. gondii* genomes per reaction tube, i.e. $\leq$ 10 parasites/mL depending on the different pre-analytical protocols used in the participating centre), with 'performance scores' varying by a 20-fold factor among laboratories. Our data stressed the fact that differences do exist in the performances of molecular assays in spite of expertise in the matter. Finally, these studies allowed proposing a PCR sensitivity threshold to diagnostic laboratories as a minimal objective to reach for an optimal molecular diagnosis of toxoplasmosis.
Another multicentric prospective study aimed at comparing the analytical performances of different PCR protocols used for Toxoplasma detection, this time at very low concentrations (2.5 and 1 T/mL). Each of eight centers received 3 lyophilized samples of a *T. gondii* suspension at both concentrations, and extracted and tested each sample. This (i) confirmed

the differences in analytical sensitivities observed between assays, and (ii) allowed to experimentally assess the minimum number of samples to be tested at these very low concentrations, in relation to the heterogeneity of distribution linked to the Poisson law of large numbers (Varlet et al. article in preparation).

3. Specificities of EQAs for *Toxoplasma* molecular detection

The specificities of this EQA are related to (i) the fragility of *T. gondii* tachyzoites, and (ii) the low parasitic loads frequently encountered in protozoal infections..

We have observed that tachyzoites are extremely and rapidly subject to degradation at ambient temperature, even in nutritive media such as in vitro cultivation media or biological fluids (unpublished). This may occur during transport to participating centers or even during manipulations on the bench. First, this implies that all manipulations be carried out on ice and any temporary storage of the samples made at +4°C. Second, this gave a strong impulse towards freeze drying as the most adapted preparation method for EQAs (see below).

The demonstration of low parasite loads in a large proportion of infected AF samples (<10 tachyzoites per mL) (Costa et al., 2001; Romand et al., 2004) makes it necessary to test sensitivity with low Toxoplasma concentrations. Moreover, this choice allows a finer and more stringent assessment of the performances of the methods used by each laboratory year by year, as diagnostic methods for pathogens are particularly fallible with low concentrations of pathogens in the biological sample (Chernesky et al., 2002; Kaiser et al., 2007; Lachaud et al., 2001; Lachaud et al., 2002; Pelloux et al., 1998). It is noteworthy that these considerations probably apply to most other parasitic diseases.

4. Different protocols for EQA of *Toxoplasma*-PCR

A large number of variables have to be taken in consideration when preparing such an EQA. All of them must be subject to a choice made on both scientific, technical and practical grounds.

The matrix has a profound impact upon the sensitivity of the PCR assay. Indeed, the presence of inhibitors such as hemoglobin, blood proteins, anticoagulants etc. (Al-Soud & Radstrom, 2001) is obviously more a problem when analyzing blood-related products than more 'simple' biological fluids such as amniotic fluid, cerebrospinal fluid or aqueous humor. Also of note, the total amount of DNA in the template preparation, which includes DNA from the pathogen and its host, is also much higher in the former than in the latter. This, in turn, directly influences the amount of free Mg^{2+} ions in the reaction mix, and hence the enzymatic activity of the DNA polymerase (Bastien et al., 2008). Thus, the observed analytical sensitivity of the assay is somehow proportional to the volume of template DNA input in the PCR reaction (Burggraf et al., 2005) and high concentrations of human DNA do decrease the sensitivity of molecular assays designed to detect infectious pathogens. Another impact of the sample matrix is its effect on extraction efficiency. Indeed, it is well-known that the paucity of cells in a biological fluid leads to inefficient DNA recovery, due to the absence of "carrier-DNA" effect during extraction. Therefore, it is not recommended to use a non-cellular fluid such as saline (NaCl 0.9%) or PBS or culture medium as a matrix. Actually, the working group for molecular diagnosis of the CNRT did use saline but with high concentrations of the parasite (> 10^5/mL), in order to precisely avoid this problem.

QCMD, a 'not for profit' European organization dedicated to EQAs in molecular diagnosis in microbiology, is also using plasma as a matrix, seemingly successfully. Yet, we believe further testing is required to assess the extraction efficiency of low concentrations of *T. gondii* in plasma.

All EQAs for *Toxoplasma*-PCR published to date used panels consisting of samples of human amniotic fluid (Bastien et al., 2007, Kaiser et al., 2007). Although this avoids tackling with the problem of PCR inhibition, it allows good extraction efficiency and perfect mimicry of a diagnostic sample. In spite of the fact that amniotic fluid drawn from hydramnios is often naturally diluted, hence not ideal because pauci-cellular, it constitutes the most common source of matrix, and is collected after informed consent has been obtained. There is no consensus as to whether the patient should be seronegative for toxoplasmosis or seropositive with proven sound immunity due to past toxoplasmosis.

Samples from several patients are pooled in order to constitute a homogeneous matrix. The presence of blood should be avoided except if intentionally added to induce inhibition. Of note is the fact that we also tested bovine amniotic fluid, which proved rather satisfactory, but was abandoned because (a) it often presented bacterial contaminations and (b) it impeded the detection of human beta-globin by participants (as an inhibition control).

The parasitic material to be tested may be previously extracted DNA or whole *T. gondii* tachyzoites. DNA can be used as such in its own extraction buffer. However, our experience demonstrated that the DNA extraction process and the PCR assay both contribute to the performance of the whole diagnostic method (Sterkers et al., 2010); thus, this form of EQA does not allow testing the DNA extraction method, and this can even lead to wrong assessments of the method used in the participating laboratories (Molecular biology working group of the CNRT, unpublished data).

Therefore, the use of whole tachyzoites is preferred. Yet, these can originate from different sources and be sent under different forms.

Between 2002 and 2004, our test material was based upon naturally infected samples kept at -20°C after primary diagnosis. Known PCR-positive and PCR-negative amniotic fluids were pooled separately, and the pools were tested for the presence of bacterial contamination and by Toxoplasma-PCR. The reference PCR assay used to control negative samples allowed the detection of 0.5-1 parasite/mL (Chabbert et al., 2004). The negative pool was used for making negative control samples, and also for diluting the positive pool in order to produce samples with definite Toxoplasma concentrations. The method was simple and had the advantages of (a) testing naturally infecting strains (b) sending non-infectious samples (thawed parasites being dead). Drawbacks included an increased risk of bacterial contamination (leading to DNA degradation during transport), a reduced accuracy in the quantification of parasites (parasite concentrations in the positive pool were estimated by quantitative PCR in two or three different laboratories before dilution and distribution), as well as, later, ethical problems in France with respect to this type of samples (even though signed consent was obtained from all women from whom the samples were drawn).

After that, the test material was based upon the classical method of seeding negative amniotic fluids with live *T. gondii* tachyzoites. Between 2005 and 2007, parasites were collected from ascitic fluid containing tachyzoites drawn from mouse infected with the RH strain of *T. gondii*. These were added at different concentrations into thawed amniotic fluid and the panels were sent either at ambient temperature or at +4°C. In spite of expertise and numerous precautions and controls (including bacteriological controls) in the collection and preparation of tachyzoites, this procedure was sometimes spoilt with problems, essentially leading to cell death and degradation of the parasite DNA upon arrival at destination. This remained unexplained and made us change to another mode of preparation of the parasites. Since 2008, tachyzoites have been collected from cultivated *T. gondii* (RH strain) maintained in vitro by serial passages in human foreskin fibroblast cell cultures (Besteiro et al., 2008). These have the advantages of being purified, easier to obtain, more standardized, easier to count in a counting chamber and perhaps less subject to degradation than ascites drawn tachyzoites.

These advantages were concomitantly combined with those of freeze-drying (see below).

The choice of the parasite concentrations depends upon the objectives of the EQA and the proficiency of the participants. In France, we opted for low concentrations of parasites (i) because it has been established that a large proportion of infected AFs contain Toxoplasma loads < 10 tachyzoites per ml (Costa et al., 2001; Romand et al., 2004), (ii) because diagnostic methods for pathogens are particularly fallible with low concentrations of pathogens in the biological sample (Pelloux et al., 1998), (Chabbert et al., 2004; Chernesky et al., 2002; Kaiser et al., 2007; Lachaud et al., 2001; Lachaud et al., 2002) and (iii) because most participating laboratories were proficient in this diagnosis. The highest concentrations that were ever sent were 100 and 50 tachyzoites per mL. This choice proved right , since these concentrations were always detected by the participants (see below). Conversely, the use of low concentrations come up against the pitfall of the Poisson law of large numbers; this predicts the probability of not having any tachyzoite in the sample tested after aliquoting, as a function of the mean predicted concentration in the initial positive pool. We have experimentally tested this probability by assessing the reproducibility of PCR results among low concentrations samples, which implies that several samples be tested at 'low' concentrations (it is important to keep in mind that the definition of 'low' concentrations is dependent on the sensitivity threshold of the method tested). Such a practice aims at preventing a misevaluation of the method performances. The heterogeneity of distribution which was observed from the PCR results confirmed the Poisson Law, implying that, even if the PCR is realized in triplicate, several samples should be sent to the participants for testing the sensitivity of the methods using concentrations below 5T/mL.

The volume of the sample to be tested should ideally get close to the volume tested in routine practice. However, due to a general lack of matrix fluid, it is difficult to provide each participant with 10-20 mL of fluid, which implies a compromise between the ideal volume and the pragmatic constraints. For this reason, in our EQAs, we have chosen to send 2-mL samples. Others have used 1 mL (Kaiser et al., 2007), which appears even further from clinical practice.

The preparation process of the panels can indeed vary widely between fresh/thawed, frozen or freeze-dried samples. In any case, the samples must be 'as identical as possible' among themselves, which implies several precautions: (i) the parasites must be precisely counted and permanently homogenized during dilution and aliquoting; (ii) the panel samples must be kept at +4°C during the whole process; (iii) all identical samples should be drawn at the same time from the same pools; (iv) after arrival at destination, if in a fluid form, they must be kept at +4°C and processed as soon as possible. It should be noted that no DNA stabiliser or additive (anticoagulants, antibiotics) is used by us at any stage. As said above, freeze-drying offers numerous advantages over the other methods, including robustness, transport at ambient temperature, long-term conservation at +4°C, non-

infectiosity, and flexibility in use, and should be preferred for any EQA using 'fragile' microorganisms. On receipt of the panel, participants store the panel at 2-8°C until processing. Full instructions for reconstitutions of freeze-dried samples are provided. Briefly, lyophilized material should be reconstituted by adding 2 mL of sterile water to each vial. After reconstitution, each cell pellet should be resuspended to the appropriate volume to allow extraction of the complete sample using the DNA extraction protocol developed in the laboratory. After testing the panel, participants may be requested to send their results back and to complete a technical questionnaire.
The composition of the EQA panel must consist of one or several samples containing *T. gondii* and at least one sample negative for *T. gondii*. This minimal requirement has increased over the years of our EQAs. The EQA organized by QCMD sends 10-12 samples using different matrices. Presently, we prefer to send three-five samples containing relatively low concentrations of *T. gondii*. Low concentrations should be sent as several replicate samples. Higher concentrations may be included as positive controls depending on the proficiency of the participants. Negative controls should also be multiplicates, as the carry-over contaminations by amplicons are by nature sporadic.

5. Controls

Due to the importance and multiplicity of samples sent to participants, the panels must be strictly controlled in a number of ways.
PCR-negative amniotic fluid matrixes are pooled and checked again by the routine PCR assay for the absence of *T. gondii* DNA , before being stored at -20°C.
The *T. gondii*-containing panels are checked both for the correct concentrations of tachyzoites and for the absence of DNA degradation. As regards the latter, this can be verified (i) by multiple PCR testing before sending the samples; and (ii) after sending the samples, by keeping an entire panel at room temperature for 5 days in the coordinating laboratory before DNA extraction and PCR testing. However, it is possible that an undetectable level of DNA degradation might contribute to the false negative results observed with very low parasite concentration (e.g. 2-6/mL). In 2006, we had noticed a major unexpected decrease of the *T. gondii* concentration, which made us cancel this edition of the EQA.
Using freeze-drying, the controls are even more stringent, but can be done once and well in advance to the EQA. Pilot experiments were performed to test the influence of freeze-drying and storage on the detectability of *T. gondii* DNA in amniotic fluid. A set of serial dilutions was prepared and aliquoted, along with a negative sample. DNA extraction without freeze-drying (one aliquot of each concentration) was immediately carried out (DNA stored at +4°C) whereas all other aliquots were then freeze-dried. Real-time PCR was simultaneously run in our laboratory on both the non-freeze-dried and the freeze-dried samples (three samples for each). The number of target copies was deduced from the crossing point (Ct) value above which the signal significantly exceeds the baseline. This showed that freeze-drying results in a moderate reduction of Ct values ranging from 0.2 to 1 Ct, whichever the concentrations of tachyzoites tested (from 100.000 to 2 per mL). These differences can then be taken into account in the concentration calculations announced to the participants. Based on these results, three panels are then shipped to four selected reference laboratories. Correct results from all four laboratories for each concentration validate the freeze-drying of the panel.

Freeze-dried samples have been shown to be stable for at least one year storage at +4°C, from PCR testing of panels after 12 months (our unpublished data).

6. Results of the French national EQAs

6.1 Qualitative results

Depending on the year, the whole panel included two to five samples, totaling 765 samples during the 7-year period. Participant results and questionnaires were analyzed on an anonymous basis. The overall results are shown in Table 1.

Date	**2002**	**2003**	**2004**	**2005**	**2007**	**2008**	**2009**	**Total**
Number of laboratories	21	22	23	25	28	29	28	176
Number of samples/panel	2	4	5	5	5	5	5	-
Negative samples	21/21	42/44	45/46	50/50	56/56	58/58	56/56	328/331
False-positive [a]	**0/21**	**2/44**	**1/46**	**0/50**	**0/56**	**0/58**	**0/56**	**3/331**
Positive samples	17/21 (81%)	41/44 (93%)	67/69 (97%)	72/75 (96%)	57/74 (77%)	75/77 (97%)	72/74 (97%)	401/434 (92%)
< 2 T/mL					13/28			13/28 (46%)
2-6 T/mL		19/22	22/23		24/28		54/56	119/129 (92%)
6-10 T/mL	17/21		22/23	23/25		27/29	28/28	117/126 (93%)
20 T/mL		22/22		24/25	26/28			72/75 (96%)
50T/mL			23/23			29/29		52/52 (100%)
100T/mL				25/25		29/29		54/54 (100%)
False-negative [b]	**4/21**	**3/44**	**2/46**	**3/50**	**21/56**	**2/58**	**2/56**	**37/434 (8.5%)**

Table 1. Overall qualitative results of the French national EQAs for molecular detection of Toxoplasma from 2002 to 2009. [a] Number of negative samples found to be positive. [b] Number of positive samples found to be negative

The number of participants in the scheme increased from 21 in 2002 to 28 in 2009 (one centre closed in 2009). All laboratories used an 'in-house' PCR assay with diverse extraction methods. The panel composition and the performance of the participants in relation with the year are shown in Table 1. Data for the entire panel showed a constant progression from the implementation of the national EQA to date. Indeed, correct results were reported in 17/21 (81%) data sets in 2002; 17/22 (77%) in 2003; 20/23 (87%) in 2004; 22/25 (88%) in 2005; 27/29 (93%) in 2008 and 26/28 (93%) in 2009; 2007 was an exception as we wanted to test very low parasite concentrations, which lead to a high number of negative results. The most concentrated positive panels (50 and 100T/mL) were identified correctly in all data sets.

Low positive concentrations (2-6 T/mL) were detected in 86%, 96%, 86% and again 96% of the samples in 2003, 2004, 2007 and 2009, respectively. And very low positives (< 2 T/mL) were detected in 13/28 (46%) of the samples.

The general level of sensitivity in French laboratories was high. The total number of false negative results reported was equivalent to 8.5% of all tests performed on all positive panel members. Except for 2 laboratories, sensitivity problems were observed only for low parasite concentrations (<10 T/ml), where two to four of the 21-29 laboratories reported false negative results. But it should be noted that a single (very low) concentration (< 2 T/ml) was responsible for about half of these false negative results; if one takes out these very low concentrations, false negative results represented 5.4% of 406 samples. Considering the 6-10 T/mL concentration which corresponds to the parasitic load in half of the amniotic fluids according to Costa et al. (Costa et al., 2001), false-negative results represented 7.1% of 126 samples.

It is interesting to note that over the whole period of our study, except in 2007, the number of false negatives decreased over the years from 6.7% in 2002-2004 (Bastien et al., 2007) to 2.7% in 2009. Considering the 6-10 T/mL concentration, the number of false negatives decreased from 19% laboratories in 2002 to 6.9% in 2008 and 0% in 2009 (Table 1). The false-negative results did not appear to be related to a particular PCR method (e.g., real-time, DNA target or primer pair), but can be attributed to technical proficiency, PCR optimization and laboratory practices. Identical conclusions were reached by Kaiser et al. 2007 (Kaiser et al., 2007).

Also noteworthy is the fact that, although we always used lower concentrations than in other EQAs, the percentage of false negative results appeared to be very low as compared with the scarce results published to date. Indeed, a study in 1998 reported 40% of false-negative results at a concentration of 10 T/mL (Pelloux et al., 1998). Another international study involving 33 laboratories (of which less than four seemed to be included in the national network) reported 19.5% false negative results) (Kaiser et al., 2007), using concentrations starting from 20 T/mL down to 5 T/mL.

Overall, according to our experience among proficient centers for this molecular diagnosis, and taking into account that all these centers can detect concentrations < 5 T/mL, we consider that a minimum threshold to reach should be defined at 5 tachyzoites per mL based upon a DNA dilution series. This should be confirmed using samples directly extracted at this concentration.

With regard to specificity, the cumulative rate of false-positive results (0.9%; 3/331 samples) was very low in comparison with other molecular EQA programmes, e.g., 2-10% (Kaiser et al., 2007; Walton et al., 2005); 11.6% (Pelloux et al., 1998); or 35% (Valentine-Thon et al., 2001). Indeed, remarkably, only two centers reported a false positive result in 2003; and no false positives have been reported since from any of the centers. Yet, their presence, as well as the fact that the last false positive result observed was from a real-time PCR assay, stress the need to constantly re-evaluate protocols for the prevention of carryover contamination. The multiplicity of replicates of negative controls may help to reveal more contaminations of this type, as they tend to be sporadic in nature.

6.2 Quantitative results

The quantitation of the *T. gondii* amounts in biological fluids is relevant in view of two publications that established a relationship between parasitic loads and either the prognosis or the severity of the infection in the fetus (Costa et al., 2001, Romand et al., 2004). However, its application in routine practice poses numerous problems and appears as a real challenge to standardization. This is shown on concrete grounds by the fact that, although most proficient centers in France are now equipped with real-time PCR apparatuses (the number of equipped centers rose from 13 to 28 out of 29 between 2004 and 2009), only half in 2008, and a third in 2009, of these centers really quantified *T. gondii* in their results to the EQAs (Sterkers et al. unpublished data).

Relative quantification should be a first step, and should be straightforward as long as the PCR assay shows good repeatability and reproducibility. This may be expressed in number of Cts and can show an increase or a reduction on parasite amounts on a rather fine scale (with an error coefficient corresponding to a delta of Cts ≥ 2). Absolute quantification is more interesting, but poses the problem of the nature and accuracy of the material used for the concentration range. Indeed, (i) ideally, due to different inhibition properties depending on the biological fluid examined, the concentration range testing should be carried out in the same matrix as the one examined (amniotic fluid, blood…); (ii) the nature of the reference material to be quantified in the range may vary widely, from a heterologous plasmid, far from the parasitic target, to plasmids containing inserts of *T. gondii* DNA sequence, and to whole tachyzoites; (iii) the assessment of the amount of targeted DNA before the DNA extraction/PCR should be highly accurate. In any case, it should be kept in mind that the less concentrated is the DNA to be quantified, the less reproducible and accurate is the quantification.

Our experience in the EQAs with respect to quantification reveals wide variations in the results for the same sample. In 2008, for example, quantitative results were reported in 17 (59%) data sets, two of them expressing data as Ct (cycle threshold) values and 15 expressing them as concentrations (tachyzoites/mL). Eleven data sets allowed us calculating the mean, median and standard deviation. Four data sets were excluded as they were considered as off the correct range (variation of at least 10 fold for one point of the range) (Table 2). In 2009, quantitative results were reported in 23 (82%) data sets. But this time, probably due to the communication of the previous year's results and to recommendations from the CNRT, the majority of centers (17) expressed the results as semi-quantitative, whereas 9 laboratories expressed them as concentrations (three laboratories reported both). It is noteworthy that among the four centers which were considered off the correct range in 2008, one (C1) under-evaluated the amounts in 2008 but over-evaluated them in 2009, two did not quantify in 2009, and one did not improve much (Table 2).

The notification of parasitic loads to clinicians remains extremely rare in France, both because of the difficulties encountered in absolute quantification and because there is a general assumption in the community that the two studies quoted above should be confirmed. The need for standardization is here critical. First, there should be a consensus about the matrix and the nature of the DNA target to be used. Second, in order to allow comparison between centers, the estimates of the amounts of targets in the concentration range should be identical for all. Actually, a standardized and accurate concentration range should be distributed to all centers, so that all can work using a robust and reproducible material allowing multicentric comparisons. This is the present project of the CNRT, with the concern of the correlation between the real estimation and the results obtained using serial dilutions of DNA or extractions of diluted suspensions of tachyzoites. Indeed, we repeatedly obtained different results when serially diluting pre-extracted DNA or when directly extracting amniotic fluid spiked with *T. gondii* at different (relatively low) concentrations. A different challenge again will be posed by the implementation of quantification in whole blood, which might be a useful tool for the follow-up of immunosuppressed patients. Our experience with whole blood shows wide variations in the cycle thresholds, corresponding to variations of parasitemiae by a 1000-fold factor.

		2008			2009		
Real concentration in the samples (tachyzoites/mL)		**100**	**50**	**10**	**10**	**5**	**5**
Centres	*C1*[a]	*10*	*4*	*1*	*1245*	*660*	*142*
	C2	20	10	4	15	1	2
	C3	47	20	3	10	5	5
	C4	65	23.7	7.2	23	7	3
	C5	86	30	12	<200 (Ct=30)	<200 (Ct=35)	<200 (Ct=33)
	C6	95	44	10			
	C7	110	50	10	42	7	13
	C8	120 (Ct=29.5)	50 (Ct=30.5)	3 (Ct=34.5)	5	1	3
	C9	126.5	80	15.7	176.5	62	34.5
	C10	135	100	20			
	C11	170	18	2	(Ct=31)[c]	(Ct=36)	(Ct=36)
	C12	225	75	15			
	C13				10	*100*	10
	C14	*350*	*250*	*100*			
	C15	*1100*	*810*	*68*	*100*	*0*	*21*
	C16	*4000*	*1600*	*40*			
	C17	(Ct=34)[b]	(Ct=36)	(Ct=38)			
Mean 1 ± SD[d]		444.0±1019.1	211.0±434.4	20.7±28.1	181.2±430.1	61.3±159.4	
Mean 2 ± SD[e]		**109.0±57.2**	**45.5±29.1**	**9.3±6.0**	**17.5±13.4**	**5.2±3.8**	
Median 1[d]		120	50	10	19	7	
Median 2[e]		**110**	**44**	**10**	**12.5**	**5**	
Max		4000	1600	100	1245	660	
Min		10	4	1	5	1	

[a] Italicized lines and figures show the data which were considered as off the correct range (variation of at least 10 fold for one point of the range)
[b] this center did not express the results as an absolute quantification
[c] this center did not express the results as an absolute quantification in 2009
[d] mean and median calculated with all data sets
[e] mean and median calculated without including the data sets considered as "off range": these were found to be very close to the initial amounts estimates

Table 2. Performances of absolute quantification (in tachyzoites/mL) by real-time PCR in French national EQAs in 2008 and 2009

7. Discussion

Our experience with the French national EQAs for the molecular detection of *T. gondii* has been the source of many lessons and improvements. Our researches for an optimal matrix and parasitic material were profitable, and lead to an efficient tool for this unique evaluation of molecular detection for a parasite.

7.1 Diversity of the methods used for molecular diagnosis

The questionnaires associated to the national EQAs for the molecular detection of *T. gondii* have permitted to document the extreme diversity, as well as the evolution, of the methods used, whether for DNA extraction or the PCR itself, as well as of the practices associated to this diagnosis. This has been the subject of a previous article (Sterkers et al., 2009) and another one is in preparation. It is important to stress that the 2002-2009 period of study has witnessed the massive progression of quantitative 'real-time' PCR technology, as opposed to 'conventional' (end-point detection) PCR, as well as, more recently, the emergence of the use of automated DNA extraction. Yet, at the time of writing, all PCR assays used in France are 'in house' (or 'laboratory-developed') assays, and we are therefore far from standardization.
As said in the Introduction, this situation, not only are the source of variations in the assays' performances which may be detrimental for patient's care, but also hamper any valuable comparison of data among different centers. Unfortunately, standardization of 'laboratory-developed' methods appears impossible in a world of rapidly changing technologies and wide commercial offer. This will eventually be made possible (i) through a homogenization of pre-analytic and analytic practices, and (ii) through the commercialization of truly highly-performing PCR kits for this pathogen, keeping in mind that these should be evaluated against 'in-house' methods (Morelle et al., submitted). While waiting for the second event, the French CNRT is working towards the first goal.
The French national EQAs have encountered great success and allowed many achievements. They have enhanced communication among proficient laboratories; they have brought the means for authorized centers to self-evaluate their performances; and finally, in part due to the return of information and recommendations attached to the results, they have allowed an improvement in the performances of the PCR assays realized in these centers. The CNRT endeavors to push these improvements further by distributing standard material and establishing recommendations about methods (including DNA primers and targets) and pre-analytic practices.

7.2 Recommendations

The return of information in a yearly report provided the opportunity for writing proposals aiming at improving laboratory performances and practices for this diagnosis. Part of these recommendations have been discussed previously (Sterkers et al., 2009). One of the recent recommendations is to reach a minimal PCR sensitivity threshold, which is being established using a standard biological material. This is not mandatory but, joint to the distribution of benchmark material, should help laboratories to carry out continuous self-evaluation and improve both the performances of their PCR assays and the associated practices. Another important recommendation remains the refined optimization of the PCR conditions (Bastien et al., 2008); indeed, although this appears less necessary with highly-performing modern real-time thermocyclers, we have seen several examples of these where relatively simple changes in the PCR conditions improved the sensitivity of the technique. In any case, the differences that have been observed among centers in the different EQAs for *T. gondii* appear to be independent of the target, primers or technology, hence to rely essentially on proficiency and care in the optimization of the PCR conditions. Another lesson which has been learnt from these EQAs is the crucial importance of evaluating the diagnostic method as a whole, meaning the DNA extraction together with the PCR method. The former must be adapted to the latter, which means that both should be optimized together (our unpublished data). Thus, even an optimized PCR assay may yield poor results when DNA has been extracted by a suboptimal method for that assay.

7.3 Perspectives

The EQAs for molecular detection of toxoplasmosis hitherto have focused on the diagnosis of congenital toxoplasmosis. In view of the increasing prevalence of toxoplasmosis in immuno-suppressed individuals, there is a clear need to evolve towards the testing of biologically relevant matrices other than amniotic fluid, such as whole blood. It is noteworthy that the molecular diagnosis of CT offers the advantage of being readily confirmed by an independent method (serological follow-up until the age of one year), according to the well defined clinical and biological criteria for CT (Lebech et al., 1996) and updated by (Remington et al., 2004; Tissot Dupont et al., 2003). This makes it possible to calculate sensitivity, specificity, and positive and negative predictive values. Moreover, amniotic fluid is a 'simple' matrix for the PCR, from which spiked mimic samples can easily be made. Conversely, the significance of a positive signal in the blood of immunocompromised patients is much more difficult to analyze, and simulated buffy coat is certainly less easy to produce. Moreover, a blood matrix inevitably yields some degree of inhibition of the PCR, which imposes a supplemental bias in the analysis. It is also important to stress that a PCR assay that has been optimized using amniotic fluid will not necessarily be so when using blood. Finally, as discussed above, another necessary evolution is the assessment of the accuracy of absolute quantification.

In total, EQAs for Toxoplasma-PCR have been clearly evolving from a relatively non-professional venture among colleagues to a highly skilled organization. They appear as a necessity for the improvement of molecular diagnosis of infectious diseases, and are becoming compulsory by law in many countries. The progression of molecular diagnosis in Parasitology and Mycology should therefore see a gradual extension of such EQAs.

8. Acknowledgments

We wish to thank Pierre Marty (CHU of Nice), Jacqueline Franck (CHU of Marseille), Hervé Pelloux (CHU of Grenoble), Nathalie Van Langendonck (CHU of Tours) and François Peyron (CHU of Lyon) for their help during the different years of EQAs. We also thank all the centers that participated to these national programs, and more precisely B. Cimon (Angers), P. Millet (Bordeaux), D. Quinio (Brest), M. Vergnaud (Caen), B. Carme and C. Aznar (Cayenne), V. Livrelli and M. Cambon (Clermont-Ferrand), S. Bretagne (Créteil), A. Bonnin and F. Dalle (Dijon), H. Pelloux and M.P. Brenier-Pinchart (Grenoble), L. Delhaes (Lille), M. L. Dardé and D. Ajzenberg (Limoges), F. Peyron (Lyon), J. Franck (Marseille), B. Fortier † (in memoriam) and M. Machouart (Nancy), F. Morio and M. Miegeville (Nantes), H. Yera (Paris-Cochin), P. Thulliez (Paris-Institut de Puériculture), S. Brun and F. Touafek (Paris-Pitié Salpétrière), P. Roux (Paris-Saint Antoine), J. Menotti and F. Derouin (Paris-Saint Louis), F. Foudrinier and I. Villena (Reims), J.P. Gangneux and F. Robert-Gangneux

(Rennes), P. Flori (Saint-Etienne), D. Filisetti (Strasbourg), M.H. Bessières and S. Cassaing (Toulouse), T.H. Duong (Tours) and O. Eloy (Versailles). We also acknowledge financial support from the French association of hospital practitioners and teachers in Parasitology-Mycology (ANOFEL) and from the Institut de Veille Sanitaire (through the National Reference Centre for Toxoplasmosis).

9. References

Al-Soud, W. A. & Radstrom, P. (2001) Purification and characterization of PCR-inhibitory components in blood cells. *J Clin Microbiol*, 39, 485-493.

Bastien, P. (2002) Molecular diagnosis of toxoplasmosis. *Trans R Soc Trop Med Hyg*, 96 Suppl 1, S205-215.

Bastien, P., Jumas-Bilak, E., Varlet-Marie, E. & Marty, P. (2007) Three years of multi-laboratory external quality control for the molecular detection of *Toxoplasma gondii* in amniotic fluid in France. Clin Microbiol Infect, 13, 430-433.

Bastien, P., Procop, G. W. & Reischl, U. (2008) Quantitative real-time PCR is not more sensitive than "conventional" *PCR. J Clin Microbiol*, 46, 1897-1900.

Beld, M. G. H. M., Birch, C., Cane, P. A., Carman, W., Claas, E. C. J., Clewley, J. P., Domingo, E., Druce, J., Escarmis, C., Fouchier, R. A. M., Foulongne, V., Ison, M. G., Jennings, L. C., Kaltenboeck, B., Kay, I. D., Kubista, M., Landt, O., Mackay, I. M., Mackay, J., Niesters, H. G. M., Nissen, M. D., Palladino, S., Papadopoulos, N. G., Petrich, A., Pfaffl, M., W., Rawlinson, W., Reischl, U., Saunders, N. A., Savolainen-Kopra, C., Schildgen, O., Scott, G. M., Segondy, M., Seibl, R., Sloots, T. P., Tang, Y.-W., Tellier, R. & Woo, P. C. Y. (2007) Experts Roundtable: Real-Time PCR and Microbiology. In: *Real-Time PCR in Microbiology: From Diagnosis to Characterization* (ed. Q. A. Ian M. Mackay Sir Albert Sakzewski Virus Research Centre). Caister Academic Press.

Berger, F., Goulet, G., Le Strat, Y. & Desenclos, J. (2008) Toxoplasmosis in pregnant women in France: trends in seroprevalence and incidence, and associated factors, 1995-2003. *Bulletin Epidémiologique Hebdomadaire,* 14-15, 117-121.

Besteiro, S., Bertrand-Michel, J., Lebrun, M., Vial, H. & Dubremetz, J. F. (2008) Lipidomic analysis of *Toxoplasma gondii* tachyzoites rhoptries: further insights into the role of cholesterol. *Biochem J*, 415, 87-96.

Burggraf, S., Reischl, U., Malik, N., Bollwein, M., Naumann, L. & Olgemoller, B. (2005) Comparison of an internally controlled, large-volume LightCycler assay for detection of *Mycobacterium tuberculosis* in clinical samples with the COBAS AMPLICOR assay. *J Clin Microbiol*, 43, 1564-1569.

Chabbert, E., Lachaud, L., Crobu, L. & Bastien, P. (2004) Comparison of two widely used PCR primer systems for detection of toxoplasma in amniotic fluid, blood, and tissues. *J Clin Microbiol*, 42, 1719-1722.

Chernesky, M., Smieja, M., Schachter, J., Summersgill, J., Schindler, L., Solomon, N., Campbell, K., Campbell, L., Cappuccio, A., Gaydos, C., Chong, S., Moncada, J., Phillips, J., Jang, D., Wood, B., Petrich, A., Hammerschlag, M., Cerney, M. & Mahony, J. (2002) Comparison of an industry-derived LCx *Chlamydia pneumoniae* PCR research kit to in-house assays performed in five laboratories. *J Clin Microbiol*, 40, 2357-2362.

Costa, J. M., Ernault, P., Gautier, E. & Bretagne, S. (2001) Prenatal diagnosis of congenital toxoplasmosis by duplex real-time PCR using fluorescence resonance energy transfer hybridization probes. *Prenat Diagn*, 21, 85-88.

Guy, E. C., Pelloux, H., Lappalainen, M., Aspock, H., Hassl, A., Melby, K. K., Holberg-Pettersen, M., Petersen, E., Simon, J. & Ambroise-Thomas, P. (1996) Interlaboratory comparison of polymerase chain reaction for the detection of *Toxoplasma gondii* DNA added to samples of amniotic fluid. *Eur J Clin Microbiol Infect Dis*, 15, 836-839.

Kaiser, K., Van Loon, A. M., Pelloux, H., Ferrandiz, J., Picot, S., Wallace, P. & Peyron, F. (2007) Multicenter proficiency study for detection of *Toxoplasma gondii* in amniotic fluid by nucleic acid amplification methods. *Clin Chim Acta*, 375, 99-103.

King, L., Villena, I., Ancelle, T., Wallon, M., Garcia, P., Thulliez, P. & Goulet, V. (2008) Congenital toxoplasmosis: implementation of a surveillance system in France. *Bulletin Epidémiologique Hebdomadaire*, 14-15, 122-123.

Lachaud, L., Chabbert, E., Dubessay, P., Reynes, J., Lamothe, J. & Bastien, P. (2001) Comparison of various sample preparation methods for PCR diagnosis of visceral leishmaniasis using peripheral blood. *J Clin Microbiol*, 39, 613-617.

Lachaud, L., Marchergui-Hammami, S., Chabbert, E., Dereure, J., Dedet, J. & Bastien, P. (2002) Comparison of six PCR methods using peripheral blood for detection of canine visceral leishmaniasis. *J Clin Microbiol*, 40, 210-215.

Lebech, M., Joynson, D. H., Seitz, H. M., Thulliez, P., Gilbert, R. E., Dutton, G. N., Ovlisen, B. & Petersen, E. (1996) Classification system and case definitions of *Toxoplasma gondii* infection in immunocompetent pregnant women and their congenitally infected offspring. European Research Network on Congenital Toxoplasmosis. *Eur J Clin Microbiol Infect Dis*, 15, 799-805.

Luft, B. J. & Remington, J. S. (1992) Toxoplasmic encephalitis in AIDS. *Clin Infect Dis*, 15, 211-222.

Nissapatorn, V. & Abdullah, K. A. (2004) Review on human toxoplasmosis in Malaysia: the past, present and prospective future. *Southeast Asian J Trop Med Public Health*, 35, 24-30.

Pelloux, H., Guy, E., Angelici, M. C., Aspock, H., Bessieres, M. H., Blatz, R., Del Pezzo, M., Girault, V., Gratzl, R., Holberg-Petersen, M., Johnson, J., Kruger, D., Lappalainen, M., Naessens, A. & Olsson, M. (1998) A second European collaborative study on polymerase chain reaction for *Toxoplasma gondii*, involving 15 teams. *FEMS Microbiol Lett*, 165, 231-237.

Remington, J. S. & Desmonts, G. (1995) Toxoplasmosis. In: *Infectiuous Diseases of the Fetus and newborn infant* (ed. J. S. R. a. J. O. K. Editors). W.B. Saunders, Philadelphia.

Remington, J. S., Thulliez, P. & Montoya, J. G. (2004) Recent developments for diagnosis of toxoplasmosis. *J Clin Microbiol*, 42, 941-945.

Romand, S., Chosson, M., Franck, J., Wallon, M., Kieffer, F., Kaiser, K., Dumon, H., Peyron, F., Thulliez, P. & Picot, S. (2004) Usefulness of quantitative polymerase chain reaction in amniotic fluid as early prognostic marker of fetal infection with *Toxoplasma gondii*. *Am J Obstet Gynecol*, 190, 797-802.

Sterkers, Y., Varlet-Marie, E., Cassaing, S., Brenier-Pinchart, M. P., Brun, S., Dalle, F., Delhaes, L., Filisetti, D., Pelloux, H., Yera, H. & Bastien, P. (2010) Multicentric comparative analytical performance study for molecular detection of low amounts of *Toxoplasma gondii* from simulated specimens. *J Clin Microbiol*, 48, 3216-3222.

Sterkers, Y., Varlet-Marie, E., Marty, P. & Bastien, P. (2009) Diversity and evolution of methods and practices for the molecular diagnosis of congenital toxoplasmosis in France : a four years survey. *Clin Microbiol Infect.*

Thulliez, P. (1992) Screening programme for congenital toxoplasmosis in France. *Scand J Infect Dis Suppl*, 84, 43-45.

Tissot Dupont, D., Fricker-Hidalgo, H., Brenier-Pinchart, M. P., Bost-Bru, C., Ambroise-Thomas, P. & Pelloux, H. (2003) Usefulness of Western blot in serological follow-up of newborns suspected of congenital toxoplasmosis. *Eur J Clin Microbiol Infect Dis*, 22, 122-125.

Valentine-Thon, E., van Loon, A. M., Schirm, J., Reid, J., Klapper, P. E. & Cleator, G. M. (2001) European proficiency testing program for molecular detection and quantitation of hepatitis B virus DNA. *J Clin Microbiol*, 39, 4407-4412.

Walton, C., Hawkey, P. M. & James, V. L. (2005) Examination of specimens for mycobacteria in clinical laboratories in 21 countries: a 10-year review of the UK National Quality Assessment Scheme for Mycobacteria Culture. *Clin Microbiol Infect*, 11, 1016-1021.

Wong, S. Y. & Remington, J. S. (1993) Biology of *Toxoplasma gondii*. AIDS, 7, 299-316.

12

Quality and Safety in PICU and NICU: Scoreboard Control and Care Network

Fuster-Jorge, P.A.[1,4], Fernández-Sarabia, J.[1], Delgado-Melián, T.[2], Pérez-Hernández, R.[1], Jiménez-Rivera J.J.[3], Montes de Oca-Afonso M.R.[5], Domenech-Martínez, E.[4] and Sierra-López, A.[2,4]
[1]*Pediatric Department and Neonatology Section-NICU and PICU,*
[2]*Microbiology and Preventive Department,*
[3]*Intensive Medicine Department,*
[4]*Canary University Hospital, La Laguna University Medicine School,*
[5]*Ofra Medical Emergencies Service,*
Santa Cruz de Tenerife,
Spain

1. Introduction

1.1 Quality and safety: EFQM application to a health care organization

To ensure that each patient receives the best feature set to achieve an optimal process, to achieve the best possible results with minimal iatrogenic (safety) and the highest satisfaction, where the analysis of economic costs should also be able (efficiency and adequacy economic), defined according to the World Health Organization (WHO, 2010) the overall quality of the institutions, services and health units. All this is included in the organization of health care providers in a quality management system, comprehensive processes that will be compared with a model of excellence based on the best available evidence (Evidence-Based Quality Improvement) and patient-led (What?, Why? and for Who?), in order to learn, innovate, and to the formall and continuous improvement of the structure, both professional and technical, and to the full satisfaction of the professionals involved, patients or clients and global society, which must be assessed through surveys of specific satisfaction.

To achieve these goals the following are required: an adequate computer systems to record all the activity information of the health maintenance organization (intranet) and the processes offered or RDG (Relationship Diagnosis Group); a broad consensus with a full continuum involvement of all professionals (to Want, to Know and Empowerment) and community (clients and society); information and time and smooth communication between everybody involved; training and recognition of all professionals, groups formation with them in order to achieve improvement missions (What?, How? and Why?), to develop the map and diagrams of each process, carrying out the surveillance and control of key indicators (self), quantifying the results and establishing the appropriate corrective in the areas of greatest impact (Benchmarking), (CgestiónE/I, 2010; EFQM, 2010; EVALUACIONEFQM, 2010; Guía de la calidad, 2010; Moracho, 2003).

The process methodology is based on the model of excellence from the European Foundation for Quality Management (EFQM, 2010), founded in 1988 with the support of the European Commission, adapted to our area´s prior references Malcolm Baldrige Model (USA) and Premium Deming (Japan), (WHIKIPEDIA, 2010). The EFQM to compare the organization with a model to learn and improve continuously, evaluating the quality of the management carried out, identify areas for improvement and establish ratings. This model includes PDCA cycles (Plan-Do--Control-Act)) with successive actions, and repeated standardized planning, implementation, evaluation and correction. For this and all levels of the health organization, an ongoing assessment of each process will be devloped, conducting regular internal controls (self) and external (audit), all of which will compare our results with the model set (Moracho, 2003), which in the field of health is achieved not only by consulting the references published in our area for evidence-based medicine, but also by participating in multicenter care networks results that include related health care services, which in our cases are pediatric and neonatal intensive care, activity that we have always considered an excellent tool in our work and support care duties, teaching and researching.

1.2 Intensive care in pediatric and neonatal medicine: the PICU and NICU

Intensive care medicine was born in the decade of the 60s in response to social and welfare needs of reducing mortality and the number and importance of disability that originated, and it is still produced at present, critical situations potentially reversible. To this end the intensive care units (ICU) were created in specific areas of hospitals, in order to focus on these human and technical resources required to ensure top-quality care, effective and efficient. The first adult ICU was created in the U.S., Baltimore (1958) and Pittsburgh (1961), while in Spain it was a pioneer who created it in 1963 by Professor Jiménez Díaz in La Concepción Clinic (Madrid).

In Europe, pediatric and neonatal intensive care since 1980 had its own society in the European Club of Paediatric Intensive Care (ECPIC), since 1998 became the current European Society of Paediatric and Neonatal Intensive Care (ESPNIC, 2010), that it supports a close relationship with the European Society of Intensive Care Medicine (ESICM). The ESPNIC is an organization dedicated to promote and advance the art and science of pediatric and neonatal intensive care based on evidence, and to meet the needs of this client group so important in the European and international health context.

The pediatric intensive care constitutes a specific area of knowledge of national and international prestige. The PICU is the main framework for action, although not the only one, with a very specialized care for critically ill patients within the pediatric departments and pediatric services. Pediatric intensive care physicians have the right and duty to participate in care, teaching and researching of critical child care wherever they are, that is in any of the various levels of the health care. Also in the early 60s the first PICU was created in parallel with the tremendous growth experienced in those years in the field of surgery and other pediatric specialties. In Spain, the first one was created in the mid-60's Children's Hospital Vall d'Hebron (Barcelona); after the development of these units, they have been extended by almost all the national territory. In our country, the Spanish Society of Pediatric Intensive Care (SECIP, 2010), was founded over 25 years ago, preceded in 1976 within the AEP by the Pediatric Intensive Care Section, joining in 1980 as the Pediatric Group Spanish Society of Intensive Care Medicine and Coronary Units (SEMICYUC). The SECIP is a nexus of scientific participation and support in the organization of pediatric intensive care in Spain, and it maintains a close relationship with other companies and networks of intensive care outcome of pediatric and adult outcomes, both domestic and international. However, even in our country, there is only the recognition and accreditation as a specialty itself by the Spanish Association of Pediatrics (AEP, 2010), although we hope to obtain recognition of the National Council of Medical Specialties, with the creation of a specific training area (STA).

The hospital neonatal intensive care units (NICU) are pioneering highly specialized units, which preceded the pediatric units. They served with the highest level of care to babies born very prematurely or whit severe disease. They have also evolved greatly over the past 30 years, gradually implementing the concept of regionalization in perinatal care planning and to introduce the concept of definitive evidence-based medicine. Gradually units have proliferated in both high-risk obstetric and the modern tertiary NICU in reference hospitals from our extensive regional and national healthcare network, establishing a perinatal medical transport system better and more specialized training for all professionals involved in maternal-fetal and neonatal assistance.

In recent decades there has been an increase in the number of premature births. Its causes are multifactorials, affecting socio-economic factors, older age of parents or in vitro fertilization techniques, among others. In parallel, there has also been a significant increase in the birth rate and survival of newborns with very low birth weight (VLBW), weighing <1500 grams, which is generally associated with subpopulation very premature (29-33 weeks gestation), or even extremely premature (23-28 weeks gestation), so his stay in the NICU is often very long, thus consuming 65% of health resources for global neonatology practice (Rogowski et al, 2001).

Simultaneously, there have been extensive advances, preventive, therapeutic, diagnostic and technology, in neonatal, perinatal and obstetric medicine. The achievements and collected welfare improvements have been numerous, some of these advances in neonatology, including the use of surfactant in the treatment and prevention of respiratory distress syndrome of the newborn and new forms of respiratory support and ventilation (high frequency ventilation, extracorporeal membrane oxygenation, inhalator nitric oxide, liquid ventilation, etc.) have also subsequently been successfully used in pediatric and adult critical care patients.

The advances that have had more influence and profit in the NICU are: the widespread use of antenatal corticosteroids to achieve fetal lung maturity prior to preterm birth <34 weeks gestation; the selective and early postnatal administration of exogenous pulmonary surfactant in very premature; the emergence and development of new forms of mechanical ventilation, invasive or noninvasive, with more protection to avoid injury in the immature lungs and that facilitate early extubation; the improvements made in nutrition and infant feeding, with emphasis on early enteric and progressive human milk nutrition; and finally, the universal implementation of the delivery process of care focused on the development of the newborn and his family. All this has greatly improved our results in the short, medium and long term time, mainly in the VLBW, reducing their stay in the NICU and the hospital, morbidity and mortality, thus obtaining a longer survival with a sufficient and satisfactory quality of life.

To improve the care given in these units, it is necessary to identify risks that have our local population, controlling the results too. This is essential to establish a registration system that allows self-management results, especially in newborns that have more biological and social risk. So, we know our trends in mortality, morbidity and disability among survivors, in short, medium and long term, in various biological areas but particularly in the infants neurodevelopment.

In addition, our results should be compared with other similar units in their provision of professional human resources, technical and level of care complexity. To this end, in 2002 the database of neonatal units SEN 1500 (SEN1500, 2010) was created. This is a multicenter network created as a specific working group of the Spanish Society of Neonatology (SEN, 2010), with the aim of improving the quality and safety of newborns and their families, through research programs, education and improvement of cares received. It consists of 92 hospitals, many with III-B level of care, similar to our NICU, and distributed throughout the Spanish geography. It has a central computerized database which identifies the risks in this population, assesses the results of morbidity and mortality in the short and long term and researches in the field of neonatology. The Spanish Society of Neonatology has its origins in the historical heart of the Spanish Association of Pediatrics (AEP, 2010), when the Section of Perinatal Medicine was first created, and it established relations with the European Society of Perinatal Medicine and obstetricians of the Spanish Society of Gynecology and Obstetrics (SEGO) interested in perinatology. In 1997 the Spanish National Commission Specialties include the Neonatology in its catalog, setting specific training accreditation (STA) through the AEP.

For many years, this type of network data systems is also available in different countries and communities, some with extensive international involvement. Of these, the most popular are Australian and New Zealand Neonatal Network, British Association of Perinatal Medicine, Canadian Neonatal Network, European Neonatal Network (EuroNeoNet), International Neonatal Network, National Institute of Child Health and Human Development (NICHD) Neonatal Research Network, and Vermont Oxford Network (VON).

In our NICU, while continuing to adapt the recommendations, protocols and standards established by the SEN and other neonatal societies and associations, mainly European and North American. For many years in our NICU conducted a monthly analysis. Our results since 2002 have been compared with annual report of Spanish network SEN1500, ad hoc working group of the Spanish Society of Neonatology; and since 2005 we can compare with the monthly report of the international Vermont Oxford Network (VON, 2010), consisting of 850 units worldwide, mainly from USA, Canada and Europe, where we participate individually and into the newest member SEN1500 Spanish Group, which encompasses 11 Spanish hospitals. Also, in January 2009 our NICU was accepted to participate in the European Neonatal Network (EuroNeoNet, 2010), belonging to the European Neonatal Information System, which includes the project and initiative EuroNeoStat and EurNeoSafe; EuroNeoNet is Affiliated to European Society for Paediatric Research (SPR) and to European Society for Neonatology (ESN). Surely our extensive involvement in all these networks has enabled us to detect and correct deficiencies identified progressively in our business, continually improving quality and safety of our newborns and infants.

The Canary University Hospital (HUC) is a health care, teaching and research tertiary reference center, located in the Northern Area of the Tenerife Island and belonging to the public hospital network of the Canary Islands in Spain. It has 665 beds and is located next to the Medicine School in the San Cristóbal de La Laguna City; and provides medical coverage to refer a general population of children with an average rate of 19.7% and a population aged <15 years of 66.986 children: dates from Canary Statistical Institute , 2000-2006 (ISTAC, 2010).

In our hospital many years ago we began the policy of functional standards, proposed in the U.S. since 1951 by The Joint Commission (The Joint Commission, 2010), to improve the quality and safety of patient-centered care. Also, in our Department of Neonatology,

encompassing the PICU and NICU, we began this new line of work. Since then, we have developed a comprehensive and progressive master plan with specific strategies and policies for processes that we consider most at risk, focusing our NICU and PICU in developing not only the patient but also in their family. It was not an easy task, but very satisfying. Even today, we want to and can improve a lot, we are constantly learning, innovating and we know that we have a long way to go to move closer to excellence. However, sometimes we are quite confused and tend to justify, when the results of ours evaluations show an imbalance between the model of excellence and the daily clinical reality. "The perfection of methods and confusion of goals seem to characterize our age" (Albert Einstein).
It would be very long to detail each of the strategies currently being carried out in our two intensive care units. Therefore we only summarize statement of two processes that we consider most relevant to our NICU and PICU: surveillance and control of nosocomial infections; and prompt use of the expiratory pressure and ventilation support noninvasively in newborn with very low birth weight.

2. Surveillance and control of nosocomial infections in ours PICU and NICU

Nosocomial infections (NI), are associated or related to health care, and represent a major public health problem in pediatric intensive care units (PICU) and neonatal (NICU) too, by causing high morbidity and mortality, a significant increase in hospital stay and generate increased health care costs (Centers for Disease Control and Prevention (CDC), 2010; Raymond, Aujard & the European Study Group, 2000; Lodha et al, 2001; Stover et al, 2001; Grohskopf et al, 2002; Urrea et al, 2003, Edwards et al, 2005). Therefore, epidemiological surveillance and prevention must be considered fundamental objectives of quality and safety of pediatric and neonatal critical patient care.
In hospitalized adult patients the NI figures reach 5-10% according to data from the CDC (Yokoe, 2008). These figures are even higher in the pediatric intensive care patients and the prospective European study of UCIP reveals an incidence reaching 23.5% (Raymond, Aujard & the European Study Group, 2000). In the neonatal intensive care these figures are even higher if we take into account the higher risk newborn, the extreme low birth weigh (ELBW <1000 grams), in who we find figures of incidence over 40%, both in Europe and USA. (Beth et al. 2001; Geffers et al 2008). Being aware of their impact, monitoring and control of nosocomial infections have been a priority in our units.
Thus, since 2000 the HUC Department of Microbiology and Preventive Medicine conducted surveillance in our PICU and NICU, following the "Specific Program for Surveillance of Nosocomial Infection in Spain" (PREVINE, 2010), launched by Spanish Society of Preventive Medicine, Public Health and Hygiene. Through surveillance we obtain various indicators, the most interesting being those that relate the days of use of devices (urinary catheter, mechanical ventilation and intravascular devices) with the same associated infections (urinary tract infection, pneumonia and bacteremia). These indicators are also used by the NNIS (National Nosocomial Infections Surveillance System Report ((NNIS), 2003), which since 2005 is integrated into the NHSN (National Healthcare Safety Network, Edwards et al, 2007). As a result of monitoring, we have obtained a set of indicators that are disseminated to the PICU and NICU staff. From this information it was established a line of action against the nosocomial infection that included preventive guidelines, which have been developed based on the best scientific evidence available for that, such as the CDC in USA (Centers for Disease Control and Prevention (CDC), 2010).

Specific recommendations for neonatal intensive care units have been issued (Figure 1), whose methodology is based on the detection of centers of excellence, identification and grading of public practice, its implementation in short cycles of "plan-do-study-act" and the psychological implications of all staff. The results obtained with its progressive implementation in neonatal units have been excellent (Kilbride et al, 2003).

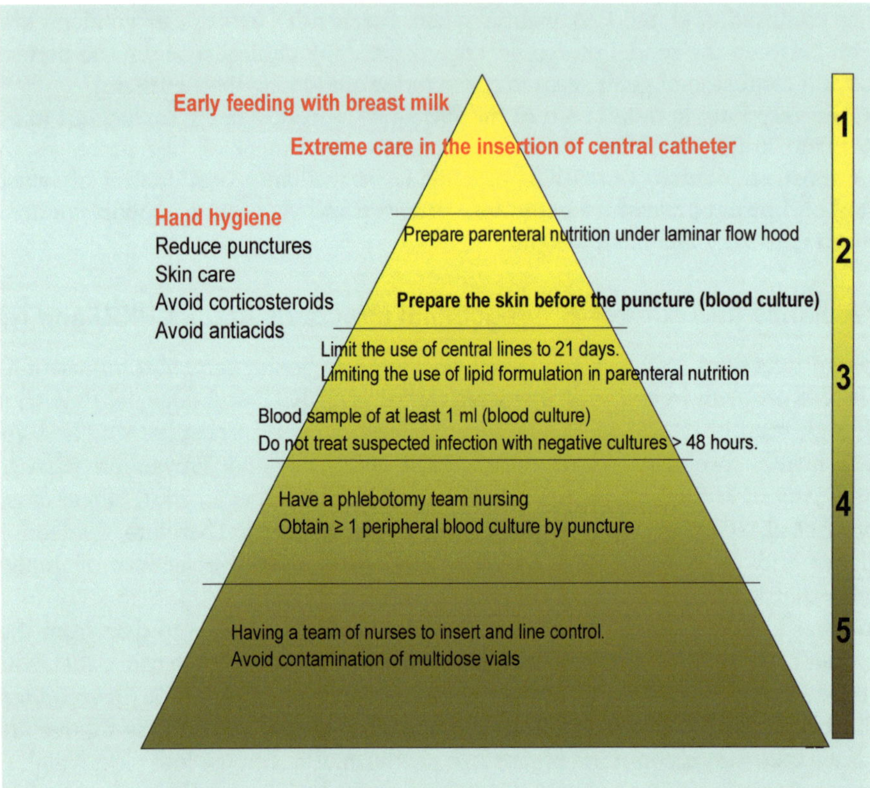

Fig. 1. Potentially beneficial practices (BPP) based on the best scientific evidence available (evidence grade 1-5), for the prevention of nosocomial infection in the NICU. Source (modified): "Evidence-Based Quality Improvement in Neonatal and Perinatal Medicine: The NIC / Q 2000 Experience" (Kilbride et al, 2003)

In addition to the NNIS, the NCID (National Center for Infectious Diseases, 2010; Edwards et al, 2007) and now the NHSN have represented over the years our most important references in infectious diseases epidemiology and preventive fields.

In our hospital we had clear since 1997 that the NI could create a problem in the new PICU of HUC, given its direct high morbidity and mortality, the increase of the stays in hospital and the economic costs at other similar units. In addition, our PICU was the first of these units opened in our island province of Santa Cruz de Tenerife, as well as our reference area, we cover for many years all our health area of influence.

The NI has been a critical priority that we controlled and analyzed continuously working with the Department of Microbiology and Preventive Medicine, relating to the proper use of invasive techniques of greater risk. We prospectively collected the main epidemiological data (number of days of hospitalization, use of intravascular devices, closed urinary catheter and mechanical ventilation) and the occurrence of related nosocomial infections (bacteremia, pneumonia and urinary tract infection).

This monitoring takes on new significance in the neonatal intensive care units. The patients seen in it, especially the lower gestational age and birth weight at high risk for nosocomial infection as already discussed. This is due to their immaturity (minor immune function, impaired skin and gastrointestinal barriers) and their NICU longer stay days (mayor bacterial and fungal colonization, central venous catheters, parenteral nutrition and invasive manipulation). In our neonatal intensive care unit there has been a cumulative incidence(CI) of nosocomial infection (number of patients with nosocomial infections x 100 / total number of patients studied) in high risk patients (defined as those with umbilical catheter, mechanical ventilation, treaties potentially serious diseases, surgically and weighing less than 1500 grams) of 46.15% in 2007, 44.30% in 2008 and 43.02% in 2009; and if we consider only newborn less than 1000 grams this figure rises to 50% in the last year; and 54% if you count the newborn less than 32 weeks in the same period of time. Of all these NI over the past year, 33.7% were bacteremias (sometimes associated with intravascular devices), 31.4% skin infections, 12.8% eye infections and 10.5% pneumonia (associated or not with mechanical ventilation). The most common microorganisms found where: 15% *Staphylococcus epidermidis*, 11.5% *Enterobacter cloacae, Enterococcus faecalis, Pseudomonas aeruginosa* and *Candida parapsilosis*.

To be our pediatric unit a new PICU created since 1997, we prioritize it at the first years of the study. It is an autonomous medical and surgical pediatric intensive care unit of level III-B; it has 4 beds in single box that generates annual revenue of more than 300 patients. During the study period we had 1031 stay days per year, with 65.93% of male patients, a mean age of 3.4 years (SD ± 4.26 and range 0-15). 73% came from the emergency unit, 7% from others areas of pediatric and neonatal hospitalization, 10% from the surgical area and the remaining 10% were outside patients. 46% had respiratory disease, 19% infections, 12% neurological diseases, 10% trauma injuries, 6% cardiovascular diseases and 2% oncohematologic diseases. The incidence of one or more organ failure was 31% (respiratory 39% and cardiovascular 34%) and of multiorgan failure was 13%; with an average severity index of 5.52 (0-26) for the PSI (Physiologic Stability Index) and 5.9 (0-41) for the PRISM (Pediatric Risk of Mortality Score), which in 26% was >10. Of all patients 39% required only monitoring and conventional treatment, 9% analgesic-sedation for invasive procedures, 64% central vascular catheter, 37% mechanical ventilation, 17% hemodynamic support and 1.8% extrarrenal depuration techniques; with a half therapeutic effort class II = 13 (range 0-50) for TISS (Therapeutic Intervention Scoring System), that also in 26% was >20 (class III-IV), according to the severity of our patients referred to PRISM. Our average stay days was 5.7 (SD ± 8.26), with an occupancy rate of 85% and a rotation rate of 42.8%. The mortality was 3.30% (33% of all deaths where organs donors); the incidence of patients with severe sequelae was 1.5%, 9% of them needed readmissions; and we had a small number of complaints (0-1/year).

During PICU surveillance conducted from 2000 to 2004 were analyzed a total of 302 patients, a NI cumulative incidence of 9.76% was found. The most frequently isolated microorganisms were bacilli Gram negative (46.60%), followed by *Candida spp* (33.30%) and cocci Gram-positive (20.1%). Predominant localization was as usual the bacteremia, with a partial cumulative incidence (PCI) of 4.09%, mainly primary face of intravascular device-related (2.99% vs. 1.10%). Tables 1-2 show our results in 2000-2004 for central catheter-associated bacteremia and their use rates, compared with data published by the NNIS (the National Nosocomial Infections Surveillance System Report (NNIS), 2003).

Year	N° of patients with CC	N° of catheter-related bacteremia	Days of CC	Tasa de incidencia del HUC	NNIS Rates, June 2003				
					Average	25%	50%	75%	90%
2000	13	2	108	18.52	7.3	3.8	5.9	8.8	11.5
2001	24	3	215	13.95					
2002	33	4	302	13.25					
2003	59	1	452	2.21					
2004	29	2	268	7.46					
Total	**158**	**12**	**1345**	**8.92**					

Table 1. Central catheter (CC)-associated bacteremia in the HUC PICU and references (*) National Nosocomial Infections Surveillance (NNIS) System Report, data summary from January 1992 Through June 2003, issued August 2003. Am J Infect Control 2003; 31:481-98. Quartils: 25%, 50%, 75% and 90%. Rate of central catheter-associated bacteremia (‰) = number of bacteremias associated with CC x 1000 / number of days of CC use

Year	N° of patients with CC	Stay Days in Hospital	CC Days use	Use Rate	NNIS Rates, June 2003(*)				
					Average	25%	50%	75%	90%
2000	31	235	108	0.46	0.46	0.30	0.41	0.53	0.60
2001	35	301	215	0.71					
2002	65	455	302	0.66					
2003	91	690	452	0.66					
2004	41	373	268	0.72					
Total	**263**	**2054**	**1345**	**0.65**					

Table 2. Use of central venous catheter (CC) in the HUC PICU and references (*) National Nosocomial Infections Surveillance (NNIS) System Report, data summary from January 1992 Through June 2003, issued August 2003. Am J Infect Control 2003; 31:481-98. Quartils: 25%, 50%, 75% and 90%. CC utilization rate = number of CC use days / number of stay days

In the same time period, the PCI of ventilator-associated pneumonia was 1.80 ‰, with a mechanical ventilation utilization rate of 0.27 days (mechanical ventilation days/total stay days) and the PCI of urinary tract infection associated with the closed urinary catheter was 2.63 ‰, with a utilization rate of 0.37 (closed urinary catheter days / total stay days).
The main conclusions obtained after this first period of the study were the following: the NI is certainly a problem in our pediatric intensive care unit, although smaller than in the neonatal unit; and bacteraemia was the predominant NI, it's related to an excessive time use of intravascular devices (central venous catheters). These comparative results helped to understand their impact on later to direct time and improvement strategies to reduce infectious morbidity: "Guidelines for the Prevention of Infections Associated with Intravascular Devices in Pediatric and Neonatal Patients" (2005); and reinforcement of the "Hygiene Program Hands ", repeating once or twice a year specific training courses for all the health personnel of neonatology and other pediatric areas too. Over the years, as shown in Figure 2, we have to improve our performance, not only in the PICU, but also and simultaneously in the NICU and other hospital neonatal units (medium and minimum cares). Nosocomial bacteremia has been steadily decreasing, but still it has a long use days rate of these invasive devices.

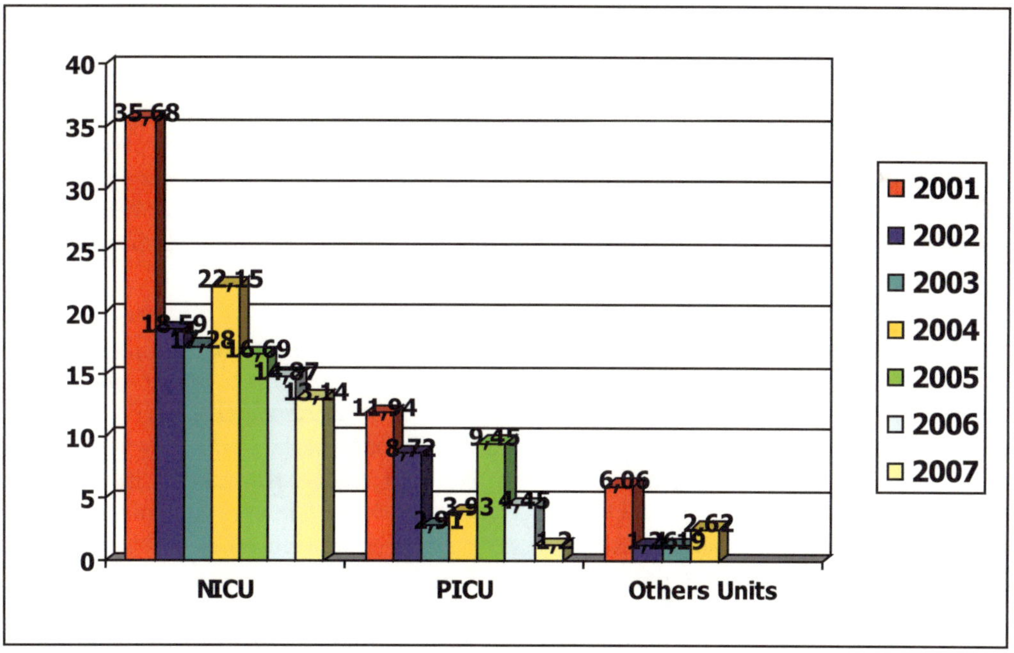

Fig. 2. Nosocomial bacteremia incidence density in the HUC neonatal units (2001-2007).
Incidence Density: DI (‰) = Nº of bacteremia x 1000 / total stays days

Currently, following the program developed by the U.S. John Hopkins Hospital and recommendations established by the WHO, the Spanish Health Ministry and the Canary Islands to the intensive care of adults, we adapted and planning an ambitious strategic plan "Bacteremia zero" (Bacteremia zero 1st edition, 2009) which attempts to reduce the incidence of bacteremia intervention is based on two main critical points to reduce related bacteremia

with central venous catheters: proper hand hygiene all the care team, medical students, family and visitors, and the proper use of intravascular catheters.
As bacteremia, whether or not the use of intravascular devices, are the leading cause of nosocomial infection in many units, both in our country (Urrea Ayala, 2007) and outside it, we should establish corrective measures to improve our results. Bacteremia associated with central venous catheters is caused mainly by coagulase-negative Staphylococci and others colonizing the skin around the insertion point of germs transmitted by the manipulation of the ICU staff. Moreover bacteremia are related to bacterial translocation from the intestinal reservoir. Thus both the care of central catheters, modification of bacterial colonization of the newborn and hygiene in the handling of this risk population have become primary goals in ours UCI.

2.1 Improve the management of central venous catheters

There are several strategies that can be taken to reduce infections associated with intravascular devices, and usually is a set of them which we must take to achieve a reduction in nosocomial infection. Therefore have been established for good clinical practice guidelines, based on the best available evidence and adapted many times for adult medicine, with proven results even in neonatal units that have achieved lower the number of infections (Aly, 2005). In this package includes not only those related to the act itself of catheter placement (choosing the right place, correct use of hand hygiene, barrier precautions such as gowns, caps, gloves and mask, a team of sufficient or proper disinfection of the skin) or maintenance (minimizing the number of entries, disinfect with alcohol keys to manipulate them, change dressings have become wet, dirty or rise, or every 7 days failure to fulfil the above, currently using chlorhexidine to clean the entry point), but should be reviewed since the indications for placement, the criteria for withdrawal or clarify the definitions of catheter-related infection to compare our results with historical series or with other institutions or networks (O' Grady, 2002).

2.2 Manipulation of bacterial and fungal colonization of the newborn

Bacterial and fungal colonization in neonatal units is related to the presence of bacteremia, sepsis and necrotizing enterocholitis in later stages. This bacterial colonization is associated with early onset and type of feeding, prematurity itself, the bacterial flora exists in each ICU and the use of broad spectrum antibiotics. For this reason the modification of the colonization of the newborn has set itself a target for the reduction of nosocomial infection. Basically there are two strategies in this regard have been introduced in recent years: the modification of intestinal bacterial colonization, with the confirmation of the importance of breastfeeding as the primary food for all newborns, and mainly for the more immature, and use of probiotics. On the other hand it's the use of antifungal prophylaxis for fungal colonization.
The feeding conditions the type of intestinal colonization. The fetal gut is sterile, whereas the adult has more than 400 bacterial species, giving it a stable environment where there is hardly room for new species (Edwards et al, 2002). The intestinal flora of breastfed infants is mainly composed of bifidobacterium and lactobacillus, while that of formula-fed infant contains more bacteroides, clostridia and enterobacterias (Balmer & Wharton, 1989; Fay & Faries, 1975). And we know that this bacterial flora it's most beneficial to the newborn and

it's associated with a reduction in mortality, sepsis and necrotizing enterocholitis. Breastfeeding achieves this in several ways: immune factors present in breast milk as IgA and lysozyme inhibits the growth of certain bacterias; the acidic pH of breast milk promotes growth of lactobacillus and bifidobacterium; breast milk has low concentrations of free iron, but contains lactoferrin, which binds iron and facilitates its absorption, thereby decreasing the amount of iron available in the colon, which favours the growth of bacteroides and enterobacterias. It also has bifidogenic factors (prebiotics) nonabsorbable oligosaccharides (lactose centre, with N-acetyl-glucosamine, galactose and sialic acid). This is why breastfeeding is undoubtedly the most physiological change colonization of the newborn intestinal flora towards a more favourable, thereby reducing the risk of nosocomial infection. It's therefore essential that in the NICU there is a training and awareness of all professionals for information, support and monitoring of breastfeeding.

As an alternative to "favourable" bacterial colonization that breastfeeding produces such good results, since 15 years there are pioneer randomized studies that pretending to demonstrate the reduction of NEC and mortality with the use of probiotics for intestinal flora modification in VLBW (Kitajima, 1997; Dani, 2002). After them a great number of meta-analysis have consolidated the results of initial studies, demonstrating that the use of probiotics reduce the incidence of necrotizing enterocholitis (Schanler, 2006; Deshpande, 2007; Deshpande, 2010) and death in newborns, with few side effects and health risks for them, although all these studies, and the latest Cochrane review (Alfaleh, 2008), conclude that further studies will be necessary in neonates, because them should clarify the use of probiotics, their class and administered doses. Although the scientific evidence seems to favour the use of probiotics, however these doubts must be clarified and by this in our unit hasn't started this prevention strategy.

Another initiative for the modification of newborn natural colonization in the NICU is the use of antifungal prophylaxis. Preterm newborn admitted to NICU are more likely than the terms neonates to be colonizationed by *Candida spp.* Colonization can occur in the gastrointestinal tract (19-30%), skin and respiratory airway (Saiman, 2006). We also know that others risk factors for invasive fungal infection are: low birth weight, use of central venous catheters, use of antacids, broad-spectrum antibiotics, parenteral nutrition, especially the lipids, and intubation for invasive ventilation (Saiman, 2000). Thus in recent years antifungal therapy has been tried to prevent colonization and invasive fungal infection in those neonates at greatest risk: the extreme low birth weigh less than 1000 grams. The invasives candidiasis have rise too much in all NICU (NNIS, 2003; NHSN, 2005). *Candida albicans* and *parapsilosis* are now the third later neonatal sepsis cause, with a highest direct mortality at ELBW. By this fluconazole universal prophylaxis is recommendable by "Clinical Practice Guidelines for the Management of Candidiasis" (Infectious Diseases Society of America, 2009) in newborn of less than 1000 grams birth weigh with rates incidence of 10-15%.

In a comparative studies with historical series (Healy & Bertini, 2005), fluconazole prophylaxis has shown evidence to reduce the incidence of fungal infection. Despite there is an A1 evidence in their favour by reduce the incidence of invasive candidiasis and enough drug safety, universal prophylaxis is not clearly established because there is some issues still unresolved (Clerihew, 2007; Kaufman, 2010): is more profitable universal fluconazole prophylaxis to control fugal nosocomial infection? ; should be it to give only when NICU has a high incidence of invasive candidiasis? Also universal prophylaxis with fluconazole may give to rise species of *Candida albicans* resistant to fluconazole, or more resistant in others species such as *Candida glabrata* and even *Candida parapsilosis* or increase the infections rate by *Aspergillus spp.* However this trend has been seen in others differences patient populations, such as cancer or transplant patients who receive prophylaxis with fluconazole, but not in the NICU.

In our NICU observational study since 2002-2008, in order to determine the incidence of invasive candidiasis and secondly the need of antifungal prophylaxis, we found that of 19198 live births in our hospital, 1.4% were VLBW less than 1500 grams (276) and 0.6% ELBW less than 1000 grams (117). All of newborns had a 0.14% (27) rates of invasive candidiasis, with 25.9% of overall direct mortality in the NICU: 22.2% (6) were neonates with a birth weigh over 1500 grams, with a 16.7% of mortality; 77.8% (21) were VLBW less than 1500 grams, with a mortality of 28.6%; and 66.7% (14) were ELBW less than 1000 grams, with a highest mortality of 35.7%. Therefore our annual incidence in ELBW population varied throughout the study time: 5.5% in 2002, 29.4% in 2003, 16.7% in 2004, 0% in 2005, 18.7% in 2006, 8% in 2007 and 0% in 2008.

By the moment we only make a strict control of the invasive candidiasis incidence, because our currently rate is around 7-8% (2009-2010); and the prophylactic fluconazole (3mg/kg once daily two times at week) is only administrated when our ELBW newborn are colonized or at 7th live days in those with more of one risk factors: venous central catheters, antibiotics, parenteral nutrition or mechanical invasive ventilation.

2.3 Hand hygiene

The importance of hand hygiene is making worldwide relevant and need the introduction of a specific hospital protocol for reduction nosocomial infection (Capretti, 2008; Won, 2004). The hands of sanitary staff are the main reservoir and vehicle for transmission of pathogens among patients admitted in the NICU. The concept of hand hygiene is a broader concept than simply disinfecting hands, includes skin care, nails, removing jewellery and artificial nails. This requires adequate use of alcohol solution (Boyce, 2002) or the hand washing with soapy water.

From 2003 the implementation in our units of the use of alcohol solution, as an alternative to hand washing, and the regular holding of workshops for hand hygiene of all health workers have attempted to improve the NI prevention. Although initially it seemed that we were improving, the most recent observational studies on the degree of hand hygiene compliance, carried out annually by the Microbiology and Preventive Medicine Department, results have not changed much and are not yet satisfactory, being similar to those already had published: not achieving a higher degree of overall compliance of 35%, when the best should be at least 65% of opportunities for hand hygiene (Table 3).

But this we hope that in the next future with the new program tailored "Bacteremia zero" (Bacteremia zero 1st edition, 2009), our two intensive care units will be strengthened definitely the training and involvement of all health workers, seeking higher quality and safety for our patients, improve our performance and achieving increasingly closer to excellence.

As a final conclusion, we believe that the creation of a multidisciplinary working group of infection control, which have involved the Department of Preventive Medicine and ours

PICU and NICU, has facilitated the implementation of effective measures to control and prevention of nosocomial infection, staff aware of our units need to achieve a steady improvement among all the results, safety and quality obtained.

	Nº oportunities	Nº hand hygiene	%
High risk	**225**	**60**	**26.66**
O1	197	53	26.90
O2	4	4	100
O3	3	0	0.00
O4	21	3	14.28
Moderate risk	**229**	**66**	**28.82**
O5	105	33	31.42
O6	64	33	51.56
Low risk	**383**	**129**	**33.68**
O7	219	44	20.09
O8	164	85	51.82
TOTAL	**777**	**255**	**32.81**

Table 3. Observational study of the completion rate (%) of hand hygiene in the PICU and NICU of HUC 2007-2009. HIGH RISK: O1 (before direct patient contact), O2 (before inserting central venous catheters), O3 (before inserting peripheral vascular catheters, urinary catheter or other invasive procedures), O4 (after switching from a contaminated site to another clean in the same patient). MODERATE RISK: O5 (after contact with intact skin of patients), O6 (after contact with body fluids, secretions, mucous membranes, broken skin, drains, bandages and dressings, etc). LOW RISK: O7 (contact with inanimate objects, medication and equipment) clinical) O8 (after removing gloves)

3. Early nCPAP and NIV in infants of very low birth weight

In recent decades, several emerging factors such as assisted reproductive techniques, improved obstetric care, the extensive knowledge of physiology and pathophysiology of fetal, perinatal and neonatal cares, and the extraordinary technological advances, diagnosis and treatment in neonatal intensive care have significantly increased the rate of prematurity (newborn less than 37 weeks gestational age). These developments, especially the use of antenatal corticosteroids for lung maturation, the postnatal pulmonary administration of surfactant and the use of invasive ventilation in the management of hyaline membrane disease or neonatal respiratory distress syndrome (RDS) have decreased early morbidity but above all have achieved better survival in the less birth weight (VLBW <1500 grams, ELBW <1000 grams) and gestational age (immature = 28-32 weeks, extreme = <28 weeks) prematures. So today the "threshold of viability" is between 400-500 grams natal weight and 22-23 weeks gestational age.

Thus there is a mayor problem for proper respiratory management of these extreme and very less weight prematures. In 25-28 weeks gestational age range the best would be to avoid invasive ventilation, oxygen therapy and prolonged endotracheal intubation; but in this subpopulation the early and elective administration of surfactant at birth in delivery room is highest recommended for its tenso-active effect in the immature lung; after this we must to try the subsequent early extubation in the NICU applying a nasal continuous expiratory pressure in the airway (nCPAP), or non invasive ventilation techniques (NIV), and administering caffeine to prevent these reintubation for the frequent premature apnoeas. This strategy (figure 3) has shown a very significant decrease morbidity and mortality, hospital stay and health costs related to the presence of neonatal RDS and the subsequent impact on his mayor respiratory after-effect: the bronchopulmonary dysplasia (BPD) or chronic neonatal lung disease (CLD), (Kilbride et al,2003).

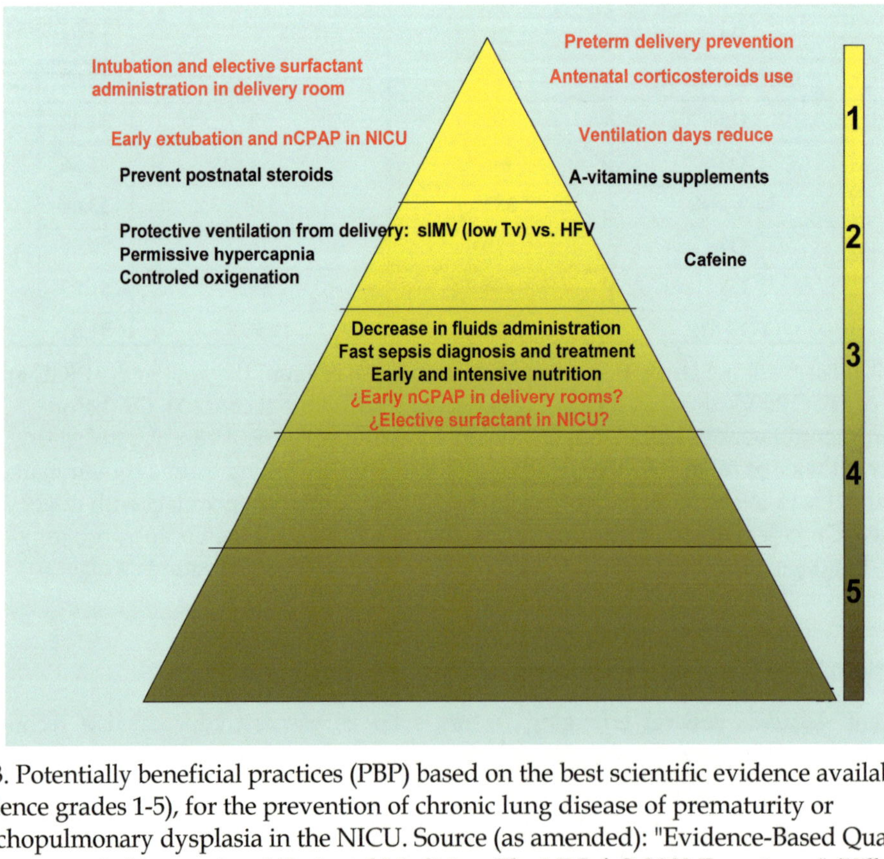

Fig. 3. Potentially beneficial practices (PBP) based on the best scientific evidence available (evidence grades 1-5), for the prevention of chronic lung disease of prematurity or bronchopulmonary dysplasia in the NICU. Source (as amended): "Evidence-Based Quality Improvement in Neonatal and Perinatal Medicine: The NIC / Q 2000 Experience" (Kilbride et al, 2003). nCPAP (nasal continuous expiratory airways pressure); sIMV (synchronized invasive mechanical ventilation); Tv (tidal volume); HFV (high frequency ventilation)

In the last ten years the early nCPAP or NIV use from the VLBW newborn birth in the delivery room, that it has been protocol in our hospital since 2003 (figure 4), have meant a change of neonatal cardiopulmonary resuscitation in delivery room and subsequent of respiratory support modality in major world NICU (Buron et al, 2006; American Heart Association, 2005). However, although it's unclear whether this early elective use has a beneficial effect, mainly by avoiding the most common squeal of extreme prematurity the BPD o CLD, or if it causes an increase in the subsequent need for mechanical ventilation and the later pneumothorax incidence, because it delay the classical treatment of RDS, that although it's more invasive, it has proven its effectiveness and efficiency in the last decades (Carlo et al, 2002; Vital & Arnold, 2005).

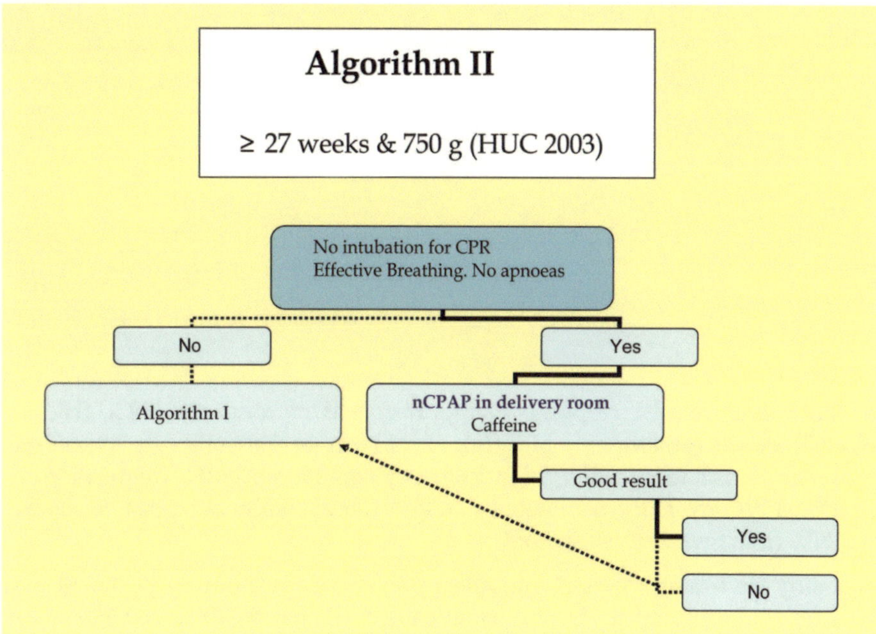

Fig. 4. Non invasive ventilator support in VLBW premature from delivery room to NICU (HUC protocol, 2003). Weeks (gestational age); g (grams birth weight); CPR (cardiopulmonary resucitation); nCPAP (nasal continuous positive airways pressure)

Do to these evidences, since 2003 we have also the "Algorithm I" (figure 5) for the extreme preterm less than 27 weeks and 750 grams, which provides for elective intubation and surfactant administered in delivery room at an early stage lung (at birth 15 minutes in delivery room or after a mayor stabilization at the first half hour of life in NICU). At the same time, the "Algorithm II" protocol includes the criteria for failure of nCPAP, or NIV if it was indicated, this is an emergency situation in which we must proceed fasting to his immediate intubation and selective surfactant administration, if there are diagnostic criterions for RDS, carried on an invasive ventilation protective synchronized support (initially assisted or controlled -A/C-, high frequency ventilation -HFV- if it's need a rescue technique, and later guaranteed volume -GV- and sometimes support pressure ventilation -SPV- too), while we maintained a strategy of "permissive hypercapnia" and a minimum "functional oxygenic ranges" through continuous monitoring of end-tidal CO_2 ($ETCO_2$) and skin crosswise oxygen saturation ($SatO_2$), respectively (Woodgate & Davies, 2001).

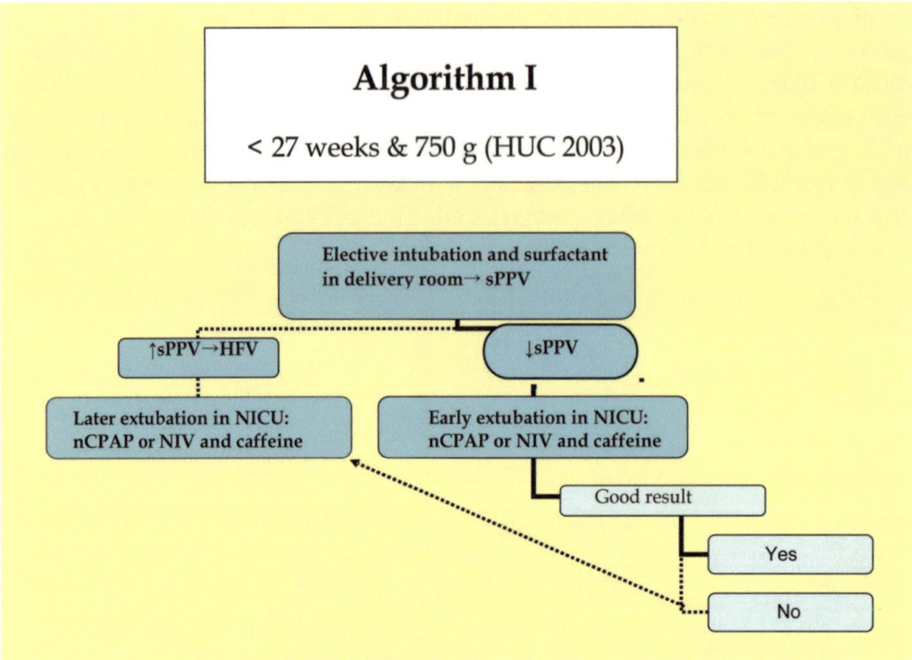

Fig. 5. VLBW infants invasive respiratory support in delivery room and NICU (HUC, protocol 2003). Weeks (gestational age); g (grams of birth weight); sPPV (synchronized positive pressure ventilation) ; HFV (high frecuency ventilation; ↑sPPV (high sPPV parameters); ↓sPPV (low sPPV parameters); nCPAP (nasal continuous positive airways pressure); NIV (non invasive ventilation)

Many currently studies are researching others preventives respiratory postnatal measures: the possibility of administering the surfactant aerosol (Down & Sinha, 2008); new synthetic surfactants; novel protective strategies for invasive ventilation (Morley et al, 2008; Stevens et al, 2007); the more safest range of oxygen saturation; the prophylactic use of inhaled nitric oxide (iNO), (Barrington et al, 2007); an appropriate pattern of postnatal corticosteroids administration to achieve early extubation (Giagliardi et al, 2007; Watterberg, 2006); the use of several pharmacologic agents as the inositol (Howlett & Ohlsson, 2003), the anti-inflammatory recombinant protein "Clara" CC10 (Levine et al 2005), the antioxidant superoxide dismutase (Davis et al 2003); an optimal electrolyte balance; the early and intensive parenteral nutrition and enteral; and a fast diagnosis and treatment of chorioamnionitis and neonatal sepsis (Fahley et al, 2008). Some of these strategies, mainly the combination of the three last measures, appear to be effective enough to reduce the incidence of BPD (Bhandari & Bhandari, 2009; Geary et al, 2008). Therefore we are implementing some of them in our NICU.

However all these postnatal actions, including the newer ventilation and protective pharmacological interventions, are probably later to be adequately effective in the prevention of premature BPD. The best would be the prenatal preventive interventions, where the prevention of preterm delivery would be more effective, but now it's hardly ever possible (Buhimschi et al, 2006). Thus the single administration of antenatal corticosteroids before premature delivery less than 34 weeks gestation, to achieve fetal lung maturity, is now the most effective strategy, but it's to discuss. Other possible preventive measures, such as the maternal administration of antibiotics in suspected chorioamnionitis, the antepartum antibiotic prophylaxis in premature prolonged rupture of the amniotic sac, or the intrapartum penicillin prophylaxis of Streptococcus agalactiae transmission, although there are highly recommended in the clinical practice guidelines, haven't getting to decrease the BPD (Bhandari & Bhandari V, 2009).

Our experiences is situated in the NICU of HUC, a health level care III-B unit (it's lacking only of ECMO -extracorporeal membrane oxygenation- and cardiac bypass surgery), with 6 open beds and 2 more in an isolation box, that generate more of 300 annual admissions, with an average stay of 7.4 days. Furthermore, we have others two neonatal hospitalization units with 8 medium neonatal cares beds and 10 minimum neonatal cares beds, both with an average of 600 annual admissions more.

We had studied the use of respiratory support techniques and the outcomes in our VLBW newborn during 2002-2008. We had compared them with the national neonatal network SEN1500 (SEN1500, 2010), a Spanish follow-up group of the Spanish Society of Neonatology (SEN, 2010) and the international Vermont Oxford Network (VON, 2010)), in which we participated since 2001 and 2005, respectively; and also with the published outcomes of the NNIS (National Nosocomial Infections Surveillance System Report (NNIS), 2003) and the SHNS (Edwards et al, 2007).

The HUC had an annual average of 2743 live births, with a total of 19198 inpatients in the seven study years. It's the north health area reference center for two level II hospitals in the Santa Cruz de Tenerife Province; both had an annual average of 1000 live births. Therefore our annual average neonatal population was of 3743, with a total of 26198 live births in them seven years. During the study we had 276 inside VLBW (1.4%), with 117 ELBW (0.6%), and 24 outside VLBW (none were ELBW); so that the total number attended in our NICU were of 300 VLBW.

We included in our study only 226 VLBW newborn (81.9%) alive in the HUC with sufficient or uncertain viability (≥23 weeks and 400 grams without malformations non compatible with life, non died in delivery room, without criteria for limitation of treatment) and that were available their parents consent. The average weight was 1146 ± 244 grams and the mean gestational age 29.5 ± 24 weeks; the 47.8% were male; 19.5% of pregnancies were in vitro fertilization and 35.4% multiple pregnancies; 74.3% were caesarean deliveries; 72.1% received prenatal corticosteroids and 32.3% antibiotics.

We collected all the variables into a computerized database, and perform a statistical analysis descriptive and comparative using the SPSS software (+) 15.0.

Our main findings and outcomes were:

1. Delivery room: 10.2% perinatal hypoxia; 78.8% CPR with O2 (oximetry preductal 85-92% at 10 minutes), 58.4% nCPAP, 46% non invasive ventilation by mask (inspiratory pressure 15-20 mm Hg and expiratory 5-6 mm Hg), 31% endotracheal intubation and invasive ventilation (tidal CO_2 35-45), 3.5% cardiac massage, 1.7% adrenaline and 8% elective surfactant (200 mg/Kg).
2. NICU: CRIB ("Babies Clinical Risk Index) 2.63 ± 2.90; 15% congenital malformation; 53.5% RDS criteria, 9.3% pneumothorax (all after a later invasive ventilation); 7% early sepsis, 20.8% nosocomial bacteremia and 4.5‰ ventilation-associated pneumonia; 8.8% BPD and 3.1% several retinopathy of prematurity (ROP grades 3-5). The 80.5% needed oxygen (FiO2> 21%), 74.8% nCPAP or NIV (30.2% exclusive from the delivery room)

and caffeine, 66.8% conventional invasive ventilation (44.6% for nCPAP or NIV failure) and 26.5% high frequency ventilation (with a total ventilation rate of 0.21 days/stay days), selective surfactant 41.1%, 1.3% iNO (for pulmonary hypertension) and 9.3% postnatal corticosteroids (to achieve his extubation and only after two weeks age).

3. At discharge: 79.2% home and only 2.8% with home oxygen-therapy and monitor; 3.5% transferred to another hospital (all for cardiac surgery); and 17.3% died (1.3% within 12 hours of life).

We found a large statistical relationship ($p<0.01$) for the early nCPAP from delivery room or NIV and the less use of oxygen in the NICU (FiO_2 <0.50 and mean of 9.7 days) and the lower incidence of BPD. Also was significant ($p<0.05$) the lower incidence of severe ROP (grades 3-5).

There was a large significant correlation ($p <0.001$) between the risk of death and a gestational age <24 weeks, the need for resuscitation at birth with adrenaline and a CRIB >9.

Besides the exclusive use of non invasive respiratory support helped the VLBW newborn to achieve faster his early progressive enteral nutrition, non-nutritive sucking at the breast with a close contact "skin to skin" with her mother and finally a mayor breastfeeding: the majority of him received first an enteral nutrition with breast milk, and 50% were maintained later beyond of his hospital discharge.

When we compared our outcomes with the VON and SEN1500 observed that: in the HUC the obstetrician administered less prenatal corticosteroids and performed a greater number of caesarean; our VLBW newborn had + 1 week gestational age, similar mean natal weight, less perinatal hypoxia, intubation and early surfactant in delivery room, but more nCPAP or NIV; less oxygen, more nCPAP and NIV, less surfactant with a lower incidence of RSD, BPD and severe ROP in the NICU, although with a higher incidence of pneumothorax associated with later invasive ventilation when nCPAP or NIV failures; our incidence rate of early sepsis were higher, also the nosocomial bacteremia but it decreased progressively, while those the pneumonia associated-ventilation and the use rate of invasive ventilation remained adequate; at discharge our VLBW newborn had lower length of stay, average weight and very few patients needed home oxygen-therapy and monitor. Furthermore and significantly, we had a high incidence of major congenital malformations.

The main conclusions were that: during the study it was a progressive implementation in our VLBW newborn of early nCPAP or NIV from the delivery room, it helped to reduce the need of oxygen-therapy, the incidence of BPD and severe ROP, but not of pneumothorax associated with a delayed invasive respiratory support and surfactant administration; as well as improving many other aspects related to quality and safety of our patients.

It was evident that the early use of non invasive respiratory support had benefits for our VLBW newborn. But at the same time, our results indicated that we should formalize and better select why patients more premature would to be candidates for elective intubation and early surfactant administered immediately after stabilization in the delivery room or in the NICU.

From then on we have made these corrective actions, which in the last two years have been quite effective and efficient, improving the quality and safety of our VLBW newborn. Although we have not even completed the measures of all indicators of this process that fully guarantee it, in 2009 year we have improved dramatically, achieving similar or even superior results to those obtained by the VON: we have increased use of antenatal corticosteroids and decreased number of caesareans; intubate and administer more early surfactant in delivery room or NICU in the smallest premature, select and better ensure early invasive respiratory support halving the incidence of delayed pneumothorax, reducing significantly the incidence of early and late sepsis. We have continued our excellent results for the BDP and ROP; also we have optimized a lot of nutrition and so the 80% of our VLBW newborn keep the breast milk at discharge.

4. References

Abramczyk, M.L. et al (2003). Nosocomial infection in a pediatric intensive care unit in a developing country. *Braz J Infect Dis*, Vol. 7, No. 6, (December 2003) 375-80, ISSN 1413-8670.

AEP (2010, September): Asociación Española de Pediatría. Retrieved from http://www.aeped.es/

AHA: American Heart Association (2005). Journal of American Heart Association: Neonatal Resuscitation Guidelines. *Circulation*, Vol. 112, (2005), 188-195.

Alfaleh, K. & Bassler, D. Probiotics for prevention of necrotizing enterocolitis in preterm infants. *Cochrane Database Syst Rev*, Vol. 23, No. 1, (January 2008), CD005496.

Aly, H. et al (2005). Is bloodstream infection preventable among premature infants? A tale of two cities. *Pediatrics*, Vol 115, (2005), 1513–1518, ISSN 0031 4005.

Bacteriemia zero 1ª edición (May 2009). Basado en el proyecto "Keystone ICU" desarrollado por la Universidad John Hopkins (Pronovost, P. et al. *N Eng J Med*, 2006; 2725: 32). Adaptado al español y publicado por el Ministerio de Sanidad y Consumo Español (actualmente Ministerio de Sanidad y Política Social), la SEMICYUC (Sociedad española de Medicina Intensiva y Unidades Coronarias) y el Departamento de Seguridad del Paciente de la WHO. Retrieved from http://www.seguridaddelpaciente.es/index.php/lang-es/component/content/article/42-sociedades-cientificas-prof/73-programa-para-reducir-las-bacteriemias-por-cateteres-venosos-centrales-en-las-uci-del-sns.pdf

Balmer, S.E. & Wharton, B.A. (1989). Diet and faecal flora of the newborn: breast milk and infant formula. *Archives of Disease in Childhood*, Vol. 64, (1989), 1672-1677.

Barclay, A.R. et al (2007). *J Pediatr Gastroenterol Nutr*, Vol. 45, No. 5, (November 2007), 569-76.

Barrington, K.J. & Finer, N.N. (2007). Inhaled nitric oxide for preterm infants: a systematic review. *Pediatrics*, Vol. 120, No. 5, (November 2007), 1088-99, ISSN 0031 4005.

Bertini, G. et al (2005). Fluconazole prophylaxis prevents invasive fungal infection in high-risk, very low birth weight infants. *J Pediatr* 2005; Vol. 147, (2005), 162-165.

Beth, H.S. et al (2001). Nosocomial infection rates in US children´s hospitals neonatal and pediatric intensive care units. *Am J* Infect *Control*, Vol. 29, (2001), 152-157.

Bhandari, A. & Bhandari, V. Pitfalls (2009). Problems and progress in bronchopulmonary dysplasia. *Pediatrics*, Vol. 123, No. 6, (2009), 1562-74, ISSN 0031 4005.

Boyce, J.M. & Pittet, D. For the Healthcare Infection Control Practices Advisory Committee and the HICPAC/SHEA/APIC/IDSA Hand Hygiene Task Force. Guideline for Hand Hygiene in Health-Care Settings. *Morb Mort Week Rep*, Vol. 51, RR16, (2002), 1-44.

Buhimschi, C.S. et al (2006). Clinical proteonomic: a novel diagnostic tool for the new biology of preterm birth, part I: proteomics tools. *Obstet Gynecol Surv*, 61 (7): 481-6.

Burón, E. & Grupo de RCP Neonatal de la Sociedad Española de Neonatología (2006). Reanimación del recién nacido. *An Esp Pediatr*, Vol. 65, No. 5, (November 2006), 470-477, ISSN 1695-4033.

Carlo, W.A. et al (2002). Minimal ventilation to prevent bronchopulmonary dysplasia in extremely-low-birth-weigt infants. *J Pediatr*, Vol. 141, (2002), 370-375.

Capretti, M.G. et al (2008). Impact of a standardized hand hygiene program on the incidence of nosocomial infection in very low birth weight infants. *Am J Infect Control*, Vol. 36, No. 6, (August 2008), 430-5.

CDC (September 2010): Centers for Disease Control and Prevention. Retrieved from http://www.cdc.gov

CgestiónE/I (September 2010): Club Excelencia en Gestión. Retrieved from http://www.clubexcelencia.org/

Clerihew, L. et al (2007). Prophylactic systemic antifungal agents to prevent mortality and morbidity in very low birth weight infants. *Cochrane Database Syst Rev*, Vol. 17, No. 4, (October 2007), CD003850.

Dani, C. et al (2002). Probiotics feeding in prevention of urinary tract infection, bacterial sepsis and necrotizing enterocolitis in preterm infants. A prospective double-blind study. *Biol Neonate*. 2002 Aug;Vol. 82, No. 2, (August 2002), 103-8.

Davis, J.M. et al (2003). Pulmonary outcome at 1 year corrected age in premature infants treated at birth with recombinant human CuZn superoxide dismutasa. *Pediatrics*, Vol. 111, No. 3, (2003), 469-76, ISSN 0031 4005.

Deshpande, G. et al (2007). Probiotics for prevention of necrotising enterocolitis in preterm neonates with very low birth weight: a systematic review of randomised controlled trials. *Lancet*, Vol. 369, No. 9573, (2007), 1614-1620.

Deshpande, G. et al (2010). Updated meta-analysis of probiotics for preventing necrotizing enterocolitis in preterm neonates. *Pediatrics*, Vol. 125, No. 5, (May 2010 - Epub 2010 Apr 19), 921-30, ISSN 0031 4005.

Down, S.M. & Sinha S.K. (2008). Aerolized lucinactant: a potential alternative to intrachacheal surfactant replacement therapy. *Expet Opin Pharmacother*, Vol. 9, No. 3 (2008), 475-8.

Edwards, A.M. & Parrett C.A. (2002). Intestinal flora during the first months of life: new perspectives. *British Journal of Nutrition* (2002), Vol. 88, Suppl. 1, (2002), S11–S18.

Edwards, A.M. et al (2005). Attributable Cost of Nosocomial Primary Bloodstream Infection in Pediatric Intensive Care Unit Patients. *Pediatrics*, Vol. 115, (2005), 868-72, ISSN 0031 4005.

Edwards, A.M. et al (2007). National Center for Infectious Diseases (NCID). Retrieved from http://www.cdc.gov/ncidod

Edwards, A.M. et al (2007). National Healthcare safety Network (NHSN) Report, data sumary for 2006, issued June 2007. *Am J Infect Control*, Vol. 35, No. 5, (June 2007), 290-301.

EFQM (2010, September): European Foundation for Quality Management. Retrieved from http://www.efqm.org/en/

EuroNeoNet: European Neonatal Network (2010, September). Retrieved from http://www.euroneostat.org/paginas/publicas/euroneo/euroNeoNet/index.html

ESPNIC (2010, September): European Society of Paediatric and Neonatal Intensive Care. Retrieved from http://www.espnic.org/

EVALUACIONEFQM (2010, September). Retrieved from http://www.evaluacionefqm.com/

Fahey, J.O. (2008). Clinical management of intra-amniotic infection and chorioamnionitis: a review of the literature. *J Mild-wifery Womens Health*, Vol. 53, No. 3, (2008), 227-35.

Fay, J.P. & Faries R.N. (1975). The inhibitory action of fatty acids on the growth of E coli. *Journal of General Microbiology*, Vol. 91, (1975), 233-240.

Finer, N.N. et al (2004) Delivery room continous positive airway pressure/positive end-expiratory pressure in extremely low birth weight infants: a feasibility trial. *Pediatrics,* Vol. 114, no. 3, (2004), 651-657, ISSN 0031 4005.

Geary et al (2008). Decreased incidence of bronchopulmonary dysplasia after early management changes, including surfactant and nasal continuous positive airway pressure treatment at delivery, lowered oxygen saturation goals, and early amino acid administration: a historical cohort study. *Pediatrics*, Vol. 121, No. 1, (2008), 89-96, ISSN 0031 4005.

Geffers, C. et al (2008). Incidence of healthcare associated infectios in high-risk neonates: results Fromm the German surveillance system for very-low-birth weight infants. *J Hosp Infect*, Vol. 68, (2008), 214-221.

Giagliardi, L. et al (2007). Antenatal steroids and risk of bronchopulmonary dysplasia: a lack of effect or case of over-adjustment? *Paediatr Perinat Epidemiol*, Vol. 21, No. 4, (2002), 347-53.

Girou, E. et al (2002). Efficacy of hand rubbing with alcohol-based solution versus standard hand washing with antiseptic soap: randomized clinical trial. *British Medical Journal,* Vol. 325, (2002), 362-6.

Guía de la calidad (September 2010) Retrieved from http://www.guiadelacalidad.com/modelo-efqm.php?op=11

Grohskopf, L.A. et al (2002). Pediatric Prevention Network. A national point-prevalence survey of pediatric intensive care unit-acquired infections in the United States. *J Pediatr,* Vol. 140, No. 4, (2002), 432-8.

Healy, C.M. et al (2005). Impact of fluconazole prophylaxis on incidence and outcome of invasive candidiasis in neonatal intensive care unit. *J Pediatr,* Vol. 147, (2005), 166-171.

Howlett, A. & Ohlsson, A. (2003). Inositol for respiratory distress syndrome in preterm infants. *Cochrane Database Syst Rev,* Vol. 4, (2003), CD000366.

Infectious Diseases Society of America (2009). Clinical Practice Guidelines for the Management of Candidiasis (2009 Update). *Clinical Infect Diseases,* Vol. 48, (January 2009), 503-35.

ISTAC: Instituto Canario de Estadística (2010). Retrieved fromhttp://www2.gobiernodecanarias.org/istac/
Kaufman, D.A. & Manzoni, P. (2010). Strategies to prevent invasive candidal infection in extremely preterm infants. *Clin Perinatol,* Vol. 37, No. 3, (September 2010), 611-28.
Kilbride, H.W. et al (March 2003). Evaluation and Development of Potentially Better Practices to Prevent Neonatal Nosocomial Bacteraemia. Evidence-Based Quality Improvement in Neonatal and Perinatal Medicine: The NIC/Q 2000 Experience. *Pediatrics,* Vol. 111, Suppl., (March 2003), e504-e518. Retrieved from http://pediatrics.aappublications.org/content/vol111/issue4/#SUPPLSE1
Kitajima, H. et al (1997). Early administration of Bifidobacterium breve to preterm infants: randomised controlled trial. *Arch Dis Child Fetal Neonatal Ed,* 1997 Mar; Vol. 76, No. 2, March 1997), F101-7.
Kugelman, A. et al (2007). Nasal intermittent mandatory ventilation versus nasal continuous positive airway pressure for respiratory distress syndrome: a randomized, controlled, prospective study. *J Pediatr,* Vol. 150, No. 5, (2007), 521-6.
Levine, C.R. et al (2005). The safety, pharmacokinetics, and anti-inflammatory effects of intrachateal recombinant human Clara cell protein in premature infants with respiratory distress syndrome. *Pediatr Res,* Vol. 58, No. 1, (2005), 15-21.
Saima, L. Strategies for prevention of nosocomial sepsis in the neonatal intensive care unit. *Curr Opin Pediatr,* Vol. 18, 101-106.
Moracho, O. (2003). Calidad hospitalaria: modelo de excelencia EFQM del Hospital de Zumarraga, Ozakidetza (Servicio Vasco de Salud). "Jornada Signo", Hospital Ramón y Cajal, Madrid, Marzo 2003.
Morley, C.J. et al (2008). CPAP or intubation at birth for very preterm infants. *New Engl J Med,* Vol. 358, No. 7, (2008), 700-8.
NHSN (September 2010): Nacional Healthcare Safety Network (CDC- NHSN). Retrieved from http://www.cdc.gov/nhsn/
NNIS (2003): National Nosocomial Infections Surveillance System (NNIS) Report 2003, data summary from january 1992 trough june 2003. *AmJ Infect Control,* Vol. 31, No. 9, (June 2003), 481-498.
NCID (September 2010): National Center for Infectious Diseases. Retrieved from http://www.cdc.gov/ncidod
O'Grady, N.P. et al (2002). Guidelines for the prevention of intravascular catheter-related infections. *Morb Mort Week Rep,* Vol. 51, RR10, (2002), 1-26.
PREVINE (2010, September): Programa específico para la vigilancia y control de la infección nosocomial en los hospitales españoles. Retrieved from http://www.mpsp.org/mpsp/Documentos/inf_nosoc/inf_hos.htm
Raymond, J. & the European Study Group (2000). Nosocomial infections in pediatric patients: an European, multicenter prospective study. *Infect Control Hosp Epidemiol,* Vol. 21, (2000), 260-3.
Rogowski, J.A. et al (2001). Economic implications of neonatal intensive care unit collaborative quality improvement. *Pediatrics,* Vol. 107, (2001), 23-9, ISSN 0031 4005.
Rowin, M.E. et al (2003). Pediatric intensive care unit nosocomial infections: epidemiology, sources and solutions. *Crit Care Clin,* Vol. 19, No. 3, (2003), 473-87.

Saiman, L. & National Epidemiology of Mycosis Survey Study Group (2000). Risk factors for candidemia in neonatal intensive care unit patients. *Pediatr Infect Dis J,* Vol. 19, (2000), 319-324.

Saiman, L. (2006). Strategies for prevention of nosocomial sepsis in the neonatal intensive care unit. *Curr Opin Pediatr,* Vol. 18, (2006), 101-106.

Schanler R.J. (2006). Probiotics and necrotising enterocolitis in premature infants. *Arch Dis Child Fetal Neonatal Ed.,* Vol. 91, No. 6, (2006), F395-F397.

Schreiber, M.D. et al (2003). Inhaled nitric oxide in premature infants with the respiratory distress syndrome. *N Engl J Med,* Vol. 349, (2003), 2099-2107.

SECIP (September 2010): Sociedad Española de Cuidados Intensivos Pediátricos. Retrieved from http://secip.blogspot.com/

SEN (September 2010): Sociedad Española de Neonatología. Retrieved from http://www.se-neonatal.es/default_principal.asp?idx=&cidioma=2

SEN1500 (September 2010): Red española de unidades neonatales de resultados y seguimiento de los recién nacidos de muy bajo peso. Retrieved from http://www.se-neonatal.es/default_principal.asp?idx=&cidioma=2

Stevens et al (2007). Early surfactant administration with brief ventilation vs. selective surfactant and continued mechanical ventilation for preterm infant with or at risk for respiratory distress syndrome. *Cochrane Database Syst Rev,* Vol. 4, (2007), CD003063.

Stover, B.H. & Pediatric Prevention Network (2001). Nosocomial infection rates in US children's hospitals' neonatal and pediatric intensive care units. *Am J Infect Control,* Vol. 29, No. 3, (2001), 152-7.

The Joint Commission (September 2010). Retrieved from http://www.jointcommission.org/GeneralPublic/Complaint/sp_qi_review.htm

Urrea, M. et al (2003). Prospective incidence study of nosocomial infections in a pediatric intensive care unit. *Pediatr Infect Dis J,* 22 (6): 490-4.

Urrea, M. et al (2007). Nosocomial infections in paediatric and neonatal intensive care units. *Journal of Infection,* Vol. 54, (2007), 212-220.

Vitaly S.H. & Arnold J.H. (2005). Bench-to-bedside review: Ventilator strategies to reduce lung injury: Lessons from pediatric and neonatal intensive care. *Critical Car,* (2005), 177-183.

VON (September 2010): International Vermont Oxford Network for improvement the quality and security of the medical assistance to the newborn and families. Retrieved from http://www.vtoxford.org/

Watterberg, K. (2006). Anti-inflammatory therapy in the neonatal intensive care unit: present and future. *Semin Fetal Neonatal Med,* Vol. 11, No. 5, (2006), 378-84.

WHIKIPEDIA (September 2010): La enciclopedia libre, web de la fundación española Wikimedia. Retrieved from http://es.wikipedia.org/wiki/Fundaci%C3%B3n_Europea_para_la_ Gesti%C3%B3n _de_la_Calidad

WHO (September 2010): World Health Organization. Retrieved from http://www.who.int/es/

Won, S.P. et al (2004). Handwashing program for the prevention of nosocomial infections in a neonatal intensive care unit. *Infect Control Hosp Epidemiol,* Vol. 25, No. 9, September 2004), 742-6.

Woodgate, P.G. & Davies M.W. (2001). Permissive hypercapnia for the prevention of morbidity and mortality in mechanically ventilated newborn infants. *Cochrane Database Syst Rev*, Vol. 2, (2001), CDOO2061.

Yokoe, D.S. & Classen, D. (2008). Introduction: improving patient safety-through Infection Control: a new health care imperative. *Infect Control Hosp Epidemiol*, Vol. 29, Suppl, (2008), S3-S11.

13

Primer Design to Sequence Analysis - a Pipeline for a Molecular Genetic Diagnostic Laboratory

Elaine Doherty[1], Renate Marquis-Nicholson[1], Jennifer M. Love[1], Clare Brookes[1], Debra Prosser[1] and Donald R. Love[1,2]
[1]*Diagnostic Genetics, LabPlus, Auckland City Hospital, PO Box 110031, Auckland 1148,*
[2]*School of Biological Sciences, University of Auckland, Private Bag 92019, Auckland 1142*
New Zealand

1. Introduction

Confirmation of a clinical diagnosis of a heritable disorder usually involves the analysis of the coding exons of a gene, plus surrounding intronic sequences. Variants that are identified using this strategy can be classed as either disease-causing or non-pathogenic polymorphisms. The variants comprise point mutations, micro insertions and deletions (indels), and copy number changes.

The detection of small variants (such as substitutions and indels) can be identified using a number of techniques including SSCP (Orita et al., 1989), DHPLC (Yu et al., 2006) and high-resolution melt-curve analysis (Garritano et al., 2009; Montgomery et al., 2010; Ngyyen-Dumont et al., 2009). These techniques can de difficult to optimise and can result in both false negative and false positive results. The accepted gold standard of mutation detection, however, is dideoxy-based sequencing. This is the most direct approach and can be the most sensitive of all techniques. However, it is also the most expensive option and can be the most time-consuming due to bottle-necks arising from designing appropriate primers, the analysis of subsequent sequence data, and the interrogation of data to filter out sequence variants that are not disease-causing. Within our laboratory, we experienced difficulties in three principal areas: standardising the primer design; objectively evaluating sequence data that can be performed in an automated manner thereby addressing the bottle-neck of trace sequence evaluation; and responding to the increasing need for in-depth analysis of sequence variants to inform referring physicians as to the clinical significance of the data.

To resolve the above issues, we developed a system that addressed each of these areas. The first involved designing primers that are assessed for unique amplification of targeted genomic regions using standard conditions. The aim here was not to tailor PCR conditions to a primer pair, but to force all primer designs to satisfy only one condition. The ancillary design feature of this module was to tail all primers such that all amplicons could be sequenced bi-directionally using only two primers (the "tail" sequences), thereby avoiding sequencing using amplicon-specific primers. The second area involved the wholesale adoption of Variant Reporter™ software (Applied Biosystems Ltd) for the automated analysis of all sequence data. This decision led to a significant initial investment in time to establish appropriate reference sequence projects for each of the genes analysed by our laboratory. But critically, these templates can be applied to any gene of interest and have been successfully implemented within our laboratory to produce quality trace data for use in a diagnostic testing. The third area comprised a streamlined approach to bioinformatic analysis of sequence variants, which has replaced *ad hoc* and superficial analysis in order to support our reports to referring physicians. This aspect of our work remains a constant challenge and should be viewed as work-in-progress as more web-based bioinformatic tools become available.

2. Materials and methods

2.1 DNA purification

DNA was extracted from peripheral blood (taken in EDTA) using the Gentra Puregene DNA Extraction kit (Qiagen Ltd). If blood spots/Guthrie spots were used then DNA was extracted using the Qiagen Miniblood kit (Qiagen Ltd).

2.2 PCR conditions

PCR was performed using the following reagents in a 25µl reaction volume: 50 ng genomic DNA, Faststart buffer (Roche Applied Science), 2 mM $MgCl_2$, forward and reverse primers at 0.8µM each, 0.4mM dNTPs (GE Healthcare Ltd), 1 unit Fast Start Taq DNA polymerase (Roche Applied Science) and +/- 5µl GC-rich solution (Roche Applied Science); most PCRs did not require GC-rich solution.

The reaction conditions for all PCRs were: 95°C for 4 minutes followed by 35 cycles of 94°C for 45 seconds, 60°C for 30seconds and 72°C for 30seconds, with a final extension of 72°C for 10 minutes.

Primers designed using the Primer3 programme of the UCSC genome browser were tested first under the conditions above. If PCR failed, a second set of primers was designed using Primer3 and manual adjustment of the region to be amplified, or the alternative software design tool, FastPCR (see later).

5 µl of PCR product was purified using ExoSAP-IT® (USB) and subjected to bidirectional sequence analysis using Big-Dye Terminator v3.0 (Applied Biosystems), and 2.5ng per 100bp of amplified product. 20µl sequenced products were purified using Clean-Seq (Agencourt). 15ul of purified sequenced product was then subjected to capillary electrophoresis using a 3130xls Applied Biosystems Genetic Analyser with a 50cm capillary array.

2.3 Sequence analysis

All trace sequences were analysed with KB™ Basecaller v1.4. Most of the sequence analysis parameters were set using the default values defined in Variant Reporter™ (Applied Biosystems). The mixed base threshold was set at the default of 25%. After base calling, the 5′ and 3′ ends were trimmed to remove sequence until fewer than 4 bases in a window of 20 had Quality values of less than 20; the resulting sequence is termed the clear range or contiguous read length (CRL). Traces were filtered with a minimum Trimmed Read Length (TRL) of fewer than 50 base pairs; the TRL is sequence that remains after trimming using the defined quality values. Traces were filtered with a Median Under Peak (PUP) of less than 4 (this is the median value within the clear range of the ratio of the signal of the highest secondary peak to the signal of the main called base). Traces were filtered if the expected TRL was less than 50% of the trace length providing a minimum trace score of 35. This is the average quality value of bases in the clear range which equates to a base-call error rate of

0.031%. A reference negative control DNA sample was sequenced for each gene in order for the algorithm to work.

3. Results

3.1 Primer design

Our main aim was to produce high-quality sequence traces spanning all coding exons plus flanking intronic sequence of any gene of interest. We arbitrarily chose to examine 20 base pairs of flanking intronic sequence in order to detect splice acceptor and donor variants. The priority for the primer design process was to produce primers that allowed for PCR amplification under standard buffer conditions using only one cycling protocol. Placing the emphasis on primer design rather than PCR optimisation was advantageous in reducing both technical time and reagent costs. Figure 1 shows the primer design workflow.

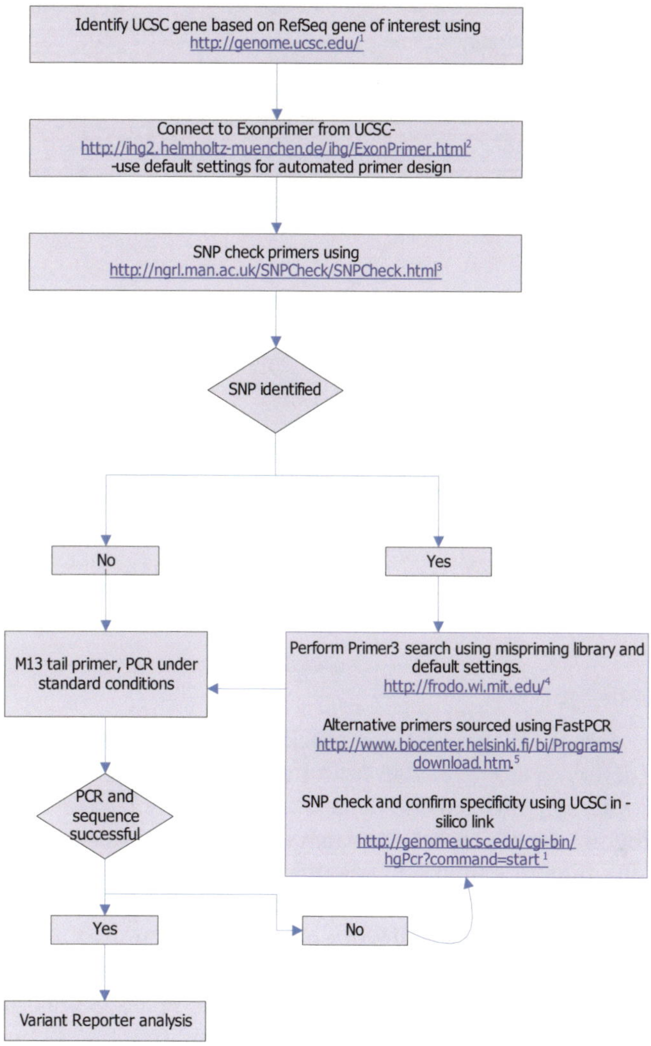

Fig. 1. Primer design workflow

The first objective involves the identification of the mRNA sequence of each gene of interest. This is achieved by interrogating the public UCSC genome browser page http://genome.ucsc.edu/ (Figure 2A) and identifying an appropriate RefSeq accession number (Figure 2B), and the relevant UCSC accession number (Figure 2C). This website also provides a direct link to ExonPrimer for design of primers specific to this mRNA sequence (http://ihg.helmholtz-muenchen.de/ihg/ExonPrimer.html); see Figure 2. ExonPrimer provides exon and amplicon size and creates additional primer sets as alternatives. We use the ExonPrimer default parameters as this achieves our objective of designing primers flanking the 20 bases of intronic sequence.

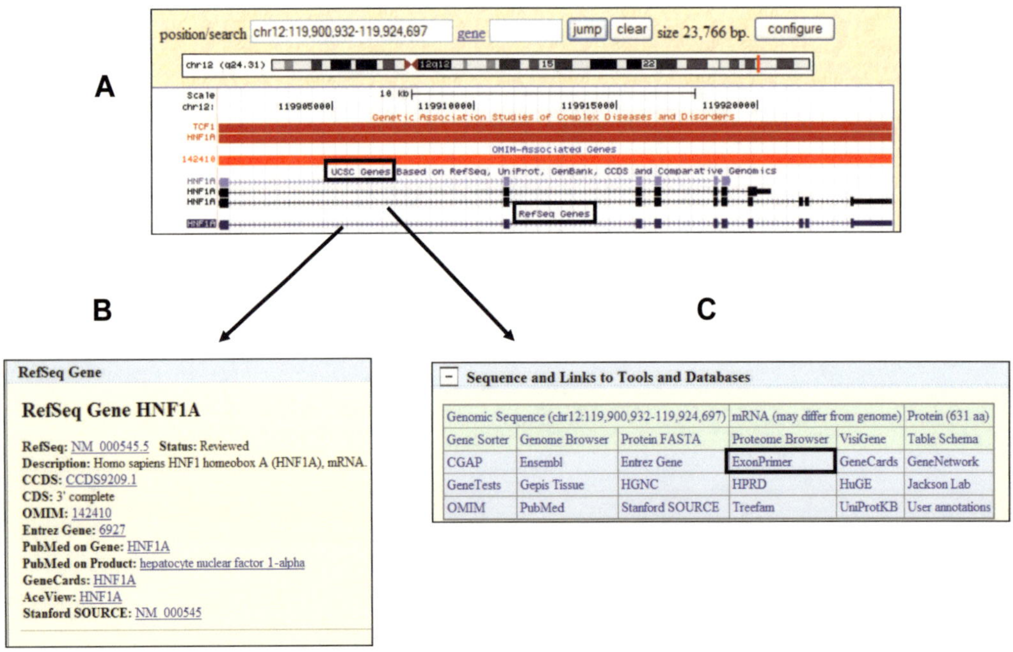

Fig. 2. ExonPrimer design

Following primer design, all primers are checked for single nucleotide polymorphisms (SNPs) using the software tool available from the National Genetic Reference Laboratory, Manchester, UK (https://ngrl.manchester.ac.uk/SNPCheckV2/snpcheck.htm). This is a bioinformatics program which allows for batch checking of primers for SNPs. It uses the current build of the human genome to identify the position in the sequence where the primers bind. The current release of dbSNP is used to detect any known SNPs at the primer binding sites. Primers that do not show SNPs are tailed with universal M13 sequences and used in PCR +/- GC-rich solution and our standard amplification conditions. Figure 3 shows a schematic of the tailed primers flanking two exons of a hypothetical gene (panel A), and the amplification products achieved after the PCR amplification of 11 unique exons (+/- genomic DNA).

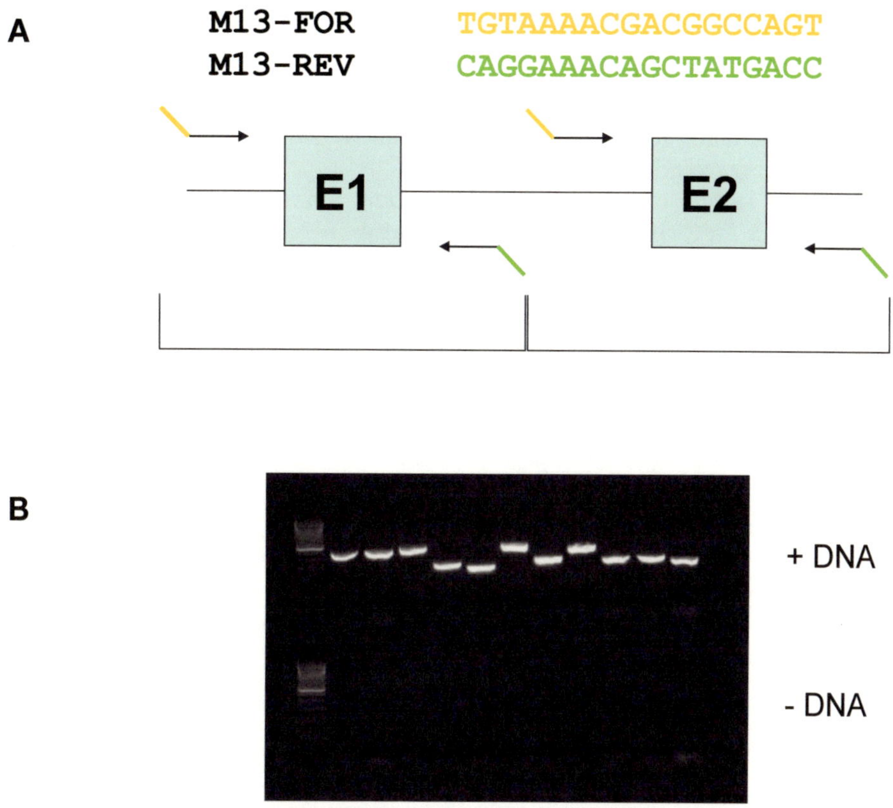

Fig. 3. Schematic of amplicons and gel electrophoresis of amplicons

If a SNP is detected then a second set of primers is designed using Primer3 (Rozen & Skaletsky et al., 2000; http://frodo.wi.mit.edu/primer3/) in which we use the default settings; however, we highlight the human mis-priming library to avoid designing primers to regions of interspersed repeats. If the need arises, alternative primers are also designed using the freely available FastPCR programme (http://www.biocenter.helsinki.fi/bi/Programs/download.html). This programme includes various bioinformatics tools that can be used for the design of a variety of PCR applications. Following identification of primers using the default settings, we select primer pairs with annealing temperatures close to 60^{0}C and amplicon lengths of approximately 400bp for optimal electrophoresis through our 50cm capillary array. Again, SNP checks of these primers are performed and all successful primers are tailed with universal M13 forward or reverse sequences. Unlike Exonprimer, these tools do not have an integrated search for cross-complementation to other areas of the genome. The UCSC website can be used for this purpose by selecting the *in silico* PCR option. This screens the primer pairs against the human genome and provides details of sequences that lie between the primers, and includes the primer pair. It also includes chromosome location and hence can be used to identify primer pairs which amplify homologous regions of the genome. The above design-on-demand pipeline has enabled us to design and implement primers for screening a number of disease-associated genes; we are in the process of writing a programme that will perform all of the above steps automatically.

3.2 Variant Reporter™ Reference sequence design

Our chosen software for the analysis of sequence traces is Variant Reporter™ developed by Applied Biosystems. This software tool uses advanced algorithms and quality metrics to automate the detection of variants and streamline the data analysis process, thereby greatly reducing the time required to analyse trace sequences. Additional features also allow for quality trace scoring over defined regions.

The first step in setting up sequence analysis using Variant Reporter™ is to produce a reference sequence for each gene, as shown in figure 4.

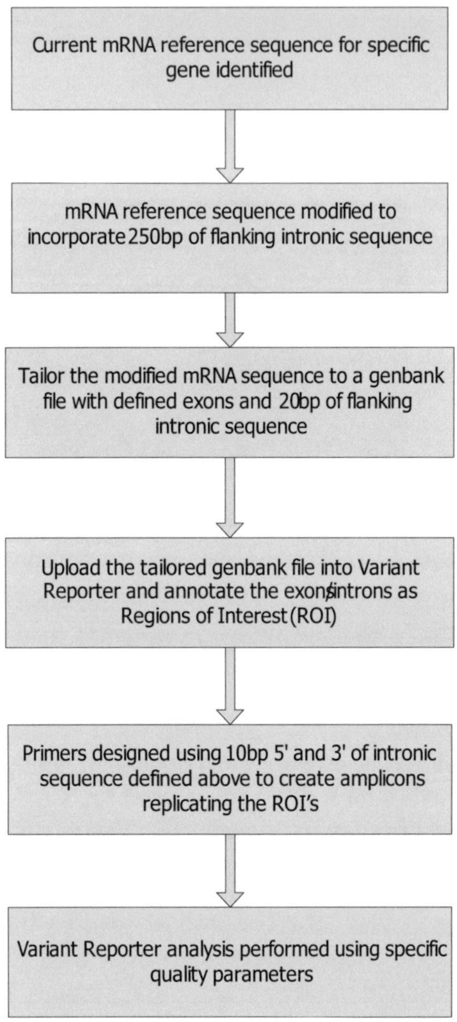

Fig. 4. General workflow in setting up a project in Variant Reporter™

In our experience, the Variant Reporter™ programme has difficulty accepting large genomic reference sequences, therefore we reduce the reference sequence to a size which can be accommodated by the software. In our laboratory the reference sequences used are those of the mRNA of the gene of interest, which can be downloaded from the UCSC genome browser. This approach is in accordance with current international guidelines, which specify the reporting of changes in relation to the mRNA accession number as specified by the Human Genome Variation Society (HGVS) Guidelines (http://www.genomic.unimelb.edu.au/mdi/mutnomen/). We modify each mRNA reference sequence using the UCSC genome browser to include enough intronic sequence to cover the maximum length of every amplicon we intend to sequence. This involves amending the mRNA reference sequence of the gene of interest to include 250bp of 5′ and 3′intronic sequence.

We then adapt the modified mRNA reference file to a tailored Genbank reference file which includes the defined exons and 20bp of flanking introns to create a "pseudo Genbank" file. Once the file has been imported into Variant Reporter™ it can be annotated to define the regions of interest (ROIs).

In the case of small genes, the workflow for constructing appropriate files for Variant Reporter™ is described below.

1. **Identify the mRNA reference sequence for the targeted gene**

The first stage is to identify the current mRNA reference sequence for the gene of interest in the UCSC genome browser. This is done by entering the UCSC genome browser and clicking on the 'Genomes' link and entering the gene name or RefSeq accession number in the search term box (Figure 5A). On the next page, click on the highlighted RefSeq (Figure 5B), and scroll to the 'Links to sequence' and click on 'Genomic sequence' (Figure 5C).

A

B

C

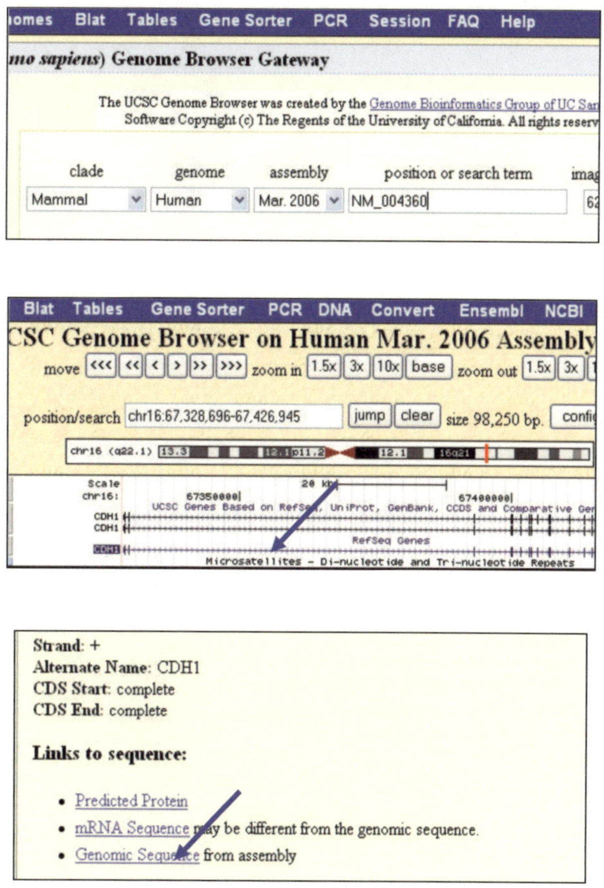

Fig. 5. Retrieving genomic sequence of gene of interest

2. **Modify the mRNA reference sequence to incorporate 250bp of flanking intronic sequence.**

The Sequence Retrieval Region Option is used to set the regions to 250bp (or to the user specification; Figure 6A). The Sequence Formatting Options allows for exon formatting in upper case and intronic regions in lower case (Figure 6B). This formatting is extremely useful for downstream visualisation of these regions in Variant Reporter™. The resulting sequence can be pasted into an appropriate program (such as notepad) in which all the RefSeq annotations are removed and the file is saved as .txt file. The lengths of each exon can be calculated easily using the word count character option. At this stage the modified mRNA sequence can be uploaded into Variant Reporter™. The exons and flanking intronic sequence can then be manually annotated.

A

B

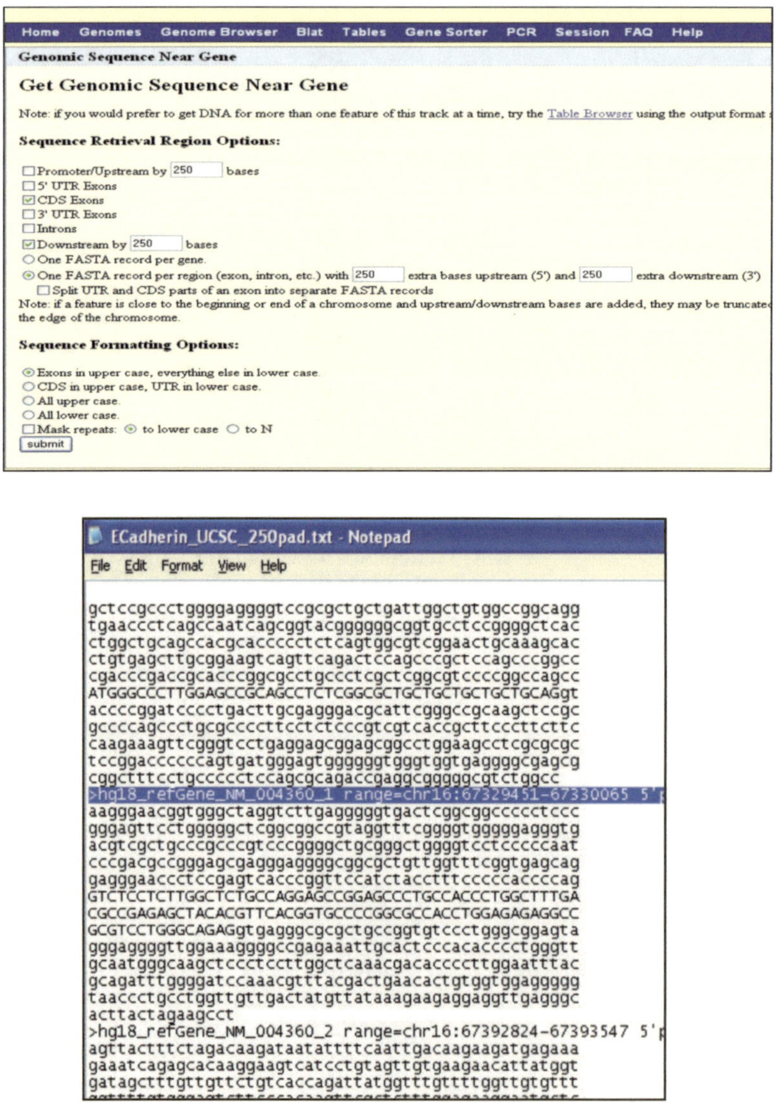

Fig. 6. Retrieving exon sequences with flanking 250bp of intron sequence

3. **Annotate exons and introns as regions of interest (ROI)**

The 'define reference' link of Variant Reporter™ is used to import the 'modified reference sequence' from step 2. At this point the user can choose to delete redundant reference sequences using the 'delete layer' option from the 'Layers' tab. The Variant Reporter™ reference sequence can now be annotated according to the user specifications using the 'Create/Edit ROI' option from the ROI tab. The ROI start for each exon is specified on the 'Cumulative Exon start co-ordinates section' of an exon formatting spreadsheet.

4. **Design primers using 10bp 5' and 3' of intronic sequence to create amplicons replicating the ROI.**

A main feature of Variant Reporter™ is that it provides quality statistics based on the amplicon, not the region of interest (ROI). Therefore, we adopted an *in silico* approach to make the ROIs coincident with the amplicons. This approach involved using the first 10 bases of each 5′ intronic region as the forward primer, and reverse complemented the last 10 bases of the corresponding 3′ intron to create the reverse primer. A freely available reverse-complement web-site (http://www.bioinformatics.org/SMS/rev_comp.html) can be used for this purpose. The primers can be easily identified by highlighting the corresponding bases in Variant Reporter™ and pasting them into a .txt file (Figure 7A).

Fig. 7. Creation of regions of interest to replicate amplicons

The reverse-complement web browser is then used to reverse complement the 3′ intron. An example of the .txt format that can be used to upload the primers into Variant Reporter™ is shown in Figure 7B. Note that the primer pair name will correspond to the amplicon name as it appears in Variant Reporter™. Once the primer sequences have been imported, the amplicons are automatically created and correspond to the exon plus flanking 20bp intron (Figure 7C). Nucleotide numbering of the coding regions starts with nucleotide 1 as A of the major initiation codon. Intronic sequences are referred to as -n or +n where –1 denotes the invariant G of the AG splice acceptor site, and +1 indicates the invariant G of the GT splice donor site, in accordance with HGVS nomenclature.

5. **Establishing specific quality parameters in Variant Reporter™**

The user can select default Analysis Parameters or customize their personal settings. We choose to set a minimum trace score of 35 equating to an average base-call error rate of 0.031%. These can be set using the Analysis Parameters tab.

The above workflow has been used to establish sequencing projects for a large number of genes, some of which are shown in Table 1.

Disease	Gene	RefSeq mRNA accession no.	Coding exons
Adrenoleukodystrophy	*ABCD1*	NM_000033.2	1-6
Gastric Cancer (E-Cadherin)	*CDH1*	NM_004360.2	1-16
Maturity Onset of Diabetes in the Young (MODY) type 2	*GCK*	NM_000162.3	1-10
MODY3	*HNF-1a*	NM_000545.4	1-10
Non-insulin dependent diabetes mellitus (NIDM) type I	*HNF-4a*	NM_000457.3	1-10
Non ketotic hyperglycinaemia	*GLDC*	NM_000170.2	1-25
Holocarboxylase synthetase deficiency	*HLCS*	NM_000411.4	4-12
Rett Syndrome	*MECP2*	NM_004992.3	1-4
Paragangliomas and Pheochromocytoma	*SDHB*	NM_003000.2	1-9
	SDHC	NM_003001.3	1-6
	SDHD	NM_003002.1	1-4
Myoclonus dystonia	*SGCE*	NM_001099400.1	1-11
Von Hippel-Lindau syndrome	*VHL*	NM_000551.2	1-3
Long QT Syndrome (LQT) type 1	*KCNQ1*	NM_000218.2	1-16
LQT2	*HERG*	NM_000238.2	1-15
LQT3	*SCN5A*	NM_198056.2	2-28

LQT5	*KCNE1*	NM_000219.3	3
LQT6	*KCNE2*	NM_172201.1	2
LQT7	*KCNJ2*	NM_000891.2	2
Myoclonus dystonia	*SGCE*	NM_003919.2	1-10
Familial adenomatous polyposis	*APC*	NM_000038.4	2-16
Multiple Endocrine Neoplasia (MEN) type 2A/2B	*RET*	NM_020630.4	10-16

Table 1. Genes for which Variant Reporter™ projects have been established

3.3 Bioinformatic analysis

Comparison of the sequence traces from a patient with those of a negative control individual in Variant Reporter™ results in the identification of sequence variants that differ from the relevant RefSeq sequences. The final step for a diagnostic laboratory is to use a method of "filtering" these variants to identify those that are benign (likely to be, but not necessarily, silent mutations at the amino acid level), and so provide a much more limited set of variants that may or may not be disease-causing. Our routine work flow is shown in Figure 8, which accesses a number of web-based analysis tools to interrogate the data in a formal manner. The reader can refer to many recent publications that highlight the work of others in classifying sequence variants in the context of their disease association (Abkevich et al., 2004; Sullivan et al., 2006; Lee et al., 2008; Arnold et al., 2009).

The work flow can be divided into a series of modules, which are summarised below:

1. Convert the locations of sequence variants in HGVS nomenclature (http://www.hgvs.org/mutnomen/) into predicted consequences on translation using the Mutalyzer web site (http://www.mutalyzer.nl/2.0/).
2. Interrogate sequence variants against SNP databases to determine if they represent known polymorphisms with associated rs identifiers. These variants are present in the general population and so would likely be non-pathogenic.
3. Interrogate sequence variants against disease-associated databases such as the Leiden Open Variation Database (LOVD; http://www.lovd.nl/2.0/) and the HGVS locus specific mutation databases (http://www.hgvs.org/dblist/glsdb.html).
4. Predict the functional consequences of predicted amino acid changes on protein function using evolutionary sequence conservation and physico-chemical considerations provided by web-based tools such as SIFT (http://sift.jcvi.org/www/SIFT_seq_submit2.html) and Polyphen-2 (http://genetics.bwh.harvard.edu/pph2/); Ng & Henikoff 2003; Tavtigian et al., 2008; Adzhubei et al., 2010; Flanagan et al., 2010.
5. Predict the effect of sequence variants on splicing (Walker et al., 2010).

The outcome of the above workflow provides laboratory staff with information that either strongly supports the conclusion that a sequence variant is a (benign) polymorphism, a disease-causing mutation, or of unknown clinical significance. The latter creates dilemmas for the laboratory, but usually leads to continuing family studies to determine if the variant co-segregates with the disease, or *in vitro*/modelling studies that necessitate research-based laboratory involvement.

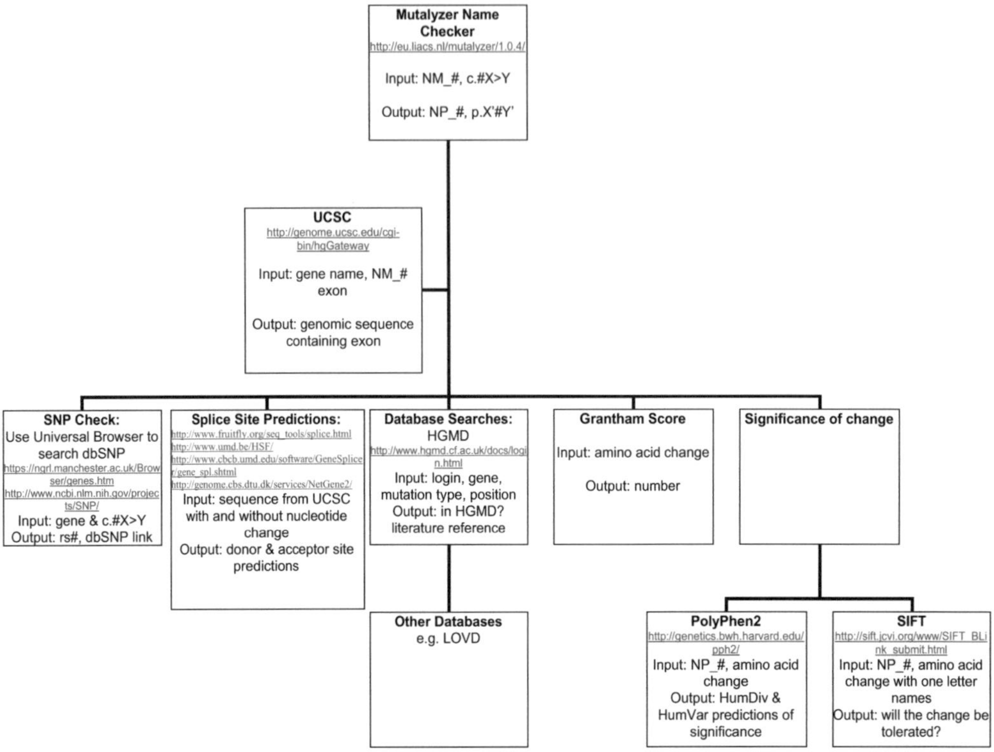

Fig. 8. Bioinformatic workflow

4. Discussion

The integration of the processes described above has allowed us to design a sequencing strategy suited to obtaining high-throughput quality controlled sequence data. To date, we have evaluated over 1000 sequence traces with the end point being high-quality data passing defined quality thresholds set in Variant Reporter™.

The first stage of the sequencing pipeline involved primer design. Here, our main emphasis was to create a structured template requiring minimal intervention from the operator. A tiered strategy consisting of the UCSC browser, Primer3 and FastPCR tools enables the selection of several primer pairs which can be M13-tailed and tested under standard conditions. By focusing initially on primer design rather than reaction optimisation we have reduced costs in technical time as well as reagents. In addition, M13 tailing of final-stage primers also decreases both operator time and error and opens the process to automation. In combination with the use of the SNP software tool and UCSC *in silico* PCR, this has enabled the rapid design of specific primers.

We chose to use primers which covered at least 20 bases of flanking introns in order to detect splice variants. Other laboratories may choose to design primers at different sites to include deeper intronic variants. However, there is a lack of international best practice guidelines defining the length of intronic sequence necessary for diagnostic reporting. Until this is resolved, each laboratory will continue to sequence differing flanking lengths affecting the detection of deeper intronic variants that potentially have pathogenic properties. As recommended by the Clinical Molecular Genetics Society (CMGS; http://cmgsweb.shared.hosting.zen.co.uk/BPGs/Best_Practice_Guidelines.htm, retired publication), it is good practice for the diagnostic laboratory to specify the length of flanking intron that is sequenced in order for the clinician to gain a clear understanding of the genomic regions that have been assessed for each patient.

The second stage of the sequencing pipeline involved establishing Variant Reporter™ reference sequences and downstream analysis. This was initially run in parallel to our established sequencing analysis system, a more manual-orientated software tool which compared test and control sequences. Here, it was necessary for the user to scan the sequences as not all changes were automatically detected by the software. These variants were not genotyped by the software, therefore the user had to manually call the variant. In addition, the software did not provide quality scoring of the sequence data.

Initial problems were encountered as Variant Reporter™ was unable to accept large genomic reference sequences. Therefore we chose to create a modified reference sequence to a size that could be accommodated by the software. This approach involved using the UCSC genome browser to incorporate 250 bp of flanking intronic sequence into the selected mRNA sequence. In turn, this modified sequence could be annotated as a Genbank file and uploaded into Variant Reporter™ to create a reference file that included highlighted regions of interest. This process alleviated the time-consuming step of creating the reference sequence manually in Variant Reporter™.

At the point of analysis it is necessary to incorporate a reference sample in the project in addition to the test sample. We routinely use a pre-sequenced negative control to compare with patient samples. Without the use of this reference sample we have found that Variant Reporter™ fails to detect pre-identified variants. As described by Applied Biosystems, this is due to the algorithm requiring a second sample for trace comparison.

Once optimisation of the reference file and project parameters was completed, we analysed over 1000 sequence traces in parallel using Variant Reporter™ and our conventional method. All substitutions identified by our conventional method were successfully detected and genotyped automatically by Variant Reporter™. In addition, the heterozygosity/homozygosity of each variant matched. All data passed our quality threshold set in Variant Reporter™ of a minimum trace score of 35 equating to an average base-call error rate of 0.031% (as now specified on all diagnostic reports produced from our laboratory). Our new system has reduced the time taken to analyse samples to roughly 10% of the original time (a reduction from 8 hours for a full multi-gene screen to 1 hour).

A major inconsistency which we have identified, however, is the failure of Variant Reporter™ to identify and correctly name indels. In order to circumvent this problem we manually evaluate all samples not reaching 100% trimmed read length (TRL) as identified through the trace quality review using the procedures above. The % TRL is the length of the sequence that remains after trimming, divided by the amplicon length. A % TRL falling beneath the 100% threshold as identified through the trace score feature can be used to indicate a potential insertion or deletion. Our in-house protocol for these samples now involves scanning of these traces in the variant review window.

Once an insertion/deletion mutation has been detected visually we manually score the result. This approach also addresses the problem of inconsistencies in genotyping data in that Variant Reporter™ uses a different nomenclature system from the current HGVS international guidelines for diagnostic laboratories (http://cmgsweb.shared.hosting.zen.co.uk/BPGs/Best_Practice_Guidelines.htm).

Both of these issues are currently being assessed by Applied Biosystems.

In our experience the Variant Reporter™ default trace score of 25 allows for trailing C shoulders indicative of a sub-optimal array. By increasing this score to 35, the threshold is raised sufficiently to exclude low-quality sequence data. At present there is no universal consensus on minimal quality settings that should be used by the genetic community. Until there is, each laboratory will continue to use its own standards of quality scoring which will result in different rates of false negatives for each laboratory. We feel that international guidelines are needed to address this concern within the diagnostic laboratory.

While we were implementing enhanced primer design and the assessment of sequence traces using Variant Reporter™, our bioinformatic analysis strategy also required attention. The approach described here is evolutionary as more programmes become available on-line to predict the biological significance of sequence variants. In this respect, it should be noted that similar challenges in biological interpretation are occurring in the field of whole genome molecular karyotyping (Marquis-Nicholson et al., 2010). The construction of international databases comprising well-curated data from diagnostic laboratories (Cotton et al., 2008) should improve our understanding of variants and improve the genetic analysis of individuals and their families with heritable disorders.

5. References

Abkevich, V.; Zharkikh, A.; Deffenbaugh, A.M.; Frank, D.; Chen, Y.; Shattuck, D.; Skolnick, M.H.; Gutin, A. & Tavtigian, S.V. (2004). Analysis of missense variation in human BRCA1 in the context of interspecific sequence variation. *J Med Genet,* 41(7), 492-507.

Adzhubei, I.A.; Schmidt, S.; Peshkin, L.; Ramensky, V.E.; Gerasimova, A.; Bork, P.; Kondrashov, A.S. & Sunyaev, S.R. (2010). A method and server for predicting damaging missense mutations. *Nat Methods,* 7(4), 248-249.

Arnold, S.; Buchanan, D.D.; Barker, M.; Jaskowski, L.; Walsh, M.D.; Birney, G.; Woods, M.O.; Hopper, J.L.; Jenkins,. M.A.; Brown, M.A.; Tavtigian, S.V.; Goldgar, D.E.; Young ,J.P. & Spurdle, A.B. (2009). Classifying MLH1 and MSH2 variants using bioinformatic prediction, splicing assays, segregation, and tumor characteristics. *Hum Mutat,* 30(5), 757-770.

Cotton, R.G.; Auerbach, A.D.; Axton, M.; Barash, C.I.; Berkovic, S.F.; Brookes, A.J.; Burn, J.; Cutting, G.; den Dunnen, J.T.; Flicek, P.; Freimer, N.; Greenblatt, M.S.; Howard, H.J.; Katz, M.; Macrae, F.A.; Maglott, D.; Möslein, G.; Povey, S.; Ramesar, R.S.; Richards, C.S.; Seminara, D.; Smith, T.D.; Sobrido, M.J.; Solbakk, J.H.; Tanzi, R.E.; Tavtigian, S.V.; Taylor, G.R.; Utsunomiya, J.& Watson, M.(2008). GENETICS. The

Human Variome Project. *Science*, 322(5903), 861-862.Orita, M.; Iwahana, H.; Kanazawa, H.; Hayashi, K. & Sekiya, T.(1989). Detection of polymorphisms of human DNA by gel electrophoresis as SSCPs. *Proceedings of the National Academy of Sciences of the United States of America (USA)*, 86(8), 2766-2770.

Flanagan, S.E.; Patch, A.M. & Ellard, S. (2010). Using SIFT and PolyPhen to predict loss-of-function and gain-of-function mutations. *Genet Test Mol Biomarkers*, 14(4), 533-537.

Garritano S, Gemignani F, Voegele C, Nguyen-Dumont T, Le Calvez-Kelm F, De Silva, D.; Lesueur, F.; Landi, S. & Tavtigian, S.V. (2009). Determining the effectiveness of High Resolution Melting analysis for SNP genotyping and mutation scanning at the TP53 locus. *BMC Genet*, 10, 5.

Lee, E.; McKean-Cowdin, R.; Ma, H.; Chen, Z.; Van Den Berg, D.; Henderson, B.E.; Bernstein, L. & Ursin, G. (2008). Evaluation of unclassified variants in the breast cancer susceptibility genes BRCA1 and BRCA2 using five methods: results from a population-based study of young breast cancer patients. *Breast Cancer Res*, 10(1), R19.

Marquis-Nicholson, R.; Aftimos, S.; Hayes, I.; George, A. & Love, D.R. (2010). Array comparative genomic hybridisation: a new tool in the diagnostic genetic armoury. *N Z Med J*, 123(1318), 50-61.

Montgomery, J.L.; Sanford, L.N. & Wittwer, C.T. (2010). High-resolution DNA melting analysis in clinical research and diagnostics. *Expert Rev Mol Diagn*, 10(2), 219-240.

Ng, P.C. & Henikoff, S. (2003). SIFT: predicting amino acid changes that affect protein function. *Nucleic Acids Res.* 31(13), 3812-3814.

Nguyen-Dumont, T., Calvez-Kelm, F.L., Forey, N., McKay-Chopin, S., Garritano, S., Gioia-Patricola, L., De Silva, D., Weigel, R., Sangrajrang, S., Lesueur, F.; Tavtigian, S.V.; Breast Cancer Family Registries (BCFR); Kathleen Cuningham Foundation Consortium for Research into Familial Breast Cancer (kConFab). (2009). Description and validation of high-throughput simultaneous genotyping and mutation scanning by high-resolution melting curve analysis. *Hum Mutat*, 30(6), 884-890.

Rozen, S. & Skaletsky, H. (2000). Primer3 on the WWW for general users and for biologist programmers. In: *Bioinformatics Methods and Protocols: Methods in Molecular Biology*, Krawetz, S. & Misener, S. (Eds), 365-386, Humana Press, Totowa, NJ.

Sullivan, L.S.; Bowne, S.J.; Birch, D.G.; Hughbanks-Wheaton, D.; Heckenlively, J.R.; Lewis, R.A.; Garcia, C.A.; Ruiz, R.S.; Blanton, S.H.; Northrup, H.; Gire, A.I.; Seaman, R.; Duzkale, H.; Spellicy, C.J.; Zhu, J.; Shankar, S.P. & Daiger, S.P. (2006). Prevalence of disease-causing mutations in families with autosomal dominant retinitis pigmentosa: a screen of known genes in 200 families. *Invest Ophthalmol Vis Sci*, 47(7),3052-3064.

Tavtigian, S.V.; Greenblatt, M.S.; Lesueur, F.; Byrnes, G.B.; IARC Unclassified Genetic Variants Working Group. (2008). In silico analysis of missense substitutions using sequence-alignment based methods. *Hum Mutat*, 29(11), 1327-1336.

Walker, L.C.; Whiley, P.J.; Couch, F.J.; Farrugia, D.J.; Healey, S.; Eccles, D.M.; Lin, F.; Butler, S.A.; Goff, S.A.; Thompson, B.A.; Lakhani, S.R.; Da Silva, L.M.; kConFab Investigators, Tavtigian, S.V., Goldgar, D.E.; Brown, M.A. & Spurdle AB. (2010). Detection of splicing aberrations caused by BRCA1 and BRCA2 sequence variants encoding missense substitutions: implications for prediction of pathogenicity. *Hum Mutat*, 31(6), E1484-505.

Yu, B.; Sawyer, N. A.; Chiu, C.; Oefner, P. J. & Underhill, P. A. (2006). DNA mutation detection using denaturing high-performance liquid chromatography (DHPLC). *Current Protocols in Human Genetics*. 7.10.1–7.10.14.

14

Innovative Approaches in Quality Management in Clinical Laboratories

Lina Souan, Maher A. Sughayer
King Hussein Cancer Center
Jordan

1. Introduction

Clinical laboratories must adhere to Good Laboratory Practices (GLP) to ensure the quality, integrity, and reliability of patient results. GLP principles include organization and personnel, a quality assurance program, facilities, equipments, reagents and materials, test systems, standard operating procedures, result reporting, and archiving of records and reports (Wikipedia,2010).
This chapter presents a review of the latest requirements in quality control and quality management in certified clinical laboratories. We will describe different stages in clinical testing procedures and highlight the critical steps that need to be monitored using quality assurance indicator tools. In addition, we will present our latest contribution to the field of quality control computer software which encompasses an innovative approach in monitoring quality control and quality assurance in laboratories.
Accumulating and analyzing quality control data as well as quality indicators help in developing the performance improvement plan in the laboratory. Performance improvement is a process which uses information from multiple sources to increase the probability of accomplishing desired outcomes to better serve the needs and expectations of patients. The most frequently used performance improvement tools are: 1. FOCUS-PDCA – framework for improvement, 2. Failure Modes and Effects Analysis (FMEA), 3. Root Cause Analysis (RCA) and 4. Control Charts. However, in this chapter we will focus on one model of quality improvement approaches which is FOCUS-PDCA and control charts and give examples on its implementation in a clinical laboratory setting.

2. Imperative

At the end of this chapter, you will be able to understand different stages of analytical testing; differentiate between internal and external controls, analyze common problems associated with invalid test results, have an overview of different QC software and our contribution to advance these programs.
Traditionally, medical laboratories used conventional paper-based systems for assessing quality of medical tests performed in the laboratory. The main drawbacks for these traditional quality control monitoring systems can be summarized in the following points.
Manual paper-based quality monitoring systems have a long lag time between tests and application of quality control procedures. The absence of a reporting system of the different

system variables hinders the long-term planning and expansion of the system. Clinical laboratories that are still using these traditional ways of monitoring the quality control results have reduced reliability and productivity as there have been manually applying and assessing quality control rules such as Westgard rules (Al-Taee,2009).
Because of these problems and the desire to deliver high quality services, most of modern medical laboratories are planning to gradually transform all administrative and communicative work to become electronic. Hence, ample efforts are put in automating the quality control process in clinical laboratories in order to increase quality of the results. Several quality control programs are commercially available; however, we created a new program that captures the results entered by each medical technologist. The program analyzes the validity of each test, suggests corrective action(s) for failed quality control data as well as reports the number of failures in performing the daily quality control per each technologist. This not only confirms validity of the implemented quality control process, but delivers higher confidence in lab testing procedures, and allows for more reliable test results. Moreover, analyzing failure reports generated by the program which are related to poor performance enables the medical director in assessing the medical technologists' competency in addition to evaluating the performance of the instrument/ clinical analyzers. Furthermore, this data can be used in setting Quality Improvement plans in the laboratory (Al-Taee,2009).

3. Quality management in clinical laboratories

3.1 Stages in performing analytical tests

All analytical tests are divided into three stages: Pre-analytical stage, analytical stage and a post analytical stage.

3.1.1 Pre-analytical stage

The pre-analytical stage is the stage that precedes the tests analysis. This stage consists of several factors such as: Written policies and procedures, requisition for analysis, preparation of patient, patient identification, effective blood collection, mixing with anticoagulants, storage until transportation, transportation collection or courier service (Sharma,2009), and keeping accurate laboratory logs to ensure complete records of specimen arrival. These steps are performed by physicians, nursing staff, doctor's assistant and laboratory staff. Recent statistics have shown that 46–68.2% of total laboratory result errors are due to pre-analytical factors (Plebani,2006). Hence, controlling these steps by monitoring them and analyzing the data reduces the infidelity in lab test results.

3.1.2 Analytical stage

Analytical factors that influence the quality of laboratory results may include factors such as: Method of analysis, reagents used for the test, equipment, calibration, controls used in the test run, analysis of the control results using statistical programs, training and education of the laboratory technologist/technician. One of the important steps in the analytical stage is running quality control material.
Quality control (QC) refers to the routine technical activities that must be included during each assay run to verify that the test is working properly (Westgard, *et al.*,2006). The main purpose of running QC material is to ensure the validity of test method and test result, to

prevent occurrence of errors by getting erroneous patient results, to identify the cause of errors and to correct errors if and when they occur.

The following steps should be followed in order to achieve the aforementioned goals of running QC material.

- Select the proper method for the analysis.
- Adopt the proper QC system
- Train the laboratory technologists on GLP practices and request that they get certification of training on the tests they perform.
- Keep all instruments calibrated and well maintained by performing the daily, weekly and monthly maintenance indicated by the manufacturer.
- Make sure that all reagents are handled and stored properly by closely monitoring the temperature in the refrigerators/ freezers and equipment during test analysis.
- Make sure that the technicians are following the written policies on properly handling and processing patient samples.

3.1.2.1 Types of Quality Control

Quality Control is divided into two categories: Internal and External Quality Control.

Internal Quality Control to the test kit is defined as control samples provided with the test kit with known reactivity.

External Quality Control to the test kit is defined as control samples not included with the test kit but provided from an external source that has been validated for use with a specific test kit.

Invalid Quality Control:

In an event that the Quality Control results were invalid, then follow one or more of the following corrective actions:

- Repeat test
- If repeatedly invalid:
    - assume a problem with test product or procedure
    - continue with alternative testing algorithm
    - Identify cause of problem
    - Inform supervisor
    - Take corrective actions

Table 1- presents examples of some of the problems encountered while performing controls of serological tests and the proposed actions that should be taken to correct theses problems. Records of internal control results must be periodically reviewed. These records should be reviewed before accepting test results. As for the external control results, they should be reviewed by test performer as well as reviewed weekly or monthly by testing site supervisor.

External Quality Control:

External Quality Control samples can be either prepared by a "Reference Laboratory" or commercially prepared.

The frequency of using test External Control samples could be as follows:

- Minimum once a week and preferably at the beginning of the week
- New shipment of test kits
- Starting a new lot number
- If environmental conditions exceeded range needed for stability of kits.

Problem	Potential Cause	Action
No control line or band present	Damaged test device or controls	Repeat the test using new device and blood sample.
	Proper procedure not followed	Follow each step of testing according to the written standard operating procedure. Re-check buffer and/or specimen volumes. Wait for the specified time before reading the test.
	Expired or improperly stored test kits or controls	Check expiration date of kits or controls. Do not use beyond stated expiration date Check temperature records for storage and testing area
Positive reaction with negative external control, i.e. false positive	Incubation time exceeded	Re-test negative control using a new device and read results within specified time limit
Extremely faint control line	The control line can vary in intensity	No action required. Any visible line validates the results.

Table 1. Potential causes of problems encountered while running QC & corresponding corrective action(s).

In summary, using internal and external quality control insures that the reagent material and solutions involved in the assay procedure did not deteriorate after validation, and they can be incorporated in the run one or more times every run of analysis. Data from running control material is plotted on control charts and interpreted following what is called Westgard Statistical rules (Thompson,2010).

3.1.2.2 Statistical QC Rules

a. Arithmetic Mean or average

The arithmetic mean of a group of numbers is found by dividing their sum by the number of members in the group (BioRad,1995). For example, the sum of the four numbers 3, 4, 5 and 20 is 32; hence, their mean is 32 divided by 4, which comes up to 8. In clinical laboratories, the arithmetic mean is often called the target value.
Note that when control material is assayed the values of the controls rarely fall on the target mean. The values rather show distribution around the mean.

b. Standard Deviation

Standard Deviation is a measure of the dispersion of a set of data from its mean. The more spread apart the data, the higher the deviation (BioRad,1995). The **sample standard deviation** is denoted by *s* and defined as follows:

$$s = \sqrt{\frac{1}{N-1}\sum_{i=1}^{N}(x_i - \overline{x})^2}, \tag{1}$$

Column A	Column B	Column C
x values	x- $\bar{x}$	(x- $\bar{x}$)^2
30	30-27.7=2.3	5.29
31	31-27.7=3.3	10.89
29	29-27.7=1.3	1.69
24	24-27.7=-3.7	13.69
28	28-27.7=0.3	0.09
33	33-27.7=5.3	28.09
20	20-27.7=-7.7	59.29
30	30-27.7=2.3	5.29
25	25-27.7=-2.7	7.29
27	27-27.7=-0.7	0.49
∑x= 277	∑(x- $\bar{x}$)= 0	∑((x- $\bar{x}$)^2)=132.1
$\bar{x}$= 27.7		

Table 2. Example illustrating the calculation of the mean and standard deviation of a set of control data points.

Where N= group size, x= variable and $\bar{x}$ is the mean. Example:

- First we calculated the arithmetic mean of the group which is (∑) the sum of the individual points 277 then divided by 10 to be equal to 27.7.
- Second, we subtracted the mean value from each control value (column B) and applied the following equation x- $\bar{x}$. As it is evident from table 1 that the difference can be a positive value or a negative value, however, the sum of the difference is always zero.
- Third, we squared the difference score to make all terms positive as seen in column C.
- Fourth, the squared difference was summed and used in the equation below to yield the predictable dispersion or standard deviation (SD or s).

$$s = \sqrt{\frac{1}{N-1}\sum_{i=1}^{N}(x_i - \bar{x})^2}, \qquad (2)$$

S= ((132.1/(10-1))$^{1/2}$= 3.83

The value of the standard deviation allows us to predict the range of control values that should be observed if the method remains stable. From the example above, we can conclude the following:

- The range of the data is from 20-33
- The mean equals to 27.7
- SD = 3.83 therefore one SD (1SD)= 3.83, 2SD=7.66, 3SD=11.49

Note that all values of data in the example above are approximately within the range of the mean ± 2SD. It is important to remember that the smaller the value of SD the better the result precision or reproducibility of the test.

c. Coefficient of variation

Coefficient of Variation (CV) is a measure of variability in a test or control level. It is expressed as a percent (BioRad,1995), and is calculated as:

$$CV(\%) = (\text{Standard deviation (s)} \div \text{Mean})(100) \qquad (3)$$

Example: If the mean value for a control group was 117.4 IU/L and the SD was 5.03 IU/L, then CV = (5.03 IU/L / 117.4 IU/L)(100) = 4.3%

Calculating the CV for a test is useful in comparing manufacturer claims, College of American Pathologists (CAP) survey results, and peer group QC reports. It can also be used as a part of the internal quality control system when performing patient precision testing if the test materials used are similar and CVs are determined under similar conditions.

d. Standard Deviation Index (SDI)

It is important to participate in external proficiency testing program. These programs calculate the standard deviation index (SDI) and use it to compare a laboratory's results to its peer group. Furthermore, it allows the lab to gauge its performance when compared to the peer group. The SDI for the mean is calculated as follows:

SDI = (mean of suspect data points - lab cumulative mean) ÷ lab cumulative standard deviation.

The target SDI is zero. This would indicate that the lab's performance is identical to the peer group average. Acceptable SDI values are between +/-1.0. Any test/method/instrument which has a SDI between +/-1.0 and 1.5 may have a problem and the lab should investigate and troubleshoot and correct any test/ method/ instrument which has an SDI of +/-2.0 or greater. The lab can use the SDI statistic to interpret proficiency testing results (BioRad,1995).

e. Median, Mode, and range

- Median: The median is the middle number of a set of values arranged in an increasing order (Reynolds,2003). For example for the variables 4, 7, 3, 6, 2, 9, 11, the median number is 6 (2, 3, 4, 6, 7, 9, 11). If you have an odd number of variables, the median is the middle number. But if you have an even number of variables, the median is the sum of the two middle values divided by 2.
- Mode: The mode is the value which occurs most frequently (Reynolds,2003). For example the mode for the variables 3,5,5,7,7,7,9,10,10, is 7.
- Range: The range is the difference between the largest and the smallest variable in a distribution. It gives an indication of the dispersion of the data points, but it does not indicate how the data within the range vary about the mean. For example, the range for the variables 6, 8, 7, 3, 10, 12, 6, 8, 7, 9,4, is 3-12.

f. Normal or Gaussian distribution

Many of the control data accumulated will follow a certain distribution, with a peak around a certain value, and a few data points that lie outside the central cluster. This curve is called the normal distribution, the bell curve or the Gaussian distribution (Schieltz,2010). A frequency distribution curve is said to be normal when the curve of histogram is symmetrical bell shape. In this case, the mean, median and mode are identical. The figure below represents the frequency distribution of a large set of laboratory values obtained by measuring a single control material. This distribution shows the shape of a normal curve. In this figure we can notice the following points.

- Mean ± 1 SD represents 68% of all values
- Mean ± 2SD represents 95% of all values (95% confidence limit)
- Mean ± 3SD represents 99.7% of all values.

At 95% confidence limit, it is obvious that 2.5% of the distribution is in the lower or left tail, and the same amount (2.5%) is present in the upper tail.

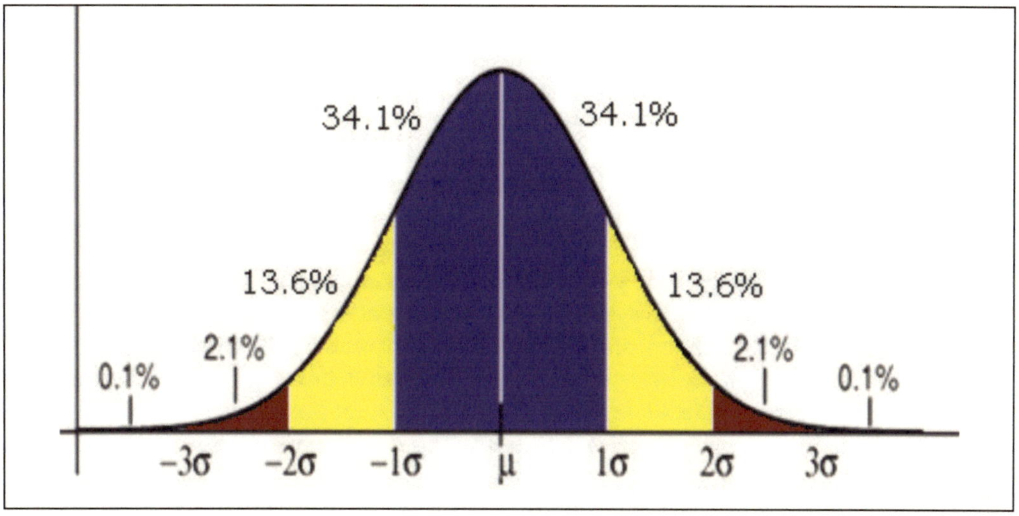

Fig. 1. Gaussian distribution illustrating, 68% of the values lie within 1 standard deviation of the mean; 95% lie within 2 standard deviations; and 99.7% lie within 3 standard deviations.

3.1.2.3 Statistical QC charts

QC data should be evaluated for its accuracy and precision and a decision has to be made either to accept or reject the results before reporting patients' results. Therefore, several methods and actions must be taken in order to reach to the correct decision such as method validation (Lumsden,2000) and participating in an External Quality Assurance program. Other measures have to be taken in the laboratory in order to assure quality of work and results such as personnel training and education, instrument calibration and maintenance, good record keeping and finally plotting and reviewing of Quality Control Charts.

Quality Control Charts are a graphical representation of QC raw data used to evaluate whether a measurement procedure is in-control or out-of-control. Control results are plotted versus time or sequential run number; lines are generally drawn from point to point to highlight any trends, systematic shifts, and random errors (Westgard,2008).

A control chart must include control limits which are lines drawn on the chart to provide graphical criteria for assessing whether a measurement procedure is in-control or out-of-control. These control limits are usually calculated from the mean and standard deviation (SD, or s) determined for a given control material. In general, interpreting the results is usually based on a specified number of results or points; when these data points are in-control; patient test-results are reported thereafter. When the control data are out-of-control, the run must be rejected and no test results can be reported before the QC is repeated and is acceptable.

The most commonly used QC chart is the Levey-Jennings chart. The Levey-Jennings chart usually has the days of the month plotted on the X-axis and the control observations plotted on the Y-axis. By observing the data plotted in the Levey-Jennings chart, we can determine if test results are in control and accurate, or if test results are not in control and consequently unacceptable. The use of the Westgard, and Cumulative Summation Rules aids in establishing an effective error-detecting plan. The Levey-Jennings chart has horizontal lines at the mean value and at the control limits, usually 2 or 3 standard deviation.

Manually constructed Levey-Jennings control charts used to be drawn by using graph paper having 10X10 or 20x20 lines per inch. One graph paper sheet is used for each level of the control. Two separate sheets must be used for a two control level because drawing both charts on a single sheet may reduce the readability of the control charts. The following steps are then are followed (Westgard,2002):

1. Label charts. Include the name of the test and the name of the control material in a top place so that this information is quickly and easily discerned when viewing the chart. The measurement unit in the following example is pg/mL, can be included in the label or included in the label for the y-axis. Other information typically included on the chart are the name of the analytical system, the lot number of the control material, the current mean and standard deviation, and the time period covered by the chart.
2. Scale and label x-axis. The horizontal or x-axis represents time and you will typically set the scale to accommodate 30 days per month or 30 runs per month. For this example, divide the x-axis into evenly sized increments and number them sequentially from 1 to 30. Label the x-axis "Day."
3. Scale and label y-axis. The y-axis represents the observed control value and you need to set the scale to cover the lowest and highest results expected. Generally the scale should cover a value as low as the mean - 4 standard deviations and a value as high as the mean + 4 standard deviations (Westgard,2002). In this example, the chart for Control 1 should be scaled to accommodate a range from 200 to 64, and 200 to 336.
4. Draw lines for mean and control limits. Mark the values that correspond to the mean and draw a black horizontal line (at 200 mg/dL for Control 1). Locate the values that correspond to the mean +1s and the mean -1s and draw blue horizontal lines (at 234 and 166 for Control 1). Locate the values that correspond to the mean +2s and the mean -2s and draw green horizontal lines (at 268 and 132 for Control 1). Locate the values that correspond to the mean +3s and the mean -3s and draw red horizontal lines (at 302 and 98 for control 1). Do the same procedure for Control 2 using the data of the mean and calculate SD as previously described and plot the data as shown in Figure 2.

Quality Control Rules

Control rule means a decision criterion for judging whether an analytical run is in-control or out-of-control. The rules for accepting or rejecting a run should be clearly stated in the quality control policy. For qualitative tests the QC rules are simple because the positive control must be positive and the negative control must be negative for the patient's results to be valid.

However, the quality control rules for the quantitative tests are set by the laboratory director. Some of the key points that the laboratory director should contemplate while setting the lab QC rules are: the medical necessity and precision of the test or assay used while setting the quality control limits, and refrain from choosing a single limit for all assays. Furthermore, the clinical significance of each test must be accounted for when evaluating QC limits and rules and the QC limits should be clearly stated in the QC policy manual and in the procedure manual. The most commonly used rule is the Levey-Jennings approach in which many laboratories choose 2 standard deviations on either side of the mean for their QC limits or rule. Since 95% of the values are covered by the 2 SD on either side of the mean in a normal distribution curve, the control will be out of limits in 1 out of 20 determinations by chance and any result exceeding 2SD values on either side of the mean is considered out of control (Westgard,2002).

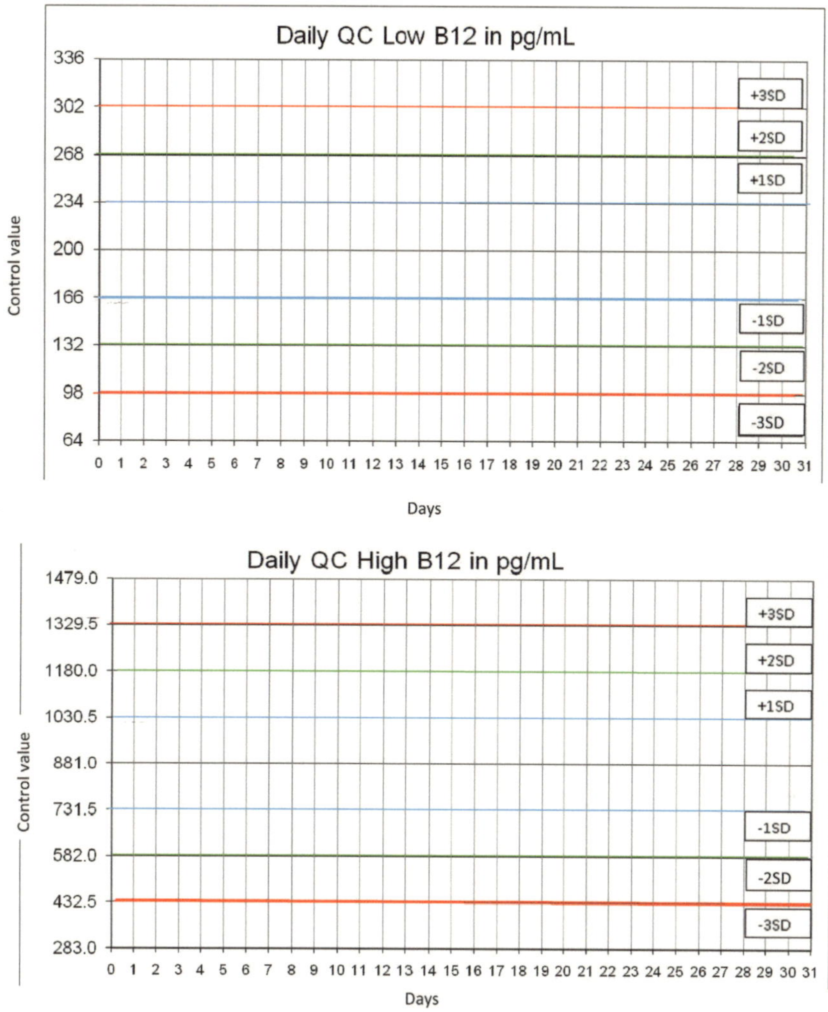

Fig. 2. A representation of Levey-Jennings chart with horizontal lines at the mean value and at the control limits, 1, 2 and 3 standard deviation.

Westgard Rules

It has been found that the false rejection rate is 5% when using one control level, and 9% when using two control levels while it goes up to 14% when using three control levels (Westgard,2002). In order to reduce the false rejection rate without compromising quality, Dr. James Westgard put some rules to analyze daily QC data.

Westgard rules are commonly used to analyse data in Levey-Jennings charts and to define specific performance limits for a particular assay. These rules can also be used to detect both random and systematic errors which will be explained later in this chapter. There are six commonly used Westgard rules of which three are warning rules and the other three are mandatory rules when two control levels are used.

The violation of warning rules should start a review of test procedures, reagent performance and equipment calibration. The violation of mandatory rules should result in the rejection of the results obtained with patients' samples in that assay (BioRad,1995).

Dr. Westgard adopted a short hand notation to abbreviate different decision criteria or control rules, e.g., 1_{2s} to indicate 1 control measurement exceeding 2s control limits. Combinations of rules are generally indicated by using a "slash" mark (/) between control rules, e.g. $1_{3s}/2_{2s}$.(Westgard,2009)

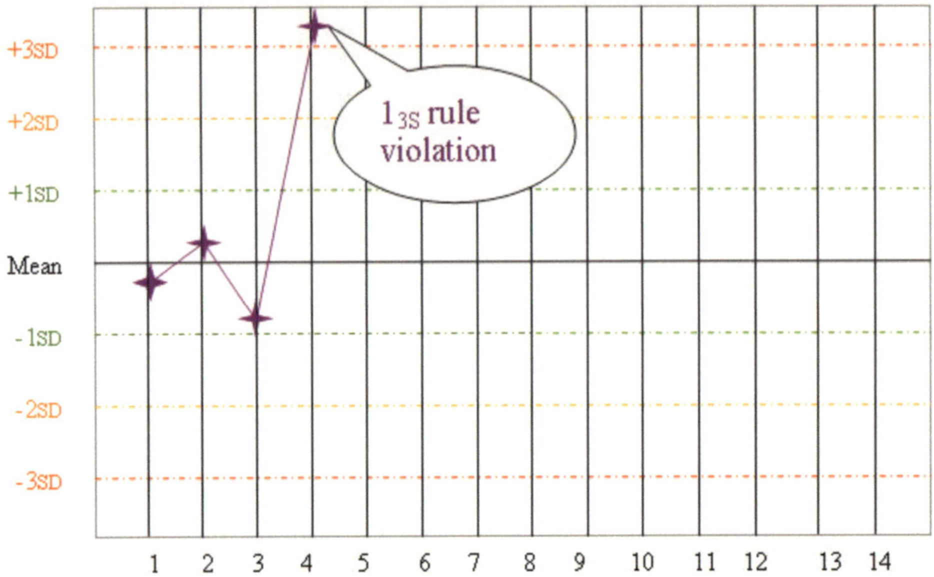

Fig. 3. Rule 1_{3s} - refers to plus/minus 3s. A run is rejected when a single control exceeds the mean ± 3s.

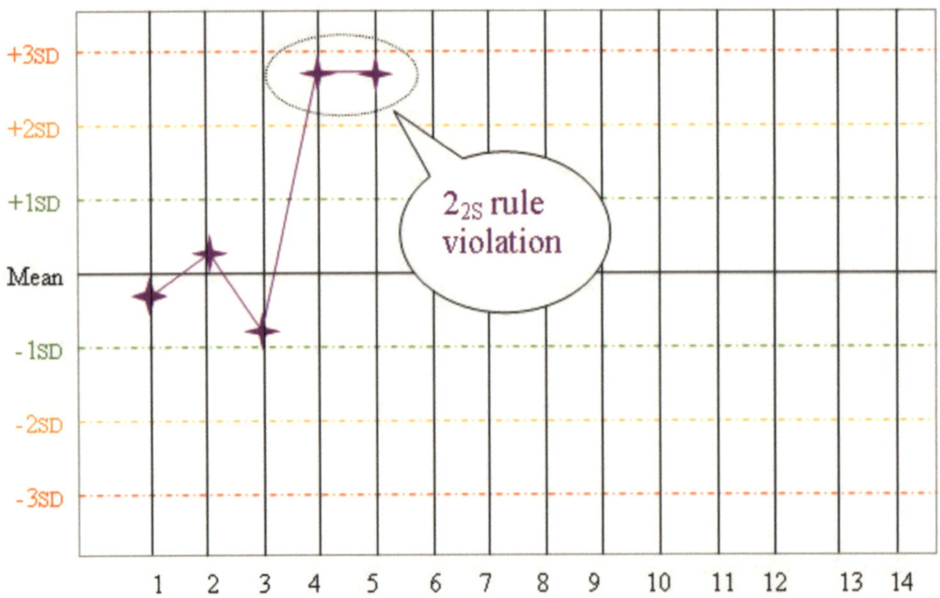

Fig. 4. Rule 2_{2s} - reject the run when 2 consecutive controls exceed the mean ± 2s.

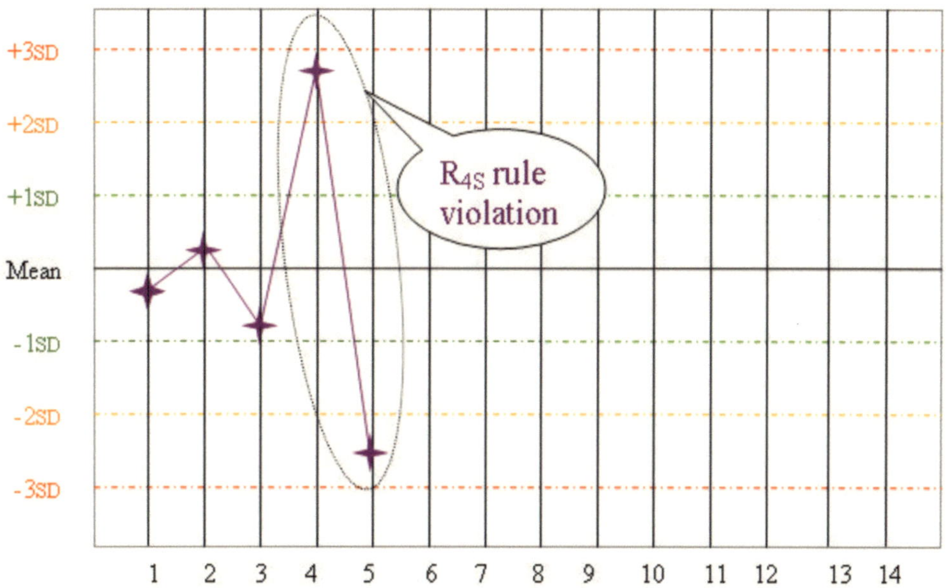

Fig. 5. Rule R_{4s} - when 1 control in a group exceeds the mean ± 2s and another control exceeds the mean in the other direction by 2s, reject the run.

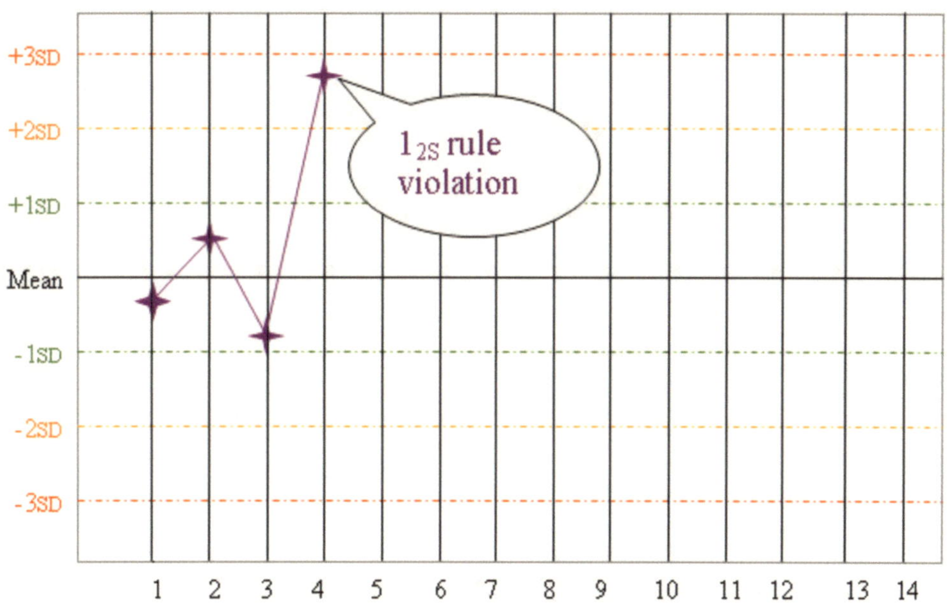

Fig. 6. Rule 1_{2s} - refers to the historical rule of plus/minus 2s from the mean, with multi-rules: a warning rule to trigger careful inspection of control data.

Fig. 7. Rule 4_{1s} - when 4 consecutive control measurements are on one side of the mean either ± 1s. Warning rule or a rejection rule depending on the accuracy of your instrument.

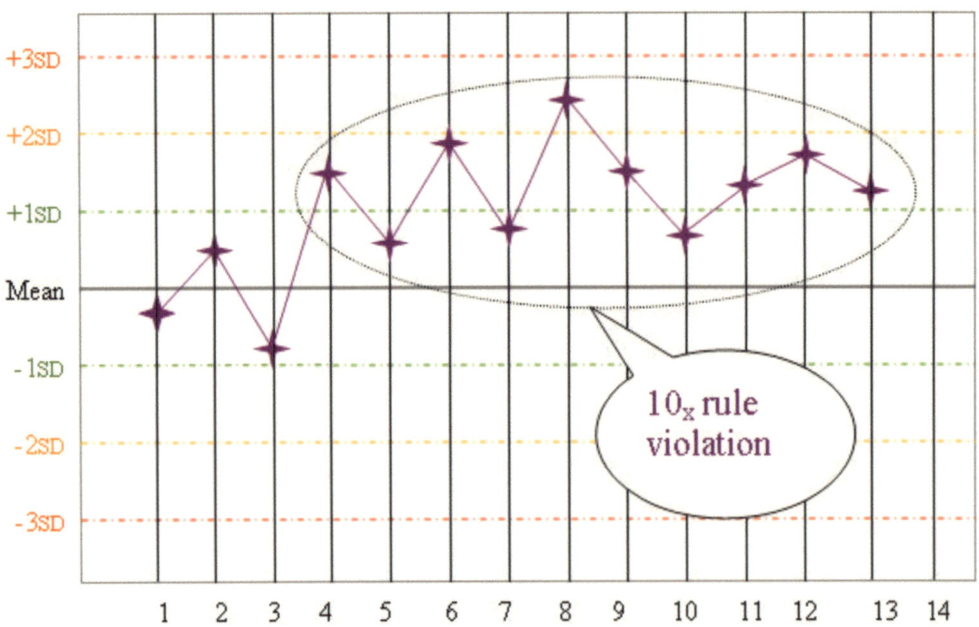

Fig. 8. Rule 10x - 10 consecutive control measurements fall on one side of the mean. If within 1 s, warning. If between 1 and 2 s, reject.

When three different control level materials are being analyzed, some other control rules fit better and are easier to apply (Westgard,2002), such as:

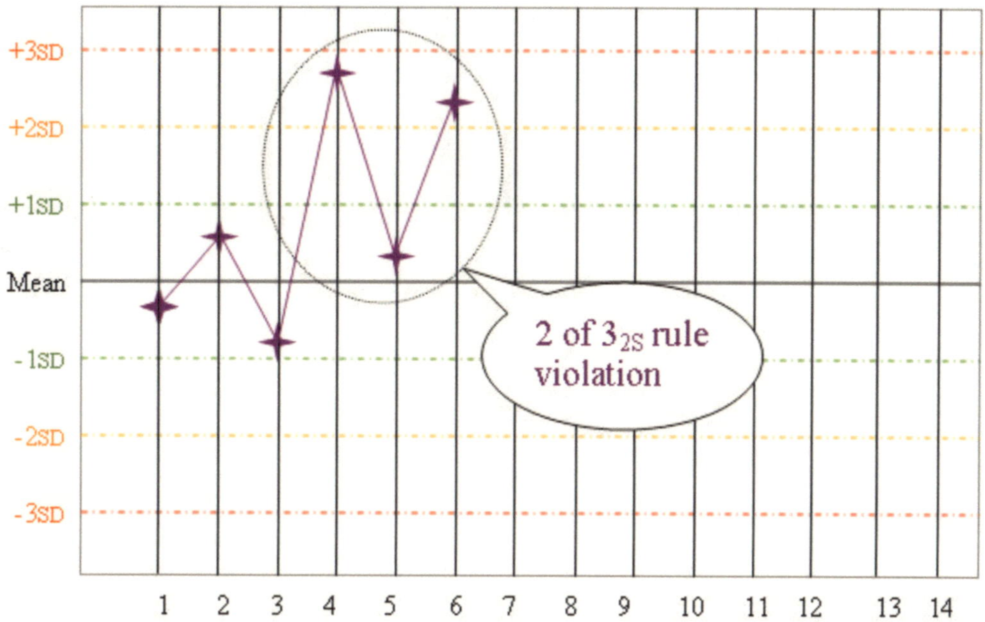

Fig. 9. Rule 2 of 3_{2s} - reject the run when 2 of 3 controls exceed the mean ± 2s

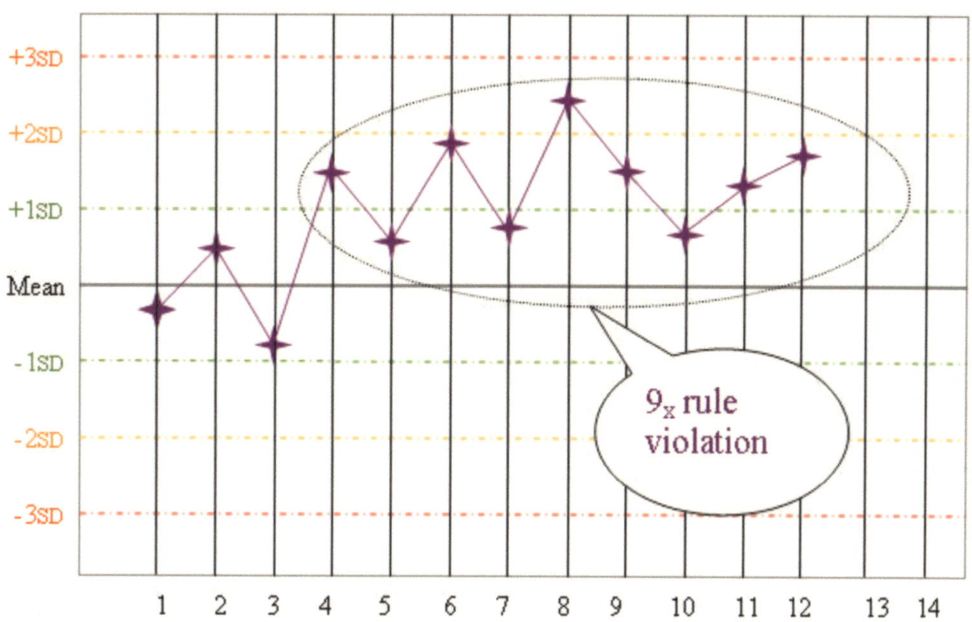

Fig. 10. Rule 9x - reject when 9 consecutive control measurements fall on one side of the mean.

Fig. 11. Rule 7T - reject when seven control measurements trend in the same direction, either higher or lower.

Examples of most commonly used "QC Rule Combinations":

- In Clinical Chemistry tests, the controls are usually run in multiples of 2, hence, the QC combination rule will be: 1_{3S} / 2_{2S} / R_{4S} / 4_{1S} / 10_X
- On the other hand in hematology, coagulation, blood gases tests, the controls are usually run in multiples of 3, hence, the QC combination rule will be: 1_{3S} / 2 of 3_{2S} / R_{4S} / 3_{1S} / 12_X

Random and systematic errors

As was mentioned earlier Levy-Jennings charts and Westgard rules help detect random and systematic errors in a run (BioRad,1995).

Random Errors are defined as those errors which affect the reproducibility or precision of a test system. They are detected by the use of 1_{3s} or R_{4s} rules and can be due to variations in line voltage, pipettes, dispensers, contamination, volume dispensed, bubbles in lines of reagents, etc.

The other type of errors is the Systematic Errors. Systematic Errors are bias or shifts and Trends in the QC data. These errors affect the accuracy of the test system. They are usually detected by the use of the 2_{2s}, 4_{1s}, or 10_x rules and can be due to calibration lot changes, temperature changes in incubator unit, light source deterioration, electronics, reagent lot changes, etc. (BioRad,1995).

Precision

Precision is the degree of agreement among repeated measurements of the same substance on the same sample, or on separate samples collected as close as possible in time and place. Precision indicates the following:

- How consistent and reproducible the laboratory method is.
- It shows how close the measurements are to each other.

- It may not mean that the sample results actually reflect the "true" value, but rather it reflects that analysis is giving consistent results under similar conditions.

Precision is verified through the use of duplicate samples or measurements (Derek-Law,2007). Duplicate samples are defined as more than one sample taken from the same source at the same time. Hence, precision is determined by calculating the Standard Deviation (SD) or coefficient of variation of the samples. The smaller the value of the coefficient of variation is (or standard deviation) the more precise the measurements.

Accuracy

Accuracy is a measure of confidence in a measurement. It indicates how close test results to the true or target value (Derek-Law,2007). The smaller the difference between the test results and its "true" or expected value; the more accurate the measurement will be. The more precise or reproducible the values are, the more reliable or accurate the result. Measurement of accuracy can be determined by comparing a sample that has a known value such as a standard reference material to the technician's measurement of that sample.

Example:

If the pH readings of a standard buffer were as follows:

7.5, 7.4, 6.7, 7.2, 6.8, 7.3, 6.5, 7.2, 6.8, 7.0, 7.4, 7.2

The mean of these results will be 7.08

Accuracy (bias) = mean value - true value (7.08-7.0) = 0.08

Therefore, the difference is biased by +0.08 units. The bias reflects the error in comparison to the mean. Figure 12 illustrates the difference between precision and accuracy (Derek-Law,2007).

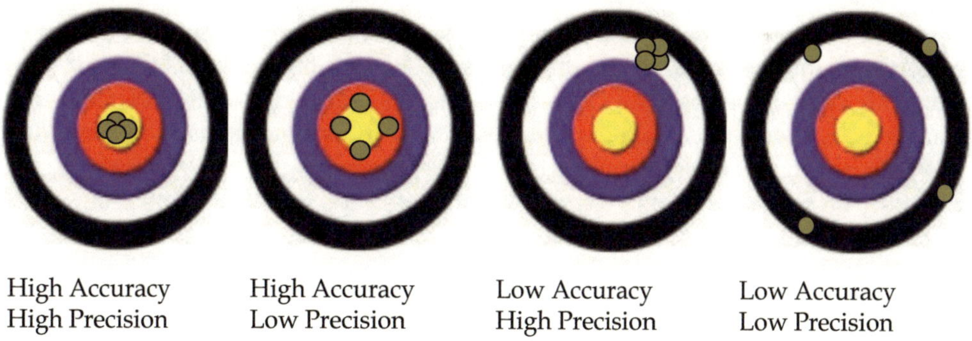

Fig. 12. A diagrammatic representation of the difference between accuracy and precision.

3.1.2.4 Computer Software for Assessing QC validation

QC programs were developed to facilitate and accelerate the implementation of QC rules before reporting patient's results. There are three commercially available programs that help in the selection of QC procedures and they are listed as follows (Westgard,2002):

- Validator® 2.0 by Westgard QC, Inc., http://www.westgard.com/lesson2.htm
- Quality Advisor by Quality Is Key, Inc., http://www.qualityiskey.com
- EZ Rules® by Westgard QC, Inc. http://www.westgard.com/ezrules.htm

Some of the currently available QC software that could be used for industrial applications and that software may be adapted for laboratory applications are summarized below (Westgard,2002).

- Bio-Rad Unity (http://www.bio-rad.com).

- MAS/Dade LabLink (http://www.mas-inc.com).
- Hematronix Real-Time QC (http://www.hematronix.com) is primarily a "real-time" peer-comparison service.
- Sigma Diagnostics Computrol on Line (http://www.sigma-aldrich.com)
- Fisher Scientific ConCurTRAK (http://www.fishersci.com
- Boston Biomedica AccuChart (http://www.bbii.com) is an internal QC program for specialized tests in infectious disease.
- Blackhawk Biosystems Virotrol QA (http://www.blackhawkbiosystems.com)
- Westgard EZ Runs ™ (http://www.westgard.com)

However, most of these programs are complex, expensive, or designed for specialized tests, or available only upon purchasing the company's control materials. Therefore, we decided to design our own QC software keeping in mind the most important features required in a QC program such as it should be straightforward and user-friendly, consists of a rapid setup for new applications, capable of easily adjusting the mean and SD according to the in-house lab values, the program should be capable of plotting the manufacturer QC mean and SD at the beginning of a new lot. Thereafter, the program must be capable of automatically updating the mean and SD values after accumulating minimum 20 points. In addition, the program must be capable of representing the QC data in charts such as the Levy-Jennings chart with the option of displaying multilevel, multistage, single chart display. Finally, the program should be equipped with an automatic flagging when a QC rule is violated and capable of generating exception or exclusion logs or reports. The best QC program will be capable of exporting the QC data and comparing it to peer-review data (Westgard,2002). The unique feature in our QC software is that in addition to the features mentioned above it is capable of monitoring work progress, report generation, and classification of system end users corresponding to access privileges. Figure 13 shows a physical layout for the client-server architecture of the proposed QC system. The main advantages of this architecture are its simplicity, low-cost and seamless compatibility with existing systems (Elmasri,2003). However, when the work environment is distributed over a wide geographical area, this architecture can be upgraded to be completely web-based application (Al-Taee,2009).

The software architecture that we developed is explained in details in our paper (Al-Taee,2009). As for the QC rules used in our newly designed program we implemented the following rules only as the program is used mainly for the serology laboratories. $\mathbf{1_{3s}}$ **/** $\mathbf{R_{4s}}$**/**$\mathbf{2_{2s}}$ **/**$\mathbf{4_{1s}}$ **/** $\mathbf{1_{2S}}$.

The reporting process in our QC program aimed at preparing and printing various types of reports. The contents of these reports vary depending on the user type. For example, the technicians are allowed to prepare and print reports related to the tests they perform while the director can prepare reports on all tests and users existing in the system database (Al-Taee,2009).

Several immediate improvements were evident through using our developed QC system. Although detailed data is not yet available; table 3 shows initial extrapolated estimates for performance compared with the original paper-based system. Furthermore, the developed QC system is capable of generating reports summarizing personnel and instrument performance by reporting the number of times the QC was repeated, the reasons for this repetition as well as how frequent the technicians was running the controls as a measure of competency and evaluation (Al-Taee,2009).

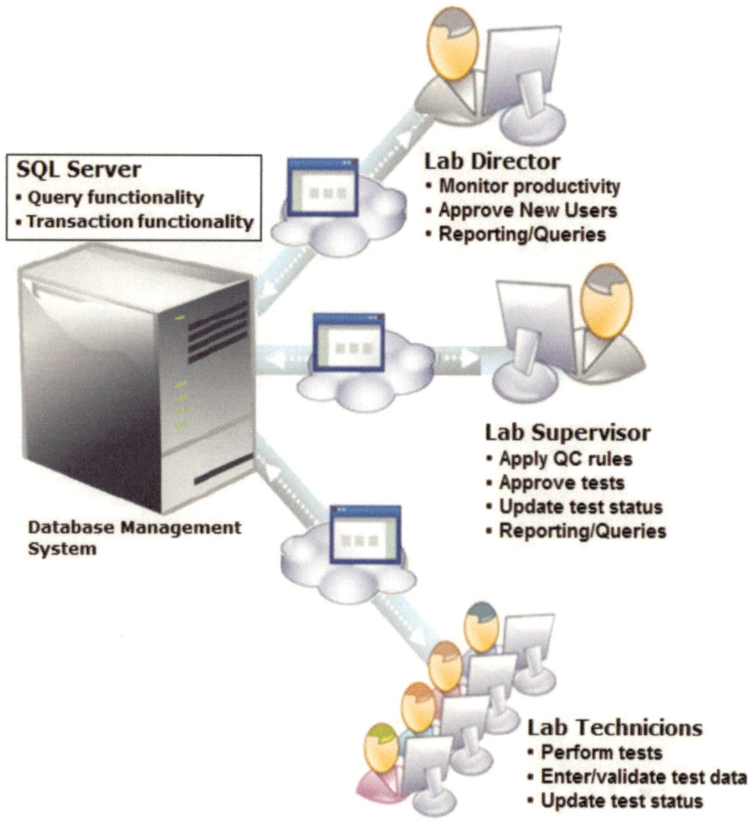

Fig. 13. The client-server architecture of the QC system.

	Estimated Processing Time	
	Web-based System	Paper-based System
Place test	< 2 min	5 min
Approve test	< 1 min	15 min - 1 hour
Apply Westgard rules	< 1 min	15 min - 1 hour
View work productivity	< 2 min	14 min - 1 hour
Prepare annual test report	< 5 min	1 - 2 days

Table 3. Comparison between the developed and paper-based QC systems.

In conclusion, the QC prototype system which we developed was tailored for the specific quality control plan of the serology tests at medical laboratories; however, it can be extended to other sections of clinical laboratories. Standardizing the QC processes on one system ensured that all clinical tests are evaluated using common quality requirements and appropriate QC rules. Furthermore, it provided better processing for the documentation of new reagent/calibrator and QC lot numbers as well as faster, enhanced and more accurate information provided to management to respond to and document QC issues. The control material for each test is processed in the lab and the results are used for monitoring the accuracy of analytical methods. Such information provides analysts with confidence that results acquired from actual samples are correct. Moreover, the developed system provides daily, weekly and monthly reporting on-line for management as well as easy tracking of proficiency testing performance and problem resolution. Finally, the adopted object-oriented development approach allows for easy maintenance and expansion and therefore more QC options can be added in the future. The innovative approach in our QC software was that it allows the medical director to monitor QC performance as well as instrument performance and technician performance due to its ability to generate reports summarizing the failures in QC classified by its origin whether it was due to instrument malfunction or technician negligence. Reviewing and carefully analyzing these reports helps keep close monitoring of clinical tests performed in the laboratory and aids in initializing quality improvement projects for failed QC. However, further work is still needed to refine this project to display the proposed corrective action(s) taken when QC values are out of control (Al-Taee,2009).

3.1.2.5 Proficiency testing

In addition to performing internal QC, it is important to participate in External Proficiency testing (Tholen,2008). Some of the advantages of participating in external quality assessment are the following:

- These programs enable laboratories to compare their performance in terms of accuracy and precision with those of their peers by analysing the same samples using their routine analytical methods, which are prepared and circulated by the scheme organisers. The results obtained by all the laboratories are analysed and compared with a reference value for the determinants by the scheme organisers and the laboratory performance is reported back to the participants.
- External Proficiency testing is a key requirement of all accreditation bodies.
- Proficiency testing helps to highlight reproducibility performance between laboratories and helps is detecting systematic errors such as bias in data reported.

3.1.3 Quality Assurance in the post-analytical stage

Quality Assurance does not end by getting the results from the instrument. The post analytical stage is very critical in assuring accurate results. This section demonstrates two types of the post-analytical checks, Technical and Administrative checks.

3.1.3.1 Technical Checks

Technical checks involve the following.

- Clinical Correlation: This is the correlation of patient's test result with the clinical diagnosis.
- Correlation with other laboratory tests: Under certain circumstances the combinations of the result of one test with other results become impossible due to discrepancy between the results. Hence, if checks for correlation between the results are made before reporting the results are made, errors can be identified and corrected before results are released to patient's file. For example, it is unusual to see 4+ bacteria by urinalysis microscopy and no growth in culture.
- Delta Checks: Delta checks refer to test results that fall outside certain clinically significant limits or critical values. Hence, the result will fail or will be flagged to alert

the technician to review the result before releasing it. Another purpose of the delta check is to compare the results of the patient with the previous result of the same test done on more than one occasion. In this case delta check or variation between the previous test results can be compared to see if they fall within the limit allowed for normal changes.

- Normal range check or limit Check: In this check the patient's results are checked if they are impossible or incompatible with the normal range of the test. Impossible results may be due to clerical errors, short sampling or clot formation, improper identification of fluid being analyzed, improper specimen identification or improper sample collection.

Technical checks usually depend on education, training, experience and technical skill of individuals performing the tests. The technologist should be qualified enough to handle and interpret test results.

3.1.3.2 Administrative Checks

Administrative Checks involve reporting results to physicians. There are several areas that should be closely monitored to ensure accurate reporting of patient's results. Some of the areas are as follows:

- Reporting Mechanism: Some problems may arise in reporting results due to: Organization and format of report, flagging of abnormal results, reporting of critical results by telephone, delay in posting results in medical records, reporting on or after the discharge of patient.
- Record Keeping: Refers to the mechanism of maintenance of records related to patient results. The importance of record keeping is to be able to refer back to previous results to check patient history. Record keeping can be accomplished by the following: Paper records, computer technology, microscopic slide from patient specimens, and tissue blocks from pathology specimens.
- Test Completion Check: This check is to verify that all tests ordered are completed before releasing the results. This step is critical in order to provide a complete picture for the diagnosis. However, some results may be delayed because the test was sent to a referral laboratory, therefore, the results must be filed in an organised secure folder until all test results are ready to be released. Another scenario if the physician desires to add another test at a later stage to verify a certain diagnosis, therefore, proper storage of the original patient-sample must be done in order to assure fast retrieval of the sample to perform the test especially if it was a precious sample such as CSF (cerebrospinal fluid) or bone marrow sample.
- Reflex Testing Protocol: This refers to the situation when a single test result may not provide the physician with enough information to determine diagnosis; hence, another test is automatically performed by the lab without the treating physician's request. For example, serum protein electrophoresis is usually accompanied by immunofixation in order to accurately diagnose the type of the multiple myeloma bands or clone.

4. Performance improvement

Performance Improvement is a process which uses information from multiple sources to increase the probability of accomplishing desired outcomes to better serve the needs and expectations of patients. The most frequently used Performance Improvement tools are: 1.

FOCUS PDCA - framework for improvement, 2. Failure Modes and Effects Analysis (FMEA), 3. Root Cause Analysis (RCA) and 4. Control Charts which we discussed earlier in this chapter. In this section of the chapter we will focus on one model of quality improvement approaches which is "FOCUS PDCA " and give examples on its implementation in a clinical laboratory setting.

Quality assurance (QA) and its component, Quality Improvement (QI), began in the engineering and manufacturing fields where statistical process control and continuous quality improvement were combined with general management methods. Since then QA and QI were combined and implemented in healthcare systems.

Quality Assurance activities are often represented as three points on a QA triangle (Figure 14). The points are: defining quality (QD), measuring quality (QM), and improving quality (QI). QD refers to developing outcome standards that the healthcare delivery system must meet in order for its patients to achieve the best health gains. QM consists of quantifying the current level of compliance with standards and finally, QI represents the use of appropriate methodologies to close the gap between the current and expected levels of quality. Quality Improvement often uses quality management tools and principles to recognize and deal with system inefficiencies (Bornstein,2001).

Fig. 14. The Quality Assurance Triangle

Approaches to conducting QI activities are numerous; one of these activities is the Performance Improvement Model FOCUS-PDCA.

The FOCUS phase helps to narrow the team's attention to a distinct opportunity for improvement (Bader,2003).

The letter "F" stands for FIND. Find a process that needs improvement and describe the process of the improvement plan.

The letter "O" stands for ORGANIZE. Choose a team who is knowledgeable in the process. Determine the team size and members and try to choose a representative from each level in the organization, select members, and be ready to document their progress.

The letter "C" stands for CLARIFY. Highlight and clarify the current knowledge of the ongoing process.

The letter "U" stands for UNDERSTAND. Understand the causes of variation. The team will be responsible for measuring the process and learning the causes of variation which will help them in preparing a plan for collecting data as well as establishing specific, measurable, and controllable variations.

The letter "S" stands for SELECT. Select the possible process for improvement and determine the action(s) that need to be taken to improve the current process.
The P-D-C-A phase allows the team to follow that opportunity and review its outcome.
The letter "P" stands for P PLAN. Plan the improvement by investigating the process, choosing possible steps for improvement, and identify which data would be helpful to collect that might aid in the improvement process.
The letter "D" stands for DO. Do the improvement/data collection/data analysis and carry out a simulation of the correction plan on a small-scale.
The letter "C" stands for CHECK. Check the data for process improvement and document the results of the change. Modify the change, if necessary and possible.
The letter "A" stands for ACT. Act to hold the gain or continue the improvement process and carry out the change if it is working. However, if the performance improvement plan fails, abandon the plan and repeat the cycle (Bader,2003).

7. Conclusion

In this chapter, we have demonstrated the latest approaches in quality control and quality management in clinical laboratories. Furthermore, we highlighted our recent work in which we constructed new quality control software which is tailored for the specific quality control plan for the immunology/ serology tests at medical laboratories. This program is capable of standardizing the QC processes in one system which ensures that all clinical tests are evaluated using common quality requirements and appropriate QC rules. Furthermore, it provides better processing for the documentation of new reagent/calibrator and QC lot numbers.
In addition, this program grants faster, enhanced and more accurate information to the management in order to respond to and document QC issues. Control material for each test is processed in the lab and the results are used for monitoring the accuracy of analytical methods. Such information provides technologists with confidence that results acquired from actual samples, are correct. Moreover, the developed system provides daily, weekly and monthly reporting on-line for management as well as easy tracking of proficiency testing performance and problem resolution. The uniqueness of our software resides in the reports generated by the program. For example, the QC failure reports are linked to the performer which gives the director a chance to assess the competency of the employees performing the tests as well as the efficiency of the analytical instruments. Furthermore, we have demonstrated how the benefits provided by the improved QC program accelerated the quality assurance process of the tests performed. Finally, the adopted object-oriented development approach that we used allows for easy maintenance and expansion and therefore more QC options can be added in the future. Further work is still needed to refine this QC software to display the proposed corrective action(s) taken when QC values are out of control.

8. References

Al-Taee, M., Souan, L., Al-Haj, A., Mohsen, A. and Muhsin, Z. (2009 of Conference). Quality Control Information System for Immunology/ Serology Tests in Medical Laboratories. 6th International Multi-Conference on Systems, Signals and Devices. ISBN, Djerba - Tunisia, 23-26 March 2009 Published,

Bader, M. K., Palmer, S., Stalcup, C. and Shaver, T. (2003). Using a FOCUS-PDCA quality improvement model for applying the severe traumatic brain injury guidelines to practice: process and outcomes. . *Evid. Based Nurs.* 6, (6-8),

BioRad. (1995). *Basic & Intermediate Systems of Quality Control for the Clinical Laboratory.*Bio-Rad Laboratories, Lit. No. Q-1102 4/05,

Bornstein, T. (2001). Quality Improvement and Performance Improvement: Different Means to the Same End? *QA Brief.* Volume 9, (6-12),

Derek-Law, E. (2007). *Accuracy Versus Precision.*Geodatabases That Are Easier to Set Up and Manage. http://www.esri.com/news/arcuser/0807/gdb_precision.html,

Elmasri, R. a. N., S. B. . (2003). *Fundamentals of database systems.*Addison Wesley, USA,

Lumsden, J. H. (2000). Laboratory test method validation. *Revue Med. Vet.* 151, (623-630),

Plebani, M. (2006). Errors in clinical laboratories or errors in laboratory medicine? *Clinical Chemistry and Laboratory Medicine.* 44, (750-759), 1434-6621

A. J. Reynolds. (2003). *Unit 5 Section 2 : Mean, Median, Mode and Range.*Mathematics Enhancement Program,

Schieltz, B., and Westover, B.; 2010). Descriptive Statistics Medialab,

Sharma, P. (2009). Preanalytical variables and laboratory performance. *Indian Journal of Clinical Biochemistry.* 24, (109-110), 0970-1915

Tholen, D. W., Berte, L.M., Boone, D.J., Cooper, W.G., Gun-Munro, J., Nobel, M.A., Sarewitz, S.J., Williams, M.L. (2008). Using Proficiency Testing to Improve teh Clinical Laboratory; Approved Guideline. *Clinical and Laboratory Standards Institute (NCCLS).* 27, (GP27-A22),

M. Thompson. (2010). *Internal quality control in routine analysis.*1757- 5958,

Westgard, J. (2009). "Westgard Rules" and Multirules.

Westgard, J. O. (2002). *Basic QC Practices.*Westgard QC, Inc., 1-886958-17-3, Madison

Westgard, J. O., Miller, W. G., Allen, K., Boone, D. J., Caines, P., Cooper, G., Garber, C., Greenberg, N., Jain, C. P., Linnet, K., Russek-Cohen, E. and Vaks, J. E. (2006). Statistical Quality Control for Quantitative Measurement Procedures: Principles and definitions; Approved Guideline-Third Edition. *Clinical and Laboratory Standards Institute.* 26, No.25, (1-33), 1-56238-613-1

Westgard, J. O., Miller, W.G., Allen, K., Boone, D.J., Caine, P., Cooper, G., Garber, C., Greenberg, N., Jain, C.P., Linnet, K., Russek-Cohen, E., Vaks, J.E. (2008). Statistical Quality Control for Quantitative Measurement Procedures: Principles and Definitions; Approved Guideline. *Clinical and Laboratory Standards Institute (NCCLS).* 26, (C24-A23),

Wikipedia; 2010). Good Laboratory Practice,

Part 2

Computer and Mathematical Approaches for Quality Control

15

Mathematically Fuzzy Approach to Quality Control

Rostovtsev A.M.
Bauman Moscow State Technical University, Moscow, Russia

1. Introduction

Romulus: Did she lay, this noble bird?
Achilles: Almost, my Emperor.
Romulus: Almost? What does that mean? A hen either lays or she doesn't.
Achilles: Not yet, my Emperor.
(The Emperor makes a decisive gesture).
Romulus: Not at all.

(F. Dürrenmatt, "Romulus the Great", 1949)

2. Quality as a mathematical function

Quality control, which initially emerged and developed in the sphere of material production and consumption, now is rapidly spreading also in nonmaterial spheres, such as development and implementation of various projects, economic programs and managerial solutions [1], which in essence are the processes which quality is considered from the point of view of operations research theory [2].
The fact that quality is functional is known since the last quarter of the twentieth century. At that time Russian researchers V. Solodovnikov and N. Filimonov have formulated the concept of quality functional [3]. Curiously enough, they came to the conclusion about necessity of mathematically fuzzy (hereinafter □ MF) approach to evaluation of acceptability of performance of complicated control systems [4]. This fully complies with the methodology of mathematical fuzziness which foundations were laid by the American scientist Lotfi Zadeh [5], who discovered that the more complicated is a system, so much the less we are capable to provide precise and at the same time practically valuable judgment about its behavior. Therefore, for the systems which complexity exceeds certain threshold level, "precision" and "significance" become almost mutually exclusive characteristics [6].
Besides, the functionality of quality is known since the times, when Japanese scientist G. Taguchi proposed "quality loss function" [7,8] for radio-electronic components and products, for which the operating performance was evaluated using binary scale (e.g., "bad" or "good"). At the same time, QFD ("Quality function deployment") [8,9] method has been developed in the USA; it is the procedure for transforming user demands into quality parameters of the processes of product planning, manufacturing, installation and enhancement (quality improvement).

MF approach to quality control is possible only in case if it is considered, figuratively speaking, from mathematically «rigorous» point of view. Therefore, it is necessary to clarify, to which class of mathematical representations the concept «quality» may be ascribed. For this purpose we must briefly deepen into terminological aspects of quality as a mathematical phenomenon. With this objective in mind, process approach [2] on the basis of standard [10] is already used.

All this has laid the groundwork of modern quality control, both for material and nonmaterial objects, which in the context of above said we will name simply "objects". As prominent Austrian-British philosopher of the 20-th century L. Wittgenstein notes in his treatise [11], "objects contain the possibility of all states of affairs".

Quality is judged by presence with the object of one (one-dimensional) or several parallel (multidimensional or vectorial) properties [2,12], hereinafter denoted as P where necessary.

They represent different aspects of the object and are determined by values which in qualitative sense are common for many objects, but in quantitative sense are individual for each object [13]. Therefore, the value is a denominated property [14] and serves as the quality index (hereinafter - QI) of the object.

The properties may be either point-wise, i.e. occurring singly in full scope which is denoted by 1 (while their absence is denoted by 0) [14], or permanent, i.e. persisting continuously during prescribed time interval. Occurrence of point-wise property or loss of permanent property must be confirmed by the result of appropriate QI check. In contrast to single occurrence of point-wise property, loss of permanency (incurring loss of quality of the object) may take place either at a point or within a finite interval.

In case of a point-wise property in nonmaterial sphere of activities, specifically, a quality management process or quality management system (hereinafter – QMS), the fact of its certification at an industrial enterprise, suppose, in year 2011, is subject to confirmation by certificate granted by a certification body.

An example of a point-wise property in the sphere of material production may be the JIT (Just-in-time) logistics concept, created in late 1950-ies in the Japanese company Toyota Motors, later accepted by other Japanese automobile manufacturers [15] and now having worldwide recognition. It suggests supply of a resource exactly at the time when it is needed, allowing reduction of stock reserves and related expenses. Examples in non-material sphere of activity may be the just-in-time fulfillment of contractual obligations and payments; an analog of this in social-behavioral aspect is just-in-time coming of a subordinate functionary to his chief, as well as paying visits.

In case of permanent property, the interval of loss of permanence represents the period (or time) of property recovery, e.g., recovery of system performance or device operability, respectively.

An example of interval loss of permanence, e.g., in the sphere of management activity, may be a short-term stoppage of work of an institution or enterprise administration for a few hours during working day for reasons which might be foreseen and timely eliminated. Exception here might be only force majeure circumstances. An example of such loss in material sphere is a temporary exceedance of allowable limits of mains voltage.

An example of point-wise loss of permanent property may be occurrence of defective products at an industrial enterprise in conditions of prescribed zero defects of production, e.g., violation of requirements to the appearance of the items (in particular, items made of plastics [16,17]).

QI may be either quantitative, if it has metric scale, or qualitative, if its values are established according to ranking scale or to classificatory scale of quality gradations [18, 19]. Point-wise property may have any of the above mentioned scales. Permanent property may have only metric scale, and only in anticipation of eventual interval loss of quality.
However, if possibility of point-wise loss of property is envisioned, then only non-graded (free) axis is conceivable.
The above said allows making some conclusions and generalizations:

- Quality is inherent to any object; in contrast to property, quality cannot appear and disappear,
- quality is functional; it may be either compliant or incompliant with the requirements raised to it by reason of presence or absence of the property with the object,
- verbal evaluations «1» and «0» correspond to presence or absence of the property,
- if the object has several properties, quality becomes multidimensional and exists only when its presence is supported concurrently by all QI,
- property may be either point-wise (occurring one-time) or permanent, i.e. persisting continuously,
- existence or absence of a property is evidenced by the values of QI, either measured or evaluated by other methods,
- these values must either lie within required (prescribed) limits of appropriate scales or be point-wise.

Now it is permissible to ask: if quality is functional [3 - 9], to which kind of mathematical functions it should be categorized, and what should be the appearance of mathematical function of quality? To get the answer, let us turn to the theory of functions [20], which allows, in case of single QI, taking it as independent variable x, to search for quality function (hereinafter - QF) in the generalized form as

$$Q =]Q(x) \qquad x \in E, \tag{1}$$

where: Q is the symbolic value «1» in case of compliance, or «0» in case of incompliance of quality with the requirements raised to it,

$Q(x)$ is the quality function (QF),

] is the symbol of a mathematical operation, decision rule or method for determining compliance or incompliance of quality with the requirements raised to it,

E means the definition domain of QF.

As is known, the concept of "function" is characterized by the following attributes:

- representation method,
- range of values,
- range of definition.

If these concepts will provide substance to the formal expression (1), then QF will be actually found.
Let us start from representation method. Four such methods are known [20]:

- analytic,
- graphical,
- tabular,
- verbal (in the form of verbal expression).

As Russian scientist N. Nazarov has rigorously proved, quality is not a quantitative value [18, 21]; hence, two methods remain applicable - graphical and verbal. Graphical representation method for function assignment is not enough; its mathematical formulation is required. Only verbal representation of a function remains available for this purpose. Addressing to [20], we find precedents of verbal representation of functions. These include:

Dirichlet function - equaling to 1, when x is rational number, and 0, when x is irrational number,

Boolean function - which has the same range of values as the previous function, though the range of its definition is different.

This gives ground to assume that QF may also be set verbally, i.e., following [2,3], to assign value 1 to quality presence, and value 0 – to quality absence, what, apropos, represents the range of values of QF sought for.

It should be noted here that by no means these values are the measure or, horrible ictum[1]), the quantity of quality. 1 and 0 here lose their numerical value [22] and quantitative essence in favor of verbal essence, accepting the sense of propositions «is present» and «is absent», respectively, i.e. become nothing more than the symbols of verbal evaluations.

Further on we will denote the verbal evaluations of quality range of values as $\rfloor$Q (quality is compliant with the requirements raised to it) and $\rceil$Q (quality is incompliant with the requirements raised to it). The symbols $\rfloor$ and $\rceil$ represent the right parts of notations «floor» - $\lfloor\ \rfloor$ and «ceiling» - $\lceil\ \rceil$ [23 - 25], located to the left of Q. The symbols $\rfloor$ and $\rceil$ are generalized into symbolic operator] appearing in the formula (1). This mnemonics corresponds to the widely known manual gestures, and the symbols are $\rfloor$ and $\rceil$contained in character sets of standard computer software.

Taking the above said into consideration, QF may be set on the basis of decision rule

$$Q =]Q(x) = \left\{ \begin{array}{l} 1, \text{ if } \rfloor Q; \\ 0, \text{ if } \rceil Q. \end{array} \right\} \tag{2}$$

The meaning of the equality sign «=» in formulas (1) and (2), as well as the essence of values 1 and 0 in formula (2), is solely symbolic, linguistic and, therefore, is not numerical, because the question is only about presence or absence of the quality P with the object being checked.

As can be seen from (2), the range of values as well as representation method of QF searched for is the same as for Boolean function. It remains to clarify the range of definition E of the function searched for. This range, which we will denote as E_Q, represents the set of values taken by QIs. The appearance of this set of values depends on the applied method of QI gradation according to the principle of preference of various ranking scales [19]. It may be single as well, i.e. may consist of one element, e.g., for already mentioned single (one-time) measurements or evaluations.

From the point of view of mathematical inequalities, functional requirements to QI may be presented either by semi-open intervals

$$x \geq x_l \qquad [x_l, \infty), \tag{3}$$

$$x \leq x_r \qquad (0, x_r] \tag{4}$$

or by segment

[1] Note: horrible to say (lat.)

$$x_l \leq x \leq x_r \qquad [x_l, x_r] \qquad (5)$$

where x_l and x_r are the so-called functional thresholds [16,26], left and right, respectively.
The range of definition E_Q of quality function is represented by the set

$$E_Q = \{x_l,\ldots,x_r\},$$

while the range of definition E_B of Boolean function is represented by the set

$$E_B = \{1,0\}.$$

If cardinalities of E_Q and E_B sets are compared, it is evident that, except for point-wise properties, when $E_Q = \{1\}$,

$$\text{card } E_Q \geqslant \text{card } E_B, \qquad (6)$$

where card means cardinality of a set.
Naturally, for point-wise properties

$$\text{card } E_Q < \text{card } E_B \qquad (7)$$

From the point of view of theory of functions, relationships (6) and (7) allow to make conclusion that QF is an extension of Boolean function in regard of range of definition.
The results of fulfillment of functional requirements (3) – (5) are presented in table 1.

Quality index, QI	Relationships of QI and functional thresholds	Functional thresholds	
		left, x_l	right, x_r
x	$\geq$	⌋Q	⌉Q
	$\leq$	⌉Q	⌋Q

Table 1. Dependence of quality Q (⌋Q or ⌉Q) on fulfillment of functional requirements

If the object possesses not one but several properties, then a certain quality function Q_j will correspond to each j-th property. In this case, assertion about the quality Q_0 of the object in whole may be made only on the basis of logical proposition

$$Q_0 = Q_1 \& Q_2 \& \ldots \& Q_j \& \ldots \& Q_n, \qquad j = \left(\overline{1,n}\right), \qquad (8)$$

where Q_j means quality of j-th property of the object,
& means logical conjunction sign.

Expression (8) testifies that ⌋Q_0 is achievable only on condition of ⌋Qj for each QI. Figuratively speaking, it may be collated (both in form and fact) with a chain which breaks if only one of its links is broken.
Thus, by means of QF, mathematical-logical formulation (8) was obtained, corresponding to the known assertion that quality is represented by the set of object properties [12, 27].

3. Mathematically fuzzy characteristic functions of quality

The graphical method of setting QF (2) may give visual presentation of the mutual relationship of the only two possible values $Q(x)$ of quality, 0 and 1, with QI values lying

along x axis; for this purpose fig. 1a and fig. 1b show graphical presentation of QF for the half-intervals (3) and (4), respectively, of its domain of definition.
Reviewing of fig.1 from the point of view of mathematical analysis leads to a conclusion that these are the plots of discontinuous piecewise-linear functions with discontinuities at points x_l and x_r. Numerical 0 of QI x at each plot coincides at the origin of coordinates with non-numerical zero relating to absence of the property with the object. Discontinuity of the function is explained by the fact that 0 and 1 along vertical axis are non-numerical. There is no mathematical fuzziness here, and cannot be in principle. However, if one looks at these plots from, so to say, the point of view of mathematical fuzziness, then, pursuant to [28, 29], it is obvious that the plot at fig. 1a resembles the rectangular mathematically fuzzy (MF) S-number (sigmoid), while the plot at fig. 1b resembles the rectangular MF Z-number (zetoid).
In order to pass from purely outward resemblance to exact matching with these MF numbers, "eine grosse Kleinichkeit"[2] is needed – vertical segments, the so-called terms[3], connecting 0 and 1 of functional thresholds – x_l and x_r.
This fact is taken into consideration at fig. 2, which, as may be ascertained, is the MF analog of fig.1 and presents the graphical method of setting QF as MF number for these thresholds [28,29]. We will denote such MF function as characteristic function (hereinafter - CF), it structurally repeats the decision rule (2) for QF and the MF characteristic of functional thresholds (3, 4), or of the combination thereof (5):

$$Q =]Q(x) = \alpha_Q(x) = \begin{Bmatrix} 1, \text{ if } \rfloor Q; \\ 0, \text{ if } \rceil Q. \end{Bmatrix} \tag{9}$$

where $\alpha_Q(x)$ is the membership function [30], in this case - of QI x to the property Q(x).
Now let us find out, which MF number corresponds to segment (5) of QF domain of definition. As shown at fig.3a, it will correspond to intersection of sigmoid 1 and zetoid 2 mentioned above. Using the terminology of the theory of sets, this intersection represents the set

$$\Pi = (S \cap Z) = \{x \mid x \in S \text{ и } x \in Z\}, \tag{10}$$

where: S, Z – the sets corresponding to S- and Z- numbers,
$\cap$ - operator of intersection of sets,
| - Sheffer stroke,
$\in$ - logical inclusion sign,
x - QI.
Fig. 3b shows the result of this intersection – the rectangular tolerant (closed) MF number, let us name it Π-number or Π-oid by analogy with S- and Z-numbers. The characteristic property of Π-number is fulfillment of conditions

$$\rfloor Q \in \Pi \text{ and } \rceil Q \notin \Pi, \tag{11}$$

where $\notin$ means logical exclusion sign.
Relations (11) mean that quality is compliant with the raised requirements ($\rfloor$Q) only within Π number, and is incompliant with the raised requirements outside it, i.e. ($\rceil$Q).

[2] Note: a big trifle (German);
[3] Note: from Latin word "termin" – borde

As these three MF numbers correspond to three possible domains of definition (3) – (5) of QF, they may be considered as the main (basic) MF characteristic numbers of quality. As far as the domains (3) – (5) are concerned, they may be named the domains of S-, Z- and П-quality, or briefly S-, Z- and П- quality, designated as Q_S, Q_Z and $Q_П$, respectively.
Thus, not only traditional mathematical methods, but also the methodology of mathematical fuzziness [5], may be applied to quality as the subject of applied mathematical research.
Basing on the main MF characteristic numbers of quality, auxiliary (or second-rank) MF numbers may be derived, which are met in course of quality control of point-wise values or permanent values.
In case of point-wise property of the object, when $x_l = x_r$, П-number degenerates into the so-called [30] singleton[4]. We will denote it as MF I-number, due to its appearance, fig. 4, and will regard it as additional characteristic number of quality.
In case of permanent property of the object, its point-wise absence may be characterized by anti-singleton, let us denote it as Ī. Geometric appearance of this singleton is shown at fig.5 as drooping vertical straight segment symbolizing point-wise absence of property (˥P). As may be seen at fig.5, its x – x axis has neither arrow orientation nor scale, and incompliance of quality (˥**Q)** with the raised requirements occurs here once and at single point.
Examples from material and non-material spheres characterized by singleton I and by anti-singleton Ī were presented above, where point-wise properties and their absence were discussed. A propos, another characteristic number related to loss (˥**P**) and recovery (˩P) of property P may be seen there.
According to the stated above, loss of property (˥**P)** leads to incompliance of quality with the requirements raised to it (˥**Q**), while recovery of property (˩P) restores such compliance to (˩Q). It is shown at fig.6 and characterizes ˥ and Q at the interval ($x_{˥Q}$, $x_{˩Q}$), where $x_{˥Q}$ and $x_{˩Q}$ denote the points of incompliance (˥) and compliance (˩Q) of quality of permanence by the property P(x) at QI axis x. Its width

$$b = x_{˩Q} - x_{˥Q}$$

in terms of MF language is the base or carrier of time (or period) of quality recovery. Appearance of this number at fig. 6 resembles the letter U, therefore we will name it MF auxiliary characteristic number U of quality absence.
For reasons of simplicity, later we will denote MF characteristic numbers S, Z, П, I in general case by common symbol Ǫ (the so-called "candle apple"[5]).
Similarly, MF characteristic numbers Ī and U in general will be represented by common symbol Ǭ, but they are not discussed here in detail because they are subjects for separate study as they refer to cases difficult for prediction and to unforeseen situations.
Therefore, further we will basically discuss the generalized MF quality characteristic numbers Ǫ or, what is the same, quality Ǫ-numbers. By the way, all the symbolic notations used above were taken from the corresponding Unicode tables of Microsoft Office Word; in particular, hexagonal numbers of symbols Ǫ и Ǭ are 01EA and 01EC, respectively.
Classification of MF characteristic numbers of quality and of quality absence is presented in table 2.

---

[4] Note: individual item existing in singular (eng.)
[5] Note: "growing up from pedicle" (contemporary folklore)

Rank	Quality presence	Name	Designation	Position at x axis of quality index
Main	Q̣	Sigmoid	S	Local
		Zetoid	Z	
		Π-oid	Π	
Auxiliary		Singleton	I	
	Q̣̄	Anti-singleton	$\bar{I}$	Free
		U-oid	U	

Table 2. Mathematically fuzzy characteristic numbers of quality

Examples from the field of production quality control characterized by MF characteristic numbers of quality are given in table 3.

MF characteristic numbers of quality	Objects and/or facilities of quality control
1	2
S	Check of dimensions of male parts by go gauges "GO". Check of dimensions of female parts by no-go gauges "NO GO". Internal diameter of male thread Functional characteristics of a micro switch: forward travel (until direct actuation) of the driving element; overtravel (after direct actuation); force of reverse actuation at the driving element; operational life (limit number of changeovers).
Z	Check of dimensions of female parts by go gauges "GO". Check of dimensions of male parts by no-go gauges "NO GO". Check of defects of geometric shape using universal measuring instruments or by amplitude sensor. Internal diameter of male thread. Functional imperfections of elastic members (e.g., nonlinearity and hysteresis of their elastic response). Functional characteristics of a micro switch: force of direct actuation; voltage drop at normally closed contacts.
Π	Check of limiting dimensions of machine parts confined by tolerance range, using universal measuring instruments and/or by double-limit contact sensor.
I	Just-in-time delivery of component parts, subassemblies, units etc. to the assembly line.
$\bar{I}$	Hang-up of contacts of micro switches in the zone of expected direct and/or reverse actuation
U	Single or intermittent failures of measuring or automatic checking facilities

Table 3. Examples of production quality control characterized by MF characteristic numbers of quality

Vectorial nature of properties of objects and of corresponding QIs [12], especially in regard of complicated systems [31], allows speaking about composite MF characteristic numbers of quality which we will denote as Q^n-numbers, where n means the number of dimensions of object properties vector. The corresponding $\bar{Q}^n$-numbers will be considered here only fragmentary, because they are more suitable for analysis of catastrophic situations (so to say, "post-flight analysis"), rather than to quality control. In this connection, QF also becomes vectorial, i.e. multi-component.

For simplicity reasons, let us consider the elementary case of a two-dimensional vector of properties which we will denote as vectorial MF characteristic two-component number Q^2 of quality. Indices 1 and 2 in the latter notation correspond to the set of two numbers, the main and auxiliary, from Q- and/or $\bar{Q}$-range. If such two-component number is composed only from Q-numbers, then, naturally, it will represent a Q^2-number. If even one of the numbers of the two-component set happens to be $\bar{Q}$-number, then, by virtue of relation (9), the resulting Q^2-number also will be $\bar{Q}$-number. This rule applies to similar composite numbers of any vectorial dimensionality. As the saying is, a scabbed sheep will mar the flock.

From the point of view of theory of sets, two-component MF numbers of quality represent a union of two heterogeneous sets, what is presented in general form as

$$Q^2 = Q_{12} = (Q_1 \cup Q_2) = \{x \mid x_1 \in Q_1 \text{ and } x_2 \in Q_2\}, \qquad (12),$$

where Q_1 and Q_2 mean the main and auxiliary MF numbers, respectively,
$\cup$ means the sign of logical uniting of sets.

As $x_1 \neq x_2$ in the area under consideration (where $\neq$ is the symbol of absence of equivalence), then numbers Q_1 and Q_2 cannot lie in the same plane. Therefore, following [32], let us create from the two planes corresponding to them the dimensionless non-numerical (verbal) axis containing the verbal segment [0,1] corresponding to its non-numerical property.

Such a coordinate system for the two-component MF characteristic number of quality Q^2 (Q_1 - main number, Q_2 - auxiliary number conjugated with Q_1) is shown at fig. 7. The main plane x_1-0-1, corresponding to Q_1 number, lies in the plane of drawing, and the auxiliary plane x_2-0-1 forms a spatial right angle with it.

Classification of the main two-component characteristic numbers of quality is given in table 4.

Type	Logical set structure	Geometric image
1	S∨Z∨Π ∪ S∨Z∨Π	Piecewise-planar surface
2	S∨Z∨Π ∪ I	Planar piecewise linear form
3	I ∪ S∨Z∨Π	-*-
4	I ∪ I	Vertical linear segment

Table 4. Classification of two-component characteristic numbers of quality

Geometric images of Ϙ²-numbers of the 1st kind from table 4 are systemized in table 5. Their coordinate axes correspond to the axes at fig. 7. In a similar way, one can systemize and represent graphically other kinds of MF numbers of this table, as well as various combinations of Ϙ- and $\bar{Ϙ}$-numbers, because it is easy to show that formula (12) also covers the case when $Q_2 = \bar{Ϙ}$.

Investigation of geometric images of Ϙ²-numbers in table 5 shows that all of them, except ПП-number, consist of only two geometric forms resembling street shelters: a shed, e.g., ZS-number, or a booth, e.g., ПS-number. It is not too difficult to note that the entire variety of images of Ϙ²-numbers obtained using the coordinate method considered above may be reproduced by sequential rotations around the vertical axis at the angles being multiple of π/2, namely:

for the "shed"-

$$SZ \looparrowleft ZZ \looparrowleft ZS \looparrowleft SS \looparrowleft SZ: \tag{13}$$

for the " booth" -

$$ПS \looparrowleft SП \looparrowleft ПZ \looparrowleft ZП \looparrowleft ПS, \tag{14}$$

where $\looparrowleft$ means the mathematical symbol of rotation at the angle π/2, in this case – counterclockwise.

Presence of rotations (13) and (14) indicates eventual group-theoretical properties of the structure formed by Ϙ²-numbers.

MF characteristic numbers of quality allow identifying and classifying multi-component (vectorial) double goals and goals of larger dimensionality stipulated by a joint action, e.g., by implementation of QMS into manufacturing activity of an enterprise. The table of kinds of QMS measurable goals [14] is convenient for this purpose, where only the due dates of achieving the goals are changed for later dates. After adding Ϙ- and $\bar{Ϙ}$-numbers to this table, we get table 6.

As one may see, the goals of this table are characterized by single-component Ϙ-, two-component Ϙ², while the latter of the goals in the list – by ternary (three-component) MF Ϙ³ -number IZS. In this connection, it seems that we deal with the structure of notations of MF numbers which may be regarded as MF symbolic language of quality control. A word in this language is a sequence of Ϙ-numbers, ordered by the priority of actions corresponding to them, namely, to organizing and implementing the measures related to quality assurance and quality control, as shown, in particular, in table 6. From the semantic point of view, this corresponds to the known statement: "the meaning of a certain word is its usage in the language" [33].

For practical application of the stated above MF approach to quality control it is necessary to know, at least in first approximation, the influence of measurement errors on the Ϙ-numbers which characterize this quality. Basing on the investigations [34, 35], it may be shown, that if one turns a priori to the probability function of MF appurtenance of this error, then for the case of most commonly encountered Gaussian normal probability law, this MF function is a symmetrical MF R-L-number. Upon completion of additive MF operations with the main

Kinds of goals			Examples of goals	Q̧ and Q̧̄ numbers of quality	
				main	auxiliary
Absolute	Index value is presented in explicit form.		To increase the number of contracts with customers to 50. To achieve time between failures of one thousand hours for the product under design.	S	
	Index value is not presented in explicit form.		To exclude cases of usage of unverified measuring instruments during acceptance tests (0 cases). To exclude cases of failures of performance (termination) of contracts with customers through the organization's fault (0 cases).	Ī	
	Range of index values (from – to, not larger than, not less than)		Availability coefficient of the product under design shall be not less than 0,95.	S	
			Number of claims lodged to the enterprise shall be not larger than 5.	Z	
			Duration of analysis of a contract with the consumers – 3-5 days	Π	
Relative	Relative value of index		To certify 100% of test equipment.	I	
			To revise 20% of enterprise standards.	S	
	Absolute increment of index value		To increase the profit per one employee of the enterprise by 5 thousand rubles. To reduce average duration of contract issuance by one day.		
	Relative increment of index value показателя		To increase the number of won tenders by 10%. To increase the average wage at the enterprise by 20%.		
	Increment of index value is not expressed in numeric form.		To increase the assortment of manufactured types (kinds) of products. To decrease the number of occupational injuries.	I	
Temporary goals (goals with prescribed deadline)	Short-term (period of achievement – not over 1 year)	Goals of the first two kinds	*All the goals stated below, set for the term not over one year*		
		The goal is not expressed in numeric form	To certify QMS in regard of compliance with GOST R ISO 9001 - 2001 during 2011 To perform training of quality service personnel during 2010. To purchase in 2011 the new climate chamber for tests.		
	Long-term (period of achievement – over 1 year)	Goals of the first two kinds (for term over 1 year)	To rewrite before 2012 all enterprise standards in the framework of preparation of QMS to certification in regard of compliance with GOST R.	Z	
			At that, to rewrite before 2011 not less than 50% of enterprise standards.		S
		Short-term for every year.	To replace, by year 2012, 100% of measurement tools used during state tests of the products by measurement tools included in the state register of measurement tools	I	Z
			At that, by year 2011, to replace not less than 70% of measurement tools.		S

Table 6. Kinds of QMS measurable goals

(S-, Z- and П-) characteristic numbers of quality, it leads to changing of location of the latter on the scale of QI x which is measured. For the simplest three-term R-L-number which is appropriate here, S-number shifts by two inter-term intervals to the right, Z-number shifts by same distance to the left, while П-number symmetrically narrows from left and from right by one inter-term interval without shifting.

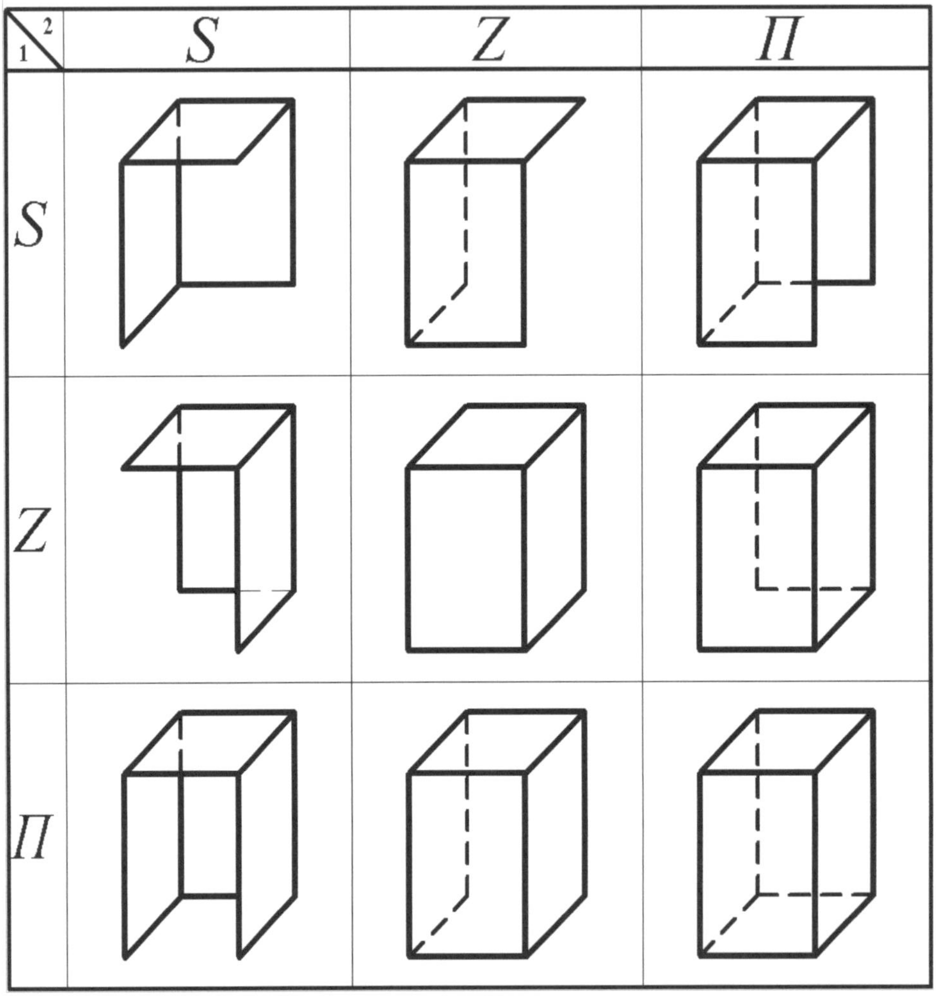

Table 5. Geometric image of mathematically fuzzy two-component Q-numbers of quality

In conclusion, I express my gratitude to Doctor of physical and mathematical sciences, Professor of Moscow State Technical University named after N.E. Bauman – M.I. Kiselev for every kind of assistance, valuable comments and proposals during work over this chapter; to the students of the chair headed by him who assisted me; and especially – to my daughter Nadezhda who undertook the labor of computer-aided making-up of the manuscript.

4. Figures

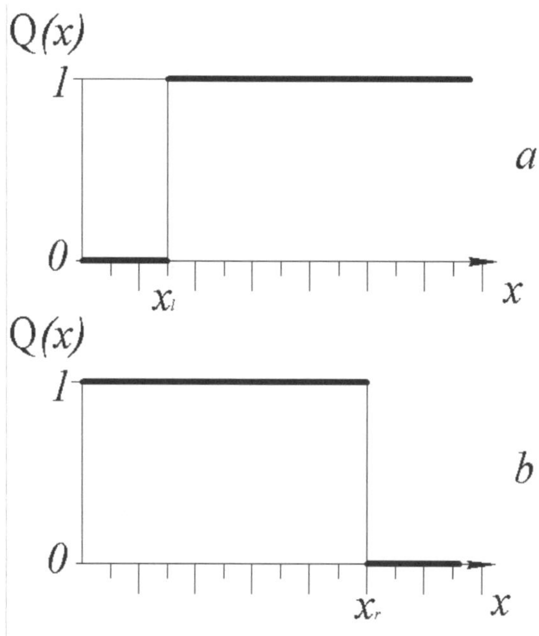

Fig. 1. Analytic quality function $Q_a = Q_a(x)$ for functional thresholds:
a) left, $x \geqslant x_l$, b) right, $x \leqslant x_r$,
x - quality index

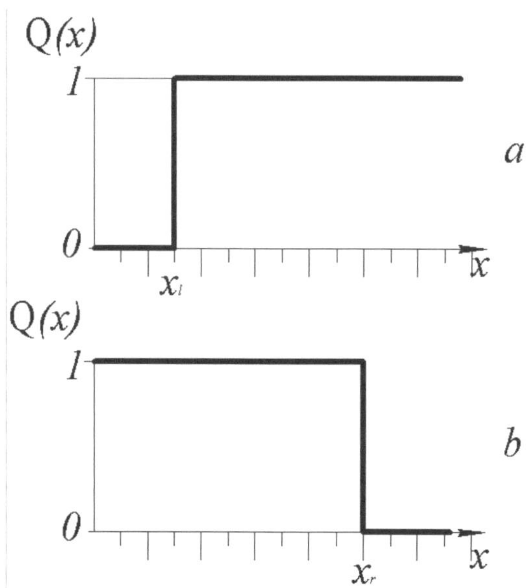

Fig. 2. Mathematically fuzzy (MF) quality function $Q = Q(x)$ for functional thresholds:
a) left (sigmoid S) $x \geqslant x_l$, b) right (zetoid Z) $x \leqslant x_r$,
x - quality index

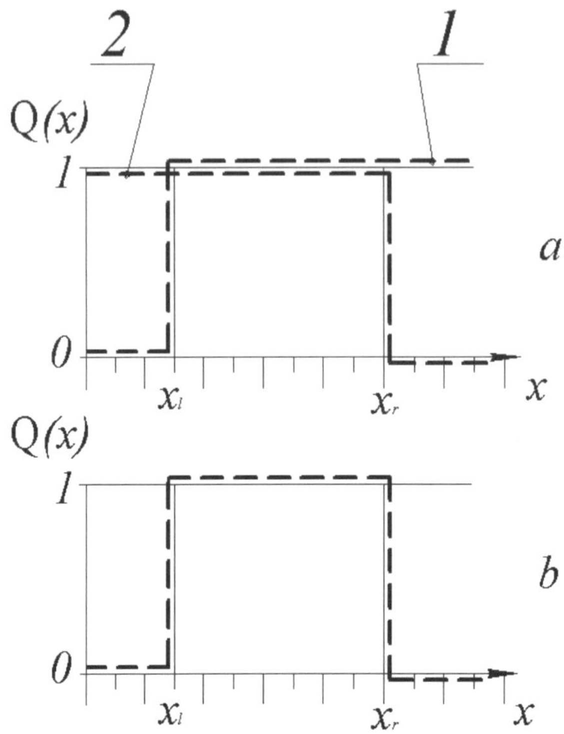

Fig. 3. Formation of MF rectangular tolerant П-number from sigmoid S and zetoid Z.
a) intersection of sigmoid (1) and zetoid (Z),
b) characteristic П-number of quality, or П-oid,
x - quality index,Q (x) - MF quality function

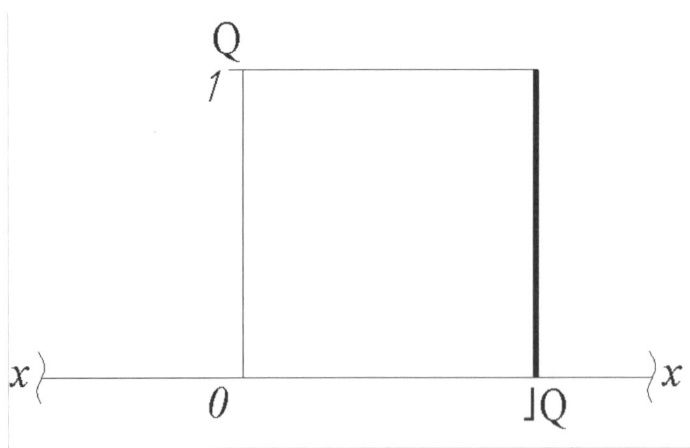

Fig. 4. Degenerate П-number or singleton I corresponding to point-wise MF quality function
$x - x$ - free axis of singleton location,
Q - axis of verbal values of quality (0 and 1)

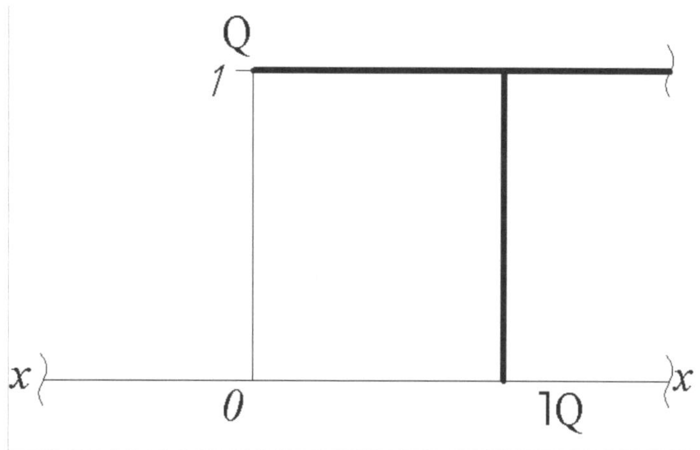

Fig. 5. Anti-singleton $\bar{\text{I}}$ of permanent quality function Q (x)
$x - x$ – free axis of anti-singleton location,
Q – axis of verbal values of quality (0 and 1)

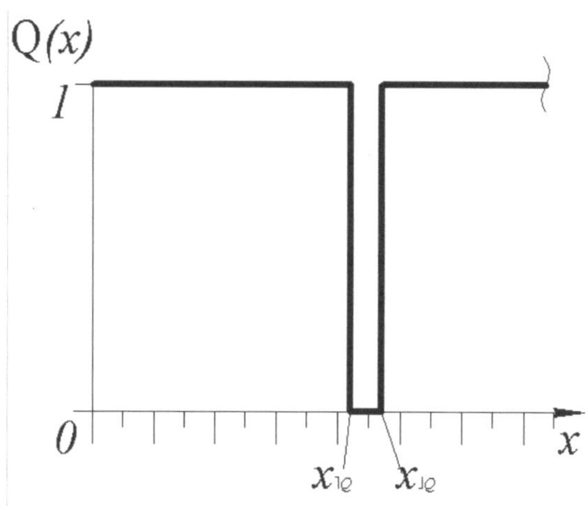

Fig. 6. MF U-number of interval absence of permanency of MF quality function Q (x)
x - quality index,
$x_{\urcorner Q}$, $x_{\lrcorner Q}$ - points of loss and recovery of permanency of the property characterized by MF quality function at the interval $(x_{\urcorner Q},\ x_{\lrcorner Q})$

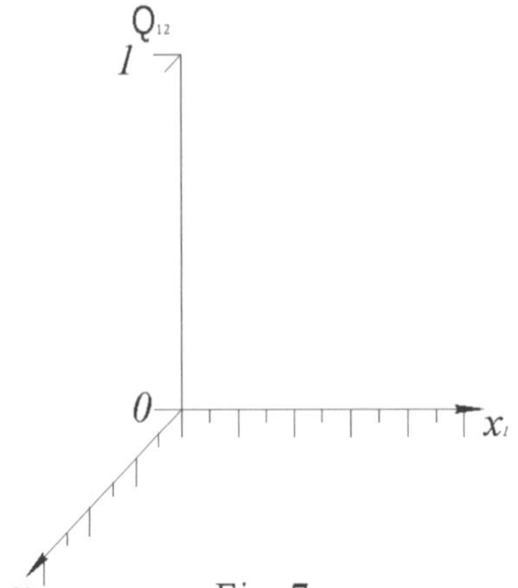

Fig. 7

Fig. 7. Coordinate system of the composite MF characteristic quality number Q^2.
a - main coordinate plane x_1 - 0 - 1, corresponding to the main number Q_1;
b - auxiliary coordinate plane x_2 - 0 - 1, corresponding to the auxiliary number Q_2;
c – plane of quality absence;
$Q_{1,2}$ - axis of verbal quality values (0 and 1) common for numbers Q_1 and Q_2

5. References

[1] Nikitin V.A., Filoncheva V.V. Quality Management Basing on ISO 9000-2000 Standards – 2nd edition [in Russian] – St. Petersburg: Piter 2004.

[2] Stepanov A.V. About Terminology and Process Approach. [in Russian] «*Standards and Quality*», 2007, No. 1, pp. 84-88.

[3] Solodovnikov V.V., Filimonov N.B. Dynamic quality of automatic regulation systems. [in Russian] *Educational aid.* – Moscow: Publishing house of Moscow State Technical University n.a. N.E. Bauman. 1987. – 84 p.

[4] Evaluation of quality characteristics of complicated systems and systems analysis. *Abstracts of reports of the VIII Interdepartmental science-technical seminar of the Academy of Sciences of the USSR.* [in Russian] – Moscow: 1978, – 298 p.

[5]. Zadeh L.A. The concept of a Linguistic Variable and its Application to Approximate Reasoning / [Translation from English into Russian] – Moscow: Mir 1976. – 165 p.

[6] Zadeh L.A. Toward a theory of fuzzy information granulation and its centrality in human reasoning and fuzzy logic // *Fuzzy sets and systems.* – Vol. 90. – 1997, N 2.

[7] Adler Yu.P. Recent Trends in Statistical Quality Control – Taguchi Method. [in Russian] – Moscow: Znaniye, 1988 – p. 56.

[8] A.N. Chekmarev, V.A. Barvinok, V.V. Shalavin. Statistical methods of Quality Management. [in Russian] – Moscow, Mashinostroenie, 1999. – 320 p.

[9] Total Quality Management: Manual for Institutions of Higher Education. O.P. Gludkin, N.M. Gorbunov, A.I. Gurov, Yu.V. Zorin: Edited by O.P. Gludkin. [in Russian] – Moscow: Radio i Svyaz. 1999 – 600 p.

[10] GOST R ISO 9000-2001. Quality Management Systems. Basic Concepts and Dictionary. [in Russian]

[11] Wittgenstein L. Logical-Philosophical Treatise / [translation from German into Russian]. General editor and author of foreword: V.F. Asmus – Moscow: Nauka, 1958 (2009). – 133 p.
[12] Nazarov N.G., Arkhangelskaya E.A. Advanced Methods and Algorithms of Processing Measurements and Product Quality Control. [in Russian] *Educational aid.* - Moscow: Publishing house of standards: 1995 - 163 p.
[13] Soviet Encyclopedic Dictionary. [in Russian] - Moscow: Soviet encyclopedia. 1980.
[14] Stepanov A.V. Objectives in the Field of Quality: Measurability and Types. [in Russian] *«Standards and Quality»*, 2007, No. 5, pp. 64-68.
[15] Ohno Taiichi, Setsuo Mito. Just-in-Time for Today and Tomorrow . - Portland, Oregon: Productivity Press, 1988. - 208 p.
[16] V.A.Kopin, V.L.Makarov and A.M.Rostovtsev, Processing Plastic Components [in Russian], Khimiya, Moscow (1988).
[17] A.M. Rostovtsev. Quality Control of Parts made of Plastics. [in Russian] Moscow: «Khimiya» Publishing house, 1974, - 112 p.
[18] N.G. Nazarov and N.T. Krushnyak. *Measur.Techn.*, 48, No.10, 969 (2005).
[19] Nazarov N.G., Bobrova I.A. Scales of Measurement in Metrology. [in Russian] - Moscow: Publishing house of Moscow State Technical University n.a. N.E. Bauman. 2001.
[20] Mathematics. *Large Encyclopedic Dictionary.* [in Russian] - Moscow, BSE. 2000.
[21] Nazarov N.G. Practical Guide for Solving Measurement Tasks Basing on Optimal Measurement Plans. [in Russian] *Educational Aid.* - Moscow: Publishing house of Moscow State Technical University n.a. N.E. Bauman, 2007 - 162 p.
[22] Mitrokhin A.N. Qualitative Value as the Element of Dimensional Analysis, or anent Dimensionality of «Dimensionless» Values. [in Russian] *«Legislative and Applied Metrology»*, 2010, No.3, p. 5-60.
[23] Graham R., Knuth D., Patashnik O. Concrete Mathematics: A Foundation for Computer Science /[Translation from English into Russian] Moscow: Mir; Binom. Laboratoria Znaniy, 2009.
[24] A.M. Rostovtsev. Extreme Choice of Tabular Data in Designing and Technologic Practice as their Dedekind section //*Fundamental and Applied Problems of Reliability and Diagnostics of Machines and Mechanisms: Materials of the 9th session of International Science school* in commemoration of V.P. Bulatov. [in Russian] St. Petersburg, 2009, p.307.
[25] A.M. Rostovtsev. Extremality in the selection and optimization of measurement devices. Measurement Techniques. Vol. 53, No. 9, (2010), pp. 870 - 875. Translated from *Metrologiya*, No. 8, pp. 8–16, August, 2010.
[26] Yu.N.Lyandon, Functional Interchangeability in Engineering [in Russian], Mashinostroenie, Moscow (1967).
[27] Germeier Yu.B. Introduction to Operations Research Theory. [in Russian] - Moscow: *Nauka*, 1973.
[28] A.M. Rostovtsev. Quality from the Point of View of Mathematical Fuzziness. *Proceedings of the 9th scientific and technical conference «State and Problems of Measurements»*. Moscow: Publication of Moscow State Technical University n.a. N.E. Bauman, 2004.

[29] A.M. Rostovtsev. Mathematically Fuzzy Approaches to Quality Control. Measurement Techniques. Vol.52, No.3, (2009). Translated from *Izmeritel`naya Teknika*, No. 3, pp.18–19, March, 2009.

[30] Matsievsky S.V. Fuzzy sets: Educational aid. - Kaliningrad: Publishing House of KGU, 2004. - 176 p

[31] Evaluation of quality characteristics of complicated systems and systems analysis. *Abstracts of reports of the VIII Interdepartmental science-technical seminar of the Academy of Sciences of the USSR*. [in Russian] - Moscow: 1978, - 298 p.

[32] Deichelmann H. Linguistische Systeme und ihre Anwendung. Darmstadt: Fachhochschule Darmstadt, 1996.

[33] Wittgenstein L. Philosophical Research // *Novoye v zarubezhnoy lingvistike* (New in foreign linguistics). Issue XVI. [in Russian] — Moscow, 1985. — p. 79—128

[34] A.M. Rostovtsev. Mathematically Fuzzy Summing of Measurement Errors. *Proceedings of the 9th All-Russian conference «State and Problems of Measurements»*. Moscow State Technical University n.a. N.E. Bauman, 23-25 November, 2004 [in Russian] Moscow: Publishing house of Moscow State Technical University n.a. N.E. Bauman, 2004, pp. 37, 38.

[35] A.M.Rostovtsev. Mathematically Fuzzy Summing of Measurement and Checking Errors. [in Russian] ibid, pp.61,62.

16

On Some Applications of Nonlinear Differential Equations in Image Processing: Concepts and Electronic Implementation

S. Morfu[1] and P. Marquié[2]
Université de Bourgogne, LE2I - CNRS UMR 5158
Dijon, France

1. Introduction

Quality control by visual inspection usually involves acquisition of images corrupted by noise and sometimes weakly contrasted. That is why image filtering is the first task commonly performed after data acquisition to take a decision in most quality control applications. Since this crucial filtering step allows to obtain a better classification in pattern recognition, the development of new processing algorithms still remains an opened research direction. In image processing field, unconventional methods of filtering inspired by the properties of nonlinear systems have recently been proposed as a convenient solution to solve problems of high computational complexity Teuscher & Adamatzky (2005). In fact, all these

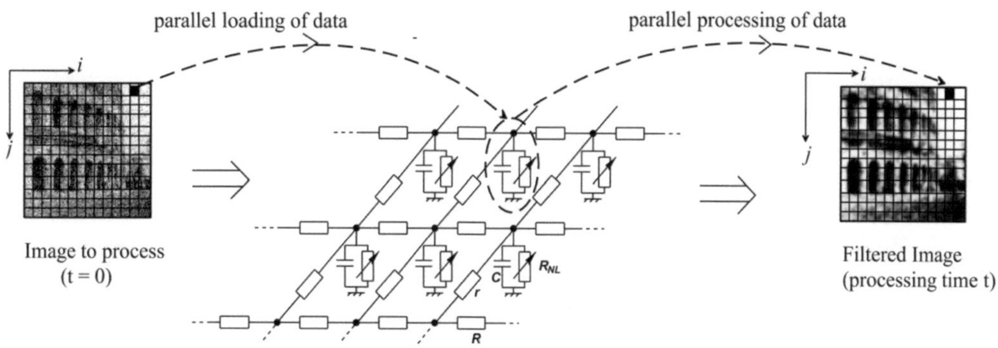

Fig. 1. The concept of image filtering using circuits ruled by nonlinear equations. The voltage at the nodes of the 2 dimensional network obeys to a set of $N \times M$ differential equations where $N \times M$ represents the image size. The initial condition loaded at each node of the network is denoted by the arrow in broken lines and corresponds to the gray level of each pixel.

unconventional algorithms are based on the properties of Nonlinear Differential Equations (NDEs) which can allow to explain various natural phenomena, such as the recurrence of ice ages, the propagation of blood pressure in arteries or the transport mechanisms in living cells to cite but a few. Another interesting feature of these NDEs is that they can also describe the voltage evolution at the nodes of a 2 dimensional electronic network. Therefore, the processing tools derived from the properties of these NDEs can potentially be implemented directly after

image acquisition. This fact represents a genuine advantage to develop intelligent embedded systems combining acquisition and processing.
Even if such embedded systems are actually far from being operational, this paradigm of nonlinear circuits ruled by NDEs has been raised since the eighties by L.O. Chua and L. Yang under the well known acronym *"CNN"* which means Cellular Neural (or Nonlinear) Network Chua (1999).
The principle of image filtering with these electronic networks is summarized in fig. 1:

- Each pixel of the initial image is considered as a cell (namely an electronic circuit) coupled to its neighbors. The processing task is set by the properties of this cell, that is by the nonlinear differential equations which rule the 2 dimensional network.
- The initial condition applied to each cell corresponds to the initial gray level of the image to process. Therefore, each cell can be viewed as a processor working in parallel.
- The dynamic of the nonlinear network is interpreted as a specific filtering process. Indeed, the image for a processing time ă"t"ă is obtained noting the state of the whole network at that time "t". Therefore, the implemented filter is recursive since the filtered image for a given processing time is deduced from the temporal evolution of the network.

This definition reveals many important features of the CNNs such as their parallel architecture, which is suitable for image processing, their strong dependence on initial conditions, offering a rich variety of potential processing and their intrinsic speed which allows real time applications Chua (1999); Chua & Yang (1988). Since their introduction as a novel class of information processing systems Chua (1999), the CNNs have been the subject of a growing interest with various applications in quality control Morfu et al. (2007); Occhipinti et al. (2001), image compression Venetianer et al. (1995) or image processing Julian et al. (2002); Morfu (2005). To summarize these two decades devoted to $CNNs$, finding new processing algorithms Grassi et al. (2006); Grassi & Grieco (2003) and designing the elementary cells of these electronic nonlinear networks Arena et al. (2003); Nagy & Szolgay (2003) constitute the two main directions of research to develop promising applications using $CNNs$.
In this chapter, the concept of processing with these CNNs ruled by NDEs will be intensively discussed. Especially, we will restrict this chapter to the study of the properties of NDEs for image processing purposes. More precisely, among the standard processing tools which can be performed with the properties of these NDEs, we propose to detail contrast enhancement, noise filtering or image segmentation, which constitute the most common pre-processing tasks performed to enhance pattern recognition and classification applications. We close this chapter by a slight discussion on a possible implementation of these NDEs.

2. Reaction-Diffusion electrical networks

2.1 Generic presentation

These kind of networks can be implemented following the electrical structure depicted in figure 2.
The voltage $U_{i,j}$ at cell (i,j) of the network obeys the standard Nagumo equation:

$$C\frac{dU_{i,j}}{d\tau} = \frac{1}{R}\left[\sum_{(k,l)\epsilon N_r}(U_{k,l} - U_{i,j})\right] - I_{NL}(U_{i,j}) \tag{1}$$

where τ corresponds to time and $I_{NL}(U_{i,j})$ represents a voltage-dependant nonlinear current. Finally, Nr denotes the set of nearest neighbors of cell (i,j).

In the following, we consider two different types of nonlinear current obtained while considering two different nonlinear resistances.

1. The first nonlinear resistor model presents a cubic nonlinear current-voltage characteristics since the nonlinear current I_{NL} obeys

$$I_{NL}(U_{i,j}) = U_{i,j}(U_{i,j} - a)(U_{i,j} - b)/(R_0 ab\beta). \tag{2}$$

In this case, setting:

$$\alpha = \frac{b}{a}, \qquad X_{i,j} = \frac{U_{i,j}}{b}, \qquad D = \frac{R_0 \alpha \beta}{R}, \qquad t = \frac{\tau}{R_0 \alpha C \beta}, \tag{3}$$

leads to the well known normalized Nagumo equation:

$$\frac{dX_{i,j}}{dt} = D\left[\sum_{(k,l)\epsilon N_r} (X_{k,l} - X_{i,j})\right] + f(X_{i,j}), \qquad \text{with} \qquad f(X_{i,j}) = -X_{i,j}(X_{i,j} - \alpha)(X_{i,j} - 1) \tag{4}$$

2. The second nonlinear resistor model presents a current-voltage characteristics approximating a sine shape for voltages belonging to the interval [-2V, 2V], that is

$$I_{NL}(U_{i,j}) = \frac{16\psi}{R_0} \sin(2\pi U_{i,j}) \simeq P(U)/R0. \tag{5}$$

In this expression, after the use of a least-square method, P(U) is a fifteenth order polynomial approximating the sine law (5) on the [-2V; 2V] voltage range. As for the first type of nonlinear resistor, a feedback resistor R_0 between the input and the output of the nonlinear source allows to obtain the desired nonlinear current by applying the Ohm law to this resistor R_0.

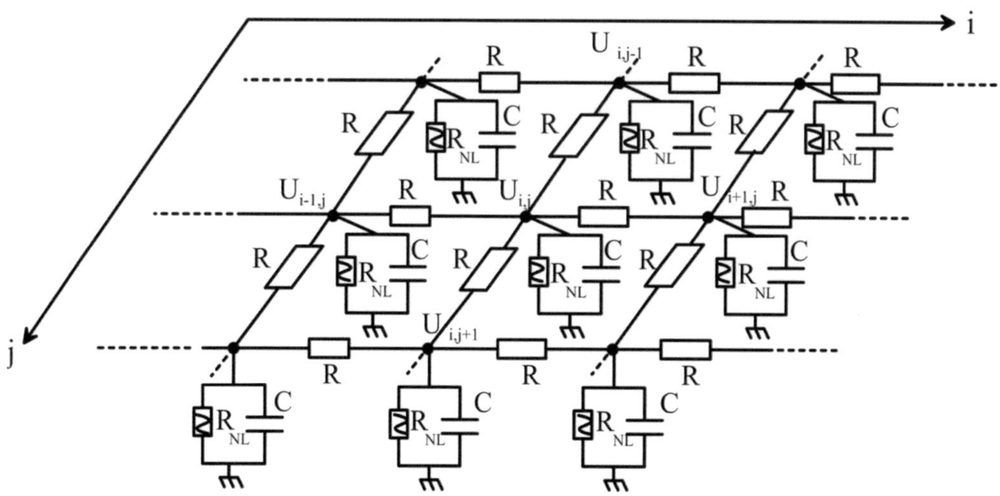

Fig. 2. The Reaction-Diffusion electrical network. R and C respectively represents a linear resistor and capacitor while R_{NL} is the nonlinear resistor whose current voltage caracteristic obeys to one of the two laws (2) or (5).

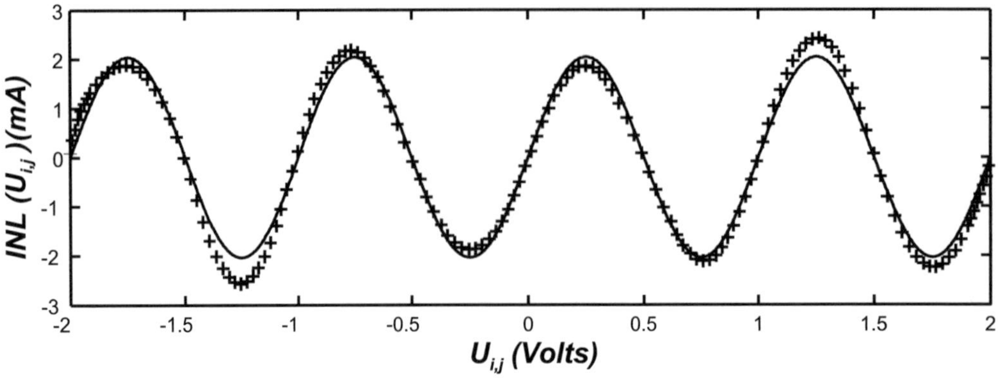

Fig. 3. Nonlinear sinusoidal current-voltage characteristics. The experimental current (+ signs) perfectly matches the theoretical law (5) on the $[-2V, 2V]$ voltage range. $R_0 = 2K\Omega$, $\psi = 0.25V$.

As shown in figure 3, the experimental nonlinear current-voltage characterisics (cross signs) perfectly fits the theoretical sine law given by (5) (solid line.)
Next, setting the following transformations

$$\tau = tR_0C, \qquad D = \frac{R}{R_0}, \qquad U_{i,j} = 4X_{i,j} - 2, \tag{6}$$

we get the normalized reaction diffusion equation

$$\frac{dX_{i,j}}{dt} = D\left[\sum_{(k,l)\epsilon N_r} (X_{k,l} - X_{i,j})\right] + g(X_{i,j}) \qquad \text{with} \qquad g(X_{i,j}) = -4\psi \sin(8\pi X_{i,j}) \tag{7}$$

2.2 Image processing applications

2.2.1 Image segmentation

For image processing application purpose, the properties of both (4) and (7) systems have been investigated. Their behavior can be understood while using the following mechanical analogy. Indeed, from a mechanical point of view, each of these two systems corresponds to a two-dimensional network of $N \times M$ particles of unit mass in the overdamped regime, coupled together with springs of strengh D. Each particle is submitted to a nonlinear force, f or g deriving from a potential $\Phi(x) = -\int_0^x f(x)dx$ or $\Phi(x) = -\int_0^x g(x)dx$, according to the considered nonlinear resistor model.
This mechanical analogy can be highlighted by applying the fundamental principle of dynamics to the network of particles presented in figure 4.
Indeed, if $X_{i,j}$ denotes the displacement of the particle (i, j), submitted to the potential Φ of figure 4.(b), the fundamental principle of dynamics writes

$$\frac{d^2X_{i,j}}{dt^2} = -\frac{dX_{i,j}}{dt} + D\left[\sum_{(k,l)\epsilon N_r} (X_{k,l} - X_{i,j})\right] - \frac{d\Phi}{dX_{i,j}}, \tag{8}$$

where $-\frac{dX_{i,j}}{dt}$ corresponds to the friction force, while $\frac{d^2X_{i,j}}{dt^2}$ represents the inertia term. Equations (4) and (7) are obtained while neglecting the inertia term versus the friction force,

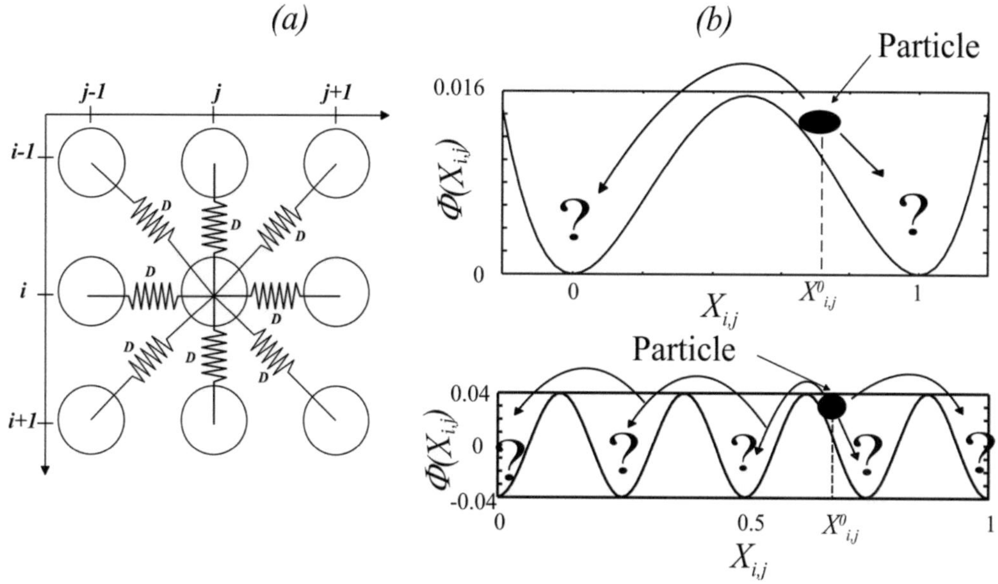

Fig. 4. Mechanical analogy of systems (4) and (7). (a) 2-D network of particles coupled by springs of strengh D and submitted to one of the two potentials represented in (b). Depending on both its initial condition and the resulting springback forces, each particle will be attracted by one of the different potential wells of figure (b).

which corresponds to the overdamped regime. The stability analysis of the two systems (4) and (7) depends respectively on the considered nonlinear forces f and g, the potentials they derive from being represented in figure 4.(b). This analysis being presented in details in Morfu et al. (2008), we just present here some remarks allowing to understand the dynamics of our system:

1. Local minima of the potential correspond to the stable states of the system, while the local maxima define the unstable states. Accordingly, in the uncoupled case ($D = 0$), a particle (or state of a cell or grey level) will evolve towards one of the stable states, depending on its initial condition versus the position of the unstable states acting like thresholds.
2. From a mechanical point of view, the term $D\left[\sum_{(k,l)\epsilon N_r}(X_{k,l} - X_{i,j})\right]$ corresponds to the neighbors-induced resulting elastic force acting on the particle (i,j). In the coupled case, ($D \neq 0$), this particle will evolve towards one of the stable states, not only depending on its initial condition but also on the resulting elastic force.

From an image processing point of view, it can be concluded that if an image is loaded as initial condition in the electrical network with cubic nonlinearity obeying to (4), the latter will evolve, for sufficiently large time, to an almost black and white picture, since only two stable states exist, corresponding respectively to 0 (black) and 1 (white).

Next, considering a sinusoidal nonlinearity (5), the system may evolve towards a larger number of stable states defined by the shape of the associate potential. It will then be possible to get a segmentation of the initial image into several regions related to the mean grey levels corresponding to the system stable states.

As an example, we present, in the context of quality control, the result of an image segmentation realized with a nonlinear network using respectively cubic and sinusoidal

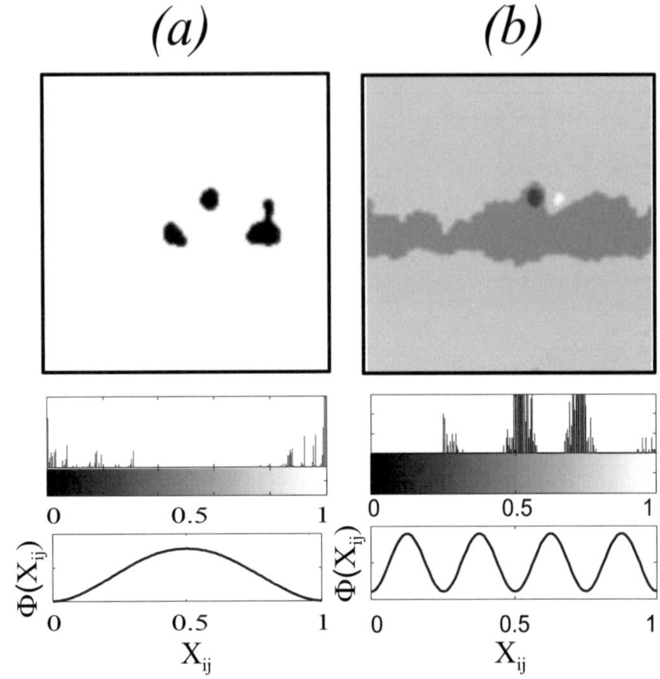

Fig. 5. Image segmentation using electrical reaction-diffusion networks. (a) The use of a cubic nonlinearity does not allow to extract the regions of interest. Indeed, as seen in the histogram and the potential under the image resulting from the segmentation process, the image grey levels are mainly localized in the two potential wells, corresponding respectively to black (0) and white (1). Parameters: $D = 0.05$, $\alpha = 0.5$, $t = 5000$. (b) The sinusoidal nonlinearity enables the extraction of the four regions of interest as observed on the image and on the grey levels repartition in four potential wells. Parameters: $D = 1.6$ and $t = 5000$.

nonlinearity. The initial noisy and weak contrasted image, considered in Morfu et al. (2008), represents a digitalized radiography of a soldering between two rods of metal. This image reveals the following four regions of interest:

1. The background of the image (light grey) which corresponds to the two rods of metal
2. The central part of the image which corresponds to the soldered joint (medium grey)
3. A dark grey spot localized in the soldered joint and corresponding to a gaseous inclusion
4. A projection of metal (white spot)

As seen in figures 5.(a) and (b), if the use of a cubic nonlinearity does not allow to extract the four specific regions, the sinusoidal nonlinearity enables to reveal the four individual components of the original noisy and weak contrasted image.

2.2.2 Noise filtering and contrast enhancement

During the last decades, a particular interest has been devoted to the development of bio-inspired processing tasks in the field of signal or image processing as seen in the previous section. In particular, it has been shown that noise filtering operations based on reaction-diffusion systems with cubic nonlinearity could be realized. Nevertheless, due to the

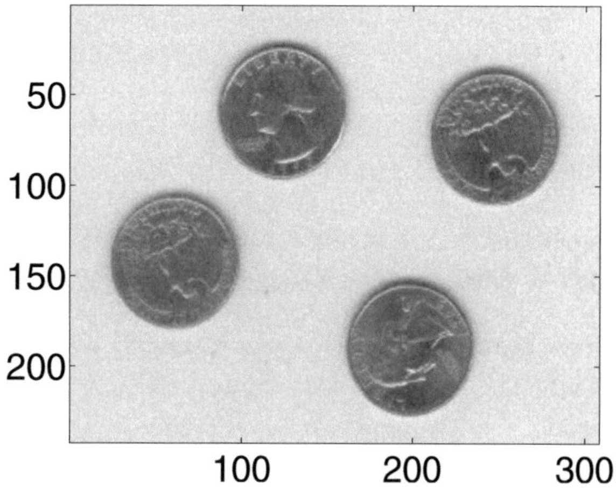

Fig. 6. Initial noisy image

homogeneous or isotropic diffusion process, this noise removal operation provides a blurry filtered image, which renders difficult the precise edge or contour location. As an example, in this section, we will consider the initial noisy image presented in figure 6.

In order to improve the noise filtering efficiency, the following diffusion process has been proposed Morfu (2009):

$$\frac{dX_{i,j}}{dt} = \frac{D}{|N_r|}\left[\sum_{(k,l)\epsilon N_r} h(\|X_{k,l} - X_{i,j}\|) \times (X_{k,l} - X_{i,j})\right] - \gamma X_{i,j}(X_{i,j} - \alpha)(X_{i,j} - 1), \quad (9)$$

where $|Nr|$ represents the total number of neighbors for the considered pixel (i, j).

The respective effects of the cubic nonlinearity on one hand, and of the anisotropy depicted by the function h on the other hand can be considered separately in equation (9).

Indeed, according to the weight γ of the cubic nonlinearity with respect to the function h, the following diffusion process may be highlighted:

1. **Isotropic and linear diffusion process.**
   When the nonlinear parameter γ is set to 0 and when the h function takes the value 1, then equation (9) corresponds to the classical linear heat equation. The resulting filtered image is represented in figure 7.(a) for the processing time $t = 6$. Furthermore, the profile of line number 50 is also shown for different processing times. It can be observed that, as time goes on, the noise effect tends to disappear but in the same time, the image details like the contours are also removed, which gives a blurry processed image.

2. **Isotropic and purely nonlinear process**
   When the h function takes the value 1, equation (9) reduces to the standard Nagumo equation (4) which describes a nonlinear process. The resulting filtered image and the profiles versus time of the line number 50 are presented in figure 7.(c) et (d). As the processing time increases, a contrast enhancement occurs (the relative amplitude of the profile increases) due to the cubic nonlinearity. However, as the process is isotropic, the contour location is not preserved since the resulting image remains blur.

3. **Purely anisotropic diffusion process**
   When $h \neq 1$ and $\gamma = 0$ (which means that there is no more nonlinearity in the system), one

get the Peronna and Malik algorithm allowing to overcome the blur effect due to isotropy Perona & Malik (1990). The main idea consists in replacing the diffusion coefficient D by a function h of the gradient amplitude, which corresponds to a modulation of the diffusion depending on the different regions of the considered image. More precisely:

- for regions containing edges or contours, diffusion is stopped by setting the h function to zero. Indeed, a contour corresponds to an important value of the local gradient. For this reason, in equation (9), the function h will depend on the local gradient modulus, which can be approximated by $\|X_{k,l} - X_{i,j}\|$ for the pixel (i,j) with neighor of index (k,l).

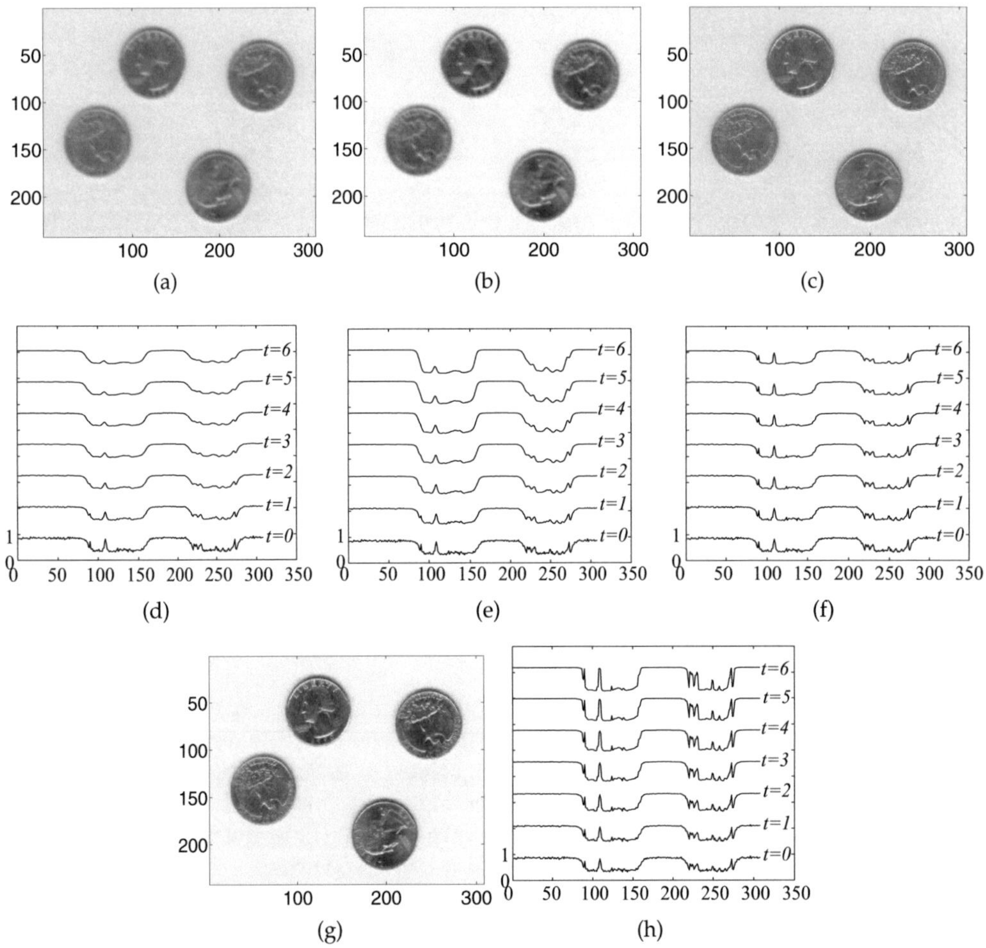

Fig. 7. : Image processing with different diffusion processes. (a) and (d) Isotropic and linear process ($h = 1$ and $\gamma = 0$). (b) and (e) Isotropic and nonlinear process ($h = 1$, $\gamma = 1$ and $\alpha = 0.5$). (c) and (f) Purely anisotropic process ($\sigma = 0.03783$, $\gamma = 0$). (g) and (h) Anisotropic and nonlinear process ($\sigma = 0.03783$, $\gamma = 1$ and $\alpha = 0.5$). In each case, the filtered image is obtained for the processing time $t = 2$ after integrating the system of nonlinear differential equations (9). Profiles of the line number 50 are also represented for different processing times.

- for homogeneous regions corrupted by noise, as the local gradient is quite weak, the diffusion process can be enhanced for eliminating the noise by the increase of the h function.

From these remarks, it comes that the h function must be a monotonic decreasing function versus the local gradient, from 1 which corresponds to the maximum of diffusion effect to 0 corresponding to no diffusion effect. As an example, Peronna and Malik have considered the following Lorentz function in their studies Perona & Malik (1990):

$$h(\|X_{k,l} - X_{i,j}\|) = \frac{D}{1 + (\frac{\|X_{k,l} - X_{i,j}\|}{\sigma})^2}, \tag{10}$$

where the parameter σ adjusts the decreasing of the function h, and then the weight of anisotropy. So it can be set in order to preserve the edges in the considered image.
According to Black et al. (1998), this parameter represents a noise estimator defined by

$$\sigma = 1.4826MAD(\nabla X)/\sqrt{2}\,, \tag{11}$$

where MAD and ∇X denote the median absolute deviation and the gradient of the image to be processed, respectively.
Considering the example of figure 6, the parameter σ has been estimated to $\sigma = 0.03783$. As observed in figures 7.(e) and (f), this diffusion process allows noise removal but without any contrast enhancement.

4. **Process combining anisotropic and nonlinear diffusion**
When $\gamma \neq 0$ and when the function h obeys to (10), the benefits of both anisotropy and nonlinearity are combined, as verified in figures 7.(g) et (h). Indeed, the obtained processing results clearly show that noise is removed while the image contrast is enhanced Morfu (2009).

3. Hamiltonian electronic processing systems

3.1 Theoretical background

As for the reaction diffusion processing system, we can use a mechanical analogy to depict the electrical network. So we consider a two-dimensional network of $N \times M$ particles of unit mass, but contrary to the previous case, the friction force is neglected versus the inertia term in equation (8). Furthermore, we will consider that all the particles are uncoupled, which corresponds to $D = 0$. Under these conditions, the oscillations of the $N \times M$ particles obey the following system of differential equations:

$$\frac{d^2 W_{i,j}}{dt^2} = f(W_{i,j}) \qquad i = 1, 2...N \text{ et } j = 1, 2...M\,, \tag{12}$$

with the the nonlinear force f which can be written under the following form

$$f(W_{i,j}) = -\omega_0^2 (W_{i,j} - m - \alpha)(W_{i,j} - m + \alpha)(W_{i,j} - m). \tag{13}$$

This force f derives from the potential Φ represented on figure 8. The roots $m - \alpha$ and $m + \alpha$ of the cubic force define the position of the two potential wells, while the root m indicates the position of the potential barrier height Δ. From a theoretical point of view, as seen in figure 8, if the potential energy of a particule (or oscillator) is greater than the potential barrier height, as no friction force is considered, this particle will oscillate without any damping that is for an infinite time.

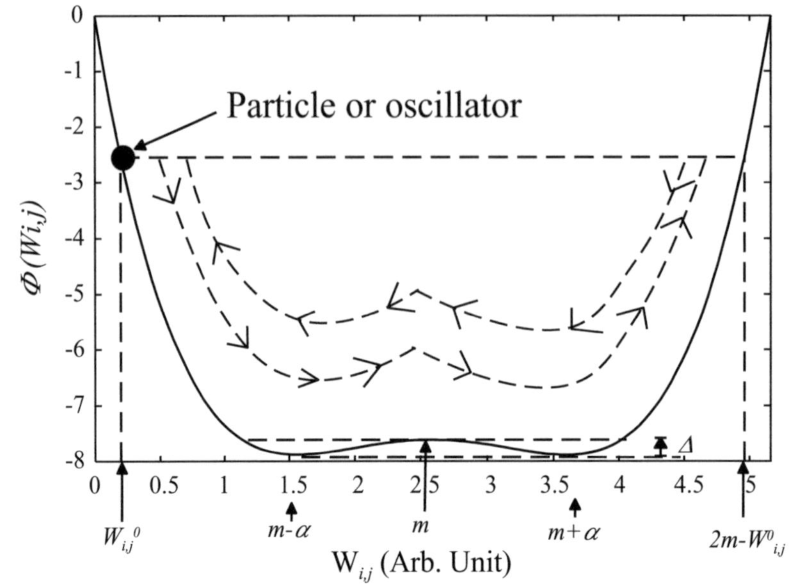

Fig. 8. Potential related to the nonlinear force (13) represented for $m = 2.58$, $\alpha = 1.02$ and $\omega_0 = 1$. As indicated by the dot line, a particle (or oscillator) which initial condition $W^0_{i,j}$ is greater than $m - \alpha\sqrt{2}$, oscillates in a symetric way, crossing periodically the potential barrier. Δ.

3.2 Electronic implementation of an oscillator

The schematic electronic diagram of this kind of elementary nonlinear oscillator is represented in figure 9. It is composed of:

- 3 external sources adjusting the three roots of the cubic nonlinearity m, $m - \alpha$ and $m + \alpha$.
- 2 AD633JNZ analog multipliers with scale factor $1/10V^{-1}$.
- 2 inverting amplifiers with gain $-K = -10$ for the compensation of the multipliers scale factors.
- a double integrator with time constant RC assuring a feedback between the input and the output of the nonlinear circuit.
- a $1N4148$ diode with threshold voltage V_T, used to apply the initial condition $W^0_{i,j}$ to the considered cell (i, j).

The diode cathode potential $W_{i,j}$ has to be expressed in order to get the nonlinear differential equation (12).
The output of the second amplifier, which can be written under the following polynomial form

$$P(W_{i,j}) = (W_{i,j} - m - \alpha)(W_{i,j} - m + \alpha)(W_{i,j} - m)\,, \tag{14}$$

is connected to the input of the circuit via the double integrator, which leads to the expression of the cathode potential:

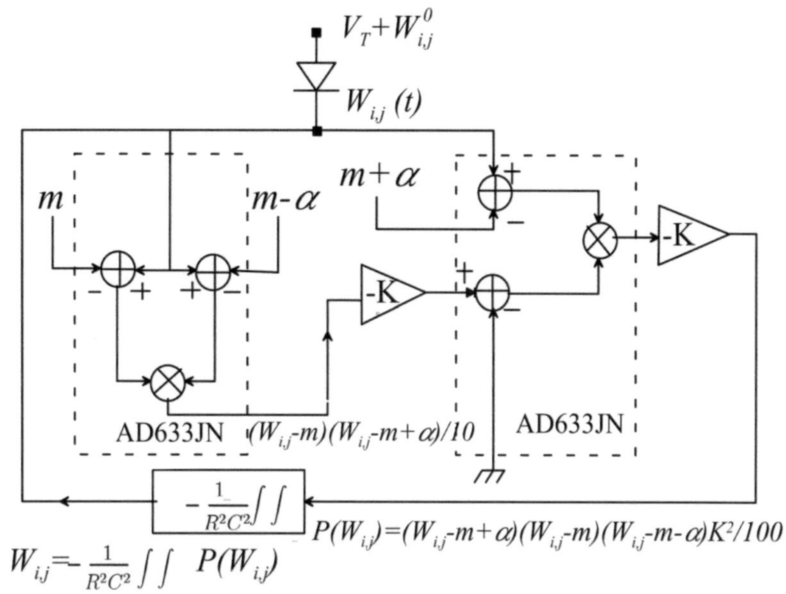

Fig. 9. Sketch of the nonlinear oscillator circuit obeying to (12). The component values are: $R = 10\ K\Omega$ and $C = 10\ nF$. $-K = -10$ stands for the scale factor of the inverting amplifiers.

$$W_{i,j} = -\frac{K^2}{100R^2C^2}\int\int (W_{i,j} - m + \alpha)(W_{i,j} - m - \alpha)(W_{i,j} - m)dt, \quad \text{with } i = 1, 2...N \text{ and } j = 1, 2...M. \tag{15}$$

Deriving twice equation 15 gives the nonlinear differential equation

$$\frac{d^2W_{i,j}}{dt^2} = -\frac{K^2}{100R^2C^2}(W_{i,j} - m + \alpha)(W_{i,j} - m - \alpha)(W_{i,j} - m). \tag{16}$$

which exactly corresponds to the normalized equation (12) provided that $\omega_0 = K/(10RC)$.

3.3 Image processing based on the properties of the nonlinear oscillators

Solutions of equation (16) can be obtained and written under the form:

$$W_{i,j}(t) = m + (W^0_{i,j} - m)\text{cn}(\omega_{i,j}t, k_{i,j}), \tag{17}$$

where $W^0_{i,j}$ corresponds to the initial condition applied to the particle (i, j) of the network, which corresponds to the grey level of the pixel (i, j) in the image processing context.
In equation (17), cn corresponds to the Jacobi elliptic function, which modulus $k_{i,j}$ and pulsation $\omega_{i,j}$ are given by

$$\omega_{i,j}(W^0_{i,j}) = \omega_0\sqrt{(W^0_{i,j} - m)^2 - \alpha^2} \quad \text{and} \quad k_{i,j}(W^0_{i,j}) = \frac{1}{2}\frac{(W^0_{i,j} - m)^2}{(W^0_{i,j} - m)^2 - \alpha^2}. \tag{18}$$

Pulsation and shape of oscillations are imposed by the modulus of the Jacobi elliptic function. They depend then on parameters α and ω_0 of the nonlinearity, but also on the initial condition $W^0_{i,j}$.
Because of the sensitivity to initial condition in this kind of nonlinear system, two oscillators with a very weak difference of initial condition will rapidly present a phase opposition. Indeed, considering respectively the initial conditions 0 and 0.2 for the oscillators of figure 10, phase opposition is attained theoretically at time $t_{opt} = 1.64\ ms$:
For this time t_{opt}, the initial weak difference between the amplitudes of the two oscillators becomes the largest one since it is equal to 5.1 (see figure 11).

The experimental results obtained using the electrical circuit present a good agreement with the theoretical predictions, despite the slight discrepancy for the optimal time (1.46*ms* instead of 1.64*ms*), mainly imputable to the components uncertainties. Let us note that the amplitude difference δ of the oscillators evoves with time, reaching periodically local minima and maxima.
The dynamics of this oscillators network can be understood in term of image processing. Considering indeed a weak contrasted image as initial condition, if the grey scale of the original image belongs to the interval [0; 0.2], then the amplitude difference between the two oscillators will follow the theoretical curve presented in figure 11. Consequently, as shown in figure 12, the particular times for which the amplitude difference vanishes will define the minima of contrast. On the contrary, the particular times corresponding to local maxima of Δ may lead to contrast enhancement with possible image inversion.

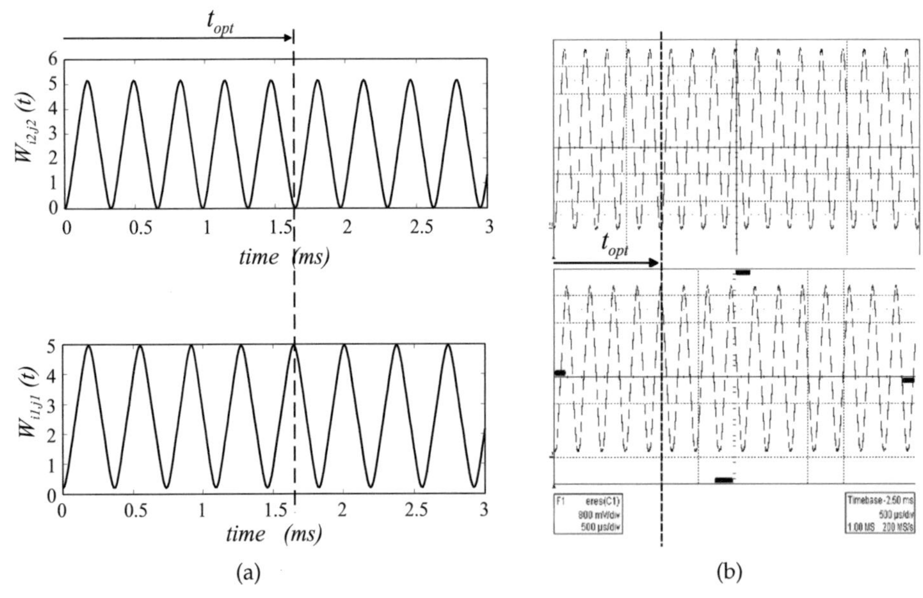

Fig. 10. Theoretical (a) and experimental (b) voltage temporal evolutions for two oscillators submitted to very narrow initial conditions. Top panel: initial condition is $W^0_{i2,j2} = 0$; Bottom panel: initial condition is $W^0_{i1,j1} = 0.2$. System parameters: $K = 10$, $R = 10K\Omega$, $C = 10nF$, $m = 2.58V$, $\alpha = 1.02V$.

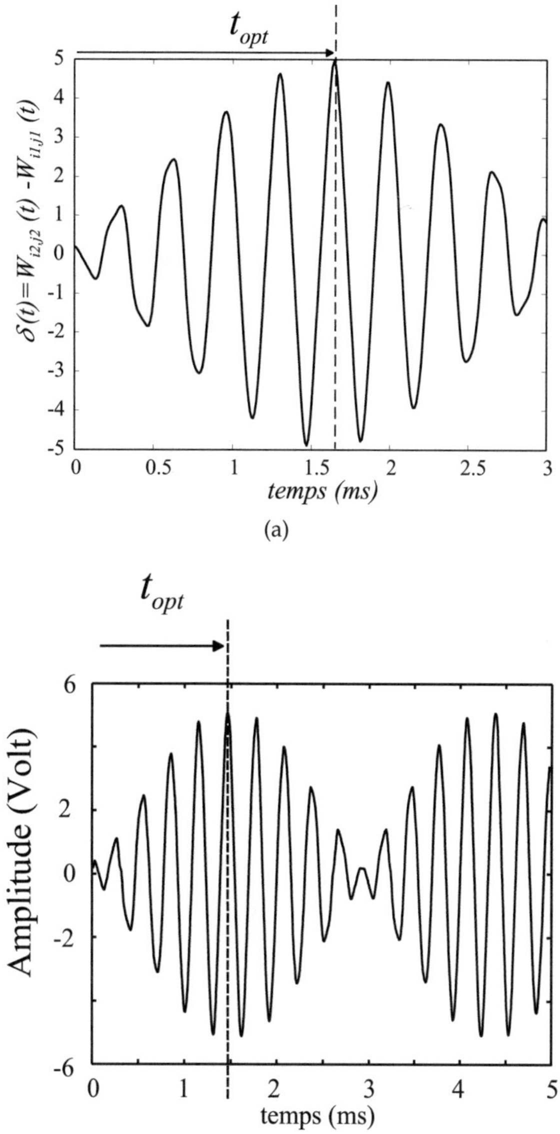

Fig. 11. Temporal evolution of the amplitude difference between the two oscillators of figure 10. (a) Theoretical evolution; (b) Experimental evolution. System parameters: $K = 10$, $R = 10K\Omega, C = 10nF, m = 2.58V, \alpha = 1.02V$.

Finally, in order to ensure the existence of elliptic Jacobi functions as solutions, modulus $k_{i,j}$ must stay in the $[0; 1]$ range. Consequently, the initial conditions applied to the lattice must be smaller than $m - \alpha\sqrt{2}$ or greater than $m + \alpha\sqrt{2}$. These conditions allow an energy greater than the potential barrier energy for all the cells of the network.

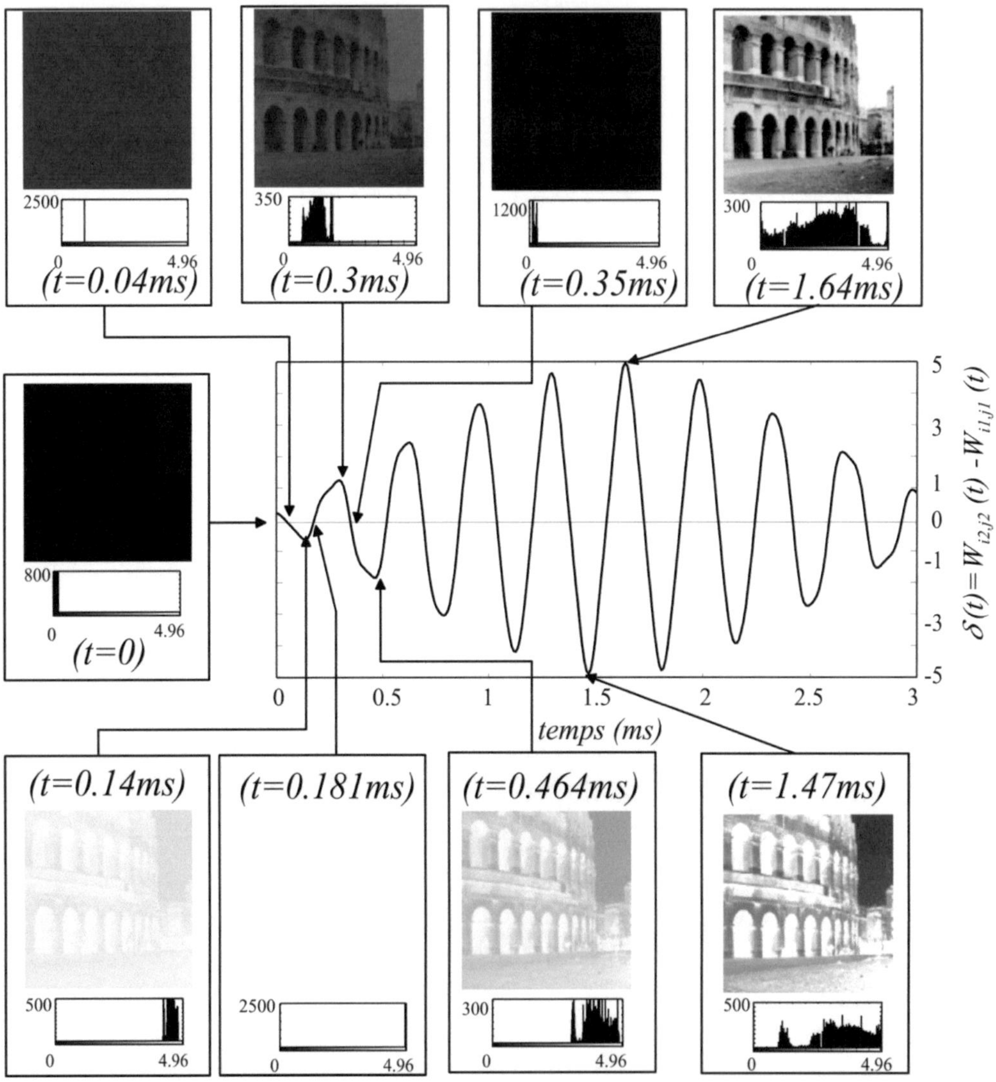

Fig. 12. Time evolution and histogram of a weakly contrasted image. The original image loaded in the network is represented with its histogram at time $(t = 0)$ on the left. The evolution of the amplitude difference shows the image dynamics for periodic particular times corresponding to local minima or maxima of contrast, with the possibility to get also image inversion.

As a result, the weakly contrasted image does not present grey levels exceeding $m - \alpha\sqrt{2}$ for its contrast to be enhanced.

4. Conclusion

In this chapter we have presented an overview of nonlinear processing inspired by Reaction-Diffusion and inertial electronic systems. Reaction-Diffusion electrical networks allow noise filtering and image segmentation while Hamiltonian electrical networks enable contrast enhancement. Each network is obtained by coupling elementary cells including classical components such as resistors, capacitors, self inductors, operational amplifiers or analog multipliers.

If the elementary cell can be easily built with such components, implementing the whole network is far from being realistic. We trust that it could be interesting to use Field Programmable Gate Array circuits (FPGA) to enable a large scale integration of the network. However, it requires to re-cast the nonlinear processing algorithms.

Lastly, as revealed by the reaction-diffusion systems, the filtering tasks are strongly depending on the nonlinearity. Therefore, we think that considering different nonlinearities could be the starting point for further development in nonlinear image processing.

5. References

Arena, P., Basile, A., Bucolo, M. & Fortuna, L. (2003). An object oriented segmentation on analog cnn chip, *IEEE Trans. on Circuits. and Syst.* 50(7): 837–846.

Black, M. J., Sapiro, G., Marimont, D. H. & Heeger, D. (1998). Robust anisotropic diffusion, *IEEE Trans. on Image Processing* 3(7): 421–432.

Chua, L. (1999). *CNN: A paradigm for complexity in Visions of Nonlinear Science in the 21st Century*, World Scientific, Singapore, chapter 13, pp. 793–826.

Chua, L. & Yang, L. (1988). Cellular neural networks: Theory, *IEEE Trans. on Circuits and Syst.* 35(10): 1257–1272.

Grassi, G., Di Sciascio, E., Grieco, L. & Vecchio, P. (2006). New object-oriented segmentation algorithm based on the CNN paradigm, *IEEE Trans. on Circ. and Syst.II: Express Brief* 53: 259–263.

Grassi, G. & Grieco, L. A. (2003). Object-oriented image analysis using the cnn universal machine: new analogic cnn algorithms for motion compensation, image synthesis and consistency observation, *IEEE Trans. on Circuits. and Syst.* 50(4): 488–499.

Julian, P., Dogaru, R. & Chua, L. (2002). A piecewise-linear simplicial coupling cell for CNN gray-level image processing, *IEEE Trans. on Circ. and Syst.* 3: 904–913.

Morfu, S. (2005). Image processing with a cellular nonlinear network, *Phys. Lett. A* 343(4): 281–292.

Morfu, S. (2009). On some applications of diffusion processes for image processing, *Phys. Lett. A* 373: 2438–2444.

Morfu, S., Marquié, P., Nofiele, B. & Ginhac, D. (2008). Nonlinear systems for image processing, *Advances In Imaging and Electron Physics, Elsevier* 152: 79–153.

Morfu, S., Nofiele, B. & Marquié, P. (2007). On the use of multistability for image processing, *Phys. Lett. A* 367: 192–198.

Nagy, Z. & Szolgay, P. (2003). Configurable multilayer cnn-um emulator on fpga, *IEEE Trans. on Circ. and Syst.I: Fund. Theory Appl.* 50: 774–778.

Occhipinti, L., Spoto, G., Branciforte, M. & Doddo, F. (2001). Defects detection and characterization by using cellular neural networks, *IEEE Int. Symposium on circuits and systems ISCAS* 3: 481–484.

Perona, P. & Malik, J. (1990). Scale space and edge detection using anisotropic diffusion, *IEEE Trans. Pattern Anal. Machine Intell.* 12(7): 629–639.

Teuscher, C. & Adamatzky, A. (2005). *Proceedings of the 2005 Workshop on Unconventional computing 2005 from cellular automata to wetware*, Luniver Press Beckington, ISBN 978-0-9551170-2-2.

Venetianer, P., Werblin, F., Roska, T. & Chua, L. (1995). Analogic CNN algorithms for some image compression and restoration tasks, *IEEE Trans. on cicuits and syst.* 42(5): 278–284.

17

Quality Control in Clinical Laboratories

Petros Karkalousos[1] and Angelos Evangelopoulos[2]
[1]*Technological Institute of Athens, Faculty of Health and Caring Professions, Department of Medical Laboratories*
[2]*Lab Organization & Quality Control dept, Roche Diagnostics (Hellas) S.A.*
Athens
Greece

1. Introduction

1.1 The automated analyzers in clinical laboratories

Nowadays, the overwhelming majority of laboratory results in clinical laboratories is being generated by automated analyzers. Modern automated analyzers are highly sophisticated instruments which can produce a tremendous number of laboratory results in a very short time. This is achieved thanks to the integration of technologies from three different scientific fields: analytical chemistry, computer science and robotics. The combination of these technologies substitutes a huge number of glassware equipment and tedious, repetitive laboratory work. As a matter of fact, the laboratory routine work has diminished significantly. Today laboratory personnel's duties have been shifted from manual work to the maintenance of the equipment, internal and external quality control, instrument calibration and data management of the generated results.

1.2 Statistical Quality control in industrial production

Quality control is an ancient procedure. For centuries manufacturers checked the quality of their products trying to find early any defect. At that time, manufacturers checked every product, one by one, without exception. Today, in industrial business, monitoring the quality of the each product is unattainable due to the large-scale production of different goods. Modern quality control is used to check the quality of a minimum number of samples from the total production. The procedure is called **statistical quality control (SQC)** or **statistical process control** (SPC). SQC is faster and more efficient than single checking.
The most general definition of SQC is: "SQC is process that minimizes the variability of a procedure" although it would be wiser to define SQC as "The process that focuses on revealing any deviations from well defined standards".

1.3 SQC in clinical laboratories' automated analyzers

SQC can be used in every automated production, like the laboratory determinations which are performed by biomedical analyzers. Unlike the industrial business where all products are similar, the laboratory determinations are totally different because of the huge biological differences among human beings. As a result, SQC can be done only for the equipment and the analytical methods and rarely for each laboratory result. SQC of automated analyzers uses as samples not the patients' results but the results of some special samples, the control samples.

The aim of this chapter is the introduction to the statistical quality control for automated analyzers in biomedical sciences such as haematology and biochemistry. The most important relevant laboratory SQC methods will be described.

2. Basic terms and definitions

2.1 Types of laboratory errors and mistakes

In laboratory practice many non-conforming results may appear. These results are divided in two major categories:

- **Errors**: Non-conforming results with "statistical meaning". This category includes all the "wrong" laboratory measures due to non-human action.
- **Mistakes**: Non-conforming results with "no statistical meaning". This category contains all the human errors e.g. mixing up samples.

Another classification of errors and mistakes is based on the time and the stage they appeared in laboratory practice.

1. The **pre-analytical stage** encompasses all the procedures which are take place before the analysis of the patients' samples on the automated analyzers (e.g. blood drawing, sample transportation, centrifugation, dilutions etc).
2. The **analytical stage** includes the analytical methods
3. The **post-analytical stage** refers to transmission of data from analyzers to the LIS, validation of results that have been produced and posting of the results to physicians or patients.

According to the previous classification, errors and mistakes are divided in three corresponding categories: *I) pre-analytical, II) analytical, III) post-analytical.* The majority of pre-analytical and post-analytical outliers are "mistakes" in contrary to analytical outliers which are considered as "errors"[1].

Tables 1-3 contain a list of the most common errors in hematology and biochemical analyzers.

Although many kinds of errors can be detected with various methods in laboratory practice, the laboratory staff focuses on detecting and eliminating the analytical errors.

The reasons are:

- The analytical errors are attributed to the laboratory staff.
- The analytical errors can be detected with SQC methods.
- Statisticians have helped in establishing certain limits for the analytical errors for every laboratory determination.

1.	Inappropriate specimen (e.g. wrong specimen-anticoagulant ratio)
2.	Wrong anticoagulant (e.g. sodium citrate in place of EDTA)
3.	Improper conservation method
4.	Inappropriate patient's preparation (e.g. wrong diet)
5.	Mistakes in patients' identification

Table 1. Common pre-analytical errors

[1] In this chapter, we will use the term "errors" for all kinds of outliers except special circumstances.

1.	Expired reagents which may lead to erroneous results
2.	Expired controls or calibrators
3.	Calibration curve time-out elapsed
4.	Failure in sampling system
5.	Failure in aspiration system of reagents
6.	Changes in analyzer's photometric unit / flow cell / measuring unit
7.	Any other analyzer's failure

Table 2. Common analytical errors

1.	Wrong matching between sample and laboratory's files
2.	Wrong copy of results from the analyzer's report to the laboratory report (in cases of manual transfer)
3.	Delay in delivering the results to the physicians, clinics or patients
4.	Loss of the results

Table 3. Common post-analytical errors

2.2 Precision and accuracy

Analytical errors influence the repeatability, reproducibility, precision, trueness and accuracy of the analytical methods. Precision and accuracy can be defined with many different ways.

1st **definition** (EURACHEM/CITAC Guide CG 4)

Precision *is the closeness of agreement between independent test results obtained under stipulated conditions.*

Accuracy *is the closeness of the agreement between the result of a measurement and a true value of the measurand.*

According to the same guide **true value**[2] is the "*value consistent with the definition of a given quantity*". In fact true value is any measurement with no errors random or systematic (see paragraph 3.4).

Measurand is a "*particular quantity subject to measurement*" or simply any substance or analyte which can be assayed in a clinical laboratory.

2nd **definition**

$$\text{Precision} = \text{repeatability or reproducibility}^3 \quad (1)$$

$$\text{Accuracy}^4 = \text{trueness} + \text{precision} \quad (2)$$

---

[2] According to the EURACHEM/CITAC Guide CG 4 the indefinite article "a" rather than the definite article "the" is used in conjunction with "true value" because there may be many values consistent with the definition of a given quantity.

[3] According to ISO 5725-3:1994 the definition of precision concludes and the "intermediate precision" (s_{zi}) with *i* denoting the number of factors. Intermediate precision relates to the variation in results observed when one or more factors, such as time, equipment and operator, are varied within a laboratory; different figures are obtained depending on which factors are held constant.

[4] :Recently the term analytical uncertainty has been used to encompass several aspects of accuracy.

Repeatability *is the degree of consensus between successive measurements which have been done on the same sample with very similar conditions (same analyzer, same user, same laboratory, same methods, same lot of reagents) in a very short time (e.g. same day).* **It is approached by within run or within day experiments and often is symbolized as s_r.**
Reproducibility *is the degree of consensus between successive measurements achieved on the same sample with different conditions (e.g. different analyzer, different user, different lot of reagents) in a long time.* **It is approached by between day experiments, can be either intra-laboratory or inter-laboratory, and often is symbolized as s_R.**
Trueness is often used to describe the classical term bias.

3rd definition

$$\text{Precision} = x_i - \overline{x} \tag{3}$$

$$\text{Trueness} = \overline{x} - \mu \tag{4}$$

x_i: a single measurement, **μ:** a true value, $\overline{x}$ **:** the average of successive measurements[5]
Based on equations 3 and 4 the equation of accuracy is transformed as follows:

$$\text{Accuracy} = (\overline{x} - \mu) + (x_i - \overline{x}) = (x_i - \mu) \tag{5}$$

2.3 Types of analytical errors

Analytical errors fall into two subcategories according to EURACHEM/CITAC Guide CG 4 as follows:

Random Errors (RE) (Fig. 1, RE)

Result of a measurement minus the mean that would result from an infinite number of measurements of the same measurand.
The mathematical definition is:

$$\Delta\text{RE} = x_i - \overline{x} \tag{6}$$

In fact random errors affect the **precision** (see paragraph 2.2) of all measurements. Random errors are attributed to either undetermined reasons (inherent error) or well defined causes. Their magnitude (**ΔRE**) is equal to the precision of a measurement and it is always greater than zero (ΔRE > 0). ΔRE can be diminished by increasing the number of measurements (it influences the $\overline{x}$). Large ΔRE increases the dispersion of the results around a true value.
The experimental standard deviation of the arithmetic mean or average of a series of observations is not the random error of the mean, although it is so referred to in some publications on uncertainty. It is instead a measure of the uncertainty of the mean due to some random effects. The exact value of the random error in the mean arising from these effects cannot be known.

---

[5] Although, accreditation bodies use the term trueness for the difference $\overline{x} - \mu$ in daily SQC practice is very common to relate the difference $\overline{x} - \mu$ with the term "accuracy" which is expressed. We will use this term for accuracy later in this chapter.

Systematic Errors (SE) (Fig. 1, SE)

Systematic error is defined as a component of error which, in the course of a number of analyses of the same measurand, remains constant or varies in a predictable way. Quite often is attributed as:
Mean that would result from an infinite number of measurements of the same measurand carried out under repeatability conditions minus a true value of the measurand and is expressed mathematically as:

$$\Delta SE = \bar{x} - \mu \qquad (7)$$

Systematic errors can be attributed to certain reasons and therefore can be eliminated much easier than random errors. ΔSE cannot by diminished by increasing the number of measurements. As opposed to the random errors their magnitude can be zero ($\Delta SE \geq 0$).
There is also another kind of analytical errors but it cannot be detected easily with SQC methods. These errors are called "**gross errors" (GE)** and their classified to the category of mistakes (Fig. 1, GE). Gross errors can result from mixing up of samples, clots in the analyzers' sampling system etc.

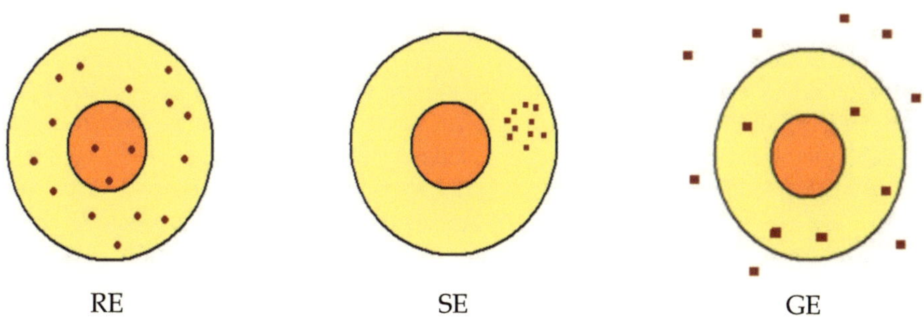

Fig. 1. Representation of the values' dispersion in random errors (RE), systematic errors (SE) and gross errors (GE)

Random and systematic errors can be detected very effectively by means of SQC methods such as Levey-Jennings charts, Westgard rules, Cusum charts e.t.c.

2.4 Total analytical error

The mathematical definition of the total analytical error (**TE**) is:

$$\text{Total analytical error} = \text{Random Error} + \text{Systematic Error} \qquad (8)$$

According to the definitions of random and systematic errors (paragraph 2.3):

$$TE = \Delta RE + \Delta SE \qquad (9)$$

Under ideal circumstances, total analytical error equals to zero, but this cannot be achieved in daily practice. Only ΔSE can be zero ($\Delta SE \geq 0$) where as ΔRE is always greater than zero ($\Delta RE > 0$) because of the existence of the inherent error.
Since $TE > 0$ is unavoidable, TE of every single determination must be lower than a specified limit. This limit is called "**allowable total analytical error**" (**aTE**) and it is different for each analyte being determined in a clinical laboratory.

2.5 Internal and External SQC

Random and Systematic errors must be detected at an early stage and then every effort should be taken in order to minimize them. The strategy for their detection consists of specific SQC methods which are divided in two categories:

Internal Quality Control (IQC). *It concludes all SQC methods which are performed every day by the laboratory personnel with the laboratory's materials and equipment. It checks primarily the precision (repeatability or reproducibility) of the method.*

External Quality Control (EQC). *It concludes all SQC methods which are performed periodically (i.e. every month, every two months, twice a year) by the laboratory personnel with the contribution of an external center (referral laboratory, scientific associations, diagnostic industry etc.). It checks primarily the accuracy of the laboratory's analytical methods. However, there are certain EQC schemes that check both the accuracy and the precision.*

Other terms for external quality control are: interlaboratory comparisons, proficiency testing **(PT) and external quality assessments schemes (EQAS).**

The metrics of internal and external quality control are based on statistical science (e.g. SDI, CV, Z-score) and they are graphically represented by statistical charts (**control charts**). Some of them are common in other industries while others specific for internal or external quality control in clinical laboratories.

2.6 Control materials

Control materials (or simply "**controls**") are all the materials which can be used for error detection in SQC methods. Although this term is considered equal to "**control samples"**, several SQC methods have been deployed based on patients' results.

Control samples are **pools** of biological fluids (serum, whole blood, urine or other materials). They contain analytes which are determined by the laboratory, ideally in concentrations/activities close to the decision limits where medical action is required. In internal and external SQC, it is common practice that laboratories use two or three different control samples which contain different quantities of analytes e.g. low, normal, high concentrations/activities. Control samples with the same analytes but different concentrations/activities are called "**levels**". Different levels check the performance of laboratory methods across all their measuring range. In most cases control samples are manufactured by analyzers' or reagents' manufacturers, but they can also be made by the laboratory personnel.

Before control samples are assayed for IQC reasons, each laboratory has to estimate their limits. **Control limits** are an upper and lower limit (see paragraph **3.2) between** which the **control values** are allowed to fluctuate.

3. Internal quality control

3.1 Normal distribution

Normal or Gaussian distribution (**N**) is the basis of SQC theory. Distribution chart is a biaxial diagram (x/y). X-axis represents the values of a variable's observations and y-axis the frequency of each value (the number of each value's appearance). It has a bell-shaped form with its two edges approaching asymptomatically the x-axis. The highest point of normal distribution corresponds to the value with the higher frequency (mode value). In any normal distribution, the mode value is equal to mean and median value of the variable.

Mode value (Mo) *is the value with the highest frequency.* It is always on the top of every distribution curve.
Median value (M) is the value which divides the variable's observations in two equal parts. It represents the "center" of the distribution.
Mean value or average value (μ or $\overline{x}$ *is equal to the value which all the observations should have if they were equal.* In normal distribution, mean, median and mode values coincide. The mean value or average can be calculated by the next formula:

$$\mu = \frac{\sum_{i=1}^{N} x_i}{N} \tag{10}$$

Where: $\mathbf{x_i}$ = Single value, $\mathbf{\Sigma x_i}$ = Sum of values, **N** = Total number of values

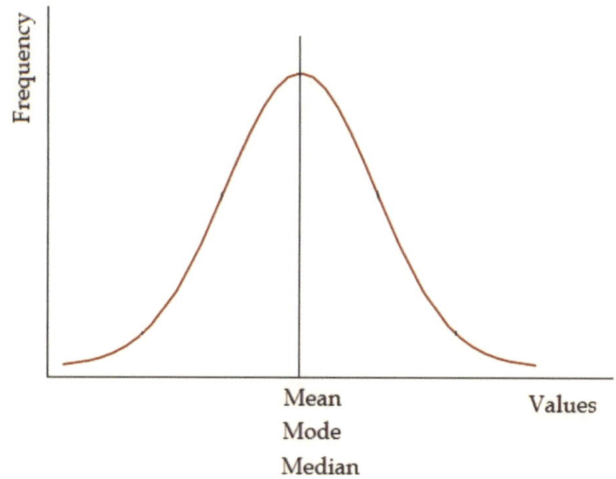

Fig. 2. The normal distribution

The length of a distribution curve defines the variance of the variable. The most common measure of variance is standard deviation (SD or s). **Standard deviation** can be calculated by the next formula:

$$s = \sqrt{\frac{\sum_{i=1}^{N} (x_i - \mu)^2}{N-1}} \tag{11}$$

The distance between the upper (UL) and lower limit (LL) of a normal distribution is six standard deviations (6s). Since mean value is in the center of normal distribution, the total range of a normal distribution is $\mu \pm 3s$ (to be more exact, not all, but nearly all (99.73%) of the values lie within 3 standard deviations of the mean). Every normal distribution can defined as N~(μ, s). For instance N~(76, 2.3) means a normal distribution with mean value = 76 and standard deviation = 2.3.
Mean value and standard deviation allow the statisticians to calculate the distance of each observation from the center (mean), using as measuring unit the s. This distance is called Z-score. It is measured in standard deviation's units by the next formula:

$$Z-\text{score} = \frac{x_i - \mu}{s} \tag{12}$$

Where: x_i = Single value, μ = Mean value, s = Standard deviation
Supposing we are looking the distance of value x_i = 80 from the mean of the normal distribution N~(100, 5). According to equation 12 Z-score is:

$$Z-\text{score} = \frac{x_i - \mu}{s} = \frac{80-100}{5} = -4$$

So the value 80 is 4 standard deviations lower than mean value 100. In Fig. 3 the location of Z-score=-4s seems to be lower than the lower limit of the distribution.

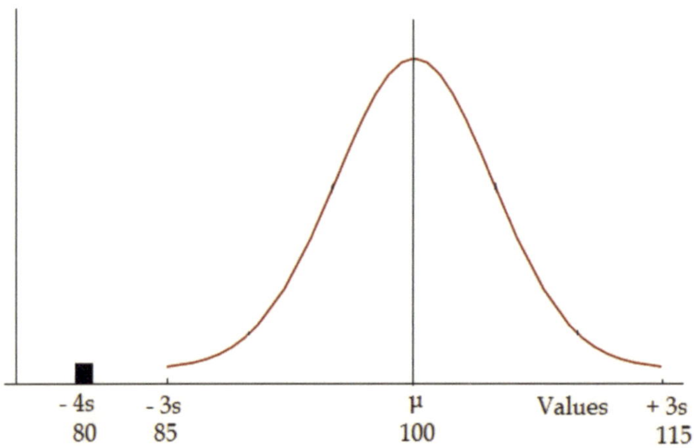

Fig. 3. The location of the value 80 with Z-score = -4

Because of the symmetric bell-shaped form of normal distribution its surface can be divided in six parts containing a specific percentage of its observations ("**empirical rule**", Table 4, Fig. 4).

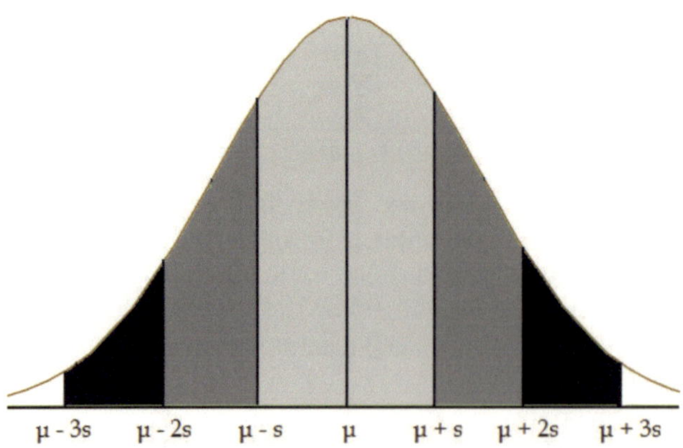

Fig. 4. The division of a normal distribution in six parts

1.	The part $\mu \pm s$ contains 68,26% of the observations.
2.	The part $\mu \pm 2s$ contains 95,46 % of the observations.
3.	The part $\mu \pm 3s$ contains 99,73 % of the observations.

Table 4. The "empirical rule" of the normal distribution

Standard deviation depicts the variation in the same units as the mean (e.g. cm, L, mol/L). So standard deviation cannot be used to compare the variance of different distributions or distributions with different mean. Therefore for comparison reasons the **coefficient of variation** (CV%) is being used. CV% is a normalized measure of dispersion of a probability distribution. It is defined as the ratio of the standard deviation to the mean and expressed as a percentage:

$$CV\% = \frac{s}{\mu}100 \qquad (13)$$

The normality (the normal distribution) of a variable is crucial for any statistical study. The "**normality test**" is the first task of the statisticians before the application of any statistic test. If the variable's values are compatible with the normal distribution then "**parametric tests**" are employed. Otherwise "**non-parametric tests**" are used. The most known normality tests are Kolmogorov-Smirnov & Shapiro-Wilk.

3.2 Calculation of control limits

Control limits are necessary for any SQC method in internal and external quality control (see paragraph 3.1). They consist of a center value (CL) and an upper and low control limit (UCL & LCL). Generally, they are created by repetitive measurements of control samples. In internal SQC two or more control samples are assayed every day and at least once per day before the patients' samples. Then the laboratorians check if all control values lie within the control limits. If at least one of the controls' measurements is outside of one of the two control limits then further actions may be required until random or systematic errors are under control.

Control limits vary depending on the control samples, the automated analyzer and the method of determination. They can (and should) be calculated by the laboratory itself, although many laboratories use the control limits established by the analyzer's manufacturer.

The steps for their calculation are the following:

1. The laboratory's staff collects 20 – 30 successive measurements from any control level.
2. Standard deviation (s) and mean value (μ) are calculated. The range $\mu \pm 3s$ is considered as "trial limits"
3. The laboratory's staff checks if any of the measurements exceed the range $\mu \pm 3s$. If so, the outlier is rejected the standard deviation and mean value are calculated once more.
4. The laboratory's staff repeats the previous procedure until no measurement exceeds the range $\mu \pm 3s$. The final μ and s are the mean value and the standard deviation of the control limits.

Control limits correspond to a normal distribution. The mean value of the control limits is symbolized as "μ" and it is considered a true value of the daily control values (see paragraph 3.4). Their standard deviation is symbolized as "s" and it is equal to the inherent error. On the contrary the mean value of the daily control values is symbolized as $\bar{x}$ and their standard deviation as "SD". SD encompasses the inherent error and any other existing random error.

3.3 Random and systematic errors in normal distribution of control values

Successive measurements derived from the same sample have a normal distribution. This is also the case with control values. Control values have always an inherent error, even if the determinations have been done under exactly the same conditions. This inherent error is the minimum random error of the process (Fig 5). If other sources of random errors exist the standard deviation of the measurements will be higher. In systematic errors the mean value of the control values has been moved further up or down from the mean value of the control limits (μ) (Fig 6).

In daily practice a measurement has both random and systematic errors. Random error may be approached by a number of concepts and statistical techniques (precision, repeatability, reproducibility, s, CV%,). Another concept for approaching the analytical variation caused by the random errors is the **imprecision (s_{meas})**. The difference $\bar{x} - \mu$, expressed as a percentage of μ, is called **$bias_{meas}$**.

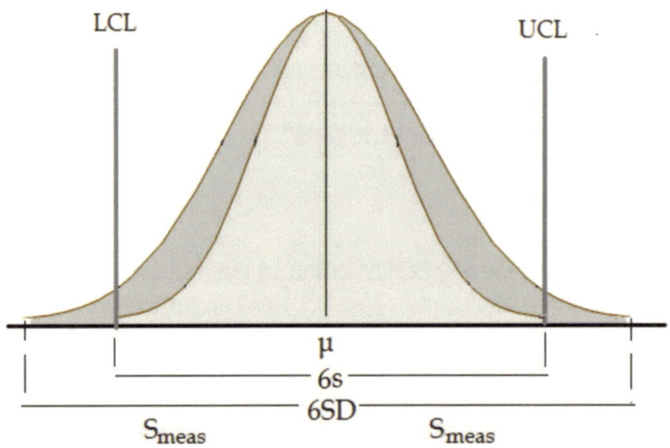

Fig. 5. The random error in normal distribution

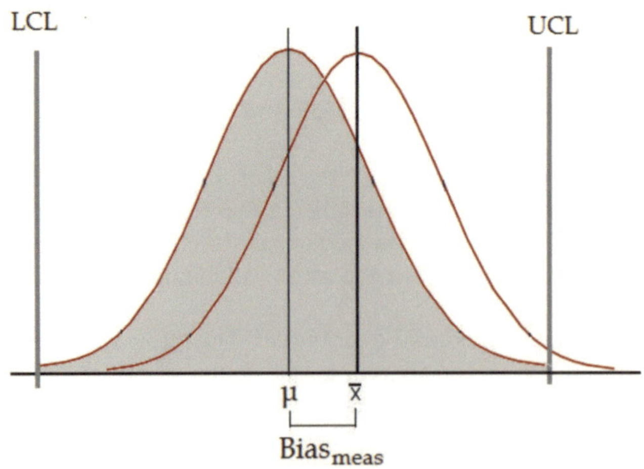

Fig. 6. The systematic error in normal distribution

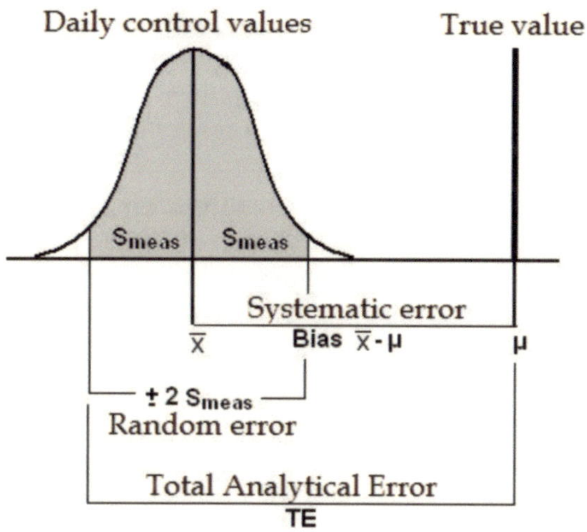

Fig. 7. The graphical display of a total analytical error. It refers to the maximum incidence of RE and SE

3.4 A true value in SQC

Although the previous procedure estimates the standard deviation and mean, with the exception of outliers, the laboratory personnel cannot be sure that the mean is the real one. The mean value of the 20-30 values will approach the true if some other procedure has come before. This procedure is the **calibration** of the automated analyzer with reference materials / calibrators. The validation of the calibration is done with "**recovery tests**" and the external quality control.

3.5 Levey-Jennings chart

Levey-Jennings chart is the most important control chart in laboratory quality control. It can be used in internal and external quality control as well. It detects all kinds of analytical errors (random and systematic) and is used for the estimation of their magnitude.
It was firstly introduced in 1952 (Levey S. & Jennings E, 1950). Stanley Levey and Elmer Jennings were inspired by Walter Shewhart charts, the most effective control charts in industrial business at that time. Shewhart and Levey-Jennings charts are based on **normal distribution** (Fig. 4).

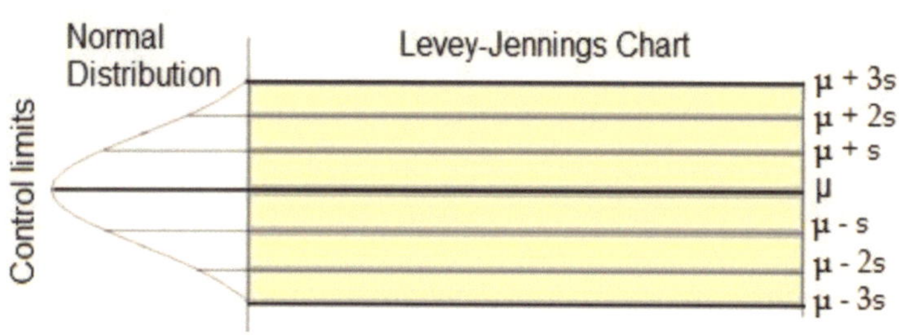

Fig. 8. Plottting of a Levey-Jennings chart from a control limits distribution curve

Plotting of a Levey-Jennings chart starts from the distribution curve of the control limits (μ±3s). The first step is to rotate clockwise the distribution curve of the control limits by 90°. The second step it to draw seven lines which start from the points μ+3s, μ+2s, μ+s, μ, μ-s, μ-2s and μ-3s. These seven lines form the Levey-Jennings chart (Fig. 7). For every different control level a different Levey-Jennings chart is being plotted.

3.6 Random and systematic errors in Levey-Jennings chart

The operator of an automated analyze,r assays on a daily basis two or three control levels with the same chemical methods and equipment as with the patient's samples. Every control value is plotted on a Levey-Jennings chart. The analyzer's operator checks if any or the daily control values exceeds certain limits. If so, there is either a random error or a systematic error or both.

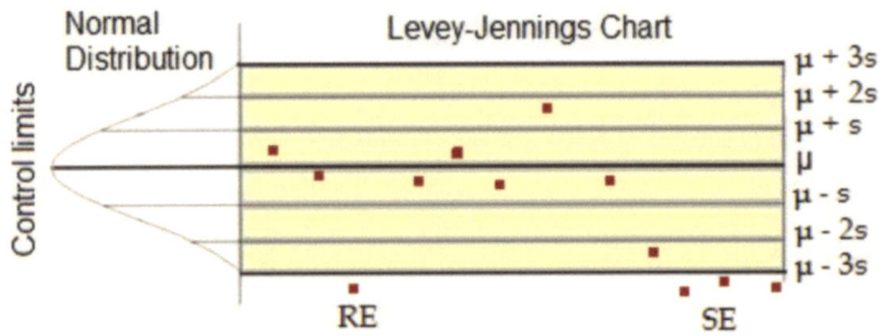

Fig. 9. The daily use of a Levey-Jennings chart with a random error (RE) and a systematic error (SE)

In Levey-Jennings chart a random error is revealed when one control value exceeds the UCL (μ+3s) or LCL (μ-3s). The detection of a systematic error is more complicated. In systematic errors two or more successive control values exceed the control limits which can be respectively μ+3s, μ+2s μ+s or μ-s μ-2s, μ-3s (Fig 9).

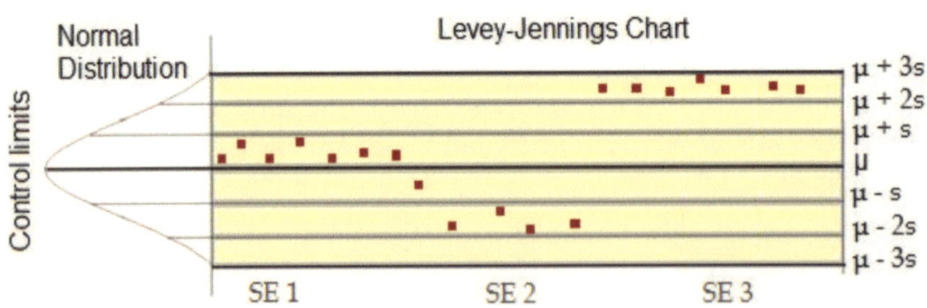

Fig. 10. Three different systematic errors on a Levey-Jennings chart. SE 1 has 7 successive control values between μ and μ + s (UCL = μ + s), SE 2 has 4 successive control values between μ - s and μ - 2s (LCL = μ - 2s), SE 3 has 7 successive control values between μ + 2s and μ + 3s (UCL = μ + 3s)

Daily values create their own normal distribution which may be different from the normal distribution of control limits. Some examples with random and systematic errors are displayed in Fig. 10 & 11.

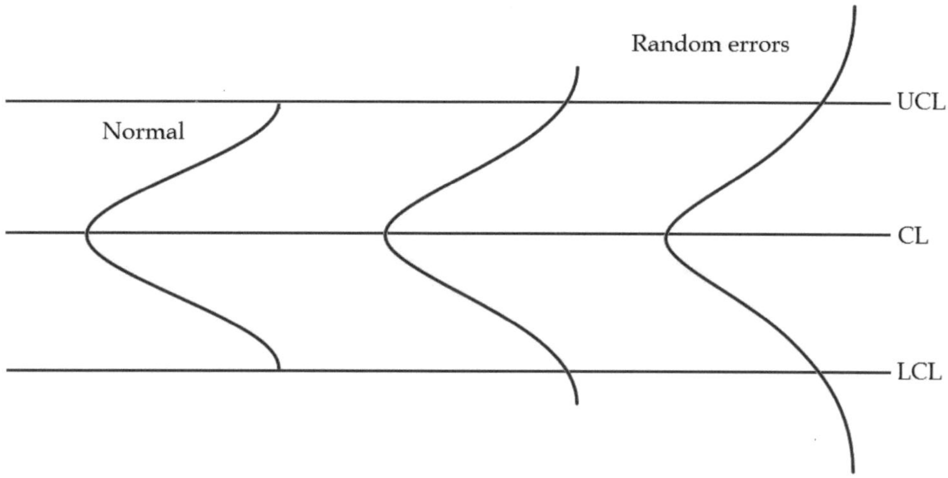

Fig. 11. The random errors on a Levey-Jennings chart

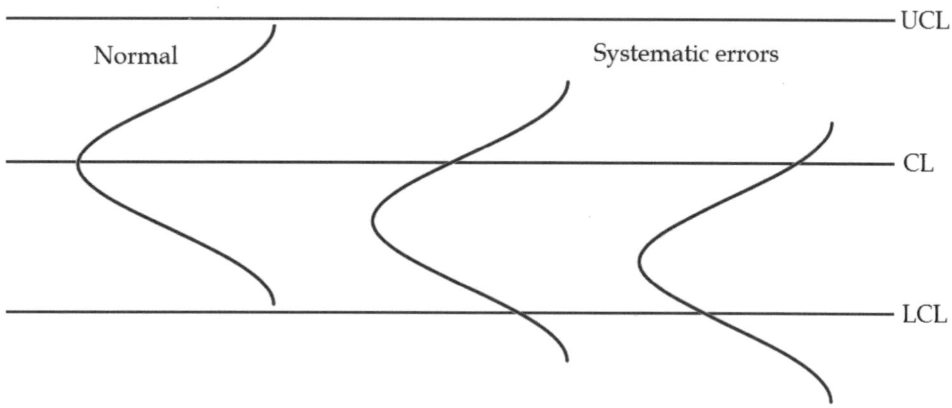

Fig. 12. The systematic errors on a Levey-Jennings chart

3.7 The Westgard rules

Error detection can be done very easily by using quality criteria. Although many of them have been proposed on the past, the most widely used are the so called Westgard rules (Westgard et al., 1981). Westgard rules (or modifications of them) are used today in almost every single biochemical, immunological or hematological automated analyzer. They are symbolized as A_L. A is the number of control values and L the control limits (Table 5).

1_{2s}. One control value lies between μ+2s/μ+3s or between μ-2s/μ-3s. It is only a warning rule.

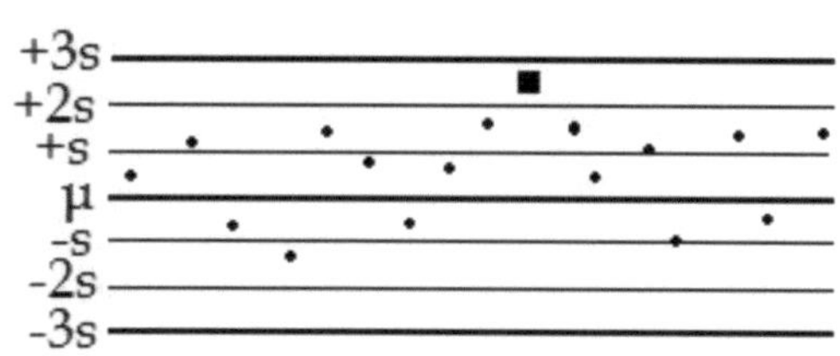

1_{3s}. One control value lies over μ+3s or under μ-3s. This criterion is sensitive to the detection of random errors. The results should be blocked and not reported to the patients. The run[6] is rejected.

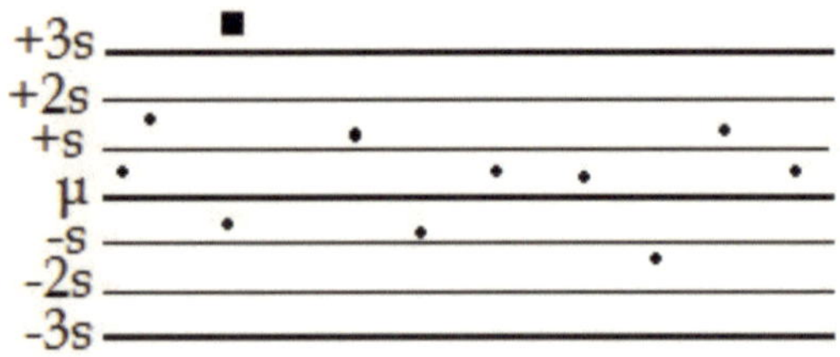

2_{2s}. Two successive control values lie between μ+2s and μ+3s or between μ-2s and μ-3s. It defines a systematic error. The results should be blocked and not reported to the patients.

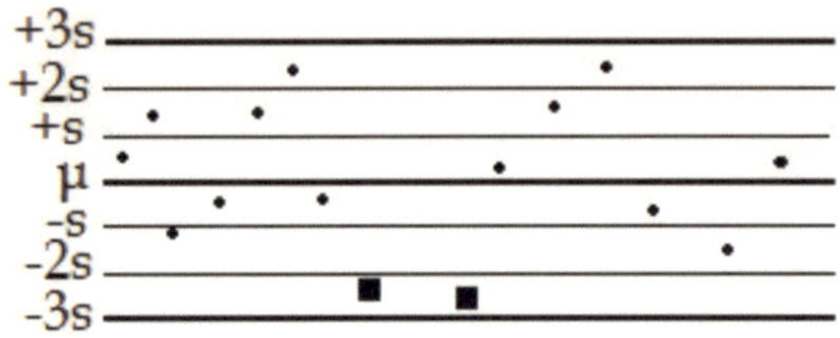

R_{4s}. The distance of two successive control values, values, is over 4s. It is a criterion sensitive to random errors. The results should be blocked and not reported to the patients. (Normally this criterion is used with two different control levels/across runs - when used with one level, as in this example, is applied for two consecutive runs).

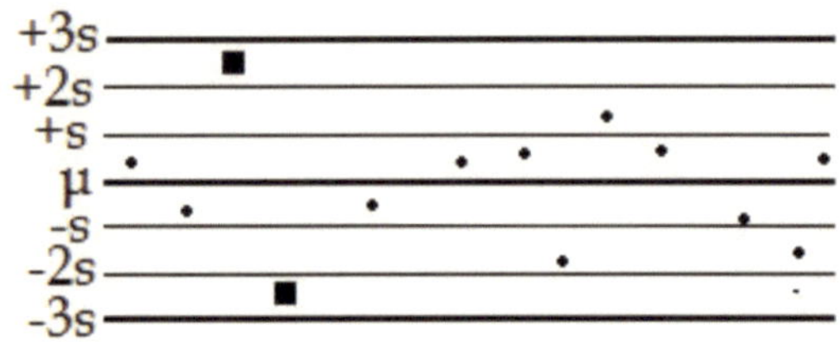

4_{1s}. Four successive control values lie between μ+1s and μ+3s or between μ-1s and μ-3s. It defines a systematic error. The results should be blocked and not reported to the patients.

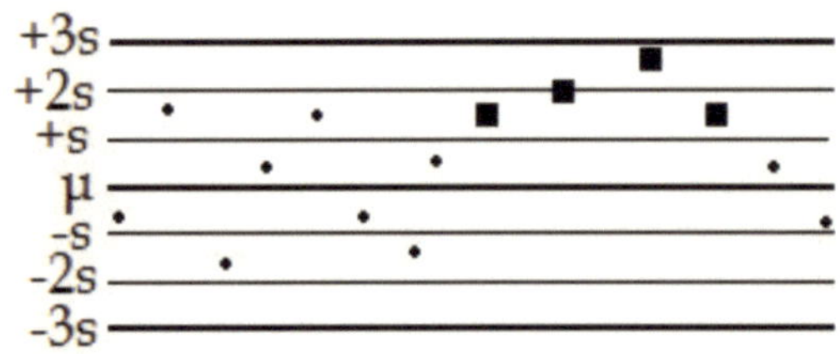

$10\bar{x}$. Ten successive control values lie between μ and μ+3s or between μ and μ-3s. It is a criterion that reveals systematic errors. The results should be blocked and not reported to the patients.

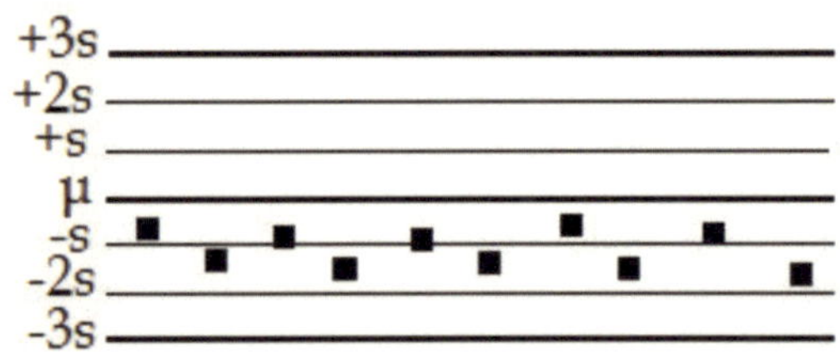

Table 5. Application of Westgard rules in one control level

In internal quality control two or even three different control samples may be used (see paragraph 2.6). In this case Westgard rules are used with a different way (Table 6).

---

[6] Analytical run or run is the analysis of patients' samples and controls during the day keeping the same analytical conditions.

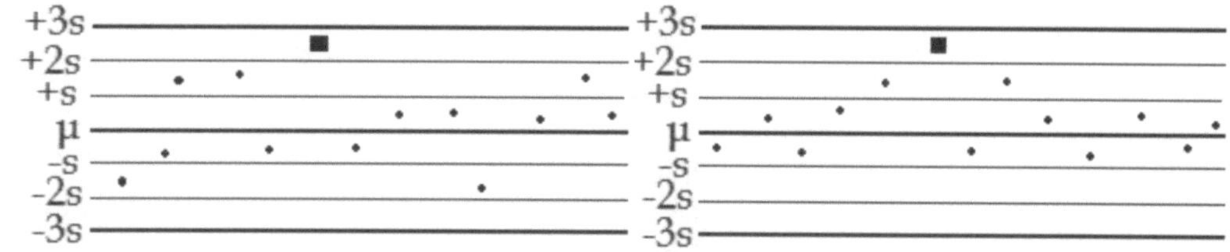

2_{2s} rule

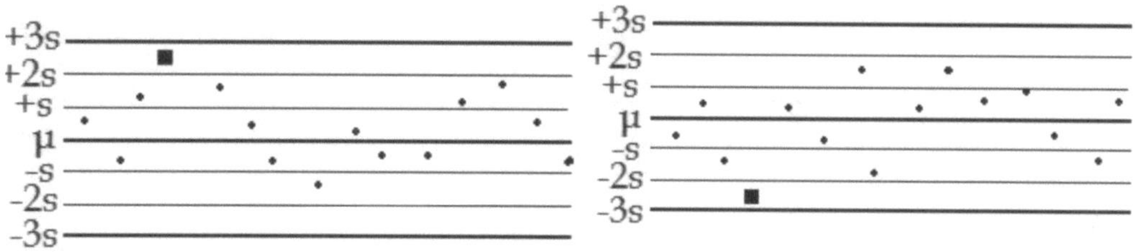

R_{4s} rule

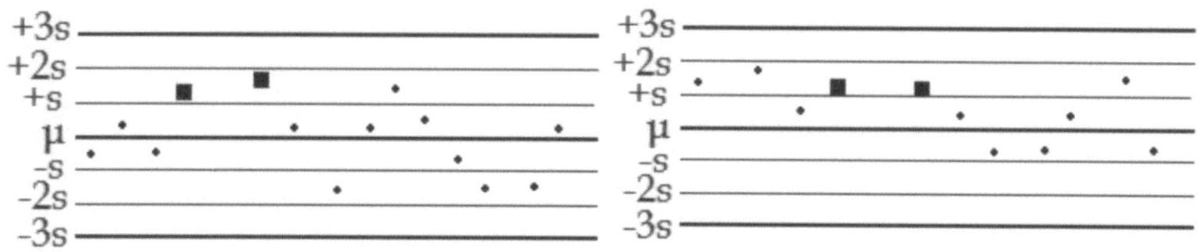

4_{1s} rule

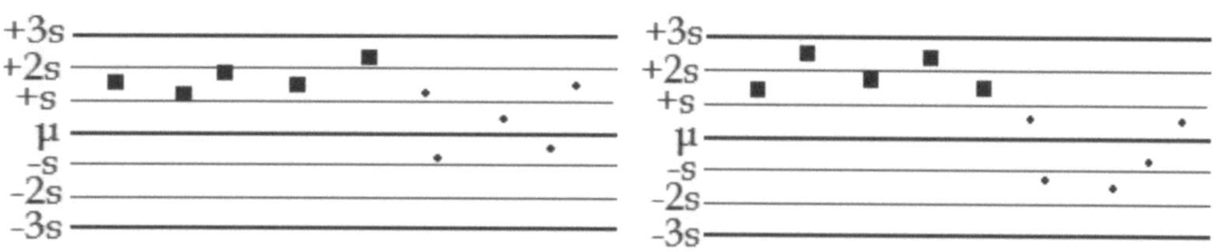

$10\bar{x}$ rule

Table 6. Application of Westgard rules in two levels of controls

3.8 The Average of Normals method

Levey-Jennings chart and Westgard rules detect random and systematic errors. This is achieved usually with the daily analysis of two different control levels. But control samples determination has some disadvantages:

- It is costly.
- It is time consuming.
- In many labs is performed once a day, as a rule before the analytical run (although several laboratories perform IQC in well defined intervals depending on the analyte or use bracketing before releasing the results).

These three disadvantages can be minimized with some others methods that use as control materials the patients' results. The most known of these is the **Average of Normals (AON)** with extensive application in biochemistry analyzers. The main disadvantage of AON method is that detects only **systematic errors**. The advantages are that it is free of charge, fast and it is done automatically throughout the day.

AON method based on the principle that the mean value of all normal results fluctuates between well defined limits (Hoffmann & Waid, 1965). The steps for implementation of AON method are:

1. The laboratory calculates the mean value and the standard deviation of the analyte reference values (RV)[7]. The laboratory can use the proposed reference values from bibliography or better estimate its own. For instance, if RV = 100 – 120 mmol/L then the mean value (μ) is 110 mmol/L. The standard deviation is (120 – 100)/6 = 3,3 mmol/L.
2. The laboratory defines which number of normal results (N) will use in AON method every day. This number of normal results will be the "control sample" of the method and it will remain steady.
3. The laboratory calculates the **standard error** (SE) of normal results with the following formula:

$$SE = \frac{s}{\sqrt{N}} \tag{14}$$

4. The laboratory calculates the confidence interval (CI) of the method with the formula:

$$\text{CI or Control limits} = \mu \pm 1.96 \times \frac{s}{\sqrt{N}} \tag{15}$$

   The confidence interval will be used for the definition of the control limits of the method.
5. Every day the laboratory calculates the mean value of N normal results. This mean value is symbolized as AON and it is calculated by the next formula (see equation 9).

$$AON = \frac{\sum_{i=1}^{N} x_i}{N} \tag{16}$$

6. If AON exceed the control limits (equation 15) then the analyte's determination has a systematic error which has to be eliminated before the analytical run.

The effectiveness of AON method depends on the number of normal results (N) which will be used daily. N is relevant to the variance of method (CV_a) and the biological variation of the analyte (CV_b). The variance of the analytical method can be calculated by the coefficient of variation of the control samples. The biological variations for most analytes are available in the scientific literature (see Appendix II & paragraph 5.1).

In Fig. 12 there is a nomogram (Cembrowski et all., 1984) which correlates number N with the ratio CV_b/CV_a. Such nomograms can be plotted by sophisticated software based on simulations theory.

---

[7] Reference values (later normal values) are the range of analyte's values of a health person.

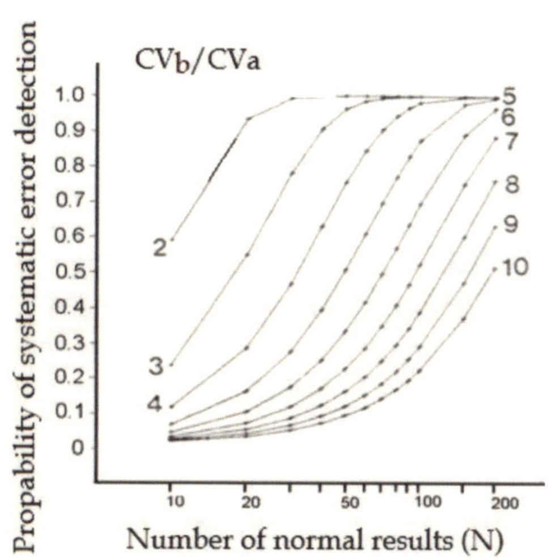

Fig. 13. Cembrowski nomogram which correlates the ratio CV_b/CV_a with the number of normals (N). The normogram detects systematic errors with ΔSE = 2s with probability of false alarm 1%

In daily practice AON method has its own control chart which detects only systematic errors. In AON chart each dot represents a daily mean value of the same analyte (Fig. 13).

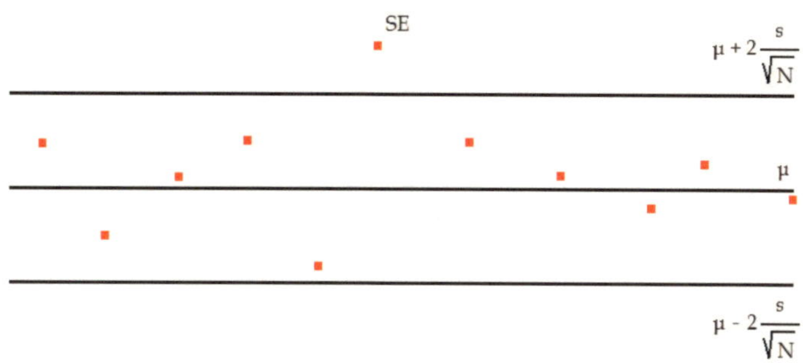

Fig. 14. AON chart. Each dot is a mean value. The point outside the upper limit represents a systematic error and not a random error like in Levey-Jennings chart

3.9 Bull's algorithm

The AON method of Hoffman & Waid is used mainly to biochemical analyzers. Hematology analyzers use another average of normals method, the Bull's algorithm (Bull, 1974). Bull had determined that some hematological parameters have very small biological variation (CVg) resulting in their mean value remaining steady. Bull applied his idea in erythrocyte indexes (MCV, MCH, MCHC) at the beginning, but today his algorithm is used for the majority of hematological parameters.

In Bull's algorithm a **moving average is calculated** instead of a mean value (equation 6). The moving average is a mean value that contains the mean value of a former moving average. In Bull's algorithm the moving average is usually calculated by a batch of 20 values, 19 of them are previous patients' values and one is the mean value of the previous batch. Bull's moving average is symbolized as $\overline{X}_B$. It is calculated by the formula:

$$\bar{X}Bi = (2-r)\bar{X}B,i-1 + rd \tag{17}$$

The constant r defines the percentage of the participation of the previous Bull's moving average to the calculation of the current one. For instance if r = 0.4 the previous Bull's moving average participate to the calculation of the current moving average with a percentage of 40%.
In contrary to AON method Bull's algorithm uses all the patient values not only the normal ones. It eliminates the outliers with the constant d (difference) which is calculated by the next formula:

$$\bar{X}Bi = (2-r)\bar{X}B,i-1 + rd$$

$$d = \operatorname{sgn}\left(\sum_{j=1}^{N} \operatorname{sgn}(X_j - \bar{X}_{B,i-1})|X_j - \bar{X}B,i-1|^P\right)\left(\frac{1}{N}\sum_{j=1}^{N} \operatorname{sgn}(X_j - \bar{X}B,i-1)|X_j - \bar{X}B,i-1|^P\right)^{1/p} \tag{18}$$

If P = ½:

$$d = \sum_{j=1}^{N} \sqrt{(X_j - \bar{X}B,i-1)}\left(\frac{1}{N}\sum_{j=1}^{N} \sqrt{(X_j - \bar{X}B,i-1)}\right)^2 \tag{19}$$

If r = 1.0 the differences $|X_j - \bar{X}_{B,i-1}|$ are all positives, then equation 17 is transformed as follows:

$$\bar{X}_{B,i} = \bar{X}_{B,i-1} + \left(\frac{\Sigma\sqrt{X_j - \bar{X}_{B,i-1}}}{N}\right)^2 \tag{20}$$

Some manufacturers have developed modified equations for Bull's algorithm. For instance, Sysmex hematology analyzers use a variation of the previous equations. Instead of the symbol $\bar{X}_B$ it uses the symbol $\bar{X}_M$.
The equation 17 is transformed as follows:

$$\bar{X}_{M(i)} = \overline{X_{(i-1)}} + \operatorname{sgn}\left\{\sum_{i=0}^{N} \operatorname{sgn}[x_{(l,i)} - \bar{X}_{M(i-1)}]\sqrt{|x(j,i) - \bar{x}_{M(i-1)}|}F \right. \tag{21}$$

The constant F is calculated by the formula:

$$F = \left\{\frac{\sum_{i=0}^{N} \operatorname{sgn}[x(_{i,j}) - \bar{x}(i-1)\sqrt{|x_{(j,i)'} - \overline{X_{M(i-1)'}}|}}{N}\right\}^2 \tag{22}$$

Bull's algorithm like AON method detects only systematic errors and it has its own control chart and its own rules. If Bull's algorithm is used for the quality control of erythrocyte

indexes the control limits of Bull's chart are $\overline{X}_B \pm 3\%$. The range ± 3% comes from the biological variation of the erythrocyte indexes which is almost 1%. The center line or target of Bull's chart is the Bull's average ($\overline{X}_B$) which has been calculated after nine successive calculations (equations 17, 21).
The three erythrocyte indexes are:

$$MCV=(Hct(\%))/(RBC(millions/\mu L))\ 1000 \quad (23)$$

$$MCH=(HgB(g/dL))/(RBC(millions/\mu L))\ 1000 \quad (24)$$

$$MCHC=(HgB(g/dL))/(Hct(\%))\ 100 \quad (25)$$

Where: MCV: Mean Corpuscular Volume, MCH: Mean Corpuscular Hemoglobin, MCHC: Mean Corpuscular Hemoglobin Concentration, Hct: Hematocrit, Hgb: Hemoglobin, RBC: Number of Red Blood Cells.
From the equations 23 - 25 is obvious that the combined study of the charts of MCV, MCH and MCHC shows which of the three hematological parameters Hct, HgB and RBC is increasing or decreasing due to a systematic error (Fig 14).
For the interpretation of Bull's charts two rules have been proposed (Luvetzksy & Cembrowski, 1987):

1. $1_{3\%}$. Bull's moving average exceeds the limit ± 3%.
2. $\overline{3}_{2\%}$. The mean value of three successive Bulls' moving averages exceeds the limit ± 2%.

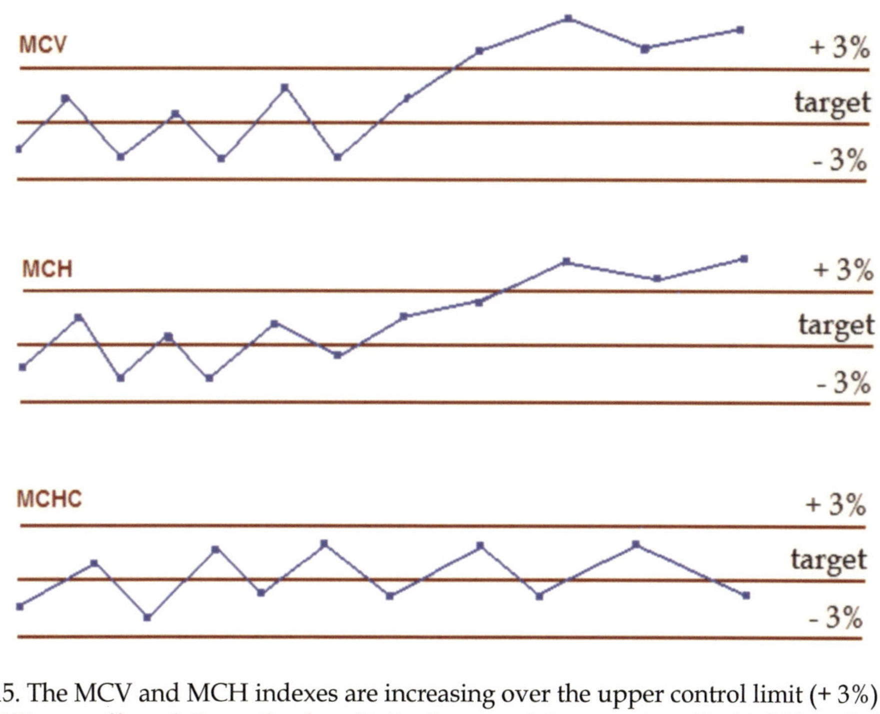

Fig. 15. The MCV and MCH indexes are increasing over the upper control limit (+ 3%) while MCHC is not affected. From the three Bulls charts and the equations 23, 24, 25 it is deduced that RBC is smaller than it has to be

3.10 Delta check method

The AON and Bull's algorithm SQC methods detect systematic errors using a group of patients (their mean value or moving average). There are two other methods that detect random errors using previous values of individual patient. The first of these is the Delta check method (Nosanchunk & Gottmann, 1974). *Delta check is the difference between the current value of one person and the previous one.*

$$\text{Delta check} = \text{Current value} - \text{Previous value} \quad (26)$$

Besides this classic equation, other approaches for Delta check calculation have been developed:

$$\text{Delta check\%} = (\text{Current value} - \text{Previous value})\,100/\text{Current value} \quad (27)$$

$$\text{Delta check} = \text{Current value}/\text{Previous value} \quad (28)$$

This difference for equation 26 should vary between two limits which are called "Delta check limits". In order to calculate them we must take into consideration the reasons of delta differences:

1. The "**intra-individual**" biological variation of the analyte (CV_I).
2. The analytical variation (s_{meas}). S_{meas} can be easily estimated by the control values.
3. The pre-analytical variation ($CV_{pre\text{-}analytical}$).
4. The improvement or deterioration of the patient's health.
5. Errors and mistakes in the pre-analytical, analytical and post-analytical stage.

The only factors which can be calculated from the list above are the first three. The equation that combines them is:

$$CV_{total} = \sqrt{CV_I^2 + S_{meas}^2 + CV_{pre-analytical}^2} \quad (29)$$

If we don't take into account the pre-analytical CV the equation 29 will become:

$$CV_{total} = \sqrt{CV_I^2 + S_{meas}^2} \quad (30)$$

Although the range $\pm\sqrt{CV_I^2 + S_{meas}^2}$ can be considered as the total variation (CV_{total}) of a result, we have to take into considerations some other points. In paragraph 3.1, the table 4 referred to the probability which have any observation of a normal distribution to be on a certain place around the mean value. For instance, the probability of an observation to be within ± 2 s from the mean value is 95%. This distance is equal to z-score (equation 11). So the range of a single result (x_i) must be:

$$x_i \pm Z - score\sqrt{CV_I^2 + S_{meas}^2} \quad (31)$$

In most cases for z-score (or simply z) the number 2 is chosen.
In delta check method, there are two results: the previous and the current one. Since total variation is the sum (with the statistical mean) of the two variations.

$$Total\ variation = \sqrt{(CV_{first\ result})^2 + (CV_{second\ result})^2} \quad (32)$$

or

$$Total\ variation = \sqrt{\left(Z\sqrt{CV_I^2 + S_{meas}{}^2}\right)^2 + \left(Z\sqrt{CV_I^2 + S_{meas}{}^2}\right)^2} \quad (33)$$

or

$$Total\ variation = \sqrt{2}\ Z\sqrt{CV_I^2 + S_{meas}{}^2} \quad (34)$$

The last calculation formula of Total variation will be the control limits of Delta check which will be calculated by the next formula:

$$\text{Control limits of Delta check} = \text{Current Value} \pm \sqrt{2}\ Z\sqrt{CV_I^2 + S_{meas}{}^2} \quad (35)$$

4. External quality control (EQC)

4.1 Calculation of EQC control limits

External quality control or external assessment scheme (EQAS) or proficiency testing program (PT)[8] refer *to the process of controlling the accuracy of an analytical method by interlaboratory comparisons.*

Its basic idea can be synopsized in the following steps:

1. The EQAS coordinator prepares and sends to the participants of the scheme one or two samples from the same pool.
2. The samples are assayed by the laboratories using the same equipment and reagents as they do in routine for the patients' determinations.
3. The EQAS coordinator gathers all the results and it groups them (**peer groups**) according the laboratories analytical methods, analyzers or any other criteria.
4. The EQAS coordinator calculates the target value (**consensus mean**) and its total variation (expressed as standard deviation) of the laboratories results.
5. If any of the laboratories has values outside of the control limits (target value ± allowable variation) then this laboratory is considered "**out of control**".
6. The "out of control" laboratories have to correct their analytical procedures.

In many countries the participation of the laboratories in EQC is mandatory. In any case, EQC is a very important part of SQC. The calculation of the consensus mean is similar to calculation of the control limits in internal quality control. Although there are many calculation methods, an EQAS coordinator normally takes following steps:

1. The EQAS coordinator calculates the mean value (μ) and the standard deviation (s) of each peer group.
2. It calculates the range μ ± 3s.
3. If there are laboratory results which are over the previous range, these results are rejected.
4. It calculates the range μ ± 2s.
5. If there are laboratory results which are over the previous range, these results are rejected.
6. The steps 4 - 5 are repeated until there will be no value outside the new range μ ± 2s.

---

[8] The term PT is used in USA in spite of EQAS which is used for the majority of the European countries.

4.2 EQAS charts and statistics
4.2.1 Standard Deviation Index
EQAS has its own charts and statistics. One of the most important statistics is the "**Standard Deviation Index**" (SDI). SDI shows the distance of the laboratory results from the consensus mean. It quantifies the inaccuracy of the analytical method. It is similar to Z-score (equation 12) and it is calculated by the formula:

$$SDI = \frac{\textit{laboratory result} - \textit{Mean value of peer group}}{\textit{Strandard deviation of peer group}} \quad (36)$$

The control limits of SDI are zero ± 2 SDI. The SDI value of each laboratory can be located on a proper chart (SDI chart) as follows:

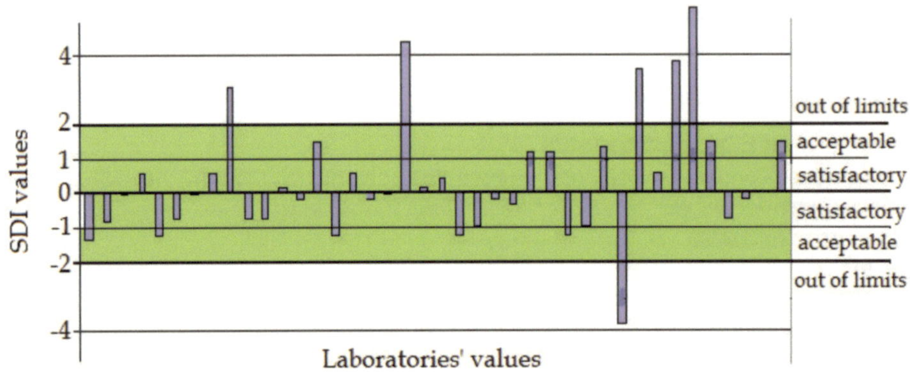

Fig. 16. SDI chart for a given analyte. Each bar corresponds to a different laboratory

Four rules are usually employed for SDI evaluation:
1. **$2/5_{1SDI}$**. Two from five successive control limits exceed 1 SDI. It is a warning rule.
2. **$\overline{X}_{1.5SDI}$**. The mean value of five SDI values exceeds the limits ±1.5 SDI. It reveals a lasting systematic error.
3. **1_{3SDI}**. One value exceeds the limits ±3 SDI.
4. **R_{4SDI}**. The range (R) between the lower and higher SDI values is over ±4 SDI.

4.2.2 Precision Index and Coefficient of Variation Ratio (CVR)
External quality assessment schemes quite often use two different statistics for the measurement of precision, the Precision Index (PI) and the Coefficient of Variation Ratio (CVR).

$$PI = \frac{\textit{Standard deviation of laboratory}}{\textit{Strandard deviation of peer group}} \quad (37)$$

The control limits of PI are < 2. Its chart is similar to SDI chart.

$$CVR = \frac{CV \textit{ of laboratory / month}}{CV \textit{ of peer group / month}} \quad (38)$$

The combined SDI/CVR chart has the ability to evaluate the total analyte's performance (precision and accuracy).

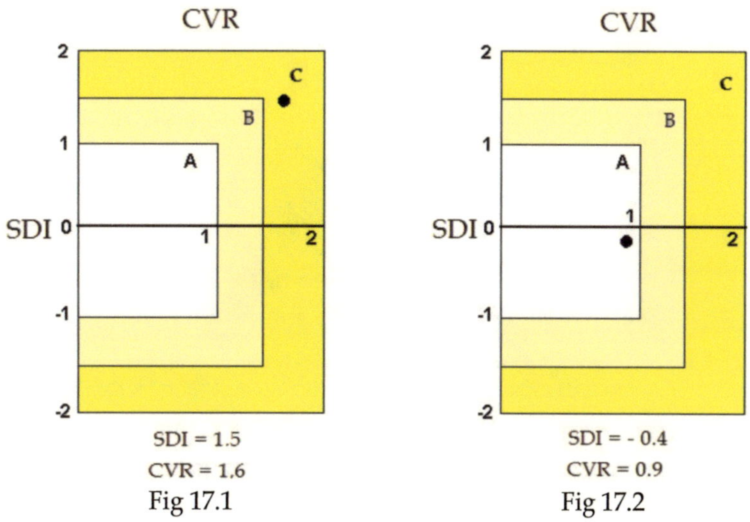

Fig. 17. CVR/SDI chart. The surface A corresponds to acceptable performance, the surface B to the grey-zone performance and the surface C to the rejected performance. In Fig. 16.1 the laboratory has unacceptable performance while in Fig. 16.2 the laboratory has acceptable performance

4.2.3 EQC normal distribution charts

More often than not, EQAS coordinators represent graphically the total performance of all the laboratories in a normal distribution chart. This normal distribution chart consists of bars (histogram) which correspond to certain groups of values. The bars may have different colors depending on the distance from the consensus mean. The bar that corresponds to each laboratory's result is marked by various methods (Fig. 17).

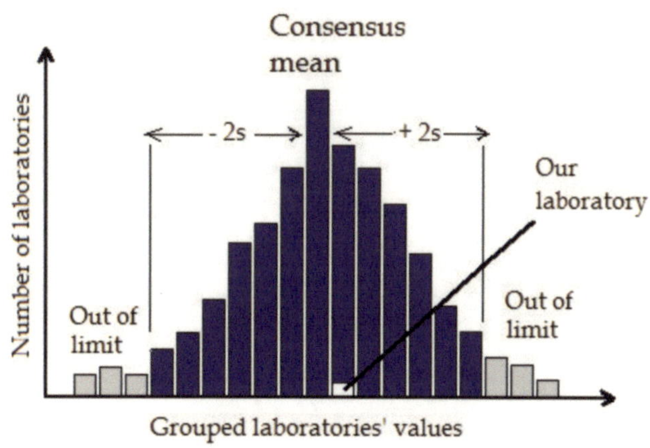

Fig. 18. An usual histogram from an EQAS report

In paragraph 4.1 we mentioned that EQAS coordinators usually group the laboratories according their analytical method and their automated analyzer. This is necessary so that the consensus mean will be more reliable and representative. Histograms containing two or more peer groups have bars with two or more different colors respectively (Fig 18).

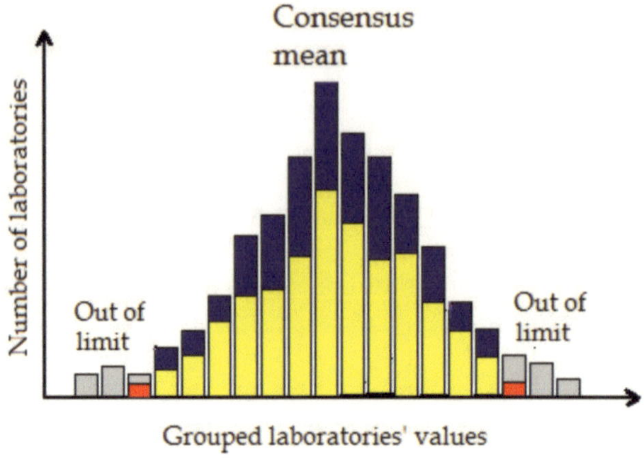

Fig. 19. A histogram from an EQAS report with two different peer groups

4.2.4 Youden plot

Most EQAS schemes use two control samples of different levels in order to check the performance of the analytical method in different concentration/activities, and preferably close to the decision limits. Youden plot is a rectangular chart of which the four angles correspond to the control limits of the two control levels [-4SD - +4SD] (Fig. 19). The acceptable part, the grey-zone and the rejected part have different colors. Each dot represents a different laboratory and therefore Youden plot describes the whole EQAS scheme. Dots (laboratories) that lie across the diagonal of the rectangular, at 45°, but are far for the center correspond to laboratories with proportional analytical error. The greater the distance from the center, the greater the proportional error. Dots restricted in the central rectangular, correspond to laboratories of which the performance is considered acceptable for this specific analyte.

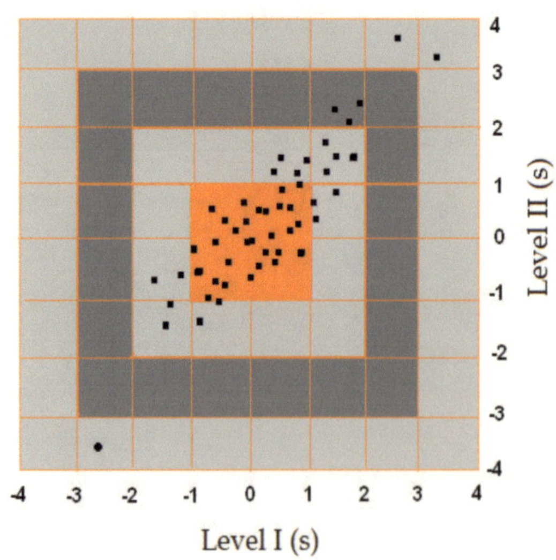

Fig. 20. A typical example of Youden plot. Each dot represents a different laboratory and its performance for the analyte in question

4.2.5 Yundt chart

Yundt chart helps to illustrate the performance of an analytical method across all its measuring range[9]. It needs at least three control levels to be plotted (Fig 20 & 21). If the line across the dots of the three levels is a straight one then the laboratory has a very good linearity. If not, there may be several issues with the linearity of the method.

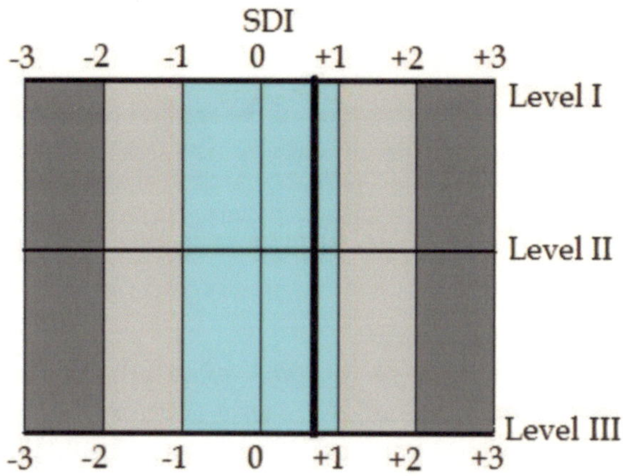

Fig. 21. A Yundt chart with three control levels. In this example the laboratory has a very good linearity

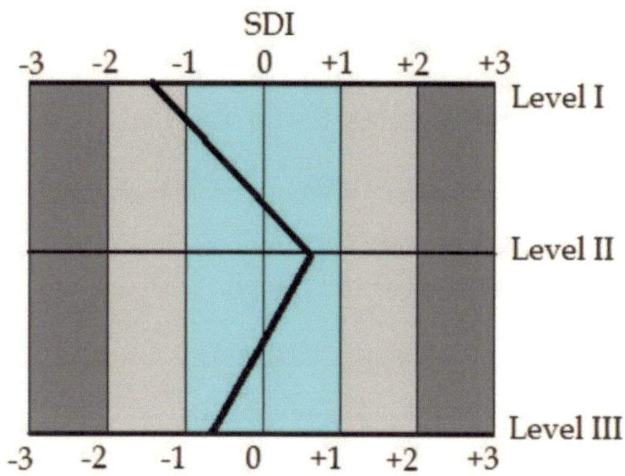

Fig. 22. A Yundt chart with three control levels. In this example the performance varies across the measuring range

5. Quality specifications

SQC's goal is to detect random and analytical errors or better, to ensure that total error is lower than the allowable total analytical error (**aTE**). aTE depends on some characteristics of the analyte and the analytical method, e.g. imprecision, bias, biological variation. According to these characteristics some analytes need more or less rigorous SQC rules than others. The

[9] Measuring range is the ability to perform within acceptable analytical error throughout the defined measuring range of an analytical method.

laboratorians choose the SQC methods (statistics, charts and rules) according to these characteristics of the analyte and the analytical method.
There are two common practices:

- aTE depending on the analytical performance of the analytical method (imprecision, bias%). This practice is extensively used in the United States. In US the laboratories are obliged to follow the restrictions of CLIA (Clinical Laboratory Amendments Act). Among others obligations their determinations must have TE less than the aTE which CLIA proposes (Appendix I).
- aTE depending on the biological variation of the analyte (intra-individual variation and within-individual variation). This is usually the European practice where most laboratories, on a voluntarily basis, choose their SQC methods so as their TE to be less than aTE (Appendix II). aTE when comes from biological variations is symbolized as $TE\%_b$.

5.1 Criteria of acceptable performance

A lot of European laboratorians have accepted some criteria for performance of the analytical methods. These criteria are called "criteria of acceptable performance" and they are based on the human biological variation (Ricos et al., 2000). Two kinds of biological variations have been calculated:

- The intra-individual variation (CV_I).
- The inter-individual variation (CV_w)

On the other hand the analytical variation can be divided in two subcategories:

- Within-day analytical variation ($CV_{w\text{-}day}$)
- Between-days analytical variation ($CV_{b\text{-}day}$)

The total analytical variation is calculated by the formula:

$$CV^2_{an\text{-}total} = CV^2_{w\text{-}day} + CV^2_{b\text{-}an} \tag{39}$$

1st Criterion of acceptable performance: The total analytical variation must be lower or equal from the half of the intra-individual variability.

$$CV_{an\text{-}total} \leq 0.5\ CV_I \tag{40}$$

2nd Criterion of acceptable performance: The within-day analytical variation must be lower or equal from ¼ of the allowable total analytical error $TE\%_b$.

$$CV_{w\text{-}day} \leq 0.25\ TE\%_b \tag{41}$$

3rd Criterion of acceptable performance: The total analytical variation must be lower or equal from 1/3 of the allowable total analytical error $TE\%_b$.

$$CV_{an\text{-}total} \leq 0.33\ TE\%_b \tag{42}$$

4nd Criterion of acceptable performance: The bias% of the analytical method must be:

$$Bias\% < 0.25\sqrt{CV_I^2 + CV_W^2} \tag{43}$$

5th Criterion of acceptable performance: The allowable total analytical error $TE\%_b$ of the analytical method must be:

$$TE\%_b \leq k0.5\ CV_W + 0.25\sqrt{CV_w{}^2 + CV_I^2} \tag{44}$$

For k = 1.65 the equation 44 is equal to:

$$TE\%_b \leq 1.65\ \Delta RE + \Delta SE \tag{45}$$

6. Appendix

Appendix I.

Allowable total errors from CLIA (USA) for some common biochemical parameters.

AST/SGOT	μ ± 20 %
ALT/SGPT	μ ± 20 %
ALP	μ ± 30 %
LDH	μ ± 20 %
AMY	μ ± 30 %
ALB	μ ± 10 %
CK	μ ± 30 %
Gluc	μ ± 10 % or μ ± 6 mg/dL
Urea	μ ± 9 % or μ ± 2 mg/dL
UA	μ ± 17 %
Creat	μ ± 15% or μ ± 0.3 mg/dL (greater)
Chol	μ ± 10 %
TRIG	μ ± 25 %
HDL	μ ± 30 %
TP	μ ± 10 %
ALB	μ ± 10 %
TBILI	μ ± 20 % or μ ± 0.4 mg/dL (greater)
Ca	μ ± 1.0 mg/dL
IRON	μ ± 20%
Na	μ ± 4 mmol/L
K	μ ± 0.5 mmol/L
Cl	μ ± 5%
Mg	μ ± 25%
pO_2	μ ± 3 SD

Appendix II.

Desirable Specifications for Total Error (TE%), Imprecision (s_{meas}), and Bias%, derived from intra- (CV_I%) and inter-individual (CV_w%) biologic variation for Common Biochemical parameters.

Analyte	CV_I%	CV_w%	s_{meas}	Bias%	TE%
TBILI	12.8	10	12.8	10.0	31.1
DBILI	18.4	14.2	18.4	14.2	44.5
ALB	1.6	1.3	1.6	1.3	3.9
TP	1.4	1.2	1.4	1.2	3.4
GPT/ALT	5.8	4.5	5.8	4.5	14
GOT/AST	6.0	5.4	6.0	5.4	15.2
γGT	6.1	10.7	6.1	10.7	20.8
LDH	3.7	5.0	3.7	5.0	11.1
ALP	3.2	6.4	3.2	6.4	11.7
Ca	1.0	0.8	1.0	0.8	2.4
Phos	4.3	3.2	4.3	3.2	10.2
Chol	3.0	4.0	3.0	4.0	9.0
Trig	10.5	10.7	10.5	10.7	27.9
Creat	2.2	3.4	2.2	3.4	6.9
Urea	6.2	5.5	6.2	5.5	15.7
UA	3.7	5.0	3.7	5.0	11.1
Cl	0.6	0.4	0.6	0.4	1.4
CK	11.4	11.5	11.4	11.3	30.3
CK MB	9.9	7.8	9.9	7.8	24.1
Gluc	2.5	2.3	2.5	2.3	6.3
HDL	3.6	5.2	3.6	5.2	11.1
LDL	4.2	6.8	4.2	6.8	13.6
Fe	13.3	8.8	13.3	8.8	30.7
Mg	1.6	1.7	1.6	1.7	4.3
K	2.4	1.8	2.4	1.8	5.8
Na	0.4	0.3	0.4	0.3	0.9
AMYL	4.4	7.4	4.4	7.6	14.6
CHE	3.5	3.1	2.7	2.9	7.4

Disclaimer: The views described here are the authors' personal interpretation and approach, which may differ from those of the institutions they work.

7. References

Bishop J & Nix A. (1993). Comparison of Quality-Control Rules Used in Clinical Chemistry Laboratories. *Clin Chem*; 39(8): 1638-49.

Bonini P, Plebani M, Ceriotti F, Rubboli F. (2002). Errors in Laboratory Medicine. *Clin Chem*; 48(5): 691-8.

Carey N, Cembrowski G, Garger C, Zaki Z. (2005). Performance characteristics of several rules for self-interpretation of proficiency testing data. *Arch Pathol lab Med*; 129(8): 997-1003.

Cembrowski G, Chandler E, Westgard J. (1984). Assessment of "Average of normals" quality control. Procedures and Guidelines for implementation. *Am J Clin Pathol*; 81(4): 492-9.

Fraser G. (1988). The application of theoretical goals based on biological variation data in proficiency testing. *Arch Pathol Lab Med*; 112(4): 404–15.

Harris E, Kanofsky P, Shakarji G, Cotlove E. (II) (1970). Biological and Analytic Components of Variation in long-term studies of Serum Constituents in Normal Subjects. *Clin Chem*; 16(12): 1022–27.

Hoffmann R & Waid M. (1965). The "average of normals" method of quality control. Am J Clin Pathol; 43:134-41.

Karkalousos P. (2007). The schemes of external quality control in laboratory medicine in the Balkans. *JMB*; 26:245–7.

Levey S & Jennings E. (1950). The use of control charts in the clinical laboratory. *Am J Clin Pathol*;20:1059-66.

Lewis S. (1998). The trend of external quality assessment in Europe. Sysmex Journal International; 8(2): 67- 71.

Lippi G, Guidi G, Mattiuzzi C, Plebani M. (2006). Preanalytical variability: the dark side of the moon in laboratory testing. *J Lab Med*; 30(3): 129–36.

Panagiotakis O, Anagnostou-Cacaras E, Jullien G, Evangelopoulos A, Haliassos A, Rizos D. (2006). ESEAP: the national External Quality Assessment Scheme for clinical chemistry in Greece & Cyprus. *Clin Chem Lab Med*; 44(9):1156-7.

Parvin C. (1991). Estimating the performance characteristics of quality-control procedures when error persists until detection. *Clin Chem*; 37(10): 1720-24.

Parvin C. (1992). Comparing the power of Quality-Control Rules to Detect Persistent Systematic Error. *Clin Chem*; 38(3): 358- 63.

Parvin C (1992). Comparing the power of Quality-Control Rules to Detect Persistent Random Error. *Clin Chem*; 38(3): 364- 69.

Parvin C. (1993). New Insight into the comparative power of quality-control rules that use control observations within a single analytical run. *Clin Chem*; 39(3), 440-47.

Plebani M & Carraro P. (1997). Mistakes in a stat laboratory: types and frequency. *Clin Chem*; 43(8): 1348–51.

Radin N. (1996). The multi-rule Shewhart Chart for Quality Control. *Clin Chem*; 30(6): 1103-4.

Ricós C, Baadenhuijsen H, Libeer J-C, Petersen PH, Stockl D, Thienpont L, Fraser C. (1996). External quality assessment: currently used criteria for evaluating performance in European countries and criteria for future harmonization. *Eur J Clin Chem Biochem*; 34: 159–65.

Ricós C, Alvarez V, Cava F, García-Lario JV, Hernández A, Jiménez CV, et al. Current databases on biological variation: pros, cons and progress. *Scand J Clin Lab Invest* 2000; 20(8): 733–5.

Ricós C, Alvarez V, Cava F, García-Lario JV, Hernández A, Jiménez CV, et al. Integration of data derived from biological variation into the quality management system. *Clin Chim Acta* 2004; 346(1): 13–8.

Ricós C, Garcia-Victoria M, de la Fuente B. (2004). Quality indicators and specifications for the extra-analytical phases in clinical laboratory management. *Clin Chem Lab Med* 2004; 42(6): 578–82.

Radin N. (1996). The multi-rule Shewhart Chart for Quality Control. *Clin Chem*; 30(6): 1103-04.

Westgard J. (1992). Simulation and modeling for optimizing quality control and improving analytical quality assurance. *Clin Chem*; 36(2): 175–8.

Westgard J, Groth T. (1981). A multirule Shewart Chart for quality control in clinical chemistry. *Clin Chem*; 27(3): 493-501.

Williams G, Young Z, Stein M, Cotlove E. (I) (1970). Biological and analytic components of variation in long-tem studied of serum constituents in normal subjects. *Clin Chem*; 16(12): 1016- 22.

Wood R. (1990). A simulation study of the Westgard multi-rule quality-control system for clinical laboratories. *Clin Chem*; 36(3): 462- 65.

18

Water Quality Control from the Perspective of Water Supply System Users' Safety

Prof. PhD. ing. Janusz Ryszard Rak
and PhD. ing. Barbara Tchórzewska-Cieślak
Rzeszow University of Technology
Department of Water Supply and Sewage Systems,
Al. Powstańców Warszawy 6, 35-959 Rzeszów,
Poland

1. Introduction

It is believed that the modern concept of quality is derived from the definition of Cicero, who in the first century BC created the Latin term "qualitas", which meant a characteristic, property of the object. In the seventeenth century AD Descartes introduced the dualistic concept of quality, distinguishing the primary quality (weight, shape) and the secondary quality that comes from the senses (smell, flavour). After the Second World War, an American scientist W.E. Deming in Japan introduced a method of quality control. In turn, J.M. Juran defined quality as the degree to which a particular product meets the needs of the buyer. In this way a system of total quality management was created. As a result of scientific research and Japanese and American practical implementations, in the years 1985-1987 the ISO series 9000 were formed.

Polish accession to the European Union caused a significant increase in the interest in national quality standards in water companies, which is received highly positively by drinking water consumers. A creator of a probabilistic methodology for safety analysis of technical objects F.R. Framer said that the risk depends not only on the severity and extent of the possible failures, but also on their likelihood

The risk may be considered tolerable (controlled) when the growing loss corresponds to strongly decreasing likelihood of a serious failure or disaster (Haimes 2009).

"Defence in Depth Philosophy" relies on the use of multiple barriers for physical, technical, procedural and organizational security. The activation of any barrier causes reactions at the local levels of system security. Directive 98/83/EC of 3 November 1998 on the quality of water intended for human consumption has committed member states to monitoring the quality of water intended for human consumption (EPA 2006). Member States should take all necessary measures to ensure regular monitoring of water quality to verify that the water available to consumers meets the requirements of current international legal norms. In 2004 in the third edition of Guidelines for Drinking-Water Quality the WHO presented guidelines for the development of so-called Water Safety Plans (WSP) (WHO 2002, 2004). which are intended for collective systems of water supply, and which meet the requirements for protection of critical infrastructure, as well as direct standards for continuous supply people with water in the required amount, under appropriate pressure and quality consistent with current standards (Rosen et al.2008).

In 2009, the project of European standard prEN 15975-1:2009 Security of drinking water supply. Guidelines for risk and crisis management. Part 1 Crisis management which will be gradually introduced as standards in individual member states, was developed.

2. Quality systems in the production of drinking water

Guarantees and drinking water consumers' safety can be assured through the following quality systems (Hellier 2000,Rak 2003):

- managing for quality - Quality Management- (QM),
- hazard analysis at critical control points - Hazard Analysis Critical Control Points - (HACCP),
- risk analysis and control of biological contamination - Risk Analysis Biocontamination Control – (RABC)
- good hygienic practice – (GHP),
- good manufacturing practice – (GMP),
- early warning for dangerous food - Rapid Alert System for Food –(RASF),

The rules of GHP and GMP are implemented before the introduction of HACCP (Hellier 2000).

Quality systems for the production of drinking water define such terms as (Johanson 2008, Rak 2009):

- threat – biological, chemical or physical factors that may occur in drinking water and cause the negative consequences to human health,
- monitoring - a system of processed observations, measurements and studies for a particular purpose, carried out on representative samples,
- risk – risk of the negative consequences to human health and the severity of such consequences as a result of the consumption of drinking water (Mays 2005),
- risk analysis – the procedure consisting of three interrelated elements, including risk assessment, risk management and information about risk,
- risk assessment – a scientifically based process consisting of three stages, including hazard identification, hazard characterization, exposure assessment and risk characteristic
- risk management – the proceedings of the competent authorities providing ways of risk prevention and control, based on risk assessment and the applicable requirements for sanitary-hygienic safety in the production of drinking water
- information about risk – it means to exchange information and opinions of threats, risk and factors connected with risk, during risk analysis, between risk managers, consumers, producers, traders, and scientists.
- procedure – it is an established pattern of conduct - a description of activities to enable the execution of certain task,
- instruction - it is an operational procedure of a lower order – it gives detailed steps in a logical sequence of implementation, describes step by step the task of the job, explains its implementation.

3. Defining the risk for the systems of collective water supply

3.1 Introduction

In the safety analysis, assuming the appearance of undesirable events according to the exponential distribution of failure free operation time , the probability of such an event (failure) is determined by the formula:

$$Q = 1 - \exp(-\lambda t) \tag{1}$$

In risk assessment $\lambda \cdot t \ll 1$, then the probability Q can by approximated:

$$Q = \lambda t \tag{2}$$

It allows to conclude that, regardless of the value of failure intensity λ = const, an increase in the risk exposure is associated with time t and always results in an increase of that risk.
The classic definition of a quantitative risk r is the product of the probability of incident P and its negative consequences C (Faber & Steward 2003, Haimes 2009, Kaplan & Garrick 1981, Kaplan 1997):

$$r = P \cdot C \tag{3}$$

The consequences can be determined in the range from zero to one, while C = 1 is attributed to decease (Guikema & Pate-Cornell 2002). In this way, limited to such case , r = P.
In epidemiology the concepts of absolute risk and relative risk have been distinguished. The absolute risk is a number of observed cases O and expected cases E in a given human population. The relative risk is calculated from the formula (Rak 2003):

$$rr = \frac{O}{E} \tag{4}$$

Often the so called excess relative risk is also calculated from the formula:

$$err = \frac{O - E}{E} \tag{5}$$

The magnitude of the risk can be also related to person-years number PY. Then the formula (4) take the form:

$$err = \frac{O - E}{PY} \tag{6}$$

For example, if rr = 1,4, this means that the excess relative risk is 0.4, representing an increase of 40% over the normal frequency of deaths due to the poisoning by the contaminated water. It is believed that the risk of fatal disease caused by the consumption of poisoned water is proportional to a dose. This is a known hypothesis resulting from the assumption of additivity of effects. However, there is a discussion of "high" and "low" dose. In the health risk assessments remains a problem of consistency of mentioned above hypothesis in the field of low-dose (Johanson 2008). An illustration of this may be the following reasoning: if the dose caused a lethal effect with a probability of 0.2, for example, 200 deaths for every 1,000 people, it is by no means clear that a dose 100 times smaller will cause two deaths for every 1,000 people. Undoubtedly it is true that this is not clear but it is not true that it can not be. Seveso II directive 96/82/EC Seveso 1996) on the control of major accident hazards involving dangerous substances is applicable from 3 February 1999. It introduces the concept of:

- major accident prevention policy (MAPP),
- safety management system (SMS),
- major accident prevention strategy (MAPS),
- plant with increased risk (PIR), plant with high risk (PHR), internal emergency plan (IEP), external emergency plan (EEP),
- safety report (SR).

In the article 3 of this directive there is a definition of *major accident*, which means " an occurrence such as a major emission, fire, or explosion resulting from uncontrolled developments in the course of the operation of any establishment covered by this Directive, and leading to serious danger to human health and/or the environment, immediate or delayed, inside or outside the establishment, and involving one or more dangerous substances". In the article 8 there is a new term - *domino effect*, meaning the sequence of events leading to a major accident. Seveso II Directive introduces risk analysis. It is postulated that the risk analysis does not only consist in demonstrating the potential likelihood of death .A product of the negative environmental and social effects and frequency of undesirable events associated with them is the basis of risk analysis and to draw appropriate conclusions. Preliminary hazard analysis (PHA) should be used as a technique for estimating the risk (Hellier 2000). If possible threats should be eliminated or reduced at source through methods that improve safety. It can be done based on past practices that has already been proved. The methods of achieving this objective are (Ezell et al. 2000, Li& et al. 2009, Macgillivray et al. 2007):

- risk should be as low as reasonably achievable (ALARA)
- risk should be as low as reasonably practicable (ALARP)
- the use of the best accessible technology - (BAT), in the manufacturing process and in major accidents prevention.

3.2 Risk valuation

Currently, the global trend is the maximum acceptable individual risk not greater than 10^{-6} per year, while the group risk 10^{-5} per year.

It is assumed that the level of individual risk between 10^{-4} and 10^{-6} per year requires risk/cost-benefit analysis (RCBA) to be carried out in accordance with the ALARP principle (risk as low as reasonably practicable).

British legislation (the Health and Safety Executive) accepts the maximum voluntary risk level 10^{-3} per year, but the imposed group risk related to industrial hazards 10^{-4} per year, for at least 10 years.

The ALARP principle is used in many different areas to analyse the risk associated with industrial activity (Faber &Steward 2003, Hrudey& Hrudey 2004 , Pollard et al.2004, Rak 2003). It is assumed that:

- the upper limit of the ALARP area is:
  - for workers 10^{-3} deaths per year
  - for the public 10^{-4} deaths per year
- the lower limit of the ALARP area is:
  - for workers and the public 10-6 deaths per year

The equivalent fatality rate , 1 death = 10 major injuries = 200 minor injuries, is applied..
The acceptance of risk by the people is the following:

- 10^{-8} (one death per year per every 100 million people) - the threat is felt incidentally,
- 10^{-6} (one death per year per every million people) - these incidents are reported, but any special procedures to counteract actively to such threat are not taken,
- 10^{-4} (one death per year per 100,000 people) - citizens are demanding organized protective action,
- 10^{-2} (one death per year per 100 people) - the risk of this kind should be reduced individually.

Historical analysis of undesirable catastrophic events in technical systems indicate that they are caused by 3 to 5 failures occurring simultaneously at the same time or in the small interval.

3.3 The concept of microrisk

The concept of microrisk associated with human activities was introduced by G. Marx.
He defined it as a risk of death as a result of performing given activity per one million people. In this way, the size of the risk can be given in units of [microrisk / year]. For example, the size of the risk for human labour in some sectors of the economy is (Haas 1999.):

- work in the electromechanical plant 100 microrisk/year, that is 10^{-4}
- work in the coal mine 800 microrisk/ year, that is $8 \cdot 10^{-4}$,
- work at high voltage lines 1,200 microrisk/ year, that is $1,2 \cdot 10^{-3}$,
- work in oil platforms 1800 microrisk / year, that is $1,8 \cdot 10^{-3}$

Selected human daily activities corresponding to 1 microrisk are:

- breathing polluted air in urban smog for 10 x 24 hrs,
- 2500 km of travelling by train,
- 2000 km of travelling by plane,
- 100 km of driving by car,
- working in the industrial plant for 5 days

Using the concept of microrisk seems to be very clear and widely accepted. The above data show that air travel is safer than driving by car.

- travelling by plane

1 microrisk = 2000 km, that is

$$\frac{1}{10^6} = 2000\,\text{km}$$

1 death = $2,0 \cdot 10^9$ km or 0,5 death per 10^8 km

- driving by car

1 microrisk = 100 km, that is

$$\frac{1}{10^6} = 100\,\text{km}$$

1 death = 10^8 km or 10 deaths per 10^9 km

3.4 The health risk in the use of public water supply-examples

Risk of death

The number of deaths 10 and individual risk of 10^{-5} were assumed as a threshold value for highly developed countries. In the case of an increase of this threshold value for a given public water supply, the risk of death should be reduced in inverse proportion to the number of injured people. The relationship of risk versus the expected number of victims is shown in Figure 1.

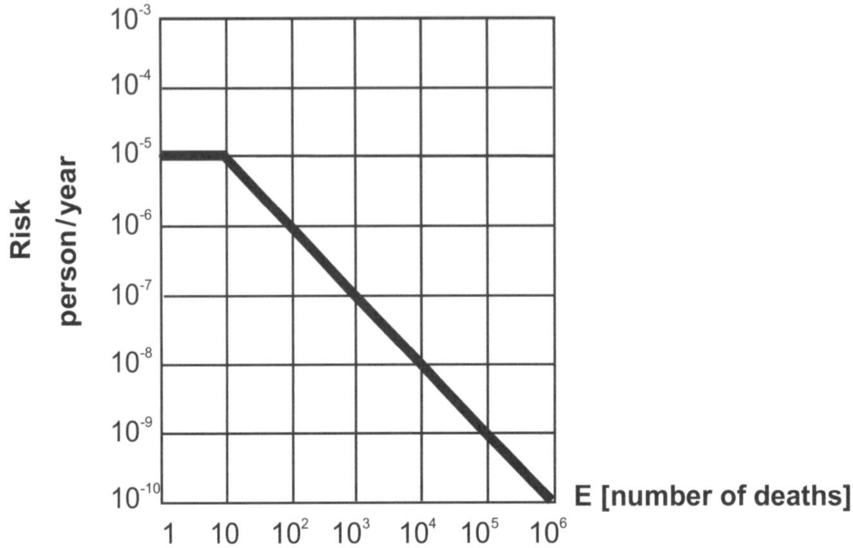

Fig. 1. Acceptable risk of death

Risk of chronic injury

The number of chronic injuries 10 and individual risk of 10^{-4} were assumed as a threshold value. The relationship of risk versus the expected number of victims is shown in Figure 2.

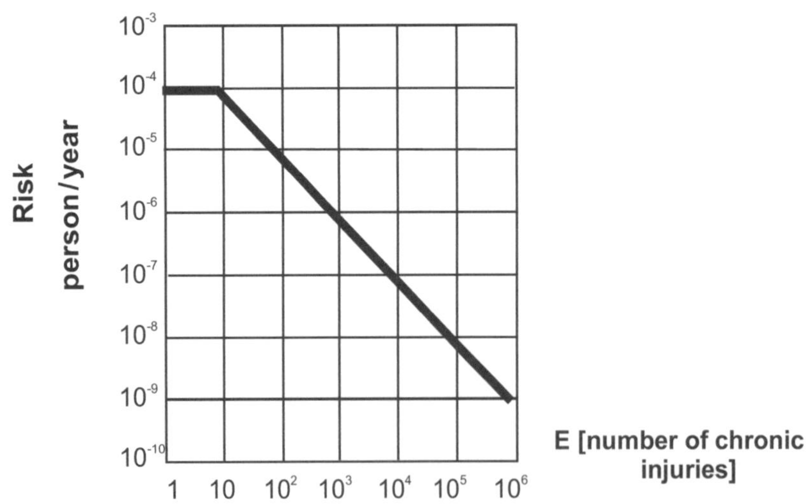

Fig. 2. Acceptable risk of chronic injury

Risk of gastric ailments

The number of gastric ailments 100 and individual risk of 10^{-3} were assumed as a threshold value. The number of chronic injuries is shown in Figure 3.

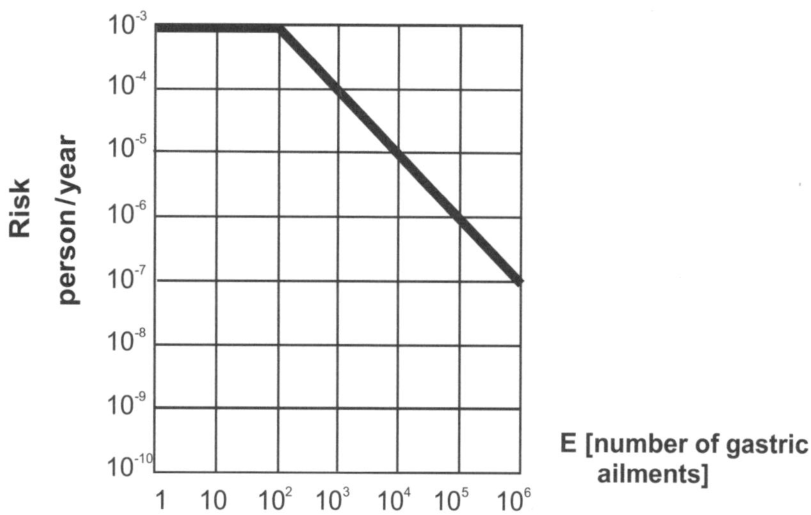

Fig. 3. Acceptable risk of gastric ailments

4. The methodology of the HACCP system for systems of collective water supply

The HACCP system is one of the systems ensuring quality and health benefits of drinking water (Havelaar 1994). The implementation of the system improves trust among the particular participants of drinking water market. The beneficiaries of the effects of the introduction of the HACCP system are manufacturers, retailers, supervisor service, and, above all, consumers. A well-functioning HACCP system allows to avoid unjustified claims, protects against the loss of trust and reliability of water supply companies in the market

External benefits from the introduction of HACPP by the company producing drinking water are:

- the increase of consumers confidence,
- the improvement of the image of the company,
- the increase of confidence at official inspection units.
- the increase of water consumers safety

The HACCP system is a change of philosophy in assessing the quality of drinking water as a product.. The assessment of the conditions of drinking high quality drinking water production, instead of only final control, ensures the increase of health safety connected with water consumption. The HACCP system is a procedure to ensure the safety of drinking water by identifying and assessing the scale of threats in terms of health quality and risk of threats during all stages of drinking water manufacturing. This system also aims to identify ways to reduce threats and to establish corrective actions.

The HACCP system for food control was created in the U.S. in the late sixties of last century, at the request of NASA - (National Aeronautics and Space Administration). The origin of HACCP was related to the scientific research on food for astronauts free from pathogenic micro-organisms.

In 1971 Pollsburgy company presented this system at the American National Conference for Food Protection. The HACCP system has been accepted by World Health Organization (WHO) and International Commission on Microbiological Specifications for Foods (ICMSF).

As a result of Polish accession to the European Union industry for the production and distribution of drinking water was obliged to apply the HACCP system. It is considered the most effective tool to ensure that food which is water will not be contaminated or polluted and will be safe for consumers health. The HACCP system is created individually for each production line and the type of distribution, taking into account the specific character of the given activity.
The main principles are:

- the identification of the possible biological, chemical and physical threats and the methods of counteraction,
- prevention, in form of a control of the particular phases of drinking water production process and distribution , not just the final product,
- applied in the entire production cycle, from water intake, through treatment, storage, distribution to consumers.

There are seven basic stages of implementing the HACCP system (Havelaar 1994):

Step 1. Threats analysis

It consists of:

- the identification of potential threats in the category of occurrence: biological, chemical, physical. The medical reports indicate that 90% of diseases resulting from poor quality water consumption is caused by its microbiological contamination,
- the establishment of a source and a reason, as well as the preventive activities, (general procedures, direct actions),
- the assessment of risk of threat.

The following definition of risk was taken:
Risk (R) is an arranged four element set (Rak&Tchorzewska 2006, Tchorzewska 2007) :

$$r = (S_i, P_{Si}, C_{Si}, O_{Si})$$

The formula used to determine the measure of risk is the following:

$$r = \frac{P_{Si}\ C_{Si}}{O_{Si}} \tag{7}$$

where:
S_i - representative emergency scenario i ,described as a series of successive undesirable events,
P_{Si} - point weight related to the probability that i representative emergency scenario S_i appears,
C_{Si} - point weight related to the size of losses caused by i representative emergency scenario S_i,
O_{Si} - point weight related to the WSS protection against i- representative scenario S_i, (protective barriers: clean water tanks, monitoring system etc.)
the WSS protection system - O - is inversely proportional to the size of the risk. The more extended system of protective barriers, the less risk of threat to water consumer safety
The proposed method is an expert method , and in such cases, the predefined values of risk assessment as descriptive measures of parameters included in the formula for risk determination, are used.. Every time some level of weight is assigned to the parameters P_{Si}, C_{Si} and O_{Si}, according to the following point scale :

- low - L = 1
- medium - M = 2
- high - H = 3

In this way the point risk scale in the numerical form, within the range [0,33 ÷ 9], has been obtained.

P = L = 1								
C								
L = 1			M = 2			H = 3		
O								
H = 3	M = 2	L = 1	H = 3	M = 2	L = 1	H = 3	M = 2	L = 1
r								
LLH 0,33	LLM 0,5	LLL 1	LMH 0,66	LMM 1	LML 2	LHH 1	LHM 1,5	LHL 3
P = M = 2								
C								
L = 1			M = 2			H = 3		
O								
H = 3	M = 2	L = 1	H = 3	M = 2	L = 1	H = 3	M = 2	L = 1
r								
MLH 0,66	MLM 1	MLL 2	MMH 1,33	MMM 2	MML 4	MHH 2	MHM 3	MHL 6
P = H = 3								
C								
L = 1			M = 2			H = 3		
O								
H = 3	M = 2	L = 1	H = 3	M = 2	L = 1	H = 3	M = 2	L = 1
r								
HLH 1	HLM 1,5	HLL 3	HMH 2	HMM 3	HML 6	HHH 3	HHM 4,5	HHL 9

Table 1. The three parametric risk matrix

Description of the parameters of risk factors according to the formula (7):

- The category of probability of failure - P
  - low probability – once in 10–100 years – P=1;
  - medium probability – once in 1–10 years – P_i=2;
  - high probability – 1–10 times a year or more often – P=3.

- The category of consequences - C
    - small – perceptible organoleptic changes in water, isolated consumer complaints, financial losses up to $5 \cdot 10^2$ EUR – C=1;
    - medium – considerable organoleptic problems (odour, changed colour and turbidity), consumers health problems, numerous complaints, information in local public media, financial losses 10^2 EUR– C=2;
    - large – endangered people require hospitalisation, professional rescue teams involved, serious toxic effects in test organisms, information in nationwide media, financial losses over 10^3 EUR – C=3.
- The category of protection - O
    - low protection level – O=1, municipal water quality standard monitoring, according to valid standards (monitoring at selected points of WSS that are equally distributed throughout the water supply system, in particular: water intake, WTP control points shown in fig.1, the places where water is fed into the network, selected points of the water network)
    - medium protection level – O=2, greater than normal monitoring of WSS (total network monitoring e.g. utilising SCADA software (Tchorzewska 2009)),
    - high protection level – O=3, special monitoring (e.g. within the framework of the multi-barrier system, including raw water biomonitoring based on test organisms, and use of industrial television with movement detectors on strategic objects).

The point weights presented above are a proposal to the preliminary risk assessment and can be modified for a given WSS. The advantage of the presented method is the possibility to compare risk Table 2 summarizes the risk categories and corresponding point scales. According to the matrix for risk assessment given in Table 1 we can analyse different undesirable events, taking the following scale of risk

Risk categories	Point scale
Tolerable	$0,33 \leq r \leq 1,0$
Controlled	$1,5 \leq r \leq 3,0$
Unacceptable	$4,0 \leq r \leq 9,0$

Table 2. Risk categories

Step 2. The establishment of critical control points (CCP)

Critical control point in a WSS is a point, step or procedure which allows for control and prevention, eliminating or reducing risks associated with the possibility of losing the safety of water consumers to an acceptable level It enables to achieve the purpose of the system through the control of drinking water health safety .The condition of CCP determination is the possibility to monitor it and control the threat.. To determine CCP we can use a decision tree method. It allows to determine CCP through a logical sequence of questions and answers concerning the possibility to eliminate or reduce threat at a given point to the acceptable level. Below you can see the exemplary decision tree questions, according to the Dutch procedures.

Question 1

Are there any preventive measures in relation to the given threat ?

Yes: go to question 2 No: go to question 1a

Question 1a

Are preventive measures necessary in terms of health safety ?

Yes: go to question 1b No: it is not a CCP

Question 1b

Is the threat controlled by standard procedures?

Yes: it is not a CCP No: modify the process or/and preventive measures

Question 2

Does the given preventive measure eliminate or reduce the threat to the acceptable level ?

Yes: go to question 4 No: go to question 3

Question 3

Can contamination by the identified threat reach the inadmissible level or can it increase to the unacceptable level ?

Yes: go to question 4 No: it is not a CCP

Question 4

Can the threat be eliminated in the further process or can it be reduced to the tolerable level?

Yes: it is not CCP No: it is CCP

Step 3. The establishment of the critical limits for every control point

After the establishment of the CCP, you must specify one or more indicators of contamination to be controlled, and the desired values, tolerances and unacceptable critical value. Index selection criterion should be the speed, ease and accuracy of measurement and the possibility of monitoring. In case of difficulties in this area the visual and/or sensor assessment should be used.

Step 4. The establishment of CCP monitoring procedures

The CCP monitoring is a base of the HACCP functioning. The results obtained from the monitoring have to be recorded. For monitoring procedures one should specify:

- a method of monitoring,
- a character, continuous or periodic,
- the frequency of periodic monitoring,
- a way of supervision,
- the principle of check and calibration of measuring devices

In Figure 1 the control points in the water production subsystem are presented.

For water distribution subsystem the scope of monitoring quality of water intended for human consumption in accordance with regulation covers the following range of indicators [5]:

- Physical and organoleptic parameters
- colour,
- turbidity,
- pH,
- conductivity,
- smell
- taste
- Chemical parameters
- ammonium ion,

- nitrite,
- nitrate,
- chlorine available
- Σ chlorates and chlorites,
- Fe and Mn for water from underground intakes
- Microbiological parameters
- Escherichia coli
- Enterococcus –
- coliform bacteria –
- Clostridium perfingens for water from surface water intakes

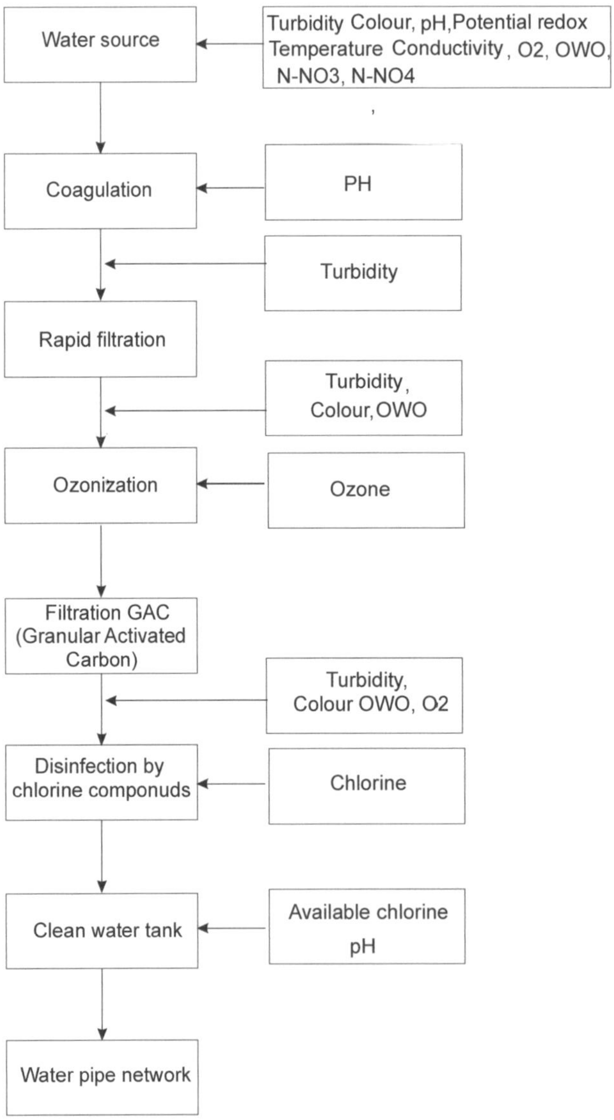

Fig. 4. Location of the control points and indicators of water treatment process quality

Step 5. The establishment of corrective actions

Corrective action must be taken when monitoring shows a trend to exceed tolerated values of indicators of contamination or such values are exceeded. It should be possible to stop the process of drinking water production to remove the causes that led to corrective action

Step 6. The establishment of system verification procedures

Verifications are performed after the implementation of the HACCP system, as its first evaluation. Then the frequency of subsequent verifications should be established. Verification is always performed after making changes in the technological production process and also when the undesirable events occur.
Effectiveness of the HACCP system can be verified through internal and external audits. There is the possibility to obtain HACCP certificate which strongly increases the confidence of current and future consumers of tap water.

Step 7. The creation of documentation

System documents should contain a plan of HACCP and the records testifying system operation. The preparation, storage and monitoring of documents should be established.

5. Conclusions

- With regard to the management the modern WSS is characterized by centralization and decentralization. As for the risk, during normal system operation centralized control is necessary, while in periods of threats decentralized action is required.
- In the risk analysis one should not put the greatest emphasis on the accuracy of the results, but, above all, on the "success" or "failure" of projects related to improving the safety. The purpose of risk analysis is to provide the information necessary to make decisions related to risk reduction. In the European Union countries, from a methodological point of view, two kinds of approach to the problem of improving safety, are being observed.. These are qualitative and quantitative methods of risk assessment.
- The proposed methods for assessing risk associated with tap water consumption are compatible with current trend in a global scale.

6. References

Ezell B. & Farr J. & Wiese I.(2000) Infrastructure risk analysis of municipal water distribution system. *Journal of Infrastructure Systems*, ASCE. vol 6(3) ,pp: 118–122

Faber M. H. &Steward M. G.(2003) Risk assessment for civil engineering facilities: critical overview and discussion. *Reliability Engineering and System Safety*, vol . 80, pp: 173-184.

Guikema S. D. & Pate-Cornell M.E.(2002) Component choice for managing risk in engineered systems with generalized risk/cost functions. *Reliability Engineering and System Safety*, vol: 78, pp: 227-238.

Haimes. Y.Y. (2009). On the Complex definition of risk: a systems-based approach. *Risk Analysis*. vol 29(12), pp: 1647-1654.

Haas C.N&Rose J.B & gerba C.P (1999).Quantitive microbial risk assessment. John Wiley & Sons. New York, USA.

Hastak H.& Baim E, (2001). Risk factors affecting management and maintenance cost of urban infrastructure. *Journal of Infrastructure Systems, ASCE*, vol 7 (2), pp:67–75

Hellier K (2000) Hazard analysis and critical control points for water supplies, *63rd Annual Water Industry Engineers and Operator's Conference*, Warrnambool

Havelaar A.H . (1994).Application of HCCP to drinking water supply.Food control. Vol 5(3), pp:145-152

Hrudey, S.E. & Hrudey, E.J. (2004) '*Safe drinking-water. Lessons from recent outbreaks in affluent nations*'. IWA Publishing, London, UK.

Johanson B. (2008). Public Views on Drinking Water Standards as risk Indicators. *Risk Analysis* vol 28(6) pp: 1515-1530

Kaplan S& Garrick BJ. (1981) On the quantitative definition of risk. *Risk Analysis*. vol1 (1), pp:11-27.

Kaplan S. (1997). The words of risk analysis. *Risk Analysis*, vol 7(4), pp:407-417.

Li, H. & Apostolakis G. E. & Gifun, J. & Van Schalkwyk, W. & Leite, S. & Barber, D., (2009). Ranking the Risk from Multiple Hazards in a Small Community. *Risk Analysis*, vol:3, pp: 438-456.

MacGillivray BH& Sharp JV& Strutt JE & Hamilton P.D & Pollard S.J.T. (2007) Benchmarking risk management within the international water utility sector. Part I: design of a capability maturity methodology. *Journal of Risk Research* vol 10 pp: 85-104

Mays L.W. (2005). The role of risk analysis in water resources engineering. *Department of Civil and Environmental Engineering, Arizona State University. www.public.asu.edu/lwmays*: 8-12.

Pollard, S.J.T. & Strutt, J.E. & Macgillivray, B.H. & Hamilton, P.D. & Hrudey, S.E. (2004). Risk analysis and management in the water utility sector - a review of drivers, tools and techniques. *Process Safety and Environmental Protection* vol. 82(B6), 1-10.

Rak J.(2003) A study of the qualitative methods for risk assessment in water supply systems. Oficyna Wydawnicza Politechniki Wrocławskiej. *Environment Protection Engineering*, vol.3 (4), pp:123-134.

Rak J. & Tchórzewska-Cieślak B. (2006). Five - parametric matrix to estimate risk connected with water supply system operating *Environment Protection Engineering*. Wydawnictwo Politechniki Wrocławskiej. Vol 2, pp: 37-47

Rak J.(2009).Selected problems of water supply safety. *Environmental Protection Engineering*. Vol 35, pp: 29-35

Rosen, L. &Lindhe, A. & Hokstad, P. & Sklet, S. & Rostum J. & Pettersson, T. J. R., (2008). Generic Framework for Integrated Risk Management in Water Safety *Plans. 6th Nordic Drinking Water Conference*, Oslo

Seveso II directive 96/82/EC *(Safety Management System)*

Tchórzewska-Cieślak B.(2007) Method of assessing of risk of failure in water supply system. *European safety and reliability conference ESREL.2007 Risk, reliability and societal safety*. Taylor & Francis, t.2, Norway, Stavanger, s.1535-1539

Tchórzewska-Cieślak B,(2009). Water supply system reliability management. *Environmental Protection Engineering*. Vol 35 , pp.29-35

Water Safety Plans (Revised Draft), (2002) Report publication WHO/SDE/WSH/02.09 (World Health Organization, Geneva,.

WHO, (2004), *Guidelines for Drinking Water Quality*, 3rd edn. (draft) (World Health Organization, Geneva).

United States Environmental Protection Agency , 2006. *Decision-Support tools for predicting the performance of water distribution and wastewater collection systems*. Washington D.C.: National Risk Management Research Laboratory Office of Research and Development U.S. Environmental Protection Agency

19

The Application Research of Intelligent Quality Control Based on Failure Mode and Effects Analysis Knowledge Management

Zhao Xiuxu
Wuhan University of Technology, Wuhan, Hubei, 430070, China

1. Introduction

SPC provides statistical techniques and graphical displays control charts to enable the quality of conformance to be monitored and special causes of process variability to be eliminated(Montgomery D.C. 2005). Combining with computer technology and the automatic control technology, SPC has realized the product quality on-line control. With online SPC system, a large amount of manufacturing process quality data can be collected and stored in SPC database. These data can be visited via the operation in SPC database, and many kinds of quality reports can be generated for specific using. However, this technology has not been all well applied in enterprises(Venkatesan G., 2003). In practical manufacturing process, after the operators getting the analyzed results from SPC system, they need to make decision to eliminate the fluctuation in manufacturing process. But it is largely dependent on the personal knowledge and experience level of the operator on-site. Under this specific situation, the decision results are uncertain, and it may also lead to the quality problems can't be solved in time. So, it is very important to find appropriate methods to support the operators' decision making, especially the techniques which can facilitate the operators make right decision in time.

Considering the existing limitations discussed above, this research is prompted by the need to have a technique that will enable the intelligent acquisition of Failure Mode and Effects Analysis (FMEA) knowledge for the decision making in the quality control process. The study specifically looks at the intelligent knowledge searching and reasoning in FMEA repository, which can facilitate the operators decision making.

In this chapter, FMEA knowledge sharing in the manufacturing process quality control will be discussed after the brief review of FMEA, and an application system named IQCS has been developed to manage the FMEA knowledge for the manufacturing process quality continual improvement.

2. FMEA and manufacturing process quality control

2.1 The origin and development of FMEA

FMEA is a technique that identifies the potential failure modes of a product or a process and the effects of the failures, and assesses the criticality of these effects on the product functionality. According to BS 5760 Part 5, "FMEA is a method of reliability analysis intended to identify failures, which have consequences affecting the functioning of a system within the limits of a given application, thus enabling proofs for action to be set." (British Standards Institution, 1991) It provides basic information for reliability prediction, product and process design (Sheng-Hsien Teng & Shin-Yann Ho, 1996).

In the early 1950s, Grumman Corporation in U.S. applied the FMEA firstly and got very well results (Hai ZHANG & Zhi-bin ZHOU, 2007). Since then, FMEA obtained widespread applications in the military system design, such as the aviation, astronautics, ships, weapon etc. Under the leadership of Ford Company, Big Three auto makers started to introduce FMEA in the product quality improvement (Daimler Chrysler Co., Ford Co. & General Motor Co., 2008). Then, FMEA had been widely applied in vehicle safety assurance. With the application value of FMEA had been widely recognized, FMEA has been applied in more and more other fields, such as machinery, medical equipment, nuclear power, food safety assurance etc. After a long time development and improvement, FMEA has become a necessary reliability analysis work which must be completed in system development (Stamatis D.H. 2003).

Traditionally, FMEA is used in hard copy or spreadsheet format to capture the potential problem of a design or process. The knowledge captured is aimed for reuse. However, as the knowledge in the FMEA grows, it gets harder and harder to be reused. The implementation of a highly manual FMEA system is a difficult task. FMEA is found to be not user friendly, hard to understand and not very flexible.As a result, many companies use FMEA merely to satisfy the contractual requirements of their customers (Dale B.G., & Shaw P., 1996). Users always find FMEA is a "tedious and time-consuming activity" (Price Chris, Pugh David R., Wilson Myra S., & Snooke Neal, 1995). It is especially true when FMEA is used in complex systems with multiple functions (Hai ZHANG, & Zhi-bin ZHOU, 2007).

2.2 The significance of FMEA knowledge acquisition in process quality control

With the increasing complexity of modern manufacturing system, there are more and more uncertain factors in manufacturing process, and the difficulties of process reliability analyzing increased greatly. Meanwhile, the quality problem dealing and analyzing are restricted enormously by the personal knowledge and experience level of the on-site operators. FMEA knowledge resources, which have been accumulated in the earlier periods, can not only facilitate the comprehensive decision making, but also improve the quality problem analyzing process. So, it is significant for the manufacturing process continual improvement to utilize the FMEA knowledge resources.

3. Intelligent quality control system

3.1 The structure and orinciple of IQCS

As shown in Fig.1, the structure model of the IQCS mainly contains three tiers:

1. Function tier: It makes the real-time data acquisition and analysis in the manufacturing process. It can output the statistical analysis results in the form of quality report, and input the data of manufacturing process into the SPC system database.
2. Data tier: It implements the FMEA process, which is conducted by the experts coming from different area, such as design, manufacture and quality management etc. The knowledge and experience of experts will be extracted by the "Brain Storming" activity. Then, the FMEA results will be transformed into FMEA knowledge according with specific method, and put into FMEA repository.

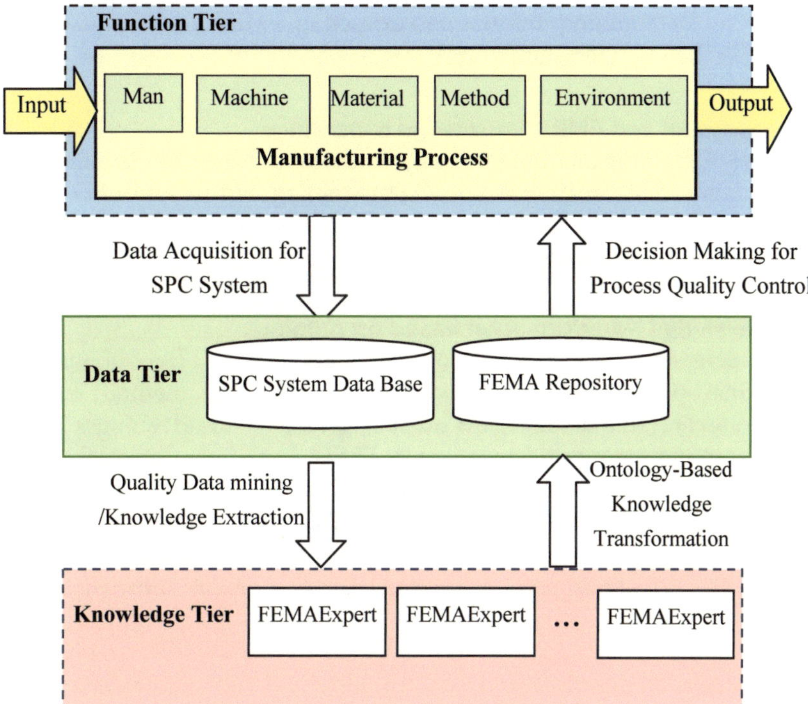

Fig. 1. The principle of IQCS based on FMEA repository

3. Knowledge tier: It is a database that is designed to store the data collected by SPC system, and the knowledge will be stored in the FMEA repository which is designed with specific structure.

In the practical manufacturing process, the function tier of SPC system will do the collecting and analysis of the real-time process data, then, the statistical analysis results will be output according to the users' specific requests, and the manufacturing process data will be put into the SPC system database. The function tier of FMEA is to complete the FMEA process. The interaction tier, which includes SPC system database and FMEA Repository, is responsibe for the dynamic interaction between FMEA and SPC system. There are two tasks: the first one is to extract information dynamically from SPC system for FMEA process, and the other one is to visit FMEA repository to provide decision supporting for quality control and system adjusting in the manufacturing process. So, it is very important to establish the interaction mechanism of the interaction tier.

3.2 The operation of IQCS system.

In manufacturing process, the function of IQCS is realized by the operation of FMEA and SPC system through data mining/ information extracting process and dealing with problems and decision process. The data mining/ information extracting process collects manufacturing process data dynamically from SPC system, and these data will be provided for the experts to conduct FMEA process. The results of FMEA will be transformed and put into FMEA repository, provide real-time decision supporting for operators to analyze the quality problems, and adjust manufacturing system in time. In this process, the most important keys are Data mining/Information extracting, FMEA knowledge transformation, and the interaction between SPC and FMEA.

3.2.1 Data processing and FMEA knowledge extracting.

To make full use of the large amount of quality data and messages in manufacturing process in SPC system database, the main function of Data mining/FMEA Information Extracting is to develop specialized algorithm of knowledge discovery and data mining based on Knowledge Discovery in Databases.

3.2.2 FMEA knowledge transformation based on ontology

Because of the diversity and complexity of FMEA knowledge, object-oriented knowledge expressive method, process model method and predicate logic method, frame described method and production rule method now often used all have disadvantages in some extent. These methods can not express the meaning of FMEA knowledge precisely. Ontology has accurate form of expression and explicit semantic which can definite the relationships between concepts, concept and object, also objects. This useful expressive form reduces the misunderstandings of relationship of concept and logic, and makes the share and reuse of knowledge possible. It is the important theory evidence of FMEA knowledge system based on ontology.

The expressive method based on ontology can depict basic knowledge system of certain domain through normal description of concept, term and the relationship between them. The method can not only depict the object-hierarchical model, such as organization, resource, product, but also denote nonfigurative affair, such as faith, target, plan, activities. Using ontology to denote knowledge can make FMEA knowledge-frame legible. FMEA knowledge may be organized systematic and constructional through knowledge-modeling based on ontology. The well-organized model can promote analysis and answer of quality. Meanwhile, it is useful to share and reuse the knowledge inner corporation even between corporations in the same domain.

3.2.3 The interaction between SPC system and FMEA repository

In order to combine FMEA with SPC dynamically, the internal relations between SPC system and FMEA repository must be established combining with specific manufacturing process, and knowledge discovery in FMEA is driven by the results of SPC system. The coordination mechanism of database and repository is introduced in the transformation of tacit knowledge which mainly come from the experience of experts(You-zhi XU, Dao-Ping WANG, & Bing-ru Yang, 2008). Based on the learning form above references, the IQCS system operation process was developed. Firstly, the structured data of product key quality characteristics value are founded according to the manufacturing process, secondly, intelligent searching and reasoning process in FMEA repository will be started through the association of product and its corresponding characteristic value, finally the FMEA knowledge responded to the key quality characteristic value will be attained.

3.3 An illustrative example Of IQCS

Combining with specific manufacturing quality control process, the quality data extracting and the abnormal quality problem analysis process in IQCS are as follows:

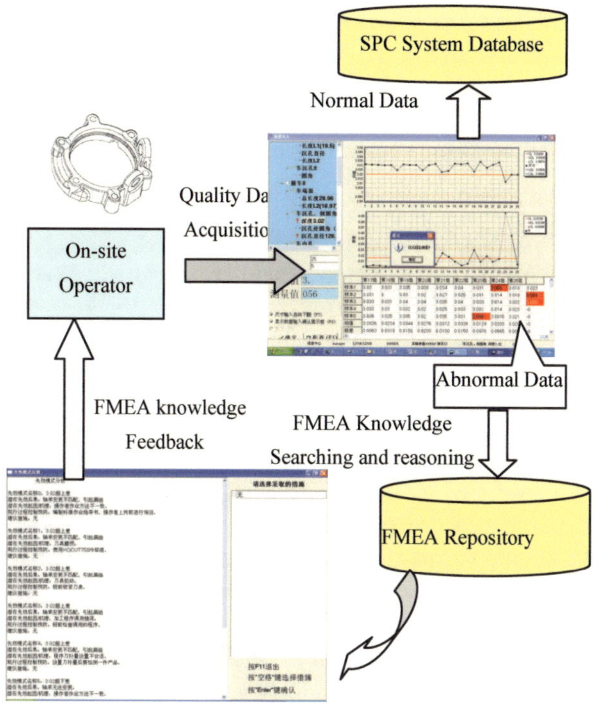

Fig. 2. A specific quality control process based on FMEA repository and SPC system

Step 1. The operator extracts the quality data according to the key quality characteristics. The regular data will be put into the SPC database, if the irregular data appear, then switch to step2;

Step 2. The system will remind the on-site operator with a pop-up message and start to the searching in FMEA repository at the same time.

Step 3. After finding out the corresponding FMEA knowledge in the FMEA repository, the results will be sent back to the operator through system interface to support the operators' decision making.

The specific flow chart of this process is shown in Fig.2.

4. Conclusion

In order to solve the problems of analyzing quality problems in manufacturing process and eliminate the uncertainty in manufacturing system adjustment decision making, this paper presents an approach which can transform the tacit knowledge in FMEA process effectively through a FMEA repository based on ontology.

Combined FMEA with SPC system, the intelligent searching and reasoning can be realized through the cooperation of FMEA repository and SPC database. And FMEA repository based on ontology can realize the FMEA knowledge sharing and reusing through the intelligent reasoning technology.

Based on the theoretic study, an application system IQCS has been developed for specific manufacturing process. Its application shows that it can not only improve the efficiency of

quality control, but also prevent the potential quality problem. Thus, the independence of quality fluctuation in manufacturing process will be enhanced.
This research provides an effective way for decision making of process quality improvement and the manufacturing system adjusting. Based on the current study, further research will focus on enable techniques development which can facilitate the dynamic coordination of FMEA repository and SPC system in manufacturing process.

5. Acknowledgment

Thanks for Shaoguan Hongda Gear Corp. to provide financial aid to support this research.

6. References

British Standards Institution, (1991). Reliability of systems, equipment and components, Guide to failure modes, effects and criticality analysis (FMEA and FMECA). BS5760-5-1991

Daimler Chrysler Co., Ford Co. & General Motor Co., (2008). Potential Failure Mode and Effects Analysis, FMEA (Fourth Edition).

Dale B.G., & Shaw P., (1990). Failure mode and effects analysis in the UK motor industry: a state-of-the-art study, *Qual. Reliab. Int.* V6, No. 3, pp. 179-188. ISSN 07488017

Hai ZHANG, & Zhi-bin ZHOU, (2007). Progress of Failure Mode and Effects Analysis, *Aeronautical Manufacturing Technology*, Vol. 8, No.65. pp.64-66. ISSN 1671-833X

Montgomery D.C. (2005). *Introduction to Statistical Quality Control* (fifth ed.), John Wiley, ISBN 9780471656319, Hoboken, NJ.

Price Chris, Pugh David R., Wilson Myra S., & Snooke Neal, (1995). Flame system: automating electrical failure mode and effects analysis (FMEA), *Proceedings of the Annual Reliability and Maintainability Symposium*, pp.90-95. ISSN 0149144X

Sheng-Hsien Teng, & Shin-Yann Ho, (1996). Failure mode and effects analysis - an integrated approach for product design and process control, Int. J. *Qual. Reliab. Mannag.* Vol. volume number, No. issue number, Vol. 13, No. 5, pp.8-26. ISSN: 0265-671X

Stamatis D.H. (2003). Failure Mode and Effect Analysis: FMEA from Theory to Execution (Second Edition). American Statistical Assoc Quality Press, ISBN0873895983, Milwaukee,Wisconsin, USA,

Venkatesan G., (2003). Process control of product quality, *ISA Transactions 42*. pp.631–641. ISSN 00190578

You-zhi XU, Dao-Ping WANG, & Bing-ru Yang, (2008). Research on Tacit knowledge Transformation Model Based on Double Bases Cooperating Mechanism, *Soft Science*, Vol. 22, No.8, pp.6-10. ISSN 1001-8409

20

Model Driven Adaptive Quality Control in Service Oriented Architectures

Tomasz Szydło and Krzysztof Zieliński
AGH University of Science and Technology
Poland

1. Introduction

Service-Oriented Architecture (SOA) enables the development of applications that are built by combining loosely coupled and interoperable services (Erl, 2004). In SOA, the centre of gravity shifts from development to integration. An application in SOA is composed from a set of services that are connected together in the integration platform. Such an application is described as an abstract composition which can be executed in containers (e.g. the Enterprise Service Bus), responsible for mapping abstract services to concrete service instances during execution. The mapping process can be performed by a user or planner prior to execution or during runtime. It should take into account Quality of Service aspects, enabling quality control to become an integral part of SOA application execution.

Automation of quality control can explore the concept of adaptive or autonomic systems (Ganek & Corbi, 2003). An adaptive software system enables software to modify its structure and behaviour in response to changes in its execution environment (McKinley et al., 2004). Adaptation can be differentiated into static or dynamic, depending on when it takes place. Static adaptation assumes that adaptive behaviour is *hardwired* or configured before the application starts. Dynamic adaptation processes enable extensions or replacement of system components during execution without stopping and restarting the system. Thus, this category of adaptation is particularly suitable for SOA applications.

Autonomic systems (Kephart & Chess, 2003) represent a more mature step in the evolution of dynamic adaptive systems. In such systems the operator does not influence the system directly, but only defines general policies which specify system behaviour. The system is able to collect new knowledge and use it in the adaptation process. In the case of SOA applications, this means that integrated components may be dynamically managed by business rules and policies, taking into account experience gathered from former application runs.

In spite of the differences between adaptive and autonomic systems, their general design follows the same MAPE (Monitor, Analyze, Plan, and Execute) (IBM, 2006) paradigm. This very well known paradigm has not yet been widely adopted by the SOA application execution environment, creating an interesting space for research. Its practical exploitation requires suitable mechanisms implementing the MAPE control process to be embedded into execution containers for SOA services. The analysis and planning steps for SOA applications require a definition of metrics and their calculation on the basis of monitoring data. The values of these metrics could be used for the SOA application execution model construction, which may subsequently be utilized in the planning step. The model represents knowledge about the SOA application execution history and service interdependency. Its construction is somewhat optional, as a planning strategy might exploit only online monitoring parameters which refer to simple services.

This chapter presents the model-driven adaptive SOA application execution environment as a more mature version of adaptive systems. Such an execution environment is a promising approach to managing complexity, leveraging software models which we refer to as *models@run.time* (Blair et al., 2009). Such a model is defined as a causally connected self-representation of the associated system that emphasizes the structure, behaviour or goals of the system from a problem space perspective. Runtime models provide "abstractions of runtime phenomena" and support reasoning. They may used for dynamic state monitoring and control of systems during execution or to dynamically observe the runtime behaviour in order to understand a specific behavioural phenomenon.

The chapter focuses on the construction of SOA application execution models and mechanisms supporting the MAPE process which allow it to be used for quality control at runtime. The proposed approach takes into account the fact that service delivery is intimately connected to Quality of Service and Quality of Experience requirements, expressed through contracts between service providers and end users. It is also essential that SOA applications - frequently composed at runtime - offer new theoretical and practical possibilities for QoS/QoE management. These aspects are very important for research on next-generation SOA (Erl, 2010) technologies, dealing with management, governance and support for composite services through lightweight orchestration.

The structure of the proposed chapter is as follows. First, the requirements of QoS/QoE control in SOA systems are presented. A clear distinction between QoS and QoE is introduced. The taxonomy of QoS metrics for SOA systems is also presented. When referring to QoS we note that this term is used to describe quality from the point of view of providers as well as clients. Paradoxically, system users might be satisfied with high QoE even if the system itself does not implement all the functionality paid for by the clients. On the other hand, fulfilment of all required quality aspects does not automatically result in user satisfaction if QoE remains low. The importance of QoE has strong temporal characteristics as it is frequently analyzed in brief intervals, whereas QoS is usually averaged over a longer period of time. It should also be noted that clients may disagree with the provider as to the overall quality of the system. Since major differences between these assessments are commonplace, they should also be taken into account when considering quality control in SOA systems.

Following this discussion, model-driven adaptive quality control of SOA applications is presented. A statistical model of SOA application execution which could be used for this purpose is introduced and its runtime updates are considered. An adaptation process based on a fitness function is formulated. The computation of this function relies on the proposed statistical model of SOA application execution. This part of the chapter is based on ongoing research by the authors (Szydlo & Zielinski,2008; Szydlo,2010). The following part of the chapter considers adaptability mechanisms for SOA systems. They exploit the fact that each SOA application is composed of a set of services linked to one another via the integration platform - ESB. Such an application can be described as an abstract composition and then executed in an ESB container responsible for mapping abstract services onto specific instances in the course of service processing. The mapping process can be performed by the user or planner prior to execution or it can take place at runtime. Such an approach allows instances to be replaced during execution if QoS/QoE requirements are to be fulfilled. This, in turn, leads to practical implementation of QoS-driven composition as a foundation of the investigated runtime adaptability concept. Sensors and effectors for adaptation activity are described and location of their installation within the ESB architecture is explained. The usability of the proposed approach is illustrated by a case study. The chapter ends with conclusions.

2. Quality management in SOA systems

Quality management is a crucial element as it ensures that systems meet their requirements with respect to specific performance metrics. Dealing with QoS is a sign that technology is progressing beyond initial experimentation to a production deployment stage - as in the case of service orientation. In order to manage quality, a set of metrics has to be defined. Clients and providers are bound by SLAs which define the terms and conditions of service quality that the provider must deliver to customers. A key aspect of SLA is QoS information which consists of several criteria such as execution duration, availability, execution time, and many others. SLA also comprises financial information such as the price for using a service, and the way in which penalties are calculated. To define quality requirements, a common understanding of offered quality between users and providers is necessary (Dobson & Sanchez-Macian, 2006).

It is important for both consumer and provider to understand metric semantics (Ludwig, 2003). Clear definitions must be associated with particular parameters (e.g. response time as average value or the maximum response time over the last ten minutes). Similarly, the point of measurement is important. Metric values gathered on the client side are often different than those gathered within the service container or the service itself.

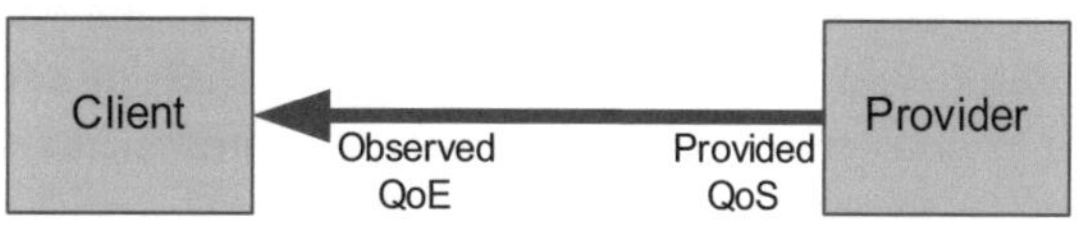

Fig. 1. Perceiving quality

Analysis of existing frameworks in terms of QoS shows that this term is used interchangeably for describing quality from the provider as well as from the client point of view. Authors have therefore decided to use term Quality of Experience (QoE) for Quality of Service observed at the client endpoint. This idea is presented in Fig. 1. Paradoxically, system users might be satisfied with system behaviour that exhibits high QoE even if the system does not implement all the functionality for which the clients actually paid. On the other hand, the system may fulfil all the required quality aspects, but users might still not be satisfied given low QoE. The importance of QoE has a temporal character as well, as it is observed over a finite interval, whereas QoS is more of an average statistic. Client may disagree with the system provider as to the quality of the system. In fact, it is very common for both assessments to differ fundamentally.

The definition and evaluation of metrics is essential for model-driven adaptive quality control of SOA applications as they could be composed at runtime with services offered by many providers. It is also a fundamental element of the MAPE process and the execution model. This is why the metrics classification used for in the context of SOA applications should be considered in more detail.

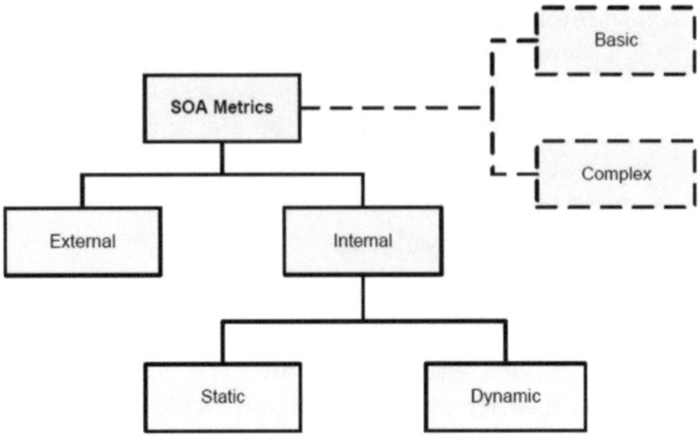

Fig. 2. Taxonomy of SOA metrics

We refer the reader to the taxonomy presented in Fig. 2 which distinguishes the following categories:

- **Internal and External Metrics** Internal metrics addresses solely those aspects of a system which are available exclusively for system architects. They might provide detailed information on source code, or the number of used software modules. External metrics are available for system users and provide information about system reliability, efficiency or response time. Evaluation of external metrics is more difficult than of internal ones, as such metrics often have subjective meaning. Additionally, metrics calculated at early stages of product development may not reflect the values of final versions;
- **Static and Dynamic Metrics** Static metrics describe invariable parameters. These might include parameters derived from software documentation or semantic description, or metrics describing source code. Examples of such metrics include configuration and security aspects of software (Hershey et al., 2007). Dynamic metrics are evaluated during runtime. Sample metrics of this kind include execution time and network throughput;
- **Basic and Complex Metrics** Basic metrics are directly gathered from system monitoring. They might provide simple information about system behaviour, e.g. the time necessary to deliver a message. Complex metrics are combinations of basic ones. For example, response time might be defined as the time needed to deliver a message to a service plus the time needed to process this message.

A number of authors have tried to organize metrics for SOA (Rud et al, 2007) (Hershey & Runyon, 2007); however, in practice, most such schemas are variations of availability, execution time and throughput.

The presented metrics could refer to simple or composite services which constitute an SOA application. Typically, computing composite service metrics requires additional information e.g. on how the execution of one service may impact the execution of another service (Menasce, 2004). The problem becomes even more complicated if services are invoked in parallel. Such additional information may be provided by the execution model which typically describes existing interdependencies. This problem is further elaborated in Section 3 where the SOA application execution model is presented.

3. Model driven adaptive quality control in SOA systems

Runtime adaptability requires detailed information on how the system behaves and how certain services influence others. This information is necessary for proper formulation and implementation of adaptation strategies. The use of models for SOA adaptability seems to be a promising approach (Blair et al., 2009). Models represent knowledge about the working system and hide unnecessary information, providing only facts related to the problem on hand. An execution model may be constructed on the basis of monitoring data collected while observing service QoS/QoE and execution infrastructure load. Updating models at runtime enables the system to respond to changes in the execution environment. The model may then be used to modify composition of the application in order to achieve high-level system management goals. Such modifications address service selection, binding protocol and interaction policy choices.

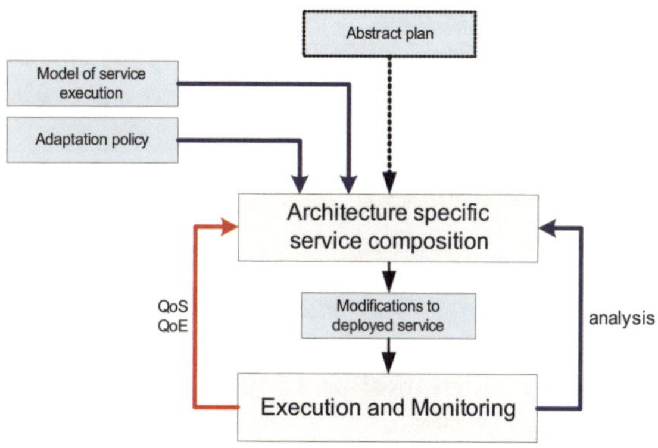

Fig. 3. Model-driven adaptation

The concept of model-driven adaptation is presented in Fig. 3. The service provider composes an application in a selected technology and provides an adaptation policy to the system along with a composite service execution model. The architecture-specific service composition layer continuously modifies the deployed service according to the provided adaptation policy. The abstract plan which acts as input for architecture-specific service composition might be hidden and used only by the programmer during application development. The knowledge of how the system behaves is represented by the service execution model. For this purpose, statistical models can be used. The use of such models for adaptive quality control requires mechanisms for model identification, updates and exploitation in the adaptation process.

3.1 Statistical models of composite service execution

Composite service execution is a collection of activities applied during each service invocation. In other words, each invocation is a particular flow in a graph of possible activities represented as a model M. By observing the composite service activity at a given point in time, these flows can be described by statistical data associated with edges of a graph. Thus, the model is composed of a graph and static information which can be provided by the system designer or identified automatically based on system activity monitoring. Once used for adaptation, the model can be updated at runtime, reflecting the current behaviour of the system.

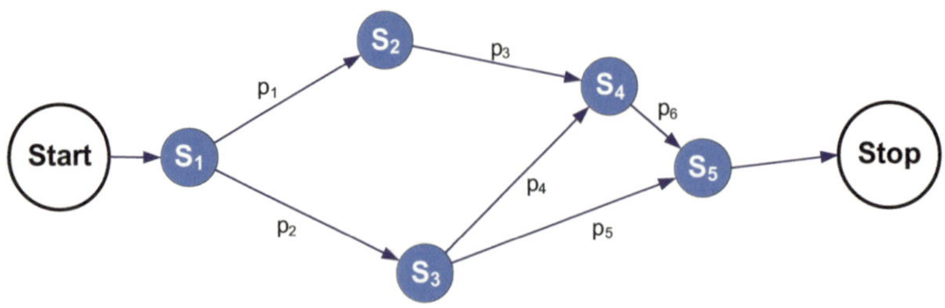

Fig. 4. Sample model of complex service execution at $Level_1$

A formal composite service execution model is represented as a *Directed Acyclic Graph (DAG)* $M = (V;E)$ and includes two special nodes: *Start* and *Stop*. At $Level_1$, nodes in this model represent abstract services - specified by description of their functionality, e.g. WSDL files, while at $Level_2$ particular *Service* instances (specific services) are considered. A sample model of complex service execution at $Level_1$ is shown in Fig. 4. Edge $(s1; s2) \in E$ means that service s_1 communicates with service s_2 with probability $p(s1; s2)$. For each vertex s_j the probabilities on edges originating in a given node add up to 1 provided that successors are not invoked in parallel:

$$\sum_{s_j \in SUCC(s_k)} p(s_j, s_k) = 1$$

Once the model is defined and statistical information is collected, it can be used for adaptation purposes. From the service provider's perspective, statistical information can be very valuable because it enables identification of the busiest services and bottlenecks, as well as optimization of resource allocation.

3.2 Model updates at run-time

Updating statistical models at runtime strongly depends on the implementation technology of integration middleware. SOA leverages loose coupling of services, which might be connected by additional infrastructural elements, e.g. Enterprise Application Integration (EAI) patterns (Hohpe & Woolf, 2003).

The composite service execution model at $Level_1$ contains abstract services, whereas in the execution environment each abstract service might be represented by several instances, used interchangeably. Hereinafter, the number of messages sent between abstract services s_1 and s_2 is understood as a sum of messages sent between all instances of these abstract services.

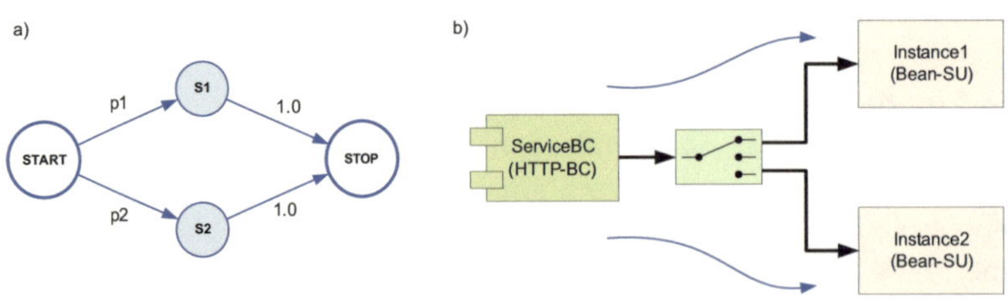

Fig. 5. Model synchronization a) complex service execution model; b) EAI model

When a model contains all the services involved in message passing, calculating the probabilities for model edges becomes straightforward - they reflect the number of messages sent between services divided by the total sum of messages originating at a source node. Nevertheless, the complex service execution model may not contain all the services deployed in ESB, as shown in Fig. 5 where the EAI content router is excluded from the model. This is justified by the fact that such an element plays a strictly technical role and does not influence the functionality of the composite service. It has also been optimistically assumed that the number of messages sent between services s_1 and s_2 is the minimum number of messages sent between services that are on the shortest path between s_1 and s_2.

3.3 Adaptation process

The proposed execution model can be used for computation of expected values of selected metrics if some service instances take part in execution. The adaptation process may analyze all the possible combinations of service instance invocation flows that might occur in the execution process and select one for execution, based on metric values.

When decisions are made locally, i.e. without analysis of how a particular service instance may affect the overall application, the results often lead to unsatisfactory solutions. The concept of statistical models presented in this chapter enables estimating global behaviour of systems when a certain invocation flow of instances is selected for execution.

Referring to the example in Fig. 4, let us assume that for services $s_1,\ldots, s_5$, instances $i_1,\ldots,i_5$ are selected. Thus, the presented model now is at $Level_2$. Let us now consider the execution time metric. Let t_k be the execution time of service i_k. If $Time_C$ is the predicted total execution time for subsequent invocations of a composite service, then for the presented model and selected instances, it can be given as:

$$\begin{aligned} Time_C = {} & p_1p_3p_6\left(t_{i_1} + t_{i_1,i_2}^{RTT} + t_{i_2} + t_{i_2,i_4}^{RTT} + t_{i_4} + t_{i_4,i_5}^{RTT} + t_{i_5}\right) \\ & + p_2p_4p_6\left(t_{i_1} + t_{i_1,i_3}^{RTT} + t_{i_3} + t_{i_3,i_4}^{RTT} + t_{i_4} + t_{i_4,i_5}^{RTT} + t_{i_5}\right) \\ & + p_2p_5\left(t_{i_1} + t_{i_1,i_3}^{RTT} + t_{i_3} + t_{i_3,i_5}^{RTT} + t_{i_5}\right) \end{aligned}$$

The composite service model at $Level_1$ may generate several models C at $Level_2$ because for each abstract service several instances may be available. Given a set of metrics Q, for each service C a QoS_C vector can be calculated. For each *i-th* metric $QoS^i_C \in [0;1]$ is the value of that metric based on historical invocations of a complex service and information stored in the composite service execution model.

There might be several ways to select the most suitable solution. For example, let us define a fitness function that determines how well a selected set of services matches user expectations. The weights of each metric express its importance in the overall fitness indicator:

$$fitness\left(QoS_C\right) = \sum_i w_i \times QoS_C^i$$

Providing exact weights for each metric is a multidimensional decision problem. Despite its inherent complexity, it can be approached through decomposition. The Analytic Hierarchy Process (Forman & Selly, 2001) is a technique based on mathematics and human psychology for prioritizing elements of a decision problem. For each pair of metrics the user specifies which one is preferred, in the form of a fraction between 1/9 and 9/1. The result of AHP is a vector of weights w_i for each metric i. From the number of possible solutions, one chooses the solution with the highest fitness factor and modifies the execution container in order to use the selected service instances. A different approach assumes the use of QoE during adaptation. For example, when the client pays for a particular level of quality, the system may constantly switch between sets of service instances so that the client finally receives what he paid for.

The presented concept of quality control requires suitable mechanisms which should extend the SOA integration to support execution of service instances selected by the adaptation policy. Contrary to other solutions (Vuković, 2006; Gubala et al., 2006; Chafle et al., 2006), the proposed approach requires the service provider to compose applications in the selected technology and then provide an adaptation policy along with a model of composite service execution. The adaptation system monitors the deployed composite services and modifies their operating parameters in accordance with the provided adaptation policy. All these stages of adaptation are described in detail in next section, referring to the MAPE paradigm.

4. Adaptability mechanisms for SOA systems

The MAPE paradigm consists of four sets of functionality: monitor, analyze, plan and execute. Analyzing and planning responses to changes in the execution infrastructure can be affected by model-driven adaptation described in the previous section. Other sets of functionality are strictly related to the managed resource. Monitor functions collect details from the managed resource and correlate them into symptoms that can be analyzed. The execute functions provide mechanisms for performing necessary changes to the system.

Currently, the Enterprise Service Bus (ESB) looms large as a promising approach for deployment of SOA applications. Most of the ESB implementations are compliant with Java Business Integration (JBI), developed under the Java Community Process (JCP) and specified as Java Specification Request (JSR) 208, as an approach to implementing SOA. ESB provides mechanisms for message normalization, routing between selected components and location transparency, making it possible to access services by name without prior knowledge of their exact location. This is why ESB is the suitable place for mechanisms required by the adaptation process.

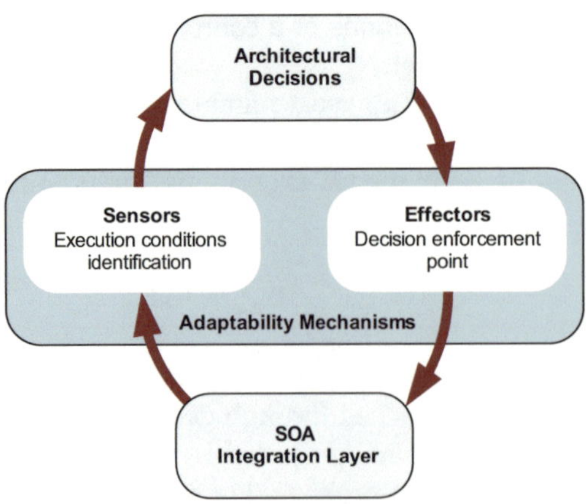

Fig. 6. Extending the integration layer with adaptability mechanisms

Fig.6 depicts a processing cycle based on the MAPE paradigm and corresponding to a standard control/adaptation loop. The Architectural Decision block continuously monitors the system and handles events whenever an action needs to be taken. It monitors the environment using *sensors*, analyses what is found and then plans and executes specific actions through *effectors* operations. *Sensors* monitor current system state while *effectors* have the ability to change the state or configuration in order to perform adaptation.

Sensors

An adaptive system acts in accordance to various goals, requiring several sorts of information gathered from ESB. In most cases, this information will be disjunctive, so one would expect to deploy specialized types of sensors rather than generic ones. It has been found that the interceptor design pattern (Schmidt et al., 2000) fulfils requirements for dynamic installation and reinstallation of sensors. This leads to the conclusion that sensors can be implemented as interceptors and installed whenever necessary.

The most appropriate location of sensors is the place where inter-service/inter-component message transmission takes place. JBI specification defines elements which participate in sending and passing messages. A message is created in a service, passes through a specialized Service Engine (SE) or Binding Component (BC) and is then sent (through a Delivery Channel - DC) to a Normalized Message Router (NMR), which reroutes it to a particular destination through the same components. Injecting interceptors into services or SEs requires knowledge of various types of services and engines. Thus, a better place to inject message exchange interceptors is the DC through which every SE is connected to NMR.

The amount of monitoring information gathered from ESB would be overwhelming. A more suitable approach is to correlate or aggregate events and send notifications only if complex events take place. For this purpose, the stream of simple events from ESB could be processed by a Complex Event Processor (CEP) (Luckham, 2002) which, in due time, would notify the Architectural Decision block.

Effectors

Implementing an adaptation policy requires action, i.e. a set of operations which modify the execution of an application. These operations are performed by effectors which have to be implemented and installed in the execution environment. It has been found that modifying message routes between service instances is crucial for implementation of the proposed SOA application adaptation concept.

Rerouting messages to other instances is justified only when these instances share the same interface and provide identical functionality. This process could be performed effectively by installing suitable effectors in ESB. Current implementations of ESBs that are compliant with the JBI specification share some common attributes used during message processing. Each invocation of a complex service is described by its CID (Correlation ID) and is constant for one invocation, even when passed among services. A particular message sent between services is described by EID (Exchange ID). Generally speaking, the invocation of a complex service is described by CID and consists of several message exchanges described by EID which complies with the execution model proposed in Section 3. Extending NMR with effectors implementing a selected routing algorithm, capable of modifying message routing, would yield an adaptable ESB.

Priority	CID	Intentional Service Name	EID	Destination Service Endpoint
1	n/a	n/a	+	Service Endpoint 1
2	+	+	n/a	Service Endpoint 2
3	n/a	+	n/a	Service Endpoint 3
4	+	n/a	n/a	Service Endpoint 4
5	n/a	n/a	n/a	Service Endpoint 5
6	n/a	n/a	n/a	Service Endpoint 6

Table 1. Examples of routing rules

Once the message reaches the NMR, the routing algorithm checks conditions of routing rules in the routing table (see Table 1). If the message matches a routing rule, that rule is fired and the Service Endpoint from the routing rule substitutes the intended destination Service Name. Routing rules are split into groups with different priorities that are analyzed in a particular order. In every message, header parameters such as Correlation ID, intended Service Name and Exchange ID, are matched to routing rules. If the message matches several rules, one of them is selected on a *round-robin* basis to provide load balancing.

5. Case study

In order to verify the presented concept of SOA application composition and execution, the Enterprise Service Bus (ESB) as an integration platform has been extended by adaptability mechanisms. The prototype design follows the architecture of Adaptive ESB presented in the previous section. Adaptability mechanisms are implemented using Aspect Oriented Programming (AOP) (Kiczales et al., 1997), where new features can be enabled and disabled without modifying original ESB container source code.

The prototype implementation of adaptive ESB is exploited in a scenario which demonstrates how the proposed concept of quality control can be applied to a situation where clients should receive variable QoS depending on some constraints. The provided composite service execution model is used to generate a possible set of service instances that can be invoked during execution. Subset selection bases on a fitness function defined in Section 3.3, which determines user preferences by associating weights with the available composite service metrics. Possible solutions are then ordered according to the values of the calculated metrics and the adaptation policy selects a particular set of service instances based on high-level goals, i.e. the cost of execution and the overall client budget.

5.1 Composite service

Let us assume that one would like to create a *GeoWeather* service that provides weather information for a given location. Location data will be provided as a zip code, GPS location, IP address of computer user or country name. Another assumption is that accuracy would be sufficient if limited only to cities.

Widely-available weather services do not provide weather information for GPS coordinates. Instead, they provide weather information for a given readable address; thus all possible input formats have to be converted to addresses and then polled for weather information.

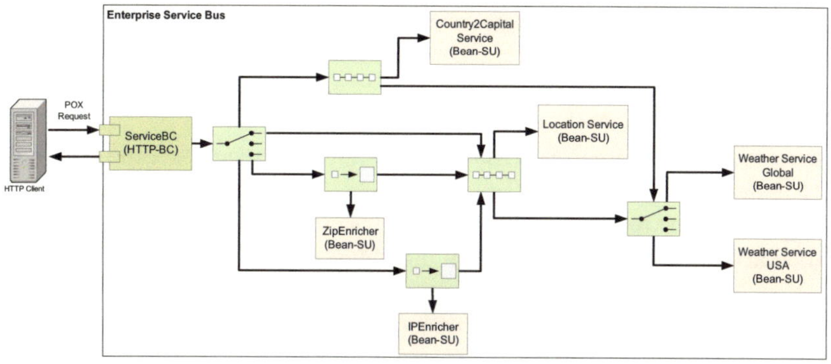

Fig. 7. EAI model of complex service used in the case study

To deploy this application into ESB, it has to be described in terms of EAI patterns. Invoking services in a sequence might be performed with the use of *RoutingSlip* integration patterns, while modifying the message content can be done using the *ContentEnricher* pattern. An EAI-based model of the presented composite service is depicted in Fig.7.

Let us assume that the client operates an Internet website that provides information for amateur pilots and that he wants to include very accurate weather forecasts for airports on the main website. The client is interested in a service which returns weather information for given geographical coordinates. The provider notices that this website receives 8,200 visits per week (on average) and, given the QoS selected by the client, it costs 54,000 credits. As the website become more popular, the number of invocations may significantly increase. The client therefore decides to buy a service with variable QoS and with the maximum budget of 55,000 credits. Detailed statistics of user invocations are presented in Fig. 8. Two different profiles can be distinguished. The first one reflects typical interest in the website, while the second one occurs when there is increased interest in the website. The increased numbers of page visits before the end of the week can be attributed to a national flying contest.

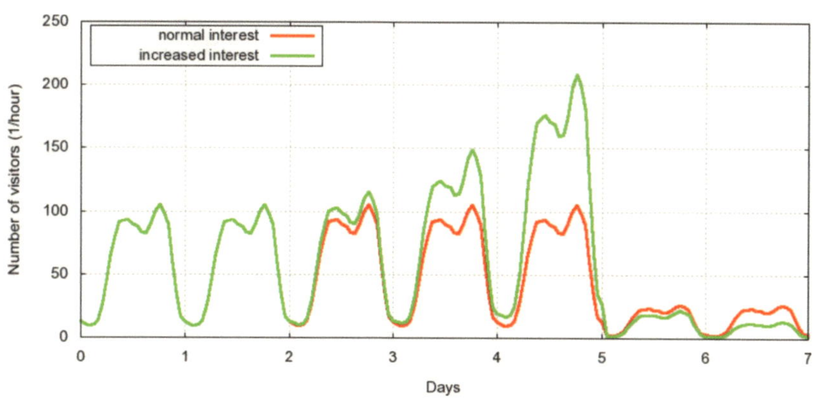

Fig. 8. Number of visitors per hour

5.2 Importance of metrics

For this type of service, it is better (from the customer's point of view) to achieve very accurate data slower than to rapidly access inaccurate information. The author therefore decides to apply the fitness function presented in Section 3, which expresses the degree of overlap between the functionality of the composite service and the user's preferences. This means that each metric has to be normalized and its value has to be in the [0; 1] range where 1 means that this metric has the best possible value.

Three metrics are defined for this scenario: availability, execution time and data quality. Availability is defined as the percentage of successful invocations. Execution time metric is normalized linearly (for response time equal to 0*ms* the value of the metric is 1, while for response time above 1000*ms* the corresponding value is 0). The data quality metric equals 0 if data is accurate to within 100*km*, and 1 if it is accurate to within 10*km*.

Weights for each metric in the fitness function are calculated using the mentioned earlier AHP method. The customer determines the relative importance of each metric:

- *availability* is three times as important as *execution time;*
- *availability* is five times less important than *data quality;*
- *execution time* is five times less important than *data quality.*

Following calculations, the fitness function used in this case can be given as:

$$fitness = 0,20 * availability + 0,097 * execution\_time + 0,702 * data\_quality$$

5.3 Composite service execution model

In order to react to changes in the execution environment, a composite service execution model and an adaptation policy have to be defined. The model of the analysed service and its projection onto instances found in the ESB is depicted in Fig. 9. This model can be derived by simply transforming the EAI-based model shown in Fig. 7. The service is only a subset of a larger, complex service; hence the model contains only these services that are used in the invocations.

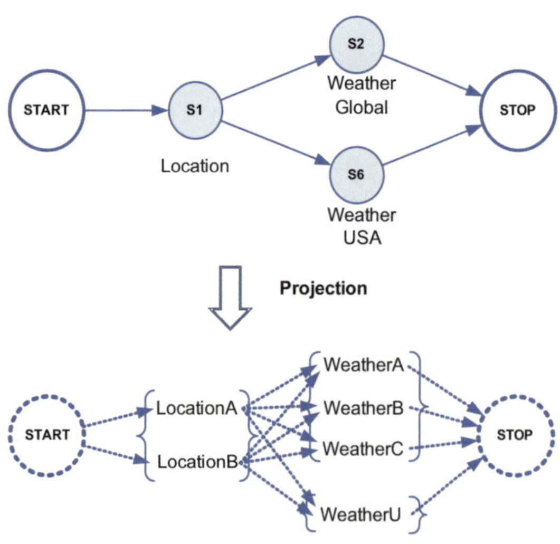

Fig. 9. Projection of the model of complex service execution at $Level_1$ onto composite services

There are six possible combinations of service instances that might be executed. The model, along with the expected values of metrics for projected composite services, is updated at runtime and used by the adaptation policy.

Composite service	Service *Location* instance	Service *Weather* instance	Cost [unit/100]	Availability [%]	Data quality [%]	Execution time [ms]	*fitness factor*
C_1	*LocationA*	*WeatherA*	650	0.81	0.95	200	0.737
C_2	*LocationA*	*WeatherB*	600	0.81	0.90	300	0.657
C_3	*LocationA*	*WeatherC*	350	0.72	0.70	500	0.535
C_4	*LocationB*	*WeatherA*	550	0.89	0.80	300	0.623
C_5	*LocationB*	*WeatherB*	500	0.89	0.80	400	0.614
C_6	*LocationB*	*WeatherC*	250	0.79	0.70	600	0.526

Table 2. Fitness factors of composite services for the variable QoS test case

Table 2 contains expected metric values for the composite services that might be invoked and the evaluated fitness factor for each composition.

5.4 Adaptation strategy

In the presented test case, a simple accounting system has been implemented which constantly monitors user invocations and updates the associated accounts. The same module calculates the expected number of invocations until the end of the accounting period. The number of invocations per week is calculated as a sliding window from the last 7 days. It is then used to calculate the expected number of invocations for the remaining days until the end of the accounting period. Given the number of credits left and the expected numbers of invocations until the end of accounting period, the module calculates the maximum price for subsequent invocations. This module is expressed as a fact and inserted into the working memory of the policy engine.

5.5 Results

Detailed statistics of user invocations are presented in Fig. 10. Increased numbers of page visits before the end of the week can be attributed to a national flying contest. The total number of invocations is not the expected 8,400 but approximately 10,200 as depicted in Fig. 10B. During the week, the system tries to estimate the total number of invocations during the accounting period - this fact is reflected in the figure. Because the actual number of invocations is greater than expected, system has to switch to a different set of instances to bring the total sum below the assumed 55,000 credits. Coloured areas in Fig. 10 show which composite service is used by the system to provide the service. One can notice that on the 4th day the system has decided to switch the composite service to the less expensive C_2 with fitness factor 0.657 instead of C_1 with 0.737 and then to C_4 with 0.623. As the expected number of invocations during the whole accounting period increases, the maximum number of credits that might be spent on each invocation keeps decreasing. This leads to further adjustments (on day 5) where the composite service is switched to C_5, C_3 and finally C_6. During the last 2 days, the number of invocations is noticeably smaller, so the system returns to its initial state i.e. to composite service C_1 with the best fitness value. Without an adaptation policy as depicted in Fig. 10A, the client would have to pay 64,000 credits in the described case; however the adaptation policy ensures that the cost is kept below the assumed limit of 55,000 credits. Execution time of provided service is depicted in Fig.10C, data quality in Fig. 10D and availability in Fig. 10E.

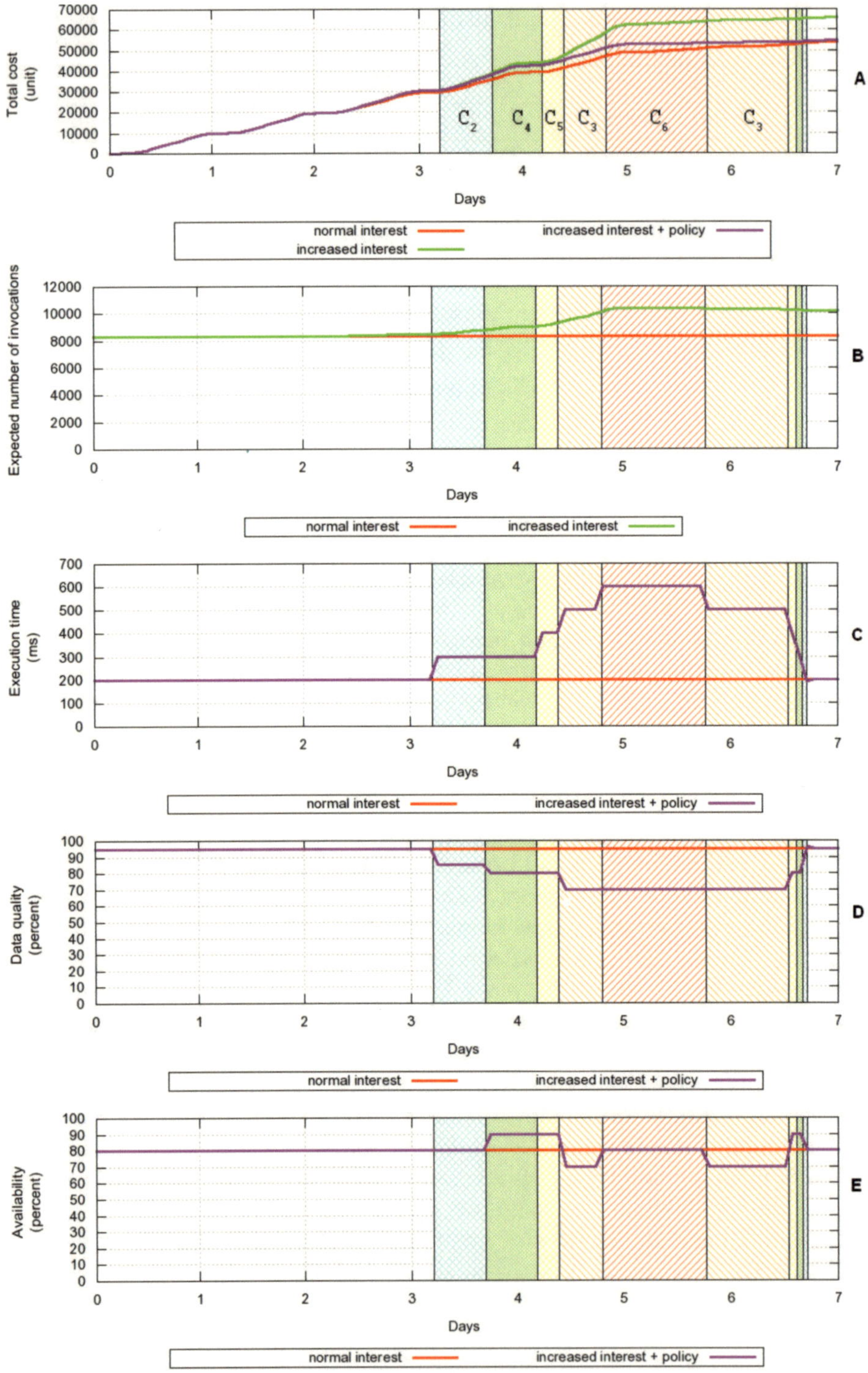

Fig. 10. Test case scenario with variable QoS

6. Summary

SOA applications which are loosely coupled and composed from self-contained, autonomous services with well-defined contracts are particularly well suited for adaptive control of QoS. Their execution environments (such as ESB) could be effectively equipped with mechanisms which allow implementation of MAPE paradigms.

A crucial element of the adaptation loop is a decision subsystem which enforces application adaptation actions, maintaining QoS/QoE guarantees under changing operational conditions and constrains. This rather complex process may be practically performed with SOA application execution model support. The presented study shows that a statistical model could be used for this purpose.

Statistical models, which can be identified at runtime, represent knowledge and experience derived from the SOA application execution history. This knowledge may be transparently used by the execution environment to control QoS according to a selected policy. The proposed approach leads to a new, important category of adaptive SOA systems which may satisfy consumer requirements under changing QoS conditions without involving a human operator.

7. References

Ali Arsanjani, Liang-Jie Zhang, Michael Ellis, Abdul Allam, and Kishore Channabasavaiah. *S3: A Service-Oriented Reference Architecture*. IT Professional, 9(3):10-17, 2007.

Gordon Blair, Nelly Bencomo, and Robert B. France. Models@ run.time. Computer, 42:22-27, 2009.

G. Chafle, K. Dasgupta, A. Kumar, S. Mittal, and B. Srivastava. Adaptation in Web Service Composition and Execution. In *Web Services*, 2006. ICWS '06. International Conference on, pages 549-557, Sept. 2006.

Glen Dobson and Alfonso Sanchez-Macian. Towards Unified QoS/SLA Ontologies. In SCW '06: *Proceedings of the IEEE Services ComputingWorkshops* (SCW'06), pages 169-174, Washington, DC, USA, 2006. IEEE Computer Society

Thomas Erl. SOA Principles of Service Design (The Prentice Hall Service-Oriented Computing Series from Thomas Erl). Prentice Hall PTR, Upper Saddle River, NJ, USA, 2007.

Thomas Erl. Modern SOA Infrastructure: Technology, Design, and Governance, Prentice Hall PTR, Upper Saddle River, NJ, USA, 2010

Ernest H. Forman and Mary Ann Selly. Decision By Objectives - How To Convince Others That You Are Right. World Scientific Publishing Co. Pte. Ltd., Singapore u.a., 2001.

Alan G. Ganek and Thomas A. Corbi. The dawning of the autonomic computing era. *IBM Systems Journal*, 42(1):5-18, 2003.

T. Gubala, D. Harezlak, M.T. Bubak, and M. Malawski. Constructing abstract workflows of applications with workflow composition tool. , *Cracow'06 Grid Workshop*, Vol 2: K-Wf Grid: the knowledgebased workflow system for Grid applications: October 15-18, 2006 Cracow, Poland

Paul Hershey and Donald Runyon. SOA Monitoring for Enterprise Computing Systems. In *EDOC '07: Proceedings of the 11th IEEE International Enterprise Distributed Object Computing Conference*, page 443, Washington, DC, USA, 2007. IEEE Computer Society.

Paul Hershey, Donald Runyon, and Yangwei Wang. Metrics for End-to-End Monitoring and Management of Enterprise Systems. In *Military Communications Conference*, 2007. MILCOM 2007. IEEE, pages 1-7, Oct. 2007.

Gregor Hohpe and Bobby Woolf. Enterprise Integration Patterns: Designing, Building, and Deploying Messaging Solutions. Addison-Wesley Longman Publishing Co., Inc., Boston, MA, USA, 2003.

IBM Corporation, An Architectural Blueprint for Autonomic Computing, June 2006

Kiczales, Gregor; John Lamping, Anurag Mendhekar, Chris Maeda, Cristina Lopes, Jean-Marc Loingtier, and John Irwin (1997). "Aspect-Oriented Programming". *Proceedings of the European Conference on Object-Oriented Programming*, vol.1241. pp. 220–242.

Jeffrey O. Kephart and David M. Chess. The Vision of Autonomic Computing. *Computer*, 36(1):41–50, January 2003.

David Luckham. The Power of Events: An Introduction to Complex Event Processing in Distributed Enterprise Systems. Addison-Wesley Professional, May 2002.

Heiko Ludwig. Web services QoS: external SLAs and internal policies or: how do we deliver what we promise? In *Web Information Systems Engineering Workshops*, 2003. Proceedings. Fourth International Conference on, pages 115–120, 2003.

P. K. Mckinley, S. M. Sadjadi, E. P. Kasten, and B. H. C. Cheng. Composing adaptive software. *Computer*, 37(7):56–64, 2004.

Daniel A. Menasce. Composing Web Services: A QoS View. *IEEE Internet Computing*, 8(6):88–90, 2004.

Aad Van Moorsel. Metrics for the Internet Age: Quality of Experience and Quality of Business. Technical report, *5th Performability Workshop*, 2001.

Brice Morin, Olivier Barais, Jean-Marc Jezequel, Franck Fleurey, and Arnor Solberg. Models@ run.time to support dynamic adaptation. *Computer*, 42:44–51, 2009.

Dmytro Rud, Andreas Schmietendorf, and Reiner Dumke. Resource metrics for service-oriented infrastructures. In Daniel L¨ubke, editor, *Proceedings of the Workshop on Software Engineering Methods for Service-oriented Architecture* 2007 (SEMSOA 2007), Hannover, Germany, pages 90–98. Leibniz Universit¨at Hannover, FG Software Engineering, May 2007.

D. Schmidt, M. Stal, H. Rohnert, and F. Buschmann. Pattern-Oriented Software Architecture, Volume 2: Patterns for Concurrent and Networked Objects. Wiley, 2000.

Tomasz Szydlo and Krzysztof Zielinski. Method of Adaptive Quality Control in Service Oriented Architectures. In *ICCS '08: Proceedings of the 8th international conference on Computational Science*, Part I, pages 307–316, Berlin, Heidelberg, 2008. Springer-Verlag.

Tomasz Szydlo, QoS driven Semantics Based SOA Applications Composition and Execution, Ph.D. dissertation, AGH-UST Krakow, October 2010.

Maja Vuković, Context Aware Service Composition. Technical report, University of Cambridge, 2006.

21

Research about New Predictive-Maintenance Methodology using VBA for Marine Engineering Applications

José A. Orosa, Angel M. Costa and Rafael Santos
University of A Coruña
Spain

1. Introduction

Currently, one of the most important activities that helps marine engineers to maintain the quality of any service is statistical process control (SPC). Statistical process control is the applied science that helps to collect, organise and interpret the wide variety of information available for engineering applications (harris et al., 1999; Kumar et al., 2008).

SPC can help to understand and reduce the variation in any business process, which thus leads to less expenditure of time and money on rework and waste.

In the first place, control charts show how consistently a process performs and whether or not we should attempt to adjust it. Second, it compares the process performance to customer requirements, providing a process-capability index that functions as an ongoing, accurate indicator for improvement of quality. Finally, the resulting process-capability index facilitates a quick evaluation of the results of quality initiatives designed to improve process consistency.

Because it involves complex mathematics, computers are ideally suited to develop the SPC charts. They easily collect, organise and store information and present results in easy-to-understand graphs, called the control charts. Computers accept information typed in manually, read from scanners, manufacturing machines or details imported from other computer databases. Furthermore, as concluded in recent research works (Vosniakos and Wang, 1997; Lee et al., 2006), a computer collecting information in real time can detect very subtle changes in a process and can even warn in time to prevent process errors before they occur.

The process model proposed in those studies included the manufacturing process, responsible for generating the product, and a measuring process, responsible for generating information about the quality.

The process evaluation leads to intelligent decision-making for quality improvement. Therefore, recent research works (Louit et al., 2009; Kumar et al., 2008) have proposed a practical procedure to discriminate between the common renewal approach towards the time-to-failure model and the use of a non-stationary model.

Currently, to automate quality-information generation, there is a certain tendency to apply private software resources furnishing quality-control charts (Statgraphics, 2009), which can solve typical engineering problems. Nevertheless, when a marine engineer tries to solve his particular working problems at the ship or in a power station, he cannot depend on licenses and private software-related problems that can stop his activity. Consequently, engineers must apply programming languages to develop their own applications in accordance with the particular conditions of each engine room (Cebi et al., 2009; Deris et al., 1999).

Therefore, programming languages, such as FORTRAN, Pascal and C, are not appropriate for marine engineers who must work on-site with limited resources, in contrast to the work spot of a general engineer.

As a result of previous research works and meetings with experienced marine engineers and companies, the researchers of the Department of Energy and Marine Propulsion have concluded that the programming software must be user-friendly and must not be too heavy for the computers of the power station.

Furthermore, it must be borne in mind that a marine engineer hardly ever stays in the same ship or power station; accordingly, he must be ready to start his work from scratch within a limited period of time.

From previous experiences in thermodynamic studies (Orosa, 2009), researchers have analysed the principal computing software resources, such as C, Matlab, Mathematica, Engineering Equation Solver or EES, Pascal, FORTRAN and Visual Basic. From their research, they conclude that Visual Basic for applications (VBA) is the best solution for these typical engineering problems.

Microsoft (MS) VBA is a very simple language to converse with Excel, a user- friendly programming language that can be easily used by engineers who are interested in and ready to spend some time in a very rewarding and enabling learning process. People who use applications that host VBA can automate and extend the application functions, thereby shortening the development cycle of customising engineering solutions (VBA, 2009; Verma, 2003; Verhoefa et al., 2008).

The only problem related with this language is that sufficient information is not available about VBA. Consequenly, researchers of the Department of Energy and Marine Propulsion, in collaboration with the Department of Informatics, began a self-learning process involving VBA to define a new methodology for marine engineers, which allows them reduce the time elapsed when they carry out maintenance studies at work.

In particular, this chapter presents a deeper study to generate a software resource to generate a control chart of attributes and, thereby, derive a research tool for a faster quality-control study, based on the results obtained and not by the calculation process.

As explained, VBA has been selected because of its macro programming system that can be implemented with MS Excel (Wu and Liua, 2008).

2. Objectives

The fixed objective of this study was to automate the control chart of attributes, combining the feasibility of working with MS Excel with the user-friendly data introduction of Visual Basic, to determine the real feasibility of VBA as a working tool with reference to other private softwares.

Finally, this software has been tested under real on-board situations, such as detection of failures related to exhaust gas-temperature sampling in the main engine.

3. Materials

3.1 Programming languages

The programming language selected was VBA of MS Office Excel 2003. Microsoft owns the VBA, whose code is compiled (Microsoft, 2010, Excel, 2010; Jelen and Syrstad, 2008 and Roman, 2002) in an intermediate language called the P-code (2010); the latter code is stored by the hosting applications (Access, Excel, Word) as a separate stream in structured storage files independent from the document streams.

The intermediate code is then executed by a virtual machine. Therefore, the obtained software resource can only be operative in a Windows-98 or above Operating System. Calculus operations and graphics are carried out in MS Excel, with a report in MS Word. Accordingly, MS Excel and MS Word must be of version 2000 or above.

3.2 Software characteristics

In a typical software application, the user specifies a target mean and the desired average run length before a deviation of that magnitude is detected. In real applications, after some working hours, control limits must be obtained from the data sampled and be compared with design conditions.

With the help of the Department of Informatics, the Department of Energy and Marine Propulsion began to solve typical problems and send the generated code to the corresponding module of MS Excel file. Therefore, the software designed must develop a control chart that will detect deviations of a specified magnitude within an acceptable time.

4. Methods

As explained before, in the present study, the exhaust-gas temperature in the engine will be analysed using control charts and related with daily events. To do it, control-chart selection and sampling of data were applied.

4.1 Control charts of attributes

As explained above, the control chart is one of the most important and commonly used among the Statistical Quality Control methods for monitoring process stability and variability. It is a graphical display of a process parameter plotted against time, with a centre line and two control limits (Jennings and Drake, 1997).

Attribute charts are a set of control charts specifically designed for data on attributes. Attribute data are also known as "count" data. Typically, the number of times a specific condition is observed is counted (usually something not preferred, such as an error) in a specific sample from the process.

This is different from measurement data in terms of its resolution. As a consequence of this, attribute data have less resolution because only if some error occurs is it counted, in contrast to obtaining a measurement to find the enormity of the condition.

For example, attribute data for a maintenance-care process might include the number of tasks of maintenance in a pump, whereas variable data for the same process might be the measurement of the pressure decrease in the same pump. Thus, attribute data generally provide us less information than what measurement data would for the same process. Thus, for attribute data, prediction is not possible if the process tends towards an undesirable state, because it is already in this condition.

To select an adequate attributable control chart, each difference must be considered.
The Np-Chart monitors the number of times a condition occurs, relative to a constant sample size, when each sample can either have this condition or not.
The p-Chart monitors the percent of samples having the condition, relative to either a fixed or varying sample size, when each sample can either have this condition or not. The c-Chart monitors the number of times a condition occurs, relative to a constant sample size. In this case, a specific sample can have more than one instance of the condition, in which case, the total number of times it occurs in the sample is counted.
The u-Chart monitors the percent of samples having the condition, relative to either a fixed or varying sample size. In this case, a specific sample can have more than one instance of the condition, in which case counting of all the times it occurs in the sample is carried out.

4.2 p-Chart calculations

In a case study, the exhaust-gas temperature sample failure in the main engine is sampled and, consequently, the p-charts were selected. In this type of charts, the calculation of the average of all samples and the upper and lower control limits are shown in Equations 1, 2 and 3.

$$\overline{p} = \frac{\Sigma p}{k} \tag{1}$$

$$UCLp = \overline{p} + 3 \cdot \sqrt{\frac{\overline{p} \cdot (1 - \overline{p})}{\overline{n}}} \tag{2}$$

$$LCLp = \overline{p} - 3 \cdot \sqrt{\frac{\overline{p} \cdot (1 - \overline{p})}{\overline{n}}} \tag{3}$$

When the sample sizes are not constant, the average sample size, $\overline{n}$, should be used for n in the control-limit calculations, as shown in Equation 4.

$$\overline{n} = \frac{\sum n}{k} = \frac{N}{k} \tag{4}$$

Where
n = the sample size, the number of individuals in a sample
k = the number of samples
N = number of individuals, the total number of observations
N = n times k if all samples are of the same size.
p = the fraction of "successes", which may be defective items, expressed as a decimal
1-p =q=the fraction of "failures", which may be the qc-pass items.
p+q =1; hence, q = 1-p.

4.3 Software design

To design the software, the methodology used for developing a control chart must be analysed.

In this section, the steps that must be considered to properly develop the programme are listed:

The user must:

1. Select the characteristics of quality.
2. Determine the size of the subset if it is going to be constant.
3. Compile the data into groups.

Once the type of control chart is selected, it must be considered that, although it is possible to plot individual samples, the control chart will be distorted if they are not normally distributed and thus will not give a clear picture of any process changes.

To avoid this distortion, a sample size of four to six is adequate to detect moderate-to-large shifts in a parameter and 15–25 samples to detect small shifts. In this case study, groups of 150 samples of temperature have been selected.

The software must:

4. Permit the insertion of the data for each group.

   This will automatically calculate the control limits. These limits, together with the average p value of all groups, will be reflected in the graph.
5. Permit the discarding of data.

   It should be remembered that only those values, which are attributable subgroups identified through the day book and can be corrected, are discarded.

   On the contrary, those groups whose causes are not attributable are retained as part of the data.

   Moreover, items that are out of control below the lower control limit are not discarded, because these represent exceptionally good quality. If the point in the out-of-control side is due to a lower error of inspection, it must be discarded.

   After eliminating the group caused by attributable reasons, the software must automatically re-calculate and re-present the control limits. Once the p-graph is under control, a comparative study can be carried out with reference to the product specifications of the previous studies. For this, the control limits must be found within the area bounded by the specifications.

   These specifications are the typical values of the product to be analysed, which are introduced into the programme the first time that data are inserted into the area above the control limits.
6. The software must let a clear output of information.

In summary, it can be concluded from these steps that the software must plot the sample data, define the control limits, delete the data related to attributable causes, and recalculate the limits of the process towards a controllable situation.

5. Results

5.1 Software

The methodology begins with the automation of the calculus of Equations 1, 2 and 3 in MS Excel, as shown in Figure 1.

In the figure, it can be observed that the engineers must automate the selection of constants in MS Excel in accordance with the problem data.

Subsequently, the main screen of the software has been developed with VBA and is represented in Figure 2.

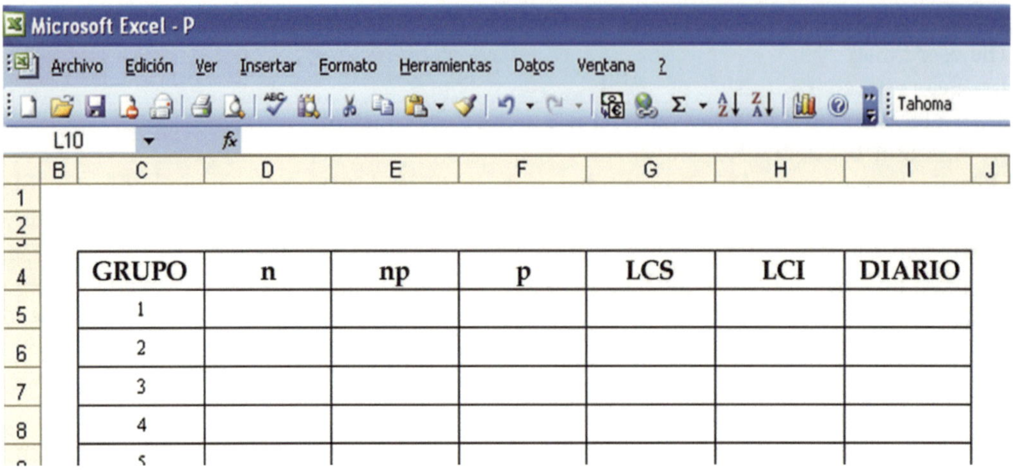

Fig. 1. Calculus automation in MS Excel

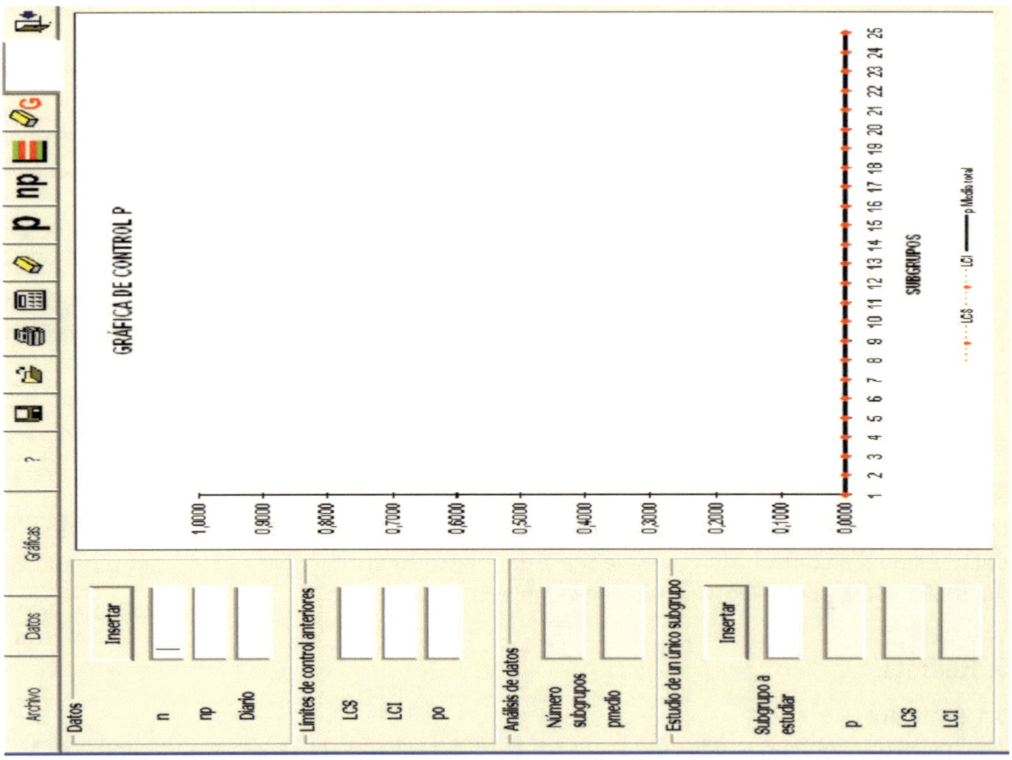

Fig. 2. Control Chart of the attributes P

5.2 Control-chart test

A test of the software developed was carried out with the sample data of the number of errors sampled in the exhaust-gas temperature of the main engine (np), figure 3.

Most ofen, they were instantaneous sampling failures and, in other cases, real sensor failure. It was proposed to characterise the process with the aim of differentiating between instantaneous and real sampling failures. For this purpose, the data was sampled in groups of a constant value of 150 samples, as shown in Table 1, and compared with other software resources, such as Statgraphics.

Fig. 3. Main engine

subgroup	n	np
1	150	6
2	150	1
3	150	4
4	150	2
5	150	0
6	150	3
7	150	3
8	150	1
9	150	4
10	150	5
11	150	1
12	150	5
13	150	4
14	150	1
15	150	0
16	150	2
17	150	3
18	150	4
19	150	8
20	150	1
21	150	2
22	150	3
23	150	0
24	150	1
25	150	1

Table 1. Number of sampling errors detected in each group of 150 samples

5.3 Test results

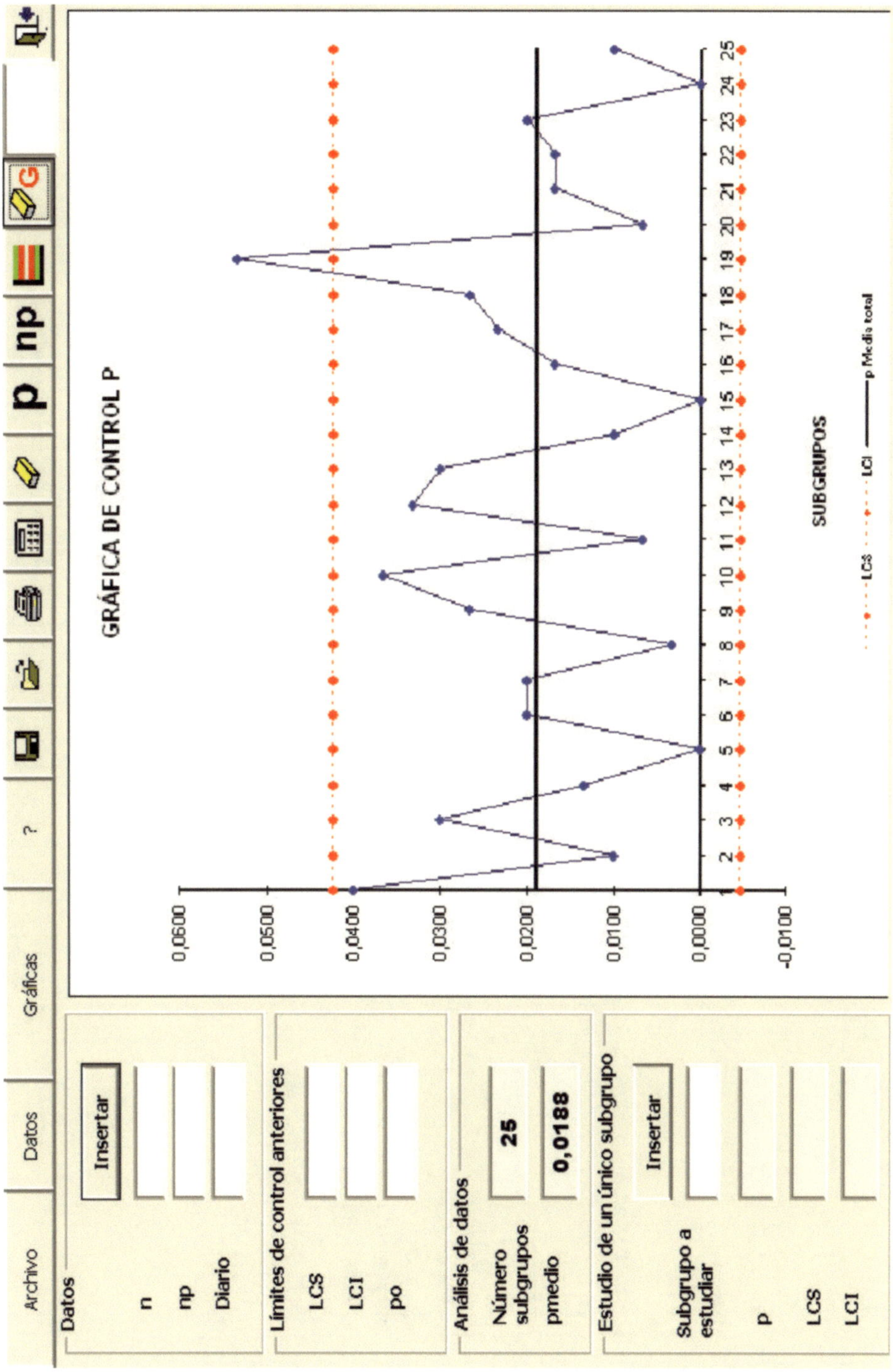

Fig. 4. Graphic units do not comply

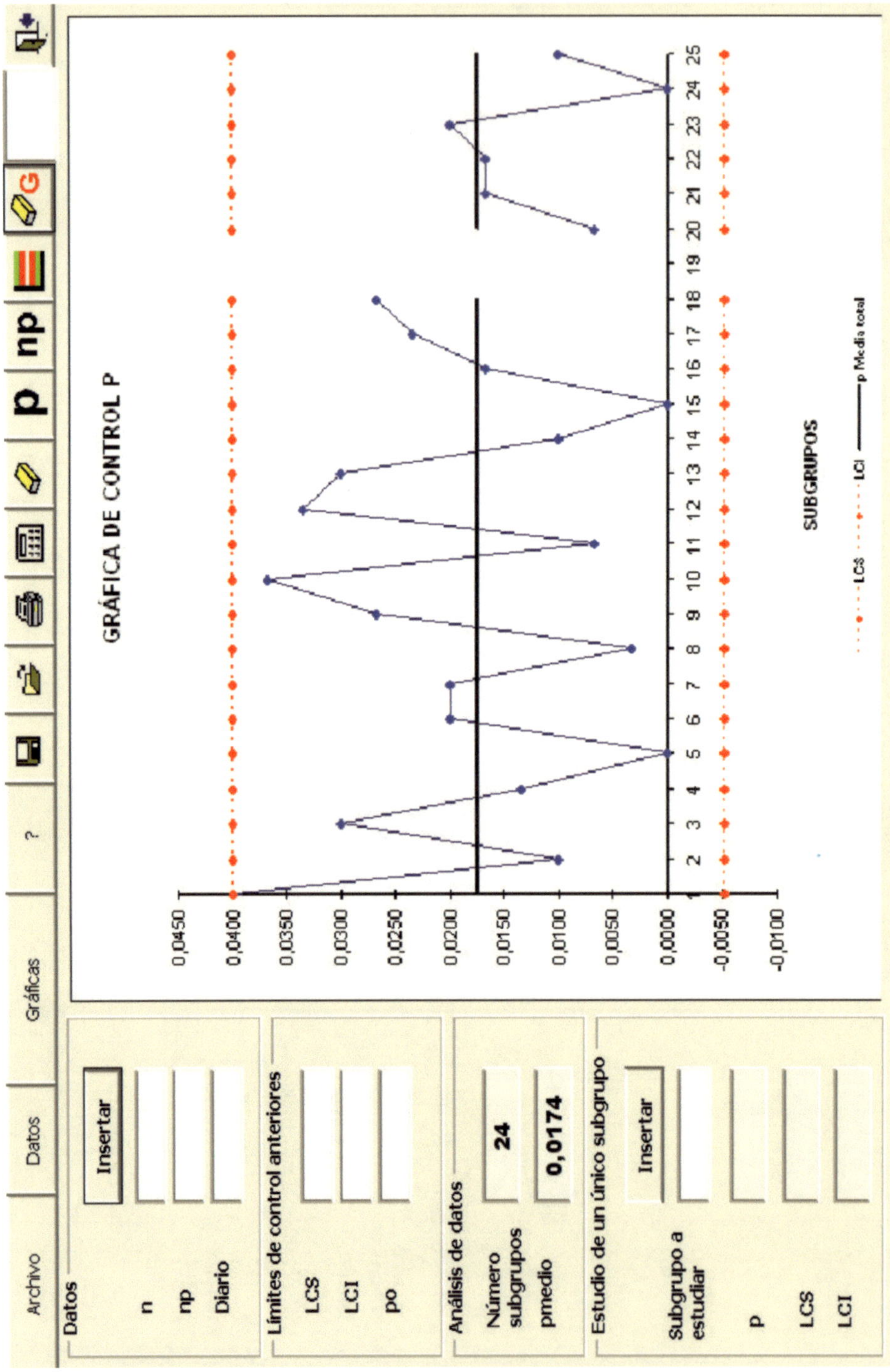

Fig. 5. Control Chart limits

6. Discussion

Statistical Process Control is the applied science that helps to collect, organise and interpret the wide variety of information available for engineering applications. This manuscript describes a study about development of the optimal methodology to develop a quality-control analysis of attributes.

As explained above, after the selection of the characteristics of quality and the size of groups, the calculus was automated to show the general solution of the problem of Figure 1. This screen will help persons who do not want to waste too much time to solve the mathematical solutions of the problem and want to improve the quality of the process.

As observed from Figure 2, when the programme is started, it displays the main menu of operation.

This screen shows the menus: file, data, graphs and help such as in other Visual Basic software. The options to save, open, print, delete, insert data and delete group are also inserted. In this screen, the number of groups, diary comments and specifications limits are added. Finally, the output data are shown as mean values and limits in the bottom of the screen.

Once all the data are inserted, each chart includes the statistically determined upper and lower control limits, indicating the bounds of the expected process behaviour (Figures 4 and 5).

The fluctuation of the points between the control limits is due to the intrinsic variation of the process. In other words, this variation is due to "common causes" that influence the process. Any points outside the control limits can be attributed to a "special cause," implying a shift in the process. When a process is influenced by only the common causes, it is stable and can be predicted. Thus, the main role of the control chart is to identify the occurrence of special causes, so that they can be removed, resulting in a reduction of variation in the overall process.

Subsequently, the process can be further improved by either relocating the process to an optimal average level or decreasing the variation due to common causes. Once the software is developed, it is tested using real case studies, which shows a clear agreement with other software resources, such as Stat graphics.

In this case study, the exhaust-gas temperatures of the main engine were sampled in groups of 150 samples and the number of sampling errors was noted. From this data, and using the herein-developed software, it was analysed and the control limits of the exhaust gas temperature were determined.

After calculating the control limits, the chart must be interpreted by looking for out-of-control points. If there are any points, their special causes must be eliminated. Conduct brainstorming sessions and experiments designed to find those process elements that contribute to sporadic changes in process location.

To use the data in hand, auto drop is turned on, which will remove the statistical bias of the out-of-control points by omitting them from the calculations of the average and control limits.

In particular, one of the 25 samples exceeded the upper limit. This value was related to an attributable reason of abnormal combustion and could be discarded, as shown in Figure 4. From this same figure, changes in the process-control limits and mean values can be observed. These values define the real conditions of the main engine temperatures.

Nevertheless, even when all the points on a control chart are within the control limits, a process can still display out-of-control behaviour due to systematic or random behaviour. This can be identified by the points showing a process shift, a trend or a recurring cycle, or the points forming two groups on either side of the process average.

These patterns can be identified by the Western Electric Control Rules:

1. One point lies outside (3 × standard deviation) (SD)
2. Two out of three consecutive points plotted are greater than 2 × SD.
3. Four out of five consecutive points plotted are greater than 1 × SD.
4. Eight consecutive points are plotted on one side of the central line.
5. An unusual or non-random pattern in the data.

In accordance with this methodology, the normalised data was processed and plotted, after deleting the attributable failure, as observed in Figure 4. The authors' effort to detect tendencies in the case study was not fruitful with reference to the ranges of SD.

It was related with a good maintenance periodicity. Because of this characteristic, this evolution must be analysed with a certain frequency to detect future failures due to wastage of elements.

To get a report of the analysis, the output information can be furnished through the data shown in the main menu by printing or generating a report in MS Word.

In particular, it is very easy to generate a report in MS Word with VBA due to the fact that it works in most of the MS resources, such as Word, Power Point and Access, among others. This report consists of three control charts and a data table showing the values obtained numerically.

Another advantage of the software is that, once developed, it permits engineers to solve a high number of control-chart problems within a few minutes and, consequently, use most of the time to understand the results and brainstorm thought the chart analysis.

Finally, it should be noted that this methodology presents a disadvantage. To obtain an adequate behaviour, the software needs an executable file that locates the files in each directory and creates the corresponding shortcuts of initialisation and uninstalls the software.

This packing has been carried out with private software due to the fact that VBA does not possess this option.

In summary we can say that results show that Visual Basic for Applications is the most adequate language that can reduce the programming time and is compatible within the available limited resources.

Furthermore, time elapsed in solving a typical maintenance study is reduced to less than fifty per cent of the time required in the classical method. In spite of this advantage, a self-executable file is lacking, which must be developed as a future improvement of the methodology.

7. Conclusions

Statistical Process Control is the applied science that helps to collect, organise and interpret the wide variety of information available for engineering applications. This manuscript describes a study about development of the optimal methodology to develop a quality-control analysis of attributes.

As shown in this report, VBA is an interesting tool to design new engineering software resources and to subsequently implement the learning methods. This new methodology for each engineering application lets students to be self-adapted to each real condition at work. Furthermore, once the software is developed, it lets students solve a high number of control-chart problems within a few minutes and spend most of the time in understanding the results through the charts study.

Nevertheless, in spite of the fact that VBA has clear advantages with reference to other typical programming languages, it has not been exploited in depth during the preceding years by professors and engineers. Consequently, this language does not have its own packing tools that generate, install and uninstall files and icons.

For the future implementation in newer MS Office versions, it will be interesting to implement these tools and evolve a definitive teaching methodology that involves less time. This methodology must be implemented along with more information, such as books and student guides (El-Khouly et al., 2000).

Finally, future research studies must be completed for quantification of the real learning-time implementation obtained with this new methodology and its advantages in accordance with the Bologna declaration.

8. Acknowledgements

We record our appreciation for the sponsorship extended by the University of A Coruña to this project under its IV Contract Programme (2008).

9. References

Cebi, S.; Celik, M.; Kahraman, C.; and Deha, I. (2009). An expert system towards solving ship auxiliary machinery troubleshooting: SHIPAMTsolver. *Expert Systems with Applications*. 36. pp.7219-7227.

Deris, S.; Omatu, S.;Ohta, H.; Shaharudin Kutar, Lt Cdr; Samat, P. A. (1999). Ship maintenance scheduling by genetic algorithm and constraint-based reasoning. *European Journal of Operational Research*. 112. pp.489-502.

Discovering Macros of Visual Basic for Applications (VBA) for Microsof Excel. Introduction to VBA. http://www.excel-vba.com/excel-vba-introduction.htm. Accessed June 2010.

El-Khouly, M. M.; Far, B. H.; Koono, Z. (2000). Expert tutoring system for teaching computer-programming languages. *Expert Systems with Applications*, Vol. 18, No. 1, pp. 27-32. ISSN:0957-4174.

Harris, T. J., Seppala, C. T. and Desborough, L.D. (1999). A review of performance monitoring and assessment techniques for univariate and multivariate control systems. *Journal of Process Control*. 9. pp. 1-17. ISNN: 0959-1524.

Jelen, B.; Syrstad, T. (2008). Visual Basic for Applicatiosn (VBA) and macros for Microsoft Office Excel 2007 (business solutions), Second edition, Indianapolis, pp. 42-52. ISBN: 0789736829.

Jennings, A.D. and Drake, P.R. (1997). Machine tool condition monitoring using statistical quality control charts. *International Journal of Mach. Tools Manufact*. Vol. 37, No. 9 pp.1243-1249. ISSN:0890-6955.

Kumar Sharma, R., Kumar, D. and Kumar, P. (2008). Predicting uncertain behaviour of industrial system using FM- A practical case. *Applied Soft Computing*. 8. pp.96-109. ISSN:1568:4946

Lee, S. Lee, J. K. Park, B. Lee, D. Kim, S. Lee, K. (2006). Development of internet-based ship technical information management system. *Ocean Engineering*. 33 1814-1828. ISSN:0029:8018.

Louit, D. M., Pascual, R. and Jardine, A.K.S. (2009). A practical procedure for the selection of time to failure models based on the assessment of trends in maintenance data. *Reliability Engineering & System Safety*. doi:10.1016/j.ress.2009.04.001. ISSN: 0951-8320.

Microsoft P-Code technology. http://msdn.microsoft.com/es-es/default.aspx. Accessed June 2010.

Microsoft Help. ACC: Visual/Access Basic Is Both a Compiler and an Interpreter. Accessed June 2010.

Orosa, J.A. (2009). Programming languages for Marine Engineers. *Computer Applications In Engineering Education*. DOI:10.1002/cae.20339.ISSN: 1061-3773.

Quality Control Charts. http://www.qualityamerica.com/knowledgecente/knowctrWhen_to_Use_an_Attribute_Chart.htm. Accessed June 2010.

Roman, S. (2002). Writing Excel Macros with VBA. Sebastopol CA: O'Reilly, Chapter 15. ISBN: 0596003595.

Statgraphics Centurion Web page. http://www.statgraphics.com/control_charts.htm#variables. Accessed April 2009.

Verhoefa, A.; Degrèvea, J.; Huybrechsa, B.; Van Veenb, H.; Pexb, P. and Van der Bruggena, B. (2008). Simulation of a hybrid pervaporation–distillation process. *Computers & Chemical Engineering*, Vol. 32, No. 6, Pages 1135-1146. doi:10.1016/j.compchemeng.2007.04.014. ISSN: 0098-1354.

Verma, M. P. (2003). Steam tables for pure water as an ActiveX component in Visual Basic 6.0. Geotermia, Instituto de Investigaciones Electricas. *Computers & Geosciences,* Vol. 29, No 9, pp. 1155-1163. DOI:10.1016/S0098-3004(03)00136-5. ISSN: 0098-3004.

Visual Basic for Applications Frequently Asked Questions. http://msdn.microsoft.com/en-us/isv/bb190540.aspx. Accessed February 2009.

Vosniakos, G & Wang, J. (1997). A software system framework for planning and operation of quality control in discrete part manufacturing. *Computer integrated Manufacturing Systems,* Vol. 10, No 1. pp.9-25. ISSN:0951-5240.

Wu, C.; Liua, Y. (2008). Development and evaluation of an ergonomic software package for predicting multiple-task human performance and mental workload in human-machine interface design and evaluation. *Computers & Industrial Engineering,* doi:10.1016/j.cie.2008.06.013. ISSN: 0360-8352.

22

Development of VBA Based Ship Technical Corrective Management System for Marine Engineers

José A. Orosa, Angel M. Costa and Rafael Santos
University of A Coruña
Spain

1. Introduction

Recent research works aim at developing an advanced tool, based on expert system principles to solve ship auxiliary machinery troubleshooting (Cebi et al., 2009). In a broad sense, expert systems are computer applications, which embody some/no algorithmic expertise to solve certain types of problems.

An expert system, also known as knowledge based system, is a computer program that contains the knowledge and analytical skills of one or more human experts, related to a specific subject. In this sense, some advantages common to most computer systems could include the following (García, 2003):

- Control over maintenance activity.
- Control over spending.
- Facility or consulting history.
- Easy to obtain ratios and indicators.

Some disadvantages would be:

- High initial investment in equipment and programs and manpower for the implementation.
- Bureaucratic system.
- Increased staff dedicated to indirectly unproductive tasks.
- The information provided is often not sufficiently reliable.

An example of expert system is the SHIPAMTsolver (Cebi et al., 2009).

This system was developed based on PRO-LOG language. It was developed utilizing a considerable amount of information and collected data on the different types of ships machinery systems, system failures, system indicators, and troubleshooting methods through textbooks, handbooks, and interviews with chief engineers who are the marine experts on ship marine system.

Nowadays, maintenance activity in ships is proposed to be developed with GMAO (computer maintenance management system) software (Louit et al., 2009; harris et al., 1999; and Vosniakos and Wang, 1999). To analyze the obtained data with this software, it is evident that for an adequate resolution of each problem, an adequate model selection to determine the time taken for a particular component or system to reach failure is essential, which can in turn lead to wrong conclusions and decisions. The gap between researchers and practitioners of maintenance has resulted in the fact that although many models rely on very specific assumptions for their proper application, these are not normally discriminated by the practitioner according to the real operating conditions of their plants or fleets. To solve this problem, this data is sent to land for analysis by experts (Macián et al., 1999; MS Excel, 2009).

Thus, earlier research works on Internet-based ship technical information management system was developed (Lee at al., 2006). In that case study, implementation of an IDE (Integrated Data Environment) that represents the CALS (Continuous data acquisition and life-cycle support) was proposed. It was an environment where each participant could access data during the entire lifecycle of a product, without limitations imposed by the geographical location, heterogeneity in hardware, software and platform.

Vast amount of documents produced during the entire lifecycle of a ship were systematically and integrated managed. This method has the advantage of being able to continuously maintain and improve the shipbuilding information system, and consequently increase the productivity and further advancements in shipbuilding technology. Further, this technology shows the advantage of implementing the integration and optimization of design, manufacturing and management system to collaborate between expert groups. Particularly, it can integrate the islands of automation prevalent among the departments and related companies.

Analysis of real situations in ships reveals whether the maintenance software is a private software or one developed by shipping companies, on their own. In spite of this, most marine engineers tend to implement its work activities in Microsoft Office, due to this private software does not let them adapt the software to suit ship characteristics and equipment in a user friendly manner.

The problem arises when a new crew boards the ship and finds a different Microsoft Office maintenance data structure from that in other ships. Consequently, they need to invest too much time to understand the data processing structure. This problem also becomes significant when this data needs to be transferred from Microsoft Office to its GMAO software for analysis on land, at the end of each day, week or month.

However, it is expected that from the corrective maintenance analysis of data on land, wrong conclusions can be drawn as most of the data must be understood on ship and, necessarily, this private software must be improved.

To solve this problem, the Department of Energy and Marine Propulsion of University of A Coruña launched an action program to research for a methodology based on MS Excel and VBA, to structure this maintenance data and form the base for future friendly GMAO applications.

In the present chapter, the first prototype of this complementary application was developed in collaboration with experienced marine engineers who improved and tested these applications during the working periods for a year, under real ship conditions.

2. Objectives

The objective of this chapter is to develop a suitable method of adaptive maintenance computer application that could be modified to suit the maintenance conditions for each ship, by the marine engineers. Once the application was developed, it had to be tested under real ship navigation conditions to test the new improvements.

3. Materials and methods

In this chapter, the first prototype of a complementary corrective maintenance application is proposed to be developed in collaboration with experienced marine engineers.

3.1 Marine engineers

Marine engineers must be the sole personnel to utilize this software and introduce the data. Thus, a more personal and bureaucratic system will not be needed, and the information will be reliable. Utilizing this information, these marine engineers will develop its statistical studies of control charts and Weibull analysis on board, to reduce its maintenance cost.

3.2 Maintenance

Maintenance is defined as a combination of all technical and administrative actions, including supervision actions intended to retain an item in, or restore it to, a state in which it can perform a required function.
It is a set of organised activities that is performed to maintain an item in its best operational condition, with minimum cost. Activities of maintenance function could include either repair or replacement activities, which are necessary for an item to reach its acceptable productivity condition; else these activities, should be carried out at minimum possible cost. Five types of maintenance are listed in Table 1.

Run to Failure Maintenance (RTF)	The required repair, replacement or restore action performed on a machine or a facility after the occurrence of a failure to bring this machine or facility to at least its minimum acceptable condition. It is the oldest type of maintenance.
Preventive Maintenance (PM)	It is a set of activities performed on plant equipment, machinery, and systems before the occurrence of a failure to protect them and to prevent or eliminate any degradation in their operating conditions.
Corrective Maintenance (CM)	In this type, actions such as repair, replacement, or restore will be performed after the occurrence of a failure to eliminate the source of this failure or reduce the frequency of its occurrence.
Improvement Maintenance (IM)	It aims at reducing or eliminating entirely the need for maintenance.
Predictive Maintenance (PDM).	It is a set of activities that detects changes in the physical condition of equipment (signs of failure) to conduct the appropriate maintenance work to maximize the service life of the equipment without increasing the risk of failure.

Table 1. Types of maintenance

3.3 Main engine combustion system

In our case study, the principal systems to be analyzed for corrective maintenance are the main engines and auxiliary engines of a merchant ship. Therefore, these services are described. On analysis of the services to operate diesel engines, there may be differences in characteristics and extent, depending mainly on the size and speed of the marine engines. The main services are cooling, lubrication, fuel, start and reverse motion, and remote control and automation (rowen, 1996; Taylor, 1996). Figure 1 shows that the fuel system is more complicated when heavy fuel is used. Due to the high viscosity of this fuel, it does not always flow by gravity, requiring heated tanks to reduce the viscosity, especially in cold waters. Although sedimentation tanks are also installed, centrifugal or similar purification systems are necessary to extract water and other waste. As sedimentation tanks are present, the fuel is pumped into the tanks via the service purification equipment. The big slow two-stroke engines and four times of high power, are suitable to support heavy fuels, and are significantly cheaper than light, though they necessitate more complex equipment (heating tank preheaters, lined pipes, etc..) plus a maintenance plan for the engine because of the high waste combustion. Therefore, the importance of this maintenance plan is clearly related to an adequate management system.

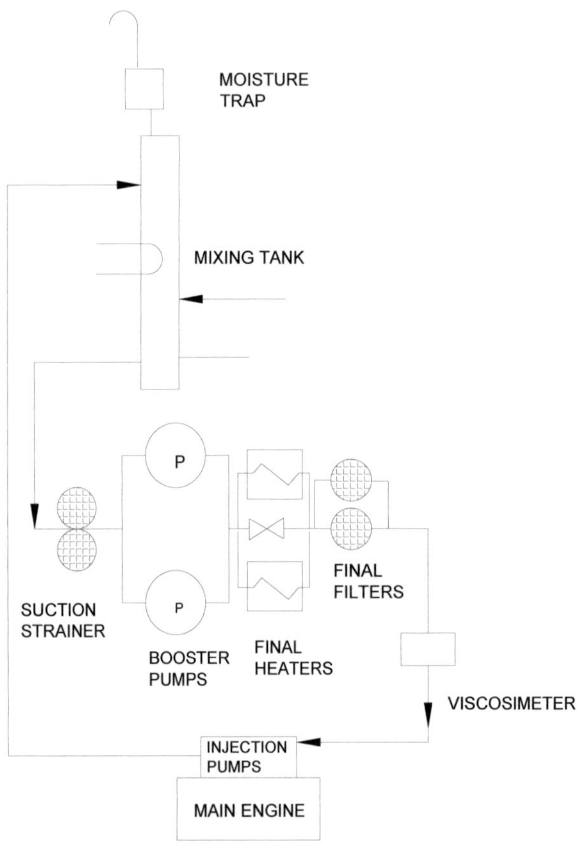

Fig. 1. Fuel treatment and service system

In Figure 2 the main engine lubricating-oil circulating system is shown. Oil, draining from the bearings and cooling passages at the bottom of the crankcase, passes into an independent sump that is built into the double bottom below the engine, from where it is drawn by the lubricating-oil circulating pump, for redistribution via a filter and a cooler. A full flow filter is provided in the pump discharge line. It could be of the duplex, basket type, but better filtration is usually afforded by a disposable-element or self-cleaning simplex unit, with a standby filter in a bypass. Filtered oil is distributed to the engine bearings, for governing and control service to the valve gear and for piston cooling.

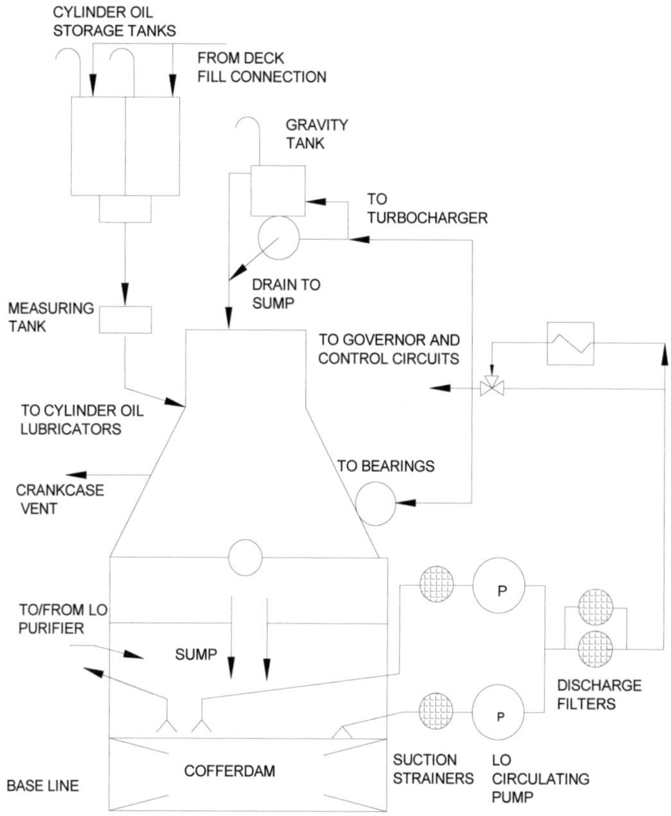

Fig. 2. Main engine lubricating-oil circulating system

3.4 Visual Basic for Applications

Due to the vast amount of documents produced during the whole life of a ship, it is proposed, based on marine engineers' experience, to structure this data in Microsoft Excel spreadsheets. The principal problem encountered when such spreadsheets are used, is to obtain a clear structure and an adequate and easy access to a comprehensive spreadsheet. To solve this problem the VBA (Visual Basic for Applications) and hyperlinks in Microsoft Excel were proposed to access an adequate structure and spreadsheets.

As explained earlier, Microsoft presents the VBA, whose code is compiled (Microsoft, 2009) in an intermediate language called P-code (Microsoft P-Code, 2009); the latter code is stored by the hosting applications (Access, Excel, Word) as a separate stream in structured storage files (eg., .doc or .xls), independent from the document streams. The intermediate code is then executed by a virtual machine (hosted by the hosting application). Despite its resemblance to many old Basic dialects, the VBA is not compatible with any of them, except Microsoft Visual Basic, where the source code of VBA Modules can be directly imported, and which shares the same library and virtual machine (Mathworks, 2009).

Another advantage of this methodology is that, with such files, most computers will not require a high memory resource, and the processor and information will be accessible for expert economists, that can help to solve the problems with the adequate control method. Further, this software can be adapted to ship equipment in a user friendly manner.

4. Results

4.1 Maintenance activities

The first step in developing our application was to study the principal maintenance activities in the main and auxiliary engines. These activities were obtained from collaboration with marine engineers, during its working periods. Results are summarized in Tables 2 and 3.

Main and auxiliary engines	
Element	**Task**
Compressor 1	Clean with water
Centrifugal filter	Clean
Turbine	Water blasting
Oil	Analysis
Combustion	Pressure
Relief valve	Check operation
Bypass valve	Check operation
Leaking fuel	Check
Air leaks	Check
Turbocharger	Cleaning air filters
Automatic alarm stops	Check operation
Rotocaps	Check operation
Air cooler	Check pressure drop
Regulator	Oil change
Mechanical over speed	Check operation
Injector	Check operation
Piston rings and cylinder	Check operation
1 cylinder injection pump	Check operation
Shirt	Check operation
Piston cylinder 1	Check the cooling gallery
Air compressor scan	Replace
Main bearing	Inspect
General review	

Table 2. Tasks in the main and auxiliary engines

Fuel centrifuge	Oil centrifuge	Compressors
Sewage pump fuel	Purifier oil pump	Oil Change
Oil Change	Oil Change	Air filter
Intermediate Maintenance	Intermediate Maintenance	Non-return valve compressor
Increased maintenance	Increased maintenance	High pressure switch valve
		Low pressure switch
		Compressor safety valves
		Refrigeration compressor Gallery

Table 3. Task in fuel treatment and lubricating-oil circulating systems

4.2 MS Excel file

The Department of Energy and Marine Propulsion, in collaboration with marine engineers, developed the structure of the prototype application showed in Figures 3 to 6. In Figure 3 it is evident that most marine engineers prefer to structure the data as preventive and in corrective data. All the existing spreadsheets were hidden. Access to each of them was done in VBA, and with hyperlinks to each main screen.

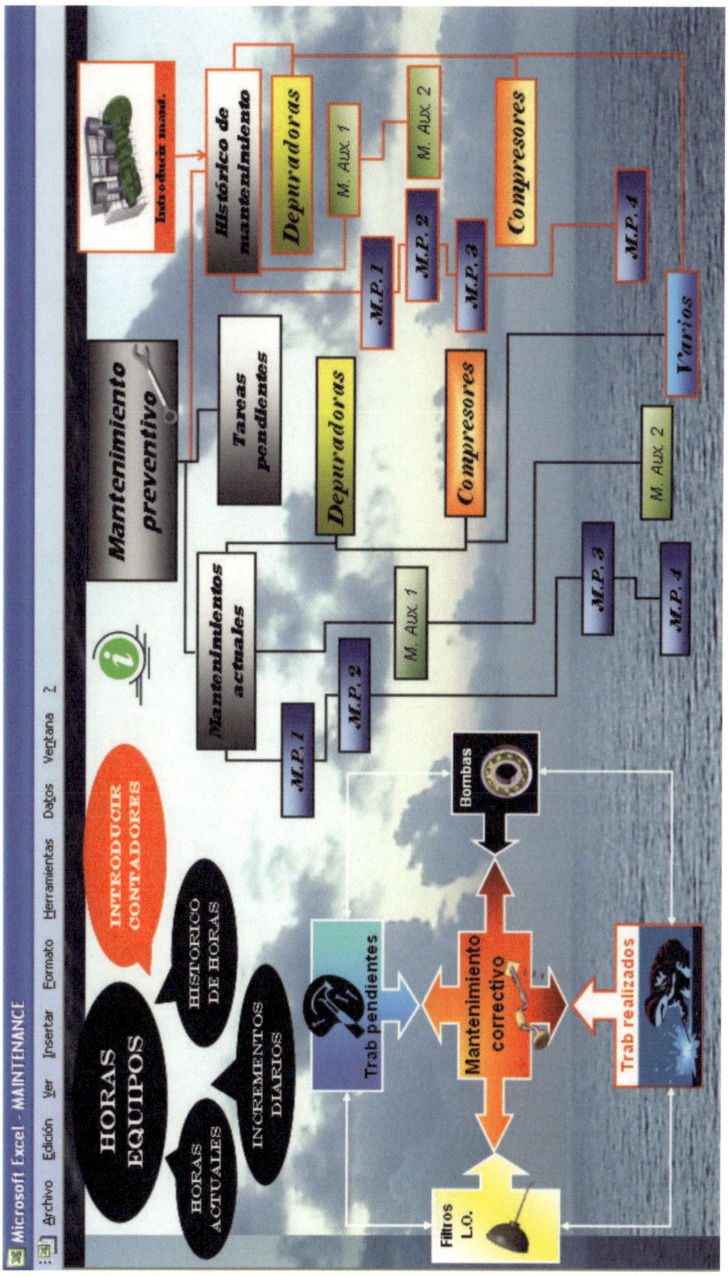

Fig. 3. Maintenance application structure

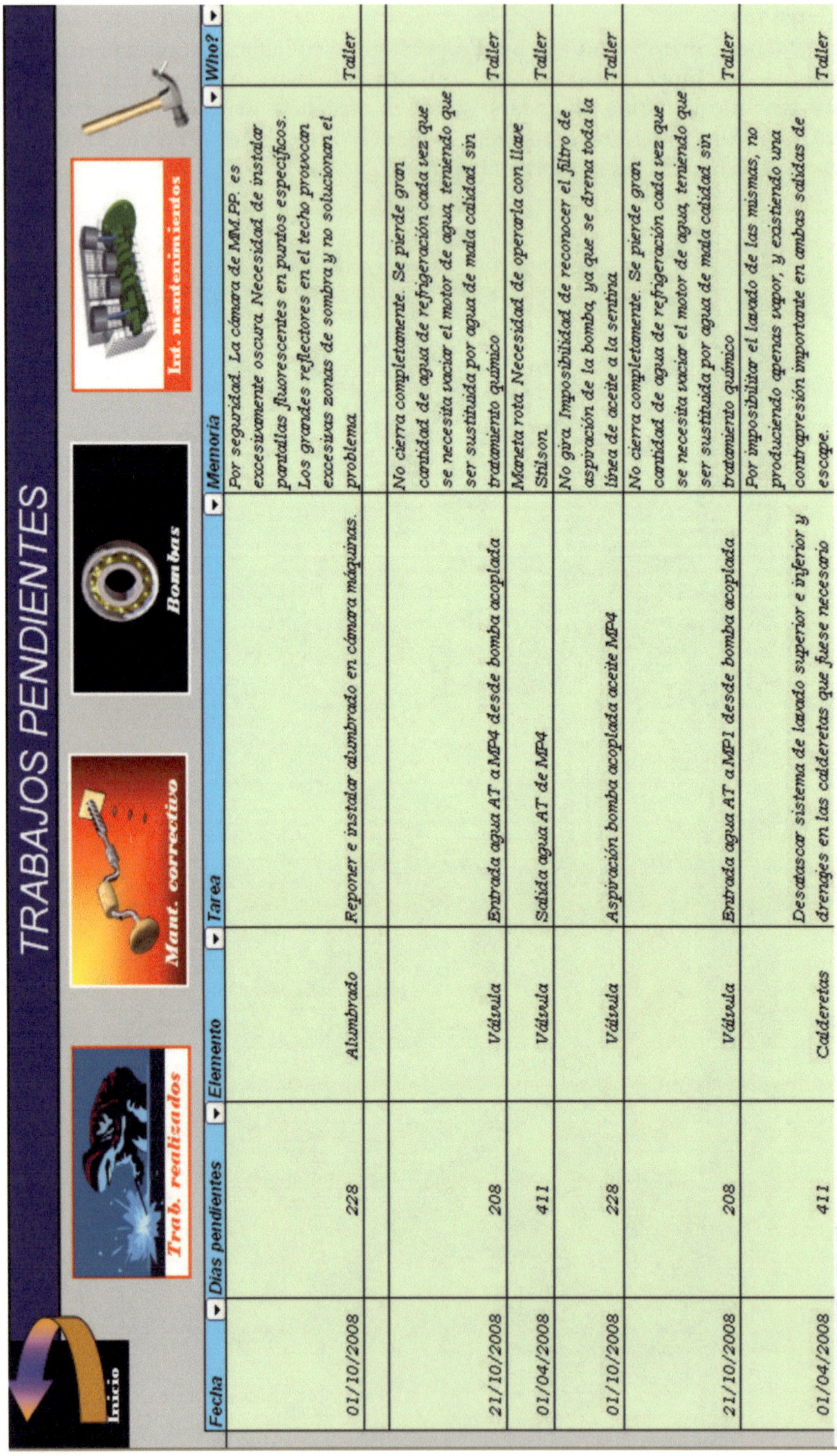

Fecha	Dias pendientes	Elemento	Tarea	Memoria	Who?
01/10/2008	228	Alumbrado	Reponer e instalar alumbrado en cámara máquinas.	Por seguridad. La cámara de MM.PP. es excesivamente oscura. Necesidad de instalar pantallas fluorescentes en puntos específicos. Los grandes reflectores en el techo provocan excesivas zonas de sombra y no solucionan el problema.	Taller
21/10/2008	208	Válvula	Entrada agua AT a MP4 desde bomba acoplada	No cierra completamente. Se pierde gran cantidad de agua de refrigeración cada vez que se necesita vaciar el motor de agua, teniendo que ser sustituida por agua de mala calidad sin tratamiento químico	Taller
01/04/2008	411	Válvula	Salida agua AT de MP4	Maneta rota. Necesidad de operarla con llave Stilson.	Taller
01/10/2008	228	Válvula	Aspiración bomba acoplada aceite MP4	No gira. Imposibilidad de reconocer el filtro de aspiración de la bomba, ya que se drena toda la línea de aceite a la sentina.	Taller
21/10/2008	208	Válvula	Entrada agua AT a MP1 desde bomba acoplada	No cierra completamente. Se pierde gran cantidad de agua de refrigeración cada vez que se necesita vaciar el motor de agua, teniendo que ser sustituida por agua de mala calidad sin tratamiento químico	Taller
01/04/2008	411	Calderetas	Desatascar sistema de lavado superior e inferior y drenajes en las calderetas que fuese necesario	Por imposibilitar el lavado de las mismas, no produciendo apenas vapor, y existiendo una contrapresión importante en ambas salidas de escape.	Taller

Fig. 4. Pending task screen

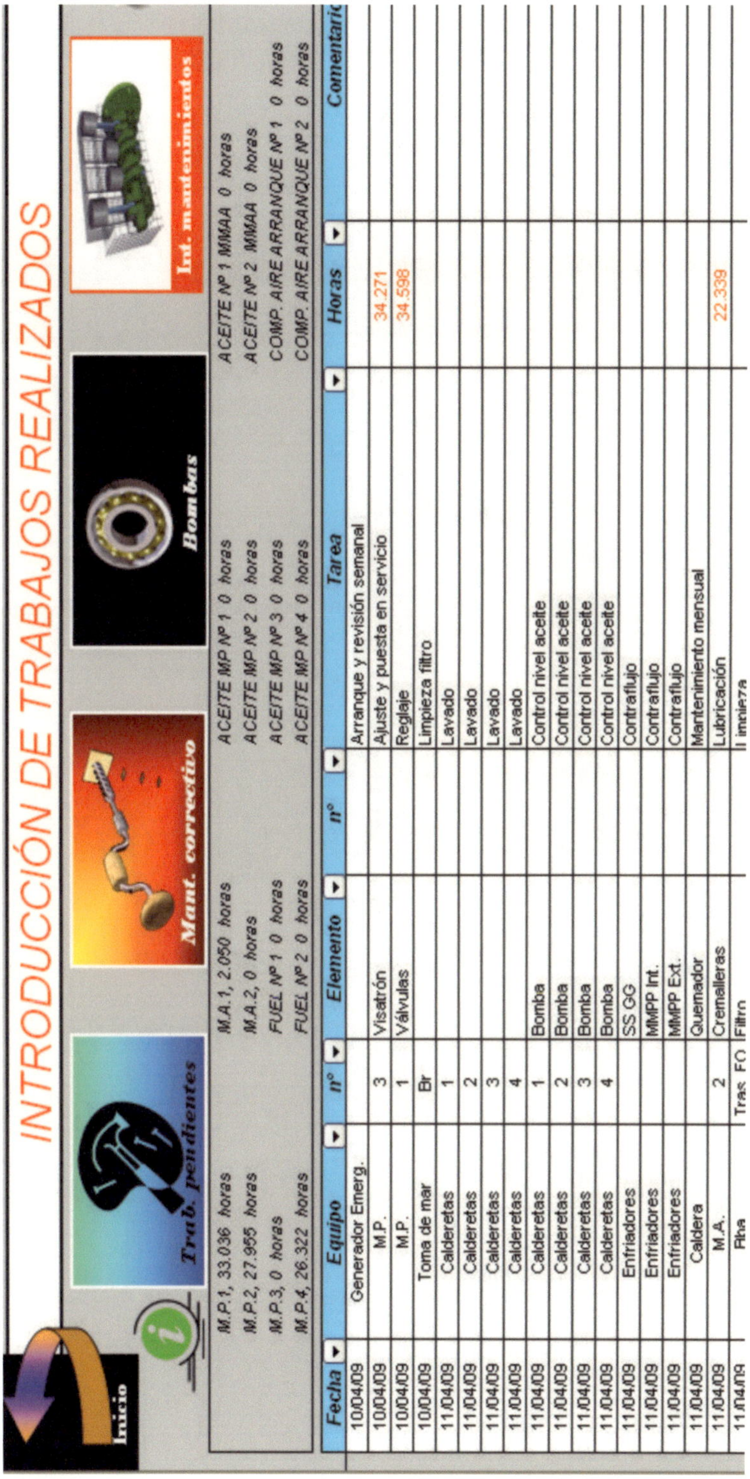

Fig. 5. Task finished screen

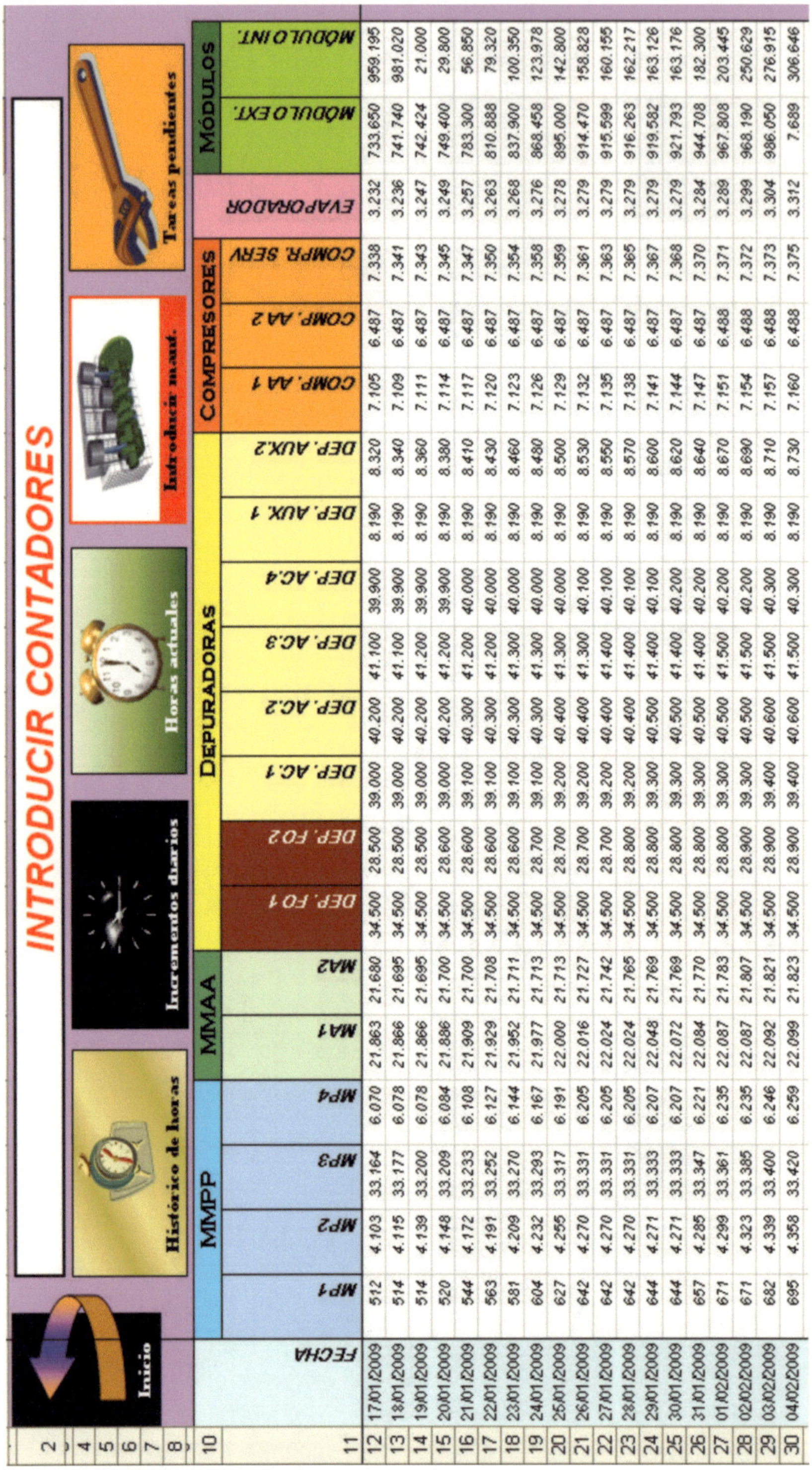

		MMPP				MMAA		DEPURADORAS								COMPRESORES				MÓDULOS	
11	FECHA	MP1	MP2	MP3	MP4	MA1	MA2	DEP. FO 1	DEP. FO 2	DEP. AC.1	DEP. AC.2	DEP. AC.3	DEP. AC.4	DEP. AUX. 1	DEP. AUX.2	COMP. AA 1	COMP. AA 2	COMPR. SERV	EVAPORADOR	MÓDULO EXT.	MÓDULO INT.
12	17/01/2009	512	4.103	33.164	6.070	21.863	21.680	34.500	28.500	39.000	40.200	41.100	39.900	8.190	8.320	7.105	6.487	7.338	3.232	733.650	959.195
13	18/01/2009	514	4.115	33.177	6.078	21.866	21.695	34.500	28.500	39.000	40.200	41.100	39.900	8.190	8.340	7.109	6.487	7.341	3.236	741.740	981.020
14	19/01/2009	514	4.139	33.200	6.078	21.866	21.695	34.500	28.500	39.000	40.200	41.200	39.900	8.190	8.360	7.111	6.487	7.343	3.247	742.424	21.000
15	20/01/2009	520	4.148	33.209	6.084	21.886	21.700	34.500	28.600	39.000	40.200	41.200	39.900	8.190	8.380	7.114	6.487	7.345	3.249	749.400	29.800
16	21/01/2009	544	4.172	33.233	6.108	21.909	21.700	34.500	28.600	39.100	40.300	41.200	40.000	8.190	8.410	7.117	6.487	7.347	3.257	783.300	56.850
17	22/01/2009	563	4.191	33.252	6.127	21.929	21.708	34.500	28.600	39.100	40.300	41.200	40.000	8.190	8.430	7.120	6.487	7.350	3.263	810.888	79.320
18	23/01/2009	581	4.209	33.270	6.144	21.952	21.711	34.500	28.600	39.100	40.300	41.300	40.000	8.190	8.460	7.123	6.487	7.354	3.268	837.900	100.350
19	24/01/2009	604	4.232	33.293	6.167	21.977	21.713	34.500	28.700	39.100	40.300	41.300	40.000	8.190	8.480	7.126	6.487	7.358	3.276	868.458	123.978
20	25/01/2009	627	4.255	33.317	6.191	22.000	21.713	34.500	28.700	39.200	40.400	41.300	40.000	8.190	8.500	7.129	6.487	7.359	3.278	895.000	142.800
21	26/01/2009	642	4.270	33.331	6.205	22.016	21.727	34.500	28.700	39.200	40.400	41.300	40.100	8.190	8.530	7.132	6.487	7.361	3.279	914.470	158.828
22	27/01/2009	642	4.270	33.331	6.205	22.024	21.742	34.500	28.700	39.200	40.400	41.400	40.100	8.190	8.550	7.135	6.487	7.363	3.279	915.599	160.155
23	28/01/2009	642	4.270	33.331	6.205	22.024	21.765	34.500	28.800	39.200	40.400	41.400	40.100	8.190	8.570	7.138	6.487	7.365	3.279	916.263	162.217
24	29/01/2009	644	4.271	33.333	6.207	22.048	21.769	34.500	28.800	39.300	40.500	41.400	40.100	8.190	8.600	7.141	6.487	7.367	3.279	919.582	163.126
25	30/01/2009	644	4.271	33.333	6.207	22.072	21.769	34.500	28.800	39.300	40.500	41.400	40.200	8.190	8.620	7.144	6.487	7.368	3.279	921.793	163.176
26	31/01/2009	657	4.285	33.347	6.221	22.084	21.770	34.500	28.800	39.300	40.500	41.400	40.200	8.190	8.640	7.147	6.487	7.370	3.284	944.708	182.300
27	01/02/2009	671	4.299	33.361	6.235	22.087	21.783	34.500	28.800	39.300	40.500	41.500	40.200	8.190	8.670	7.151	6.488	7.371	3.289	967.808	203.445
28	02/02/2009	671	4.323	33.385	6.235	22.087	21.807	34.500	28.900	39.300	40.500	41.500	40.200	8.190	8.690	7.154	6.488	7.372	3.299	968.190	250.629
29	03/02/2009	682	4.339	33.400	6.246	22.092	21.821	34.500	28.900	39.400	40.600	41.500	40.300	8.190	8.710	7.157	6.488	7.373	3.304	986.050	276.915
30	04/02/2009	695	4.358	33.420	6.259	22.099	21.823	34.500	28.900	39.400	40.600	41.500	40.300	8.190	8.730	7.160	6.488	7.375	3.312	7.689	306.646

Fig. 6. Screen of working hours

4.3 Time to develop the application

Time utilized to develop the application was much less than the time taken to configure most of GMAO applications, because most of the data is similar, in the same kind of service in different ships. It took about 35 hours to develop the basic application, according to calculations.

Another advantage was that it was not necessary to undertake a prior course of study on how to employ the spreadsheet. However, if necessary, a tutorial was added, as shown in Figure 7. Finally, it becomes very easy for engineers on board to develop a graphical representation of data, and to do statistical data analyses that will help solve the different operational problems.

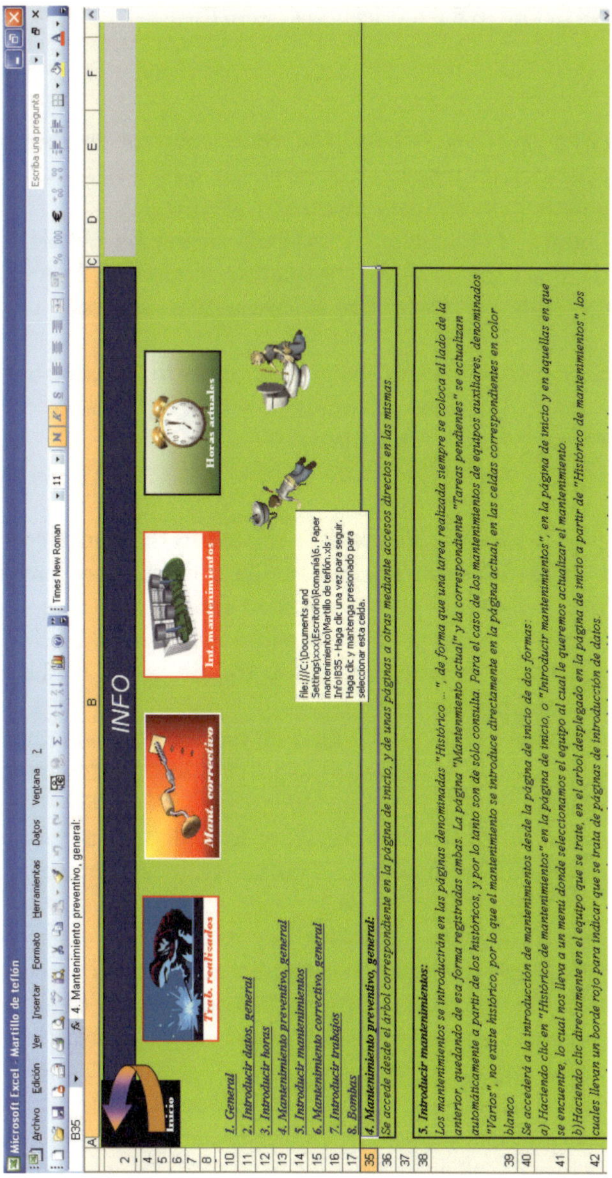

Fig. 7. Information screen

4.4 Application test on ship

A practical case study was done in a merchant ship during its sea lane of activity for a year. During this period, two expert marine engineers trained in this application employed it. As a result of this application, parameters like fuel oil filter mean life were implemented because of a greater importance of time between filter changes and centrifuge operation.

Particularly, in its application, during working periods, an excessive waste of filter was detected, that was caused by the lack of maintenance of the earlier automatic systems.

In other cases, an adequate stock allowed the same replacements to be located in different equipment, to solve a failure in the lubrication system (bearings).

5. Discussion

As obtained in earlier research works, the complexity of marine systems and time constraints on the operational processes require effective corrective actions to be performed, to manage and troubleshoot in auxiliary machinery aboard ships.

Due to the vast amount of documents produced during the whole life of a ship, it is proposed, based on the experience of the marine engineers, to structure this data in Microsoft Excel spreadsheets. Once the software was selected, the structure of the information was defined.

In Figure 3 it is evident that most marine engineers prefer to structure the data as preventive and corrective data. All the designed spreadsheets were hidden. Access to each of them was done in VBA and with hyperlinks t each main screen.

In corrective data, the filters mean life, pending task, and finished task are stored, as shown in Figures 4 and 5.

The preventive section, as seen in Figure 6, is related to the number of working hours of each equipment such as the main engine, auxiliary engine, compressors and purifiers, besides others. Related cells easily permit a way to consider the time remaining to begin a maintenance task. Further, it was possible to add a very simple alarm that warns anyone who is near a maintenance process. Therefore, the proposed methodology ensures the required supervision by shipboard personnel.

However, the time elapsed to design this kind of application was much less than the time taken to configure a typical GMAO software as most of the data is similar, used in the same kind of service in different ships.

Another advantage was that it was not necessary to undertake a prior course of study on how to employ the spreadsheet; however, if necessary, a tutorial was added, as shown in Figure 8.

Finally, it is very easy to develop a graphical representation of data, and to do statistical data analyses that will help solve different operational problems on board, and export this to another application like Microsoft Word. Further, due to the small size of the obtained files, they can be accessed anywhere in the world by the Internet, to allow maximum availability of the information.

Finally, on testing this methodology in real navigation, an excessive waste of filter was detected due to the lack of maintenance of the earlier automatic systems. In other cases, an adequate stock inventory file allowed the location of replacements in different equipment to solve a failure in the lubrication system.

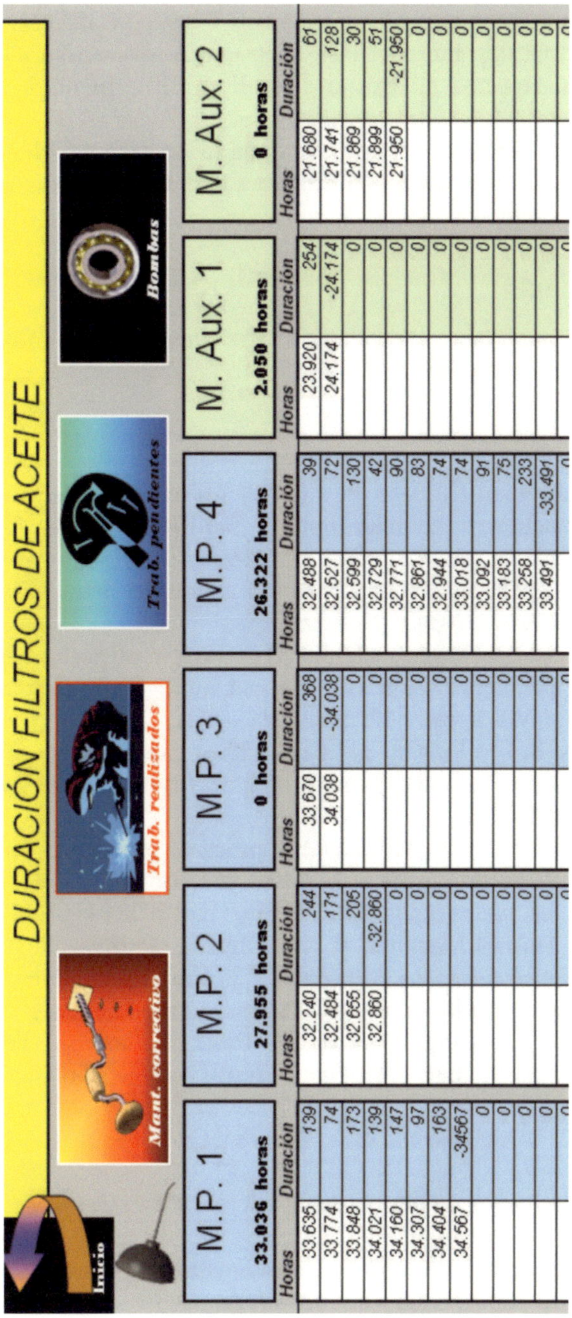

Fig. 8. Corrective maintenance of filters in the main and auxiliary engines screen

6. Conclusions

As evident, this practical case study, based on the marine engineers' maintenance process enabled an improvement in the methodology.

Particularly, the methodology based on Microsoft Excel spreadsheets and VBA permitted easy implementation of ship maintenance. For example, corrective maintenance problems such as, detecting unexpected failures and locating replacements in the storeroom were observed.
Finally, research work in the future must be done to develop a GMAO that would reduce the time elapsed to develop such applications, and to adapt the software to suit the characteristics of each ship.

7. Acknowledgements

We would like to express our appreciation to the University of A Coruña for its sponsor support in its IV contract programme (2008).

8. References

Cebi S, Celik M, Kahraman C, Deha Er I (2009) An expert system towards solving ship auxiliary machinery troubleshooting: SHIPAMT solver. *Expert systems with Applications*. No. 36, pp. 7219-7227. ISSN:0957-4174.

García Garrido S (2003) Organización y gestión integral del mantenimiento. Editorial Díaz de Santos. ISBN:8479785489. España.

Harris TJ, Seppala CT, Desborough LD (1999) A review of performance monitoring and assessment techniques for univariate and multivariate control systems. *Journal of Process control*. Vol. 9, pp. 1-17. ISNN: 0959-1524.

Lee S, Lee J. K, Park B, Lee D, Kim S, Lee K. (2006) Development of internet-based ship technical information management system. *Ocean Engineering*. No. 33, pp. 1814-1828. ISSN:0029:8018.

Louit DM, Pascual R, Jardine A. K.S (2009) A practical procedure for selection of time to failure models based on the assessment of trends in maintenance data. *Reliability Engineering and System Safety*. Doi:10.1016/j.ress.2009.04.001. ISSN: 0951-8320.

Macián Martínez V. Tormos Martínez, B. and Olmeda González P. (1999) *Fundamentos de ingeniería del mantenimiento*. Editorial U.P.V. Universidad Politécnica de Valencia.

Mathworks. http://www.mathworks.com/, Accessed February 2009.

Microsoft Excel. Using MS Excel for Weibull analysis. http://www.qualitydigest.com/jan99/html/body_weibull.html. Accessed June 2009.

Microsoft Help. ACC: Visual/Access Basic Is Both a Compiler and an Interpreter. Accessed June 2009.

Microsoft P-Code technology. http://msdn.microsoft.com/es-es/default.aspx. Accessed June 2009.

Rowen, A. L. (1992). Marine engineering. *Diesel Engines*. Published by the society of Naval Architects and Marine Engineers. New Jork.

Taylor, D A. (1996) *Introduction to Marine Engineering*. Published by Butterworth-Heinemann.. Elsevier Ltd. ISBN: 0 7506 2530 9. Great Britain.

Vosniakos G. and Wang J. (1999) A software system framework for planning and operation of quality control in discrete part manufacturing. *Computer Integrated Manufacturing Systems*. Vol. 10, No. 1. pp. 9-25. ISSN:0951-5240.

23

Quality Assessments of Seafloor Mapping with Multibeam Echo Sounders

Jan Terje Bjørke[1,2] and Stein Nilsen[1]
[1]*Norwegian Defence Research Establishment,PO Box 115, N-3191 Horten,*
[2]*Norwegian University of Life Sciences, N-1432 Ås*
Norway

1. Introduction

Seafloor topography is often mapped by the use of multibeam echo sounders, see Figure 1 for illustration of the principle on measurement with multibeam echo sounders. These systems evolved from single beam systems twenty five years ago (Glittum et al., 1986) and are still in development in the direction of increased measurement precision, higher density of the sample points and higher number of measured data points per second. To ensure high quality of the measurements, the survey system must be regularly verified and calibrated, statistics about measurement errors derived, measurement quality monitored, spot checks of the surveyed area performed and identification of spikes, suspicious or unexpected observations carried out. Measurement errors can be divided into three classes: Systematic errors, random errors and outliers. Methodologies to perform the quality assessments considered, will be described and illustrated in some case studies. The final product in seafloor mapping is often a digital elevation model (DEM). From a DEM different cartographic products as contour line maps, perspective views, shaded relief maps or maps with coloured depth intervals can be derived. Digital elevation models can also be used in navigation of vessels, ships and boats or for the construction of technical installations in oil and gas winning. The range of applications of DEMs is therefore large. Since the construction of DEMs can be based on different interpolation techniques, an assessment of common methods will be presented.
Quality issues of spatial data are discussed in Guptill and Morrison (1995): *Spatial data quality is a key issue in an era where current electronic technologies and communications networks such as Internet allow easy access to digital data in spatial form.* Elements of spatial data quality are: (1) lineage, (2) positional accuracy, (3) attribute accuracy, (4) completeness, (5) logical consistency and (6) semantic accuracy. In the present chapter we will not cover all these aspects of spatial data quality, but we will illustrate some quality parameters with examples from seafloor mapping. These quality aspects are: (1) systematic and random measurements errors, (2) identification of unexpected observations and (3) accuracy of digital elevation models. The topics we will cover, belong to the following quality elements of Guptill and Morrison (1995): Positional accuracy, attribute accuracy and semantic accuracy.

2. Systematic measurement errors

Systematic errors in the measurement system should be detected and corrected before the survey is carried out. For this purpose we have developed a patented program system, SeaCal,

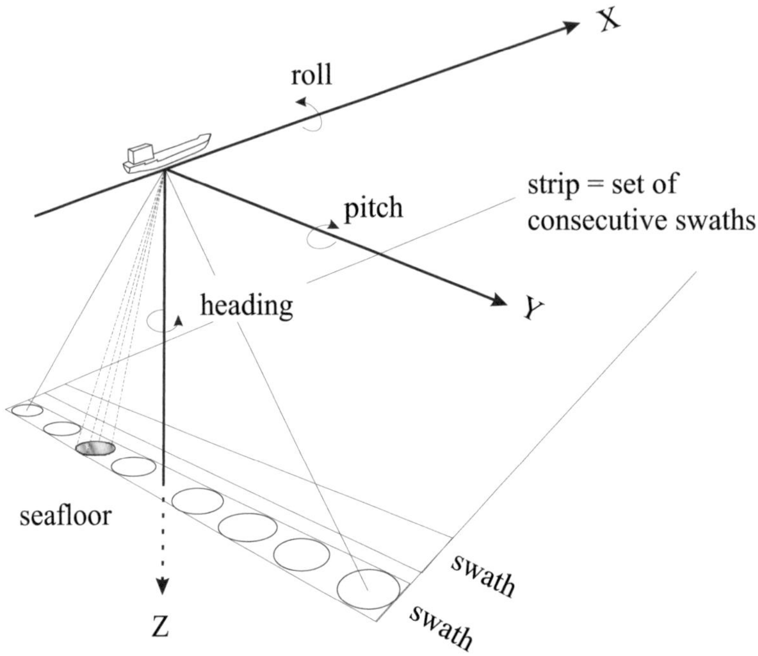

Fig. 1. Principle on measurement and coordinate axes of multibeam echo sounders. A fan of beams measures depth values on the seafloor. The coordinates of the measured points can for simplicity be explained as polar coordinates, i.e., they are computed from information about distance and direction from the centre of the measurement system to the seafloor.

that computes the calibration parameters from field measurements, see Bjørke (2004) and Bjørke (2005).

2.1 SeaCal

Systematic errors can be detected from field measurements by running several overlapping survey lines in different directions, see Figure 2. Static offsets in the three rotation angles—roll, pitch and heading—horizontal shift of the transducer relative to the reference point of the positioning system and systematic errors in the measured sound speed are derived simultaneously from the application of least-squares adjustment (Bjørke, 2005) as:

$$V = (A^T P_1 A)^{-1} A^T P_1 L_1, \tag{1}$$

where V is a vector of the unknown parameters, A a coefficient matrix, P_1 a weight matrix and L_1 a vector of observations. Since the stability of the system of equations depends on the relief of the seafloor and the configuration of the calibration lines, restrictions are put on the unknown parameters by

$$V = (A^T P_1 A + P_2)^{-1} (A^T P_1 L_1 + P_2 L_2), \tag{2}$$

where P_2 is a weight matrix of the unknown parameters and L_2 a vector of *a priori* parameter values; see for example Leick (1995). For example, in a flat area a small (differential) offset in the parameters often have no effect on the measured depth values and several parameters cannot med determined. Therefore, the stability conditions in Equation (2) are convenient to prevent numerical problems in the computations.

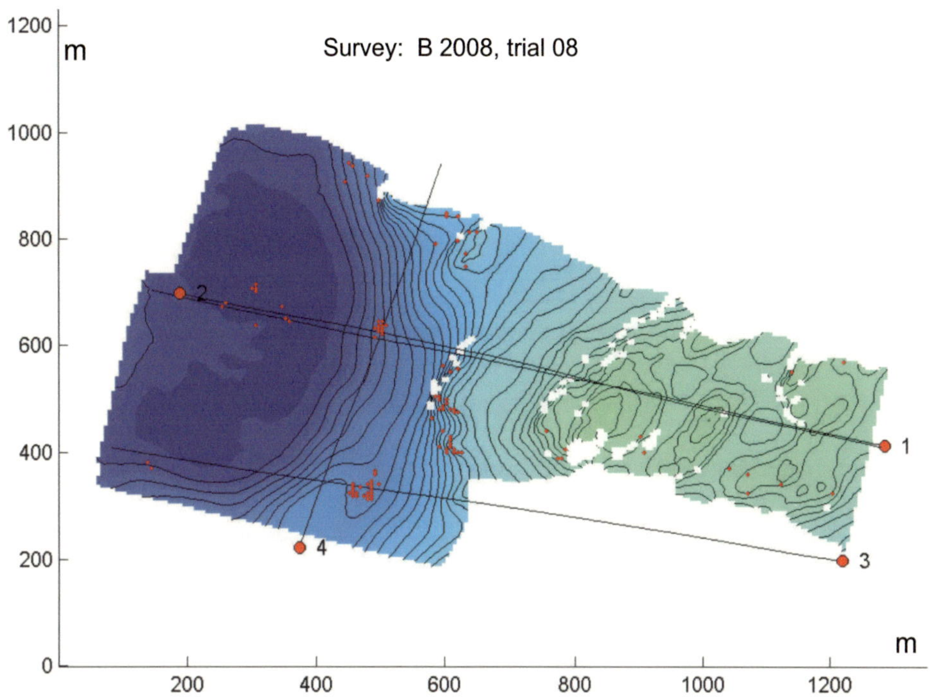

Fig. 2. Field calibration of multibeam echo sounder. The map shows the selected area, named B 2008, trial 08, with four calibration lines. There are three parallel lines and one crossing line. The sea depth varies from 85m to 244m.

The variance-covariance matrix of the unknowns can be computed from

$$Q_V = \sigma_0^2 (A^T P_1 A + P_2)^{-1}, \tag{3}$$

where σ_0^2 is the variance of the unit weight. Here, the diagonal elements define the variance of the parameters and the other elements their covariance. The variance $\sigma_{V_i}^2$ of parameter V_i is derived as

$$\sigma^2{}_{V_i} = Q_{V_{ii}},$$

and the covariance σ_{V_i,V_j} of the two parameters V_i and V_j is computed as

$$\sigma_{V_i,V_j} = Q_{V_{ij}}.$$

The correlation coefficient ρ_{V_i,V_j} measures how separable two parameters V_i and V_j are and is derived from Q_V as

$$\rho_{V_i,V_j} = \frac{\sigma_{V_i,V_j}}{\sigma_{V_i}\sigma_{V_j}}.$$

SeaCal utilizes the relation between horizontal movement and the corresponding change of the seafloor depth. This calibration methodology requires seafloor with some relief if all the static offset parameters of the transducer are to be computed.

The coefficient matrix A is defined in Table 1. Here, we suppose that the fan of beams is approximately in a vertical plane which are approximately perpendicular to the strip direction. The terrain slope is used to compute the change of the seafloor depth that corresponds to a certain horizontal movement. Rollscale is computed on the basis of Snell's law—-the law about refraction. Since ray tracing is not applied, it is assumed that a correction of the beam direction can be explained from changing the beam direction at the passage between the transducer and the water.
Snell's law can be written as

$$\phi_1 = \arcsin \frac{c_1}{c_0} \sin \phi_0 = \arcsin(k \sin \phi_0), \text{ where } k = \frac{c_1}{c_0};$$

c denotes the sound speed and ϕ the beam angle. We assume that $c_0 = 1500\text{m/sec}$ and get the relation between a small change dc_1 in sound speed and k as

$$dc_1 = 1500dk.$$

How a small change in k effects ϕ can be found from differentiation as

$$d\phi_1 = \frac{\sin \phi_0}{\sqrt{1 - k^2 \sin^2 \phi_0}} dk \approx \frac{\sin \phi_0}{\sqrt{1 - \sin^2 \phi_0}} dk \text{ when } k \approx 1. \tag{4}$$

We define

$$\Phi(\phi) = \frac{\sin \phi_0}{\sqrt{1 - \sin^2 \phi_0}} \tag{5}$$

and write Equation 4 as

$$d\phi_1 = \Phi(\phi) dk. \tag{6}$$

Factor dk will be termed rollscale factor.

name of parameter	small change of parameter	change in sea depth
roll	$\Delta\phi$	$[\partial H_y h - y]\Delta\phi$
pitch	$\Delta\omega$	$[-\partial H_x h]\Delta\omega$
heading	$\Delta\kappa$	$[\partial H_x y]\Delta\kappa$
translation in X	ΔX	$[\partial H_x]\Delta X$
translation in Y	ΔY	$[\partial H_y]\Delta Y$
relative tide correction	Δh_i $\forall$ strips	Δh_i
rollscale factor	Δk	$[\partial H_y h - y]\Phi(\phi)\Delta k$
horizontal scale factor	m_y	$[\partial H_y y]m_y$

Table 1. Definition of the calibration parameters in SeaCal. ∂H_x and ∂H_y are the terrain slope along the two horizontal axes, y the distance from the nadir point of the transducer to the measured point and h the vertical distance from the transducer to the measured point.

2.2 Field survey

The inspection of systematic errors of the multibeam echo sounder should be run on a regularly basis. For example, at the start and at the end of a survey, or as often as once a month. The length of the time period used to run the field measurements for the verification, is typically thirty minutes to one hour. Figure 2 shows measurements to control the calibration of a multibeam system. In this case four lines are measured in an area where the sea depth varies from 85m to 244m. The survey lines are each of length 1000m, approximately. The results of the calibration is shown in Figure 3. The table shows the corrections to be added to the different parameter settings. For example, roll should be corrected $+0.107°$, the eccentricity of the transducer relative the GPS-antenna 1.84m in X-direction and -0.54m in Y-direction. The standard deviation of the depth measurements after the calibration is $\pm 0.11m$. The correlation between pitch and X is high, which says that in the present case the two parameters cannot be well separated. The reliability factor indicates how well the parameters can be computed. A red lamp signalises that the parameter considered should not be changed. In order to get a reliable value of this parameter, another calibration should be run—for example, find an area with a relieff that is more suited for field calibration of multibeam echo sounders.
Figure 4 illustrates the difference between a well calibrated multibeam echo sounder system and a system whithout a fine tuned calibration. The two graphs show the standard deviation of the depth measurements before and after the calibration is fine tuned. The calibration in Figure 2 is used for the construction of the two graphs. In the fine tuned case the standard deviation of the depth measurements is ± 0.11m against ± 0.49m in the case without fine tuning.

2.3 Error budget

From the contribution of the error sources defined in Table 1 an error budget can be set up. We assume the different error components are independent and compute their joint contribution σ by the law of variance propagation of random measurements. The error budget in Table 2 is computed for different beam angles and terrain slopes. The budget assumes sea depth 100m and certain values of the standard deviation of the different error sources. The important aspect of the computations is the exponential growth of σ as the beam angle increases. Therefore, for seafloor mapping with multibeam echo sounder the beam angle should be restricted to for example maximum $60°$.

3. Computation of random measurement errors

Random measurement errors in multibeam echo sounder data are functions of the kind of measurement equipment used; the sea depth, the terrain slope and the weather conditions during the survey. Since the magnitude of the random measurement error may vary from one survey to another, efficient methods to derive this error component is attractive in quality control of the depth measurements. The method we have developed, makes it possible to derive the random measurement error from one single survey. This makes it possible to monitor the random measurement error during a survey, and random measurement errors can be derived without measuring the same area several times, which saves field work. Traditional methods, as for example proposed by Maas (2002), requires that the same area is measured twice. The method we launch, is based on the application of semivariogram modelling in geostatistical theory, see for example Cressie (1993); Olea (1999) or Wackernagel (1998) for an overview of geostatistical methods.

3.1 Application of theory from geostatistics

In geostatistics we consider terrain surfaces as realizations of random functions. The height Z in position u can then be defined as $Z = (u, \omega)$, where $u \in D \subset \mathbb{R}^2$. The letter ω denotes the

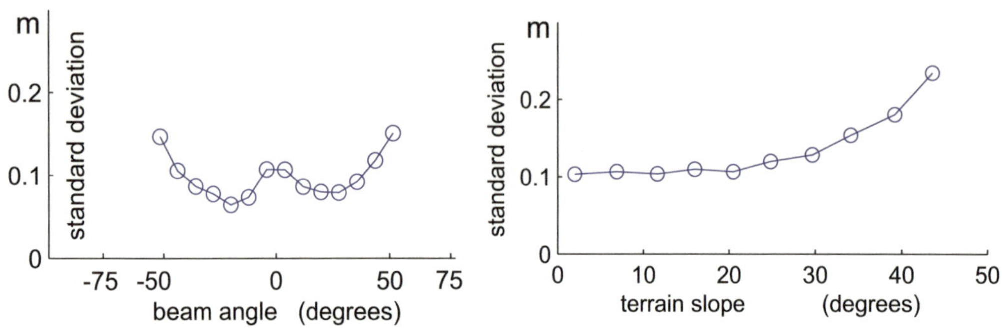

Calibration results

parameter	value	maximum correlation	reliability	fixed
roll	0.107	0.11	1.00	
pitch	0.033	0.94	0.94	
heading	0.003	-0.06	0.90	
x0	1.84	0.94	0.94	
y0	-0.54	-0.28	0.91	
h0	0.00	0.00	0.00	x
roll-scale	0.0007	0.11	0.84	
y-scale	0.0013	-0.28	0.95	

Standard deviation of the depth measurements after the calibration: 0.11

Legend

OK Evaluate Evaluation strongly recommended

Fig. 3. Field calibration of multibeam echo sounder from the survey shown in Figure 2. The graphs show how the precision of the measurements varies with beam angle and terrain slope. The table gives the results of the calibration computation, i.e, the parameter values, information about correlation and reliability. A red sign indicates that evaluation is recommended before the calibration values are entered into the system.

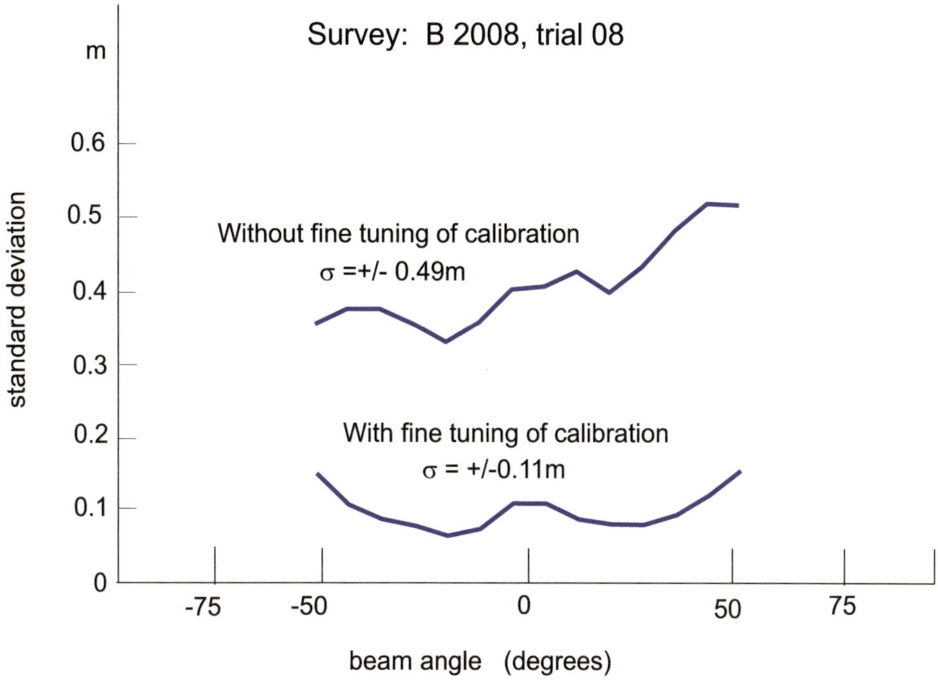

Fig. 4. Demonstration of the influence of not well calibrated multibeam echo sounder. Data used for the computation is the same as the data as used in Figure 3, i.e., survey B 2008, trial 08. The graphs show how the precision of the measurements varies with beam angle in the two cases: (1) the system is well calibrated and (2) the calibration is not fine tuned. The standard deviation of the residuals in the two cases is ±0.11m and ±0.49m, respectively.

particular realization, or outcome. For a fixed position $u', Z(u',\omega)$ is a random variable. For a fixed $\omega' \in \Omega$, $Z(u,\omega')$ is a deterministic function of u, called a realization of the random function.

Let $f(u) = Z(u,\omega')$ denote the exact terrain surface we are trying to model. We can write:

$$f(u) = m(u) + z(u),$$

where m is the *drift* and z a realization of a zero mean random function. The computation of the drift represents a challenge in geostatistics, see for example Olea (1999), page107, Cressie (1993), page 169 or Wackernagel (1998), page 213. Later on we will show how development of the mathematical theory of wavelets offers methods to compute approximations to the drift.

The values we observe can be seen as samples of another surface

$$\tilde{f}(u) = m(u) + z(u) + w(u), \tag{7}$$

where w is assumed to be a zero mean white noise component due to measurement errors. In order to obtain an estimate of the random measurement errors of the multibeam echo sounder, $w(u)$ is the component we in this section want to compute; but first, we will deal with how to compute the drift.

The drift is defined as the expected value of a random function:

$$m(u) = E(Z) = \int_{\Omega} Z(u,\omega)\, dP(\omega), \tag{8}$$

	σ in meter					
Beam angle	Terrain slope					
	0°	10°	20°	30°	40°	50°
0°	± 0.10	± 0.11	± 0.13	± 0.17	± 0.23	± 0.30
5°	± 0.10	± 0.11	± 0.13	± 0.17	± 0.23	± 0.30
10°	± 0.10	± 0.11	± 0.13	± 0.17	± 0.23	± 0.31
15°	± 0.10	± 0.11	± 0.14	± 0.18	± 0.23	± 0.31
20°	± 0.11	± 0.11	± 0.14	± 0.18	± 0.24	± 0.32
25°	± 0.11	± 0.12	± 0.14	± 0.18	± 0.24	± 0.32
30°	± 0.11	± 0.12	± 0.15	± 0.19	± 0.25	± 0.33
35°	± 0.12	± 0.13	± 0.16	± 0.20	± 0.26	± 0.35
40°	± 0.13	± 0.14	± 0.17	± 0.21	± 0.28	± 0.37
45°	± 0.15	± 0.16	± 0.19	± 0.23	± 0.30	± 0.39
50°	± 0.17	± 0.18	± 0.21	± 0.26	± 0.32	± 0.43
55°	± 0.21	± 0.22	± 0.25	± 0.30	± 0.37	± 0.48
60°	± 0.27	± 0.28	± 0.31	± 0.36	± 0.43	± 0.55
65°	± 0.37	± 0.38	± 0.41	± 0.46	± 0.54	± 0.67
70°	± 0.57	± 0.58	± 0.60	± 0.65	± 0.74	± 0.88
75°	± 0.99	± 1.00	± 1.03	± 1.08	± 1.17	± 1.32
80°	± 2.20	± 2.21	± 2.24	± 2.29	± 2.39	± 2.56
85°	± 8.77	± 8.78	± 8.81	± 8.86	± 8.96	± 9.14

Table 2. Standard deviation of measured sea depth derived from the error budget. The sea depth is defined to 100m. The following error components are considered: Roll ($\sigma_\phi = \pm 0.05°$), pitch ($\sigma_\omega = \pm 0.1°$), heading ($\sigma_\kappa = \pm 0.1°$), error in measured sound speed at the transducer ($\sigma_v = \pm 1.0\text{m/sec}$) which corresponds to rollscale $\sigma_k = \pm 0.00067$, error in the measured position of the transducer in the three axis directions X ($\sigma_x = \pm 0.1\text{m}$), Y ($\sigma_y = \pm 0.1\text{m}$) and Z ($\sigma_z = \pm 0.1\text{m}$).

where $P(\omega)$ is the distribution function of ω. The procedure we will apply to approximate the drift, uses spatial averaging over subsets of D as proposed by Bjørke and Nilsen (2005) and Bjørke and Nilsen (2007).

We start the drift computation by dividing the geographical area into grid blocks, making sure a suitable number of measurements fall within each block; emphirical tests show that an average of approximately ten values per block gives good results. Based on these values, we estimate the average depth a_i inside each block and approximate the drift by generating the sum

$$m(u) = \sum_{j=1}^{n} a_j \phi_j(u), \tag{9}$$

where n is the number of blocks. Each function ϕ_j is continuous and has integral 1 over block j and integral zero over all other blocks. The method we use to construct these functions is based on average interpolating subdivision; for an introduction see Sweldens and Schröder (1996). Note that we are not interpolating the observations—the surface we construct has the same average values over the grid blocks as the original estimated values—and the surface does not in general pass through any of the measured data points. In the Appendix a C-implementation of the 2D-average interpolating algorithm is given.

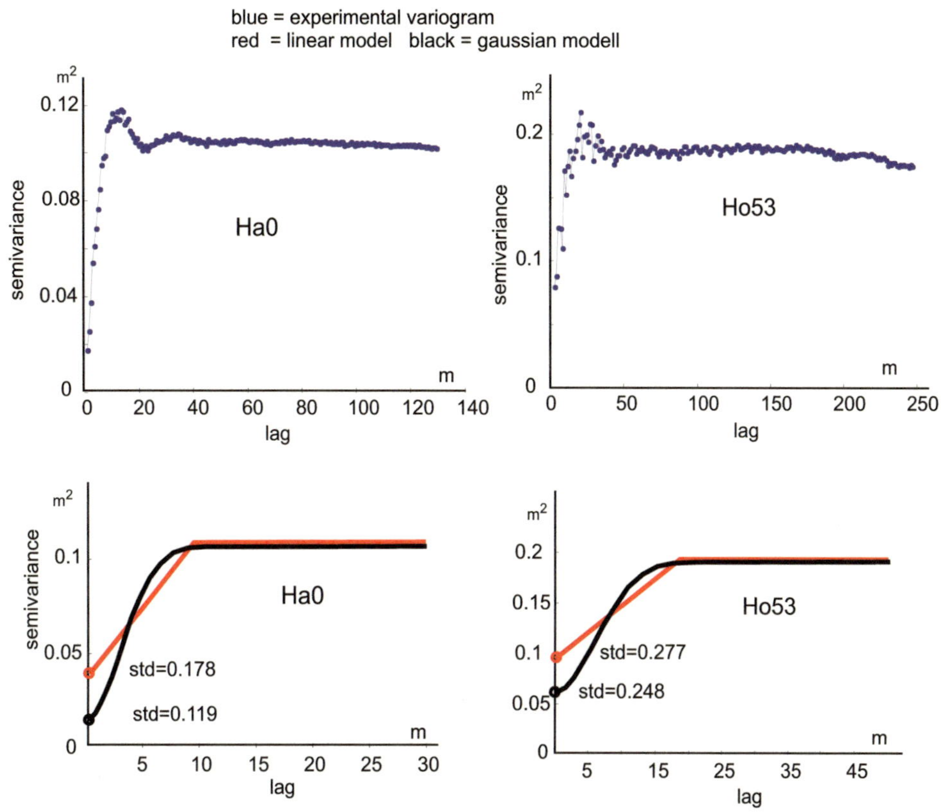

Fig. 5. Examples of experimental semivariogram and the fit of a linear and a Gaussian variogram model; note the parabolic form of the Gaussian variogram near the origin. Visually, one may expect the piecewise liner variogram to follow the linear part of the gaussian variogram more close. A reason for the discrepancy is that the curve fitting is computed from weighted least squares and that the number of observations in the lower part of the experimental variogram is small compared with the number of observations in the lag classes near the break point of the variogram.

When we have a model of the drift, we can compute the residuals between the drift and the observations as

$$r(u) = \tilde{f}(u) - m(u) = z(u) + w(u). \tag{10}$$

The two components $z(u)$ and $w(u)$ cannot be separated from these equations, but from the semivariogram we can derive $w(u)$. In the semivariogram $w(u)$ appears as the so-called nugget effect, see e.g. Olea (1999) or Cressie (1993).
The semivariogram is defined as

$$\gamma(h) = \frac{1}{2}E[\{r(u) - r(u+h)\}^2], \tag{11}$$

where u and $u + h$ are any two locations in the sample domain. The following expression is an unbiased estimator for the semivariogram

$$\tilde{\gamma}(h) = \frac{1}{2n(h)} \sum_{i=1}^{n(h)} \{r(u) - r(u+h)\}^2], \tag{12}$$

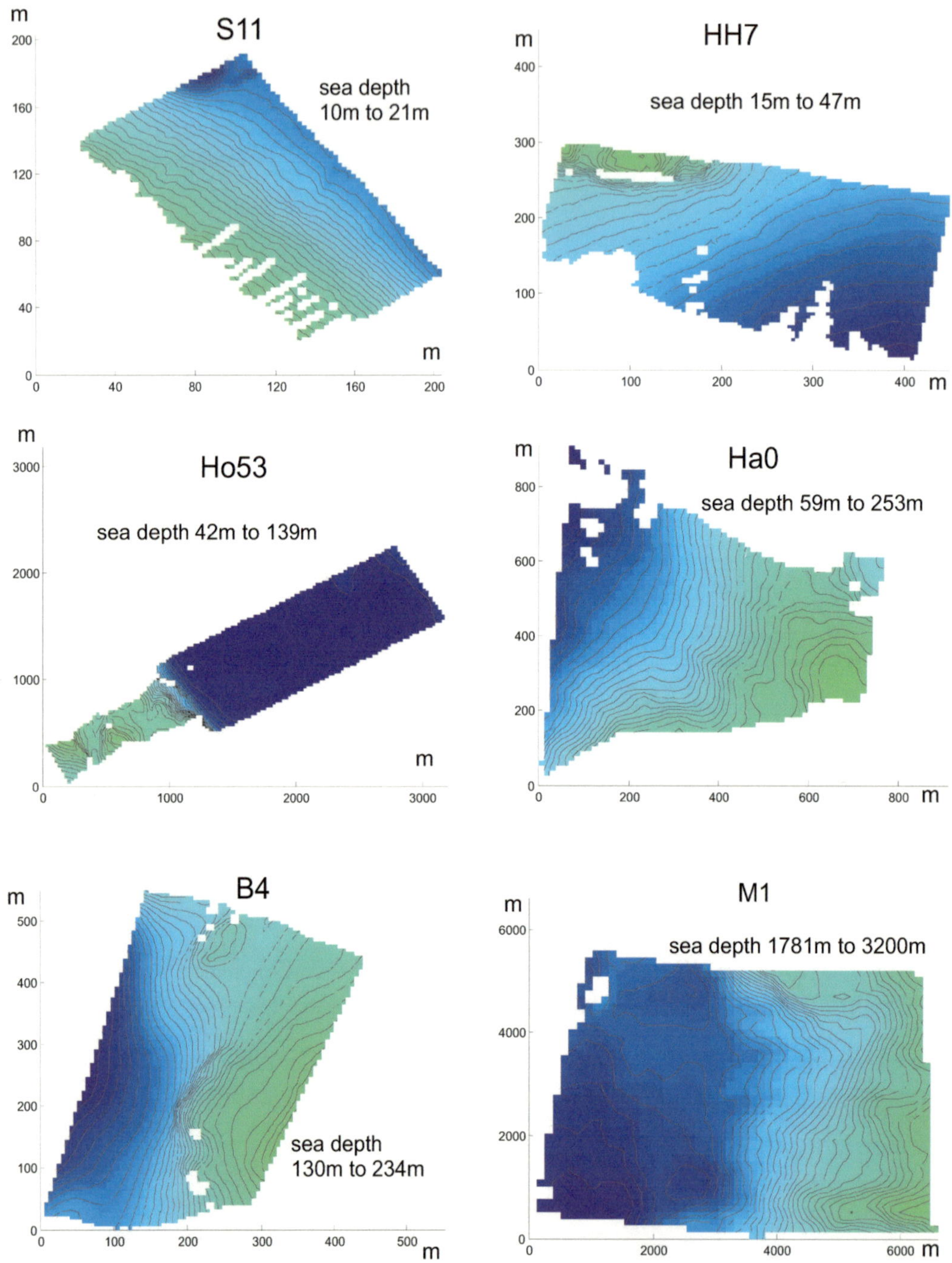

Fig. 6. Six maps of the twelve survey areas used for the study of random measurement errors of the multibeam echo sounder and for comparing different DEM-methods. Map S11 is also representative for survey S1, map HH7 representative for survey HH0 and so on.

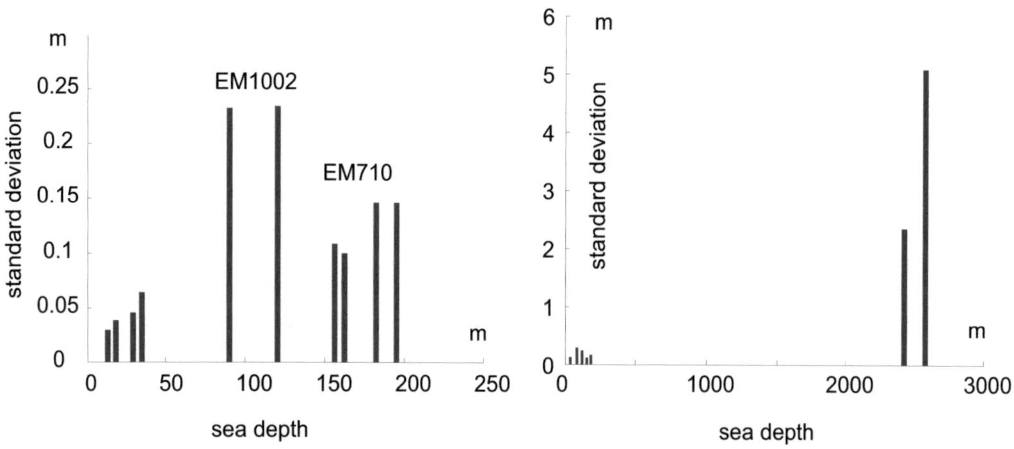

Fig. 7. Random errors of multibeam echo sounders used in the study areas. The values of σ_w are derived from the linear variogram as presented in Table 3. The kind of multibeam echo sounder used for the measurements is indicated, i.e., EM1002 and EM710; EM710 is a more modern system than EM1002.

where $n(h)$ is the number of pairs of variables at distance class h. Figure 5 shows two examples of experimental variograms. The number of permissible functions to be used for variogram fit is infinite, but in practice the options is reduced to a handful (Olea 1999, page 76). Some popular models are spherical, exponential, Gaussian, power and linear semivariogram models.

In the forthcoming experiments a linear and a Gaussian model will be used. The linear model is composed of two pieces. The first fits the part of the experimental variogram with small lags, and the next the rest of the variogram. Let h be the lag. Then the piecewise linear semivariogram is the model

$$\gamma(h) = \begin{cases} w + Ch \ , 0 < h \leq a \\ w + Ca \ , h > a, \end{cases} \tag{13}$$

where w is the noise and a the range, i.e., the lag at which the semivariogram reaches a constant value.

Let h be the lag. Then the Gaussian semivariogram is the model

$$\gamma(h) = w + C\left(1 - e^{-3(\frac{h}{a})^2}\right), \tag{14}$$

where w is the noise and a the range. The Gaussian semivariogram has a parabolic form near the origin. This property is useful when the noise of the measurements is to be computed. In many applications the noise is set to zero, but in our use of the variogram, the noise is a component we want to derive. Therefore, in the case of quality assessment of multibeam measurements, the ability of the variogram model to detect the noise, is important.

The standard deviation of the measurement errors can be derived from the variogram as

$$\sigma_w = \sqrt{\gamma(0)}, \tag{15}$$

see for example the explanation in Bjørke and Nilsen (2007) and the noise component as indicated in Figure 5. The variogram models are fitted by the use of least squares method

Survey	Sea depth	$\gamma(0)$	$\sigma_w = \sqrt{\gamma(0)}$		a	$\gamma(a)$	$\sigma_z = \sqrt{\gamma(a)}$
		linear	linear	Gaussian	linear	linear	linear
	m	m^2	m		m	m^2	m
S1	12-21	0.00181	±0.043	±0.038	1.8	0.00252	±0.050
S11	10-21	0.00121	±0.035	±0.029	1.6	0.00162	±0.040
HH0	18-43	0.00584	±0.076	±0.064	1.8	0.00717	±0.085
HH7	15-47	0.00398	±0.063	±0.045	1.7	0.00507	±0.071
Ho53	43-197	0.08879	±0.298	±0.235	17.4	0.19108	±0.437
Ho83	42-139	0.07188	±0.268	±0.233	12.0	0.09913	±0.315
Ha0	59-253	0.03719	±0.193	±0.108	9.3	0.10530	±0.325
Ha5	62-261	0.04562	±0.214	±0.100	11.8	0.16232	±0.403
B4	130-234	0.02791	±0.167	±0.146	3.6	0.03214	±0.179
B6	145-244	0.02938	±0.171	±0.146	4.7	0.04102	±0.203
M1	1781-3200	24.72150	±4.972	±4.964	84.8	32.76960	±5.725
M10	1780-3092	7.74608	±2.783	±2.473	85.8	17.94542	±4.236

Table 3. Computation of random measurement errors derived from the linear and the Gaussian semivariogram. In the table $\gamma(0)$ represents the variance of the random measurement errors and σ_w its standard deviation, i.e., $\sqrt{\gamma(0)}$. The value of the lag when the variogram reaches a constant value, is termed the range and denoted a. The corresponding value $\gamma(a)$ of the semivariogram is termed the sill. The square root of the sill approximates the standard deviation σ_z of the residuals.

where the number of observations in each lag class is used as weights such that more observations gives higher weight. When the Gaussian model is computed, one should watch for invalid solutions; visual presentation of the variogram model is recommended. One should also test for invalid semivariograms, for example, if the least squares curve fitting gives negative semivariance at the origin.

3.2 Demonstration of the computation of random measurement errors

Computation of random measurement errors from the experimental semivariogram will be derived from the twelve surveys illustrated in Figure 6. By applying Equation 15 to the two variogram models in Figure 5, we find that in area Ha0 σ_w is ±0.178 for the linear variogram and ±0.119 for the Gaussian variogram. In area Ho53 the linear and the Gaussian variogram give σ_w is ±0.277 and ±0.248, respectively. This indicates that the Gaussian variogram may give lower values for σ_w than the linear variogram. Note the parabolic form of the Gaussian model near the origin.

Table 3 shows the results from applying the linear and the Gaussian variogram models to the data in the twelve test areas. The standard deviation of the random measurement errors increases with increasing sea depth. In the shallow water areas σ_w is some centimeters and is of the magnitude several meters in the deep water areas. The range a represents the correlation distance. For lags less than a, the residuals are correlated. The value of a varies

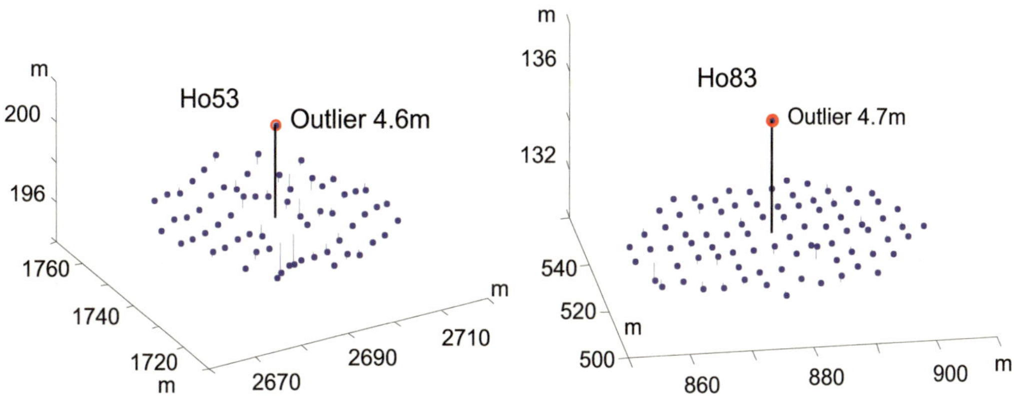

Fig. 8. Identification of isolated outliers

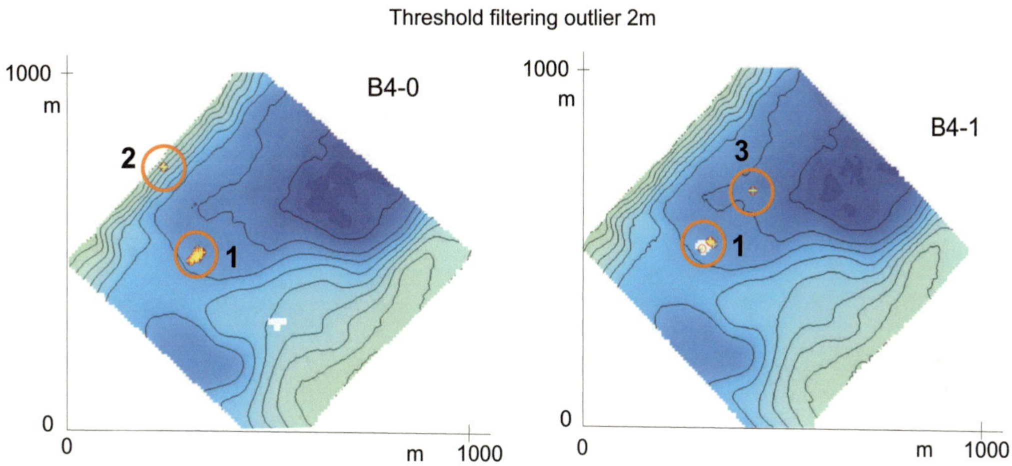

Fig. 9. Outliers identified in two surveys of the same area. In the red circle numbered 1 there are unexpected observations in both the surveys. The vertical distance between the contour lines is 5m

from approximately two meter in the shallow water area and up to almost one hundred meter in the deep water area. We notice that σ_w systematically is lower for the Gaussian model than the linear model. In a coming section on quality of digital elevation models the discrepancies between the two models will be discussed. The conclusion from this evaluation is that the Gaussian model seems to give more reliable semivariograms than the linear model.

Figure 7 shows how the random measurement error varies with the sea depth. The main impression is that σ_w increases with increasing sea depth, but at depth 100m there are two surveys—Ho53 and Ho83—that show higher value of σ_w than surveys at depth 150m to 200m. This can be explained from the survey equipment used. For Ho53 and Ho83 multibeam echo sounder EM1002 is used, and in the other case EM710 is applied. Since EM710 belongs to a newer generation multibeam echo sounders than EM1002, the measurement precision of EM710 is improved compared with the precision of EM1002. This explains why Figure 7 does not show a monotonic increasing relation between σ_w and increasing sea depth.

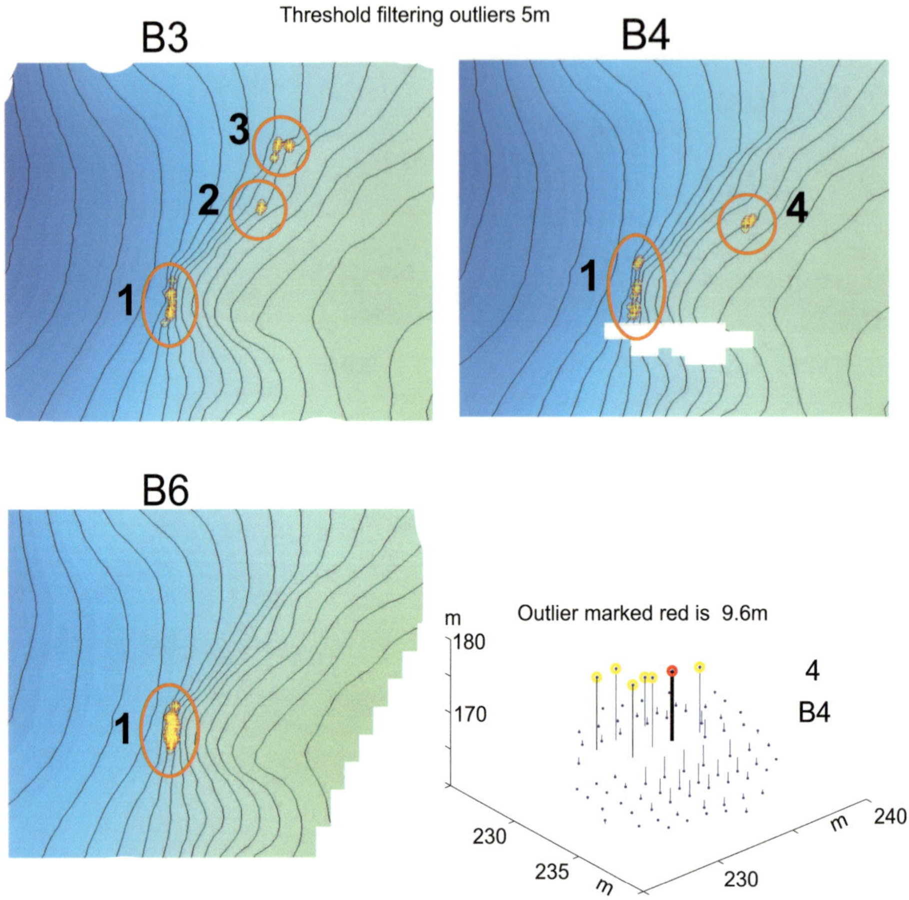

Fig. 10. Outliers identified in three surveys of the same area. In the red circle numbered 1 there are unexpected observations in all the surveys. The vertical distance between the contour lines is 5m. A perspective view of outliers in circle 4 in survey B4 is shown.

4. Detection of outliers in the measurements

Some case studies will illustrate the detection and the interpretation of outliers in depth measurements with multibeam echo sounder. We define an outlier as an unexpected large residual between the drift surface and the measured depth values. Unexpected large may be set to a threshold derived from the standard deviation σ of the residuals, for example 6σ. In the selected test areas the portion of the residuals greater than 6σ is less than 0.1% of the number of measured depth points. Average interpolation wavelets algorithm will be used for the drift computation, see the Appendix for description of this method. The average interpolation wavelets algorithm is well suited for detection of outliers, because average values are resistant against large measurement errors.

Figure 8 illustrates the detection of isolated outliers in the two surveys Ho53 and Ho83. In both cases the outlier is of magnitude 4m-5m. A question is: Can these unexpected observations be regarded as measurement errors?. For example, is there a human construction on the seafloor that have caused the spike or have we detected a ship wreck? In principle, we need more information to classify an outlier as a huge measurement error. Knowledge of properties of the

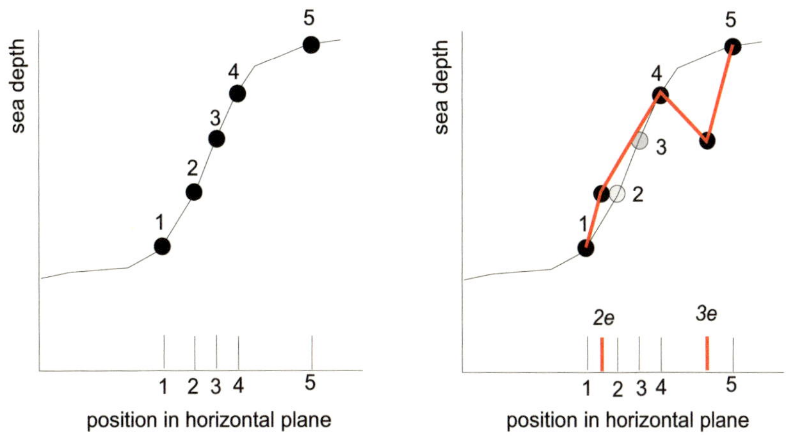

Fig. 11. Demonstration on how a horizontal position error may effect a measured terrain profile. In the first case we assume no position error. In the second case point 2 is associated to position 2e and point 3 to position 3e. The other terrain points in the second case are associated to their correct positions. The red profile line shows how the position errors have changed the profile from a monotonic increasing profile to a profile with a sharp local minima.

measurement system, weather conditions during the survey and other additional information will be of value when outliers are to be evaluated.

In Figure 9 the same area is measured twice. This gives us more information about the outliers compared with information from one single measurement of the seafloor. The red circles in Figure 9 mark areas with outliers. As we can see, the area numbered 1 occur in both surveys. Therefore, this indicates that there may be some terrain features in this area that cause the unexpected observations.

Figure 10 shows three surveys of the same area. Also in this case there are areas with outliers that are marked in all the tree surveys, see the circle numbered 1. The outliers in circle 4 in survey B4 are illustrated in the perspective view. A group of outliers of size 10m, approximately, occurs in an area of size 8m×8m.

Figure 11 demonstrates how a horizontal position error may effect a measured terrain profile. In the first case we assume no position error. In the second case point 2 is associated to position 2e and point 3 to position 3e. The other terrain points in the second case are associated to their correct positions. The red profile line shows how the position errors have changed the profile from a monotonic increasing profile to a profile with a sharp local minima.

The identification of outliers in the depth measurements raises several questions of how these measurements should be treated:

1. Should the measurments classified as outliers be removed from the data set?;
2. should the area with the outliers be inspected by new measurements?;
3. should the trend surface be used as a substitute for the measured data points?;
4. how to implement a system for automated detection and elimination of outliers?;
5. and how to design the user interaction in an automated system for handling of outliers.

Therefore, there are research question to be answered when an automated system for elimination of outliers are to be realized.

5. Quality of interpolation methods in digital elevation models

A product from seafloor mapping is often digital elevation models (DEMs). From a DEM different cartographic products as contour line maps, perspective views, shaded relief maps or maps with coloured depth intervals can be derived. Digital elevation models can also be used in navigation of vessels, ships and boats or for the construction of technical installations in oil and gas winning. The range of applications of DEMs is therefore large. Since the construction of a DEM can be based on different interpolation methods, an assessment of common interpolation methods will be presented. The methods we will investigate is average interpolation wavelets, triangle methods (TIN), moving surface and kriging. A brief description of these methods will be presented before we run the experiment.

5.1 Average interpolating wavelets

Average interpolating wavelets was presented in Section 3; see this section and the Appendix where a C implementation of the 2D-average interpolating wavelets algorithm is given..

5.2 TIN

Triangle models are often used in digital terrain models. The method is frequently referenced as triangulation irregular network (TIN). The construction principle of the network follows the so-called Delaunay principle. There exist several methods for the computation of a Delaunay triangulation of a set of points, see for example Kreveld (1997). When the triangles are computed, interpolation of depth values in a point with coordinates (x, y) is carried out from the linear interpolation

$$z(x, y) = a_1 + a_2x + a_3y,$$

where the coefficients are computed on the basis of the depth value in the three corners of the circumscribing triangle of the point considered.

5.3 Moving surface

Moving surface interpolates the depth z in position (x, y) from a polynomial. In our application we select coefficients from the bi-cubic polynomial

$$\begin{aligned} z(x, y) = {} & a_1 + a_2x + a_3y + a_4xy + \\ & a_5x^2 + a_6y^2 + a_7x^2y + a_8xy^2 + a_9x^2y^2 + \\ & a_{10}x^3 + a_{11}y^3 + a_{12}x^3y \\ & + a_{13}xy^3 + a_{14}x^3y^2 + a_{15}x^2y^3 + a_{16}x^3y^3; \end{aligned} \tag{16}$$

which can be written as

$$z(u) = \sum_{j=1}^{k} a_j b_j(u) \tag{17}$$

where k is the number of coefficients of the polynomial and $b_j(u)$ a function of the (x, y) coordinates of point u.

The coefficients a_j of the polynomial is derived by applying least squares approximation to the data points in a neighbourhood of the point to be interpolated. This requires the following system of equations to be solved:

$$\mathbf{v} = \mathbf{f} - \mathbf{B}\boldsymbol{\Delta},$$

where

$$\mathbf{v} = \begin{bmatrix} v_1 \\ v_2 \\ \vdots \\ v_n \end{bmatrix}; \mathbf{f} = \begin{bmatrix} z_1 \\ z_2 \\ \vdots \\ z_n \end{bmatrix}; \mathbf{B} = \begin{bmatrix} b_1(u_1) & b_2(u_1) & \cdots & b_k(u_1) \\ b_1(u_2) & b_2(u_2) & \cdots & b_k(u_2) \\ \vdots & \vdots & \vdots & \vdots \\ b_1(u_n) & b_2(u_n) & \cdots & b_k(u_n) \end{bmatrix} \text{ and } \mathbf{\Delta} = \begin{bmatrix} a_1 \\ a_2 \\ \vdots \\ a_k \end{bmatrix};$$

$$\text{such that } \sum_{i=1}^{n} p_i v_i v_i = \min. \tag{18}$$

Here,

p_i is the weight of data point u_i;

n is the number of data points;

k is the number of coefficients to be computed;

$\mathbf{f}$ is a vector of observed depth values;

$\mathbf{\Delta}$ is a vector of the coefficients to be computed;

and $\mathbf{B}$ is a matrix of variables defined in Equations 16 to 17 such that $b_j(u_i)$ is variable j for data point i.

Weight p_i is computed from the euclidian distance $d(\cdot)$ between point q_0 to be interpolated and its neighbouring point q_i as

$$p_i = \frac{1}{\epsilon + (d(q_0, q_i))^2},$$

where ϵ is a small positive number to prevent division by zero. By defining P as a $n \times n$ weight matrix as

$$\mathbf{P} = \begin{bmatrix} p_1 & 0 & 0 & \cdots & 0 \\ 0 & p_2 & 0 & \cdots & 0 \\ 0 & 0 & p_3 & \cdots & 0 \\ \vdots & \vdots & \vdots & \vdots & \vdots \\ 0 & 0 & 0 & \cdots & p_n \end{bmatrix},$$

the solution for $\mathbf{\Delta}$ is

$$\mathbf{\Delta} = \mathbf{N}^{-1}\mathbf{t}, \text{ where } \mathbf{N} = \mathbf{B}^t\mathbf{P}\mathbf{B} \text{ and } \mathbf{t} = \mathbf{B}^t\mathbf{P}\mathbf{f}. \tag{19}$$

5.4 Kriging

The roots of kriging goes back to pioneering work seventy years ago. Kriging is a form of generalized regression for the formulation of a minimum square error sense. There are several forms of kriging: Simple kriging, ordinary kriging, universal kriging, block kriging and other forms. In our experiment we will apply simple kriging and universal kriging. Details of the kriging equations are well covered in geostatistical literature, for example, Olea (1999) gives a presentation well suited to be read by engineers. Therefore, only a summary of the kriging equations will be given in the present paper.

5.4.1 Simple kriging

Simple kriging requires knowledge of the drift $m(u)$ and the semivariogram $\gamma(h)$ of the residuals $r(u)$ between the drift and the observations $\tilde{f}(u)$, see Equationes 10 and 12. The simple kriging equations take the form

$$\mathbf{MA} = \mathbf{S}, \tag{20}$$

where

$$\mathbf{M} = \begin{pmatrix} \gamma(u_1,u_1) & \gamma(u_1,u_2) & \cdots & \gamma(u_1,u_k) \\ \gamma(u_2,u_1) & \gamma(u_2,u_2) & \cdots & \gamma(u_2,u_k)) \\ \vdots & \vdots & \vdots & \vdots \\ \gamma(u_k,u_1) & \gamma(u_k,u_2) & \cdots & \gamma(u_k,u_k) \end{pmatrix}$$

and

$$\mathbf{A} = (\lambda_1\ \lambda_2\ \cdots \lambda_k)^t \text{ and } \mathbf{S} = \gamma(u_1,u_0)\ \gamma(u_2,u_0)\ \cdots \gamma(u_k,u_0)^t.$$

The interpolated value in u_0 is derived from

$$\hat{f}(u_0) = \sum_{j=1}^{k} \lambda_j r(u_j) + m(u_0),$$

which expresses that that the interpolated value is computed from a linear kombination of the residuals in the neighbouring data points.

In simple kriging the variance in the interpolated point is computed as

$$\sigma^2 = \mathbf{A}^t\mathbf{S}. \tag{21}$$

5.4.2 Universal kriging

Universal kriging does not require knowledge of the drift, but the semivariogram of the residuals between the drift and the observations must be known. Therefore, determination of the semivariogram may in universal kriging also require knowledge of the drift. The equations in universal kriging represents a generalization of the equations of simple kriging and takes the form

$$\mathbf{CV} = \mathbf{F} \tag{22}$$

where

$$\mathbf{C} = \begin{pmatrix} \gamma(u_1,u_1) & \gamma(u_1,u_2) & \cdots & \gamma(u_1,u_k) & b_1(u_1) & b_2(u_1) & \cdots & b_n(u_1) \\ \gamma(u_2,u_1) & \gamma(u_2,u_2) & \cdots & \gamma(u_2,u_k) & b_1(u_2) & b_2(u_2) & \cdots & b_n(u_2) \\ \vdots & \vdots & \vdots & \vdots & \vdots & \vdots & \vdots & \vdots \\ \gamma(u_k,u_1) & \gamma(u_k,u_2) & \cdots & \gamma(u_k,u_k) & b_1(u_k) & b_2(u_k) & \cdots & b_n(u_k) \\ b_1(u_1) & b_1(u_2) & \cdots & b_1(u_k) & 0 & 0 & \cdots & 0 \\ b_2(u_1) & b_2(u_2) & \cdots & b_2(u_k) & 0 & 0 & \cdots & 0 \\ \vdots & \vdots & \vdots & \vdots & \vdots & \vdots & \vdots & \vdots \\ b_n(u_1) & b_n(u_2) & \cdots & b_n(u_k) & 0 & 0 & \cdots & 0 \end{pmatrix}$$

and

$$\mathbf{V} = [\lambda_1\ \lambda_2\ \cdots \lambda_k\ \mu_1\ \mu_2\ \cdots \mu_n] \text{ and } \mathbf{F} = [\gamma(u_1,u_0)\ \gamma(u_2,u_0)\ \cdots \gamma(u_k,u_0)\ b_1(u_0)\ b_2(u_0)\ \cdots b_n(u_0)].$$

The parameters λ_i are Lagrange multipliers to ensure that universal kriging becomes an unbiased estimator. The drift is represented by the coefficients $b(u)$. In our implementation we have taken them from the bi-cubical ploynomial in Equation 17. How many coefficients that are used, is decided from the number of data points in the neighbourhood of the point to be estimated.

The interpolated value in u_0 is derived from

$$\hat{f}(u_0) = \sum_{j=1}^{k} \lambda_j \tilde{f}(u_j).$$

In universal kriging the variance in the interpolated point is computed as

$$\sigma^2 = \mathbf{V}^t\mathbf{F}. \tag{23}$$

When solving kriging equations, one should be aware of possible *numerical problems*, see for example Davis and Morris (1997). For example, if two points have identical coordinates, they will produce equal kriging equations, and the system of equations will become singular.

5.5 Evaluation of the interpolation methods

Computation of interpolation errors of different interpolation methods will be derived from the twelve surveys as indicated in Figure 6. A portion of the sample points are selected as control points. This subset C is subtracted from the original data set A, and a new set B is derived as $B = A \setminus C$. Set B is used for the interpolation of depth values in the points of C. The residuals between the measured and the interpolated depth values of set C is computed. Interpolation methods to be investigated are simple and universal kriging, TIN, moving surface and wavelets. The results of the computations are shown in Table 4. In nine cases the variation of the standard deviation in the control points for the different interpolation methods is less than 10%. In one case the variation is of magnitude 30% and in two cases 100%. In seven cases universal kriging performs better than TIN and equal in five. Moving surface works better than TIN in six cases, equal in four and weaker two cases. Average interpolating wavelets performs better than TIN in two cases, equal in one and weaker in nine cases. Ranking the methods on basis of the standard deviation in the control points gives:

1. Universal kriging,
2. simple kriging,
3. moving average,
4. TIN,
5. and average interpolating wavelets.

An attractive property of kriging is its ability to estimate the variance of the interpolated depth values. From Equation 23 the average kriging variance σ_u^2 in the control points of the twelve survey areas can be derived. The results of these computations are presented in Table 5. The realism of σ_u^2 depends on how well the variogram is modelled. An evaluation of σ_u^2 will be carried in the following. The residual $r(u)$ between the measured value $\tilde{f}(u)$ and universal kriging value $g(u)$ in point u is

$$r(u) = \tilde{f}(u) - g(u).$$

From the law of variance propagation of independent measurements we can compute the variance $\hat{\sigma}_r^2$ of $r(u)$ as

$$\hat{\sigma}_r^2 = \sigma_w^2 + \sigma_u^2,$$

Survey	Sea depth m	Simp. kr. m	Univ. kr.. m	TIN m	Mov. surf. m	Wavelets m
S1	12-21	±0.045	±0.044	±0.045	±0.044	±0.051
S11	10-21	±0.041	±0.041	±0.041	±0.040	±0.044
HH0	18-43	±0.052	±0.053	±0.073	±0.067	±0.078
HH7	15-47	±0.057	±0.058	±0.060	±0.059	±0.074
Ho53	43-197	±0.316	±0.304	±0.317	±0.277	±0.381
Ho83	42-139	±0.275	±0.285	±0.301	±0.259	±0.304
Ha0	59-253	±0.152	±0.144	±0.146	±0.178	±0.324
Ha5	62-261	±0.155	±0.151	±0.148	±0.196	±0.388
B4	130-234	±0.175	±0.180	±0.204	±0.175	±0.183
B6	145-244	±0.168	±0.166	±0.175	±0.165	±0.191
M1	1781-3200	±6.259	±0.6.084	±6.201	±5.098	±5.494
M1	1780-3092	± 3.302	±3.245	±3.437	± 2.876	± 4.138

Table 4. Standard deviation for different interpolation methods. The standard deviation is computed from the residuals between measured depth value and interpolated value in the control points.

Survey	σ m	Survey	σ m
S1	±0.043	Ha0	±0.194
S11	±0.036	Ha5	±0.222
HH0	±0.078	B4	±0.169
HH7	±0.063	B6	±0.174
Ho53	±0.306	M1	±5.124
Ho83	±0.265	M10	±2.759

Table 5. Average standard deviation of depth values computed from universal kriging variance, see Equation 23

where the noise component σ_w^2 of the measurements can be derived from the semivariogram and the universal kriging variance σ_u^2 by the application of Equation 23. The variance of $r(u)$ can also be computed from the residuals in the control points as

$$\sigma_r^2 = \frac{1}{n}\sum_{i=1}^{n}(\tilde{f}(u) - g(u))^2,$$

where $\tilde{f}(u)$ is the measured depth value in u and the $g(u)$ the value computed from universal kriging and n the number of control points. From the standard deviation of the two variances the relation

$$q = \frac{\hat{\sigma}_r}{\sigma_r} \tag{24}$$

is computed and presented in Table 6.

Survey	Gaussian	linear	Survey	Gaussian	linear
S1	1.30	1.38	Ha0	1.54	1.90
S11	1.12	1.22	Ha5	1.61	2.04
HH0	1.90	2.05	B4	1.24	1.32
HH7	1.33	1.54	B6	1.37	1.47
Ho53	1.27	1.41	M1	1.17	1.17
Ho83	1.24	1.32	M10	1.14	1.21
mean value				1.35	1.50

Table 6. Factor computed to evaluate the realism of the kriging variance in the control points, see Equation 24. The factor is computed for both the Gaussian and linear semivariogram given in Table 3. The closer the factor is to 1, the more realistic the semivariogram is modelled.

From Table 6 we can see that the Gaussian and the linear semivariogram on average give q-factor 1.35 and 1.50, respectively. The closer the factor is to 1, the more realistic the kriging variance will be. Therefore, the Gaussian semivariogram gives slightly more realistic values of the kriging variance than the linear semivariogram. The Gaussian variogram seems therefore to be preferable when applied to seafloor mapping from multibeam echo sounder measurments.

6. Conclusions

Quality issues of spatial data are illustrated by examples from seafloor mapping with multibeam echo sounder. The methodologies to compute systematic errors and random errors of the multibeam echo sounders represent rather new approaches—although, well known techniques as least squares adjustment and geostatistical tools are applied. The average interpolating wavelets algorithm shows its strength in the computation of the drift of seafloor models, and we have demonstrated how the drift is used to derive residuals for the computation of random errors, outliers and the semivariogram used in kriging interpolation. We have shown how the three error components—random measurement errors, systematic errors and outliers—can be identified from depth measurements with multibeam echo sounders. Since these error sources are important to the quality of seafloor maps and digital elevation models, their parameters should be documented and follow the measurements as so-called *metadata*.

Different interpolation methods are compared. These methods can be used to compute digital terrain models with regular sampling patterns from scattered data. Grid models, for example, have regular sampling pattern and are very popular computer representations of digital elevations models. The interpolation methods investigated, are selected from the class of models often used in seafloor mapping. The general impression of the comparison is kriging interpolation shows slightly higher precision than the other methods. For practical purposes we recommend that factors as *speed of computation and numerical stability* should be considered when interpolation methods are to be implemented.

7. Acknowledgements

This research is supported by FFI-projects 113602 and 113002.

8. References

Bjørke, J. T.: 2004, En metode for feltkalibraring av systemparametere i et multistråleekkolodd-system (Eng. a method for field calibration of system parameters in a multibeam echo sounder system), Norwegian patent no. 315766, Norwegian Patent Office, PO Box 8160 Dep. 0033, Oslo, Norway.

Bjørke, J. T.: 2005, Computation of calibration paramters for multibeam echo sounders using least squares method, *IEEE Jornal of Oceanic Engineering* 30(4), 818–831.

Bjørke, J. T. and Nilsen, S.: 2002, Efficient representation of digital terrain models: compression and spatial decorrelation techniques, *Computers & Geosciences* 28, 433–445.

Bjørke, J. T. and Nilsen, S.: 2003, Wavelets applied to simplification of digital terrain models, *International Journal of Geographical Information Science* 17(7), 601–621.

Bjørke, J. T. and Nilsen, S.: 2005, Trend extraction using average interpolating subdivision, *Mathematical Geology* 37(6), 615–634.

Bjørke, J. T. and Nilsen, S.: 2007, Computation of random errors i digital terrain models, *Geoinformatica* 11, 359–382.

Bjørke, J. T. and Nilsen, S.: 2009, Fast trend extraction and identification of spikes in bathymetric data, *Computers & Geosciences* 35, 1061–1071.

Cressie, N. A. C.: 1993, *Statistics for Spatial Data, Revised Ed.*, Probability and Mathematical Statistics, John Wiley & Sons, Inc., New York. 900 pp.

Davis, G. J. and Morris, M. D.: 1997, Six factors which affect the condition number of matrices associated with kriging, *Mathematical Geology* 29(5), 669–683.

Glittum, M., Heier, O., Pohner, F. and Vestgard, K.: 1986, Multibeam echo-sounder system for sea bed mapping, *Oceanology: Proceedings of an international conference, March 4 - 7, 1986, Brighton, UK*, Advances in Underwater Technology and Offshore Engineering. Oceanology, Volume 6, The Society for Underwater Technology. ISBN 0-86010-772-8.

Guptill, S. C. and Morrison (editors), J. L.: 1995, *Elemenst of spatial data quality*, Elsevier Science Ltd., Oxford, U.K.

Kreveld, M. J. v.: 1997, Digital elevation models and TIN algorithms, *in* M. J. Kreveld, J. Nievergelt, T. Roos and P. Widmayer (eds), *Algorithmic foundations of geographical information systems*, Vol. 1340 of *Lecture Notes in Computer Science*, Springer-Verlag, London,UK, pp. 37–78. ISBN 3-540-63818-0.

Leick, A.: 1995, *GPS: Satellite surveying*, second edn, John Wiley & Sons. Inc., USA.

Maas, H.-G.: 2002, Methods for measuring height and planimetry discrepances in airborne laserscanner data, *Photogrammetric Engineering & Remote Sensing* 68(9), 933–940.

Olea, R. A.: 1999, *Geostatistics for Engineers and Earth Scientists*, Kluwer Academic Publishers, Boston. 303 pp.

Sweldens, W. and Schröder, P.: 1996, Building your own wavelets at home, *Wavelets in Computer Graphics*, ACM SIGGRAPH Course Notes, pp. 15–87.

Wackernagel, H.: 1998, *Multivariate geostatistics*, second edn, Springer, Paris.

9. Appendix: 2D average interpolating wavelets algorithm

The average interpolating wavelets algorithm used in our experiments, will presented as a C-implementation The implementation follows the descriptions in Bjørke and Nilsen (2002) and Bjørke and Nilsen (2003), see also applications of the metod in Bjørke and Nilsen (2007; 2009) and Bjørke and Nilsen (2005). The calling syntax of the algorithm is **2dforward(** A, i, j, q **)** where A is a $2^n \times 2^n$ matrix of cell average values, i and j are the number of rows and columns of A, and q the number of levels in the expansion. The following example will clarify the role of expansion. Assume matrix A_0 as

$$A_0 = \begin{bmatrix} 2.1 & 4.3 & 1.8 & 2.9 \\ 3.5 & 5.8 & 7.1 & 3.2 \\ 8.7 & 3.2 & 7.8. & 3.6 \\ 4.1 & 6.9 & 4.4 & 6.7 \end{bmatrix}$$

with spatial resolution 10m. From this matrix we want to make an expansion to a matrix with spatial resolution 2.5m. This means that we must increase the size of A_0 from 4×4 cells to size 16×16 cells. From A_0 we construct a new matrix A with size 16×16 cells by filling out the extra cells with zeros as

$$A = \begin{bmatrix} 2.1 & 0 & 0 & 0 & 4.3 & 0 & 0 & 0 & 1.8 & 0 & 0 & 0 & 2.9 & 0 & 0 & 0 \\ 0 & 0 & 0 & 0 & 0 & 0 & 0 & 0 & 0 & 0 & 0 & 0 & 0 & 0 & 0 & 0 \\ 0 & 0 & 0 & 0 & 0 & 0 & 0 & 0 & 0 & 0 & 0 & 0 & 0 & 0 & 0 & 0 \\ 0 & 0 & 0 & 0 & 0 & 0 & 0 & 0 & 0 & 0 & 0 & 0 & 0 & 0 & 0 & 0 \\ 3.5 & 0 & 0 & 0 & 5.8 & 0 & 0 & 0 & 7.1 & 0 & 0 & 0 & 3.2 & 0 & 0 & 0 \\ 0 & 0 & 0 & 0 & 0 & 0 & 0 & 0 & 0 & 0 & 0 & 0 & 0 & 0 & 0 & 0 \\ 0 & 0 & 0 & 0 & 0 & 0 & 0 & 0 & 0 & 0 & 0 & 0 & 0 & 0 & 0 & 0 \\ 0 & 0 & 0 & 0 & 0 & 0 & 0 & 0 & 0 & 0 & 0 & 0 & 0 & 0 & 0 & 0 \\ 8.7 & 0 & 0 & 0 & 3.2 & 0 & 0 & 0 & 7.8. & 0 & 0 & 0 & 3.6 & 0 & 0 & 0 \\ 0 & 0 & 0 & 0 & 0 & 0 & 0 & 0 & 0 & 0 & 0 & 0 & 0 & 0 & 0 & 0 \\ 0 & 0 & 0 & 0 & 0 & 0 & 0 & 0 & 0 & 0 & 0 & 0 & 0 & 0 & 0 & 0 \\ 0 & 0 & 0 & 0 & 0 & 0 & 0 & 0 & 0 & 0 & 0 & 0 & 0 & 0 & 0 & 0 \\ 4.1 & 0 & 0 & 0 & 6.9 & 0 & 0 & 0 & 4.4 & 0 & 0 & 0 & 6.7 & 0 & 0 & 0 \\ 0 & 0 & 0 & 0 & 0 & 0 & 0 & 0 & 0 & 0 & 0 & 0 & 0 & 0 & 0 & 0 \\ 0 & 0 & 0 & 0 & 0 & 0 & 0 & 0 & 0 & 0 & 0 & 0 & 0 & 0 & 0 & 0 \\ 0 & 0 & 0 & 0 & 0 & 0 & 0 & 0 & 0 & 0 & 0 & 0 & 0 & 0 & 0 & 0 \end{bmatrix}.$$

By calling **2dforward(A,4,4,2)**, the function returns a matrix A where the empty cells are filled in with interpolated values.

```
int forward2d(double *matrix,int size_i,int size_j,int numlevels)
{
   int i,j,step,smallstep,level,maxlev,tmp,Tmp,minsize;

   if(size_i<size_j){
      tmp=size_i;
      Tmp=size_j;
   }else{
      Tmp=size_i;
      tmp=size_j;
   }
   minsize=3;
   maxlev=0;
   step=1;
```

```
    while(((tmp%2)==0) && ((Tmp%2)==0) && ((tmp/2)>=minsize) &&
    (maxlev<numlevels)){
       tmp/=2;
       Tmp/=2;
       maxlev++;
    }
    for(level=1;level<=maxlev;level++){
       smallstep=step;
       step*=2;
       for(i=0;i<size_i;i+=smallstep){
          for(j=0;j<size_j;j+=step){
             matrix[size_j*i+j+smallstep]-=matrix[size_j*i+j];
             matrix[size_j*i+j+smallstep]*=0.5;
             matrix[size_j*i+j]+=matrix[size_j*i+j+smallstep];
          }
          matrix[size_j*i+smallstep]-=
             -0.375*matrix[size_j*i]
     +0.500*matrix[size_j*i+step]
     -0.125*matrix[size_j*i+2*step];
          for(j=step;j<size_j-step;j+=step){
             matrix[size_j*i+j+smallstep]-=
                +0.125*matrix[size_j*i+j+step]
       -0.125*matrix[size_j*i+j-step];
          }
          matrix[size_j*(i+1)-smallstep]-=
             +0.125*matrix[size_j*(i+1)-3*step]
     -0.500*matrix[size_j*(i+1)-2*step]
     +0.375*matrix[size_j*(i+1)-step];
       }
       for(j=0;j<size_j;j+=smallstep){
          for(i=0;i<size_i;i+=step){
             matrix[size_j*(i+smallstep)+j]-=matrix[size_j*i+j];
             matrix[size_j*(i+smallstep)+j]*=0.5;
             matrix[size_j*i+j]+=matrix[size_j*(i+smallstep)+j];
          }
          matrix[size_j*smallstep+j]-=
             -0.375*matrix[j]
             +0.500*matrix[size_j*step+j]
             -0.125*matrix[size_j*2*step+j];
          for(i=step;i<size_i-step;i+=step){
             matrix[size_j*(i+smallstep)+j]-=
                +0.125*matrix[size_j*(i+step)+j]
                -0.125*matrix[size_j*(i-step)+j];
          }
          matrix[size_j*(size_i-smallstep)+j]-=
             +0.125*matrix[size_j*(size_i-3*step)+j]
             -0.500*matrix[size_j*(size_i-2*step)+j]
             +0.375*matrix[size_j*(size_i-step)+j];
       }
    }
    return(maxlev);
}
```

Part 3

Quality Control in the Industry

24

Electrochemical Methods in Analysis of Biofuels

André L. Santos[1], Regina M. Takeuchi[1],
Paula G. Fenga[2] and Nelson R. Stradiotto[2]
[1]*Faculdade de Ciências Integradas do Pontal - UFU - Ituiutaba*
[2]*Departamento de Química Analítica - Instituto de Química - UNESP - Araraquara*
Brazil

1. Introduction

Energy is vital for modern society since commercial, agricultural and industrial activities are highly energy-consuming. Moreover, the marvels of the modern technology have enhanced the aspirations of people for an improved quality of life, which is more and more dependent on energy. It is a consensus that this very high energy demand can not be adequately supplied by traditional energy sources such as fossil-based fuels. In addition, there are many concerns about the usage of these fuels in future, mainly taking into account that they are obtained from a finite source and that their production and utilization are processes highly pollutant. These environmental concerns are widely justified taking into consideration, for example, that 56.6 % of anthropogenic green house gases (GHG) emissions of 2004 were associated to fossil fuel use [1]. Thus, for a sustainable future, it is essential to develop integrated energy models linking both traditional and modern renewable energy sources in order to decrease the global dependence on petroleum and to minimize environmental problems associated to its use as energy supply. A very complete discussion focused on the development and evaluation of integrated energy models can be found in the review written by Jebaraj and Iniyan [2].

Among renewable energy sources, biomass has the highest potential to adequately respond to energy requirements from modern society of both developed and developing countries [3]. Biomass is a term for all organic material originated from plants which convert sunlight into plant material through photosynthesis. The term includes all land- and water-based vegetation, as well as all organic wastes. Therefore, biomass resource is the organic matter, in which the energy of sunlight is stored in chemical bonds [4]. Renewable energy is of growing importance in responding to concerns over the environment and the security of energy supplies. In this context, biomass provides the only renewable source of fixed carbon, which is an essential ingredient of our fuels [3]. Biomass can be converted in different forms of useful energy and the literature presents some works especially dedicated to the technologies of biomass conversion [5] including thermochemical conversion processes [6]. The conversion process and its efficiency are influenced by the composition of biomass. Therefore its knowledge is important to select the more appropriated conversion process. A complete discussion about biomass composition and its variability is presented by Vassilev et al [7].

In despite of the advantages of biomass as energy source, some discussions have been done about the potential environmental damages associated with mega-scale biomass production and there are some concerns about its competition with food production [8-11]. These discussions are very important and must be encouraged in scientific community, because they have an important role for the development of responsible, efficient and environmentally correct energetic policies. In the same direction of these discussions, some authors have worked aiming to decrease NO_x emissions of oxygenated biofuels such as biodiesel and ethanol being this subject reviewed by Rajasekar et al [12]. In despite of the concerns associated with global use of biomass to produce energy, there is a consensus in scientific community that if biomass is efficiently produced and processed and if responsible energetic policies are adopted, biomass will represent a very important global source of energy in the future [13]. There are three main political concerns driving bioenergy production today: energy security, global climate change trends and rural incomes [1]. Therefore, bioenergy production involves important sectors of our society such as agricultural, commercial, environmental and industrial. Thus, the efficient and environmentally compatible development of bioenergy requires a stronger emphasis on inter-disciplinary research as stated by Petersen [1].

Bioenergy is especially attractive for tropical or subtropical developing countries because they present ideal climatic conditions to produce biomass. Besides environmental advantages, the use of bioenergy in these countries has also economical advantages because it minimizes their needs of imported petroleum. Some authors present revisions about the use of bioenergy in some developing countries such as Taiwan [14] and Mexico [15] and an interesting discussion about regional added value provided by regional biomass sources is presented by Hoffmann [16]. Bioenergy can be used in three ways: production of liquid transportation fuels, heating and electricity generation. It presents some advantages when compared with other renewable energy sources. For example, bioenergy can be produced continuously in contrast to intermittent wind and solar sources. Moreover, biomass is the only source of high grade renewable heat. However, the possibility of to obtain renewable transportation liquid fuels is the most attractive feature of biomass and a very important application of bioenergy nowadays [17].

Transportation is one of the most energy-consuming activities in modern society. According to International Energy Agency statistics, this sector is responsible for 60% of the world's total oil consumption [18]. Transportation is also almost completely dependent on petroleum-based fuels such as gasoline and diesel [19]. Therefore, transportation sector is very sensible to the several economic limitations associated to petroleum such as its geographically reduced availability and instability of price. In addition, petroleum-based fuels are highly pollutant and nowadays petroleum-based motor vehicles are responsible for more than 70 % of global carbon monoxide emissions and 19 % of global dioxide carbon emissions [18]. These facts clearly show the importance of decreasing the usage or replacing petroleum-based fuels in a near future. Liquid biofuels for transportation have recently attracted attention of different countries due their characteristics such as renewability, sustainability, common availability, regional development, rural manufacturing jobs, reduction of greenhouse gas emissions and biodegradability [20].

The term biofuel is referred to solid, liquid or gaseous fuels that are predominantly produced from biomass. Biofuels include bioethanol, biobutanol, biodiesel, vegetable oils, biomethanol, pyrolysis oils, biogas and biohydrogen. Among biofuels, bioethanol and biodiesel are by far the most used nowadays and many scientists agree that they have potential for replacing petroleum-based fuels [20]. Therefore, it is expected that the demand for biofuels will rise drastically in a near future. Some governmental actions in different countries support this statement. For example, nowadays bioethanol derived from corn responds by approximately 2 % of the total transportation fuel in USA and another 0.01 % is based on biodiesel. The USA Department of Energy has set ambitious goals to replace 30 % of the liquid petroleum transportation fuel with biofuels by 2025 [21]. In this context, there are no doubts that biofuels, mainly bioethanol and biodiesel, will play a very important place in the worldwide energetic policies.

This text is dedicated to the quality control of liquid automotive biofuels by using electroanalytical techniques. Presented discussions will be restricted to bioethanol and biodiesel, the most used transportation liquid biofuels nowadays. More general discussions about biofuels can be found in the literature [22] including the production of other biofuels [23,24], technologies for biofuel production [25,26] economic [27] and environmental [28,29] aspects related to biofuels.

1.1 Bioethanol and biodiesel

Bioethanol is by far the most widely used biofuel for transportation worldwide. Production of bio-ethanol from biomass is one way to reduce both consumption of crude oil and environmental pollution [18]. Besides these advantages, bioethanol has a higher octane number, broader flammability limits, higher flame speeds and higher heats of vaporization, which lead to theoretical efficiency advantages over gasoline in a combustion engine [18].

USA and Brazil are, respectively, the two largest producers of bioethanol. USA have surpassing Brazil since 2005 and in 2008 there were 136 ethanol plants in USA which are able to produced 7.5 billion gallons of ethanol annually [30]. Brazil was pioneering in the usage of bioethanol for transportation in large scale, starting in 1975 when the National Alcohol Fuel Program (ProAlcool) was initiated. ProAlcool was a response from Brazilian government to petroleum crises occurred in 1973. Ethanol fuel was firstly used in Brazil in its anhydrous form as an additive for gasoline and its use has banned lead from Brazilian gasoline. Nowadays, besides the use of anhydrous ethanol as additive, hydrated ethanol is used as fuel in its pure form or blended with gasoline in any proportion in flex-fuel engines. The large scale use of bioethanol has brought several advantages to Brazilian society, including reduced dependence of imported petroleum, jobs generation, acceleration of rural economy, developing of production technologies besides environmental advantages such as substantial reduction of GHG emissions. The interior of São Paulo State is one of the most rich and developed areas in Brazil. In this region, ethanol production is the main agricultural and industrial activity which demonstrates the positive impact of ethanol usage in Brazilian economy.

Most of the bioethanol produced in the world derives from plant juice containing sucrose from sugarcane in Brazil and starch from corn in the United States [31]. Ethanol can also be produced from sugars derived from lignocellulosic material by hydrolysis of the cell wall using enzymes, physical and chemical treatments. Ethanol obtained for this process is called second generation ethanol. The efficiency of the lignocellulose conversion process has not yet proven to be economical, but second generation ethanol is a highly desired goal because it could significantly broaden the choice of feedstock which would contribute to avoid competition between ethanol and food production. Several aspects of bioethanol production are discussed in literature including technologies to produce anhydrous ethanol [32], biotechnological procedures [33], alternative feedstocks [34], pretreatments for efficient enzymatic hydrolysis [35] and works totally devoted to second generation bioethanol [36-39].

Biodiesel is the second most used liquid biofuel worldwide. It is obtained from renewable sources such as vegetable oils and animal fats. Its biodegradability, nontoxic nature and low emissions profile are the main attractive features of this biofuel [40]. The performance of biodiesel as transportation fuel is very similar to petrodiesel [41], which makes biodiesel the only biofuel potentially able to replace petrodiesel. Some characteristics of biodiesel such as its low sulfur content and its lubricating property make it a fuel even better than petrodiesel [41]. However, before the use of biodiesel as a fuel in its pure form some of its properties must be improved such as its viscosity, oxidation resistance and crystallization temperature [42].

Germany and France are the largest producers of biodiesel [43] being Europe responsible for 70 % of the biodiesel produced in the world. European biodiesel production has drastically increased from 1998 to 2008. In this period, biodiesel production grew from 475 thousand tons to 16 million tons [42]. In USA the production of biodiesel has increased from 2 million gallons to 700 million gallons in the period 2000-2008. A significant increase on biodiesel fuel market is expected not only in developed but also in developing countries such as China, Brazil, Indonesia and Malaysia [42]. Nowadays the main usage of biodiesel is as an additive to petrodiesel forming blends containing 5-10 % of biodiesel [42]. The percentage of biodiesel depends on regulatory legislation of each country and there is a global tendency in to increase biodiesel contents in these blends.

Biodiesel is produced by a transesterification reaction between triacylglycerides from renewable biological resources, such as plant oils or animal fats and a short-chain alcohol such as methanol or ethanol. Therefore, biodiesel is a mixture of monoalkyl esters of fatty acids (mostly methyl or ethyl esters) [42]. The main side product in the production of biodiesel is glycerin which after additional purification could be used in perfumery, medicine, and in the microbiological synthesis of ethanol, succinic acid, etc. [42]. There are three main transesterification processes that can be used to produce biodiesel: base-catalyzed, acid-catalyzed and enzyme-catalyzed transesterification [43]. The raw material used as oil source for biodiesel production includes edible and non-edible oils, algae, waste cooking oil, etc. [41]. There are several vegetable oils used as raw material for biodiesel production. In the United States and Brazil soybean oil is a source that is already scaled up for biodiesel production. Nevertheless, other sources, such as rapeseed (in Europe), sunflower, peanut, cotton, palm oil, coconut, babassu, and especially castor oil, may also be used in different parts of the world once their cultivation can achieve an economic up-scaling [43]. The use of edible oil and arable food land to produce biodiesel is very controversial and it has been considered the main drawback of biofuels. Thus there is a huge interest in the use of alternative raw materials for biodiesel production. Several scientists have worked to develop new producing processes in which algae can be used as raw material [44,45]. However, several challenges need to be overcome in order to make possible the commercial production of biodiesel from algae at a scale sufficient to make a significant contribution to our transport energy needs.

1.2 Quality control of biofuels

According to discussed in previous sections, the importance of biofuels for world energetic future is evident. Therefore, the assurance of the properties and quality of biofuels is a key aspect to their successful commercialization and market acceptance. Besides economic aspects, the quality of biofuels has implications over environmental aspects since some contaminants can lead to severe operational problems which can increase the emissions levels of engines using biofuels or reduce the efficiency of the biofuel. Moreover, some contaminants can accelerate undesirable reactions leading to storage instability. The composition of biofuels is highly variable because of several factors such as variability of raw materials, influence of climate and soil, transportation and storage method, etc. This fact reinforces the importance of rigid international technical specifications able to guarantee uniform quality parameters for biofuels produced in different regions. However, technical specifications are not the only requirement for an efficient quality control. The quality control of biofuels depends also on reliable official analytical methods and certified reference materials. Reliable official analytical methods are the tool responsible to guarantee that technical specifications are obeyed while certified reference materials are extremely important to validate and check the performance of both official and new analytical methods [46].

Quality control of biofuels is a complex and interdisciplinary issue that represents a real challenge to our society. In order to respond to this challenge there are necessary integrated actions involving scientific and political sectors. International integrated actions seem to be the best way to assure the homogeneity of the quality of biofuels produced in different regions of the world. Fortunately, these actions are in progress since 2007 when Brazil, European Union (EU) and USA, the three largest biofuels producers, elaborated a White Paper on internationally compatible biofuel standards issued by a tripartite task force. USA, EU and Brazil agree that there are different standards for biofuels which were known to be an obstacle to the free circulation of biofuels among the three regions. It is stated on the White Paper that the participants agree to promote, whenever possible, the compatibility of biofuels-related standards in their respective regions. Such compatibility would not only facilitate the increasing use of biofuels in each of the regional markets, but also would support both exporters and importers of biofuels by helping to avoid adverse trade implications in a global market [47].

The subsequent subsections of this work will be focused in the contribution of analytical chemistry for quality control of bioethanol and biodiesel. Especial attention will be addressed to official analytical methods and technical specifications adopted by EU, USA and Brazil. Some attention will be also addressed to alternative non-electrochemical analytical methods. Despite this work is centered on electroanalytical methods, some references about other analytical techniques will be included in order to provide additional material for readers interested in more general aspects of analytical chemistry of bioethanol and biodiesel.

1.2.1 Bioethanol quality control

The main bioethanol producers have important regulatory agencies which are responsible to implement technical specifications and official norms for quality control of bioethanol. In Brazil, the agency responsible for the inspection and commercialization rules of all fuels is ANP (National Agency of Petroleum, Natural Gas and Biofuels). ANP resolutions establish criteria of quality control using Brazilian Norms (NBR) which are defined by ABNT (Brazilian Association of Technical Norms). In USA and EU the quality of bioethanol is regulated, respectively, by ASTM (American Society for Testing and Materials) and ECS (European Committee for Standardization). ECS resolutions are ruled by European Norms (EN). Technical specifications adopted in EU, USA and Brazil regulate some physical properties of bioethanol, such as appearance and density and also the content of some organic and inorganic species. The complete list of technical specifications adopted by the three main producers of bioethanol is presented in Table 1.

Quality parameter	USA		Brazil		EU
	ASTM D4806-10	ASTM D4806-10 Undenatured	ANP36/2005 Anhydrous	ANP36/2005 Hydrous	EN15376-07
Appearance	Clear & Bright	Clear & Bright	Clear & no impurities	Clear & no impurities	Clear & Bright
Color	Dye allowed, but not mandated	Dye allowed, mandated but not mandated	Dye mandated for in country, but not for export.	Dye prohibited for in country	Dye allowed, but not mandated
Maximum density at 20 °C, kg/m^3	----	----	791.5	807.6	----
Maximum electrical conductivity, µS/m	----	----	500	500	----
Maximum acidity, mg/L	0.007	0.0074	0.0038	0.0038	0.007
pHe	6.5-9.0	6.5-9.0	----	6.0-8.0	Dropped
Minimal ethanol Content, vol.%	92.1	93.9	99.6	----	[96.8]
Minimal ethanol + C3-C5 saturated alcohols content, vol.%	----	[98.4]	----	----	98.8
Minimal total Alcohol content, vol.%	----	[98.95]	99.6	95.1	[99.76]
Maximum saturated C3-C5 alcohols content, vol.%	----	[4.5]	----	----	2.0
Maximum water content, vol.%	1.0	1.05	[0.4]	[4.9]	0.24
Maximum methanol content, vol.%	0.5	0.53	----	----	1.0
Minimal/Maximum denaturant content, vol%	1.96/5.0	No denaturant	No denaturant	No denaturant	Set by country 0/1.3
Maximum Hydrocarbons content, vol.%	----	----	3	3	----
Maximum solvent-washed gum content, mg/100 mL	5.0	5.3	----	----	----

Maximum gum or residue by evaporation, mg/100mL	5 (washed gum)	5.3 (washed gum)	----	5 (unwashed)	10 (unwashed)
Maximum sulfate content mg/kg	4	4.2	----	4	----
Maximum chloride content mg/kg	40	42.1	----	1	25
Maximum sodium content mg/kg	----	----	----	2	----
Maximum copper content, mg/kg	0.1	0.105	0.07	----	0.1
Maximum iron content, mg/kg	----	----	----	5	----
Maximum phosphorus content, mg/L	----	----	----	----	0.5
Maximum sulfur content, mg/kg	30	5	----	----	10

Numbers in [ ] are calculated estimates and not specified limits

Table 1. Ethanol specifications adopted in USA, Brazil and EU [47]

Table 1 shows that technical specifications are very similar in the three regions and therefore they do not constitute serious commercial barriers. On the White Paper, technical specifications were classified in three groups: Category A - similar specifications; Category B - specifications with significant difference but which can be aligned and Category C - specifications very different which can not be aligned in short term. Among the ethanol technical specifications, only water content was classified in category C, thus if Brazil and USA exporters wish to supply EU market additional drying will be required [47]. The similarity on technical specifications will be improved even more, since these three regions are working together in this direction since 2007. This will significantly contribute to homogenize the quality of bioethanol produced in different regions which is very important to the consolidation of this fuel as an important product in the global market.

Regulated parameters are those able to seriously compromise the general quality of bioethanol causing economical or environmental damages. For example, high values of acidity, conductivity, water content and extreme pH values can intensify the corrosion power of ethanol. Therefore, if these parameters are not properly regulated, ethanol fuel could cause fast deterioration of metallic components present in the production process, transportation, storage and engines using ethanol fuel. Some chemical species such as sodium, chloride and sulfate are also able to intensify the corrosion power of ethanol and, therefore, they must be regulated as well. Copper and iron can catalyze polymerization reactions in blends ethanol-gasoline; these reactions promote the formation of gum and sediments which seriously affects the performance and cause several damages in the engine and fuel injection system. Sulfur constitutes an example of contaminant able to increase the emission level of ethanol fuel, since after combustion it is converted in SO_x which are released for atmosphere, thus the regulation of sulfur content is important from environmental concerns.

Regulated Specie	Official Analytical Method	Technical Norm
Chloride	Ion Chromatography	ASTM D7319-09 NBR 10894-07 EN 15492-08
Sulfate	Ion Chromatography	ASTM D7319-09 NBR 10894-07 EN 15492-08
Sodium	Flame Photometry	NBR 10422-07 EU and USA have no specification
Copper	Atomic Absorption Spectroscopy (AAS)	ASTM D1688-07 NBR 11331-07 EN 15488-07
Iron	Atomic Absorption Spectroscopy	NBR 11331-07 EU and USA have no specification
Sulfur	X-ray Fluorescence Spectrometry (USA)	ASTM D2622-10
	Ultraviolet Fluorescence* (USA)	ASTM D5453-09
	X-ray Fluorescence Spectrometry (EU)	EN 15485-07
	Ultraviolet Fluoresce (EU)	EN 15485-07
		Brazil has no specification
Phosphorous	Spectrophotometry (EU)*	EN 15487-07 Brazil and USA have no specification
Methanol	Gas Chromatography (EU and USA)	ASTM D5501-09 EN 13132-00 Brazil has no specification
C3-C5 alcohols	Gas Chromatography (EU)	EN 13132-00 Brazil and USA have no specification

* After reaction with ammonium molybdate

Table 2. Official analytical methods adopted in Brazil, USA and EU for quantification of some inorganic and organic contaminants of bioethanol [47,49]

To ensure that required technical specifications are fulfilled there are necessary reliable official analytical methods and their development constitutes the main contribution of Analytical Chemistry community to bioethanol quality control. Official analytical methods are established by technical norms which specify all conditions of analyses including analytical technique, sample pretreatment, calibration procedure, etc. The main analytical official methods used to regulate the content of inorganic and organic contaminants in bioethanol fuel are presented in Table 2.

From Table 2 it can be observed that official analytical methods adopted for bioethanol quality control are predominantly instrumental ones, which are usually characterized by high analytical performance showing high sensitivity, selectivity and analytical frequency. Most of technical norms are very recent and they present updated analytical methods. In the last five years it was observed an impressive evolution in some specifications. For example, sulfate specification was initially based on a gravimetric method, afterwards a volumetric method was adopted and more recently the official analytical method adopted by Brazil, EU and USA is based on ion chromatography, a very modern and reliable analytical technique. Chloride analytical official method has also evolved from volumetric for an ion chromatography method which is also adopted by the three regions [46]. The use of modern instrumentation and analytical techniques is important to ensure that precise and exact analytical data will be obtained which is the basis for a reliable quality control. On the other hand, the use of modern and more expensive analytical instrumentation usually faces resistance from producers who believe these investments are unnecessary. Regulatory agencies has fulfilled they role to overcome such resistances by adopting high performance and modern analytical method for the quality control of bioethanol. This is very important in order to guarantee that ethanol will take its place as a key global product in a very near future.

Although bioethanol technical specifications are modern, they still have some problems which have attracted the attention of several researches. Spitzer et al [48] have presented a very interesting discussion about pH and conductivity determination in ethanol fuel including metrological problems associated with these parameters. High ethanol and low water containing solutions typically lead to difficulties in pH determination with unstable readings, long response times and device depending results. These are indications for low conductivity of the sample, dehydration of the glass electrode membrane and large junction potentials. Due to a lack on ethanol based reference buffer solution, the pH electrode in all test methods is calibrated in aqueous standard buffer solutions which is still far from being satisfactory. Therefore, worldwide harmonization in pH and conductivity measurement practices is urgently needed for reliable results as stated by Spitzer et al. [48].

Another limitation of bioethanol technical specifications is that they contemplate only a few numbers of chemical species. For example there is no specification for acetaldehyde which is a well known ethanol fuel contaminant. Higher aldehydes which can decrease ethanol storage stability are not regulated as well. This fact has encouraged many scientists to develop alternative analytical methods for quantification of several chemical species in ethanol fuel. Most of these alternative analytical methods are instrumental ones and they bring interesting refinements in bioethanol analysis. The most complete work about alternative analytical methods for bioethanol fuel analysis was written recently by Oliveira et al [49]. This work revises the main advances in the quantification of organic and inorganic species in ethanol fuel by using modern analytical techniques. From this work it can be concluded that gas chromatography is predominant for the quantification of organic contaminants in bioethanol followed by high performance liquid chromatography (HPLC). Atomic spectroscopy techniques are the most used for the quantification of cationic inorganic contaminants of bioethanol. For analysis of anionic inorganic it is observed a large variety of analytical methods such as electrophoresis, UV-spectroscopy, besides potentiometric and conductometric methods. Since Oliveira's work provides a very comprehensive and recent revision about the use of modern alternative analytical methods for analysis of bioethanol, this topic will not be covered here.

1.2.2 Biodiesel quality control

The same regulatory agencies responsible for bioethanol specifications in Brazil, EU and USA are also responsible for specifications of biodiesel in these three regions. As stated previously, biodiesel is produced from a transesterification reaction between a vegetal oil or fat and an alcohol, mainly methanol or ethanol in presence of a catalyst, usually a strong base such as sodium or potassium hydroxide. The transesterification reaction plays a crucial role for the final quality of biodiesel because side products, residues of catalyst and reagents are the main sources of contamination of the final product. Besides glycerol, the main side product of transesterification reaction, mono-, di- and triacylglycerol can be also formed since this reaction is a stepwise process [50]. Therefore, the content of glycerols in biodiesel must be limited by specification since these species brings several operational problems causing loss of efficiency and damages in the engines. Besides glycerols other contaminants are introduced in the transesterification step such as Na, K, free fatty acids and alcohols and all of them have deleterious effect on final quality of biodiesel. Therefore these species are regulated by technical specifications. The complete list of specifications of biodiesel adopted in Brazil, EU and USA is presented in Table 3.

Biodiesel specifications adopted in Brazil and USA are applicable for both fatty acid methyl esters (FAME) and fatty acid ethyl esters (FAEE), whereas the current European biodiesel standard is only applicable for FAME. Also, the specifications for biodiesel in Brazil and USA are used to describe a product that represents a blending component in conventional petroleum based diesel, while the European biodiesel standard describes a product that can be used either as a stand-alone diesel fuel or as a blending component in conventional petroleum based diesel [47]. Taking into consideration these facts it is not surprising that there are some significant differences among the sets of specifications adopted in Brazil, EU and USA as can be observed in Table 3.

Table 3 clearly shows that biodiesel specifications are not so closely aligned in the three regions as bioethanol specifications. Contrasting with bioethanol, biodiesel is not a single compound but it is a mixture of fatty acids esters which makes biodiesel more complex than bioethanol. In addition, biodiesel can be produced via methylic or ethylic route, originating two chemically different kinds of biodiesel: FAME and FAEE mixtures. These features associated with the fact of biodiesel is produced from several types of feedstock on different regions also lead to significant variations in the performance and characteristics of the final product making a very hard challenge to develop a common set of specifications for biodiesel produced in different regions. On the White Paper, the biodiesel tripartite task force has identified ten specifications classified in category C [47], which demonstrates a significant divergence in specifications and properties of biodiesel produced in the three regions. This could be seen as a strong impediment to trade. However, in most of the cases, it is possible to meet different regional specifications by blending various types of biodiesel to achieve the desired quality and specifications [47] which drastically minimize trade problems.

Quality parameter	Units	Limits: USA ASTM D6751/2003	EU EN14214/2003	Brazil ANP 07/2008
Ester Content	% mass	----	96.5 min[a]	96.5 min
Density at 15 °C	Kg/m³	----	860-900	----
Density at 20 °C	Kg/m³	----	----	850-900
Kinematic viscosity at 40 °C	mm²/s	1.9-6.0	3.5-5.0	3.0-6.0
Distillation Temperature, 90 % Recovered	°C	360 max[b]	----	360 max
Oxidation Stability, 110 °C	hours	3.0 min	6.0 min	6.0 min
Flash Point	°C	130.0 min	120 min	100 min
Cetane Number		47 min	51 min	Report
Cloud Point	°C	Report	----	----
Cold Filter Plugging Point	°C	----	5 max (Grade A) 0 max (Grade B) -5 max (Grade C) -10 max (Grade D) -15 max (Grade E) -20 max Grade F)	19 max
Total Contamination	mg/Kg	----	24 max	24 max
Carbon Residue (on 100 % sample)	% mass	0.050 max	----	0.050 max
Acid Number	mg KOH/g	0.50 max	0.50 max	0.50 max
Water and Sediment	% volume	0.050 max	----	0.050 max
Water Content	mg/Kg	----	500 max	500 max
Methanol or Ethanol Content	% mass	0.20 max	0.20 max	0.20 max
Free Glycerol	% mass	0.02 max	0.02 max	0.02 max
Monoacylglycerol Content	% mass	----	0.80 max	Report
Diacylglycerol Content	% mass	----	0.20 max	Report
Triacylglycerol Content	% mass	----	0.20 max	Report
Total Glycerol	% mass	0.24 max	0.25 max	0.25 max
Linolenic Acid Methyl Ester	% mass	----	12.0 max	----
Polyunsaturated Methyl Ester	% mass	----	1 max	----
Iodine Value	$g\ I_2/100\ g$	----	120 max	Report
Sulfated Ash	% mass	0.020 max	0.02 max	0.02 max
Na + K	mg/kg	5 max	5 max	5 max
Ca + Mg	mg/kg	5 max	5 max	5 max
Copper Strip Corrosion	Rating	Class 3	Class 1	Class 1
Phosphorous Content	% mass	0.001 max	0.0010 max	0.0010 max
Sulfur Content	mg/Kg	15/500 max	10 max	50 max

[a]Minimal value necessary, [b]Maximum value allowed

Table 3. Biodiesel specifications adopted in USA, Brazil and EU [47]

Another consequence of the higher complexity and variability of the chemical composition of biodiesel is the need of a large number of specifications for chemical contaminants. This is necessary because biodiesel presents more chemical species together which increases the possibility of chemical processes able to compromise the quality of this biofuel. For example, free glycerol may separate from biodiesel and to accumulate on the bottom of storage containers or vehicle fuel tank attracting other polar components such as water, monoglycerides and soaps. These agglomerates can result in damage to the vehicle fuel injection system due to impairment of fuel filters. Alkali and alkali-earth are also able to damage engines and fuel injection systems because they can promote the formation of ash and soaps. Alkali metals are introduced in biodiesel composition from alkali catalyst residue while alkali-earth metals can be introduced from hard washing water [47]. The content of ethanol or methanol is another very important specification because these alcohols can significantly accelerate corrosion of metallic components of engines; they can cause adverse effects on injectors due to its high volatility and low lubricity [47]. The corrosion power of biodiesel can be also increased by the presence of excess of free acids, which is expressed by the acid number. This parameter is also useful to indicate the degree of fuel ageing during storage because the acid number increases due biodiesel degradation promoted mainly by esters hydrolysis reactions, originating free fatty acids. Thus, acid number is a very important quality parameter which is limited by technical specifications in Brazil, EU and USA. The ability of biodiesel to cause corrosion of copper, zinc and bronze parts of the engine and the storage tank is evaluated by the copper strip corrosion. To determine this quality parameter, a copper strip is heated to 50°C in a biodiesel bath for three hours, and then compared to standard strips to determine the degree of corrosion. This corrosion resulting from biodiesel might be induced by some sulfur compounds and by acids, so this parameter is correlated with acid number [47]. Water is other important chemical contaminant of biodiesel because high water contents promote biological growing, causing slime formation compromising fuel filters and lines in production systems. Moreover, high water contents promote biodiesel hydrolysis reaction originating free fatty acids, increasing the acid number and the corrosion power of biodiesel. Therefore, water content is directly related to storage stability of biodiesel since water is able to degrade biodiesel by hydrolysis reaction. A review of the test methods used to evaluate the stability of biodiesel was recently written by Jain and Sharma [51]. Other chemical contaminants whose contents are limited by specifications are phosphorus and sulfur. Phosphorus can reduce the efficiency of exhaust catalytic system of vehicles and sulfur increases the SO_x emissions [47]. Therefore, both contaminants cause environmental damages and their contents must be regulated.

Some of the specifications of biodiesel (Table 3) are indicative of its purity such as ester content, density, distillation temperature and total contamination [47]. Other specifications are directly related to biodiesel performance as fuel. For example, the cetane number describes biodiesel propensity to combust under certain conditions of pressure and temperature. High cetane number is associated with rapid engine starting and smooth combustion. Low cetane number causes deterioration in this behavior and causes higher emissions of hydrocarbons and particulate. In general, biodiesel has slightly higher cetane number than fossil diesel. Cetane number increases by increasing length of both fatty acid chain and ester groups, while it is inversely related to the number of double bonds [47]. The number of double bonds can be evaluated by the iodine number, linolenic acid methyl ester and polyunsaturated methyl ester content and also by oxidation stability. Kinematic viscosity is another important quality parameter for fuels because high viscosity compromises the spray injected in the engine and at low temperatures high viscosity can damage fuel pumps. Flash point is a measure of flammability of a fuel and therefore it is an important safety criterion in transport and storage. Flash point values for biodiesel are usually double those observed for petroleum-based diesel which represents a safety advantage of biodiesel [47]. Thus, the set of biodiesel specifications adopted in Brazil, EU and USA are able to guarantee both the purity and the suitable performance of biodiesel. More detailed discussion about ASTM and EN technical specifications can be found in the literature [50,52].

Analytical official methods used for the determination of the content of the main inorganic and organic contaminants of biodiesel are presented in Table 4.

From Table 4 it can be observed that official analytical methods adopted for biodiesel quality control are also predominantly instrumental ones. The complexity of biodiesel requires highly selective and sensitive analytical methods, because there are several chemical species together in the sample and the contaminants are minor components of biodiesel, therefore they must be reliably quantified even at low concentrations. These requirements are filled only by instrumental analytical methods. Analogously to the observed for bioethanol, technical specifications of biodiesel are very modern and they are based on updated instrumental analytical methods. From Table 4 it is possible to observe that gas chromatography is the only analytical technique used for quantification of organic contaminants of biodiesel. For the quantification of inorganic contaminants, there is predominance of atomic spectroscopic methods, which are characterized by extremely high selectivity and sensitivity. Despite the extremely high analytical performance of ICPOES, the high instrumental costs and the needing of highly specialized operators are strong limitations for its use in routine analysis. In order to minimize problems associated with very expensive analysis, there are still active technical norms for inorganic contaminants based on AAS, a high sensitive, selective and instrumentally simpler technique than ICPOES. Therefore, for quantification of Na, K, Ca and Mg technical regulation adopted in Brazil, EU and USA allows the use of both techniques.

There are two very important works about analysis of biodiesel: the review written by Knoth in 2006 [50] and the review written by Monteiro in 2008 [52]. A more recent review about quality and analytical methods for biodiesel analysis was written by Lôbo et al [53] in 2009. All these works bring comprehensive discussions about both official and alternative analytical methods employed for biodiesel analysis. From these works it is possible to conclude that organic contaminants of biodiesel are quantified mainly by gas chromatography (GC), which is the base not only of official but also of alternative analytical methods. Alternative GC-based methods have been also used for monitoring the transesterification reaction [52]. Flame ionization detector (FID) is the most used detection system for GC-biodiesel analysis; however the use of coupled mass spectrometer (MS) has increased lately. This detection system significantly increases the identification power of GC, avoiding ambiguities about the chemical nature of the eluting species [52]. Recent approaches for development of GC-based alternative analytical methods include the use of ethyl oleate as internal standard for glycerol determination [54], the use of ionic liquid as stationary phase for determination of FAME in diesel blends [55] and the determination of polyunsaturated FAMEs with wax capillary column using methyl tricosanoate as internal standard. [56]. Two-dimensional gas chromatography has also been successfully used for analysis of diesel blends [57] and for determination of fatty acid mixtures composition [58]. These methods have brought interesting advances on determination of organic contaminants of biodiesel.

Regulated Specie	Analytical Official Method	Technical Norm
Water + Sediment	Centrifugation	ASTM D2709-96
Water	Karl Fisher (Coulometry)	EN 12937-01 ASTM D6304-07
Methanol or Ethanol	Gas Chromatography	EN 14110-03 NBR 15343-08
Free Glycerol	Gas Chromatography	ASTM D6584-10 EN 14105-03 EN 14106-03 NBR 15341-06 NBR 15771-09
Total Glycerol	Gas Chromatography	ASTM D6584-10 EN 14105-03 NBR 15344-10
Linolenic Acid Methyl Ester	Gas Chromatography	EN 14103-03
H^+ (acid number)	Potentiometric Titration	ASTM D664-09 NBR14448-09
	Visual indicator Titration	EN 14104-03
Na + K	Atomic Absorption Spectroscopy	EN 14108-03 EN 14109-03 NBR 15554-08 NBR 15555-08 UOP* 391
	Inductively Coupled Plasma Optical Emission Spectroscopy (ICPOES)	EN 14538-06 NBR 15553-08
Ca + Mg	Inductively Coupled Plasma Optical Emission Spectroscopy	EN 14538-06 NBR 15553-08
	Atomic Absorption Spectroscopy	ASTM D4951-09 EN 14107-03 NBR 15556-08

Phosphorus Content	Inductively Coupled Plasma Optical Emission Spectroscopy	ASTM D4951-09 EN 14107-03 NBR 15553-08
Sulfur Content	Ultraviolet Fluorescence Spectrometry	ASTM D5453-09 EN 20884-10 EN 20846-10

* Universal Oil Products

Table 4. Official analytical methods adopted in Brazil, USA and EU for quantification of some inorganic and organic contaminants of biodiesel [47]

Liquid chromatography (LC) methods are less common than GC for quantification of organic contaminants of biodiesel. However, according to some authors, HPLC methods are able to provide short time analysis with no need of chemical derivatisation. HPLC is very useful to study the transesterification reaction because it is compatible with gradient elution, which is necessary to the efficient separation between mono-, di- and triacylglycerols [50,52]. Santori et al [59] have proposed a liquid chromatographic method which successfully enables the monitoring of transesterification reaction. Other recent examples of biodiesel analysis by LC methods include determination of acylglicerols, free fatty acids and FAME by size exclusion chromatography [60] and determination of ester in glycerol phase after transesterification by a RP-LC method [61].

Besides chromatographic methods, some spectroscopic techniques have been also used to study the transesterification reaction and to determine organic contaminants in biodiesel. Among these methods, nuclear magnetic resonance (NMR) and Fourier transform infrared (FTIR) spectrometry are the most commonly employed [52]. NMR was successfully used by Nagy et al [62] to evaluate both the yield of transesterification and the overall quality of biodiesel by a procedure based on glycerol hydroxyl phosphitylation followed by ^{31}P-NMR analysis. Other applications of NMR to biodiesel analysis include determination of biodiesel levels in blends [63]; determination of free fatty acids [64] and FAME [65] in biodiesel samples and also evaluation of the biodiesel preparation process [66]. An analytical method using FTIR spectroscopy has been developed by Mahamuni and Adewuyi [67] to determine biodiesel content in the reaction mixture, which allows to monitor the transesterification reaction. Moreover, this method could be successfully employed to determine the biodiesel content in petrodiesel-biodiesel blends. These authors have also shown that small modification in the analytical method enables its application to detect biodiesel adulteration with soy oil. Chuck et al [68] have performed an interesting study by using three analytical techniques: UV, FTIR and refractive index to analyze the chemical identity of biodiesel and its content in blends. These authors have concluded that FTIR is suitable to determine biodiesel content in biodiesel-petrodiesel blends while refractive index is very useful to determine the biodiesel identity. These authors also concluded that UV Spectrophotometry can be successfully used for determination of total amount of polyunsaturated esters in samples of B100 biodiesel.

An interesting advance in biodiesel analysis was introduced by Abdelnur et al [69]. These authors have described a very simple and efficient alternative method for biodiesel typification by using mass spectrometry. This method allowed the analysis of biodiesel without necessity of any sample pretreatment which is a great advantage over most of official and alternative analytical methods. Another example of using mass spectrometry for biodiesel analysis can be found in the work by Prates et al [70]. These authors have combined mass spectrometry with multivariate calibration to analyze biodiesel-petrodiesel blends. The proposed method was a simple, fast and reliable way for quantification of biodiesel in fuel blends.

Some authors have used Flow Injection Analysis (FIA) systems to analyze biodiesel. Pinzi et al [71] have proposed a FIA method with spectrophotometric detection based on determination of free and bound glycerol to monitor biodiesel production. This method was based on liquid–liquid extraction of glycerol from the biodiesel to an aqueous ethanolic phase in which glycerol was oxidized to formaldehyde with meta periodate with subsequent reaction with acetylacetone. The reaction product was photometrically measured at 410 nm. The analytical performance of the proposed method was compared with EN 14105 technical norm at a 95 % confidence level and no statistical difference between the results obtained by both methods was observed. The main advantages of the proposed method are: high analytical frequency, low cost (compared with GC and HPLC) and minimizing of sample handling. Rio et al [72] have used a Sequential Injection Analysis (SIA) based flow titration to determine the total acidity of biodiesel samples. A diode array spectrophotometric detector was used linked to chemometric tools. The results obtained with the proposed method were in good agreement with those obtained by official analytical methods, demonstrating the reliability of the method. Moreover, the use of a SIA method helps to reduce the amounts of sample, reagents and time consumed.

Analysis of inorganic contaminants of biodiesel has been performed mainly by atomic spectrophotometric methods and several recent works about it are available in the literature. Lyra et al [73] have used flame atomic absorption spectrometry (FAAS) to develop an analytical method for quantification of Na, K, Ca and Mg in biodiesel samples. In order to overcome problems associated with the high viscosity of biodiesel, the authors have prepared a microemulsion as sample pretreatment procedure. This procedure has allowed the reliable quantification of the metallic species in biodiesel, which was demonstrated by the concordance between the results obtained by the proposed method and those obtained by the official NBR 15556 technical norm. The preparation of a microemulsion was also successfully used by Jesus et al [74] for determination of Ca and Mg in biodiesel samples by FAAS. The use of microemulsions is a very interesting way to make biodiesel compatible with sample introduction systems used in FAAS because microemulsions are prepared in a very fast, simple and inexpensive way. This sample pretreatment strategy was adopted also by Chaves et al [75] for determination of K and Na in biofuel samples by FAAS. Another interesting sample pretreatment procedure is the dry decomposition in muffle furnaces. This procedure was successfully employed by Oliveira et al [76] for determination of Na in biodiesel by flame atomic emission spectrometry. Graphite furnace atomic absorption spectrometry (GF-AAS) has been also successfully used for biodiesel analysis. A GF-AAS method for direct determination of phosphorus in biodiesel samples was proposed by Lyra et al [77]. These authors have used an automatic solid sampling accessory which allowed P determination in biodiesel without any pretreatment or sample dilution. GFAAS has been also used for quantification of heavy metals such as As [78], Cd, Pb and Tl [79] in biodiesel samples. Although the content of these metals is not limited by specification, their high toxicity justifies the interest on their quantification in biodiesel samples. Aranda et al [80] have proposed an analytical method for determination of Hg in biodiesel samples. The samples were introduced directly as oil-in-water emulsions in a flow injection manifold followed by cold vapor generation coupled to atomic fluorescence spectrometry (FI-CV-AFS). The authors were able to quantify both inorganic and organic mercury (as methyl mercury) after irradiation of the sample with a UV source.

2. Electroanalytical methods

Electrochemical techniques are versatile and powerful analytical tools, which are able to provide high sensitivity, low detection limits, associated to the use of inexpensive instrumentation which presents as an additional advantage relatively low operator training requirements. These features make electrochemical techniques very attractive not only for routine analytical applications but also for fundamental physicochemical research such as molecular interactions, or interface processes [81]. Electroanalytical methods involve the measurement of an electrical property such as current, potential, conductivity or electric charge which are directly related to the analyte concentration. Therefore, electroanalysis is based on the fundamental relationships existing between electricity and chemistry and it has been used in several areas such as biomedical analysis, environmental monitoring, industrial quality control, etc. [82]. The most common routine applications of electroanalysis include potentiometric determination of pH, amperometric oxygen and glucose probes, conductometric control of ultra purified water, coulometric quantification of water in non-aqueous materials, etc. [81]. Besides these well established routine applications, the intensive research on electroanalysis has brought new and impressive potential applications for electroanalytical methods. A very interesting example of a new using of electroanalytical methods is found in the work written by Doménech-Carbó [83]. This author discusses the application of electroanalytical methods to quantify analytes from works of art demonstrating that electroanalysis is potentially useful to provide information about conservation and restoration of cultural goods. Another impressive and modern application of electroanalytical methods is the development of amperometric electronic tongues for food quality analysis [84].

Researches on electroanalysis are not limited to develop new applications for electroanalytical methods but they are also interested on the improvement of the analytical performance of electrodes and electrochemical sensors. Thus, special attention has been devoted to the use of new materials for electrode construction. These researches have produced electrochemical devices highly sensitive and selective which have allowed the use of electroanalytical methods even for analysis of very complex samples such as biofuels. These electrodes with superior analytical performance are based on modern materials and their development was achieved mainly due the incorporation of nanotechnology on the electrode production processes. Therefore, several nanomaterials have contributed to developing modern electrodes such as carbon nanotubes [85,86] and several kinds of nanoparticles [87]. The use of zeolites, silicas and nanostructured porous electrodes has also significantly contributed to development of electrodes with improved analytical performance [88,89].

The main electroanalytical techniques are potentiometry, conductometry, coulometry, voltammetry and amperometry. Recent advances and fundamental aspects of these techniques will be briefly discussed here. More complete and deeper fundamental discussions can be found in classical electroanalytical literature [82,90,91].

Potentiometry is based on thermodynamic relationships between the electrical potential measured at equilibrium conditions and the activity of some specie in solution. In electrochemical sense, equilibrium conditions mean that there is no current flowing across to the electrochemical cell. Therefore, in potentiometry, information about composition of a sample is obtained by measuring electrical potential between two electrodes in absence of appreciable current. Analytical applications of potentiometry include the detection of endpoint in potentiometric titrations and direct potentiometry in which the potential measured is used to provide the analyte concentration. The main applications and advances in potentiometric methods involve direct potentiometry, thus only this modality will be discussed here. The whole electrochemical system required for potentiometric measurements includes a potentiometer which allows potential measurements and prevents the flowing of current between the electrodes; a reference electrode whose potential is constant and insensitive to the sample composition and an indicating electrode which potentiometrically responds to the analyte activity [92]. Potentiometry allows accessing activities (or effective concentration) data instead analytical concentration. This is extremely advantageous for fundamental physicochemical studies however, for analytical purposes this is unfavorable since in this case information required is about analytical concentration and not activities. To overcome this, calibration procedures in direct potentiometry must ensure that the activity coefficient of analyte is the same in both sample and calibration standards. Standard addition method is also very useful for direct potentiometric analysis. Thus, if calibration is properly performed, the electrical potential measured between reference and indicating electrodes (with $I = 0$) can be directly related to analyte concentration (in logarithmic form). Membrane based electrodes are the most common indicating electrode used in direct potentiometry. They are also called Ion Selective Electrodes (ISEs) because their high selectivity. The potential developed across the ISE/sample/reference electrode system is not related to redox reactions but it is a junction potential which it is related to specific chemical equilibrium that takes place in the solution/membrane interface such as ion exchange. Useful membranes should present non porous nature, insolubility in the sample solution, little conductivity, high mechanical stability and selective reactivity towards the analyte. The most known membranes are the glass membrane for pH measurements and the crystalline LaF_3 membrane for determination of fluoride. ISEs are very attractive analytical devices because they present fast responses and wide linear range; they are not affected by color or turbidity of sample; they are inexpensive and finally they are not sample destructive [82]. A recent and very complete review about electrochemical theory related to potentiometry was written by Bobacka et al [93]. Modern ISE research is focused mainly in developing new ion recognition species, also called ionophores. Thus several new ionophores have been developed which have produced more stable and selective ISEs. Another trend in ISE research is to develop more robust and practical devices, thus there is significant interest on the development of all-solid state ISEs and miniaturized systems [93]. A very interesting discussion about advances in potentiometric sensors was written by Privett et al [94]. These authors present an impressive number of references discussing the main advances in potentiometry covering ISEs development and analytical applications in several areas. ISEs can be successfully used as detection system in flow analysis combining the high selectivity of these devices with all advantageous features of FIA systems such as low reagent and sample consumption, better repeatability, minimized sample manipulation and consequent lower risk of sample contamination, high analytical frequency, and low instrumentation cost. Trojanowicz [81] has presented a very instructive discussion about the use of potentiometry as detection system in flow analysis and capillary electrophoresis.

The electrical conductivity of a solution can be used for analytical purposes and this is the base of conductometric methods. The solution conductivity depends on the ion concentration and the characteristic mobility of the ions in solution. Therefore, the conductivity of one-solute solution can be used to directly provide information about the concentration of ionic species in solution. However, in multiple-solute solutions the contribution of single ionic specie to the total solution conductivity can not be done and in these situations conductometry is not useful for analytical purposes. This lack of selectivity has discouraged the development of conductometry as a widespread electroanalytical technique. However, conductometry is one of the most sensitive techniques available and therefore its modern applications exploit this high sensitivity in situations in which selectivity is not required. One example of application of conductometry is the monitoring of the quality of ultra purified water. The conductivity in this case is very low and the presence of even extremely low concentrations of ionic impurities causes a drastic increase in conductivity indicating the deterioration of the water. Another modern application of conductometry is as detection system in ion chromatography. In this case, selectivity is not required since ionic species are previously separated by the chromatographic column. Therefore ion chromatography enables the quantification of ionic species at extremely low concentration due the high sensitivity of conductometry. All these applications are based on direct conductometry measurements; however conductometry can be also used to detect endpoint in conductometric titrations. General considerations about conductometric titrations can be found in the literature [92,95]. Conductometry uses a simple, robust and inexpensive instrumentation, which includes an alternating voltage source and a conductance cell composed by two Pt electrodes sealed in only one piece. Alternating voltage is necessary to avoid electrolysis. Conductometric systems are very versatile and they can be constructed in several sizes and shapes which makes conductometric detectors totally adaptable to flow analyses systems.

Coulometric methods are based on the electrolytic reduction or oxidation of the analyte for a time sufficient to ensure its quantitative conversion to a new state of oxidation. Electrolysis can be performed at controlled potential or controlled current conditions. Controlled current electrolysis leads to two kinds of coulometric techniques: electrogravimetry and coulometric titrations. Only this last modality will be discussed here because it represents the main used coulometric technique, general discussions about other coulometric techniques can be found in the literature [91,92,95]. All coulometric methods do not require any calibration procedure because the measured property (electrical charge or mass) is related to the quantity of analyte by fundamental relationships [9091-92,95]. This is advantageous because makes analysis faster and more reliable since errors associated to the preparation of chemical standards are eliminated. Moreover, coulometric methods are the most precise and exact analytical methods available in analytical chemistry [95]. In coulometric titrations the titrant is generated in situ by electrolysis at controlled current. The relationship between electrical charge and the amount of titrant generated is based on Faraday's law of electrolysis [90]. To be used in a coulometric titration, the electrode reaction must satisfy the following requirements: it must be of known stoichiometry and it must occur with 100 % current efficiency [91]. In the same way of classical titrations, coulometric titrations also require methods to detect the endpoint, which can be visual or instrumental [95]. However, coulometric titrations offer a number of advantages over classical titrations: (a) very small amounts of substances can be determined without the use of micro volumes. (b) Standard solutions and standardizations procedures are unnecessary. (c) Substances that are unstable or inconvenient to use because of volatility or reactivity can be employed as titrant since they are generated and consumed instantaneously. (d) Coulometric titrations are easily automated. (e) Dilution effects are absents, making endpoint location simpler [91,95]. Instrumentation required for coulometric titrations is also simple and inexpensive. It consists of inert electrodes, a constant-current source which can be simply a high voltage power supply and a resistor, a coulometer or a timer [91]. Modern applications of coulometric titrations include the determination of uranium [90] and several metallic ions, halides, oxygen and unsaturated organic compounds. However the most common use of coulometric titration is the quantification of water content in non-aqueous liquids, procedure known as Karl-Fischer titration [91,95]. In this method I_2 is electrolytically generated in an appropriated reaction medium in which I_2 suffer a redox reaction involving also water as reagent. Thus, there is a stoichiometric relationship between electrogenerated I_2 and water. The endpoint is detected by using other pair of detector electrodes in an independent circuit. The detector circuit maintains a constant current between the two detector electrodes. At the equivalence point, excess I_2 is produced and it is electrochemically reduced to I^- causing a drastic voltage drop in detector circuit which marks the endpoint. The content of water is an important quality parameter for several products such as edible and lubricating oils, pharmaceutical raw material, food products, industrial solvents and also automotive fuels. Thus, Karl-Fischer analysis is currently widely employed and there are several commercial automated equipments to reliably perform these analyses at a relatively low cost.

Voltammetry is by far the most used electrochemical technique in academic and scientific fields and it is clearly the most useful technique for low-level quantifications. These techniques are based on continuous variation of the potential applied across the electrode-solution interface and the resulting current, which is the property related to the analyte concentration, is recorded. The applied potential, therefore, acts as the drive force for the electrochemical reaction, i.e. reduction or oxidation of the analyte. Cyclic voltammetry (CV) is the most widely used voltammetric technique for acquiring qualitative information about electrode processes. The success of cyclic voltammetry derives from its ability to provide thermodynamic and kinetics data about heterogeneous electron-transfer reactions [82]. CV is usually the first experiment performed in an electroanalytical study, because it offers a fast location of redox potentials of the analyte. Moreover, CV enables the convenient evaluation of the effect of the media, i.e. pH, supporting electrolyte, etc. upon the redox processes of analyte [82].

For analytical purposes, linear sweep voltammetry (LSV), differential pulse voltammetry (DPV) and square wave voltammetry (SWV) are the most widely used techniques. DPV and SWV are usually preferred due their better discrimination between faradaic and capacitive currents which allows achieving detection limits as low as 10^{-8} mol L^{-1} [90,95]. These different modalities of voltammetric techniques are classified according to the adopted potential-time programs. The basic principles of voltammetric techniques and their respective potential-time programs are extensively discussed on electrochemical books [90-92], articles [96] and in some books specially devoted to voltammetry [97]. Contrasting to coulometric techniques, voltammetry involves extremely low scale electrolysis; therefore, the fraction of analyte oxidized or reduced is negligible. Thus the sample composition remains unchanged after the voltammetric analysis. Another unique feature of voltammetric techniques is that they involve electrolysis in conditions of total limitation by mass transport, thus diffusion of analyte from the body of the solution to the electrode/solution interface plays a crucial role in voltammetric analysis [90].

Comparatively to previously mentioned electrochemical techniques, voltammetry requires the most sophisticated instrumentation, which consists in a potentiostat and a three-electrode electrochemical cell. A potentiostat is necessary to maintain a controlled potential between reference and working electrode without allowing current passage. The potentiostat also presents a secondary circuit connecting working to auxiliary electrode and this circuit is responsible for all current flowing through the electrochemical cell. Auxiliary electrode, therefore, is used just to make possible current flowing. Auxiliary electrodes are constituted by inert and highly conductive materials, such as Pt, stainless steel, etc. Reference electrode is the key to achieve controlled potential conditions and therefore it is a very important component in voltammetric analysis. The literature presents some works totally devoted to reference electrodes including discussions about practical problems associated to them [98] and new techniques for the fabrication of more robust all solid state and miniaturized reference electrodes [99].

Working electrode is the electrode in which the analyte suffers electrochemical reduction or oxidation providing the current that is proportional to its concentration. Several materials can be used as working electrode in voltammetric techniques and they can be divided in two main groups: mercury electrodes, for example: dropping mercury electrode (DME) and hang dropping mercury electrode (HDME) and solid electrodes such as: platinum, gold, glassy carbon, carbon paste electrodes (CPEs), etc. DME has historical importance in electrochemistry, since it was directly involved in the development of polarography by Heyrovsky [49] being this technique the precursor of the modern voltammetric techniques. In 2009 it was celebrated the 50th anniversary of Heyrovsky's Nobel Prize for polarography and a celebrative review was written by Barek [100] this author was also involved in a very important and complete historical review about polarography written in 2002, year in which was celebrated eighty years of the discovery of polarography [101]. Despite its historical importance, the use of DME has decreased lately due mainly to environmental reasons and practical limitation for on-line and *in situ* analyses. Gradually, DME has been replaced by solid working electrodes. Among solid electrodes, glassy carbon (GC) is the most employed in electroanalysis because of its excellent mechanical and electrical properties, wide useful potential range and chemically inert nature. However, other kinds of carbon, such as: graphite, reticulated glassy carbon, pyrolytic carbon and more recently carbon nanotubes [86] also occupy an important place in electroanalysis. Therefore, carbon-based working electrodes are the most widely used in modern voltammetry and a review about the contributions of carbon materials for the development of electrochemical sensors was written by Qureshi et al [102]. Another very important review about advanced carbon electrode materials was written by McCreery [103].

Other very important kind of carbon-based electrodes widely used in voltammetry are the CPEs. These electrodes are constituted by a mixture between carbon powder and a water immiscible organic liquid, which acts as binder agent (pasting liquid). CPEs have become one of the most popular electrode materials used for the laboratory preparation of various electrodes, sensors, and detectors. Therefore, CPEs are used in practically all areas of fundamental and applied electrochemistry including modern electroanalysis. The success of CPE can be attributed to its very attractive physicochemical and electrochemical properties.

The advantageous features of CPEs include low residual current, wide useful potential window, chemical inertness, ease surface renewal, low cost and easiness of prepare. However, the most attractive feature of CPE is the easiness to perform a chemical modification of the electrode. A CPE can be modified simply adding the chemical modifier to the mixture carbon powder/binder agent with no need of specific interactions between the chemical modifier and the electrode material. The chemical modification of a CPE modulates its chemical and/or physical properties establishing specific interactions between the analyte and the electrode surface leading to an electrode with improved analytical performance, i.e. more sensitive and/or selective. In the year 2008, it was exactly fifty years since Professor Ralph Norman Adams has introduced the CPEs. Svancara and a team of very important experts in electroanalysis/electrochemistry have written a very instructive celebrative review [104]. These authors are also responsible for the most complete and recent work about CPEs in which the main advances, trends on CPEs development and application are extensively discussed [105].

Chemically modified electrodes (CMEs) are particularly attractive kind of electrode. CMEs are prepared by immobilization of chemical species onto electrode surface aiming to modulate its properties in order to establish specific interactions between analyte and electrodic surface. Electrode surfaces are usually chemically modified by one of these methods: adsorption (chemical or physical), covalent bond, polymer film coating or via composite preparation. CMEs correspond to a powerful and versatile approach to develop electrodes with analytical performance superior to conventional ones. These electrodes have been bringing impressive advances in electroanalysis and reviews about the preparation and application of these devices are available in the literature [106].

The sensitivity of voltammetric techniques can be significantly improved by using a preconcentration step previously to the voltammetric scan, this procedure characterizes the called stripping voltammetric techniques. The preconcentration step makes use of specific interactions between analyte and electrode surface, which can involves non-electrochemical processes such as adsorption or electrochemical ones such as electrodeposition. Stripping voltammetric techniques are characterized by very low detection limits, extremely high sensitivity, multielement and speciation capability, suitability for on site and in situ applications and minimal sample treatment requirements. The high sensitivity and consequently the low detection limits obtained with stripping analysis techniques are consequences of the preconcentration step, which enhances the analyte concentration at electrode surface. More general and fundamental discussions about stripping voltammetric techniques can be found in the electrochemical literature [82,90,91].

An important advance in voltammetric techniques is the use of microelectrodes which usually present dimensions not greater than 25 μm. Compared with conventional dimension electrodes, microelectrodes present several attractive features, such as improvement of mass transport due to radial diffusion, enhancement of signal-noise ratio, low background currents and immunity to ohmic drop. This last characteristic is very attractive, since it enables the use of microelectrodes in high resistive media without the necessity of supporting electrolyte, minimizing the possibility of chemical contamination of samples. A limitation of using microelectrodes is the extremely low currents recorded with require very sensitive equipments and very powerful amplification systems. In order to overcome this limitation, microelectrodes arrays has been extensively studied and used for development of electrochemical sensors. In these arrays, currents recorded at each individual microelectrode are additive, therefore, microelectrode arrays combine the attractive properties of microelectrodes with macro scale current values leading to very high sensitivity. Xie et al [107] have presented a very interesting review about the use of microelectrodes for quantification of metallic species in water samples and a recent review about the use of microelectrode arrays to develop amperometric sensor was written by Ordeig et al [108]. Classical literature of electrochemistry also brings information about microelectrodes and microelectrode arrays [82,90,91].

Another very interesting electrochemical technique based on potential-current relationships is amperometry. In this technique, the working electrode potential is kept at a controlled and invariable value which is just sufficient to promote the electrochemical reduction or oxidation of analyte. In these conditions, the current flowing through the electrochemical cell will be directly proportional to the analyte concentration. The instrumentation and the electrodes used in amperometric techniques are exactly the same used for voltammetry. Therefore, amperometry also involves electrolysis of minimal quantity of the analyte under condition of total limitation by mass transport. The combination of amperometry with hydrodynamic conditions is a very interesting alternative to develop electroanalytical methods. Hydrodynamic conditions strongly improve the mass transport due to the forced convection and as a consequence, higher current values are obtained allowing the development of more sensitive electroanalytical methods. Another attractive feature presented by amperometry in hydrodynamic condition is the shorter time of analysis.

A great characteristic of all electroanalytical techniques is their total compatibility with flowing conditions which makes them extremely attractive to be used in detection systems in FIA, liquid chromatography and capillary electrophoresis. The combination of electrochemical detection with liquid chromatography (LC-ED) is a powerful analytical strategy which has proven to be one of the more efficient options when attempting to characterize complex matrices. LC-ED is, therefore, widely used for the trace determination of easily oxidizable and reducible organic and inorganic compounds in complex matrices, providing detection limits as low as 0.1 pmol for a number of electroactive compounds. The electrochemical cell and electrodes in LC detector system can be constructed in many different designs according to the convenience. All electroanalytical technique can be employed in LC detection, however a large prevalence of amperometry is observed mainly due its operational simplicity. Additional fundamental information about LC-ED can be found in classical electroanalytical literature [90] and a review about the use of carbon electrodes as detectors in LC is also available [109]. Capillary electrophoresis is a very powerful technique used for separation of charged species. The required apparatus consists of a fused silica capillary filled with a buffer solution. A very high potential (usually 10 - 40 kV) is applied across the capillary and the sample is introduced in an extremity of the capillary while the detection system is located in another extremity. Due to the ionized silanol groups on the capillary walls an electroosmotic flow is produced which causes the elution of all compounds at the cathode, regardless their charge. Compounds are separated based on their differences in electrophoretic mobility; thus, usually the order of elution is faster cations, slower cations, neutral species and finally anions. Electrochemical detection is very advantageous in capillary electrophoresis because it is based on a reaction at electrode surface, therefore the electrochemical cell volume can be extremely small without loss of sensitivity [90]. Modern aspects of electrochemical detection coupled to capillary electrophoresis are discussed in the work written by Trojanowicz [81].

3. Electroanalytical techniques applied to bioethanol and biodiesel quality control

Electroanalytical techniques are adopted in both official and alternative methods used for bioethanol and biodiesel quality control. Official analytical methods exploit mainly those electroanalytical techniques whose instrumentation is simple and inexpensive in which there is no necessity of highly specialized operators. This is expected since official analytical methods are used as routine analyses which must be simple, practical and easily performed. On the other hand, alternative electroanalytical-based methods usually adopt advanced electroanalytical techniques and sophisticated instrumentation. Thus, these methods bring the main advances in electroanalytical field. They are characterized by superior analytical performance, presenting extremely low detection limits associated with very high sensitivity and selectivity. In the next sections, official and alternative electroanalytical-based methods used for bioethanol and biodiesel analysis will be revised and discussed.

3.1 Official electroanalytical-based methods

Important quality parameters of both bioethanol and biodiesel are regulated by technical norms based on potentiometry, conductometry and coulometry. These electroanalytical techniques are preferred due their simplicity, low cost, reliability associated with a low consume of time and reagents. Currently, there are no official analytical methods based on voltammetry or amperometry.

The pH of bioethanol is a very important quality parameter associated to its corrosion power. This quality parameter is regulated by technical norms based on direct potentiometry by using a glass electrode. In Brazil and USA technical norms describing the potentiometric determination of pH of bioethanol are, respectively, NBR 10891-06 and ASTM D6423-08. According to previously discussed, there are metrological problems associated to the determination of pH of bioethanol, since the glass electrode calibration is performed with aqueous buffers while bioethanol is predominantly a non-aqueous system [48]. It is interesting to observe that NBR 10891-06 explicitly describes that there is no correlation between pH values obtained for bioethanol and pH of aqueous solutions. However, this same norm sets that the electrode calibration must be performed by using aqueous standard buffer solutions which is very questionable. Some researches have worked in order to overcome these metrological problems by performing extensive studies about the behavior of glass electrodes in ethanol media [110]. These studies have allowed the introduction of correction factors which is extremely useful to improve the reliability of pH values determined in bioethanol contributing to the development of new and more appropriated technical norms. Direct potentiometry is also the base of a European technical norm (EN 15484-07) that specifies the content of chloride in ethanol fuel by using a chloride-ISE. Potentiometric titration is also used as official analytical method adopted for determination of acid number of biodiesel in Brazil, EU and USA. Technical norms that regulate this quality parameter are NBR 14448-09, EN 14104-03 and ASTM D664-09. In potentiometric titration, the calibration of glass electrode is not critical, because this technique uses relative and not absolute potential values, therefore, all those problems associated with calibration in direct potentiometry are absent in potentiometric titration. This makes technical norms based on potentiometric titration more reliable than those based on direct potentiometry.

As previously discussed, the main modern application of coulometry is for quantification of water content in non-aqueous media by Karl-Fischer titration. This electroanalytical technique is the base of the technical norms adopted in EU (EN 15489-07) and USA (ASTM E1064-08) for determination of water content in bioethanol. Currently, Brazil does not regulate water content in this biofuel. Karl-Fischer titration is also used as official analytical method to determine the water content in the biodiesel produced in Brazil, EU and USA. The respective technical norms adopted in EU and USA are, respectively, EN 12937-01 and ASTM D6304-07. Brazil, currently adopt the European norm to regulate water content in its produced biodiesel.

Conductometry is used as official analytical method for specification of three quality parameters of bioethanol and one quality parameter of biodiesel. Therefore, conductometry is the most used electroanalytical technique for bioethanol and biodiesel quality control. Conductivity itself is a specified property of bioethanol regulated in Brazil by technical norm NBR 10547-06. Currently EU and USA have no specification for bioethanol conductivity but they are considering adding a specification for this property [47]. This is a very important specification because conductivity can easily and quickly determine if bioethanol is contaminated by ionic species, including those that are not specified. In addition, conductivity values are also related to corrosion power of bioethanol, i.e. high conductivity values are associated with high corrosion power. Therefore, conductivity is a very useful property able to provide information about the overall quality of bioethanol. Moreover conductivity measures require a very simple and inexpensive instrumentation and the test is very ease to perform. The other two specified quality parameters of bioethanol regulated by official analytical methods based on conductometry are chloride and sulfate content. Brazil, EU and USA have specifications for the content of these anions; the adopted technical norms are, respectively, NBR 10894-07, EN 15492 and ASTM D7319-09. All these technical norms use ion chromatography with conductometric detection as official analytical method. This is an example of the success of coupling chromatographic separation with electrochemical detection.

A very interesting use of conductometry is in the evaluation of the oxidation stability of biodiesel. This property is specified in Brazil, EU and USA and these three regions adopted the European technical norm EN 14112-03. The oxidation stability of biodiesel is a crucial parameter to ensure correct fuel performance in the vehicles, as well as in storage and distribution. EN 14112-03 adopts the rancimat procedure to determine oxidative stability of biodiesel. Rancimat is a totally automated accelerated oxidation test carried out at elevated temperature and under exposure of air. In rancimat procedure, a stream of purified air is passed through the heated sample (usually 3-6 g of biodiesel) and it is subsequently bubbled through a vessel containing deionized water in which there is a conductivity cell continuously monitoring the system. The resulting oxidation products, i.e. volatile organic acids are carried by the air stream from the sample to the deionized water. When the produced organic acids reach deionized water, their dissociation equilibria are established producing carboxylate anions and H^+. The presence of these ionic species causes an increase in conductivity of the water. Generally, oxidation is slow at first stages and it quickly accelerates after its initiation. The point in which the maximum change of rate of oxidation is obtained is called induction time and it is detected by a large and abrupt increase on the conductivity. EN 14112-03 defines a minimal induction time of 6 h at 110 °C. The oxidative stability obtained at 110 °C can be extrapolated to provide stability data under storage and transportation temperature. Rancimat procedure constitutes a very smart use of conductometry and this method is also used to evaluate oxidative stability of raw material and products from pharmaceutical and food industry. Rancimat procedure has been also used to evaluate the performance of antioxidant agents. Therefore, conductometry is a very important electroanalytical technique widely used for quality control of several materials.

3.2 Alternative electroanalytical-based methods

The number of alternative electroanalytical-based methods used for bioethanol and biodiesel analysis is higher than the official electronalytical-based methods as shown in Table 5.

From Table 5 it can be observed that the number of chemical species analyzed by alternative electroanalytical-based methods is higher than those whose contents are regulated by official technical norms. This demonstrates that electroanalytical community is not only concerned about impurities that constitute a problem for biofuels quality today but it is also concerned about species which may become a future problem for the quality and commercialization of bioethanol and biodiesel. Table 5 also shows the prevalence of voltammetry and stripping voltammetric techniques as base for alternative electroanalytical methods which is expected since modern voltammetric techniques are very suitable for low-level quantifications. Therefore, most of alternative electroanalytical methods are characterized by very low detection limits.

Direct potentiometry is especially attractive to determine chemical species which can not be easily oxidized or reduced such as alkali metals. Thus, some recent works have described the use of direct potentiometry to determine potassium in biodiesel samples. Two strategies have been adopted in these works: the use of commercial K^+-ISE which is protected from direct contact with biodiesel sample and the development of new indicating electrodes. The first strategy was used by Rapta et al [112] which have employed a PVC membrane-based commercial K^+-ISE to quantify potassium in FAME samples. In order to prevent electrode swelling promoted by FAME, two compartments were used one of them was filled with distilled water and the other was filled with FAME sample. Both compartments were separated by a semi-permeable membrane, constituted by a 20 μm thickness cellulose acetate film. Indicating and reference electrode were immersed in the compartment containing distilled water. The cellulose acetate membrane is permeable to K^+ allowing its extraction from FAME compartment to the aqueous compartment in which K^+ was potentiometrically quantified. This procedure was very efficient to avoid electrode deterioration from exposure to FAME sample which was demonstrated by the good reproducibility and concordance between the results obtained by the proposed method and those obtained by official AAS analytical method. Castilho and Stradiotto [111] have used the second strategy to determine K^+ in biodiesel samples. These authors have developed a new K^+-ISE based on a nickel-hexacyanoferrate film which was deposited onto glassy carbon by cyclic voltammetry in presence of both ions $Fe(CN)_6^{3-}$ and Ni^{2+}. These electrodes were successfully used for quantification of K^+ in biodiesel samples after a liquid-liquid extraction using HCl 0.1 mol L^{-1} as aqueous phase. The developed electrode has presented a near Nernstian behavior and a very satisfactory selectivity with selectivity coefficient for Na^+, NH_4^+ and Li^+ equal to 0.01; 0.5 and 0.0089, respectively. In both works, direct potentiometry has allowed K^+ quantification in biodiesel samples in a very reliable, fast, simple and inexpensive way. These features make direct potentiometry very useful, for example, to evaluate the efficiency of biodiesel final washing procedure for the elimination of KOH used as catalyst in transesterification reaction.

Biofuel	Technique	Analyte	Detection Limit	Reference
Biodiesel	Direct potentiometry	K^+	19 µmol L^{-1}	[111]
Biodiesel	Direct potentiometry	K^+	0.4 ppm	[112]
Ethanol	Direct potentiometry	Cu^{2+}	----	[113]
Ethanol	Conductometric titration	Total Acidity Cl^-	3.0 mg L^{-1} 0.88 mg L^{-1}	[114]
Ethanol	Conductometry/Ion chromatography	Cl^- and SO_4^{2-}	----	[115]
Biodiesel	Conductometry/Ion chromatography	Alkaline Earth and Alkaline metal-ions	----	[116]
Ethanol	Conductometry/Capillary electrophoresis	Inorganic cations and anions	----	[117]
Ethanol	SWV	SO-7 dye	0.09 µmol L^{-1}	[118]
Ethanol	SWV	SB-14 dye	0.29 µmol L^{-1}	[119]
Ethanol	SWV	SB-14 dye	93 nmol L^{-1}	[120]
Ethanol	DPV	Acetaldehyde	0.81 µmol L^{-1}	[121]
Ethanol	SWV	Acetaldehyde	0.24 µmol L^{-1}	[122]
Ethanol	DPV	Acetaldehyde	2.0 µmol L^{-1}	[123]
Ethanol	CV	Furfuraldehyde	0.70 mmol L^{-1}	[124]
Biodiesel	LSV	Glycerol	25 µmol L^{-1}	[125]
Biodiesel	LSV	tert-butylhydroquinone	71 nmol L^{-1}	[126]
Ethanol	Stripping voltammetry-DPV	Cu^{2+}	31 nmol L^{-1}	[127]
Ethanol	Stripping voltammetry-LSV	Cu^{2+}	22 nmol L^{-1}	[128]
Ethanol	Stripping voltammetry-DPV	Cu^{2+}	39 nmol L^{-1}	[129]
Ethanol	Stripping voltammetry-SWV	Cu^{2+} Pb^{2+}	120 ng L^{-1} 235 ng L^{-1}	[130]
Ethanol	Stripping voltammetry-DPV	Zn^{2+} Cu^{2+} Pb^{2+}	0.24 µmol L^{-1} 0.88 µmol L^{-1} 0.66 µmol L^{-1}	[131]
Ethanol	Stripping voltammetry -DPV	Pb^{2+}	7.2 nmol L^{-1}	[132]
Ethanol	Stripping voltammetry-DPV	Ni^{2+}	2.0 nmol L^{-1}	[133]
Ethanol	Stripping voltammetry-DPV	Ni^{2+}	2.7 nmol L^{-1}	[134]
Ethanol	Stripping voltammetry-SWV	Fe^{3+}	2.4 µmol L^{-1}	[135]
Ethanol	Stripping voltammetry-LSV	Zn^{2+}	0.26 µmol L^{-1}	[136]
Ethanol	Stripping voltammetry-LSV	Cl^-	0.13 mg L^{-1}	[137]
Ethanol	Stripping voltammetry-DPV	Sulphur	15 ng g^{-1}	[138]
Biodiesel	Potentiometric stripping	Cu^{2+}	200 ng g^{-1}	[139]
Biodiesel	Amperometry-enzymatic	Free and total glycerol	$1x10^{-5}$ (w/v)	[140]
Ethanol	Amperometric detection-HPLC	Aldehydes Ketones	1.7 ng mL^{-1} 2.0 ng mL^{-1}	[141]
Ethanol	Amperometric detection-HPLC	Acetaldehyde	3.80 µg L^{-1}	[142]
Ethanol	Amperometric detection-HPLC	SB-14 dye	6.4 nmol L^{-1}	[143]
Ethanol	Amperometric detection-FIA	Cl^-	3.7 µmol L^{-1}	[144]

Table 5. Alternative electroanalytical-based methods used for bioethanol and biodiesel analysis

Conductometry coupled to ion chromatography has been successfully used for simultaneous quantification of several inorganic ions in both bioethanol and biodiesel [115,116] which demonstrates the versatility and analytical power of combining chromatographic separation and conductometric detection. Conductometric detection in ion chromatography and capillary electrophoresis represents a very advanced use of conductivity measures. The main approach currently adopted in these detection systems is contactless conductivity detection (CCD) in which electrodes are electrically isolated from the sample being the measurements performed at high frequency conditions. One of the advantageous features of CCD is the total absence of electrode fouling effect since electrode is not in direct contact with sample solutions. Deeper discussions about CCD are beyond of the scope of this work. Additional information about the principles and application of CCD can be found in the original works introducing CCD as detection system in capillary electrophoresis [145,146]. CCD coupled to capillary electrophoresis was used by Muñoz et al [117] for the successful and simultaneous quantification of several cations (Na^+, K^+, NH_4^+, Ca^{2+} and Mg^{2+} using Li^+ as internal standard) and anions (SO_4^{2-}, Cl^- and NO_3^- using lactate as internal standard) directly in commercial ethanol fuel samples. The authors have used a homemade capillary electrophoresis equipment which has presented many advantageous features such as low cost, shorter time analysis and high analytical performance with low detection limits and high sensitivity. Classical conductometric methods have been also employed for ethanol fuel analysis. Avelar et al [114] have used conductometric titration for determining chloride content and free acidity in hydrated ethanol fuel samples. The authors have observed that conductometric titration allows the reliable quantification of total acidity and chloride content in hydrated ethanol fuel being the obtained results in good agreement with those obtained by official analytical methods. Therefore conductometric titration could be successfully used for a simple, inexpensive and fast evaluation of free acidity and chloride content which are corrosive agents representing very important quality parameters for ethanol fuel.

From Table 5 it is possible to observe a large number of voltammetric and stripping voltammetric-based analytical methods which are used mainly for ethanol fuel analysis. Voltammetric based analytical methods are used mainly for determination of organic compounds while stripping voltammetric-based methods are used mainly for quantification of transition metal ions. Working electrodes used in voltammetric methods are mainly based on carbon materials, such as glassy carbon [118,119,121,124], screen-printed carbon electrode [120] and carbon paste electrode [126]. In stripping voltammetric-based methods there is a prevalence of chemically modified carbon paste electrodes which are characterized by a superior ability to preconcentrate the analyte at electrode surface.

Organic species determined in ethanol fuel by voltammetric methods are mainly dye markers and aldehydes. Dye markers are usually purposely added to fuels in order to guarantee their identity and discourage adulterations, thus these dyes are important to quality control of fuels. Two very important dyes marker are solvent blue 14 (SB-14) and solvent orange 7 (SO-7) and the literature has shown that both can be successfully quantified by voltammetry in ethanol fuel samples. Trindade and Zanoni [119] have developed an analytical method for quantification of SB-14 in ethanol fuel samples based on SWV using a glassy carbon working electrode. The authors have performed all voltammetric measurements in a mixture Britton-Robinson (B.R) buffer: dimethylformamide (1:1 v:v).

Under the optimized experimental conditions, SB-14 has presented a very intense and defined cathodic voltammetric peak at -0.4 V vs. $Ag/AgCl/KCl_{saturated}$ which has allowed the sensitive and reliable quantification of this dye in commercial ethanol fuel samples. In order to improve even more the analytical performance of this analytical method, these same authors have replaced the glassy carbon working electrode by a screen-printed carbon electrode [120], making the voltammetric method more practical and inexpensive. The main advantages of screen-printed electrodes are their modest cost, portability and above of all their simplicity of use since only one small arrangement contains working, auxiliary and reference electrodes. Another improvement introduced in the SWV-quantification of SB-14 was the use of an anionic surfactant which has increased the solubility of SB-14 minimizing the quantity of dimethylformamide required in electrochemical cell from 50% to 30% (v:v). This is extremely advantageous since dimethylformamide is a relatively toxic and expensive reagent. The use of the surfactant has brought other benefits to the analytical method, since its presence has tripled peak current values. Moreover, according to the authors, the surfactant forms a charged film on the electrode surface acting as an anti-fouling agent. All these positive factors have allowed lowering detection limit from 2.9×10^{-7} mol L^{-1} (using glassy carbon in absence of surfactant) to 9.3×10^{-8} mol L^{-1} (using screen-printed electrode in presence of surfactant). The most impressive advantageous feature of these methods is their ability to quantify SB-14 directly in ethanol fuel samples without any sample pretreatment which is extremely attractive because it avoids time-consuming pretreatment procedures making the analytical method faster, simpler, less expensive and more reliable. This same group has yet contributed to the field of electroanalysis of dye markers by developing a voltammetric method for quantification of SO-7 directly in ethanol fuel samples [118]. This voltammetric method has also presented excellent analytical performance associated with high simplicity and low cost.

Ethanol fuel is produced by a distillation process; thus its main organic contaminants are those ones more volatile than it such as acetaldehyde. Whereas that acetaldehyde is a toxic air contaminant able to cause lung irritation and that it is a possibly carcinogenic agent, the quantification of acetaldehyde has attracted much attention. Several analytical methods have been proposed for quantification of acetaldehyde in ethanol fuel including voltammetric ones. However, voltammetric determination of aldehydes is difficult due the extremely high potential values required for promote their electrochemical reduction or oxidation. The strategy used to make possible to quantify acetaldehyde by using voltammetric techniques is to adopt a chemical derivatisation procedure before voltammetric analysis. The chemical derivatisation step is responsible to form a more easily oxidizable or reducible product which can be detected voltammetrically. Thus, acetaldehyde has been quantified by voltammetry as imine derivatives [123] and after derivatisation with hydrazine [122] and 2,4-Dinitrophenylhydrazine [121]. In all these cases there is the formation of the group RR´C=N- which is easily reduced to the correspondent amine [147]. All these voltammetric methods allow the quantification of acetaldehyde in a very sensitive way with low detection limits. The only working proposing a direct voltammetric method for quantification of an aldehyde without chemical derivatisation was presented by Saczk et al [124]. These authors have studied the voltammetric reduction of furfuraldehyde in ethanol media using a glassy carbon electrode. The authors have observed that furfuraldehyde presents a well defined cathodic peak which has allowed its reliable quantification in ethanol.

Only two works involving voltammetric analysis of biodiesel were found in the literature. The first one is the work by Lourenço and Stradiotto [125] which have used a Pt electrode to quantify free glycerol in biodiesel samples. The analytical method is based on electrocatalytic oxidation of glycerol promoted by PtO_x electrogenerated onto electrode surface. Before the voltammetric analysis the samples were submitted to a simple pretreatment consisting of a liquid-liquid extraction with water followed by elution of aqueous phase on a C-18 cartridge for elimination of interfering species. Voltammetric scans were recorded in $HClO_4$ solutions and very well defined voltammetric peaks associated to glycerol electrooxidation were obtained. The proposed method has presented a linear range from 56 to 560 mg Kg^{-1}, which is appropriated for biodiesel analysis since its maximum free glycerol content allowed is 200 mg Kg^{-1}. The reliability of the proposed method was demonstrated by addition-recovery experiments performed at a glycerol-free biodiesel sample and the recovery percentages were always close to 100 %. The second voltammetric method used for biodiesel analysis was proposed by Araújo et al [126] for determination of the antioxidant tertbutylhydroquinone (TBHQ). The poor oxidative stability of biodiesel is considered the main drawback to its use in large scale as alternative fuel. Thus several studies have been conducted in order to increase oxidative stability of biodiesel by adding antioxidant agents and TBHQ have yielded good results. Thus, in future the addition of TBHQ in biodiesel may become mandatory, and its content would be controlled by technical specifications. In this context, the development of analytical methods for quantification of TBHQ in biodiesel samples is quite important. Araújo et al [126] have developed a voltammetric method to quantify TBHQ by using a carbon paste electrode and performing voltammetric measurements in B.R. buffer solutions containing the surfactant cetyltrimethylammonium bromide (CTAB). They have observed that the presence of CTAB significantly intensifies the voltammetric peak associated to TBHQ oxidation increasing the sensitivity of the voltammetric method. Addition-recovery experiments were performed in spiked biodiesel samples and recovery percentages were always close to 100 % showing that the proposed method is very reliable. Moreover, the method has allowed TBHQ quantification directly in soybean biodiesel samples by simply diluting them in methanol without any additional pretreatment procedure. Therefore, this method can be seen as a potential analytical approach for future routine quantification of TBHQ in biodiesel samples.

Stripping voltammetric-based methods are the most sensitive electroanalytical methods employed for bioethanol and biodiesel analysis. These methods have been used mainly for quantification of metal ions in ethanol fuel samples and they have presented extremely low detection limits which usually are in the range of nmol L^{-1}. Most of stripping voltammetric methods employed for bioethanol analysis use chemically modified CPEs, which have significantly contributed for electroanalysis of biofuels. Among chemical modifiers used for constructing CPEs, organofunctionalized silica has received especial attention due mainly to the extremely advantageous features presented by silica such as high adsorption capacity, chemical and mechanical stability, insolubility in virtually all solvents and possibility of functionalization with a large variety of functional groups. CPEs chemically modified with organofunctionalized silica have been successfully used for quantification of Cu^{2+} [127,129] and Ni^{2+} [133] in ethanol fuel samples by using anodic stripping voltammetry. In these works CPEs were chemically modified with silica organofunctionalized with aminothiol groups which are able to efficiently preconcentrate metallic ions by complexation with their

S and N atoms. These works brought another important advance for development of CPEs for ethanol fuel analysis since they have produced more rigid and stable composites by replacing mineral oil, the binder agent used in conventional CPEs, by alternative binders. Conventional CPEs are totally inadequate to be used directly in ethanol fuel samples because they are instable in non-aqueous media due dissolution of binder agent which leads to CPE disintegration. Solid paraffin and castor oil-derived polyurethane have been successfully used as agent binder for the quantification of metal ions in ethanol fuel samples [127,129,133]. Both binder agents are very attractive because they are able to produce carbon composites with better mechanical resistance. Moreover these binder agents present low cost and they are easily manipulated allowing preparing the electrodes in a fast, simple and inexpensive way. Solid paraffin-based CPEs were totally stable in ethanol media allowing the quantification of Cu^{2+} [127] and Ni^{2+} [133] directly in the ethanol fuel samples without any pretreatment or dilution step. Polyurethane-based CPE was successfully used by Cesarino et al [129] for quantification of Cu^{2+} in ethanol fuel samples after mixing them with 30% (v:v) of a KNO_3 aqueous solution. Detection limits for all these works are in the range of nmol L^{-1} and their results were in good agreement with those obtained by official analytical methods demonstrating that they are able to provide Cu^{2+} and Ni^{2+} quantification in ethanol fuel samples in a very sensitive and reliable way. Other chemical modifiers used to prepare CPEs for determination of metal ions in ethanol fuel samples include: the anionic-exchange resin Amberlite® IR120 for quantification of Pb^{2+} [132], dimethylglyoxime for quantification of Ni^{2+} [134] and 1,10-ortophenantroline for quantification of Fe^{3+} [135]. All these examples use conventional CPEs; therefore their use was restricted to mixtures ethanol fuel: aqueous electrolyte solution which however has no seriously compromised their overall analytical performance.

Metal-based and metallic electrodes have been also used for quantification of metal ions in ethanol fuel samples by stripping voltammetric methods. Oliveira et al [131] have used a mercury film electrodeposited onto a glassy carbon electrode for simultaneous determination of Zn, Cd, Pb and Cu in ethanol fuel samples. The procedure adopted to prepare the electrodes was very simple consisting in to submit a rotating disc glassy carbon electrode to -0.9 V vs. SCE for 20 min at 200 rpm in a solution containing 2 x 10^{-5} mol L^{-1} of Hg^{2+}. This electrode was successfully used to determine Zn, Cd, Pb and Cu in ethanol fuel after diluting the samples in aqueous LiCl solution in the ratio 20:80% (v:v), respectively. The authors have used anodic stripping voltammetry and they performed a comparative study between the analytical performance of LSV, DPV and SWV modes. The results have shown that DPV was the most appropriated voltammetric modality which was used to simultaneously quantify Zn, Cd, Pb and Cu in three commercial ethanol fuel samples by using 10 min of electrodepositing time. The obtained results were in good agreement with those obtained by FAAS method, indicating that the proposed method is very reliable. Muñoz and Angnes [130] have used a commercial gold electrode to develop an anodic stripping-SWV method for simultaneous determination of Pb^{2+} and Cu^{2+} in ethanol fuel. The voltammetric measures were performed in a mixture ethanol:water (75:25% v:v) by using 900 s of electrodepositing time. Recovery studies were performed in spiked commercial ethanol fuel samples and recovery percentages were always close to 100 %. A gold microelectrode was used by Takeuchi et al [128] for determination of Cu^{2+} in commercial ethanol fuel samples. These authors have developed a linear sweep anodic stripping voltammetric method using 300 s of electrodeposition time. Not only the preconcentration step but also the voltammetric scan was performed directly in the ethanol sample without any pretreatment. The authors have found that copper in commercial ethanol fuel samples seems to be distributed in its labile and organic complexed forms which can not be efficiently electrodeposited in electrode surface. As a result, the stripping voltammetric method has provided copper contents lower than those obtained by FAAS. The authors have observed that acidification of samples was a very efficient way to recover copper from its complexed forms. After sample acidification, the developed analytical method was successfully employed for Cu^{2+} quantification in six commercial ethanol fuel samples and the obtained results were concordant with those obtained by official FAAS method. This was the first work in literature in which both preconcentration and detection steps were performed directly in the commercial sample by simply adding 1.0 mmol L^{-1} of H_2SO_4. Therefore, this method has allowed the quantification of Cu^{2+} in ethanol fuel samples in a very fast, simple and sensitive way with detection limit of 22 nmol L^{-1}.

The literature also brings stripping voltammetric methods for quantification of non metals in ethanol fuel samples. Ferreira et al [137] have used a HMDE for develop a cathodic stripping LSV method for quantification of chloride in ethanol fuel samples. The accumulation step was performed at +0.15 V vs. Ag/AgCl/$KCl_{saturated}$ potential in which mercury is oxidized forming Hg_2Cl_2. After the accumulation step the potential was swept in negative direction in order to promote the electrochemical reduction of Hg_2Cl_2 which produces a cathodic voltammetric peak providing the analytical signal proportional to chloride concentration. The developed method was employed for chloride quantification in five commercial ethanol fuel samples without any pretreatment or supporting electrolyte addition. The obtained results were concordant with those obtained by a potentiometric comparative method. The proposed method has allowed chloride determination in a very fast way since only 10 s of accumulation time was required. Aleixo et al. [138] have developed a cathodic stripping voltammetric method for determination of total sulfur in ethanol. The analytical method was based on the reduction of sulfur compounds to sulphite using Raney nickel. After that, sulphite was converted to H_2S which was detected by DPV-cathodic stripping voltammetry at a HMDE. The calibration graph covered the concentration range of 1-40 ppb, and the method was successfully employed for determination of sulfur in ethanol.

Almeida et al [139] have presented the only work in which a stripping-based method is used to analyze biodiesel samples. These authors have used constant current potentiometric stripping at a gold electrode to determine Cu^{2+} in biodiesel. In potentiometric stripping analysis (PSA) of a metal ion, the preconcentration step is the same as for voltammetric stripping, that is, the metal is electrodeposited onto electrode surface at constant potential conditions. After the preconcentration step an anodic current (or a chemical oxidant) promotes the metal stripping from electrode surface. As the metal is removed from electrode surface, the working electrode potential varies according to the Nernst equation. The plot of potential as a function of time gives a curve analogous to a normal redox titration curve due the logarithmic term on Nernst equation. A sudden change in the potential ("endpoint") occurs when all electrodeposited metal has been removed from electrode surface and the time necessary to reach this point is proportional to the bulk concentration of the metal ion [90]. A more convenient peak-shaped response for analytical purposes is obtained by plotting dt/dE vs. E which produces analytical signs in form of peaks which are more convenient. The constant current PSA method used by Almeida et al [139] has allowed the determination of Cu^{2+} in biodiesel without sample decomposition. The analyses were performed after a simple dilution of biodiesel samples in a mixture ethanol-HCl aqueous solution, producing a homogeneous mixture in which Cu^{2+} was directly quantified. The authors have used an electrodeposition potential of 0 V for 300 s and a stripping current of 0.5 μA. These experimental conditions were used to perform Cu^{2+} determination in three spiked commercial biodiesel samples. The obtained results have shown recovery percentages close to 100 % for all analyzed samples, thus PSA has allowed Cu^{2+} quantification in biodiesel samples in a very simple, fast inexpensive and reliable way. Therefore, this work represents a great contribution to electroanalysis of biodiesel, since it is the first one using a stripping method for biodiesel analysis with no need of sample decomposition or extraction procedures.

Amperometry-based analytical methods have been also used for develop alternative analytical methods mainly for determination of organic compounds in bioethanol. The only work describing an amperometric method for biodiesel analysis was recently written by Luetkmeyer et al [140]. These authors have developed an electroenzymatic methodology using an oxygen Clark-type electrode for the determination of free and total glycerol in biodiesel samples. The enzymatic conversion of glycerol consumes oxygen, which is measured amperometrically in a Clark-type electrode and is correlated with the concentration of glycerol in the sample. The analysis of biodiesel samples were performed after a simple liquid-liquid extraction procedure in which glycerol was extracted from biodiesel samples by a mixture containing distilled water, ethanol and heptane. The authors have found a very good correlation between the results obtained by the developed enzymatic amperometric method and the official gas chromatographic method. The proposed method was shown to be promising for the analysis of glycerol in biodiesel samples, with a simpler and inexpensive methodology compared with the gas chromatography technique.

Other examples of using amperometry for bioethanol analysis involve amperometric detection in flow analysis and amperometric detection in liquid chromatography. Paula et al [144] have developed an amperometric-FIA method for quantification of chloride in ethanol fuel samples. These authors have used a glassy carbon electrode chemically modified with a poly-aniline film immobilized onto electrode surface by electrochemical polymerization carried out by cyclic voltammetry. This electrode is able to detect chloride because redox processes of poly-aniline create positive charges inside the polymeric film, these charges must be balanced by negative counter ions from solution. As anions with large hydrated radius can not penetrate in the polymeric film some selectivity is achieve towards small anions such as Cl^-. In the proposed method, the authors have kept work potential at +0.5 V vs. $Ag/AgCl/KCl_{saturated}$ in glycine without adding supporting electrolyte. In these experimental conditions, the obtained anodic current is close to zero, since poly-aniline can not be oxidized in absence of anions as counter ions. When the sample containing chloride reaches the amperometric detector this anion enables the poly-aniline oxidation giving rise to an anodic current proportional to chloride concentration. This FIA-amperometric method was successfully used for determine chloride content in six commercial ethanol fuel sample and the obtained results were in good agreement with those obtained by a comparative potentiometric method. All determinations were conducted without any sample pretreatment procedure which is a very attractive feature.

HPLC coupled to amperometric detection has been also used for quantification of organic species in ethanol fuel samples. The combination of separation ability of chromatography with high sensitivity of amperometric detection is a very useful strategy to analyze complex matrices such as biofuels, minimizing sample pretreatment requirements. Combination of LC with EC detection is a selective and sensitive tool for determination of a wide variety of organic compounds in several kinds of samples, including bioethanol and biodiesel. Trindade et al [143] have developed an analytical method based on HPLC with amperometric detection for the simultaneous quantification of SB-14 and solvent red 24 (SV-24) in ethanol fuel. The authors have observed that these dyes were better separated on a C18 column, using a mobile phase composed of acetonitrile and ammonium acetate (90:10, v/v). Detection was performed amperometrically using a glassy carbon electrode kept at +0.85V vs. $Ag/AgCl/KCl_{saturated}$. Recovery experiments were carried out in spiked commercial ethanol fuel samples which were analyzed after a fast filtration step in a 45 μm membrane. This analytical method has allowed the simultaneous detection of SB-14 and SV-24 in a very fast and reliable way since recovery percentages close to 100 % were obtained. HPLC coupled to amperometric detection was also used by Saczk et al [141] for quantification of carbonyl species in ethanol fuel samples. This chromatographic method was used to quantify several aldehydes and ketones after chemical derivatisation with 2,4-dinitrophenylhydrazine. The proposed method has enabled the simultaneous quantification of 5-hydroxymethylfurfural, 2-furfuraldehyde, butyraldehyde, acetone and methyl ethyl ketone in ethanol fuel in a fast and reliable way. The amperometric detection was performed by using a glassy carbon electrode kept at +1 V vs. $Ag/AgCl/KCl_{saturated}$. The chromatographic separation was carried out by using a C-18 column under isocratic conditions with a mobile phase containing a binary mixture of methanol/1.0 mmol L^{-1} $LiClO_{4(aq)}$ (80:20 v/v) and a flow-rate of 1.1 mL min^{-1}. The quantification of these carbonyl compounds is important to evaluate the ethanol oxidation process, since different aldehydes contents can be related to unsuitable condition of fermentation, transport and storage of fuel. HPLC with amperometric detection was also successfully used for quantification of acetaldehyde in ethanol fuel samples [142].

4. Conclusions and perspectives

We can observe that electroanalytical techniques are significantly contributing to bioethanol and biodiesel quality control by means of official analytical methods which are characterized by high simplicity and reliability associated with low cost and no need of specialized operators. We can also conclude that electroanalytical techniques will certainly continue contributing for bioethanol and biodiesel quality control by providing modern and efficient analytical methods which may be adopted by technical specifications in the future. It is interesting to observe that most of alternative electroanalytical-based methods are developed for determining species whose contents are not currently regulated. Therefore, the electroanalytical community is fulfilling its scientific role in anticipating solutions by developing analytical methods to quantify species which may become a future problem for the quality and commercialization of bioethanol and biodiesel.

We can also observe that the most sensitive electroanalytical methods employed for bioethanol and biodiesel analysis usually are based on CMEs. The development of this kind of electrode is a trend in electroanalytical chemistry since them present superior analytical performance comparatively to non-modified electrodes. Bioethanol and biodiesel are complex matrices, thus any analytical method employed to their analysis must be highly selective and sensitive. The use of CMEs coupled to modern electroanalytical techniques is able to satisfactorily respond to these requests. The use of analyte preconcentration coupled to high selective chemical modifiers has allowed measurements to be performed directly on biofuel or in slightly diluted biofuel samples. This brings a very important advance for biofuel analysis since it avoids time-consuming pretreatment procedures making the analytical method faster, simpler and more reliable. Thus researches on development of new highly selective chemical modifiers must be intensified in the next years.

Researches on electrodes that exploit the unique properties of nanomaterials such as nanoporous structures, nanoparticles, nanotubes, etc. have presented an impressive expansion lately. The use of nanostructured electrodes has brought impressive advances in stripping voltammetric analysis, since nanostructured modifiers present extremely attractive features such as high surface area, strong adsorptive nature resulting in lowered detection limits and high electrocatalytic activity. Thus nanostructured materials have an important role to play in the field of electroanalysis and also in biofuel electroanalysis. Therefore development and application of nanostructured electrodes for biofuel analysis will continue to expand.

Another major trend in electroanalysis is the miniaturization of electrodes and electrochemical cells because this makes electroanalytical methods highly portable allowing measures *in situ* and on line. Moreover, miniaturized electrochemical systems can be used as detectors in chromatographic methods combining the high sensitivity of electroanalytical techniques with the separation power of chromatography which is an excellent approach for analysis of complex samples such as biofuels.

We believe that the research on electroanalytical methods will continue to expand, focusing mainly on improving detection limits, selectivity and stability of CMEs. Research in the field of electrochemical sensors has not presented any sign of slowdown due mainly to the incorporation of nanomaterials into electrochemical sensors platforms. The stability of these modern electrochemical sensors has increased significantly which has enabled their use in very complex matrices. Thus, modern electrochemical sensors have potential to significantly contribute to quality control of biofuels in a very near future.

5. Acknowledgments

The authors are grateful to FAPEMIG (process APQ-00291-09), FAPESP, CAPES and CNPq (project 552843/2007-5) for the financial support.

6. Bibliography

[1] J.-E. Petersen Energy production with agricultural biomass: environmental implications and analytical challenges, *European Review of Agricultural Economics* 35 (2008) 385–408.

[2] S. Jebaraja, S. Iniyan A review of energy models, *Renewable and Sustainable Energy Reviews* 10 (2006) 281-311.

[3] T. Bridgwater Biomass for energy, *J. Sci. Food Agric.* 86 (2006) 1755–1768.

[4] P. McKendry Energy production from biomass (part 1): overview of biomass, *Bioresource Technology* 83 (2002) 37–46.

[5] P. McKendry Energy production from biomass (part 2): conversion technologies, *Bioresource Technology* 83 (2002) 47–54.
[6] M-K Bahng, C. Mukarakate, D. J. Robichaud, M. R. Nimlos Current technologies for analysis of biomass thermochemical processing: A review, *Anal. Chim. Acta* 651 (2009) 117-138.
[7] S. V. Vassilev, D. Baxter, L. K. Andersen, C. G. Vassileva An overview of the chemical composition of biomass, *Fuel* 89 (2010) 913–933
[8] C. B. Field, J. E.Campbell, D B. Lobel Biomass energy: the scale of the potential resource, *Trends in Ecology and Evolution* 23 (2008) 75-82.
[9] B. Logan Microbial fuels for the future, *Nature* 454 (2008) 943-944.
[10] K. Johansson, K. Liljequist, L. Ohlander, K. Aleklett Agriculture as Provider of Both Food and Fuel, *AMBIO* 39 (2010) 91–99.
[11] A. Evans, V. Strezov, T. J. Evans Sustainability considerations for electricity generation from biomass, *Renewable and Sustainable Energy Reviews* 14 (2010) 1419–1427.
[12] E. Rajasekar, A. Murugesan, R. Subramanian, N. Nedunchezhian Review of NO_x reduction technologies in CI engines fuelled with oxygenated biomass fuels, *Renewable and Sustainable Energy Reviews* 14 (2010) 2113–2121.
[13] P. Luckow, M. A.Wise, J. J. Dooley, S. H. Kim Large-scale utilization of biomass energy and carbon dioxide capture and storage in the transport and electricity sectors under stringent CO_2 concentration limit scenarios, *International Journal of Greenhouse Gas Control* xxx (2010) xxx–xxx.
[14] F. Chen, S.-M. Lu, E. Wang, K.-T. Tseng. Renewable energy in Taiwan, *Renewable and Sustainable Energy Reviews* 14 (2010) 2029–2038.
[15] I. V.-Vazquez, J. A. A.-Benítez, C. H.-Santiago Distribution and potential of bioenergy resources from agricultural activities in Mexico, *Renewable and Sustainable Energy Reviews* 14 (2010) 2147–2153.
[16] D. Hoffmann Creation of regional added value by regional bioenergy resources, *Renewable and Sustainable Energy Reviews* 13 (2009) 2419–2429.
[17] A. J. Dunnett, N. Shah Prospects for Bioenergy, *J. Biobased Mater. Bioenergy* 1 (2007) 1–18.
[18] M. Balat, H. Balat Recent trends in global production and utilization of bio-ethanol fuel, *Applied Energy* 86 (2009) 2273–2282.
[19] A. Demirbas Political, economic and environmental impacts of biofuels: A review, *Applied Energy* 86 (2009) S108–S117.
[20] A. Demirbas Biofuels securing the planet's future energy needs, *Energy Conversion and Management* 50 (2009) 2239–2249
[21] A. J. Ragauskas, C. K. Williams, B. H. Davison, G. Britovsek, J. Cairney, C. A. Eckert, W. J. Frederick Jr., J. P. Hallett,. D. J. Leak, C. L. Liotta, J. R. Mielenz, R. Murphy, R. Templer, T. Tschaplinski The Path Forward for Biofuels and Biomaterials, *Science* 311 (2006) 484-489.
[22] G. Schaub, A. Vetter Biofuels for Automobiles – An Overview, *Chem. Eng. Technol.* 31 (2008) 721–729.
[23] L. C. M. das Neves, A. Converti, T. C. V. Penna Biogas Production: New Trends for Alternative Energy Sources in Rural and Urban Zones, *Chemical Engineering & Technology* 32 (2009) 1147-1153.

[24] A. Röttig, L. Wenning, D. Bröker, A Steinbüchel Fatty acid alkyl esters: perspectives for production of alternative biofuels, *Appl. Microbiol. Biotechnol.* 85 (2010) 1713-1733.
[25] L. Brennan, P. Owende Biofuels from microalgae—A review of technologies for production, processing, and extractions of biofuels and co-products, *Renewable and Sustainable Energy Reviews* 14 (2010) 557-577.
[26] R. C. Saxena, D. K. Adhikari, H. B. Goyal Biomass-based energy fuel through biochemical routes: A review, *Renewable and Sustainable Energy Reviews* 13 (2009) 167-178.
[27] G. W. Festel Biofuels - Economic Aspects, *Chem. Eng. Technol.* 31 (2008) 715-720.
[28] J. C. Escobar, E. S. Lora, O. J. Venturini, E. E. Yáñez, E. F. Castillo, O. Almazan Biofuels: Environment, technology and food security, *Renewable and Sustainable Energy Reviews* 13 (2009) 1275-1287.
[29] R. Hoefnagels, E. Smeets, A. Faaij Greenhouse gas footprints of different biofuel production systems, *Renewable and Sustainable Energy Reviews* 14 (2010) 1661-1694.
[30] G. Sivakumar, D. R. Vail, J. Xu, D. M. Burner, J. O. Lay Jr., X. Ge, P. J. Weathers Bioethanol and biodiesel: Alternative liquid fuels for future generations, *Eng. Life Sci.* 10 (2010) 8-18.
[31] A. J. Waclawovsky1, P. M. Sato, C. G. Lembke, P. H. Moore, G. M. Souza Sugarcane for bioenergy production: an assessment of yield and regulation of sucrose content, *Plant Biotechnology Journal* 8 (2010) 263-276.
[32] S. Kumar, N. Singh, R. Prasad Anhydrous ethanol: A renewable source of energy, *Renewable and Sustainable Energy Reviews* 14 (2010) 1830-1844.
[33] O. J. Sanchez, C. A. Cardona Trends in biotechnological production of fuel ethanol from different feedstocks, *Bioresource Technology* 99 (2008) 5270-5295
[34] D. R. Keshwani, J. J. Cheng Switchgrass for bioethanol and other value-added applications: A review, *Bioresource Technology* 100 (2009) 1515-1523.
[35] P. Alvira, E. Tomás-Pejó, M. Ballesteros, M. J. Negro Pretreatment technologies for an efficient bioethanol production process based on enzymatic hydrolysis: A review, *Bioresource Technology* 101 (2010) 4851-4861.
[36] E. Gnansounou, A. Dauriat Techno-economic analysis of lignocellulosic ethanol: A review, *Bioresource Technology* 101 (2010) 4980-4991
[37] S. Brethauer, C. E. Wyman Continuous hydrolysis and fermentation for cellulosic ethanol production, *Bioresource Technology* 101 (2010) 4862-4874.
[38] F. M. Gírio, C. Fonseca, F. Carvalheiro, L. C. Duarte, S. Marques, R. Bogel-Łukasik Hemicelluloses for fuel ethanol: A review, *Bioresource Technology* 101 (2010) 4775-4800.
[39] C. Weber, A. Farwick, F. Benisch, D. Brat, H. Dietz, T. Subtil, E. Boles Trends and challenges in the microbial production of lignocellulosic bioalcohol fuels, *Appl Microbiol Biotechnol* 87 (2010) 1303-1315.
[40] M. Fangrui, M. A. Hanna Biodiesel production: a review, *Bioresource Technology* 70 (1999) 1-15.
[41] Y. C. Sharma, B. Singh Development of biodiesel: Current scenario, *Renewable and Sustainable Energy Reviews* 13 (2009) 1646-1651
[42] E. P. Feofilova, Ya. E. Sergeeva, and A. A. Ivashechkin Biodiesel-Fuel: Content, Production, Producers, Contemporary Biotechnology (Review), *Applied Biochemistry and Microbiology* 46 (2010) 369-378.

[43] K. Narasimharao, A. Lee, K. Wilson Catalysts in Production of Biodiesel: A Review, *J. Biobased Materials and Bioenergy* 1 (2007) 19-30.

[44] E. C Francisco, D. B Neves, E. Jacob-Lopesb, T. T. Franco Microalgae as feedstock for biodiesel production: Carbon dioxide sequestration, lipid production and biofuel quality, *J. Chem. Technol. Biotechnol.* 85 (2010) 395–403.

[45] S. A. Scott, M. P. Davey, J. S. Dennis, I. Horst, C. J. Howe, D. J. Lea-Smith, A. G. Smith Biodiesel from algae: challenges and prospects, *Current Opinion in Biotechnology* 21 (2010) 277–286.

[46] Sugarcane bioethanol R&D for productivity and sustainability. Edited by L. A. B. Cortez, Blucher-FAPESP, 2010 Chap. 21.

[47] White paper on internationally compatible biofuel standards. Tripartite task force: Brazil, European Union & United States of America, Dec. 31, 2007. Available on www.**inmetro**.gov.br/painelsetorial/biocombustiveis/whitepaper.pdf.

[48] P. Spitzer, P. Fisicaro, S. Seitz, R. Champion pH and electrolytic conductivity as parameters to characterize bioethanol, *Accred Qual Assur* 14 (2009) 671–676.

[49] M. F. Oliveira, A. A. Saczk, L. L. Okumura, N. R. Stradiotto Analytical Methods Employed at Quality Control of Fuel Ethanol, *Energy Fuels* 23 (2009) 4852–4859.

[50] G. Knothe Analyzing Biodiesel: Standards and Other Methods, *JAOCS* 83 (2006) 823-833.

[51] S. Jain, M. P. Sharma Review of different test methods for the evaluation of stability of biodiesel, *Renewable and Sustainable Energy Reviews* 14 (2010) 1937–1947.

[52] M. R. Monteiro, A. R. P. Ambrozin, L. M. Lião, A. G. Ferreira Critical review on analytical methods for biodiesel characterization, *Talanta* 77 (2008) 593-605.

[53] I. P. Lôbo, S. L. C. Ferreira, R. S. Cruz Biodiesel: parâmetros de qualidade e métodos analíticos, *Quim. Nova* 32 (2009) 1596-1608.

[54] M. V. Marques, F. F. Naciuk, A. M. S. Mello, N. M. Seibel, L. A. M. Fontoura Determinação do teor de ésteres graxos em biodiesel metílico de soja por cromatografia gasosa utilizando oleato de etila como padrão interno, *Quim. Nova* 33 (2010) 978-980.

[55] C. Ragonesea, P. Q. Tranchidaa, D. Sciarronea, L. Mondelloa Conventional and fast gas chromatography analysis of biodiesel blends using an ionic liquid stationary phase, *Journal of Chromatography* A 1216 (2009) 8992–8997.

[56] S. Schober, I. Seid, A. Meinitzer, M. Mittelbach A method for the determination of polyunsaturated fatty acid methyl esters in biodiesel: Results of an interlaboratory study, *Eur. J. Lipid Sci. Technol.* 111 (2009) 829–834.

[57] F. Adam, F. Bertoncinia, V. Couparda, N. Charona, D. Thiébautb, D. Espinata, M.-C. Hennion, Using comprehensive two-dimensional gas chromatography for the analysis of oxygenates in middle distillates: I. Determination of the nature of biodiesels blend in diesel fuel, *Journal of Chromatography* A 1186 (2008) 236–244.

[58] L. Hejazi, D. Ebrahimi, M. Guilhaus, D. B. Hibbert Determination of the Composition of Fatty Acid Mixtures Using GC × FI-MS: A Comprehensive Two-Dimensional Separation Approach, *Anal. Chem.* 81 (2009) 1450-1458.

[59] G. Santori, A. Arteconi, G. Di Nicola, M. Moglie, R. Stryjek Quantitation of Compounds in Biodiesel Mixtures with Reversed-Phase Liquid Chromatography, *Energy & Fuels* 23 (2009) 3783–3789.

[60] K. Kittirattanapiboon, K. Krisnangkura Separation of acylglycerols, FAME and FFA in biodiesel by size exclusion chromatography, *Eur. J. Lipid Sci. Technol.* 110 (2008) 110 422–427.

[61] M. Hájek, F. Skopal, J. Kwiecien, M. Cernoch, Determination of esters in glycerol phase after transesterification of vegetable oil, *Talanta* 82 (2010) 283–285.

[62] M. Nagy, B. J. Kerr, C. J. Ziemer, A. J. Ragauskas Phosphitylation and quantitative 31P NMR analysis of partially substituted biodiesel glycerols, *Fuel* 88 (2009) 1793–1797.

[63] M. R. Monteiro, A. R. P. Ambrozin, L. M. Lião, A. G. Ferreira Determination of biodiesel blend levels in different diesel samples by ^{1}H NMR, *Fuel* 88 (2009) 691–696.

[64] J. K. Satyarthi, D. Srinivas, P. Ratnasamy Estimation of Free Fatty Acid Content in Oils, Fats, and Biodiesel by 1H NMR Spectroscopy, *Energy & Fuels* 23 (2009) 2273–2277.

[65] V. M. Mello, F. C. C. Oliveira, W. G. Fraga, C. J. Nascimento, P. A. Z. Suareza Determination of the content of fatty acid methyl esters (FAME) in biodiesel samples obtained by esterification using ^{1}H-NMR spectroscopy, *Magn. Reson. Chem.* 46 (2008) 1051–1054.

[66] M. Nagy, M. Foston, A. J. Ragauskas Rapid quantitative analytical tool for characterizing the preparation of biodiesel, *J. Phys. Chem.* A 114 (2010), 3883–3887.

[67] N. N. Mahamuni, Y. G. Adewuyi Fourier transform infrared spectroscopy (FTIR) method to monitor soy biodiesel and soybean oil in transesterification reactions, petrodiesel-biodiesel blends, and blend adulteration with soy oil, *Energy & Fuels* 23 (2009) 3773–3782.

[68] C. J. Chuck, C. D. Bannister, J. G. Hawley, M. G. Davidson Spectroscopic sensor techniques applicable to real-time biodiesel determination, *Fuel* 89 (2010) 457–461.

[69] P. V. Abdelnur, L. S. Eberlin, G. F. Sa, V. Souza, M. N. Eberlin Single-shot biodiesel analysis: nearly instantaneous typification and quality control solely by ambient mass spectrometry, *Anal. Chem.* 80 (2008) 7882-7886.

[70] R. G. D. Prates, R. Augusti, I. C. P. Fortes Electrospray Ionization Mass Spectrometry and Multivariate Calibration Analysis: A Combined Protocol To Quantify Biodiesel in Blends with Petrodiesel, *Energy Fuels* 24 (2010) 3183–3188.

[71] S. Pinzi, F. P. Capote, J. R. Jiménez, M. P. Dorado, M. D. L. Castro Flow injection analysis-based methodology for automatic on-line monitoring and quality control for biodiesel production, *Bioresource Technology* 100 (2009) 421–427.

[72] V. Río, M. S. Larrechi, M. P. Callao Sequential injection titration method using second-order signals: determination of acidity in plant oils and biodiesel samples, *Talanta* 81 (2010) 1572–1577.

[73] F. H. Lyra, M. T. W. D. Carneiro, G. P. Brandão, H. M. Pessoa, E. V. Castro Determination of Na, K, Ca and Mg in biodiesel samples by flame atomic absorption spectrometry (FAAS) using microemulsion as sample preparation, *Microchemical Journal* 96 (2010) 180–185.

[74] A. Jesus, A. V. Zmozinski, J. A. Barbara, M. G. R. Vale, M. M. Silva Determination of Calcium and Magnesium in Biodiesel by Flame Atomic Absorption Spectrometry Using Microemulsions as Sample Preparation, *Energy Fuels* 24 (2010) 2109–2112.

[75] E. S. Chaves, T. D. Saint'Pierre, E. J. Santos, L. Tormen, V. L. A. F. Bascuñana, A. J. Curtius Determination of Na and K in Biodiesel by Flame Atomic Emission Spectrometry and Microemulsion Sample Preparation, *J. Braz. Chem. Soc.* 19 (2008) 856-861.

[76] A. P. O., R. D. Villa, K. C. P. Antunes, A. Magalhães, E. C. Silva Determination of sodium in biodiesel by flame atomic emission spectrometry using dry decomposition for the sample preparation, *Fuel* 88 (2009) 764–766.
[77] F. H. Lyra, M. T. W. D. Carneiro, G. P. Brandão, H. M. Pessoa, E. V. R. Castro Direct determination of phosphorus in biodiesel samples by graphite furnace atomic absorption spectrometry using a solid sampling accessory, *J. Anal. At. Spectrom.*, 24 (2009) 1262-1266.
[78] M. A. Vieira, L. C. C. Oliveira, R. A. Gonçalves, V. Souza, R. C. Campos Determination of As in Vegetable Oil and Biodiesel by Graphite Furnace Atomic Absorption Spectrometry, *Energy Fuels* 23 (2009) 5942–5946.
[79] J. S. A. Silva, E. S. Chaves, E. J. Santos, T. D. Saint'Pierre, V. L. A. Frescura, A. J. Curtius Calibration Techniques and Modifiers for the Determination of Cd, Pb and Tl in Biodiesel as Microemulsion by Graphite Furnace Atomic Absorption Spectrometry, *J. Braz. Chem. Soc.*, 21 (2010) 620-626.
[80] P. R. Aranda, P. H. Pacheco, R. A. Olsina, L. D. Martinez, R. A. Gil Total and inorganic mercury determination in biodiesel by emulsion sample introduction and FI-CV-AFS after multivariate optimization, J. *Anal. At. Spectrom.* 24 (2009) 1441–1445.
[81] M. Trojanowicz Recent developments in electrochemical flow detections—A review Part I. Flow analysis and capillary electrophoresis, *Anal. Chim. Acta* 653 (2009) 36–58.
[82] J. Wang *Analytical Electrochemistry*, Wiley-VCH, New York, 2nd Ed., 2001.
[83] A. Doménech-Carbó Voltammetric methods applied to identification, speciation, and quantification of analytes from works of art: an overview, *J. Solid State Electrochem.* 14 (2010) 363–379.
[84] M. Scampicchio, D. Ballabio, A. Arecchi, S. M. Cosio, S. Mannino Amperometric electronic tongue for food analysis, *Microchim. Acta* 163 (2008) 11–21.
[85] L. Agui, P. Yánez-Sedeño, J. M. Pingarrón Role of carbon nanotubes in electroanalytical chemistry A review, *Anal. Chim. Acta*, 622 (2008) 11-47.
[86[A. J. Saleh Ahammad, J.-J. Lee, M. A. Rahman Electrochemical Sensors Based on Carbon Nanotubes, *Sensors* 9 (2009) 2289-2319.
[87] F. W. Campbell, R. G. Compton. The use of nanoparticles in electroanalysis: an updated review, *Anal. Bioanal. Chem.* 396 (2010) 241–259.
[88] A. Walcarius Electroanalytical Applications of Microporous Zeolites and Mesoporous (Organo)Silicas: Recent Trends, *Electroanalysis* 20 (2008) 711 – 738
[89] A. Walcarius Template-directed porous electrodes in electroanalysis, *Anal. Bioanal. Chem.* 396 (2010) 261–272.
[90] P. T. Kissinger, W. R. Heineman *Laboratory techniques in electroanalytical chemistry*, Marcel Dekker, New York, 2nd Ed., 1996.
[91] A. J. Bard, L. R. Faulkner *Electrochemical methods: fundamentals and applications*. John Wiley & Sons: New York, 2nd Ed., 2001.
[92] D. T. Sawyer, A. Sobkowiak, J. L. Roberts Jr. *Electrochemistry for chemists*. John Wiley & Sons: New York, 2nd Ed., 1995.
[93] J. Bobacka, A. Ivaska, A. Lewenstam Potentiometric ion sensor, *Chem. Rev.* 108 (2008) 329–351.
[94] B. J. Privett, J. H. Shin, M. H. Schoenfisch Electrochemical Sensors, *Anal. Chem.* 82 (2010) 4723–4741.

[95] D. A. Skoog, F. J. Holler, S. R. Crouch Principles of instrumental analysis, *Harcourt Inc*, 5th Ed., 1998.

[96] J. G. Osteryoung, R. A. Osteryoung Square wave voltammetry, *Anal. Chem.* 57 (1985) 101A.

[97] R. G. Compton, G. E. Banks. *Understanding voltammetry*, Word Scientific Publishing Co: Singapore, 2009.

[98] A. W. Bott Practical Problems in Voltammetry 3: Reference Electrodes for Voltammetry, *Current separations* 14 (1995) 64-68.

[99] M. W. Shinwari, D. Zhitomirsky, I. A. Deen, P. R. Selvaganapathy, M. J. Deen, D. Landheer Microfabricated Reference Electrodes and their Biosensing Applications, *Sensors* 10 (2010) 1679-1715.

[100] J. Barek 50 Anniversary of Nobel Prize for Polarography, *Polarography Lounge* (2009) 27-31.

[101] J. Barek, J. Zima Eighty Years of Polarography - History and Future, *Electroanalysis* 15 (2003) 467-472.

[102] A. Qureshi, W. P. Kang, J. L. Davidson, Y. Gurbuz Review on carbon-derived, solid-state, micro and nano sensors for electrochemical sensing applications, *Diamond & Related Materials* 18 (2009) 1401–1420.

[103] R. L. McCreery Advanced carbon electrode materials for molecular electrochemistry, *Chem. Rev.* 108 (2008) 2646-2687.

[104] I. Svancara, K. Vytras, K. Kalcher, A. Walcarius, J. Wang Carbon paste electrodes in facts, numbers and notes: A review on the occasion of the 50-years jubilee of carbon paste in electrochemistry and electroanalysis, *Electroanalysis* 21 (2009) 7-28.

[105] I. Svancara, A. Walcarius, K. Kalcher, K. Vytras Carbon paste electrodes in the new millennium, *Cent. Eur. J. Chem.* 7 (2009) 598-656.

[106] J-M. Zen, A. S. Kumar, D-M Tsai Recent updates of chemically modified electrodes in analytical chemistry, *Electroanalysis*, 15 (2003) 1073-1087.

[107] X. Xie, D. Stueben, Z. Berner The application of microelectrodes for the measurements of trace metals in water, *Anal. Lett.* 38 (2005) 2281-2300.

[108] O. Ordeig, J. Campo, F. X. Munñoz, C. E. Banks, R. G. Compton Electroanalysis Utilizing Amperometric Microdisk Electrode Arrays, *Electroanalysis* 19 (2007) 1973-1986.

[109] M. M. Villalba, J. Davis New directions for carbon-based detectors: exploiting the versatility of carbon substrates in electroanalysis, *J. Solid State Electrochem.* 12 (2008) 1245–1254.

[110] R. M. da Silva, P. P. Borges, I. C. S. Fraga, N. R. Stradiotto New methodology for pH measurements in fuel ethanol using glass electrode, *J. ASTM International* (2010). In analysis.

[111] M. S. Castilho, N. R. Stradiotto Determination of potassium ions in biodiesel using a nickel(II) hexacyanoferrate-modified electrode, *Talanta* 74 (2008) 1630-1634.

[112] P. Rapta, J. Paligova, H. Rotheneder, J. Cvengros Determination of potassium in fatty acid methyl esters applying an ion-selective potassium electrode, *Chem. Pap.* 61 (2007) 337-341.

[113] G. Oliveira Neto, S. H. P. Serrano, E. F. A. Neves A New Type of Precipitating Agent for Copper Membrane Electrode, *Anal. Lett.* 20 (1987) 1363-1377.

[114] H. M. Avelar, P. J. S. Barbeira Conductometric determination of total acidity and chloride content in automotive fuel ethanol, *Fuel* 86 (2006) 299-302.
[115] H. Herrmann, K. Chassiniol, D. Thomas Ion chromatography for ethanol analysis, *LaborPraxis* 31 (2007) 46-47.
[116] A. Steinbach, A. Willie Ion chromatography in biofuels analysis. Anions, antioxidants, alkaline earth and alkali metal-ions in biodiesel, *GIT Spezial Separation* 28 (2008) 54-56.
[117] R. A. A. Munoz, E. M. Richter, D. P. Jesus, C. L. Lago, L. Angnes Determination of Inorganic Ions in Ethanol Fuel by Capillary Electrophoresis, *J. Braz. Chem. Soc.* 15 (2004) 523-526.
[118] D. C. Romanini, M. A. G. Trindade, M. V. B. Zanoni A simple electroanalytical method for the analysis of the dye solvent orange 7 in fuel ethanol, *Fuel* 88 (2009) 105-109.
[119] M. A. G. Trindade, M. V. B. Zanoni Square-wave voltammetry applied to the analysis of the dye marker, Solvent blue 14, in kerosene and fuel alcohol, *Electroanalysis* 19 (2007) 1901-1907.
[120] M. A. G. Trindade, M. V. B. Zanoni Voltammetric sensing of the fuel dye marker Solvent Blue 14 by screen-printed electrodes, *Sens. Actuators* B 138 (2009) 257-263.
[121] V. S. Rodgher, L. L. Okumura, A. A. Saczk, N. R. Stradiotto, and M. V. B. Zanoni Electroanalysis and Determination of Acetaldehyde in Fuel Ethanol Using the Reaction with 2,4-Dinitrophenylhydrazine, *J. Anal. Chem.* 61 (2006) 889-895.
[122] V. S. Rodgher, N. R. Stradiotto, M. V. B. Zanoni Eletroanálise de derivados acetaldeído com hidrazina e determinação em amostras de álcool combustível por voltametria de onda quadrada, *Quim. Nova* 29 (2006) 662-665.
[123] G. L. P. da Silva, O E. S. Godinho Polarographic determination of free acetaldehyde by direct reduction of its imino-derivates in aqueous alcoholic media, *Anais - ABQ* 46 (1997) 256-262.
[124] A. A. Saczk, M. F. Oliveira, L. L. Okumura, N. R. Stradiotto Comportamento voltamétrico da redução de 2- furfuraldeído em etanol utilizando eletrodo de carbono vítreo, *Eclet. Quím.* 27 (2002), 141-151.
[125] L. M. Lourenço, N. R. Stradiotto Determination of free glycerol in biodiesel at a platinum oxide surface using potential cycling technique, *Talanta* 79 (2009) 92-96.
[126] T. A. Araújo, A. M. J. Barbosa, L. H. Viana, V. S. Ferreira Voltammetric determination of tert-butylhydroquinone in biodiesel using a carbon paste electrode in the presence of surfactant, *Colloids and Surfaces B: Biointerfaces* 79 (2010) 409–414.
[127] R. M. Takeuchi , A. L. Santos, P. M. Padilha, N. R. Stradiotto Copper determination in ethanol fuel by differential pulse anodic stripping voltammetry at a solid paraffin-based carbon paste electrode modified with 2-aminothiazole organofunctionalized silica, *Talanta* 71 (2007) 771-777.
[128] R. M. Takeuchi, A. L. Santos, M. J. Medeiros, N. R. Stradiotto Copper determination in ethanol fuel samples by anodic stripping voltammetry at a gold microelectrode, *Microchim. Acta* 164 (2007) 101-106.
[129] I. Cesarino, G. Marino, É. T. G. Cavalheiro A novel graphite–polyurethane composite electrode modified with thiol-organofunctionalized silica for the determination of copper ions in ethanol fuel, *Fuel* 89 (2010) 1883–1888.

[130] R. A. A. Munoz, L. Angnes. Simultaneous determination of copper and lead in ethanol fuel by anodic stripping voltammetry, *Microchem. J.* 77 (2004) 157-162.

[131] M. F. Oliveira, A. A. Saczk, L. L. Okumura, A. P. Fernandes, M. Moraes, N. R. Stradiotto Simultaneous determination of zinc, copper, lead, and cadmium in fuel ethanol by anodic stripping voltammetry using a glassy carbon-mercury-film electrode, *Anal. Bioanal. Chem.* 380 (2004) 135-140.

[132] M. F. Bergamini, S. I. Vital, A. L. Santos, N. R. Stradiotto Determinação de chumbo em álcool combustível por voltametria de redissolução anódica utilizando um eletrodo de pasta de carbono modificado com resina de troca iônica Amberlite IR 120, *Eclet. Quím.* 31 (2006) 45-52.

[133] R. M. Takeuchi , A. L. Santos, P. M. Padilha, N. R. Stradiotto A solid paraffin-based carbon paste electrode modified with 2-aminothiazole organofunctionalized silica for differential pulse adsorptive stripping analysis of nickel in ethanol fuel, *Anal. Chim. Acta* 584 (2007) 295-301.

[134] F. O. Tartarotti, M. F.Oliveira, V. R. Balbo, N. R. Stradiotto Determination of nickel in fuel ethanol using a carbon paste modified electrode containing dimethylglyoxime, *Microchim. Acta* 155 (2006) 397-401.

[135] C. S. Mattos, D. R Do Carmo, M. F. Oliveira, N. R. Stradiotto Voltammetric determination of total iron in fuel ethanol using a 1,10 fenantroline/nafion carbon paste-modified electrode, *Int. J. Electrochem. Sci.* 3 (2008) 338-345.

[136] M. F. Oliveira; A. A. Saczk; L. L. Okumura; N. R. Stradiotto Determinação de zinco em álcool combustível por voltametria de redissolução anódica, *Eclet. Quím.* 27 (2002) 153-160.

[137] H. B. P. Ferreira, T. A. F. Lima, P. J. S. Barbeira Voltammetric determination of chloride ion in automotive fuel ethanol, *Electroanalysis* 20 (2008) 390-395.

[138] L. M. Aleixo, S. Rath, O. E. S. Godinho, A. U. Ivaska Determination of total sulfur in ethanol by reduction using raney-nickel and detection by cathodic stripping voltammetry, *Analyst* 113 (1988) 1427-1429.

[139] E. S. Almeida, M. A. N. A. Monteiro, R. H. O. Montes, R. Mosquetta, N. M. M. Coelho, E. M. Richter, R. A. A. Muñoz Direct Determination of Copper in Biodiesel Using Stripping Analysis, *Electroanalysis* 22 (2010) 1846-1850.

[140] T. Luetkmeyer, R. M. Santos, A. B. Silva, R. S. Amado, E. C. Vieira, E. D'Elia Analysis of Free and Total Glycerol in Biodiesel Using an Electrochemical Assay Based on a Two-Enzyme Oxygen-Electrode System, *Electroanalysis* 22 (2010) 995-999.

[141] A. A. Saczk, L. L. Okumura, M. F. de Oliveira, M. V. B. Zanoni, N. R. Stradiotto Determination of aldehydes and ketones in fuel ethanol by high-performance liquid chromatography with electrochemical detection, *Chromatographia* 63 (2006) 45-51.

[142] A. A. Saczk, L. L. Okumura, M. F. de Oliveira, M. V. B. Zanoni, N. R. Stradiotto Determination of acetaldehyde in fuel ethanol by high-performance liquid chromatography with electrochemical detection, *Anal. Sci.* 21 (2005) 441-444.

[143] M. A. G. Trindade, D. Rinaldo, W. Vilegas, M. V. B. Zanoni Determinação de corantes marcadores do tipo azo e antraquinona em combustíveis por cromatografia líquida com detecção eletroquímica, *Quim. Nova* 33 (2010) 146-150.

[144] D. T. Paula, H. Yamanaka, M. F. Oliveira, N.R. Stradiotto Determination of chloride in fuel ethanol using polyaniline-chemically modified electrode in flow injection analysis, *Chem. Technol. Fuels Oils* 44 (2008) 435-440.
[145] J. A. F. da Silva, C. L. Lago An oscillometric detector for capillary electrophoresis, *Anal. Chem.* 70 (1998) 4339-4343.
[146] A. J. Zemann, E. Schnell, D. Volgger, G. K. Bonn Contactless Conductivity Detection for Capillary Electrophoresis, *Anal. Chem.* 70 (1998) 563-567.
[147] H. Lund, O. Hammerich. *Organic Electrochemistry*, 4th Ed., Marcel Dekker, New York, 2001.

25

Quality Control Implementation in Manufacturing Companies: Motivating Factors and Challenges

Hairulliza Mohamad Judi, Ruzzakiah Jenal and Devendran Genasan
Industrial Informatic Programme, Faculty of Information Science and Technology, Universiti Kebangsaan Malaysia 43600 Bangi, Selangor Darul Ehsan, Malaysia

1. Introduction

The pressure from globalisation has made manufacturing organisations moving towards three major competitive arenas: quality, cost, and responsiveness. Quality is a universal value and has became a global issue. In order to survive and be able to provide customers with good products, manufacturing organisations are required to ensure that their processes are continuously monitored and product quality are improved.

Manufacturing organisation applies various quality control techniques to improve the quality of the process by reducing its variability. A range of techniques are available to control product or process quality. These include seven statistical process control (SPC) tools, acceptance sampling, quality function deployment (QFD), failure mode and effects analysis (FMEA), six sigma, and design of experiments (DoE).

The purpose of this chapter is to present the implementation of quality control in four manufacturing companies and identify the factors that influence the selection of quality control techniques in these companies. The paper discusses the reasons for applying quality control techniques, the techniques used, and problems faced by them during the implementation. The paper begins with an overview of quality control and its implementation in organisations. This is followed by the description of four selected companies in this study including their products and company backgrounds. The application of quality control in each company is then presented. The motivating factors for the companies to apply quality control and challenges faced by companies in implementing quality control are discussed.

2. Quality control

Quality can be defined as fulfilling specification or customer requirement, without any defect. A product is said to be high in quality if it is functioning as expected and reliable. Quality control refers to activities to ensure that produced items are fulfilling the highest possible quality. Most of tools and techniques to control quality are statistical techniques. Quality control techniques can be classified into basic, intermediate, and advance level, but there is no consensus among researchers in the classification. For example, Xie and Goh (1999) consider

DoE as an intermediate level technique whereas Antony et al (1998) classified the technique as advanced. Nevertheless, the content is more important than the classification.

Among the basic techniques are SPC. SPC is a statistical approach for assisting operators, supervisors and managers to manage quality and to eliminate special causes of variability in a process (Oakland, 2003). The initial role of SPC is to prevent rather than identify product or process deterioration, but Xie and Goh (1999) suggest for its new role to actively identifying opportunities for process improvement. The main tools in SPC are control charts. The basic idea of control charts is to test the hypothesis that there are only common causes of variability versus the alternative that there are special causes By continuously monitoring the process, the manufacturing organisation could prevent defect items to be processed in the next stage and to take immediate corrective action once a process is found to be out of control (Hairulliza et al., 2005).

DoE and Taguchi methods are powerful tools for product and process development. Taguchi methods, for instance, aim at making product or process that robust to undesirable disturbances such as environmental and manufacturing variations. However, the application of these two methods by industries is limited (Antony and Kaye, 1995). Antony et al (1998) explore the difficulties in the application including improper understanding and fear of statistical concepts in the methods, thus propose a methodology for the implementation.

Process capability study is an efficient method to examine the capability of a process to produce items that meet specifications. The method gains rapid growing interest due to increased use of quality system QS9000, where use of process capability studies is requested (Deleryd et al, 1999). The findings from capability study might require adjustment of process using other statistical technique such as SPC or DoE. Capability studies conducted by Motorcu and Gullu (2004) and Srikaeo et al (2005) show that the machine tool and process capability and production stability was evaluated and necessary steps to reduce poor quality production was carried out using other statistical techniques.

FMEA is a powerful method to detect where exactly problems can occur and to prioritise possible problems in the order of their severity (Dale et al., 2003). The tool is useful to identify problems in product, i.e. design FMEA, as well as to trouble shoot problems in process, i.e. process FMEA (Xie and Goh, 1999). Six sigma is also a statistical tool for ensuring defect free products through process continuous improvement. The term six sigma originated at Motorola and many inspired worldwide organizations have set goal towards a six sigma level of performance (Breyfogle and Cupello, 2001). The application of six sigma has been mainly used in manufacturing industry. An example of the use of six sigma in non-manufacturing industry is in software development (Mahanti and Antony, 2005).

Acceptance sampling is another statistical techniques to make a decision whether to accept or reject a lot based on the information from sample. The application of acceptance sampling allows industries to minimise product destruction during inspection and testing, and to increase the inspection quantity and effectiveness. The application of acceptance sampling has been mainly used in manufacturing industry. Similarly, its application in non-manufacturing industry is widely reported such as Thorpe et al. (1994), Gardiner and Mitra (1994) Bathika (2003) and Slattery (2005).

3. Research method

The study aims at exploring the implementation of quality control in Malaysian organisations. Four industries were participating in the study. They will be identified as Company A, Company B, Company C, and Company D. These are among listed operating manufacturing companies in the Bandar Baru Bangi Industrial Area. Initially, 10 companies were contacted to be invited to participate in this study. As a result, only four agree to cooperate. An appointment has been made with the Quality Manager of each company for an interview. The meeting enables the researchers to get as much information about quality control implementation in the company. A list of questions has been prepared prior to the company visit as a guidance for the researcher to get the information. The interview with the Quality Manager and factory visit take about two to three hours for each company.

4. Quality control implementation

This section will describe the application of quality control techniques in each company. The discussion will present some examples of quality control techniques in each company.

4.1 Company A

Company A is a subsidiary of locally-owned National Power Limited. It produces various cables ranging from low to high voltage capacity. This company is the only local producer of fiber optic. The mission of Company A is to become the main local producer and distributer of cable, and to enter the global market soon.

Company A is striving to be the main supplier of electric cable in the country and consequently in the region. One of the ways to achieve the company mission is by ensuring high quality cables produced by them. The produced cables need to be inspected in order to ensure only good cables will be delivered to customers.

Therefore, acceptance sampling is applied by selecting a random sample from a production lot of cables and checking for defect cables in the sample (if any) to later decide whether the whole lot should be delivered to customer or not. The information is also useful for management to make any adjustment in production to ensure the cable meet the specifications. Each cable type has a set of standards or specifications comprising of for example thickness diameter, breaking load etc.

For a big-size cable such as XXX cable, the thickness diameter is measured manually. About 5 or 6 cables are randomly selected from a production lot of for example 100 cables, and tested for their outer covering diameter, to ensure that they are within the specification. Caliper is used to measure the thickness diameter. If the measured diameter is within the minimum and maximum specification, the cable is considered passed for the inspection.

Each selected cable will be measured once, and the reading figure will be recorded in a Checksheet form. Usually 5 selected cables represent a production lot, therefore a checksheet will comprises of five measurement figures. If at least two measurements are beyond the maximum specification, improvement will be made in the manufacturing department. The measurement for selected cable is recorded in a checksheet as in Table 1.

There were seven quality variables recorded in the sheet and five measurements need to be taken for each variable. The specification readings are printed in the sheet as a guideline. The quality engineer will be able to make a suitable decision whether to allow cables for shipment or not. The record in Table 1 is useful for production planning. Analysis on those data enable planning unit to monitor the production performance weekly or monthly.

Company A seems to be satisfied with the applied quality control techniques. With enough trained workers and tools to perform quality control tasks, the Quality Manager in Company A agrees that the current quality control implementation is not facing much difficulties. The only problem that they experience for current implementation is regarding setting up a new set of specifications that involves various measurements each time Company A get new customers. The procedure takes tremendous amount of time to prepare the new specifications due to various tests are needed to inspect the new cable required by new customers.

	Description	Unit	Measurment						Min	Ave	Spec
			M1	M2	M3	M4	M5	M6			
1	Aluminum H68 Condition	mm									Min:18.2 Max:18.6
2	Conductor screen R. Thickness	mm									Min. Spot 0.5
3	XLPE Insulation Diameter	mm									Min: 26.0 Max: 28.0
4	Core Diameter	mm									Min: 27.0 Max: 32.2
5	Min. Breaking Load	N									2708
6	Wrapping Test	-									No Crack
7	Black MDPE Sheath	mm									Min: 32.1 Max: 36.2

Table 1. Checksheet to inspect the selected cable

Company A has some planning to apply other quality control techniques and tools in future. The new techniques and tools are hoped to solve the current problems and to increase the efficiency and effectiveness in ensuring the products meet customer requirements.

4.2 Company B

Company B is a producer of soft drinks. This company produces drinks with a famous brand and other new brands. There are altogether about 10 branches under the parent company of Company B all over Malaysia. It produces drinks according to the specified standard that involves only mixing the specified ingredients and packing them.

The parent company of Company B produces a well-known soft drink brand and the drinks are marketed worldwide. It has a reputable name and makes every effort to maintain the status. Company B follows the quality control procedure as prescribed by the parent company. The drinks produced by Company B need to be monitored to ensure not only customers taste it good but also they will keep drink it.

Various variables are used to monitor the drink quality including mix amount, temperature, bottle condition, acid amount, etc. Among important ingredient in producing the drinks are a kind of sugar known as Brix or Soliber Solid and carbon dioxide (CO^2). The specification for Brix amount is between 10.7° to 11.0°. At least 10 tests are conducted each day in this company to ensure that Brix amount is under control. As for carbon dioxide pressure, the specification value is between 29 to 30 Newton Pascal.

Control chart is used to monitor whether the ingredients are stable. A random sample is selected from the drink production for the analysis. With the help of a special QC machine, the measurement is conducted automatically from the connection to the big mixing machines. The QC machine is able to record, analyse, and produce graphical presentation from the data. The presented control charts enable workers to take appropriate actions once the process is out of control. Among the improvement strategy is monitoring and examining the ingredient mixing process, adjusting the machine, monitoring the responsible worker for the ingredient mixing process.

Some of the graphical presentations of the analysis are as in Figure 1. The graphs show that the company needs to find out root causes that hinder them from achieving monthly and daily specification target.

Among the difficulties experienced by Company B in QC implementation is when the QC machine breaks down. Workers need to conduct control charting manually including taking the measurement reading for Brix and CO^2. Currently, the QC machine is connected to a few production machines. In future, the company intends to have an integrated system with a connection of QC machine to all production machines and monitoring computers. This enables an efficient data sharing and quality monitoring.

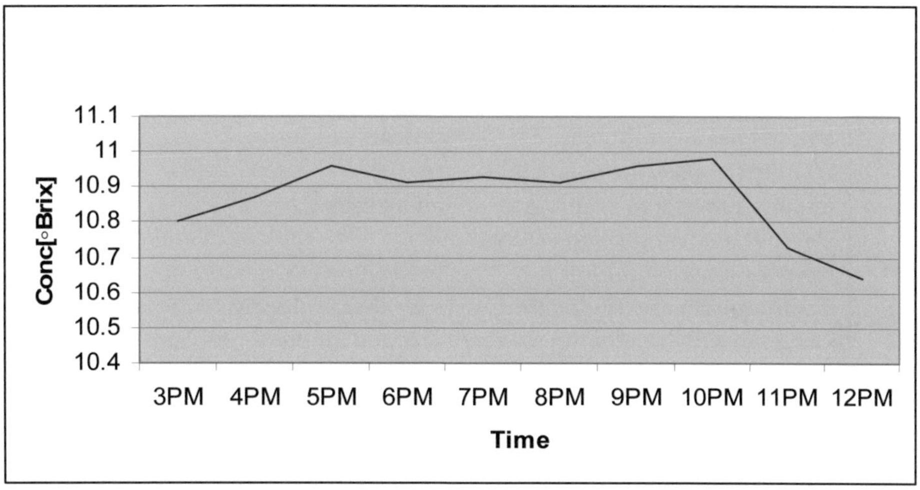

Fig. 1. Hourly Brix Amount

4.3 Company C

Company C produces plastic items. The production of items applies injection plastic mould. Among the produced plastic items are TV housing benzet and TV rear cover. Besides, the products are used as components in audio system, video frame, VCD frame, and air-conditioning frame. Company C produces plastic components for industrial usage. The company supplies raw material for many other big companies locally and abroad. This situation makes them give particular attention for quality control. The main reason for implementing quality in this company is due to customer requirement.

SPC tools are applied in various production stages. For example, workers collect data on product and process performance such as defect items and machine malfunctioning using checksheet. Suitable control chart such as np is used to identify whether the production process for video frame is stable. If there is any data above the upper control limit, corrective actions will be taken. Next, the root causes for the special variations will be identified using Cause-and-Effect diagram involving many workers in a brainstorming session.

Besides SPC, Company C also practices FMEA. The application enables Company C to identify, define and take preventive action in dealing with expected difficulties in the production process. The steps in FMEA application for producing video frame MRY are presented in Table 2. The purposes for the application are to identify the main cause for frame failure and to analytically classify the frame failure.

The application of FMEA results in systematic problem solving for various problems in the production process including Printing NG, Function NG, Cosmetic NG, Paint NG and Moulding Defects. The effectiveness of QC using FMEA was proven by analysing the performance prior to and after FMEA application. Tables 3 and 4 present these defect statistics.

No.	Step
1	Select MRY model and set the target index to be achieved.
2	Set target and date to achieve.
3	Identify problems, failures, and potential difficulties in the process.
4	Define and describe one of the failures experienced by the team.
5	Identify the known cause for the failures.
6	Identify the possible cause for the failures.
7	Identify the classification of failures model.
8	Identify preventive action to avoid failure causes.
9	Develop method for implementation.
10	Describe other failures or problems until all members satisfied.
11	Continue with identifying the cause for other problems.
12	Continue with identifying prevention action for other problems.

Table 2. Steps in FMEA application

Total Input	40070 Unit				
Defect Location	A	B	C	Total	Defect %
Printing NG	415	156	2	573	1.43
Function NG	138	53	1	192	0.48
Cosmetic NG	1520	570	7	2097	5.23
Paint NG	533	209	4	766	1.91
Moulding Defects	139	57	1	197	0.49

Table 3. Defect statistics prior to FMEA application

Total Input	53800Unit				
Defect Location	A	B	C	Total	Defect %
Printing NG	354	163	1	517	0.96
Function NG	119	53	0	172	0.32
Cosmetic NG	1296	596	2	1894	3.52
Paint NG	469	215	1	685	1.27
Moulding Defects	115	52	1	168	0.31

Table 4. Defect statistics after FMEA application

The analysis has shown that the defect rates have decreased for all defect types after applying the method. Consequently, about a thousand product items is secured from being defect items in Company C owing to FMEA application.
Among the difficulties experienced by Company C in quality control implementation is regarding the hardship for workers in working with SPC especially in using control charts. A training on basic statistics and quality concepts to conduct QC techniques is the critical action for the company.

4.4 Company D

Company D is a subsidiary of a conglomerate company in Japan. It offers an extensive range of room air conditioning units and compressors. The company is supported by over 1,000 employees and the latest technology. Company D grows to provide products, systems and services that fulfil their local and international customers, based on the latest advances in technology.
Company D also intends to be the catalyst for a changing society. This spirit has been disseminated among the workforce, and various trainings on quality and technical skills were conducted to ensure the workforce have the required knowledge and capability. In order to make sure that their products meet the high standards from customers, Company D applies rigorous quality control procedure. Company D aims that its air-conditioners offer these characteristics: comfort, energy saving, and silent.
The manufacturing process in Company D involves many stages, beginning with raw material until a complete item is produced. Each production line conducts its main function such as painting, machining, welding, and testing. One of the critical processes in Company D involves pump assembly that consists of various small processes including cuprum pipe pressing. The process takes place once the assembly of air-conditioner components completed, in which chlorofluorocarbon gas is filled into the item. The cuprum needs to be pressed so that the gas is not spreading out. If the process was not conducted properly, the cuprum pipe is leaking and the gas is released gradually from the components.
The emptiness of cuprum pipe from gas deteriorates the chilling effect thus users find it takes long time to cool down. Company D gives enough attention to the pressing process to maintain the comfort and cooling power of its air-conditioning items. To avoid this critical quality problem, ongoing monitoring of the process is conducted by taking at least five measurement of the pressed cuprum at one time, every regular hour. The readings are monitored by plotting an individual control chart to identify if the process is stable.
One of the features that users associate the brand of Company D with is silent. This distinguish characteristic is designed in each air-conditioning item produced by Company D to ensure that users could enjoy the cool environment peacefully without any disturbance. The new twin turbo fan delivers powerful air flow while the streamlined, drag-reducing construction, the air-conditioning items reduce noise with great efficiency. Company D determines that only air-conditioners of the best possible quality produced by the factory. Thus, the practice of quality assurance is of the utmost importance.
All items should fulfill the strict requirements of the international certification. These items are subject to quality inspection to verify that they are free from defects. A number of products will be picked at random from the production line to be thoroughly inspected if any noise is produced by the item. Decision will be made whether to accept or reject the whole item and consequently whether or not distribute them to the markets.

Another characteristic that the users like the most about the brand produced by Company D is its capability to save energy and this means a lower electricity bills to them. To achieve a high performance in the market, Company D utilises various recent technologies that support energy conservation and cost reduction. Low energy consumption plays an important role in ensuring items produced by Company D is at highest possible quality and providing an environment that is sustainable for the future.

Techniques such as FMEA and six sigma are exploited to help the organisation identify, diagnose, and define difficulties and problems with the current products and consequently take tremendous action in improving and radically change them. Innovation such as air purification has made items produced by Company D exceed the needs of customers. With the innovation, the air-conditioners take in fresh air with sufficient oxygen thus keeping the same level with the outside atmosphere, so indoor air is balanced as in the natural environment.

A powerful ventilation system takes away foul odours and bacteria, scrubs out and expels pollutants such as carbon dioxide and formaldehyde. Bacteria and odours are promptly eliminated, as a result of the combination of the Nano Titanium catalyst with silver and zinc. In Company D, enough investment has been made on manufacturing technology such as the application of Scroll Involute Finish Cutting and 3-D Coordinate Measuring Machines. This ensures the fast and precise inspection of components of compressors with little tolerance for errors and high reliability.

Company D has set an example that to be dominant in a fierce global competition, quality must not be compromised. New idea must be identified to accommodate the unpredictable, turbulent, and ever-changing market, due to factors such as changing customer needs, technological advances, and shorter product life cycle.

5. Issues in quality control implementation

This section will discuss important issues in quality control implementation. These include motivating factors, challenges and future trends in QC implementation.

5.1 Motivating factors in QC implementation

The implementation of quality control in four Malaysian companies is reported in this study. As shown in the case study, the reasons for the implementation might come from the company itself, the parent company, or from customer. On the current world market, both manufacturers and consumers require guarantees for the quality of products and services. One of the ways to ensure that the required quality is obtained at appropriate cost and time is by applying quality control in the organisation. The aim of these companies is the same i.e. to produce high quality and reliable products, meet customer expectation, fulfill ISO condition and compete in the market.

Among various applied quality control techniques in the companies, SPC is the most popular applied techniques. Control chart, checksheet, and cause and effect diagram were used in these companies to analyse and interpret the data related to product quality. SPC is not only easy and simple to be used, it provides a collection of powerful problem-solving techniques to achieve process stability and reduce variability, and can be used in any process (Montgomery, 2005). Consistent with previous research (Srikaeo et al, 2005; Mason and Antony, 2000; Antony and Taner, 2003; Xie and Goh, 1999), control chart is the most widely used SPC techniques in industries especially to monitor production process.

Acceptance sampling is also applied together with SPC in some companies to get better results of process and product improvement. The case study has shown that quality control implementation in Company A relies heavily on acceptance sampling. In company like this one, acceptance sampling plays an important roles to decide whether to accept or reject particular lot that contains tremendous amount of products. Conviniently, only small amount of products will be inspected to represent the whole lot. Though acceptance sampling technique receives many critics including using detection instead of prevention approach, expensive activity, provides unreliable information, acceptance sampling still has its function in industry. The finding supports Mital et al. (1998) that inspection remains an essential and critical activity for most companies even if there are some inspection-free companies due to their process stability and small process variability. Ahmed and Hassan (2003) also found that among the most used quality control techniques in the Malaysian SMIs was inspection-sampling.

FMEA is applied in one of the companies to particularly help the company from financial loss due to product scraps and defects. Though the procedure is a bit complex, the production team in Company C works together to accomplish the FMEA activity that results in defect rates reduction. Workers knowledge about the process and products were harnessed throughout FMEA activities and their participation in QC activities has boosted the moral to contribute for company excellence. On the other hand, FMEA is utilised in Company D to innovate and radically change the current procedure of products are manufactured.

The four investigated industries show some pattern in the applied QC techniques. Company A relies only on acceptance sampling and checksheet to analyse and interpret the quality data. The Quality Manager seems to satisfy with the techniques since the workers are competent to conduct the related tests and the company shows capability to manufacture products that conform to specification. In the case of Company B, the selected quality control techniques and machines are recommended by the parent company, be able to monitor that the production process is stable. Company C and D takes a step forward from the other two companies in using FMEA to improve the productivity problem, besides using SPC.

The study concludes that the selection of quality control technique in these companies is influenced by three factors. First, ease of use of the technique. A simple but powerful technique will be easily adopted by workers. A more advance technique might require employee knowledge, and might risk employee acceptance to learn new methods. Second, technique that measures product specification fulfillment. Since customer and producer agree with product specification, the selected technique should provide some measurement of specification fulfillment. Third, technique that improves current critical quality and productivity problem. Quality and productivity measures the competitiveness of the company. If company is facing serious problems in these matters, certain quality control techniques will be applied to improve the situation.

The application of quality control technique shows similar pattern of usage in the production stage. Many industries report QC application in the post-production stage, meant for finished products. The application of combination of techniques as early in the process as possible will result in better product and process performance. Xie and Goh (1999) mention that companies will get maximum benefit by improving both the product and process if they can use the approach right from the design stage.

Some of QC techniques such as six sigma, DOE, Taguchi methods, and capability studies are left behind from being used in these four industries. The finding has proven Antony and Kaye's (1995) report that the application of DoE and Taguchi methods by industries is limited. One of the reasons is due to lack of knowledge in the technique. These industries might review the benefit of designing the product and process using DoE and Taguchi especially in Company C since it produced a wide variety of finished products. Capability study is also a useful technique these companies might consider to measure organization's ability to produce goods that fulfill tolerance limit. Some improvement in control chart (Xie et al, 1999) also need to be considered by these companies. The new technique based on the cumulative count of items produced before nonconforming ones are observed allows monitoring the process automatically.

The application of information technology and internet facilities in QC is among recent development in this field (Besterfield, 2004). Real-time QC would be a helpful solution for companies facing issues around lack of quality awareness, passive response, machine problem, unreliable incoming materials quality, and tight customer schedule (Chung et al., 2007). A computer collecting information in real time can detect very subtle changes in a process, and gives warning in time to prevent process errors before they occur (Woo and Law, 2002; Chin et al., 2004).

Research finding in this study shows that each company has shown capability or potential ability to practice QC effectively and efficiently using these facilities. Companies that use computer numerically controlled (CNC) machines enjoy automatic QC. The machine enables not only precise data reading but also controllable data error and efficient report preparation. The application of statistical packages and softwares to analyse data has been performed in these companies. It helps workers to analyse data and make graphical presentation to monitor process performance. Besides, some company is moving towards integrated QC system to allow all departments conduct an efficient QC.

5.2 Challenges and future trends in QC implementation

The study found that each company has its strengths and weaknesses in applying QC. For example, in Company A, though the applied method seems to be simple, it involves tedious procedure to conduct the inspection test. To overcome the complex measurement problem, Company A might consider using a more efficient technique to conduct the inspection test. This might involve using a better tool such as digital caliper, so that a precise and fast reading can be collected from the selected sample. Furthermore, the use of coordinated electronic inspection tools such as digital caliper allows Company A to integrate manual inspection with electronic inspection (Conci and Proenca, 2002).

Manual approach in conducting quality control could be found in some of these companies. For example, to monitor the cuprum pipe pressing process in Company D, individual control chart is used to detect the presence of specific variations in the process. The chart is produced by hand, thus require vast amount of time for its preparation. The workers seem to be comfortable with hard copy files and manual method in producing document for the produced items. The lack of confident in soft copy database was believed to be contributed by the fear that anyone might intrude and manipulate the data that could deteriorate the reputation of the company.

Previous researches that have been conducted show similar problems in implementing quality control among local manufacturing organisations. Among the critical problems are regarding poor process monitoring, inability to perform data analysis and the implementation of control chart only on the finished products and not in real-time manner (Hairulliza and Teoh, 2007; Hanida et al., 2009).

With the ability of computers to accept information that were typed in manually, read from scanners or manufacturing machines, or imported from other computer databases, the resulting control charts can be examined in greater detail, incorporated into reports, or sent to users across the internet. Real-time SPC would be a helpful solution for companies in facing issues such as lack of quality awareness, passive response, machine problem, unreliable incoming materials quality, and tight customer schedule (Anthony et al., 1998). Real-time quality control also provides great efficiency to the management as it takes time to prepare manual control charts and the time allowed to perform meaningful data analysis is limited (Hairulliza et al., 2010).

Among the reported benefits from real-time SPC were the ability of the production to provide all the charts they needed and responded flexibly to the demands of its employees, and the corporate staff in quality uses it to analyze monthly reports sent electronically by several divisions (Anthony et al., 1998). SPC tools serve not only in their technical aspects, but build an environment in an organization in which all individuals seek continuous improvement in quality and productivity.

Mohd Nizam et al (2009a, 2009b) and Rosmaizura et al (2008) study the barriers in developing an online SPC system, the findings show that factors related to top management support, costs to develop the system, inter-departmental relationship, education and training on SPC are impeding manufacturing organisations from conducting real-time process monitoring. Strong commitment from top management is crucial for a successful implementation of SPC (Jafri and Chan, 2001; Putri and Yusof, 2008; 2009).

The research found that worker training need to be given particular attention by management especially in Company B, C and D. The expected stumbling blocks in implementing quality control come from the technical aspect of the tools (Grigg dan Walls, 1999). Therefore the application of quality control techniques requires knowledge and training. Sufficient exposure to quality concepts and technical abilities need to be considered to ensure that QC activities are really efficient. This is particularly important if the industry intends to try new QC or implement a quite complex and tedious QC. The training does not only give important information and knowledge to workers but also build confidence and acceptance from them.

Support from management in funding the programme is necessary to ensure its effectiveness. However, one of the constraints faced by local companies in implementing quality control is limited investment to improve current techniques. The management might consider partnership with local agencies in enhancing quality and productivity in their company, for instances National Productivity Corporation (NPC) and Standard and Industrial Research Institute of Malaysia (SIRIM). Various training and seminar series were organized by these institutions particularly involving small and medium enterprises (SMEs) to help many organisations to enhance their competencies in market through quality and productivity achievement.

In future, manufacturers are expected to face an increasingly uncertain external environment with a cumulative effect of changes in customer requirements, global competition, and technological advancement. Quality, flexibility, cost and time are regarded as among the most important competitive weapons, in order for manufacturing companies

to survive. Manufacturers face the challenge of improving efficiency and lowering costs. QC techniques would be continuously exploited to help organizations to improve and innovate their products and process in order to be accepted by customers.
Because of the escalating concern on sustainable place and resource for future generation, manufacturers are expected to give more consideration on the environmental effect from their operations. Hence, implementation of environment conservation, atmosphere friendly manufacturing practices, and green technology appear to be dominant. The QC technique will integrate these environmental issues as its important elements. Simplicity and readiness for use would be the expectation for QC techniques of the future. New research is under way to design new QC approach that combine the conventional and model-free system to keep up with the advancing technology, expanding manufacturing process and growing product varieties.

6. Conclusion

The study finds that the motivating factors for these companies to apply quality control come internally from the management and parent company or externally from customer. SPC and acceptance sampling are used widely by the companies. Six sigma, DOE, Taguchi methods, and capability studies are left behind from being used in these four industries, due to lack of knowledge in the technique. The selection of quality control technique in these companies is influenced by three factors: ease of use of the technique; ability to measure product specification fulfillment; and ability to improve critical quality and productivity problem.

7. References

Ahmed, S. & Hassan, M. (2003). Survey and case investigations on application of quality management tools and techniques in SMIs. *International Journal of Quality & Reliability Management*, 20(7), pp. 795-826.

Antony, J. & Kaye, M. (1995). Experimental quality. *Manufacturing Engineer*, 74(4), pp.178-181.

Antony, J. & Taner, T. (2003). A conceptual framework for the effective implementation of statistical process control. *Business Process Management Journal*, 9(4), pp.473-489.

Antony, J., M. Kaye, & Frangou, A. (1998). A strategic methodology to the use of advanced statistical quality improvement techniques. *The TQM Magazine*, 10(3), pp.169-176.

Battikha, M.G. (2003). Quality management practice in highway construction, *International Journal of Quality & Reliability Management*, 20(5), pp. 532-550.

Besterfield, D. H. 2004. *Quality Control.* Prentice Hall, New Jersey.

Breyfogle, F.W.III & Cupello, J.M. (2001). *Managing Six Sigma: A Practical Guide to Understanding, Assesing and Implementing the Strategy that Yield Bottom-Line Success.* John Wiley and Sons, New York.

Chin, S., K. Kim, & Kim, Y.-S. (2004). A process-based quality management information system, *Automation in Construction*, 13, pp. 241–259.

Chung, W.W.C., K.C.M. Wong & Soon, P.T.K. (2007). An ANN-based DSS system for quality assurance in production network. *Journal of Manufacturing Technology Management.* 18(7), pp. 836-857.

Dale, B.G., H.S., Bunney, & Shaw, P. (2003).Quality management tools and techniques: an overview. In Dale, B. G. (ed): *Managing Quality* (4th Edition). Blackwell, Oxford.

Deleryd, M., R. Garvare, & Klefsjo, B. (1999). Experiences of implementing statistical methods in small enterprises. The *TQM Magazine*, 11(5), pp.341-350.

Gardiner, S.C. & Mitra, A. (1994). Quality control procedures to determine staff allocation in a bank. *International Journal of Quality & Reliability Management*, 11(1), pp.6-21.

Grigg, N. P. & Walls, L. (1999). The use of statistical process control in food packing: Preliminary findings and future research agenda. *British Food Journal*, 101(10), pp. 763-784.

Hairulliza, M. J., Noraidah, S.A. & Teoh, K.F. (2010). The Design for Real-Time Paper Perforation Quality Control. Lecture Notes in Engineering and Computer Science. IMECS 2010, 3, pp 1849-1851

Hairulliza, M. J., & Teoh, K.F. (2007). Output Confirmation Test System for Cigarette Paper Perforation Quality Measurement, The 5th International Conference on Quality and Reliability, Nov 5-7, Chiang Mai, Thailand, pp. 288-292.

Hairulliza, M. J., Hazura, M. & Erna, B.N. (2005). Aplikasi Carta Kawalan Dalam Industri Produk Minuman, Seminar Kebangsaan Komputeran Industri IComp 2005, Putrajaya, 24 September 2005, pp. 43-47.

Hanida, A. S., Norazlin, K., Noraidah S.A., Hairulliza, M.J. (2009). Statistical Process Control in Plastic Packaging Manufacturing: A Case Study, 2009 International Conference on Electrical Engineering and Informatics, 5-7 August 2009, Selangor, Malaysia, 199-203.

Mahanti, R. & Antony, J. (2005). Confluence of six sigma, simulation and software development. *Managerial Auditing Journal*, 20(&), pp. 739-762.

Mason, B. & Antony, J. (2000). Statistical process control: an essential ingredient for improving service and manufacturing quality. *Managing Service Quality*, 10(4), pp. 233-238.

Mohd Nizam A.R., Rosmaizura Jaharah A.G., Zulkifli M.N., Ahmad R.I. & Suriani A.R.. (2009a). Barriers to Implementing an Online SPC System in Malaysian Automotive Manufacturing Companies. European Journal of Scientific Research, 30(2), pp.315-325.

Mohd Nizam A.R., Rosmaizura M.Z., Zulkifli M.N., Jaharah A.G., Baba, M.D., Nurhamidi M. & Ahmad R.I. (2009b). The Implementation of SPC in Malaysian Manufacturing Companies. European Journal of Scientific Research, 26(3), pp.453-464.

Montgomery, D.C. (2005). *Introduction to Statistical Quality Control* (5th edition). John Wiley, New York.

Motorcu, A. R. & Gullu, A.K. (2004). Statistical process control in machining, a case studdy for machine tool capability and process capability. *Materials and Design*, 27, pp. 364-372.

Oakland, J.S. (2003). *Statistical Process Control*, 5th ed., Oxford: Butterworth-Heinemann.

Putri, N.T. & Yusof, S. M. (2008). Critical success factors for implementing quality engineering in Malaysian's and Indonesian's automotive industries: a proposed model. *International Journal of Automotive Industry and Management*, 2(2), pp 1 - 16, 2008

Putri, N.T. & Yusof, S.M. (2009). Critical Success Factors for Implementing Quality Engineering Tools and Techniques in Malaysian's and Indonesian's Automotive

Industries: An Exploratory Study. Proceedings of the International MultiConference of Engineers and Computer Scientists 2009 Vol II. IMECS 2009, March 18 - 20, 2009, Hong Kong

Rosmaizura M.Z., Mohd Nizam A.R., Zulkifli, M.N., Jaharah A.G., Baba M.D., & Nurhamidi M. (2008). Case studies on the Implementation of Statistical Process Control in Malaysian Manufacturing SMEs. Seminar on Engineering Mathematics, pp 70-79.

Slattery, J. (2005). Sampling for quality assurance of grading decisions in diabetic retinopathy screening: Designing the system to detect errors. *International Journal of Health Care Quality Assurance*, 18(2), pp.113-122.

Srikaeo, K., J. E., Furst, & Ashton, J. (2005). Characterization of wheat-based biscuit cooking process by statistical process control techniques. *Food Control*, 16, pp. 309-317.

Thorpe, C.D., W.D. Torrence, & Schniederjans, M. (1994). Quality assurance in human resource management for Computer-Integrated Manufacturing. *International Journal of Quality & Reliability Management*, 14(2), pp.18-30.

Woo, T.M. & Law, H.W. (2002). Modeling of a quality control information system for small- to medium-sized enterprise. *Integrated Manufacturing Systems*, 13(4), pp. 222-236.

Xie, M. & Goh, T.N. (1999). Statistical techniques for quality. *The TQM Magazine*, 11(4), pp. 238-241.

Xie, M., X.S. Lu, T.N. Goh, & Chan, L.Y. (1999). A quality monitoring and decision-making scheme for automated production processes. *International Journal of Quality and Reliability Management*, 16(2), pp.148-157.

26

Quality Control Methods Based on Electromagnetic Field-Matter Interactions

Ognyan Ivanov and Mariana Kuneva
Institute of Solid State Physics – Bulgarian Academy of Sciences
Bulgaria

1. Introduction

Field-matter interactions are very attractive for practical applications in contactless and rapid analysis. Here, two types of effects, induced by irradiation of solids with electromagnetic waves, are considered: acoustoelectric effect and surface photo-charge effect (SPCE) together with the possibilities for their application in a large area of quality control activities.

The transverse acoustoelectric voltage (TAV) effect is discussed in details as a typical example of an acoustoelectric effect. Application of TAV for control of semiconductor substrates and devices is considered.

The presented paper offers also a detailed review of our recent studies on SPCE and its possible applications for surface structures visualization, semiconductor characterization, sensors for fluids, composition analysis, food control, etc. The emphasis is put on the nearly universal nature of the SPCE. Experimental approaches and setups for detecting the studied effects are also described as well as the main experimental results, highlighting the advantages and disadvantages of the quality control methods based on TAV and SPCE.

2. Acoustoelectric phenomena

The acoustoelectric effect is due to interaction of the quasi-static electric field, induced by the propagation of an acoustic wave in a piezoelectric material, with the charge carriers of the irradiated area. Acoustic wave (volume or surface type), propagating in a solid, produces a mechanical deformation, which, in case of piezoelectric solid, creates an electric field. The surface acoustic waves (SAW's) are more interesting and closer to the subject of this paper.

The next considerations concern SAW propagating along the surface of a piezoelectric material, on top of which, at a distance shorter than the SAW wavelength, a semiconductor plate is placed (Fig. 1). In this case, the interaction between the SAW and the charge carriers is induced by the electric wave, accompanying the acoustic wave, which extends away from the piezoelectric.

Such a structure has several advantages: piezoelectric substrates with a high value of the electro-mechanical coupling coefficient can be used; various semiconductor materials can be pressed against the piezoelectric providing a small mechanical perturbation for the SAW propagation, etc. By using special supports placed parallel to the SAW propagating

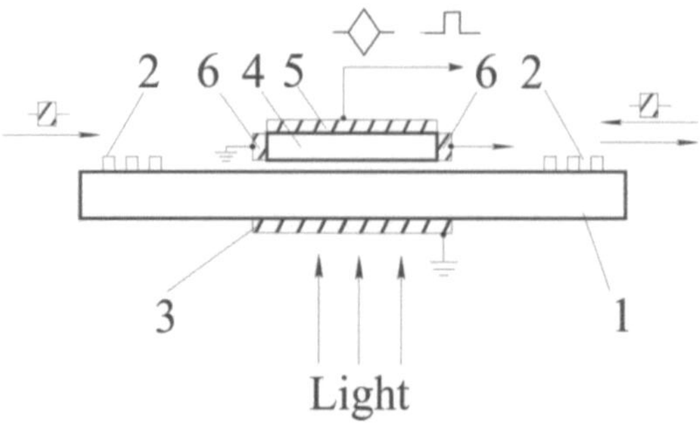

Fig. 1. Testing structure for acoustoelectric interaction: 1 - piezoelectric; 2 - SAW transducer; 3 - bottom electrode; 4 - semiconductor; 5 - top electrode; 6 - electrodes for detection of the longitudinal acoustoelectric effect

direction, the air gap between the semiconductor and the piezoelectric can be controlled and made as small as 50 - 500 nm. For achieving high power and large region for the acoustoelectric interaction, the width of the acoustic path is usually about 2 mm. A metal electrode, usually grounded, is deposited on the bottom surface of the SAW delay line (a piezoelectric with deposited input and output SAW transducers). In combination with the electrodes, deposited on the top surface and/or the front and back sides of the semiconductor, the bottom electrode is used for measurements of the parameters of various acoustoelectric effects. In order to monitor a constant air gap and to use this setup for spectral analysis or device design, the bottom electrode should be optically transparent. A semi-transparent thin gold layer is commonly used for this purpose, but it has some disadvantages. The separated medium structure requires a high-quality polishing of both surfaces as well as a high plane parallelism between them.

Since in a layered structure (Fig. 1) the electric field penetrates away from the piezoelectric structure, an effective interaction between the SAW and the free carriers in a semiconductor, pressed against the piezoelectric surface, is possible. If the semiconductor surface is not homogeneous because of non-uniform distribution of the impurities in the active area, the efficiency of the acoustoelectric interaction will change. This way, information about the processes, taking place on the surface or in the volume, can be obtained.

The studies of such layered piezoelectric–semiconductor structures could be very important for analysis of various physical properties (e.g. non-destructive characterization of the electronic properties of semiconductors) as well for the design of various devices such as converters from optical images into electrical signals, convolvers, acoustoelecric memory - an effect described firstly in [Korshak et al.,1976], etc.

When electric field penetrates into the semiconductor, it leads to a redistribution of the charge carriers and a periodic spatial variation of their density, which, in turn, causes acoustoelectric interaction, i.e. an interaction between the acoustic wave and the electron system of the semiconductor. The field induced by SAW has its longitudinal and transverse components. It should be noted that the acoustoelectric effects cannot be fully understood in the frame of the linear acoustoelectric interaction. The acoustoelectric phenomena, generated by SAW propagation, are due to both linear and non-linear components of the acoustoelectric interaction. The non-linear acoustoelectric interaction can cause acoustoelectric effects of combined frequencies as a result of non-linear interaction between two SAWs. One of the effects provoked by this interaction is the convolution of the two signals. Closely related to the convolved signal is the transverse acoustoelectric voltage.

TAV represents the potential difference across the electrodes 3 and 5 in Fig. 1 induced by the transverse component of the electric field, generated by the SAW. The signal from the convolution of the two signals applied to the SAW transducer 2 is measured on the same electrodes. The voltage induced by the longitudinal component of SAW's electric field could be measured across the electrodes 6. This component is responsible for charge carrier transfer. The experimental results correlate well with the theoretical model, based on the assumption that both polarity and magnitude of the voltage are determined by the carriers trapped in the surface states.

When acoustic power increases, the traps get filled and the surface under the contact charges negatively, so the polarity of the "trapping" TAV is opposite to the one of the "principle" effect. Such simultaneous occurence of the two types of TAV has been observed for the structure $LiNbO_3$-Si and in CdS. At a certain acoustic power, the sum of the two effects will become zero and further increase of the acoustic power will lead to a sign reversal of the voltage.

The TAV effect provides a large area for investigations. The obtained results on the kinetics and polarity of TAV are contradicting [Kunigelis, 1978; Fritz, 1981]. These problems are discussed and clarified in [Konstantinov, L. et al., 1985; Ermolova, O. et al., 1985,;Ivanov, O.,1989] The effects of TAV relaxation versus the rise and fall times of the optical excitation pulse at constant SAW parameters were investigated in [Ivanov & Germanova, 1988].

TAV – application for qualitative semiconductors characterization

The TAV-based method is considered to be very promising for characterization of the electrical properties of semiconductors. It is contactless, very sensitive and the measuring equipment is relatively simple. In addition, there are no special requirements to the exciting electrode (5 in Fig. 1), deposited on top of the semiconductor. Controlling the penetration depth of the electric field into the crystal is possible by appropriately adjusting the distance between the semiconductor and the piezoelectric according to the Debay length and the SAW wavelength. The method allows rapid comparative qualitative analysis of semiconductor surfaces (the relative distribution of the surface states concentration could be monitored); characterization of semiconductors with high defect density by TAV spectroscopy (the influence of the trapped charges on the transverse acoustoelectric voltage), etc. Because of TAV's high sensitivity, the method can be used for the analysis of samples with high resistivity, which cannot readily be characterized with other methods.

When a radio frequency (RF) pulse is applied to the input transducer of the delay line, a relaxation response is generated in a direction normal to the SAW propagation plane in piezoelectric-semiconductor structure. The amplitude and the relaxation time of this response, as well as their dependence on the duration and the amplitude of SAW pulse depend on the interaction between the wave and the free carriers in the space charge layer, so they can provide information on some surface and volume electronic properties and parameters of the semiconductor. Usually, apart from the interaction with the SAW pulse, optical excitation and/or DC pulses are applied to the semiconductor simultaneously. This allows more dependences to be established experimentally - for example, the dependence of TAV on the wavelength and the intensity of the incident radiation, on the pulse magnitude and polarity, etc. These measurements could provide information about the energy distribution of the recombination centers and traps in the band gap of the semiconductor, their coefficients of trapping and relaxation for the majority and minority carriers, the cross-section for optical excitation, etc. Some difficulties in the assessment of these characteristics exist due to the complicated nature of the processes involved [Ivanov, 1989].

Concerning experimental difficulties, it has to be noted that the amplifier used for the TAV measurements should have high input impedance, providing conditions, similar to those of an open circuit. Most probably, a lot of the published results were obtained when an amplifier with inappropriate input parameters had been used, so spurious circuit effects were added to the effective relaxation processes in the structure or even dominated in the output signal. It causes some doubt in the reliability of some of the experimental data, reported in the literature. Another problem is that the kinetics of the transverse acoustoelectric effect was not profoundly studied, although it may contribute significantly to the explanation of the charge trapping effect and the determination of the trap parameters.

There are some improvements in the measured structure which could be introduced. It was established [Borissov et al., 1985] that the use of a transparent conductive layer of SnO_2 as a bottom electrode in the structure presented in Fig. 1 has a number of advantages and is of crucial importance for the studies of acoustoelectric interactions and for the registration of the spectral dependences.

In spite of the difficulties discussed above, the possibilities for many practical applications of the TAV-based method for semiconductor characterization are indisputable, which was confirmed by the experiments described below.

A TAV-effect-based new method for a rapid analysis of semiconductor surfaces was developed [Strashilov et al., 1987; Ivanov, 1989] which includes scanning of the studied surface with a narrow beam of monochromatic light and detection of the generated TAV signal. The analysis was qualitative and comparative, i.e. only the relative distribution of the surface states' concentration could be monitored. Fig. 2 shows a TAV amplitude profile, obtained by scanning the surface of a n-type Si plate (specific resistance of 10^3 Ω.cm), in the middle of which a circular region with p-type conductivity was created. The location of the p-zone is clearly observed in both horizontal and vertical directions.

The influence of the trapped charges on the TAV at the Si- SiO_2 interface has been investigated in order to extend the application of TAV spectroscopy for characterization of semiconductors with high defect density. The effect of compression on the electronic state was investigated and a pressure threshold was observed, above which the TAV magnitude decreases and a relevant number of stress-induced traps are created on the Si - SiO_2 interface. This threshold determines the values of the applied voltage for non-destructive measurements - they should be below the threshold.

Since the TAV-based method is very sensitive, it can be used for the analysis of samples with high resistivity, which cannot readily be characterized with other methods. An example for this application is the characterization of the interface of high-resistivity ZnSe-GaAs heterostructure by single and double TAV spectroscopy. The method is used for studying of the interface band and the impurity transitions of the heterostructure at room temperature. From the TAV spectrum, a conduction band offset of 0.059 eV has been determined.

The TAV effect could be applied for modification and analysis of semiconductor surfaces with specific properties via acoustic wave stimulated absorption - an application of great interest. The contactless control provides conditions for continuous monitoring and/or in-time termination of the process.

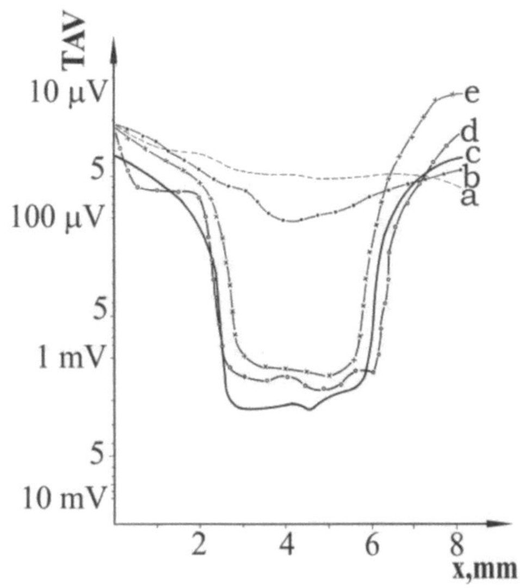

Fig. 2. Variations of TAV amplitude on laser beam scanning of the surface of a p-type area in a n-type Si plate. Curves (a - e) correspond to a laser beam shift of 1 mm in vertical direction

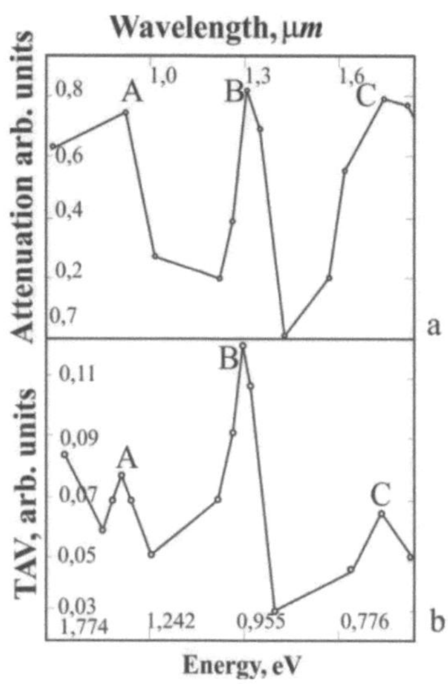

Fig. 3. Spectral dependence of the SAW attenuation (a) and the TAV amplitude (b)

A number of works are devoted to the spectral dependence of the TAV effect. Such measurements could provide information about the surface conductivity spectrum as well. The peaks in the surface photo-conductivity spectrum could be attributed to optically-induced electron transitions which were also registered by other methods. These peaks can also be used to derive the value of the band gap energy as a function of the temperature.

In cases requiring more detailed information, simultaneously obtained spectral dependences characteristic for several acoustoelectric effects could be used. Such experiments are performed for CdS samples where the dependences of the SAW attenuation coefficient, the convolution signal and the transverse acoustoelectric voltage on the wavelength of the incident light are studied. An interesting feature is the peak at 480 nm wavelength (2.6 eV), which was assigned to the intrinsic absorption edge of the material and gives the band gap energy. The experiment showed that the transverse effect is more sensitive than the SAW attenuation and the convolution signal.

The relaxation of the transverse acoustoelectric effect was studied in a number of papers as a prospective technique for characterization of semiconductor properties. A combination of relaxation dependences and spectral investigations has been also used since it provides some additional possibilities for analysis. It is established that the incident light wavelength strongly influences the transverse effect and causes SAW attenuation (α). The positions of the characteristic minima in the spectral dependences of the relaxation time and the peak positions in the α-dependences coincide.

A plot of the TAV amplitude for an epitaxial layer of GaAs as a function of the wavelength of the incident light as well as the spectrum of the SAW attenuation due to acoustoelectric interaction are presented in Fig. 3. The two spectra exhibit many identical features, which is an evidence for the common physical nature of their origin. Some differences which could be seen are due to the different impact of the semiconductor parameters on the two effects. The peaks in these spectra are also attributed to optical excitation: band-to-band electron transitions (peak A) and surface levels-to-conduction band electron transitions (peaks B, C).

The TAV technique is applied for determining the band gap energy as well as the energies of the impurity level transitions in a GaAs layer (1.4 μm thick), deposited on a vicinal Si (100) substrate by molecular beam epitaxy. It is shown that they are slightly lower compared to the bulk GaAs. This might be attributed to the influence of the residual strain on the band structure of GaAs, deposited on Si. It was also pointed out that at low temperatures the dependence of the TAV on bias voltage considerably changes.

The investigation of the transverse acoustoelectric effect is a fascinating field in solid state physics [Ivanov, O., 1989]. A complete solution of the aforementioned problems and the establishment of the exact relationships between the amplitude and the kinetics of the transverse effect and the semiconductor parameters will extend the field of its applications.

3. Surface Photo-Charge Effect

A new type of electromagnetic field-matter interaction effect was observed during our studies on the TAV effect. It was established that when the sample 4 from Fig. 1 was illuminated with modulated light, an alternating voltage was generated across the electrodes 3 and 5 in absence of acoustic wave when the sample is a metal one. Russian and American scientists took part in this study as well. The wavelength of the radiation was selected appropriately to exclude the contribution of the external photoelectric effect. It was observed later that the effect takes place under exposure of each solid to light and RF radiation. The next part will be completely devoted to our own studies in this field.

The SPCE represents generation of an alternating voltage when a solid interacts with a modulated electromagnetic field. The voltage frequency is equal to that of the incident radiation. The potential electric difference between the irradiated sample and the common ground of the system is measured experimentally. The SPCE is a very fast effect: for example, an irradiation with 20 ns laser pulse results in a signal response which reproduces precisely the waveform of the incident pulse [Pustovoit et al., 1989a]. The rapidity of SPCE is

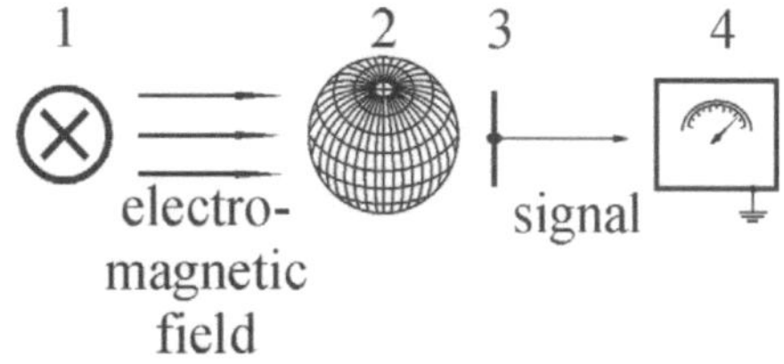

Fig. 4. A general scheme of the SPCE observation: 1- electromagnetic field source, 2 - irradiated solid, 3 - signal-measuring electrode, 4 - measuring device

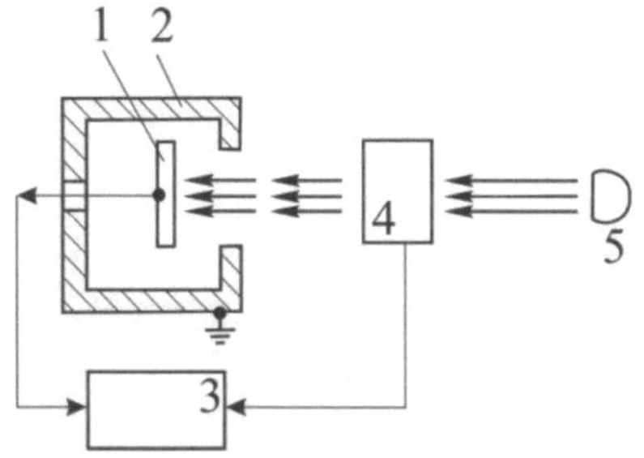

Fig. 5. Experimental setup: 1-specimen, 2-screen, 3-measuring device, 4-modulator, 5-light source

of great importance for many practical applications. The SPCE-based interaction takes place between arbitrary source of electromagnetic radiation and any solid. Unlike the case of acoustoelectric interactions where the presence of a piezoelectric layer or crystal is required, the excitation of SPCE does not need additional structures or elements on the sample. A basic scheme for observation of the SPCE is presented in Fig. 4.

Possible mechanisms causing SPCE are proposed and analyzed, but they could not be considered complete or determinative.

As the experiments have shown, the arising of a signal from a solid after electromagnetic irradiation is a universal feature of solids. SPCE has been registered in the frequency range 1Hz-1Gz, infrared, visible and the beginning of ultraviolet. Our hypothesis is that the effect exists in the whole frequency range of the electromagnetic radiation. SPCE can be induced only by modulated incident radiation. The lack of response upon non-modulated field is evidence that the detected signal is induced by the SCPE.

An important feature of the SPCE is the dependence of the signal created by it on the specific characteristics of the irradiated object. This creates numerous opportunities for fast and contactless analysis of solids, liquids and gases.

3.1 SPCE in conductors and semiconductors

It is well known that under appropriate energy of the photons, optical radiation, incident on the surface of a conductor, may lead to emission of electrons. The theory of the photoelectric effect in metals was developed by Tamm [Tamm, 1975]. According to the theory, the simultaneous conservation of the energy and the momentum of free electrons is impossible, thus the photoelectric effect is due solely to the presence of a sharp non-uniformity on the conductor surface leading to an exponential decay of the wave functions of the electrons. Since in the near-surface region of a conductor the distribution of charges is non-uniform, a so-called double layer is formed on the conductor–vacuum interface. This layer is mainly responsible for the photoelectric effect in metals. The electrostatic field of this layer repels the free charges away from the surface.

In the case of SPCE, the incident radiation creates a force acting on the medium. The magnitude of that force in the simpliest case is proportional to the gradient of the permittivity. This force causes a redistribution of the electrons near the surface of the conductor and a corresponding change in the electrostatic potential of the double layer, i.e. the incident radiation induces a macroscopic dipole momentum on the surface of an isolated conductor. If the conductor is electrically connected to another solid conductor with a zero potential for a certain period of time, determined by the time constant of the system, an electric current will flow between the conductors until a new stationary distribution of the surface charges and potentials is established [Pustovoit et al., 1989b]. This charge flow or the macroscopic polarization of the conductor, induced by illumination, can be experimentally measured. The change of the double layer's potential depends on both the surface features and the intensity of the incident radiation, being proportional to the intensity or to the square root of the intensity when the latter is low or high respectively.

The SPCE in conductors was studied in details [Pustovoit et al., 1989; Pustovoit & Ivanov, 1989]. A large number of various conductors and semiconductors: metals Zn, Cd, Al, Pb, Mg, Au, Cu and copper alloys, etc., semiconductors: Si, GaAs, GaP, ferrites and ceramic superconducting materials were experimentally investigated. A good qualitative agreement between the experimental data and the theoretical predictions was observed. It was established that the amplitude of the voltage varied from sample to sample, i.e. it depends on the composition and some other parameters of the material under study. The dependence of the voltage amplitude on the modulation frequency of the incident radiation was studied, as well as the generated voltage as a function of the intensity of the incident radiation. Dependence of the SPCE on the incident angle or the polarization of the radiation was investigated but not observed. It was established that in some cases, the history of the sample (e.g. previous illumination) affects the SPCE.

A scheme of the experimental setup for SPCE studies in conductors is shown in Fig. 5.The illuminated sample 1 is placed on a grounded shield 2 and is illuminated by the light from source 5, amplitude- modulated by mechanical modulator 4. The voltage due to the surface photo-charge effect is registered by device 3, synchronized with the modulator 4. Both a voltmeter (selective or phase-sensitive) and an oscilloscope were used as detectors. A measuring device with high input impedance (100 MΩ) was used. At high intensity of the incident radiation and a strong SPCE (for example in Si plates), the input impedance of the detector could be lower at some measurements.

It should be pointed out that the measurements are contactless, i.e. there is no need of special electrodes or even a physical contact between the conductor under study and the exciting electrode. The measured signal was registered due to the mutual electrostatic induction with an electrode, placed in a close proximity.

The logarithmic dependence of the SPCE voltage on the laser beam intensity at λ=633 nm for Cu samples is shown in Fig. 6, when the output laser power was varied in the range 50 μW - 17 mW. The signal was proportional to the intensity I^{α}. Such a relationship was also obtained for a Cu sample of the same dimensions, but with the output laser power varying in the range (3- 400) mW at λ=514.5 nm. [Pustovoit et al., 1989]. At low laser intensities, α is about 0.8. As the intensity increases, the value of α approximates 0.5. Similar results were derived for various metals (Al, Pb, Au, Cd, Mg) and semiconductors (Si and GaAs).

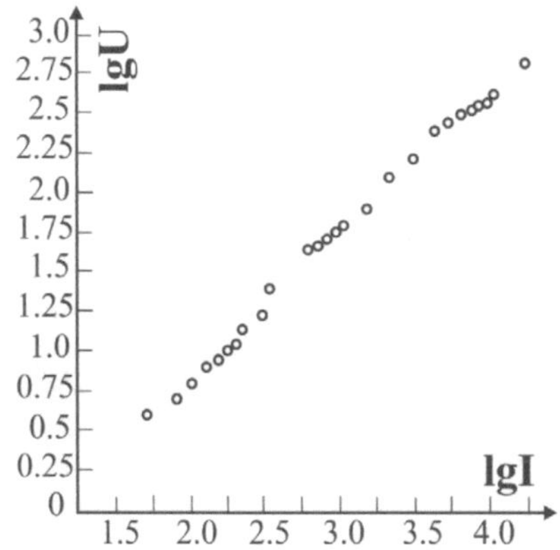

Fig. 6. Dependence of the SPCE signal on the optical irradiation intensity for a copper plate

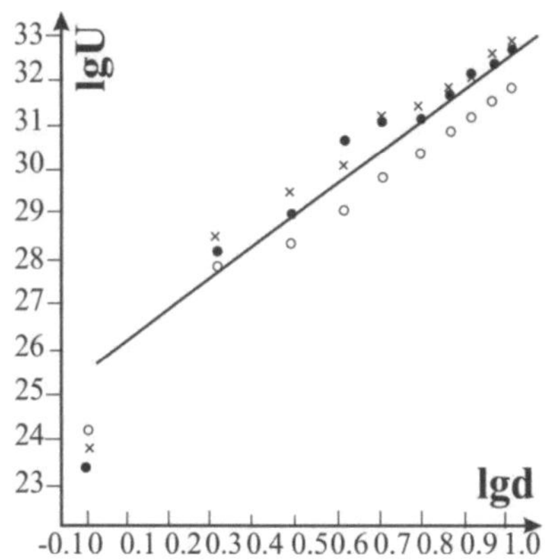

Fig. 7. Dependence of the SPCE signal on the diameter of the irradiated area for three different measurements under the same conditions

In some cases the SPCE signal significantly depends on the area of the illuminated spot. An example is shown in Fig. 7 for a Cu sample exposed to light with wavelength λ=633 nm.

When the illuminated solid is a semiconductor, the incident light can induce generation of carriers, provided that the photon energy is higher than a certain value (higher than the bandgap energy in case of electron hole generation due to band-to-band excitation, for example). High intensities could cause a high electron hole density.

As in the case of conductors, the SPCE in semiconductors is also due to the interaction of incident radiation with the free carriers in the medium, but the voltage induced by incident photons of arbitrary energy is due to the redistribution of charges already accumulated in the near-surface region. According to some theories, the photo-conductivity and the photo-voltage effect are the main contributors to the SPCE in semiconductors [Ivanov et al., 1995a]. The voltage change of the surface due to charge redistribution is measured capacitively, so there is no net charge current across the sample. Thus, the photo-induced charge is restricted within the space charge region of the semiconductor sample, resulting in total charge redistribution, compensated by potential variation within the semiconductor space charge region. In equilibrium, the optical generation of free charge carriers is balanced by recombination at the surface and by diffusion in the bulk.

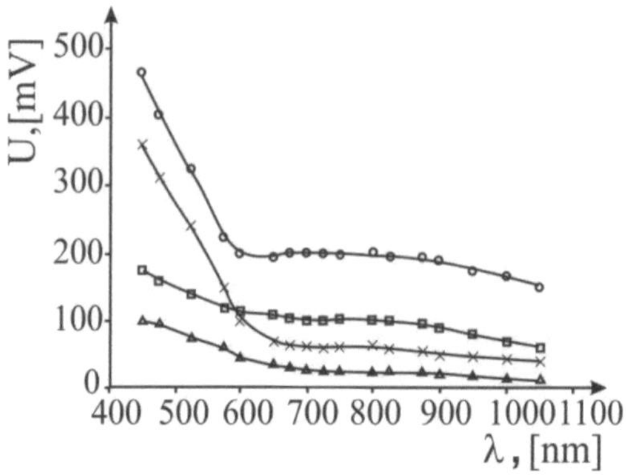

Fig. 8. SPCE spectral dependences for copper - O, lead - x, gold -Δ, brass

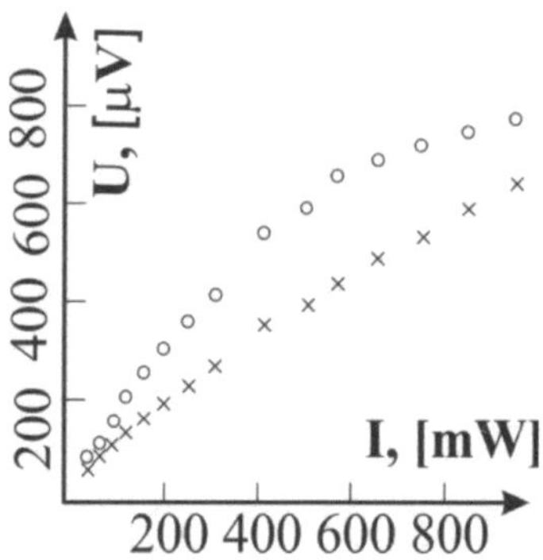

Fig. 9. SPCE signal amplitude as a function of incident light intensity for pegmatite (x) and Teflon (o)

Experimental investigations of the SPCE in a large number of various conductors were reported also in [Vankova et al., 1997]. A generation of voltage at illumination with non-monochromatic light was detected in all measured samples. The amplitude of the voltage varied from sample to sample, i.e. it depends on the composition of the material under study.

Another series of studies on spectral dependences of the SPCE at conducting surfaces was performed as well which proved that the SPCE and the external photoelectric effect in metals are two different phenomena. The SPCE spectral dependencies for Cu, Pb, Au and brass in the range 400-1100 nm were studied [Ivanov et al., 2000]. They are shown in Fig. 8. A strong decrease of the voltage at λ~600 nm is observed which is probably due to the cut-off of the external photoelectric effect. At larger wavelengths (λ>600 nm), the voltage is approximately constant; since in this spectral region the external photoelectric effect does not contribute to the signal, the response is determined by the surface photo-charge effect and thus both effects are separated. It is clear that the SPCE contributes to the signal at λ<600 nm as well, but it is screened by the much stronger external photoelectric effect. The existence of signals in the beginning of the infrared region is another indicator that the effect observed is not the external photo-effect.

The spectral dependence of the surface photo-charge effect for a photoelement in the range 450-3200 nm was also measured. A sharp decrease of the signal was observed between 450

and 700-800 nm. At λ>800 nm, the signal remained nearly constant up to 3200 nm. This dependence was similar to the one shown in Fig. 8.
The results discussed above show that a clear distinction between the SPCE and the well-known external photoelectric effect in metals exists and the two effects can be separated.

3.2 SPCE in dielectrics

The SPCE was induced in all studied dielectric samples [Ivanov et al., 1994; Ivanov et al., 1995a] when they were exposed to amplitude-modulated non-monochromatic or, in some cases, to He-Ne laser light in the visible region. Since the irradiated structure had very high output impedance, as in the case of conductors, a measuring device with high input impedance was required. For avoiding voltage generation due to the illumination of the measuring electrode itself, it was wrapped in black photographic paper.
It was established that the generated SPCE voltage depends on the type of the illuminated material. Figure 9 shows the voltage amplitude as a function of the intensity of the incident light beam for two dielectric samples: Teflon and pegmatite. The influence of the sample thickness on the voltage amplitude was also investigated for different dielectric materials. The samples thickness varied in the range 3-30 mm. It was observed that the SPCE signal sharply decreases with increase of the thickness up to 10 mm. Further increase of the thickness makes the signal diminish more gradually.
Since most of the measured materials were good thermal insulators, it is very unlikely that the generated voltage is due to thermal electromagnetic field, induced by heating of the measuring electrode. Besides, in some cases, the measured signal from the sample was considerably stronger than that obtained at direct illumination of the electrode.
Possible reason for the appearance of potential difference, generated in dielectric samples, could be the presence of dipole molecules on the dielectric surface, caused by absorption or other phenomena. The force created by the incident light is proportional to the gradient of the electrical permittivity of the medium, as in conductors, and leads to redistribution of the absorbing molecules along the surface of the dielectric. As a result, the latter becomes electrically charged [Ivanov et al., 1994].
It is quite possible that the mechanism of the SPCE differs for various types of materials, but nevertheless it should be emphasized that the effect can be induced in any solid upon illumination.

3.3 Megahertz excitation of the SPCE

When the frequency range of the SPCE exciting signal varied from 1 Hz to 1 GHz a response signal was observed after taking all necessary precautions to avoid spurious circuit effects. A large number of practically all types of solid samples with suitable dimensions were subjected to irradiation and SPCE was induced in each one of them. The results for some of them are presented in Table 1.
No amplitude modulation was applied. The amplitude of the generated voltage was different for each sample. Stronger SPCE signal was observed for conductive samples. The phase of the signal also varies from sample to sample.
It was also observed that a slight change in the sample position leads to a considerable change of the generated voltage. For more precise quantitative measurements a special holder, firmly fixing the sample, should be designed.

A systematical study of the SPCE was performed with incident radiation frequencies in the spectral range up to 1000 MHz. The tested materials were textolite, Si, poly-oxy-ethylene, glass, cardboard and quartz. For all the samples, the SPCE voltage was generated in the whole measured range. The fact that the SPCE can be induced by irradiation with electromagnetic field with frequency up to 1 GHz, and in the visible and nearby spectral range, led to the hypothesis that SPCE exists in the whole frequency range of the electromagnetic waves since it is very unlikely that there is an "interruption" in any spectral range.

Material	Size, mm	Signal, mv	Irradiating voltage U_{PP}, V
Silicon	0,48X20X70	5	0,15
Copper	0,9X15X70	6,8	0,15
Molybdenum	0,6X20X68	5,7	0,15
Piezoceramics	2,1X16X65	3,6	0,15
Potato (freshly cut sample)	2,7X20X70	8,4	4
Cardboard	1,4X20X70	4,5	40
Walnut (wood)	0,7X20X70	1,4	40
Oak (wood)	0,5X20X70	1,6	40
Teflon	2,3X20X70	1,6	40
Epoxy	1,5X20X70	3,8	40
Glass	2,4X20X70	8,6	40
Textolite	1,6X20X70	2,8	40
Quartz	1X20X70	1,4	40
Teflon	2,3X20X70	0,8	40
Poly-oxy-ethylene	2,4X20X70	3	40
Plexiglass	19X20X70	1,4	40

Table 1. SPCE signal values for some materials at incident radiation frequency of 141 kHz

Another group of experiments was performed on the relationship between the amplitude of the signal and the thickness and width of samples of the same composition. They showed that the amplitude of the signal grows with increase of the sample's thickness and width, i.e. the amplitude grows with the increase of the irradiated volume. An equation describing the relationship between the volume of the irradiated object and the generated voltage would be of great interest.

The temperature dependence of the voltage amplitude was studied for a poly-oxy-ethylene sample in the range from room temperature to 90°C. This material has a first order phase transition at 72°C. Initially, there was a slow signal increase with the temperature rise, followed by a sharp increase as the temperature reached ~70°C. This way the possibility to establish a first-order phase transition in this material by following the temperature dependence of the SPCE voltage amplitude was demonstrated as well as the fact that the SPCE signal amplitude depends not only on the composition and the volume of the irradiated material, but also on the processes taking place within the bulk of the investigated object.

A distinctive feature of the SPCE, induced by radiation in the MHz frequency range, is that no amplitude modulation of the incident electromagnetic field was required contrary to the visible range case. Up to now, SPCE has not been registered at illumination with non-modulated light. The experiments in the MHz range showed that in fact, an alternating voltage with frequency equal to the one of the incident radiation was generated. However, there are no electronic devices for signals with frequency in the range of 10^{15} - 10^{13} Hz. The amplitude modulation of the incident light enables the measurement of the low-frequency envelope of the signal.

The explanation of the SPCE will probably require new ideas. Density functional theory (DFT) including external electromagnetic fields effects is one of the possible approaches considered at present for the description of SPCE physical mechanisms.

Summarizing all experimental results on the SPCE described above, it could be concluded that the effect takes place in all solids under irradiation with electromagnetic field in the whole spectral region. The frequency of the generated potential difference coincides with the frequency of the incident electromagnetic wave. The amplitude of the SPCE voltage depends on the specific parameters of the irradiated solid. The process is very fast and allows contactless measurements from a distance.

4. Practical applications of SPCE

As discussed above, the SPCE is an inherent feature of all solids, subjected to irradiation with electromagnetic fields with frequency in various spectral ranges. The specific features of the SPCE are a prerequisite for many attractive applications including quality control.

4.1 Visualization of surface structures

Since the SPCE is sensitive to the specific features of the illuminated surface, various structures (mechanical structures, irregularities, defects, impurities, ion-implanted regions, etc.), formed on the surface of each solid, can be visually observed on a monitor. For a better resolution, the intensity and the wavelength of the incident radiation should be selected carefully in each particular case, for obtaining a maximal change in the measured SPCE voltage at a minimal variation of the parameter under study. A depth profile of the distribution of a certain structure can also be obtained by varying the wavelength of the incident laser beam.

An automated system for measurement of the surface density of current carriers and the electrical permittivity of conducting materials was developed [Ivanov et al., 1995b]. It could be used for analysis of the electron topography of conducting surfaces (semiconductors, metals, ferrites or conducting ceramics). The sample (together with an appropriate support) is mounted on a small positioning stage providing X-Y motion in its own plane by two computer-controlled electric motors. It allows the illumination spots to be positioned and the whole surface of the studied sample to be scanned. The phase and the amplitude of the generated voltage depend on the surface properties at the illuminated spot. The sample is scanned by a normally incident pulse-modulated He-Ne laser beam with 40 μm diameter of the focal spot. The scanning area is 30x30 mm with a step of 10 μm. The size of the scanning area and the scanning step could be easily improved by using more precise optical stage and finer step motors. The voltage difference arising in result of irradiation of the sample is measured by a selective voltmeter synchronized to the chopper frequency. The measured signal is applied to an analog-digital converter and to the computer, which drives the step motors for X- and Y-translations. The signal may be taken from an electrode placed near the scanned sample, but not directly connected to it. All measured data and graphics go to the computer memory. A computer program processes the data. The results are presented as a three-dimensional plot. The amplitude of the SPCE voltage is plotted along the Z-axis, normally to the surface.

The main problem of such setup is that the system producing the signal has a high-impedance output, so the measuring device must have high input impedance. With illuminating power of 1W, the signal from metals, ceramics and ferrites is about some microvolts, for semiconductors the signal is stronger and could reach some millivolts at much lower irradiation intensities.

The described system possesses high flexibility since it can be applied for investigations of all kinds of conducting surfaces. It was found out that besides the amplitude, the signal phase also changes, as a function of the surface condition. This is going to be used to derive additional information by measuring simultaneously the phase and the amplitude of the signal. Also, it is intended to extend the automation of the system by adding computer control of the illuminating beam wavelength. Since the penetration of the light in depth varies with the wavelength, this will be equivalent to a depth scan. The results will enable the application of the described device for identification of any solid surface. Such a device can be used as a microscope as well, because the minimal area of the observed region is determined only by the diameter of the laser beam and the scanning step. The analysis is quite fast, no complex equipment is required and no metal contact has to be deposited on the sample. The combination of all these advantages makes such a device very interesting.

The results from the investigation of GaP sample with sulphur impurities with a specific resistivity of 10^{-1} Ω.cm are shown in Fig. 10. The tested plate was a round wafer with a diameter d=43 mm, cut perpendicularly to the growth axis of the crystal. A part of the crystal surface with dimensions 21x29 mm, including the central area, was scanned with a step of 1 mm by laser beam (wavelength 633 nm, focal spot 50 μm, modulation frequency 170 Hz, intensity 1.8 mW). It was known that an accumulation of impurities exists in the central region of the plate, which is recognized from the sharp increase in the SPCE voltage on the 3D plot in Fig. 10.

The dependence of the SPCE voltage on the surface properties was also studied in a Si plate, implanted with boron ions [Pustovoit et al., 1989a]. The dependence, shown in Fig. 11 reveals that the signal depends dramatically on the surface properties – it describes precisely the form of the implanted region.

A Si plate with specific resistivity of 50 Ω.cm and dimensions 12x12x0.35 mm was also studied. He-Ne laser beam with λ=633 nm was used for illumination. The diameter of the illuminated spot on the surface of the plate was 40 μm. There was a 2.5 mm wide mechanical scratch within the central region of the plate. A 4 mm wide area, including the scratch, was scanned. The results, presented in Fig. 12 (a), reveal once again that the amplitude of the SPCE voltage significantly depends on the surface features. It should be pointed out that the phase of the signal was different for the smooth and the scratched surface.

An interesting fact was observed during the measurements: the contrast of the 3D pattern response depends substantially (inversely proportional) on the intensity of the laser beam (Fig. 12 (b) and (c)). This was confirmed by a series of experiments with various values of the laser beam intensity.

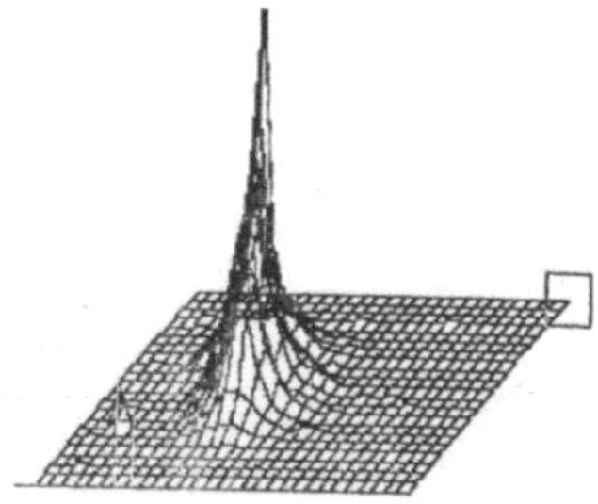

Fig. 10. Scan of a GaP plate

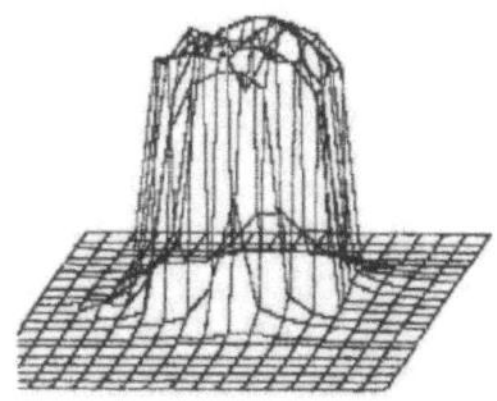

Fig. 11. A silicon plate zone pattern for a: λ=633 nm, light spot diameter 100 μm

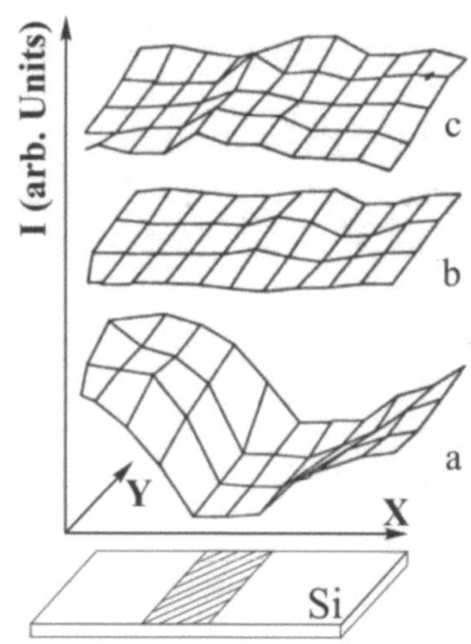

Fig. 12. Response distribution for a silicon plate. The scanned area along the x-axis is ca. 4 mm and the width of the scratched (hatched) area is ca. 2.5 mm; laser beam intensity: a-30μW, b-300μW, c-700μW

The experiments described above showed that an optical contactless method based on the SPCE can extensively be used for approximate analysis of the surface topology. Besides, such a method is universal since it can be applied to characterize any kind of surfaces.

4.2 Contactless characterization of semiconductors

Different methods have been used for investigation of defect states at semiconductor surfaces and interfaces [Berglund, 1966; Lang, 1974; White, et al., 1976; Hurtes et al., 1978, Balland et al., 1986, Tapiero et al., 1988].

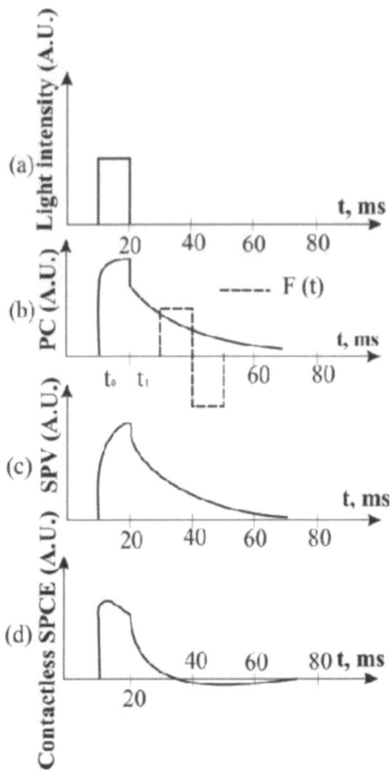

Fig. 13. Time dependence of the exciting light intensity (a); rise and decay transients of PC (b), SPV (c); and contactless SPCE (d) in the GaAs single crystal at 300 K

The SPCE temperature and spectral dependences provide information about the surface electron states and could be used for characterization of semiconductors. This technique is, to a large extent, similar to the contactless TAV technique for semiconductor characterization. In the SPCE-based method, the energy and the relative density of the trap levels can be estimated from the derivative of the SPCE voltage dependence on the energy of the incident photons.

The SPCE and the photo-current (PC) induced by excitation with short light pulses with high intensity and wavelength in the spectral range of the semiconductor's intrinsic absorption were studied for GaAs [Davydov et al., 1994]. No direct electric contact to the sample is necessary for the registration of the deep level (DL) spectrum, provided that a double pulse integrator (lock-in, double-boxcar) and a computer convolution of the SPCE transients are applied. This spectrum can be received in a wide temperature range. If the crystal surface is scanned with a light beam at a temperature corresponding to a maximum in the DL spectrum, this procedure may be applied for contactless determination of the DL density distribution profile along the scanning direction.

Figure 13 illustrates the PC transient (b), the SPV (surface photo-voltage) transient for a crystal with ohmic contact (c) and the SPCE transient for a crystal without any electrodes (d) at a temperature of 300 K. Three regions in the PC transients are observed: 1) a rapid rise to a stationary state (the generation of non-equilibrium carriers is balanced by recombination and trapping); 2) a rapid decay (recombination of free carriers); 3) a slow relaxation (recombination of trapped carriers, thermally released into the semiconductor's bands).

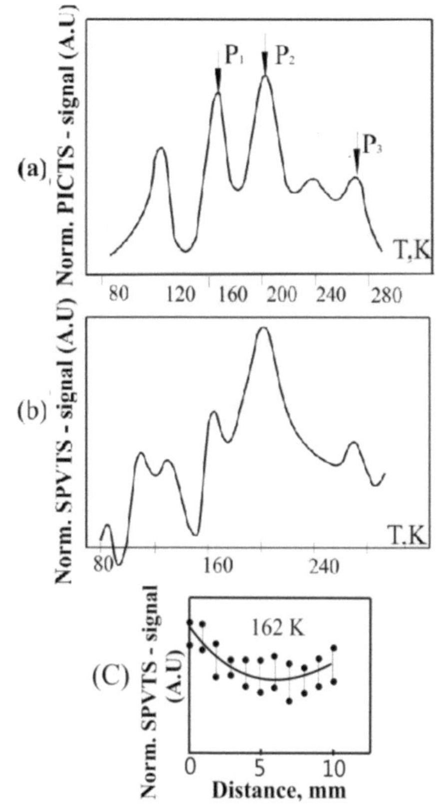

Fig. 14. DL spectra in GaAs single crystal, derived from the PC (a) and SPCE transients (b), linear distribution of P1-deep level density (c)

In order to obtain the DL spectrum of a GaAs crystal, the PC, SPV and SPCE transients were registered in the temperature interval 77–300 K and the data were stored in a computer. The slow relaxation range of each transient was found. The thermal emission from the traps to the semiconductor bands, determined by the charge carriers, was obtained and a correlation procedure was applied for the numerical analysis.

Some of the mathematical considerations used for data processing of the Photo-Induced Current Spectroscopy (PICTS) were applied to the temperature dependence of the SPCE decay transient, which allowed contactless obtaining of information on defects and impurities present in the semiconductor [Davidov et al., 1994]. The results are similar to the ones derived from the DL spectrum and the spatial distribution of the DL density along the semiconductor surface can also be obtained.

Figure 14 displays the PICTS spectrum (a) and the DL spectrum derived from the SPCE relaxation transients (b) for GaAs sample. The latter spectrum is considered as a surface photo-voltage transient spectrum (SPVTS). The comparison of these spectra shows that the same DL (P1, P2, P3) dominate markedly the PC decay transients and the SPCE relaxation transients in the same temperature intervals. The considerable differences in the depicted spectra (both in the relative intensities of the DL bands and in the disappearance of one and appearance of another DL) could be related to strong trapping into surface electronic states that reflected in the SPVTS spectrum.

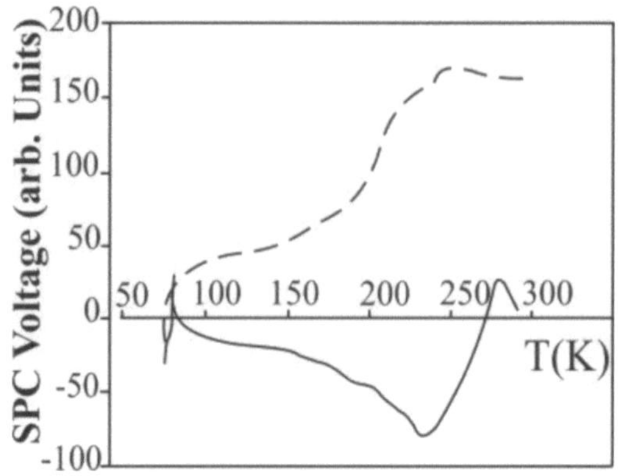

Fig. 15. SPCE voltage temperature dependencies at 633 nm (dashed line) and 488 nm (solid line)

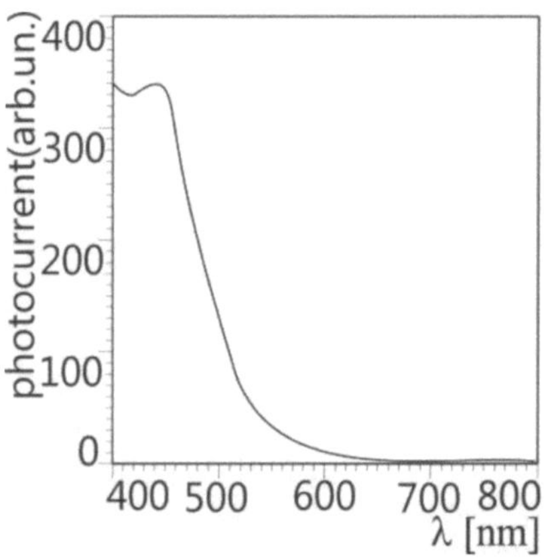

Fig. 16. Spectral dependence of the DC photo-current normalized to a constant light intensity

However, the PICTS spectrum significantly depends also on the depth of the light-excited semiconductor layer. The latter is determined by the light absorption depth and the diffusion length of the charge carriers. Obviously, such dependence may be attributed to a non-uniform DL distribution along the crystal depth.

To demonstrate the contactless SPVTS method, the surface of a GaAs crystal was scanned in order to investigate the P1 - DL density distribution profile along the scanning direction. The P1 centre parameters were not determined from the SPVTS data, since no relevant mathematical procedure exists at present. The variations of the SPVTS signal (Fig. 14 c) registered during a crystal surface scan at 162 K (P1 band maximum in the DL spectra) indicate a V-shaped distribution of the P1 - DL density. This is in agreement with the results obtained by PICTS in a sandwich-structured Ge(Au)-GaAs-Ge(Au). GaAs samples passivated with a thin ZnSe film of variable thickness were also investigated by SPCE [Abbate et al., 1995].

The temperature dependence of the magnitude and polarity of the surface photo-charge effect in CdS was experimentally studied for optical excitation in both the subbandgap and the intrinsic absorption spectral range [Ivanov & Konstantinov, 1999]. CdS single crystals with high dark resistivity (about 10 GΩ) and prismatic "real" surface were studied. The dimensions of the samples were 2 cm x 1 cm x 0.5 mm. The samples were illuminated by an Ar-ion or He-Ne laser beam (laser power of ~5 mW/cm^{-2}), mechanically chopped at 183 Hz. The results are shown in Fig. 15.

The spectral dependences of the measured DC photo-current for the same samples are presented in Fig. 16. The experiments were conducted at λ=488 nm and λ=633 nm, because the photon energy at 488 nm is higher, while at λ=633 nm it is smaller than the CdS bandgap (~2.41 eV). Excitation of electrons from the valence to the conduction band via direct band-to-band transitions is induced and, in result, the concentration of free electrons increases. This leads to a significant increase in the photo-conductivity in this spectral range. As a consequence, a negative SPCE voltage is generated at low temperatures. The magnitude of the voltage increases with the temperature (T) rise.

An example for characterization of semiconductor devices was reported in [Das et al., 1992]. In particular, MOS devices on Si were experimentally studied. The depletion, accumulation and inversion regions of the device were clearly distinguished. These curves can be used for characterization in a similar manner as the CV plots.

4.3 Sensors for fluids

It was observed that any variations in the fluid characteristics induce a corresponding change in the electron properties of the irradiated solid-liquid interface, and in result - a change in the registered SPCE signal. Therefore, provided that all other conditions remain constant, changes in fluid properties can be detected. A sketch of a setup for such measurements is presented in Fig. 17 [Ivanov & Konstantinov, 2000].

The illuminated substrate 1 should be very sensitive to SPCE (e.g. a semiconductor) for a higher voltage to be generated. The exposing radiation could be non-monochromatic or monochromatic, the latter being preferred. Since the sensitivity of the SPCE depends on both the type of the surface states at the interface 2 and the changes taking place in the studied fluid 3, appropriate wavelength, providing optimal response, should be used. The SPCE voltage is measured on the electrode 4, coupled to suitable equipment. In the case of gas or vapour sensors, a solid with maximal adsorption capacity to the respective fluid should be used as a substrate.

It should be noted that to achieve a reliable result, it is necessary to select carefully many parameters: the type of the structure that generates the signal, the wavelength and intensity of the incident radiation, the parameter of the electrical signal which has to be measured, the material of the irradiated surface, the method for signal registration, the measurement conditions, the additional influences, etc. There are many parameters that need to be optimized. The goal is a minimal change in the controlled variable to cause maximum changes in the measured electrical signal. When it is once made for a specific task, all experimental conditions should be kept constant.

Since the surface states are very sensitive to charge redistribution upon illumination, high sensitivity can be achieved in a certain spectral range. SPCE can be induced in any solid and therefore, a suitable solid together with an appropriate spectral range of the incident radiation can be assigned to any fluid in order to avoid undesirable effects as hysteresis, slow relaxation, etc.

Ar-ion laser with a power of 250 mW at the liquid-surface interface was used as radiation source. The illumination spot was the same in all the experiments in order to avoid setup

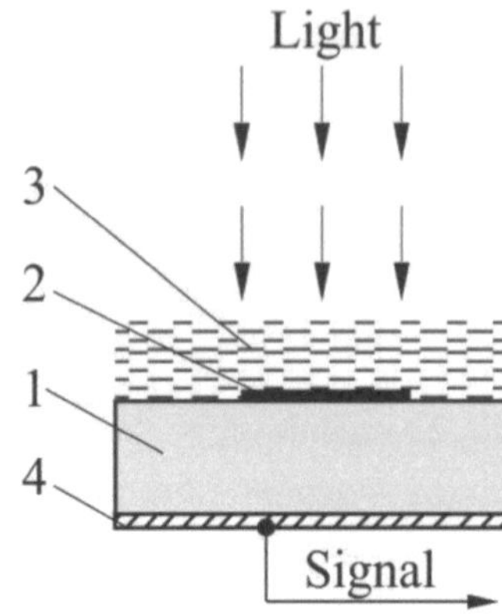

Fig. 17. Setup for the observation of SPCE at liquid-sold interface: 1-solid, 2-solid-liquid interface, generating the signal, 3-fluid under study, 4–electrode

effects. 15 ml of liquid were used in each measurement. The experimental data showed that the SPCE signal is highly sensitive to any changes in the composition and the properties of the studied fluid.

A characteristic feature of the SPCE in liquids is that the potential difference, created by SPCE, has a unique value for each liquid. For example, the amplitudes of the measured signal for several, arbitrarily selected liquids, are: tap water - 190 μV; filtered water - 209 μV; distilled water - 175 μV; alcohol - 115 μV; ammonia solution - 65 μV; acetone - 50 μV; coffee (used as an opaque liquid) - 15 μV.

The experimental results, concerning the SPCE in liquids, showed that this effect is very attractive for prospective practical applications. Under certain conditions, the SPCE signal is a function of the composition or other properties of the liquid and could be implemented as an analytical method, similar to the potentiometric methods in analytical chemistry.

As an example, a gasoline quality control device could be developed since the SPCE signal differs for leaded or unleaded gasoline with various octane grades: A-86 - 29 μV; A-93 - 54 μV; A- 93/unleaded/ - 63 μV; A-96 - 46 μV; A-95 /unleaded/ - 64 μV; A-91 /unleaded/ - 59 μV. These samples (30 ml each) were supplied by the Laboratory of Fuel. The amplitudes of the SPCE signal for a series of fuels, purchased from regular gas stations were: diesel - 48 μV; A-86 - 31 μV; A-93 - 52 μV; A-95 /unleaded/ - 68 μV.

These experiments demonstrate the possibility for qualitative detection of different liquids or variations in some liquid characteristics. Further development and calibration of the proposed method would make possible a quantitative analysis as well.

Variations of the SPCE signal were also registered upon mixing two liquids. For example, if only one drop of ammonia solution was added to 200 ml of drinking water, a 15% decrease in the amplitude of the signal was detected, whereas the signal increased by 10 % when one drop of acetone was added in the same amount of water.

Another group of experiments was performed showing that by varying the type of electrode and the illumination conditions, the SPCE signal may become dependent on the volume of the liquid. This provided an opportunity for another device - a new type level-meter for liquids to be designed.

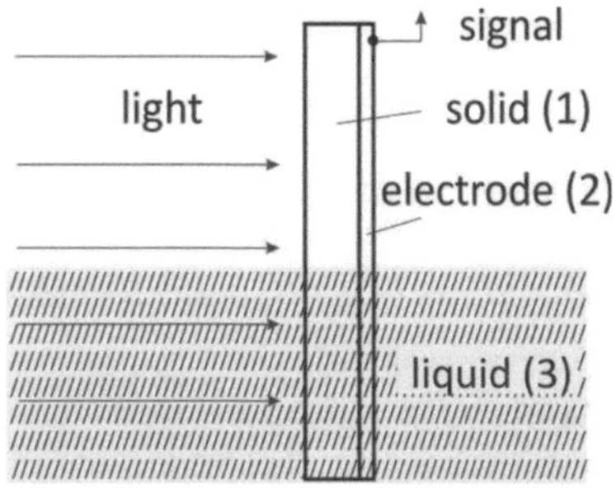

Fig. 18. Level-meter scheme: solid stick with deposited measuring structure on it (1); electrode (2); liquid (3)

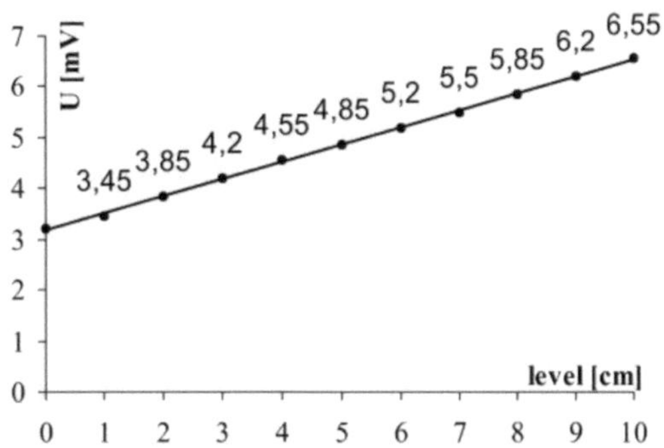

Fig. 19. SPCE response versus the petrol level

It was found that, in case of a suitably chosen structure, proportionality between the potential difference, caused by the SPCE, and the level of a certain liquid of interest can be achieved.

The essence of the idea for this type of level-meter is to put irradiated solid surface in contact with the controlled liquid. Since the electron properties at a solid surface influence essentially the interface with an overlying fluid layer, one can expect that optically-excited changes in such a system will induce measurable signals. This way, with all other conditions constant, it is possible to detect changes in the fluid. It is reasonable to expect that any kind of changes in the fluid will cause corresponding changes at the exposed interface, thus changing the SPCE signal.

The basic scheme of the device is shown in Fig. 18: (1) is a solid, in the shape of a stick. Its length is larger or at least equal to the height of the tank in which the liquid (3) is stored. A measuring structure is deposited on the stick. The stick was irradiated with modulated electromagnetic field of ultrahigh frequency. The detected SPCE signal was measured with an electrode (2) which is also placed along the length of the stick. For some device structures, the signal could be very sensitive to small changes (of the order of 0.1 mm) in the level of the liquid. Since the SPCE is characteristic for all kinds of solids, there is a large enough variety of materials which can be used to design working structures suitable for the mentioned purposes. The problem is to select the best materials for each particular case: type of liquid under control, working conditions, required accuracy, depth of the tank, etc.

As far as the measurement of liquid fuel levels is the most widespread case, a small laboratory model was constructed. It can be used to measure levels up to 10 cm of unleaded petrol 95H. The structure employed shows about 100% change of the electric signal when the liquid level varies from 0 to 10 cm. The measurement accuracy is 0.4 mm and can be increased simply by technical improvement. As it can be seen from Fig. 19, there is a linear dependence between the SPCE signal and the gasoline level. During the measurement, a selecting technique and amplification of 20 dB were used. Since the SPCE is a quite fast effect, the analysis is in real time, i.e. the measurement and the reading are performed at the moment of the level variation.

Since electric signals are measured, the question about the possible hazards of electric flash and fire arises. It must be noted that the generated potential differences are very small - of the order of nanovolts and microvolts. Such differences exist, generally, in any solid, no power supplies are required and the measuring technique does not pose additional risks. The device can be designed with electrical wiring separated from the liquid fuel.

Usually, when liquid fuels are concerned, a requirement arises to measure the level of the water at the bottom of the tank. It should be noted that even the laboratory model described can indicate the presence of water there. Both the levels of the fuel and the water can be measured simultaneously. The investigations performed so far did not reveal any difficulties of principle for the implementation of such device. However, many problems should be solved before the device leaves its laboratory-prototype stage. For example, a possible problem, which could lead to the appearance of additional errors, is the condensation of vapours on the measuring structure. Some preliminary tests in this respect were carried out. Hot water vapour was supplied to a closed chamber, in which the level-meter had been placed, until saturation was achieved. At these extreme conditions, the experimental error was about 1 mm. One way to avoid such problems is to improve the measuring structure in order to minimize the influence of the condensation. Otherwise, precautions should be taken to control the condensation.

The proposed level-meter has a simple design, requires small investments for production and, in addition, is sensitive to very small variations in the liquid level. An important advantage, in comparison to the level-meters for liquid fuels commercially available at present, is that no moving parts are involved. Hence, the results are not influenced by variations in the relative weight of the liquid and there is no chance for micro-particles to hamper the free movement of the float, used in other level-meters. The device has many advantages and can be competitive to the technical solutions currently available. The flowing velocity of a liquid can also be monitored by SPCE, since, at certain conditions, the generated voltage depends on this factor as well.

The experiments showed also that the precipitation of a substance in a solution on the surface could also be monitored. It is obvious that the precipitated substance affects the surface properties and, as a result, changes in the SPCE signal should be observed. For example, variations of the SPCE signal after precipitation of $CaCO_3$ from water on a metal surface were detected [Ivanov & Konstantinov, 2002].

The SPCE can prospectively be used for a complementary analysis together with the already existing methods for liquid control. The main advantage of such an analytical method is its universal nature - since SPCE voltage can be generated for all types of liquids, the method can be applied for characterization of any liquid. Various practical applications may be developed. The combination of optical probing of the sample and electrical detection of the generated signal is also of great importance. Provided that a suitable measuring structure is selected, minimal amounts of liquid (even a small drop), would be sufficient for the analysis. The proposed method is universal - it can be used to study various fluids under different conditions. The measurements are rapid, real-time and contactless. No complex equipment and serious financial investments are required.

4.4 Chemical composition test

The dependence of the SPCE signal on the chemical composition of the illuminated solid could find application in qualitative monitoring of the composition of various samples [Ivanov et al., 2008]. SPCE for example was used to analyze the composition of 50-cent coins (50 stotinki BGN), since they were subjected to widespread counterfeiting according to the information provided by the Bulgarian National Bank (BNB). It should be noted that there is a limited number of methods for real-time and contactless control of coins.

The development of a device for this was based on the standard type scheme for SPCE observation (Fig. 4). The sample 2 was irradiated with a modulated electromagnetic field 1. The voltage, induced by the SPCE, was measured on the electrode 3, which is in proximity to the sample 2. The signal was registered with a suitable device - a lock-in nanovoltmeter in this case. The frequency of the irradiating electromagnetic field was 144 kHz.

Initially, a large number of genuine pieces, which were in circulation, were measured in order to determine the range of the generated signals. Next, various counterfeits were tested. The comparison of the obtained results revealed that identification of the genuine from the counterfeit coins was completely possible. It was even possible to identify the types of counterfeits. A number of tests were conducted, some of them in the presence of experts from the BNB. 100% identification of the fake from the genuine coins was demonstrated. The amplitudes of electrical signals caused from SPCE and measured on counterfeits were significantly different from these for original coins. A comparison of the results showed that identification of real and counterfeit coins by the proposed method is clearly detectable, primarily due to the chemical composition. This is shown in Fig. 20, which is a one-axis graph. On the axis, the amplitudes of electrical signals are set caused by SPCE measured over 100 investigated coins. Three of them are counterfeits and they are easily found on the graph because they are out of the area of signals of the genuine coins. It is even possible to identify several different types of counterfeits, this way revealing their origin. Since the measured signals are instantaneous, the method basically finds application where an instantaneous identification is necessary. Generally, this is a real-time analysis. However, the effect is very sensitive and the whole device together with all the necessary electronic sections has to be placed in a dense metal shield. We believe that without any difficulties, the detector could be designed relatively economically and small in size, so as to be installed in any coin-operated machine.

The SPCE cannot be used to determine the exact amount of various components in an unknown sample. It can only be applied to detect whether or not the sample meets a certain standard. It is possible to trace the change of a certain impurity in a series of samples provided that the rest of the components are kept constant. This requires preliminary calibration. Theoretically, it is possible that a combination of different alloys generates a signal coinciding with the one of a genuine coin. However, such a coincidence is very unlikely and is of no practical significance. Yet, if the counterfeiters have at their disposal high-tech equipment enabling them to create and reproduce such a sample, they would rather create an ideal replica (in composition and weight) of the original coins.

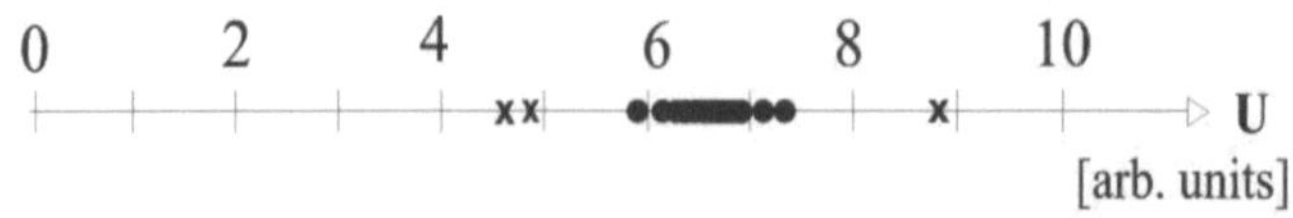

Fig. 20. The SPCE signal amplitudes, measured over 100 investigated coins: × - counterfeit coins; • - genuine coins

Presently, the commercially available detectors for real-time identification of fake coins most often operate on the principle of weight or dimensions measurements. Detectors based on composition control should be more accurate and reliable. An appropriate approach could be the building-in the proposed device into the detectors, used at present, controlling this way two parameters simultaneously. It has to be noted, however, that for successful implementation of the proposed method, the genuine coins should meet a certain standard. Otherwise, the SPCE-based identification could be considerably affected. Here, a new application emerges - analysis of the composition of coin billets prior to their stamping. The ones not complying with the standards should be returned for melting. Improvement of the standardization will undoubtedly enhance the quality and security of the coins in circulation.

The proposed device may find application in many areas - banks, shops, coin-operated machines, etc.

Besides the protection of the national currency, the SPCE-based composition tests may find many other applications. For example, various types of absorbing filters for gases and liquids can be monitored in order to determine when they have to be replaced. The absorption of the filtered substances causes changes in the chemical composition of the filter. The latter could be detected by using a SPCE-based device.

The SPCE could be used for monitoring of environmental pollution as well. This non-destructive, continuous and real-time test could also find application in areas such as ecology, industry, car production, military equipment, etc.

As far as security is concerned – analysis of drinking water (tap water or the one supplied to the army in field operations) is a very attractive possibility. Any change in the chemical composition, due to contamination with harmful substances, can be detected by the corresponding variation in the SPCE signal.

4.5 Milk and food quality control

Since milk is of great importance as a food source, supplement or ingredient, a large number of analytical methods have been developed for its investigation, including biochemical, microbiological, serological methods, etc. Principal disadvantages of commonly utilized methods are related to their slowness, problems with consumables, and the need of laboratory conditions. These methods usually require reagents, some of which are expensive, extremely unstable, and very toxic or carcinogenic. Instantaneous tests, not requiring any consumables, are still in great demand. An appliance based on the SPCE could be developed to be portable, easily operated without expensive consumables, providing fast results, and not requiring a laboratory environment [Ivanov & Radanski, 2009].

A scheme of the experimental setup is similar to that in Fig. 17. A continuous wave diode laser generating 25 mW at λ=655 nm was used as a source (L) of incident radiation. The incident light beam was chopped into periodic pulses with a modulation frequency of 800

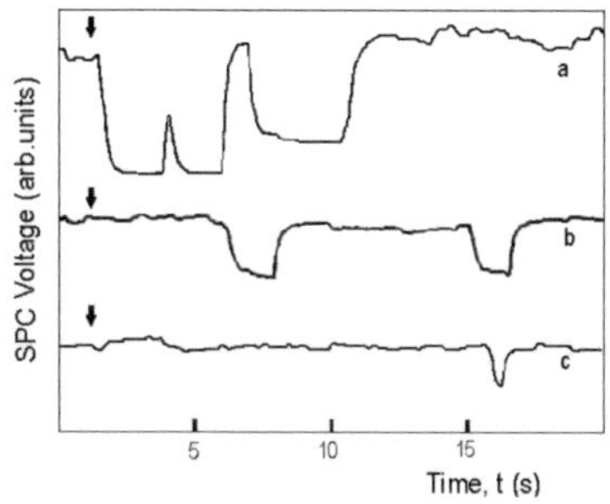

Fig. 21. Signal variation for 20 s after dripping of concentrated acetic acid in the incident spot of the laser beam: a) after the first drop; b) after the second drop; c) after the third drop. The milk was well stirred after each drop for homogenization

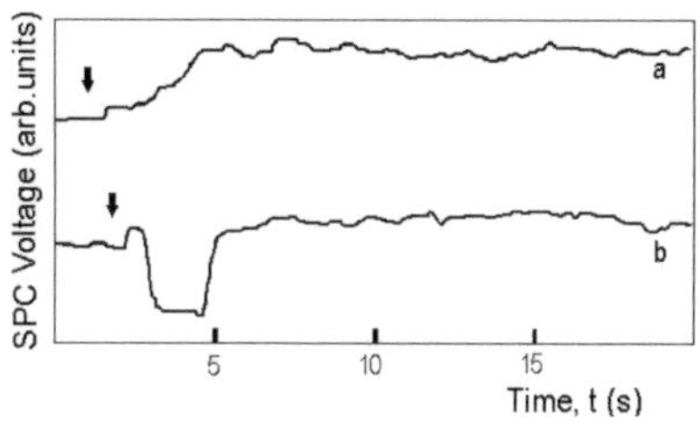

Fig. 22. Influence of the antibiotic kanamycin dripped in the incident spot of the laser beam on the signal amplitude for 20 s: a) at addition of 10 drops of a 30% solution of H_2O_2 in milk (being stored 2.5 hours before the test) to 200 ml of milk; b) pure milk signal

Hz by using a modulator (M). This value was chosen to be far from the network frequency of 50 Hz where the most parasitic signals could appear and, at the same time, to be in the working range of the optico-mechanical modulator. A pulsed laser or a pulsed LED could also be used instead of such a modulator. The milk sample was placed in the measuring arrangement. The latter is a small vessel in which two electrodes are mounted. A setup capable to perform test on only a drop of liquid [Ivanov & Konstantinov, 2002] is also possible. The signals measured are in the nano- and microvolt scale and were amplified by 20 dB by the preamplifier. The detected signal had very low amplitude, thus a lock-in nanovoltmeter, capable of extracting the signal from the background noise, was utilized. The nanovoltmeter was coupled to a recording device to follow the time evolution of signals. The reference signal to the lock-in was supplied by the modulator. The measured signal is formed by the liquid-solid interface as discussed above when the irradiated solid surface is put in contact with the milk under investigation. Most probably, the presence of a liquid affects strongly the SPCE at the illuminated solid-milk interface and thus - the capacity of the measuring structure. Correspondingly, any minor changes in the milk (concentration, contaminations, pretreatment, etc.) change the SPCE signal, indicating that the processes at the interface are strongly dependent on its properties (Fig. 21).

The required information can be obtained ether by direct observation of the signal amplitude or by using appropriate reagents (testing liquids) modifying the signal. This was demonstrated by the investigations in the case of hydrogen peroxide as an admixture (Fig. 22 a).

The milk samples (10 ml each) were characterized by measuring the amplitude of the generated potential difference and its variation after a droplet (0.04 ml) of testing liquid was added to the milk. The testing liquid was introduced both in the incident spot of the laser beam and at a certain distance (1 - 3 cm) away from it. Figures 21 and 22 demonstrate the opportunity for detecting changes in milk due to the presence of extraneous substances. The arrow signs indicate the moment of time at which the testing liquid was dripped in the milk sample. Variations in the signal response, similar to those presented in Fig. 22, were also observed in the presence of antibiotics/inhibitors in the milk samples.
The results described above demonstrate the opportunity for quality control of milk as well as for recognition of milk samples produced by different animal species since they generate signals with different amplitudes. The following types of milk were analyzed: cow, sheep, goat and buffalo. The amplitude of the signal generated by sheep milk was reduced by 15% compared to the signal, generated by cow milk. The amplitude of the signal, generated by the buffalo milk sample, was 50% smaller than the one generated by cow milk.
It should be noted that at the present stage of work the investigations are mostly qualitative ones. They are unable to identify the admixture type and its quantity in the milk. Considerable work is necessary to make possible quantitative estimations. Nevertheless, a qualitative evaluation of milk purity could be very useful for practical applications, if performed in a fast way, in real working environment, and without considerable expenses; these features are typical for the techniques proposed here.
Besides milk, SPCE can be applied for characterization of other foods (e.g. bee honey) as well.

5. Conclusion

The transverse acoustoelectric effect is considered promising for characterization of the electric properties of semiconductors. The TAV method is contactless, very sensitive and the measuring equipment is relatively simple. In addition, there are no special requirements to the exciting electrode, deposited on top of the semiconductor.
The most important advantages of SPCE-based methods include: high accuracy, low cost, instantaneous results, and possibilities for field measurements in real time, no complex equipment is required and no metal contact has to be deposited on the sample. No power supplies are required and the measuring technique doesn't pose any additional risks. This analytical method gives also great possibilities because it combines optical probing of the sample with electrical detection of the generated signal. The main advantage of SPCE is its universal nature.
The variety of possible applications of TAV and SPCE, described in this section, is promising enough to attract the attention of experts to the design and construction of electromagnetic field-matter- interactions-based devices for quality control.

6. References

Abbate, A.; Rencibia, P.; Ivanov, O.; Masini, G.; Palma, F. & Das, P. (1995). Contactless characterization of semiconductor, using laser-induced surface photo-charge voltage measurements. *Materials Science Forum,* Vol. 173-174, 221- 226, ISSN 0255-5476

Balland, J., Zielinger, J., Noguet, C. & Tapiero, M. (1986). Investigation of deep levels in high-resistivity bulk materials by photo-induced current transient spectroscopy. I.

Review and analysis of some basic problems. *Journal of Physics D,* Vol. 19, No. 1, 57-69, ISSN 0022-3727 (paper); 1361-6463 (online)

Berglund, C. (1966). Surface states at steam-grown silicon-silicon dioxide interfaces. *IEEE Transactions on Electron Devices,* Vol. 13, No. 10, 701-705, ISSN 0018-9383

Borissov, M.; Germanova, K.; Ivanov, O. & Komitov, L. (1985). Use of thransparent conductive layer for observing transverse acoustoelectronic effects. *Comptes Rendus de l'Academie Bulgare des Sciences,* Vol. 38, No. 7, 843- 846, ISSN 0366-8681

Das, P.; Mihailov, V.; Ivanov, O.; Georgiev, V.; Andreev, S. & Pustovoit, V. (1992). Contactless characterization of semiconductor devices using surface photo-effect. *IEEE Electron Device Letters,* Vol. 13, No. 5, 291- 293

Davydov, I.; Ivanov, O.; Svircov, D.; Georgiev, G.; Odrinsky, A. & Pustovoit, V. (1994). Contactless spectroscopy of deep levels in semiconducting materials: GaAs. *Spectroscopy Letters,* Vol. 27, No. 10, 1281-1288, ISSN 0038-7010 (paper); 1532-2289 (online)

Ermolova, O., Ivanov, O., Naidov-Zhelezov, O. & Solodov, I. (1985). Transverse acoustoelectric effect with bulk wave reflection in pizoelctric-semicondictor structures. *Acoustic Letters,* Vol. 9, No. 3, ISSN 1790-5095

Fritz, I. (1981). Transverse acoustoelectric effect in the separated medium surface wave configuration. *Journal of Applied Physics,* Vol. 52, No. 11, 6749-6756, ISSN 0021-8979

Hurtes, C., Boulon, M., Mitonneau, A. & Bois, D. (1978). Deep level spectroscopy in high-resistivity materials. *Applied Physics Letters,* Vol. 32, No. 12, 821-823, ISSN 0003-6951(paper + online); 1077-3118 (online only)

Ivanov, O. & Germanova, K. (1988). A study of the relaxation dependency of the transverse acoustoelectric effect in the $LiNbO_3$-CdS separated-medium configuration. *Acoustics Letters,* Vol. 11, No. 8, 148-150, ISSN 1790-5095

Ivanov, O. (1989). A review of the major investigations and problems associated with the transverse acoustoelectric effect. *Acoustic Letters,* Vol. 13, No. 4, 56-63, ISSN 1790-5095

Ivanov, O.; Mihailov, V.; Pustovoit, V. & Das, P. (1994). Surface photo-charge effect in dielectrics. *Comptes Rendus de l'Academie Bulgare des Sciences,* Vol. 47, No. 6, 21-24, ISSN 0366-8681

Ivanov, O.; Mihailov, V.; Pustovoit, V.; Abbate, A. & Das, P. (1995a). Surface photo-charge effect in solids. *Optics Communications,* Vol. 113, No. 1, 509-12, ISSN 0030-4018

Ivanov, O.; Svircov, D.; Mihailova, Ts.; Nikolov, P. & Pustovoit V. (1995b). Automatized system for measuring the surface density of current carriers and electrical permittivity of conducting materials. *Spectroscopy Letters,* Vol. 28, No. 7, 1085-1094, ISSN 1532-2289

Ivanov, O. & Konstantinov, L. (1999). Temperature dependence of the surface photo-charge effect in CdS. *Applied Surface Science,* Vol. 143, 101-103; ISSN 01169-4332

Ivanov, O.; Mihailov, V. & Djulgerova, R. (2000). Spectral dependences of the surface photo-charge effect at conducting surfaces. *Spectroscopy Letters,* Vol. 33, No. 3, 393-398, ISSN 0038-7010 (paper); 1532-2289 (online)

Ivanov, O. & Konstantinov, L. (2000). Application of the photo-induced charge effect to study liquids and gases. *Surface Review and Letters,* Vol. 7, No. 3, 211-212, ISSN 0218-625X

Ivanov, O. & Konstantinov, L. (2002). Investigation of liquids by photo-induced charge effect at solid-liquid interfaces. *Sensors and Actuators B*, Vol. 86, 287-289, ISSN 0925-4005

Ivanov, O.; Vaseashta, A. & Stoichev, L. (2008). Rapid, contactless and non-destructive testing of chemical composition of samples. *Proceedings of the NATO Advanced Study Institute on Functionalized Nanoscale Materials, Devices and Systems for Chem.-bio Sensors, Photonica, and Energy Generation and Storage*, pp. 331-334, ISSN 1874-6500, Sinaia, Romania, June 2007, Springer, Germany

Ivanov, O. & Radanski, S. (2009). Application of Surface Photo Charge effect for Milk Quality Control. *Journal of Food Science*, Vol. 74, No. 7, R79-83; ISSN 0022-1147 (paper); 1750-3841 (online)

Konstantinov, L., Strashilov, V. & Ivanov, O. (1985). Kinetics and polarity of the transverse acoustoelectric effect in the separated–medium surface acoustic wave configuration. *Journal of Physics D*, Vol. 18, No. 7, L79-L85, ISSN 1361-6413

Korshak, B.; Lyamov,V. & Solodov, I, (1976). Acoustic "after-action" accompanying surface wave interaction in a layered structure piezoelectric-semiconductor. *Journal of Experimental and Theoretical. Physics. Let*ters, Vol. 23, No. 8, 438-441, (in Russian) ISSN 0021-3640 (paper); 1090-6487 (online)

Kunigelis, V (1978). A simple theory of TAV-effect in piezoelectric-semiconductor layered structure. *Fizika Tverdogo Tela*, Vol. 20, No. 11, 3420-3423 (in Russian), ISSN BD 28 Pustovoit, V.; Borissov M. & Ivanov, O. (1989a). Photon charge effect in conductors. *Physics Letters A*, Vol. 135, No. 1, 59-61, ISSN 0375-9601

Lang, D. (1974). Fast capacitance transient apparatus: application to ZnO and O centers in GaP p-n junctions. *Journal of Applied Physics*, Vol. 45, No. 7, 3014- 3022, ISSN 0021-8979

Pustovoit, V.; Borisov, M. & Ivanov, O. (1989). Surface photo-charge effect in conductors. *Solid State Communications*, Vol. 72, No. 6, 613- 619

Pustovoit, V. & Ivanov, O. (1989). Surface charge redistribution effect in a conductor subjected to electromagnetic radiation. *Comptes Rendus de l'Academie Bulgare des Sciences*, Vol. 42, No. 4, 39-42, ISSN 0366-8681

Strashilov, V., Konstantinov, L. & Ivanov, O. (1987). Topographical study of conducting surface by using a combined photo-acoustoelectrical methods. *Appied Physics B*, Vol. 43, 17- 21

Tamm, I. (1975). *Collected Scientific Works, Vol. 1,* Nauka Publishing House, Moscow (in Russian)

Tapiero, M., Benjelloun, N., Zielinger, J., El Hamd, s. & N oguet, C. (1988). Photoinduced current transient spectroscopy in high-resistivity bulk materials: instrumentation and methodology. *J. Appl. Phys.*, Vol. 64, No. 8, 4006-4014, ISSN 0021-8979

Vankova, N., Ivanov, O. & Yordanova, I. (1997). Experimental investigations of surface photo-charge effect in different materials. *Spectroscopy Letters*, Vol. 30, No. 2, 257-266, ISSN 1532-2289

White, A., Dean, P. & Porteous, P. (1976). Photocapacitance effects of deeps traps in epitaxial GaAs. *Journal of Applied Physics*, Vol. 47, No. 7, 3230-3239, ISSN 0021-8979

27

Process Monitoring of Polymer Systems by using Near-Infrared Spectroscopy

Wei Li
School of Materials Science and Engineering, Shenyang Aerospace University
People's Republic of China

1. Introduction

Polymers are a very important class of organic materials. Nowadays, polymers have been widely used in many fields due to their specific and excellent properties. They have become increasingly significant in our lives. With the growing demands on the quality of polymer products as well as reduction of costs in production and processing, the fast, accurate and reliable monitoring methods become more and more important.

The traditional off-line monitoring methods have many limitations, one of which is lag effect. This is mainly because sampling and lab testing can be very time-consuming and for some tests a long time is required before the results are obtained. It is obviously that there is a large time gap between sampling and obtaining the results. Meanwhile, the reaction or processing of unknown quality polymer continues. Hence, the sample and end product don't have the same properties, which can result in large amounts of wasted material (Rohe et al., 1999). Another problem is that the sample analysed is small and as a result might not have all the properties that the bulk possessed (Coates et al., 2003). Due to these limitations and problems, the off-line measuring methods are clearly unsuitable for the process monitoring.

Undoubtedly, it is exciting to observe that the spectroscopic techniques can meet the above demands required by the process monitoring of the polymer systems. The vibrational spectroscopy is ideally suited to qualitative and quantitative analysis of polymers. Moreover, it can also give valuable information at the molecular level about the reactions extent and the products quality. It deserves to be specially noted that Near-infrared (NIR) spectroscopy is a promising tool among the spectroscopic techniques.

Near-infrared spectroscopy lies between the visible and mid-infrared regions of the electromagnetic spectrum, whose region extends from about 12,500-4000 cm^{-1}. This region encompasses bands that result from the harmonic overtones of fundamental and combination bands associated with hydrogen atoms, which is why compounds containing CH, OH and NH bonds lend themselves favourably to analysis by NIR spectroscopy. However, NIR spectroscopy is comparatively weaker than mid-infrared, and contains molecular structure information seriously overlapped which resulted in the silence of its research in the early stage.

The modern earliest analytical applications of NIR spectroscopy were reported in the early 1960's (Butler & Norris, 1960; Norris & Butler, 1961). Nowadays, NIR spectroscopy

technology has been extensively and effectively employed in a variety of fields, including areas of agricultural, food, chemical, petrol, pharmaceuticals, textiles, cosmetics and medical, etc. The rapid growth of the technology can be attributed to the emergence of high precision spectroscopic instruments with very high signal to noise ratios to detect minute differences in the spectra and high-speed computers to carry out the complex calculations involved in multivariate analysis (Norris 1989). On the whole, NIR spectroscopy has several advantages compared to traditional laboratory techniques, such as rapid speed and reliability of determinations, rapid and easy sample preparation, multiplicity of analysis with one operation, simple operation, and non-destructive analysis (Schultz & Burns, 1990).
The wide applicability and potential of NIR spectroscopy for polymer analyses have been recognized (Lachenal & Ozaki, 1999; Lee & Kradjel, 2007). NIR spectra contain information related to polymer properties (Lachenal, 1995) such as composition, conformation and crystallinity, therefore NIR can be widely applied for various polymer analyses. Especially, the recent progresses of modern fiber-optic-based NIR spectrometers opened up new perspectives for process monitoring in the polymer fields. The probes can be placed at very harmful working environments, while the spectrometer locates still in a process monitoring room. The use of probes connected to spectrometer via optical fibers allows direct measurements in the product line without interference in the production process. Due to the diversity of available probes, such as transmission mode and diffusive reflectance mode, the NIR technique can be used to analyze almost any kind of polymer, including melts, solutions or solid powders, etc.
The main objective of this chapter is to introduce the use of NIR spectroscopy as an alternative technique for process monitoring of various polymer systems. Firstly, some successful applications of NIR spectroscopy used in the polymers polymerization, polymers processing and polymers curing are described. Then, the quality monitoring of polymer prepregs in the production is presented.

2. Monitoring the polymerization process

In the polymerization process, the minimal changes of the process parameters may cause the great property differences of polymer products. Thus, the polymerization process could not be regulated precisely by only controlling a few process parameters, and the product quality is likely unable to meet the requirements. The ideal solution method is to measure the polymer properties continuously and adjust the processing parameters promptly by the measuring results. Among the developed techniques, the applications of NIR have been reported in the bulk, solution, emulsion, and suspension polymerizations, including monitor the compositions, monomer conversion, molecular weight averages, and average particle sizes, etc.

2.1 Bulk polymerization

A short-wavelength NIR spectroscopy was utilized for in-situ monitoring the monomer conversion during the bulk polymerization of methyl methacrylate in a mold (Aldridge et al., 1993). Using a retroreflecting array, the illumination beam was modified to pass through the sample twice in order to reduce lensing effects of the mold and Schlieren patterns caused as the reaction progressed. It was observed that the disappearance of the third overtone stretch of the CH of the vinyl group at 868nm and the concomitant appearance of the CH of the methylene group at 890nm were attributed to the monomer conversion. The spectral data at these two wavelengths showed an excellent correlation to the monomer percent conversion.

The monomer conversion of a styrene bulk polymerization was monitored on-line with NIR spectroscopy (Lousberg et al. 2002). A flow cell was utilized for sampling. The influence of the temperature of the reaction mixture on the predicted conversion was minimized by building a global multivariate calibration model. The results showed that on-line conversion monitoring with NIR offered overall an excellent accuracy (~0.32% conversion), which is comparable to the known accuracy of the at-line size-exclusion chromatography reference method. Their study also proved that the method is insensitive to either changes of the molar mass distribution or of the temperature (changes of at least 10℃) of the reaction mixture and is able to detect abnormal behavior during a batch quickly.

2.2 Solution polymerization

A fiber-optic near-infrared (NIR) spectroscopy was used to monitor the monomer conversion and the weight-average molecular weight of the polymer produced during solution polymerization of methyl methacrylate (MMA) (Cherfi & Fevotte, 2002; Cherfi et al., 2002). NIR spectra were recorded during batch and semi-continuous reactions using an in situ transmission probe. Off-line gravimetry and gel permeation chromatography (GPC) were used as reference methods to provide the conversion and the average molecular weight data set required for the calibration procedure. The calibration model was generated using PLS method. The model predictions were then validated for different operation conditions in both batch and semi-batch modes. The conversion was predicted during three validation experiments with an average standard error of prediction (SEP) of 2.1%, and the on-line evaluation of weight-average molecular weight was obtained with an average relative SEP of 5.7%.

The NIR spectroscopy used for in-line and in situ monitoring and control of monomer conversion and average molecular weight of polymer resins was presented in the styrene solution polymerization system (Fontoura et al., 2003). By using a Kalman filter state estimator and an accurate first-principle model, the control loop could be successfully closed to track desired monomer values and average molecular weights. The experimental results showed that it was feasible to use NIR spectroscopy for the simultaneous control of monomer conversion and polymer average molecular weight.

2.3 Emulsion polymerization

The NIR spectroscopy was used for the simultaneous in-line and in situ monitoring of monomer (methyl methacrylate and butyl acrylate) and polymer concentrations in the reaction medium during seeded semibatch emulsion copolymerizations (Vieira et al., 2002a). The NIR spectra were collected by a dispersive instrument equipped with a transflectance probe. The developed PLS calibration models were successfully for prediction of the major constituents of the emulsion. Based on the use of the NIR signal as process feedback, a control approach was constructed (Vieira et al., 2002b). The results showed that the control of copolymer composition and average molecular weights were achieved simultaneously.

The semi-continuous styrene/butyl acrylate emulsion copolymerization reactions were monitored by NIR spectroscopy (Reis et al., 2004). A transflection probe was immersed in the reaction medium. A set of nine reactions with slightly different formulations were carried out. The results of five reactions were used to fit a SIMPLS model, and the four remaining reactions were monitored in order to evaluate the model performance for estimation of important variables and latex properties. The study showed that simultaneous monitoring of monomer conversion, copolymer composition, particle number and average number of radicals per polymer particle was performed with success. Meanwhile, it was found that the calibration model for polymer particle size estimation was more susceptible to failure. The poor prediction of average particle sizes was also obtained in other study (Gossen et al., 1993 & Vieira et al., 2001). Obviously, the prediction of average particle sizes in emulsions systems from NIR spectra is a very controversial issue. Reis et al. think that there is a correlation between the average polymer particle size and NIR spectra, but it is not fully understood how the polymer particle size affects NIR spectra (Reis et al., 2004).

2.4 Suspension polymerization

NIR spectroscopy was used for in-line evaluation and for control and design of average particle sizes during styrene suspension polymerizations carried out in the batch mode (Santos et al., 1998 & Santos et al., 2000). It was found that the NIR spectra were sensitive to changes of the average particle size. The calibration models developed for the final average particle sizes had high prediction accuracy, and then they were used as references for implementation of an in-line procedure for control of average particle sizes. The study showed that this NIR technique allowed the successful design and real-time control of particle sizes in lab-scale styrene suspension polymerization reactors.

The morphological characteristics of PVC resins such as bulk density, cold plasticizer absorption and average particle diameter was first in-line and in real time monitored by NIR spectroscopy during suspension polymerization processes (De Faria et al. 2010a). The NIR spectra obtained at different experimental conditions were sensitive to changes of the PVC properties. The calibration models developed by PLS regression was capable to predict the morphological parameters accurately. It was shown that the modification of operation variables during the batch leaded to modification of the final morphological properties of the powder. Their study indicated that the morphology of PVC grains could be manipulated along the batch and that advanced NIR-based control procedures could be implemented for control of the morphological properties of PVC resins (De Faria et al. 2010b).

3. Monitoring the polymer extrusion processes

Extrusion is a major process for manufacture and fabrication of polymers. In the extrusion process, the resin is gravity fed from a top mounted hopper into the barrel of the extruder. Additives such as fillers, colorants and UV inhibitors are often used and can be mixed into the resin prior to arriving at the hopper. Standard chemical engineering measurements such as temperature, pressure, motor speed, and torque are used to monitor and control extruders. However, these process parameters can not ensure the product quality sufficiently due to the lack of information about polymer properties and composition. The extreme process conditions, which include high temperatures, high pressures, and harsh chemical environment, have restricted some analytical methods suitable for extrusion monitoring. The unique features of NIR spectroscopy make it adapted for such operation process. The applications of NIR spectroscopy have been realized in the polymer extrusion

processes for monitoring the polymer composition, additive concentration and flow properties, etc.
The MIR, NIR and Raman spectroscopy was used for monitoring of single screw extrusion of high density polyethylene and polypropylene blends (Coates et al., 2003). These vibrational spectroscopic techniques were then compared with in-line ultrasound velocity measurements. The in-line transmission NIR used a specifically designed stainless steel extruder section, just prior to the die. The NIR measurements for sequential extrusion showed that the absorption bands within 5150-6100cm^{-1} were changed obviously when the blend compositions were changed from 100 wt% PP to 50 wt% PP:50 wt% PE, then to 100 wt% PE. Analysis for the 10 wt% steps in blend concentration, using partial least squares regression analysis of the predicted concentration of each polymer, gave a standard deviation of prediction of ±0.39 wt%. Among the vibrational spectroscopy techniques explored, in-line NIR provided the most accurate assessment of blend ratio change for the experimental conditions.
A measurement system for in-line NIR spectroscopy for polymer extrusion processes was developed (Rohe et al., 1999; Rohe et al., 1998), which consisted of a measurement head adapted to a single screw extruder with integrated fibre optic probes (Fig.1) and an AOTF-NIR spectrometer. This system was capable of measuring different polymer compositions as demonstrated for PE/PP blends. Transmission measurements at typical extrusion conditions, where the temperature was up to 300°C and the pressure was up to 35 MPa, yielded good results with respect to mechanical and optical properties of the probes. The deviations between predicted and actual polymer composition were quite low. Fischer et al. (1997) have applied NIR spectroscopy for continuous monitoring the composition of polypropylene/ethylene vinyl acetate copolymer.

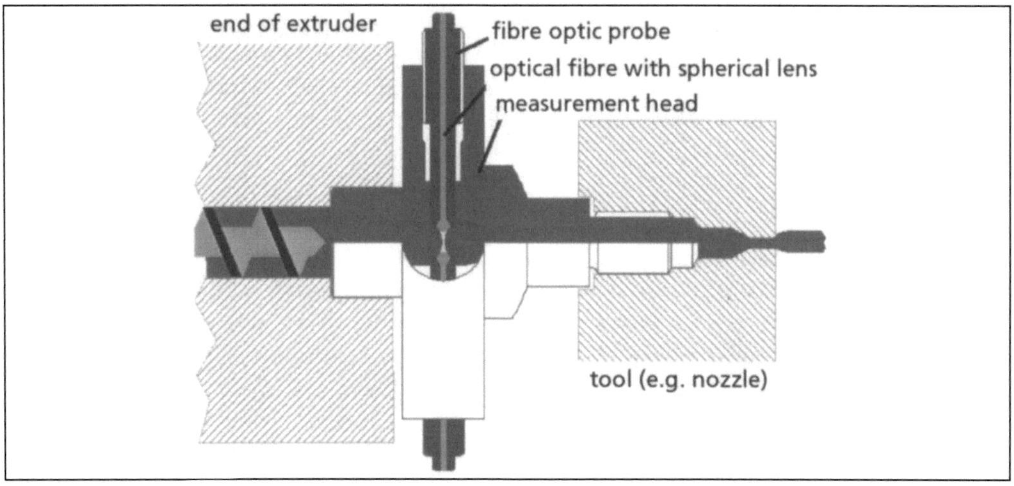

Fig. 1. Scheme of measurement head with integrated fibre optic probes for in-line measurement (Rohe et al., 1999)

Nagata et al. (2000a) developed an on-line sensor using NIR spectroscopy for monitoring CO_2 concentration in polymeric extrusion foaming processes. NIR absorption spectra were acquired by a probe installed at the foaming extruder die. The calibration curve relating the absorbance spectrum at 2019 nm to the dissolved gas concentration was derived so as to infer dissolved CO_2 gas concentration on-line from measured NIR spectra. Experimental results showed that the dissolved CO_2 concentration was successfully estimated in the molten polymer. The similar NIR system was also used for in-line monitoring the polyethylene density in the extrusion process (Nagata et al., 2000b).

The monitoring of additives content has also been studied. Batra et al. (1994) used NIR spectroscopy for monitoring the concentration of titanium dioxide (TiO_2) in molten PET. The NIR spectra were collected in-line by using a flow cell, housing two fiber-optic probes, and mounted downstream of the extruder. Because the presence of TiO_2 caused the scattering of light that resulted in a systematic baseline shift, the baseline shifts were correlated to the TiO_2 concentration data. PLS regression was applied to build models. Standard error of prediction values was about 1%. During the PP extrusion, Fischer et al. (1997) utilized NIR spectroscopy for monitoring the content of pulverized chalk in the range of weight concentrations from 0 to 40%.

In an extrusion process, the rheological properties of molten poly(ethylene vinyl acetate) (EVA) copolymers were in-line monitored by using NIR spectroscopy (Hansen & Vedula, 1998; Vedula & Hansen, 1998). The NIR transmission spectra of molten polymer flow stream were collected in a flow cell attached to a single-screw extruder. The calibration models were built on spectra in 1620-1840nm wavelength region. The SEP was 0.46 wt% for melt flow index (MI). The dynamic linear viscoelastic properties were also estimated.

4. Monitoring the polymer curing process

The final properties of thermosetting resins not only depend to a large extent on the chemical nature of the monomers utilized, but also depend on the curing process. The reason is that the properties of thermosetting resins depend on the features of the network such as the sort of hardeners, cross-link density and residual monument content. Therefore, monitoring of the curing process is benefit for the control of the structural of the resulting three-dimensional network. The applications of NIR spectroscopy for monitoring the curing of two important types of thermosetting resins, epoxy resin and bismaleimide resin, are illustrated herein.

4.1 Epoxy resin

The use of NIR spectroscopy to characterize the cure reactions of epoxy resin began in the late 1950's. The two characteristic bands for the terminal epoxide group were identified, which located in 6061 and 4546 cm^{-1} (Goddu & Delker, 1958). The epoxy/amine cure reactions were investigated by using NIR spectroscopy (Dannenberg, 1963). The epoxide group and hydroxyl group absorptions were quantified, and the effects of hydrogen bond formation on the spectra obtained were addressed. Xu et al. (1994) used in-situ NIR spectroscopy to investigate the cure mechanisms of epoxy/aromatic amine. The mechanistic models and the rate determining step were confirmed. The reaction rate constants obtained over a temperature range from 383 to 443 K demonstrated an Arrhenius temperature dependence and the corresponding activation energies and pre-exponential factors have been determined. Their study concluded that the cure rates were determined by the autocatalytic behavior of epoxy/aromatic amine system.

The remote fiber optic near-infrared spectroscopy was used for in situ real-time monitoring of the reaction kinetics of various epoxy/amine formulations (Mijović et al., 1995). In the curing reaction, the systematic changes in major absorption peaks of epoxy (4530 and 6080 cm^{-1}), amine (5056 and 6670 cm^{-1}) and hydroxyl (7000 and 4800-4900 cm^{-1}) were testified. The extent of reaction at any time *t* was calculated from the initial areas of epoxy and reference peaks, $A_{epoxy,0}$ and $A_{ref,0}$, respectively, and their corresponding values at time *t*, $A_{epoxy,t}$ and $A_{ref,t}$, according to the following equation:

$$\alpha=1-[(A_{epoxy,t})(A_{ref,0})]/[(A_{epoxy,0})(A_{ref,t})] \quad (1)$$

The peak at 4530 cm^{-1} was utilized to monitor the epoxy groups, and the peak due to the CH stretching vibration of the benzene ring at 4673 cm^{-1} was used as a reference. The effect of chemical structure and functionality of the formulation components on the reaction rate was shown. Moreover, the characteristic changes in the NIR spectra of multifunctional formulations were observed near the gel point and it was suggested that these could be used to identify gelation.

Park et al. (2000) studied the cure behavior and the reaction kinetics of epoxy resin which cured with an anhydride hardener and catalyzed by a cationic thermal latent initiator, N-benzylpyrazinium salts. The conversions of epoxy and hydroxyl groups were monitored by changes in the intensity of each band at 4530 cm^{-1} and 7000 cm^{-1}, respectively measured at different temperatures. The results revealed that the time dependence of the band intensity exhibits a sigmoidal-shape consistent with the autocatalytic reaction mechanism. From the extent of reaction which calculated from NIR absorption band at 4530 cm^{-1}, it was concluded that the cure reactions of epoxy resins were highly exothermic in nature and the cure kinetics was dependent on the temperature, resulting in the activity of catalyst.

The simultaneous dielectric/NIR spectroscopy was used to monitor the curing reaction of PMMA-modified epoxy resins (Kortaberria et al., 2004). The optical fiber NIR spectra generated in the remote mode was used for the quantitative evaluation of reaction kinetics and the extent of reaction was calculated from the absorption intensities of characteristic peaks (at 4530 cm^{-1} and 5056 cm^{-1}). The primary amines showed greater reactivity than secondary ones, being the vitrification process controlled by the secondary amine addition to the epoxy ring. The effect of the modifier as well as that of the cure temperature on the extent of reaction curves was studied, showing that the amount of modifier retarded curing reaction.

4.2 Bismaleimide resin

Mijović J. & Andjelić S. (1996) used NIR spectroscopy to investigate the mechanism and kinetics of cure of a two-component bismaleimide formulation, which composed of 4,4′-methylenebis[maleimidobenzene] and 2,2′-diallylbisphenol A. In-situ real time study of the progress of reaction was conducted in the temperature range from 140 to 250°C. The principal reaction observed was an alternating copolymerization involving maleimide and allyl double bonds. Maleimide homopolymerization was detected only in the initial stages of reaction at temperatures above 200°C. The extent of self-condensation (or etherification) of hydroxyl groups on the allyl component, which leads to cross-linking, was observed to vary with reaction temperature, suggesting a path to tailor-making networks with desired morphology and physical/mechanical properties.

The cure of bismaleimide monomer, 1,1′-(methylenedi-4,1-phenylene) bismaleimide (MDP-BMI) with aromatic diamines, such as 4,4′-diaminodiphenylmethane (DDM), was studied by using in-situ NIR spectroscopy (Hopewell et al., 2000). The concentrations of the various functional groups in the resins with respect to time were measured during isothermal cure. For MDP-BMI/DDM resins, the secondary amine was found to be less reactive than the primary amine in the nucleophilic addition reaction with maleimide double bonds. The homopolymerization of maleimide was moderately rapid in the early stages of cure in maleimide-rich formulations, and the rate slowed dramatically at higher conversion. Differences in the reactivities of different resin formulations were observed when the diamine was changed from DDM to 4, 4′-diaminodiphenylsulphone (DDS).

5. Quality control of prepreg manufacture

Fiber reinforced thermosetting matrix composites have been employed in military and aerospace applications due to their high performance. Many of these composites are made from the prepregs. Prepreg is the important intermediate of polymeric composites, and its quality will greatly influence the properties of final composite products. Thus it is important to control the quality in the manufacture process of prepreg. The volatile content, the resin content and the pre-curing degree of resin, i.e. three factors, are important to ensure the quality of the prepreg. Traditional analysis methods such as solvent extraction, burn-off and acid digestion are often used to measure the three factors. However, these methods are unsuitable for continuous measurement during the manufacture of prepregs due to the excessive time required and the destructive analysis of the specimen (Huang et al., 1997). So far, a series of process monitoring techniques have been developed to measure the prepregs quality.

5.1 Status of process monitoring technique for the prepreg quality

1. β-ray transmission technique (Huang et al., 1997; Huang et al., 2003)

The theoretical basis of β-ray montioring the prepreg quality is that the energy is attenuated during the penetration of the prepreg by β-particles. The attenuation of transmitted radiation is related to the thickness of prepreg:

$$I=I_0e^{-\mu\rho x} \tag{2}$$

where, I_0 and I are the energy before and after β-particle penetration of the prepreg, respectively; μ is the absorption coeffecient of the prepreg; ρ is the density of the prepreg; x is the thickness of the prepreg. Because the resin content is related to the prepreg thickness, the resin content is obtained by determining the thickness.

Fig.2 is the schematic of monitoring resin content of prepreg by β-ray technique (Huang et al., 1997). The prepreg specimen was passed through the middle between the β-ray source and the detector. In the test, a series of output singal of prepreg which had different rein content were obtained firstly. Then the relationship beween resin content obtained from chemical method and output singal was established. Finally, the unknown specimen was determined according to the above relationship.

The studies indicated that the β-ray technique always holded high accuracy in determining the resin content with broad range (Huang et al., 1997; Huang et al., 2003). With the nature of rapid and non-contact features, the β-ray technique can be used in the product line to continuously monitor the resin content of prepreg. It was necessary to take the protective measures because the radioactive element was used.

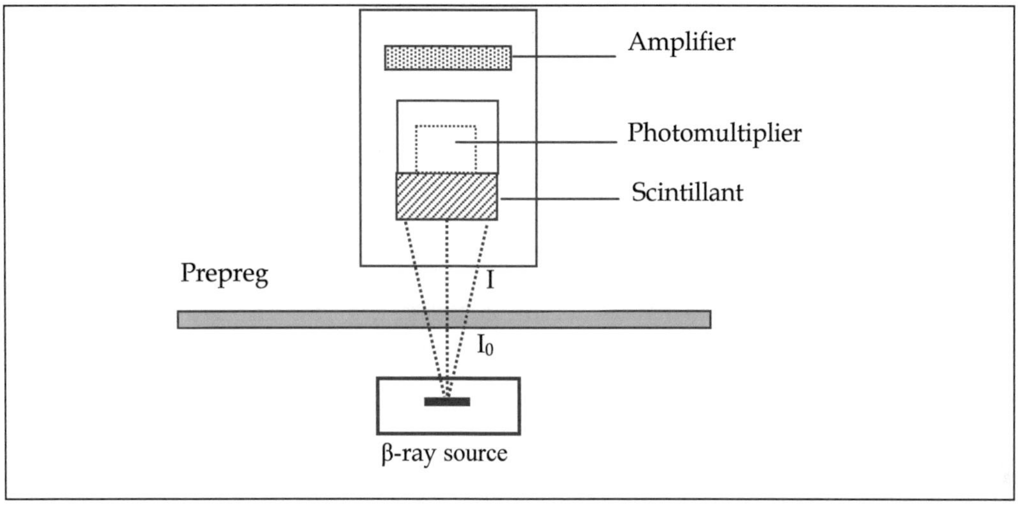

Fig. 2. Schematic of monitoring resin content of prepreg by β-ray technique (Huang et al., 1997)

2. γ-ray transmission technique (Teagle, 1983)

The principle of monitoring the resin content by γ-ray transmission is similar to the one by β-ray technique. When γ-ray penerate the prepreg, the photoelectric effect is occured accompanied with Compton Effect and elecron effect. The ray intesity is attenuated due to these three effects, the attenuation rule is accord with the exponent rule. Co60, Cs137 and Cd190 are usually used as the radiographic ray. The scintillation counter and the geiger counter are often used as dector. Due to the low resin density and the low upper measuring limit, the high precision of monitoring the resin content is difficult to obtain by using γ-ray transmission technique.

3. Photometry monitoring method (Smithgall, 1990)

The principle of monitoring the resin content by photometry technique is based on the light transmission. When the light passes through the prepregs, the transmitted light intensity is varied because the resin content of prepregs is different. Thus, the brightness received by the camera is different. Moreover, the change of resin content is judged by the different image. In the operation process, the image of prepreg which has the certain resin content is regarded as the standard image. Comparing the measured image with the standard one, the resin content is controlled by adjusting the processing parameters automatically.

The photometry technique is an on-line and non-contact monitoring method. Because its detector system is not limited to the running speed, this method is adaptable for the continuous production of the prpreg. However, the resin content is determined by receiving the transmitted light, the method is not used for the nontransparent carbon prepreg. For the transparent glass prepregs and semitransparent aramid prepregs, the ideal measurement accuracy is obtained.

4. Ultrasonic monitoring technique (Ma et al., 1991; Smith et al., 1990)

When the ultrasonic waves propagate in the prepregs, the ultrasonic delay is a function of prepreg within a definite range:

$$S_c=(S_m-S_f)V_m+S_f \quad (3)$$

where, S_c, S_m and S_f are the ultrasonic delay of prepreg, resin and fibre, respectively; V_m is the volume fraction of resin. If the ultrasonic delay of resin and fibre were known, the resin content could be calculated by the measured ultrasonic delay of prepreg.

In order to eliminate the effects caused by the air medium and the voids of prepreg, it is necessary that the ultrasonic probe and prepreg contact intimately under the force. Apparently, the ultrasonic monitoring method belongs to the contact measurement. It has the features of rapid and accurate when this method is used to determine the rather wide and thick prepreg, and the on-line monitoring can be realized in the production process. However, the accuracy of result is decided by the contact situation of the prepreg specimens and the ultrasonic probe, the partial failure on the testing prepreg may be occurred. In addition, the monitoring results are possibly occasional due to the semi soft feature of prepreg.

5. Infrared spectroscopy

The resin content and curing degree of the glass/epoxy prepreg has been continuously determined and controlled by using infrared transmission spectroscopy (Amy & Jiri, 1995). The wavenumbers at $4529cm^{-1}$ (absorption band of epoxy group) and $4055cm^{-1}$ (absorption band of methyl group) were generated by the filter. The prepreg was shined by using pulse mode, and the intensity of transmission signal was detected. The resin content of prepreg is calculated by the linear relation between the resin content and the logarithm of the transmission signal intensity at $4529cm^{-1}$. The curing degree was obtained by using the ratio of the transmission signal intensity at $4529cm^{-1}$ to that at $4055cm^{-1}$. By using the IR technique in the production, the variations of resin content and curing degree can be controlled within the ±3% in the 366 meters length range.

From the above depiction, it is apparently known that the volatile content could not be determined although the resin content and the pre-curing degree have been monitored successfully by the above mentioned methods. In the opened literatures (Li et al., 2005; Li et al., 2006a; Li et al., 2008), NIR spectroscopy in conjunction with chemometrics method has been applied to the monitoring of the three quality factors of the prepreg cloth.

5.2 Quality monitoring of the prepreg by NIR spectroscopy

The application of NIR spectroscopy for monitoring the production of glass/phenolic prepreg (Li et al., 2005; Li et al., 2006a; Li et al., 2008), carbon/phenolic prepreg (Li et al., 2006b) and carbon/epoxy prepreg (Jiang et al., 2007a; Jiang et al., 2007b) has been realized. In the production process of prepreg cloth with solution impregnation method, a FT-NIR spectrometer was assembled between the take-up mechanism and the drying tower, as shown in Fig. 3. (Li et al., 2006a; Li et al., 2008). The spectrometer could perform non-contact analysis of the product. In the test, the light from the sources was focused onto the prepreg cloth, and then the diffuse reflectance spectra from the prepreg cloth were recorded by the spectrometer. A gilded metal plate was placed under the prepreg cloth in order to enhance the diffuse reflectance effect.

The partial least square regression algorithm was used to build calibration models. The models of resin content, volatile content and pre-curing degree were developed respectively. The correlations between the predicted values and the actual values for the models are listed in Fig.4. (Li et al., 2006a). The models had determination coefficients (R^2) of 98.29 for the resin content, of 99.50 for the volatile content and of 97.66 for the pre-curing degree, respectively. The root mean square errors of prediction (RMSEP) for the resin content, the volatile content and the pre-curing degree were 0.376%, 0.169% and 0.105%, respectively.

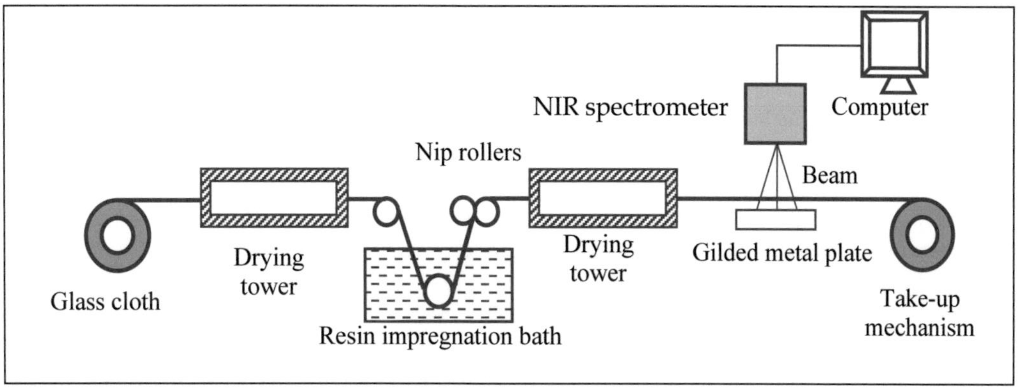

Fig. 3. Schematic of process for the manufacture of glass/phenolic prepreg cloth (Li et al. 2006a; Li et al., 2008)

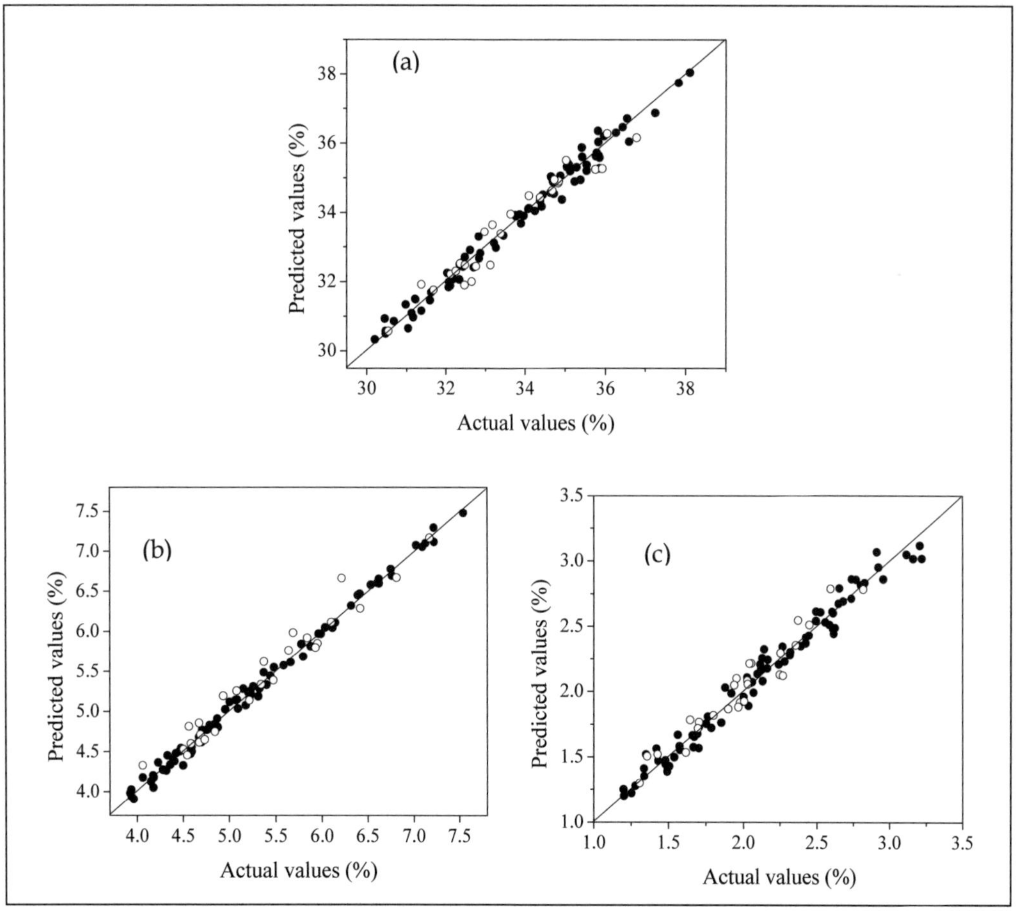

Fig. 4. Scatter plots showing correlation between chemical methods and NIR analyses for: (a) resin content; (b) volatile content; (c) pre-curing degree. Filled and open circles correspond to calibration and prediction data, respectively (Li et al. 2006a; Li et al., 2008)

In the manufacturing process, the prepreg quality is affected by processing parameters such as the concentration of resin solution, distance apart of the nip rollers, temperature of the drying tower, etc. In order to investigate the effects of processing parameters on prepreg quality, the NIR spectroscopy was used in-line for monitoring the change rules of resin content, volatile content and pre-curing degree when the processing parameters were adjusted deliberately (Li et al. 2008).

The NIR monitoring results indicated that the resin content increased with the increase of the resin concentration and the nip rollers separation. The pre-curing degree was hardly changed with these two adjustments. Meanwhile, it was also found that the volatile content increased with any increase of resin content. The reason was that the volatilisation started on the prepreg surface, the inner volatile component needed to overcome resistance when it diffused from the inner part to the surface. When there were much resin existed in the prepreg, the escape of volatile was difficult. The monitoring results also revealed that the pre-curing degree increased with the increase of the drying temperature and decreased with the increase of the production speed, and the change of volatile was opposite to that of pre-curing degree. The resin content was hardly affected by the change of drying temperature and production speed.

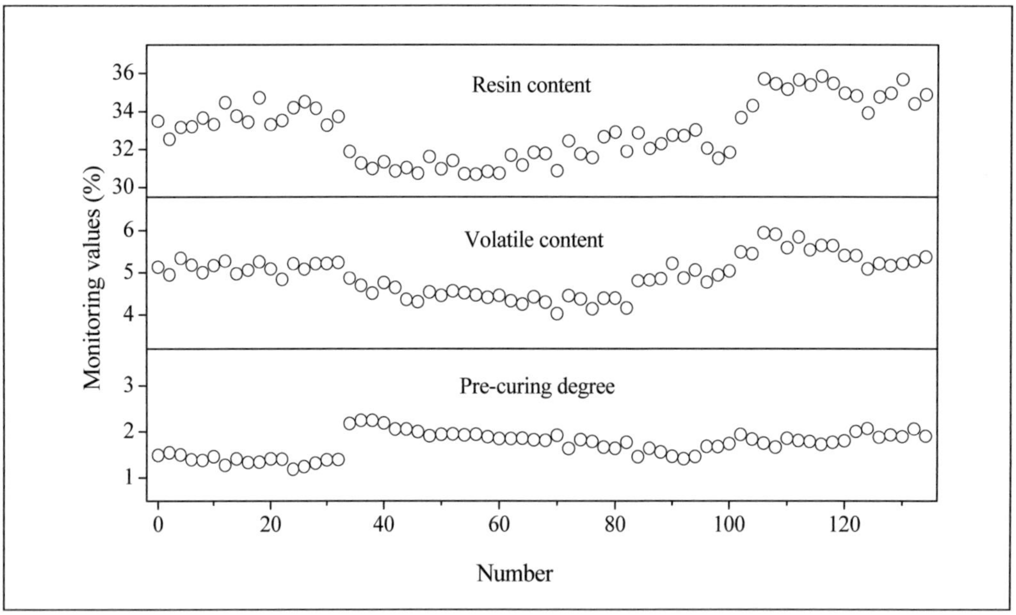

Fig. 5. Actual change of the glass/phenolic prepreg quality in the usual manufacture process (Li et al. 2008)

In the usual manufacturing process of prepreg cloth, the off-line traditonal analysis method are often used to control the prepreg quality. The actual change of the prepreg quality obtained from NIR monitoring results is illustrated in Fig. 5. It was shown that the prepreg cloths with low resin contents were produced over a long period, and the range of resin contents was rather wide. Obviously, the unqualified product was probably produced due to the delayed analysis results. In order to obtain high quality products, it was necessary to adjust appropriate processing parameters real time.

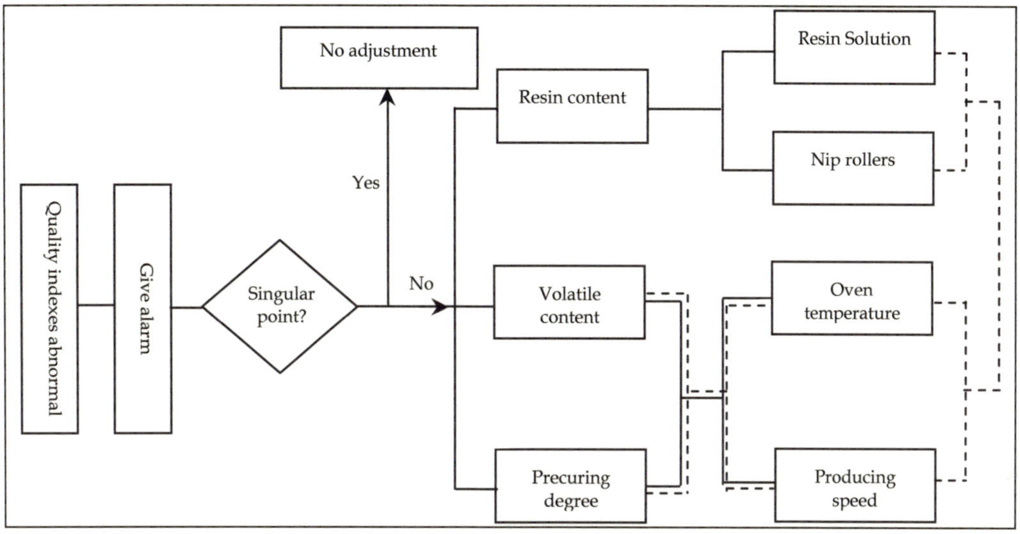

Fig. 6. Schematic of control of the prepreg quality

Fig. 7. Application of NIR for quality monitoring in the production of glass/phenolic prepreg cloth

Fig. 6 shows the schematic of control of prepreg quality. In the production process, the alarm was given when the NIR monitoring result of any quality index exceeded the preset ranges. At this moment, it was need to judge whether adjusted the processing parameters or not according to the change tendency of prepreg quality indexes. If it was only the individual singular point which appeared accidentally, there was no need to make adjustment. Otherwise, the adjustment must be carried on the processing parameters when the running tendency of prepreg quality index was changed. In case the resin content was not existed in the normal range, the general rule was to adjust the resin concentration and

the distance between nip rollers. As for the volatile content and pre-curing degree, it was required to change the oven temperature and production speed. However, it was note that the volatile content would be changed with the change of resin content by adjusting the resin concentration and the nip rollers separation. If the volatile content were changed too large, the adjustment of oven temperature and production speed was essential. In this case, the change of pre-curing degree would be occurred inevitably. It was apparently that the control of prepreg quality indexes was a complex process. The corresponding change of other two qualitative indexes should be monitored while one quality index was adjusted.
The photograph of NIR application in the production of glass/phenolic prepreg cloth is shown in Fig. 7. Based on the NIR on-line monitoring results, the quality controlling results obtained by adjusting the processing parameters are illustrated in Fig. 8.

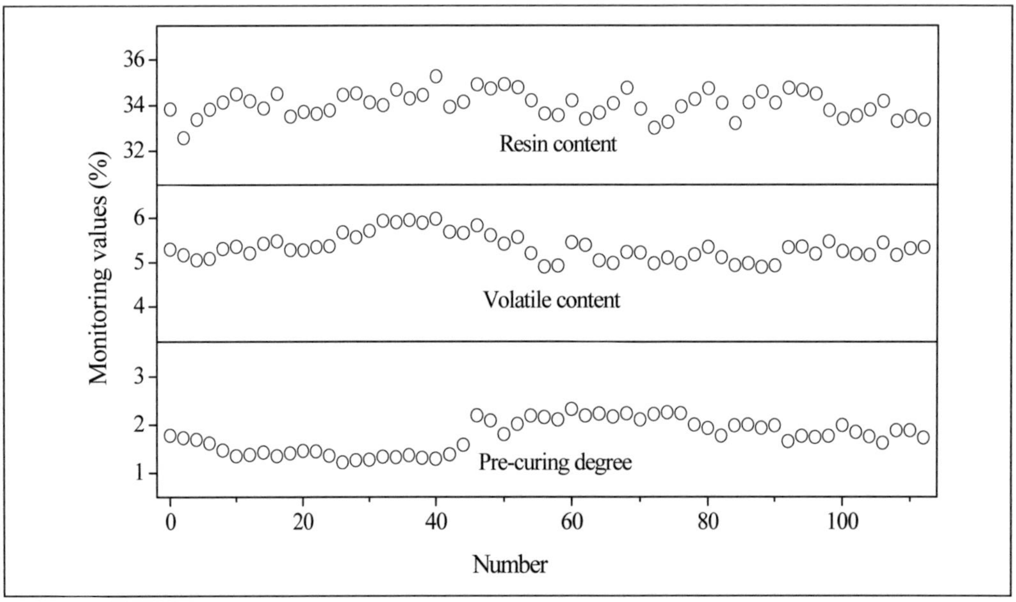

Fig. 8. Quality control of the glass/phenolic prepreg cloth by NIR method (Li et al., 2008)

From the Fig.8, it was known that the resin content, volatile content and pre-curing degree were varied within the relative narrow ranges. The results revealed that NIR spectroscopy was a suitable process monitoring technique for the production of prepreg.

6. Conclusion

Applications of the NIR technology for process monitoring of polymer systems have been summarized, which included polymerization process, extrusion process, curing process and manufacture process of prepregs. It is shown that there is a growing interest in utilization of NIRS as a tool for process monitoring in the polymer fields. The introduction of NIR technique is really helpful to improve the product quality. The polymer industry really needs rapid, reliable, non-invasive and cost effective analytical methods for process monitoring, such as NIR spectroscopy technique. Due to its intrinsic features, it is predicted that application of NIR technique in the polymer process monitoring will be become broader in the future.

7. References

Aldridge, P. K.; Kelly, J. J.; Callis, J. B. & Burns, D. H. (1993). Noninvasive Monitoring of Bulk Polymerization using Short-Wavelength Near-Infrared Spectroscopy. *Analytical Chemistry*, Vol. 65, No.24, 3581-3585, ISSN 0003-2700

Amy, M. L. & Jiri, K. D. (1995). Process for measurement of the degree of cure and percent resin of glass-fiber reinforced epoxy resin prepreg. *U S Patent*, 5457319

Batra, J.; Khettry, A. & Hansen, M. G. (1994). In-Line Monitoring of Titanium Dioxide Content in Poly(Ethylene Terephthalate) Extrusion. *Polymer Engineering and Science*, Vol. 34, No. 23, 1767–1772, ISSN 0032-3888

Butler, W. L. & Norris, K. H. (1960). The Spectrophotometry of Dense Light-Scattering Material, *Archives of Biochemistry and Biophysics*, Vol. 87, No.1, 31-40, ISSN 0003-9861

Cherfi, A. & Fevotte, G. (2002). On-line Conversion Monitoring of the Solution Polymerization of Methyl Methacrylate using Near-Infrared Spectroscopy. *Macromolecular Chemistry and Physics*, Vol. 203, No.9, 1188-1193, ISSN 1022-1352

Cherfi, A.; Fevotte, G. & Novat, C. (2002). Robust on-line measurement of conversion and molecular weight using NIR spectroscopy during solution polymerisation. *Journal of Applied Polymer Science*, Vol. 85, No.12, 2510-2520, ISSN 0021-8995

Coates, P. D. ; Barnesa, S. E. ; Sibleya, M. G. ; Browna, E. C. ; Edwardsb, H. G. M. & Scowenb, I. J.(2003). In-Process Vibrational Spectroscopy and Ultrasound Measurements in Polymer Melt Extrusion. *Polymer*, Vol. 44, No. 19, 5937–5949, ISSN 0032-3861

Dannenberg, H. (1963). Determination of functional groups in epoxy resins by near-infrared spectroscopy. *Polymer Engineering and Science*, Vol. 3, No. 1, 78-88, ISSN 0032-3888

De Faria, J. M. Jr.; Machado, F.; Lima, E. L. & Pinto, J. C. (2010a). In-Line Monitoring of Vinyl Chloride Suspension Polymerization with Near-Infrared Spectroscopy, 1-Analysis of Morphological Properties. *Macromolecular Reaction Engineering*, Vol. 4, No. 1, 11-24, ISSN 1862-832X

De Faria, J. M. Jr.; Machado, F.; Lima, E. L. & Pinto, J. C. (2010a). In-Line Monitoring of Vinyl Chloride Suspension Polymerization with Near-Infrared Spectroscopy, 2-Design of an Advanced Control Strategy. *Macromolecular Reaction Engineering*, Vol. 4, No. 8, 486-498, ISSN 1862-832X

Fischer, D.; Bayer, T.; Eichhorn, K.-J. & Otto, M. (1997). In-Line Process Monitoring on Polymer Melts by NIR-Spectroscopy. *Journal of Analytical Chemistry*, Vol. 359, No. 1, 74-77, ISSN 1061-9348

Fontoura, J. M. R.; Santos, A. F.; Silva, F. M.; Lenzi, M. K.; Lima, E. L.; Pinto, J. C. (2003). Monitoring and Control of Styrene Solution Polymerization using NIR Spectroscopy. *Journal of Applied Polymer Science*, Vol.90, No. 5, 1273-1289, ISSN 0021-8995

Goddu, R. F. & Delker, D. A. (1958). Determination of Terminal Epoxides by Near-Infrared Spectrophotometry. *Analytical Chemistry*, Vol. 30, No. 12, 2013-2016, ISSN 0003-2700

Gossen, P. D.; MacGregor, J. F. & Pelton, R.H. (1993). Composition and particle diameter for styrene/methyl methacrylate copolymer latex using UV and NIR spectroscopy. *Applied Spectroscopy*, Vol. 47, No. 11, 1852-1870, ISSN 0003-7028

Hansen, M. G. & Vedula, S. (1998). In-Line Fiber-Optic Near-Infrared Spectroscopy: Monitoring of Rheological Properties in an Extrusion Process. Part I. *Journal of Applied Polymer Science*, Vol. 68, No. 6, 859-872, ISSN 0021-8995

Hopewell, J.L.; George, G.A. &Hill, D.J.T. (2000). Quantitative Analysis of Bismaleimide-Diamine Thermosets using Near Infrared Spectroscopy. *Polymer*, Vol. 41, No. 23, 8221–8229, ISSN 0032-3861

Huang, Y. D.; Liu, L.; Zhang, Z. Q. & Wang, Y. (1997). On-line Monitoring of Resin Content for Film Impregnation Process. *Composites Science and Technology*, Vol. 58, No. 9, 1531-1534, ISSN 0266-3538

Huang, Y. D.; Sun, Y. F. & Liu, L. (2003). Beta Ray Monitoring Technique for Control of Resin Content in Continuous Fibre Prepreg Materials. *Material Science and Technology*, Vol. 19, No. 6, 815-818, ISSN 0267-0836

Jiang, B.; Huang, Y. D.; Li, W. & Liu, L.(2007a). Non-Contact Quality Monitoring of Laid Fabric Carbon/Epoxy Resin Prepreg using Near Infrared Spectroscopy. *Journal of Near Infrared Spectroscopy*, Vol. 15, No. 5, 299-305, ISSN 0967-0335

Jiang, B.; Huang, Y. D.; Li, W. & Liu, L.(2007b). Non-Destructive and Rapid Analysis of Resin and Volatile Contents in Carbon Fiber/Epoxy Resin Prepreg Cloth by Near-Infrared Spectroscopy. *Iranian Polymer Journal*, Vol. 16, No. 5, 319-326, ISSN 1026-1265

Kortaberria, G. ; Arruti, P. ; Gabilondo, N. & Mondragon, I. (2004) . Curing of an epoxy resin modified with poly(methylmethacrylate) monitored by simultaneous dielectric/ near infrared spectroscopies. *European Polymer Journal*, Vol. 40, No. 1, 129-136, ISSN 0014-3057

Lee, K. A. & Kradjel, C. (2007). NIR Analysis of Polymers. In: *Handbook of Near-Infrared Analysis*, Burns & Ciurczak, (Ed.), 3rd Ed., 529-568, CRC Press, ISBN 978-0-8493-7393-0, New York

Lachenal G. (1995). Dispersive and Fourier Transform Near-Infrared Spectroscopy of Polymeric Materials. *Vibrational Spectroscopy*, Vol. 9, No. 1, 93-100, ISSN 0924-2031

Lachenal, G. & Ozaki, Y. (1999). Advantages of Near Infrared Spectroscopy for the Analysis of Polymers And Composites. *Macromolecular Symposia*, Vol. 141, No. 1, 283-292, ISSN 1022-1360

Li, W.; Huang, Y. D.; Liu, L. & Chen, N. T.(2005). Rapid and Nondestructive Analysis of Quality of Prepreg Cloth by Near Infrared Spectroscopy. *Composites Science and Technology*, Vol. 65, No. 11-12, 1668-1674, ISSN 0266-3538

Li, W.; Huang, Y. D.; Liu, L. & Jiang, B.(2006a). The Application of Near Infrared Spectroscopy in the Quality Control Analysis of Glass/Phenolic Resin Prepreg. *Journal of Materials Science*, Vol. 41, No. 21, 7183-7189, ISSN 0022-2461

Li, W.; Huang, Y. D.; Liu, L. & Jiang, B. (2006b). On-Line Monitoring of Resin Content and Volatile Content in Carbon/Phenlic Resin Prepreg Cloth by Near-Infrared Spectroscopy. *Polymers and Polymer Composites*, Vol. 14, No. 5, 537-543, ISSN 0967-3911

Li, W.; Huang, Y. D. & Chen, P. (2008). NIR Spectroscopy: A Useful Tool for Quality Control of Glass/Phenolic Prepreg Manufacture. *Polymers and Polymer Composites*, Vol. 16, No. 1, 55-62, ISSN 0967-3911

Lousberg, H. H. A.; Boelens, H. F. M.; Le Comte, E. P.; Hoefsloot, H.C. J.; Smilde, A.K. (2002). On-Line Determination of the Conversion in A Styrene Bulk Polymerization Batch Reactor using Near-Infrared Spectroscopy. *Journal of Applied Polymer Science*, Vol. 84, No. 1, 90-98, ISSN 0021-8995

Ma, L. G. & Campbell, M. D. (1991). An Instrument for the Destructive Determination of Fiber/Resin Ratio in Fiber-Reinforced Composite Prepreg Materials. *Proceedings of the 7th Annual ASTM/ESD Advanced Composite Materials Conference*, pp. 377-381, ISBN 0871704315, Detroit, Octomber, 1991, ASM International, Detroit

Mijović, J. & Andjelić, S. (1996). Study of the Mechanism and Rate of Bismaleimide Cure by Remote in-Situ Real Time Fiber Optic Near-Infrared Spectroscopy. *Macromolecules*, Vol. 29, No.1, 239–246, ISSN 0024-9297

Mijović, J. ; Andjelić, S. & Kenny J. M. (1998). In situ Real-time Monitoring of Epoxy/Amine Kinetics by Remote Near Infrared Spectroscopy. *Polymers for Advanced Technologies*, Vol. 7, No. 1, 1-16, ISSN 1042-7147

Nagata,T. ; Tanigaki, M. & Ohshima, M. (2000a). On-Line NIR Sensing of CO_2 Concentration for Polymer Extrusion Foaming Processes. *Polymer Engineering and Science*, Vol.40, No.8., 1843-1849, ISSN 0032-3888

Nagata,T. ; Ohshima, M. & Tanigaki, M. (2000b). In-Line Monitoring of Polyethylene Density using Near Infrared Spectroscopy. *Polymer Engineering and Science*, Vol. 40, No.5, 1107–1113, ISSN 0032-3888

Nogueira, E. S.; Borges, C.P. & Pinto, J. C. (2005). In-Line Monitoring and Control of Conversion and Weight-Average Molecular Weight of Polyurethanes in Solution Step-Growth Polymerization Based on Near Infrared Spectroscopy and Torquemetry. *Macromolecular Materials and Engineering*, Vol. 290, No.4, 272–282, ISSN 1438-7492

Norris, K.H. & Butler, W.L. (1961).Techniques for Obtaining Absorption Spectra on Intact Biological Samples. *IRE Transactions on Biomedical Electronics*, Vol. 8, No. 3, 153, ISSN 0096-1884

Norris, K. H. (1989). Introduction, In *Near Infrared Reflectance Spectroscopy (NIRS): Analysis of Forage Quality*. 6, Martens, Shenk & Barton (Ed.), US Government Printing Office, Washington, D.C.

Park, S. J.; Kwak, G. H.; Sumita M. & Lee J. R. (2000). Cure and Reaction Kinetics of an Anhydride-Cured Epoxy Resin Catalyzed by N-benzylpyrazinium Salts Using Near-Infrared Spectroscopy. *Polymer Engineering and Science*, Vol. 40, No. 12, 2569-2576, ISSN 0032-3888

Reis, M. M.; Araújo, P. H. H.; Sayer, C. &and Giudici R. (2004). In Situ Near-Infrared Spectroscopy for Simultaneous Monitoring of Multiple Process Variables in Emulsion Copolymerization. *Industrial and Engineering Chemistry Research*, Vol. 43, No. 23, 7243-7250, ISSN 0888-5885

Santos, A.F.; Lima, E.L. & Pinto, J.C. (1998). In-line Evaluation of Average Particle Size in Styrene Suspension Polymerizations using Near-Infrared Spectroscopy. *Journal of Applied Polymer Science*, Vol. 70, No. 9, 1737-1745, ISSN 0021-8995

Santos, A.F.; Lima, E.L. & Pinto, J.C. (2000). Control and Design of Average Particle Size in Styrene Suspension Polymerization using NIRS. *Journal of Applied Polymer Science*, Vol. 77, No. 2, 453-462, ISSN 0021-8995

Schultz, T.P.; Burns, D.A. (1990). Rapid Secondary Analysis of Lignocellulose: Comparison of Near Infrared (NIR) and Fourier Transform Infrared (FTIR). *Tappi Journal*, Vol. 73, No. 5, 209-212, ISSN 0734-1415

Smith, D. R. Jr.; Monson, J.; Reynolds, B. & Johnsen, R. (1990). An Evaluation of an Ultrasonic Technique for Measuring Resin Content in Prepreg Material. *22nd*

International SAMPE Technical Conference, pp. 237-250, ISBN 0938994549, Boston, November, 1990, Society for the Advancement of Material and Process Engineering, Boston

Smithgall, F. B. (1990). Automatic Inspection of Advanced Composite Tape, *35th International SAMPE Symposium and Exhibition*, pp. 1880-1891, ISBN 0938994522, Anaheim, April, 1990, Society for the Advancement of Material and Process Engineering, Anaheim

Teagle, P. R. (1983). The quality control and non-destructive ebaluation of composite aerospace components. *Composites,* Vol. 14, No. 2., 115-128, ISSN 0010-4361

Rohe, T, ; Becker, W. ; Kölle, S. ; Eisenreich, N. & Eyerer, P. (1999). Near Infrared (NIR) Spectroscopy for In-Line Monitoring of Polymer Extrusion Processes. *Talanta,* Vol. 50, No. 2, 283–290, ISSN 0039-9140

Rohe, T, ; Becker, W. ; Krey, A. ; Nägele,H. , Kölle, S. & Eisenreich, N. (1998). In-line monitoring of polymer extrusion processes by NIR spectroscopy. *Journal of Near Infrared Spectroscopy,* Vol. 6, No. 1, 325-332, ISSN 0967-0335

Vedula, S. & Hansen, M. G. (1998). In-Line Fiber-Optic Near-Infrared Spectroscopy: Monitoring of Rheological Properties in an Extrusion Process. Part II. *Journal of Applied Polymer Science,* Vol. 68, No. 6, 873-889, ISSN 0021-8995

Vieira, R. A. M.; Sayer, C.; Lima, E. L. & Pinto, J. C (2001). Detection of monomer droplets in a polymer latex by near-infrared spectroscopy. *Polymer,* Vol. 42, No. 21, 8901-8906, ISSN 0032-3861

Vieira, R.A.M.; Sayer, C.; Lima, E.L. & Pinto, J.C. (2002a). In-line and In-situ Monitoring of Semi-Batch Emulsion Copolymerizations using Near-infrared Spectroscopy. *Journal of Applied Polymer Science,* Vol. 84, No. 14, 2670-2682, ISSN 0021-8995

Vieira, R.A.M.; Sayer, C.; Lima, E.L. & Pinto, J.C. (2002b). Closed-loop Composition and Molecular Weight Control of A Copolymer Latex using Near-infrared Spectroscopy. *Industrial and Engineering Chemistry Research,* Vol. 41, No.12, 2915 - 2930. ISSN:0888-5885

Xu L. ; Fu J. H. & Schlup J. R. (1994). In Situ Near-Infrared Spectroscopic Investigation of Epoxy Resin-Aromatic Amine Cure Mechanisms. *Jouranl of American Chemical Society,* Vol. 116, No. 7, 2821-2826, ISSN 0002-7863

28

Digital Image Processing for Quality Control on Injection Molding Products

Marco Sasso, Massimo Natalini and Dario Amodio
Università Politecnica delle Marche
Italy

1. Introduction

The need to increase quality of products, forces manufacturers to increase the level of control on finished and semi-finished parts, both qualitatively and quantitatively.
The adoption of optical systems based on digital image processing is an effective instrument, not only for increasing repeatability and reliability of controls, but also for obtaining a large number of information that help the easily management of production processes.
Furthermore, the use of this technology may reduce considerably the total amount of time needed for quality control, increasing at the same time the number of inspected components; when image acquisition and post-processing are feasible in real time, the whole production can be controlled.
In this chapter we describe some quality control experiences carried out by means of a cheap but versatile optical system, designed and realized for non contact inspection of injection moulded parts.
First, system architecture (both hardware and software) will be showed, describing components characteristics and software procedures that will be used in all the following applications, such as calibration, image alignment and parameter setting.
Then, some case studies of dimensional control will be presented. The aim of this application is to identify and measure the main dimensional features of the components, such as overall size, edges length, hole diameters, and so on. Of particular interests, is the use of digital images for evaluation of complex shapes and dimension, where the distances to be measured are function of a combination of several geometric entities, making the use of standard instrument (as callipers) not possible. At the same time, methods used for image processing will be presented. Moreover, a description of system performances, related to quality product requirements will be presented.
Second application examines the possibility to identify and quantify the entity of burrs. The case of a cylindrical batcher for soap, in witch its effective cross-sectional area have to be measured will be showed, together with a case study in witch burrs presence could bring to incorrect assembly or functionality of the component. Threshold and image subtraction techniques, used in this application will be illustrated together with the big number of information useful to manage production process.
Third, it will be presented a possible solution to the problem of identifying general shape defects caused by lacks of filling or anomalous shrinkage. Two different approaches will be

used, the former that quantifies the general matching of whole images and the latter that inspects smaller areas of interest.
Finally, an example of colour intensity determination of plastic parts for aesthetic goods will be presented. This application has the aim to solve the problem of pieces which could appear too dark or too light with respect to a reference one, and also to identify defects like undesired striation or black point in the pieces, depending on mixing condition of virgin polymer and masterbatch pigment. Use of pixel intensity and histograms have been adopted in the development of these procedures.

2. System description

In this section, the hardware and software architecture of the system will be described.
The hardware is composed by a camera, a telecentric zoom lens, two lights, a support for manual handling of pieces, and a PC.
The camera is a monochromatic camera with a CCD sensor (model AVT Stingray F201B ®). The sensor size is 1/1.8", and its resolution is 1624 x 1234 pixel, with a pixel dimension of 4,4 µm, a colour depth of 8 bit (256 grey levels) and a maximum frame rate of 14 fps at full resolution. The choice of a CCD sensor increases image quality, reducing noise in the acquisition phase and the high resolution (about 2 MPixel) leads to an acceptable spatial resolution in all the fields of view here adopted.
The optics of the system is constituted by a telecentric zoom lens (model Navitar 12X telecentric zoom.®); the adopted lens is telecentric because it permits to eliminate the perspective error, which is a very useful property if one desires to carry out accurate measurements. The zoom system permits to have different field of view (FOV), so the FOV can vary from a minimum of 4,1 mm to a maximum of 49,7 mm. The zoom is moved by a stepper driving, and the software can communicate and move it automatically to the desired position. The utility of this function will be clear later, when the start-up procedure is explained.
The depth of focus varies with the FOV, ranging from 1,3 mm to 38,8 mm; the mentioned camera and lens features bring to a maximum resolution of the system of 0,006 mm (FOV 4,1 mm) and a minimum resolution of 0,033 mm (FOV 49,7 mm).
To lights the scene, a back light and a front light were adopted. Both have red light, to minimize external noise and to reduce chromatic aberration. Moreover they have a coaxial illumination, that illumines surface perpendicularly and increases contrast in the image highlighting edge and shapes and improving the general quality of the image. In figure 1.a the complete system is showed. All the components are managed by a PC, that use a specific software developed in LabView®.
Its architecture is reported in figure 1.b. It has a user interface, that guides step by step the operator through the procedures for the different controls. So, the operator is called only to choose the application to use, then to load pieces in the work area and to shot photos. Every time he shots, image is acquired and processed (and stored if necessary), so the result is given almost immediately. If the operator needs to control a production batch, it is also possible to acquire several images consecutively and then post-process all of them exporting global results in an excel file.
All this operations are possible thank to the background software, that answer to the operator input. In fact, when the operator select the kind of control, the software load all the parameters necessary to the analysis. For each application, the software load all parameters

for camera, zoom and lights (that have been stored earlier), and all the information about the analysis, like calibration, templates for image alignment and so on.

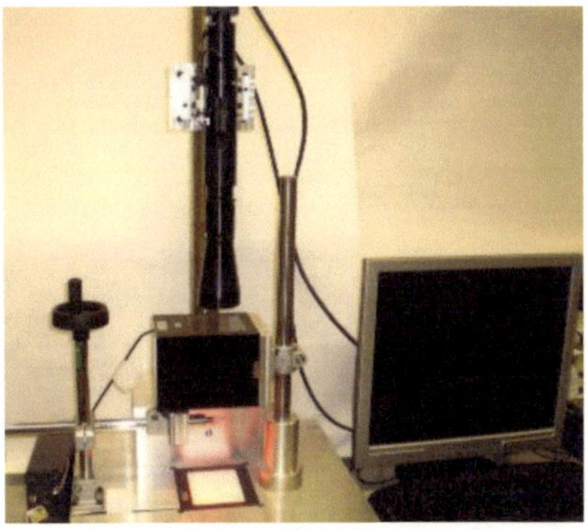

a)

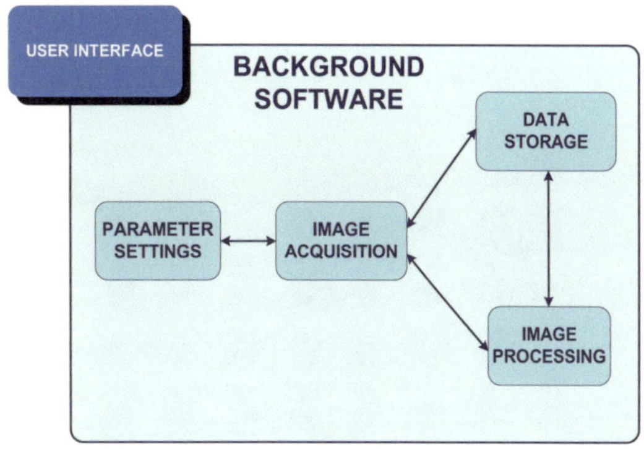

b)

Fig. 1. a) Hardware architecture; b) Software architecture

With regard to camera settings, the software controls all parameters like time exposure, brightness, contrast and so on, so when the best mix of parameters had been determined for a given application, for the analysis is enough to call it back. The same happens for the zoom control; in fact the software loads the position stored for the selected application and commands to the zoom driver to move in that position. This is useful because, if necessary, the system can acquire images of a large FOV to get information about certain features, and then narrows the FOV and acquire images with higher spatial resolution for capturing smaller details in the observed object. Each position utilised has been calibrated sooner. When a position is called, the software restores related parameters, and pass them to the following step for elaboration. The software also control light intensity, in the same way of previous components. So, all the information are passed to the acquisition step and then stored in the acquired images.

After this overview of the system, it's proper to describe two functions that are used in all applications before any other operation: image calibration and image alignment.

2.1 Image calibration

Using digital images, two kind of calibration are necessary: spatial calibration (always) and illumination and acquisition parameters calibration (depending on material and shape of the pieces to be analysed).

Spatial calibration convert pixel dimensions into real world quantities and is important when accurate measurements are required. For the application described below, a perspective calibration method was used for spatial calibration (NI Vision concept manual, 2005). So, the calibration software requires a grid of dots with known positions in image and in real world. The software uses the image of the grid and the distance between points in real world to generate a mapping function that "translates" the pixel coordinates of dots into the coordinates of a real reference frame; then the mapping can be extended to the entire image. Using this method is possible to correct perpendicularity error of camera and scene which is showed in figure 2a. This effect is actually rather reduced in the present system, as the support has been conceived to provide good mechanical alignment by means of a stiff column, that sustains camera in perpendicular position with respect to the scene.

This method of calibration is however useful and must be used to correct also alignment error (or rotation error) between image axis and real world axis (fig. 2b). It is also possible to define a new reference system in the best point/pixel for the studied application (for example, at the intersection of two edge of the piece analysed).

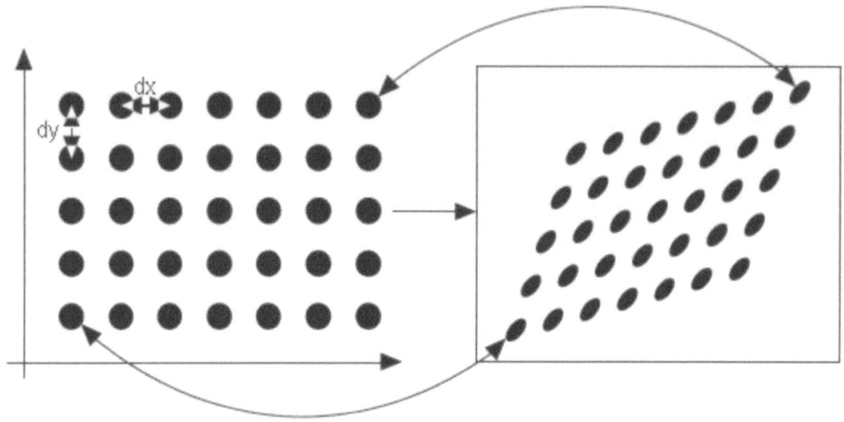

a) Perpendicularity error

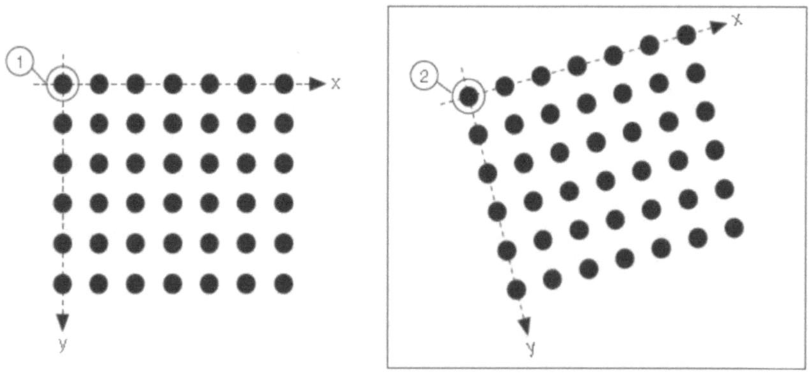

b) Rotation error

Fig. 2. Errors to correct in image calibration

About parameter setting, the problem is represented by transient materials because the light passes through the material and the dimension measured changes with intensity of illumination. So a simple method was used to correct this effect. A master of the pieces to analyse was manually measured with calliper, and the acquisition parameters (particularly brightness, contrast ant time exposure of the camera) were selected to obtain the same results with digital image measurement. All this parameters are stored for each application.

2.2 Image alignment

The other operation computed by the software before each analysis is the image alignment. This operation simplifies a lot the pieces positioning and make the analysis much easier and faster for the operator. In fact the pieces have to be positioned manually in the FOV so it is very difficult to put them always in the same position to permit the software to find features for measurement. In order to align every image with the master image that was used to develop the specific analysis tool, it could be even possible to put a given piece in any position within the field of view, and let then the software to rotate and translate the image to match the reference or master image thus to detect features and dimensions. However, for the sake of accuracy and repeatability, the positioning is aided by centring pin and support that permit to place the objects in positions that are similar to the master one used for calibration and parameters setting.

The alignment procedure is based on pattern (or template) matching and uses a cross-correlation algorithm. So, first is necessary to define a template (fig. 4b) that the software consider as a feature to find in the image. From this template, the software extracts pixels that characterize the template shape, then it looks for the pixel extracted in the new image, using an algorithm of cross-correlation; so is possible to consider the template as a sub image $T(x,y)$ of size K x L in a bigger image $f(x,y)$ of size M x N (see fig. 4), and the correlation between T and f at the pixel (i,j) is given by (J. C. Russ, 1994):

$$C(i,j)=\frac{\sum_x\sum_y T(x,y)\cdot f(x+i,y+j)}{\sqrt{\sum_x\sum_y f^2(x+i,y+j)\cdot\sum_x\sum_y T^2(x,y)}} \tag{1}$$

Correlation procedure is illustrated by fig.3. Correlation is the process of moving the template $T(x,y)$ around the image area and computing the value C in that area. This involves multiply each pixel of the template by the image pixel that it overlaps and then summing the results over all the pixels of the template. The maximum value of C indicates the position where T best matches f.

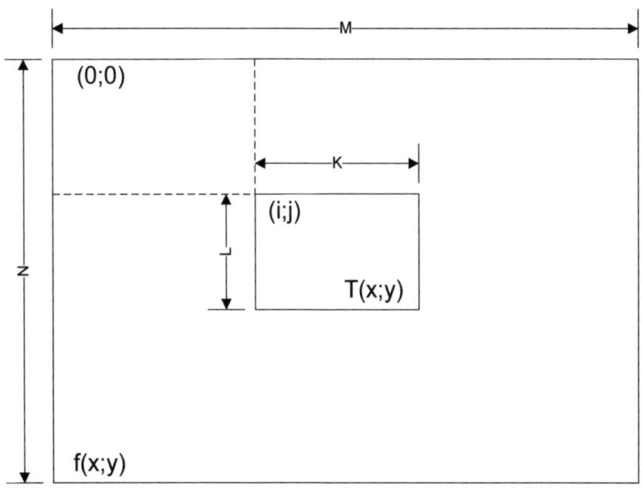

Fig. 3. Correlation process

This method requires a big number of multiplications, but is possible to implement a faster procedure: first, the correlation is computed only on some pixels extracted from the template, to determines a rough position of the template; then, the correlation over all pixels of the template can be executed in a limited area of the entire image, allowing to reduce processing time.

In fig. 4 the method applied to the first case of study is showed. Fig. 4.a reports the master image, from which the template (fig. 4b) has been extracted. It's clear, that only a part of the master image has been extracted and this part is considered as the feature to be searched in the entire image. In fig. 4.c is showed an image of a piece to align with the master, in fact, in this image, the piece has been moved behind and rotated. The software, first search the template and determines his rotation angle, rotates the image of the same entity to align the new image to the master. Then it finds the new position of the template and determines the translation vector from the master, moves the image of the fixed quantity and the new image is now aligned to the master. Black areas in fig. 4.d, are the results of translation and rotation of image; they are black because these areas have been added by the process, and a part of the image has been deleted to keep the same size of the original image. A double pattern matching is necessary, because of the image reference system is located on the left top of the image and not in the template centre. So, first pattern matching determines the rotation angle and the translation vector that have to be applied but uses only the first to rotate the image. Performing this operation in fact, the new image has the same alignment of the master image, but the translation vector changes because the rotation is performed respect to the left up corner of the image. Second pattern matching determines yet the angle (that now is zero) and the new translation vector that is used to move the image to the same position of the master. Image can be now processed.

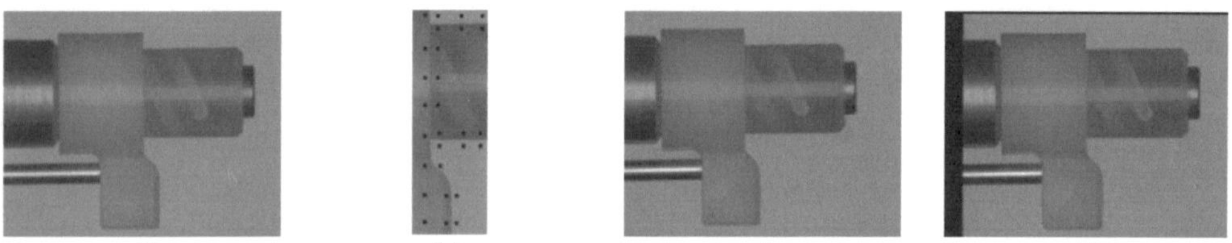

a) b) c) d)

Fig. 4. The alignment process applied to the first case of study

3. Dimensional measurement

This section describes some examples of dimensional measurement on injection moulded parts.

In fig.5.a it is shown a 3D simplified model of the part to analyse. It is constituted by two coaxial cylinders, with a ribbing on the left side; the material is Polyamide 66 (Zytel 101). In fig. 5.b, all the features that have to be measured in quality control process are represented. It's important to notice that there are two kind of feature to measure. The dimensions denoted by numbers 1, 3 and 4 require the identification of two simple feature (edges) of the part, so this will be a "simple feature" to measure, and a comparison with standard instrument is possible. Instead, feature number 2, requires the identification of an axis, by identification of two edges, and the identification of third element (edge of the ribbing) to measure the distance from the latter to the previous axis. So, three features are required and is impossible to get this measurement with standard instruments like gauges or callipers. This represent a second kind of measurement, that will be called "composed feature".

Both cases pass through the identification of edges, so the procedure used for their detection will be now described.

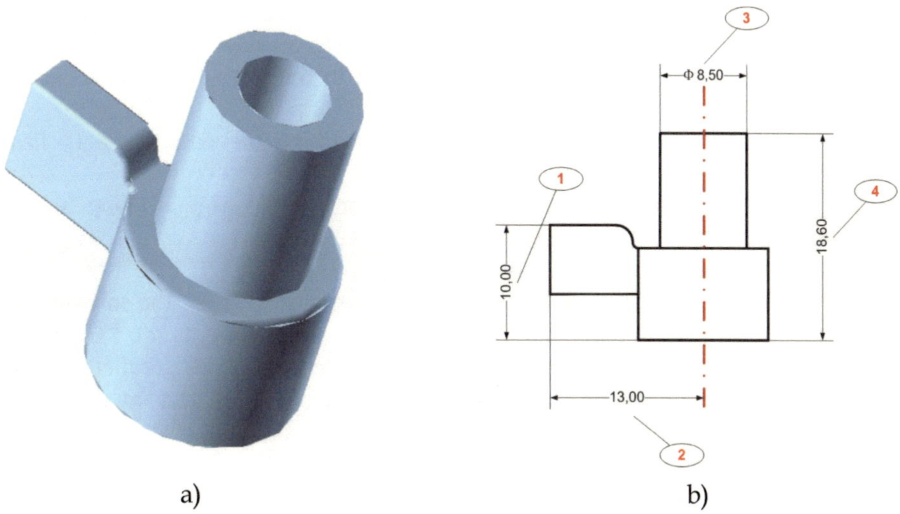

Fig. 5. a) the piece to analyse, b) requested features

More techniques are based on use of pixel mask that runs through the image, computing the sum of products of coefficients of pixel mask with the gray level contained in the region encompassed by the mask. The response of the mask at any point of the image is (Gonzales & Woods, 1992):

$$R = w_1 z_1 + w_2 z_2 + ... + w_9 z_9 = \sum_{i=1}^{9} w_i z_i \quad (2)$$

where: w_i = coefficient of the pixel mask;
z_i = gray intensity level of the pixel overlapped.

Using different kind of mask, different features can be detected. All of them, are detected when:

$$|R| > T \quad (3)$$

where T is a nonnegative threshold. Fig. 6 shows different masks for detection of different features.

a)Point

-1	-1	-1
-1	8	-1
-1	-1	-1

b)Horizontal edge

-1	-1	-1
2	2	2
-1	-1	-1

c)Vertical edge

-1	2	-1
-1	2	-1
-1	2	-1

d)+45° edge

-1	-1	2
-1	2	-1
2	-1	-1

e)-45° edge

2	-1	-1
-1	2	-1
-1	-1	2

Fig. 6. Different pixel mask for different features

In this application, an easier method, based on pixel value analysis along a pixel line, has been used. It is a simplification of derivative operators method (Gonzales & Woods, 1992) that uses gradient operators and analyses gradient vector to determine module and direction of the vector.

Usually, an edge is defined as a quick change in pixel intensity values, that represents boundaries of an object in the FOV. It can be defined by four parameters:

1. Edge strength: defines the minimum difference in the greyscale values between the edge and the background;
2. Edge length: defines the maximum distance in which the edge strength has to be verified;
3. Edge polarity: defines if the greyscale intensity across the edge is rising (increase) or falling (decrease);
4. Edge position: define x and y location of an edge in the image.

In picture 7.a is represented a part of an edge (black rectangle); first, the software requires the input of a user-defined rectangle (green in fig 7.a) that it fixes as the ROI in which to look for the edge. Then the software divides the rectangle using lines parallel to a rectangle edge, in the number specified by the user (red arrows in fig. 7.a), and analyses greyscale value of each pixel line defined, moving from the start to the end of the arrow if a rising edge is expected, vice versa if a falling edge is expected. Now It defines a steepness parameter, that represents the region (the number of pixels) in which the edge strength is expected to verify. Then, the software averages the pixel value of determinate number (width of filter) of pixel before and after the point considered. The edge strength is computed as the difference between averaged value before and after edge steepness. When it finds an edge strength higher than expected edge strength, it stores the point and continues with the analysis until the maximum edge strength is reached. Now, the point found in this way is tagged as edge start and the steepness value is added to find the edge finish. Starting from edge finish, the first point where the greyscale value exceeds 90% of starting greyscale value is set as edge position. Figure 7.b show the determination process of the edge and fig. 7.a shows the edge position determinate (yellow points).

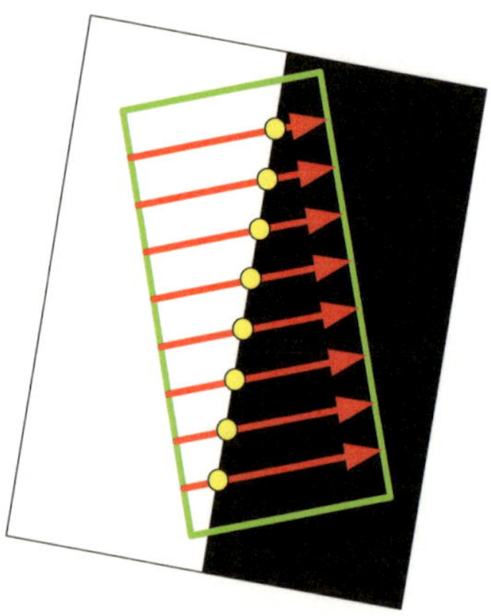

ROI used for edge detection

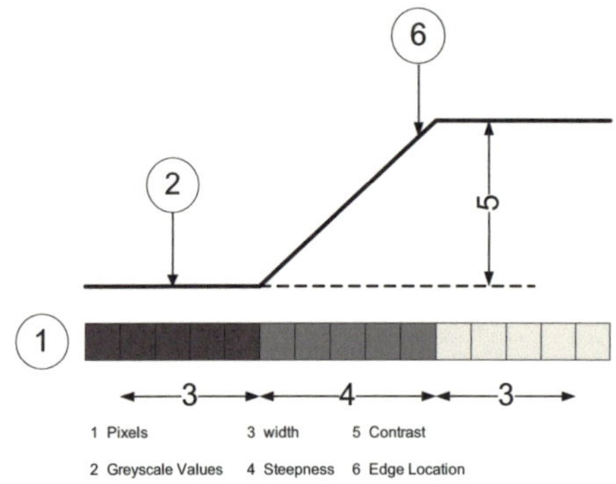

Edge position determination

Fig. 7. Determination process of edge position

It's clear that parameter settings influence edge position, especially when translucent material are analysed. In fact, translucent materials are characterised by lower greyscale intensity changes and high steepness values. If external noise cannot be eliminated appropriately, it could be difficult to get high repeatability in edge positioning in the image.

This can be avoided applying filters to the image before edge determination process. In this application, a filter to increment contrast was applied. This allow to reduce the steepness to two pixels only, and the filter width also to four pixels only. In fig.8 filter effects on the image are shown; fig. 8.a represents the original image acquired, fig. 8.b reports filtered image with increased contrast.

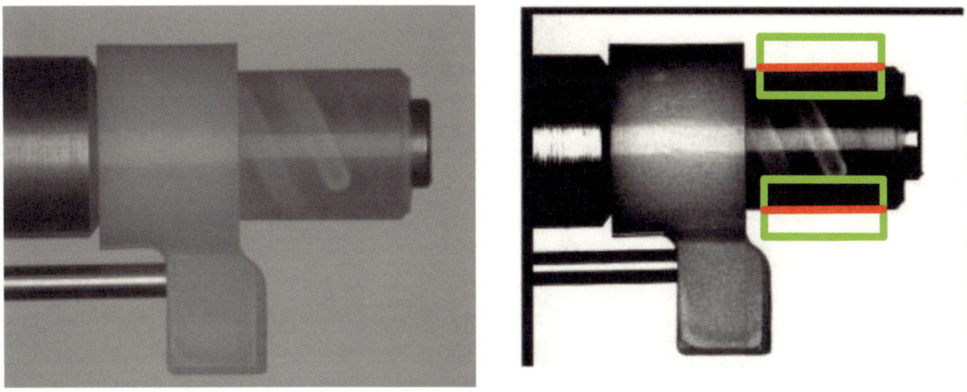

a) b)

Fig. 8. a) original image, b) aligned and filtered image

To increment contrast in the image, it's also possible to increment illumination intensity but in this case the light passes through the material in a huge way and the dimension of the part will be undervalued, so this method have to be used carefully.
This process have to be done confronting image results with contact methods or other suitable techniques; in particular, standard caliper measurement was used here.

3.1 Simple feature measurement

In this section, measurement process of simple feature will be illustrated. A simple feature is a feature that requires the identification of two elements, then some dimensional or geometric information are extracted by them. Features number 1, 3 and 4 of fig. 6.b are representative of this case. Feature number 1 will be treated as example. The objective is to determine the diameter of the part. As previously explained, two search areas have to be determined. The region may have the desired shape (circular, rectangular, square, polygonal) depending on the feature to inspect. In this case, because a line is searched, two rectangular ROI are defined (green rectangles in fig. 8.b). The ROI has fixed size and fixed position in the image; hence, alignment process is important in order to bring the part always in the same position in the image, permitting the software to find the edge at all times, with high reliability and robustness. A rake division of five pixels has been used (distance between each inspection line), it means that a point every 0,085 mm is stored (about 40 points per line). When points are determined, they are used to fit a line with a linear regression method and to determine the equation of the line in the image.
Once the equation has been determined, the process is repeated for the second edge, to determine the second equation. The rectangular search area is the same and has the same x position; this is important to define distance in following steps.
Now, using the equations is possible to find all the information required analytically. Defining first line as reference line, maximum and minimum distance between the lines has

been computed (the distance between extreme points) and has been averaged, to find medium distance between edge; also parallelism error can be evaluated as the difference between maximum and minimum edge distances. For each piece, two information have been obtained: medium distance and geometric error.
The edge determination is applicable also to other feature, such as holes; indeed, in the following case study the problem of a threaded hole is treated, where the inner diameter has to be measured.

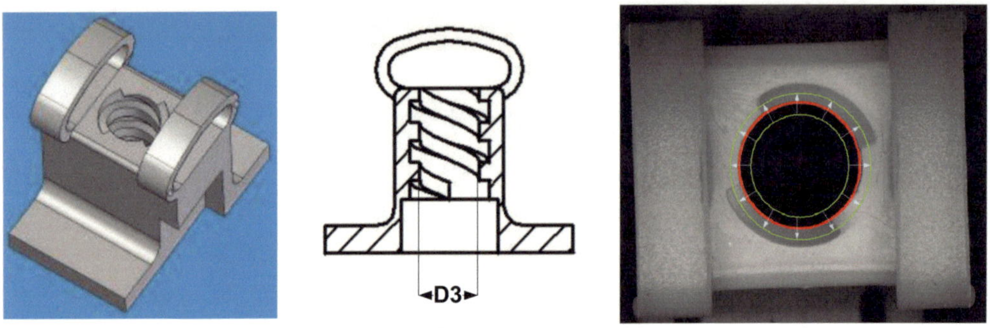

a) b) c)

Fig. 9. A problem of inner diameter determination

Traditional methods used for control process employs go-no-go gauges for diameter verification. So it is possible to say that inner diameter is comprised between established tolerance range (minimum and maximum dimension of gauges) but it's impossible to get an exact value of the diameter. The use of a standard gauge is difficult because there are not plane surfaces. Digital images offers a good solution. It is possible to determine a circular edge with the same method explained before, having care of changing the search area which now has to be annular (green circles in fig. 9.c), with inspection lines defined by their angular pitch along the circumference (white arrows in fig. 9.c). The result is the red circle reported again in fig. 9.c, which is the circle that best fits the detected circumference points.

3.2 Composed feature measurement

In this section, the measurement of feature 2 (fig. 5.b) will be explained briefly, as it can be carried out by a simple extension of the procedures already developed.
Now, the aim is to determine the distance between the ribbing edge and the axis of the hollow cylinder identified before and whose diameter has been measured. So the feature involved are three. The procedure can be similar to the previous: starting from the cylinder edges already determined (see fig.8.b), the equation of their medium axis can be easily calculated (the axis is reported in fig.10.a). Also the ribbing edge is determined with the procedure illustrated before (red in fig. 10.a), and the distance between the axis and the rib edge can be evaluated easily and quickly by analytical computations.
The same method can be applied in all that situation when it's necessary to locate any feature in the image. For example, locate the position of an hole respect to the reference system or a reference point, given by intersection of two edges; figure 10.b shows the process. The measurement can be composed as many time as desired, including all the feature that can be identified in the FOV.

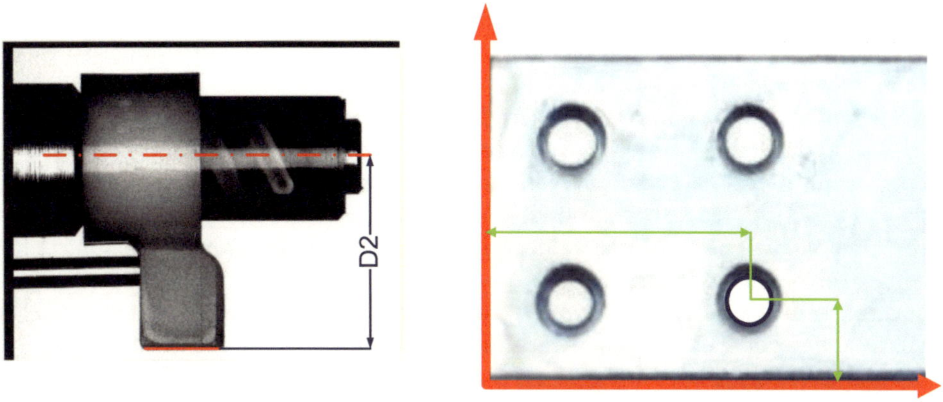

a) Ribbing distance measurement b) Hole position determination

Fig. 10. Process of composed feature measurement

4. Area measurements

Digital images are suitable also to measure 2D features, such as areas. In this section two cases of study will be illustrated. In the first example, the problem is represented by a burr caused by non perfect adherence between mould parts; the aim is to identify and measure the entity of the burr (fig. 11.a). The second problem is again about burr occurrence on the edge, as marked with a blue line in fig. 11.b, but the aim is to measure the area free from burr, available for fluid flow. The pictures also report the real image of burrs in the parts. In the first case burr covers only a part of the edge, while in the second example it extends all along the edge circumference.

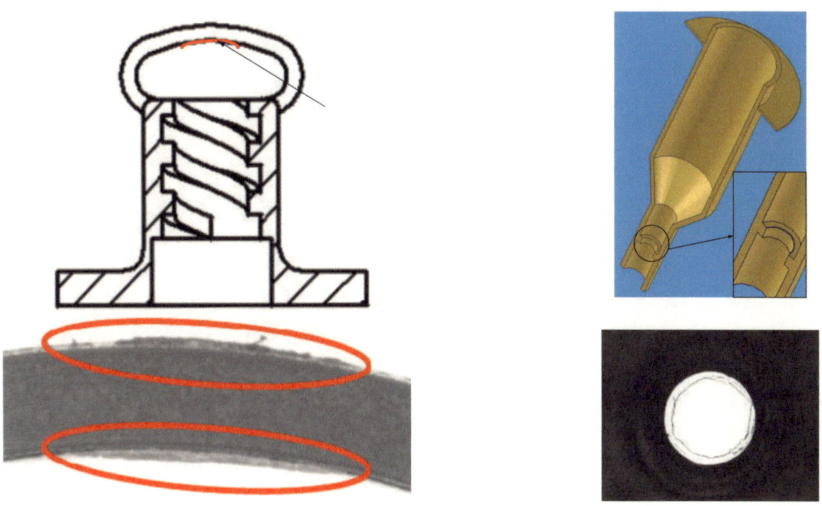

a) b)

Fig. 11. Burrs identification and quantification on two different parts

To quantify the entity of the burrs, instruments like contour projector are used. The only information available from the analysis is the maximum height of burr and it is inadequate to determine precisely the free area of the hole, or the extension on the edge. Digital images solve the problem using a combination of threshold and binary operation, so these two techniques will be now explained.

In a few words, in the digital images here presented, parts are characterised by an intensity range of gray-level much different from the background. Pixels of the part have a gray-level intensity comprised in a given interval, while all other pixels not included in this intensity range can be considered as background. Threshold operation sets all pixel that belong to the desired interval to a user-defined value, typically 1, and all other pixel of the image to zero. The image obtained is a binary image.

First case presents a burr, that has a gray-level between the free-area (gray-level ≥ 250) and the part (gray-level ≤ 120). The software, after alignment operation, extracts the zone of the image interested by analysis (fig 12.b); then, it performs a first binarization on this sub-image with a threshold value of 250, settings to 255 all pixels with a gray-level lower than 250 in order to obtain the area of the part and the burr together (fig. 12.c). To the same image extracted, a new binarization is applied, with a gray-level value of 138, and all pixel with a lower gray-level value are set to 255 to obtain the area of the part only (fig. 12.d). Finally, image in fig. 12.d is subtracted by fig 12.c to separate the burr areas. Now, with procedures similar to those used in edge detection, it is possible to determine search areas in which ROI lines are defined to identify edges. In this application an edge every 5 pixel has been determined and measured, and the software extracts the maximum value within them and returns it as measurement result. In this way, only and always the maximum burrs height is considered in measurement results. This method is conservative, but gives the certainty that the defects entity will not exceed the part requirement limits.

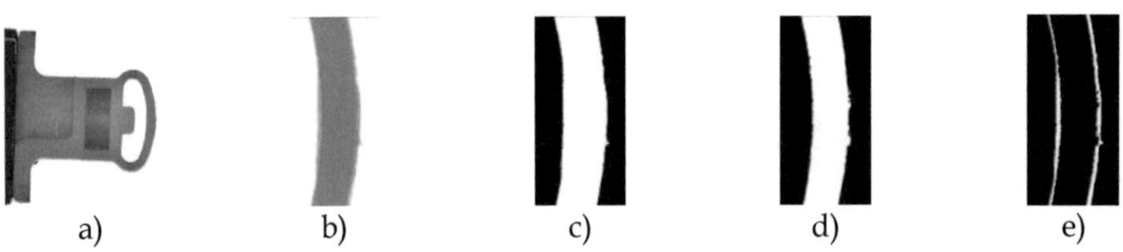

Fig. 12. Burr determination process

It's important to underline that is possible to obtain other information on burrs, that cannot be extracted with traditional methods. For example, one can measure the area (this procedure will be explained later) to understand the entity of the problem; a thin burr along all the edge indicates a defect solvable by proper tuning of injection process parameters, while big area localized on a small part of the edge indicates a damage on the mould. Furthermore, it is also possible to determine x and y mass centre position to understand where the mould is damaged. So, much more parameters are available to control production process and it is easier to define a set of them that indicates the need to repair the mould.

In the case of fig. 11.b, a different procedure that uses threshold method has been applied. Now the problem is to measure correctly the free area for fluid flow. With standard methods is only possible to have an approximation, measuring the burr height and computing the area of the annular region occupied by burr. But if this area is not annular or presents an irregular form, then it is almost impossible to get a precise measurement of its extension.

Using digital images, it is possible to implement a series of operations that compute the area with a good precision and automatically.

Since the free area has the highest gray-level of the image, the software developed computes the mass centre of the image (substituting mass with pixel gray-level values, obviously) which will always fall in the free area. From this point, the software begins to select all the pixels having a gray-level comprised in a properly defined range. Since a free area is looked for, a value of 255 has been set, with a tolerance of 5 gray-level. The algorithm stops at the burr edge, when it find a gray-level lower than 250. At the end of selection process, the software counts the number of pixel selected and, applying calibration information determines the area available for fluid flow. In fig. 12.a is reported the image before processing, and fig. 13.b shows the processed image.

a) b)

Fig. 13. The process of free area extraction

A brief explanation about threshold level must be given: smaller is the tolerance range, smaller will be the area extracted; using a tolerance range of zero, only the pixels with a gray-level of 255 will be selected and the resulting area will be the smaller as possible. In this application, the fluctuation is negligible, but this must be verified for each case and the most appropriate range has to be selected.

5. Shape defects

Digital images can also be used not only in measurement applications, but also for all controls that involve aesthetic and functionality aspects of moulded parts. This kind of controls is generally executed by expert operators, that have a depth knowledge of the process and of the part. A big experience is required and a proper training for operators is necessary. In spite of all, the work of qualified operators is not always enough for getting high product quality, because often fixed tolerance ranges are not given, and the entity of defects can be judged differently from different operators. Moreover, this kind of defects are frequently not measurable with standard instruments, and it is not easy to define a parameter to measure, check and certify part quality.

In this section, some examples of shape defects detection by digital images will be showed.

First will be explained the case of a thread termination damage.

The part of fig. 9.a has a thread in his centre, realized by a core that in his terminal part contacts the fixed plate of the mould (interested zones are highlighted by red lines in fig. 14.a). These parts of the core are very thin, and because of their low resistance property, they are the parts of the mould to undergo damage in production process earlier. If this part of the core is broken, the injection moulded component will have a defected shape that not allows the coupled screw to slide in him.

Fig. 14.b shows a correct thread termination, while fig. 14.c shows a damaged thread termination. The difference is quite small and difficult to see for a human eye that is performing many checks repeatedly. Nevertheless, carrying out a quantitative measure of such damage is basically impossible with standard instruments. Instead, using digital images it is possible to establish a procedure to determine a quantitative measure of the defect, that is a number which should be compared with admissible tolerance limits.

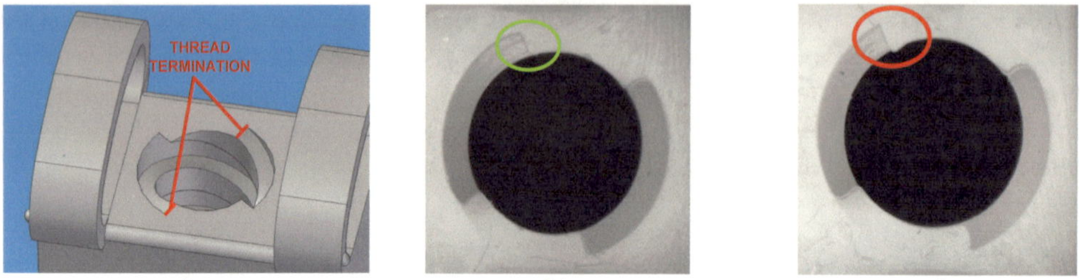

a) b) c)

Fig. 14. a) the particular; b) Good thread termination; c) Damaged thread termination

First step is always constituted by image alignment. In this case the part is aligned with image reference system, using external edges of the part, then the inner diameter of the thread is determined together with his centre (fig. 15.a); the thread termination edge is determined (green in fig. 15.b, indicated by green arrow) and the horizontal distance (indicated by P) between edge centre and thread centre is determined (fig. 15.c). This parameter has been selected because of its great sensitivity with respect to even small damages of the thread. Now is also possible to set a tolerance range within which the part can be accepted.

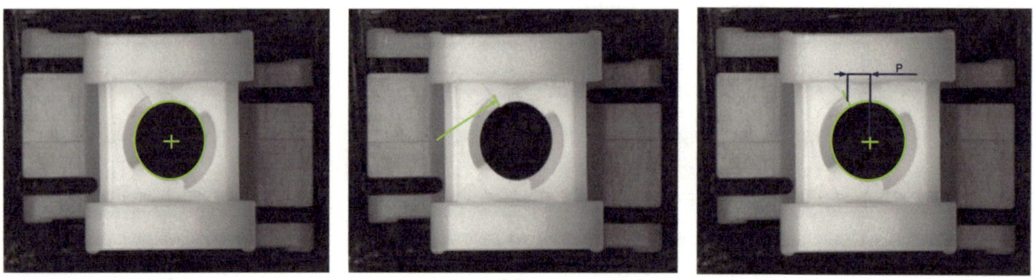

a) b) c)

Fig. 15. Measurement process of damaged thread

Here, the maximum distance accepted is 1 mm, which has been chosen analyzing more than 100 good pieces, and assuming the greater dimension as the maximum tolerance limit. Over this limit, the software alerts the operator with a warning message suggesting to reject the part. It's proper to note that expert operators are not required, because now the measurement process is totally automated and the interpretation is not subjective; however, this theme will be discussed more deeply in next section.

Next case presented below, is about individuation of process defects. To do this, part in fig 8.a will be considered. In moulding process, this particular often presents problems as short shot (fig 16.a, absence of ribbing) and air bubbles on the ribbing and on the cylinder (fig 16.b and 16.c respectively).

Different inspection methods are used, depending on what is the defect looked for. In fact, for short shot defect that is generally easy to detect, a simple pattern matching could be enough. In this case an image of the entire particular has been used as template and is looked for in inspected image. The score of searching procedure is used as parameter to control; if the parameter is lower than a threshold fixed to 95%, the piece is rejected.

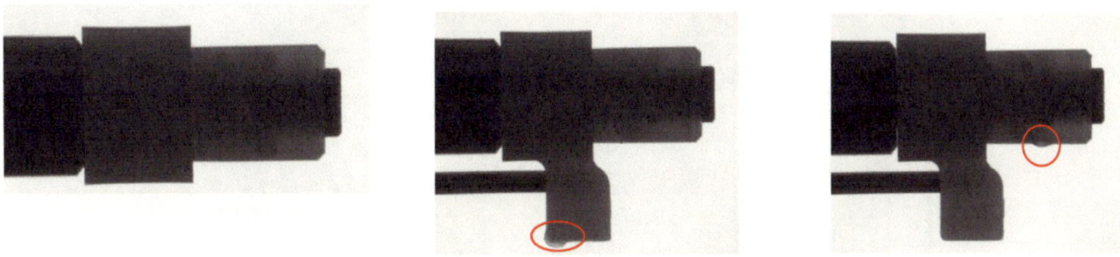

a) b) c)

Fig. 16. Moulding injection process defects

For smaller defects, that give no significant changes in score of matching operation, it is necessary to adopt methods that analyze only limited parts of the image. For example, to find air bubble of fig. 15.c, the software search and identify the edge of the cylinder, then transforms points in a line with a linear regression and determines the residual of the operation as (Labview 8.2 User Manual, 2005):

$$r = \sqrt{\sum_i e_i^2} = \sqrt{\sum_i (y_i - y)^2} \quad (4)$$

where, y_i represents the true point identified and y represent the point of the fitted line. If this parameter is higher than a fixed value, about 0.25 mm (corresponding to 15 pixel), it means that some discontinuity or irregularity is observed on the boundary surface (it doesn't matter if outwards or inwards) and part is rejected. The threshold value has been fixed analysing a sample of 100 good components and taking the bigger value founded.
The same method has been used for the air bubble on the ribbing, but in this case the edges identified and inspected are two, the same that can present the defects.

6. System performances

In this section, performances of the system, in terms of measurement precision and time saving, will be showed.
As example, the part of fig. 9.a has been used. In fig. 17, all the features to be determined are illustrated. There are four dimensional measurements, indicated as "D", a burr determination indicated as "Burr" and a shape defect identification, indicated by "screw end".
The description starts with dimensional measurement performance. Table 1 shows nominal dimension of requested features, and respective acceptable ranges.
To understand if the system is suitable to measure the required features and has a proper precision, repeatability and reproducibility, 100 images have been acquired from a component used as template, moving each time the part in the FOV to simulate a real measurement process, and resolution and standard deviation of results have been evaluated and compared with tolerance ranges.
About resolution, the system uses an image in which a pixel corresponds to 0,016 mm while the minimum tolerance range is 0,15 mm so, being their ratio approximately 10, the resolution can be considered suitable for the application.

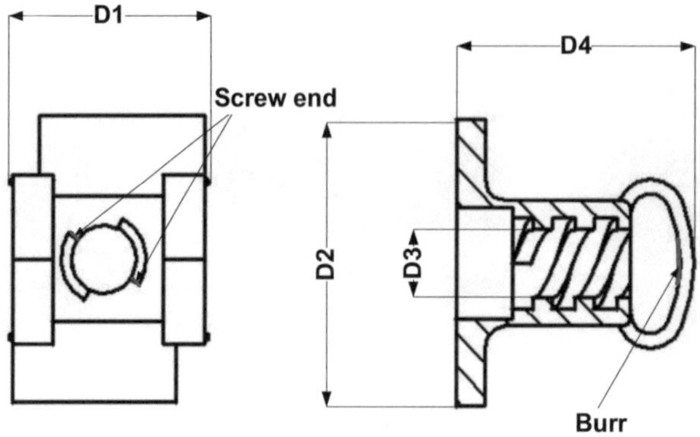

Fig. 17. The second component and the features to measure

	Nominal Dimension	Acceptable range
D1	14,00 mm	13,82÷14,18 mm
D2	19,50 mm	19,29÷19,71 mm
D3	4,80 mm	4,75÷4,90 mm
D4	17,00 mm	16,82÷17,18 mm
Burr		< 0,20 mm
Screw end		< 1,00 mm

Table 1. Feature to inspect

About repeatability, the results of 100 image measurements, on feature D1 as example, evidence a standard deviation of 0,0093 mm. Better results have been obtained with feature D2 and D4, respectively with standard deviation of 0,0055 mm and 0,0033 mm. The best result in repeatability is on D3 feature, with standard deviation of 0,0014 mm. These values include all possible disturbances, that is pixel noise and realignment errors, and can be considered as the accuracy of the instrument, because of the bias has been deleted choosing acquisition parameter settings that gives the same results as gauge measurements. About the process, results are reported in table 2.

With the exception of feature D3 (for which a comparison is not possible) it is evident that the average measures present minimum differences. This can be expected, because the image was calibrated on results of manual measurements. Standard deviation also are very similar.

Moreover, once the data has been stored, all statistical process evaluations are possible.

Assuming that the manufacture processes produce pieces whose dimensions or features are statistically scattered according to normal, or Gaussian, distributions, the process capacity can be automatically determined as (A. V. Feigenbaum, 1983):

$$Process\ Capacity = 6\sigma = \sqrt{\frac{\sum\left(u-\overline{u}\right)^2}{n-1}} \quad (5)$$

where: u = measurement of i^{nt} piece;

$\overline{u}$ = average value of measurements;

n = number of pieces evaluated.

The capacity ratio C_p and the performance index C_{pk} can be evaluated. For C_p evaluation the following equation has been used (J.M. Juran's, 1994):

$$C_p = \frac{USL - LSL}{6\sigma} \tag{6}$$

in which: ULS = Upper specification Limit;
LSL = Lower specification Limit;
6σ = Process capacity under statistical control.
The expression used for C_{pk} instead is (J.M. Juran's, 1994):

$$C_{pk} = \min(C_{up}; C_{low}) \tag{7}$$

where

$$C_{up} = \frac{USL - \bar{u}}{3\sigma} \tag{8}$$

$$C_{low} = \frac{\bar{u} - LSL}{3\sigma} \tag{9}$$

As example, table 2 reports process performances. In this case study the mould has 8 cavity, and each cavity has been measured 12 times. Totally, 96 pieces have been measured. Observing C_p results, it is noted that all measured features have an index superior to 1, so the process capability is adequate to meet established tolerance range. Moreover, the process has high repeatability, and feature D1, D2, and D4 have a C_p that ensure that no pieces can exceed tolerance range (in fact C_p is bigger than 1,63 that corresponds to 1 part per million). Only feature D3, because of a narrow tolerance range, have a C_p a bit lower alerting that a certain dispersion in production is occurring.
A further evaluation of results is possible by observing C_{pk}, in fact, C_{pk} values are lower, evidencing that the process is not centred with respect to the tolerance range. Worst situation is yet in feature D3, where C_{pk} is lower than 1; it means that a considerable percentage of pieces could exceed the tolerance limits. It's important to say, that presented data are about 8 different cavities of the mould; perhaps, dividing data per each cavity, only 1 cavity could present low C_p and C_{pk} indexes. Modifying dimensions of that cavity, indexes of the entire process could considerably increase.

		Average	Std. Dv.	C_p	C_{pk}
D1	Software	13,8960	0,0132	4,5455	1,9192
	Operator	13,9000	0,0135	4,4610	1,9827
D2	Software	19,4118	0,0153	4,5632	2,7814
	Operator	19,4110	0,0085	8,2440	4,7501
D3	Software	4,8570	0,0197	1,269	0,7276
	Operator				
D4	Software	17,0060	0,0290	2,0690	2,0000
	Operator	17,0050	0,0232	2,5896	2,5176

Table 2. Process performances

Also same consideration about reproducibility of the measurements are possible. To do this, two operators measured the same sample of pieces, once with the system and once with standard instruments. Table 3 reports these results.

		1 operator		2 operators	
		Software	Gauge	Software	Gauge
D1	Average	13,8960	13,9000	13,8970	13,8990
	Std. Dv.	0,0132	0,0135	0,0123	0,0124
D2	Average	19,4180	19,411	19,4130	19,4020
	Std. Dv.	0,0153	0,0085	0,0154	0,0134
D3	Average	4,8570		4,8540	
	Std. Dv.	0,0197		0,0194	
D4	Average	17,0060	17,0050	17,0050	16,9710
	Std. Dv.	0,0290	0,0230	0,0276	0,0397

Table 3. Reproducibility of measurements

It is possible to see that nothing is changed. Negligible differences are there between average values and standard deviation in both cases and consideration about process performances are also valid. On the contrary, measured values normally change passing from 1 to 2 operators with manual measurements, especially in feature D4 where is possible to notice a significant difference in average values (about 0,03 mm). Moreover, standard deviation increases in all cases, especially in D4, passing from 0,0230 to 0,0397. This is due to burr presence on the pieces (see fig. 11.a), as the software is able to avoid it from the measure, but this is not possible with standard instruments and the influence of the burr is different depending on the operator that is measuring. So it is possible to say that measurement process, doesn't change its precision passing from 1 to 2 operators, vice versa manually measuring, it loose precision and repeatability as higher is the number of operator. It means also, that Cp and Cpk decrease and the process could results out of control.

In fig. 18 are reported system performances about reproducibility for each feature.

Finally, once the data has been acquired, more and more quality control instruments can be automatically implemented, as example quality control chart (D. C. Montgomery, 1985).

About time saving, before explain the results, it is proper to describe the operations carried out both in manual and in optical controls.

In manual control, the operator have to execute the following steps in the order:

1. Prepare sheet for registration;
2. Control screw termination by a microscope;
3. Measure burr entities by a contour projector;
4. Measure feature D1, D2 and D4;
5. Execute Functionality test (try to screw the coupled particular);
6. Measure feature D3 with go-no-go gauge.

The order of the operation is important, because the entity of some feature can be changed during control. For example, go-no-go gauge control can damage the particular and functionality test can be compromised, as also screw termination can be damaged and incorrect consideration can be done by the operator.

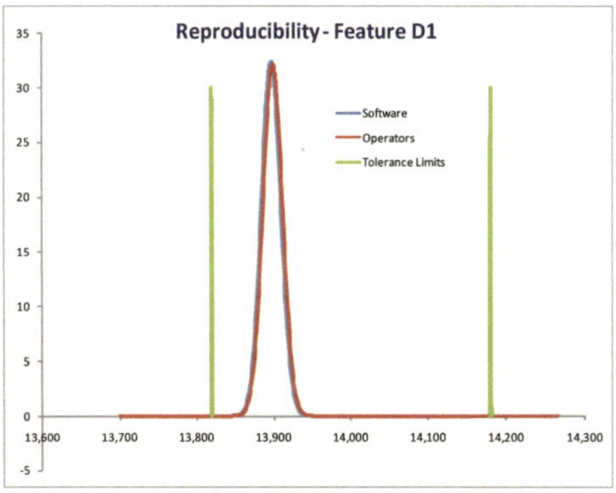

a) Reproducibility feature D1

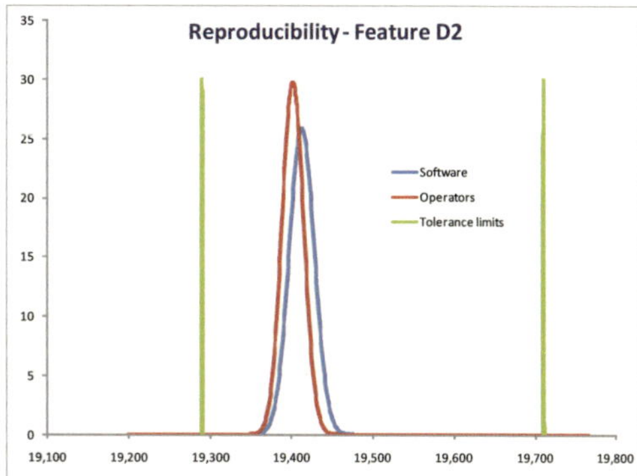

b) Reproducibility feature D2

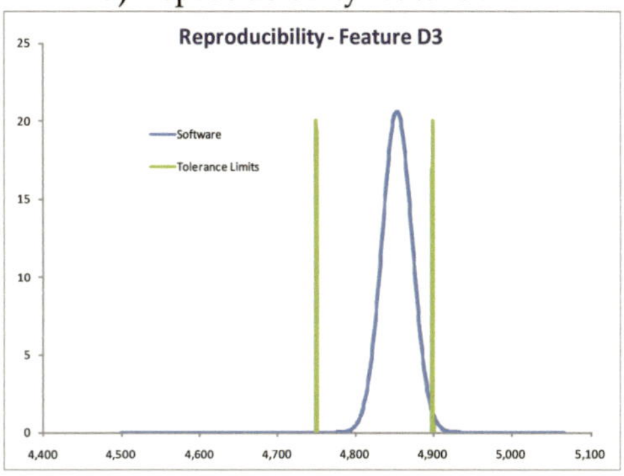

c) Reproducibility feature D3

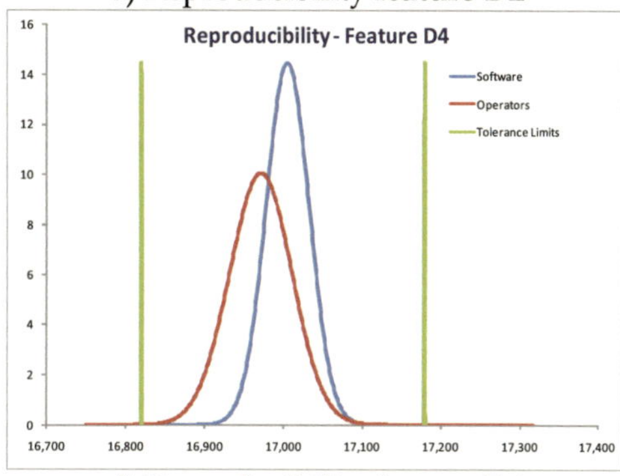

d) Reproducibility feature D4

Fig. 18. Reproducibility of the measurements

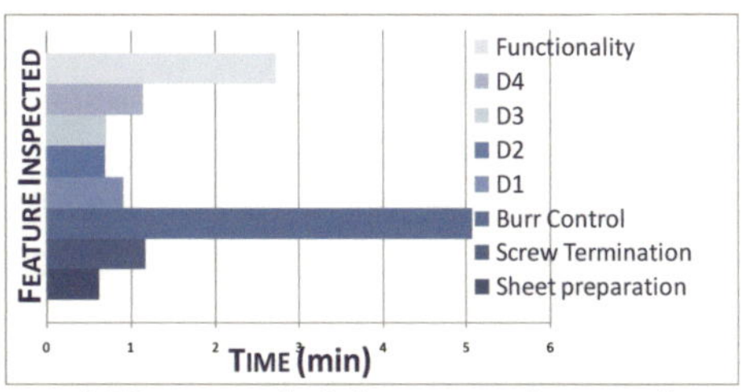

a) Time needed for manual control

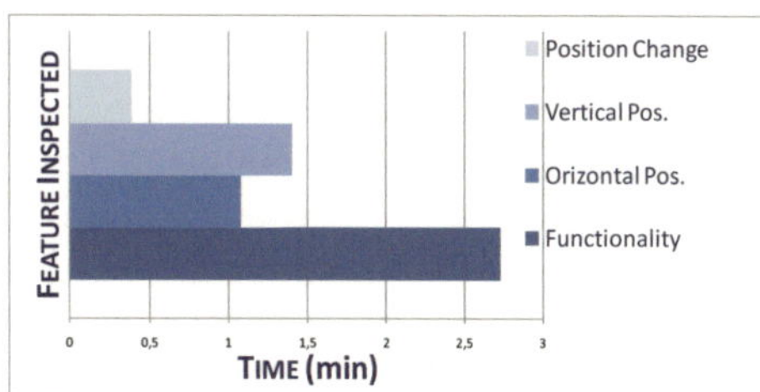

b) Time needed for automatic control

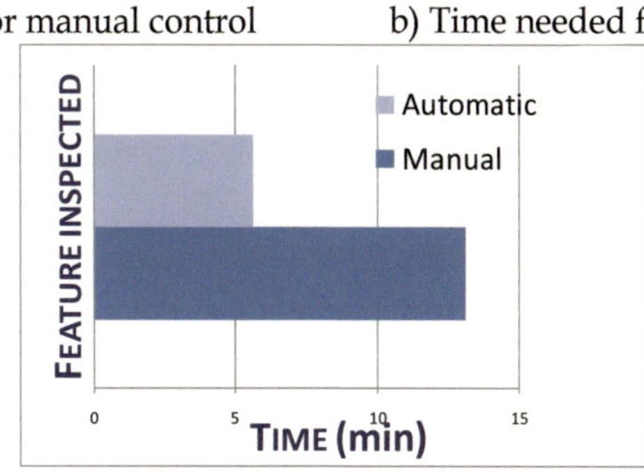

c) Difference of time amount

Fig. 19.

Time needed for each operation is reported in fig 19.a; the values are referred to 8 pieces control.
Each bar represents the amount of time needed to measure the relative feature. It's clear from the picture, that the longest control is burr determination with over 5 minutes. Then, functionality test requires about 3 minutes and all the other feature requires about 1 minutes. The amount of time needed is reported in fig. 19.c (dark blue), and is over 13 minutes.
In automatic control instead, the procedure used is different: the particular is positioned first in a vertical position in the FOV of the system (figure 17 left) and the software determines feature D1, D2, D3 and screw termination. Then the particular is moved in an horizontal position (fig. 17 centre) and the software determines feature D4 and burr entity. The time for change particular position has also been considered. Finally, functionality tests have been executed as in the previous case.
Now, the longest control is functionality test, while horizontal and vertical position requires respectively 1 minute and 1 minute and a half. Change position requires about 40 seconds. The amount of time needed for the control is about 6 minutes (fig 19.c, light blue bar).
The big time reduction (near 54%) is due to different factors.
First of all, more than a feature can be measured at the same time, using the same instrument. So, as example, measure features D1, D2, D4 and screw termination (vertical position), that requires three different instruments in a time of about 3 minutes and a half, can be measured with only one instrument in only 1 minute.
Second, the reduction of time amount also depends on the net time reduction in single feature inspection. For instance, burr determination in manual process requires over than 5 minutes, while in automatic process requires only 1 minutes and a half (together with feature D4).
Finally, time for sheet preparation and registration has been deleted, because data are automatically stored and organized in a report that can be printed if necessary.
To close this section, some limits of the system will be listed.
First of all, only particulars (or features) with dimension smaller than the maximum FOV can be inspected. So larger components have to be controlled with a different system.
Second, precision of the system depends of FOV dimension. In fact, as bigger is FOV, as lower is precision of measurement because of the number of pixel of camera sensor is constant. Third, only feature that lies in a plane or with a depth lower than depth of focus can be inspected.

7. Aesthetic applications

Digital image processing can be used also for aesthetic control.
Possible application with the developed system are three:

1. Intensity colour determination;
2. Uniformity colour determination;
3. Detection of small black-dots.

All these application are based on histograms obtained by the image.
The histogram consists in counting the total number of pixel at each level of the greyscale. The graph of an histogram is a very quick way to distinguish if in the image there are regions with different grey-level value. Histograms are useful also to adjust the image and to optimize image acquisition conditions.

The histogram is the function H defined on the grayscale range *[0,…,k,…,255]* such that the number of pixels equal to the grey-level value k is (NI Vision Concept Manual, 2005):

$$H(k) = n_k \tag{10}$$

where: k is the gray-level value,

n_k is the number of pixels in an image with a gray-level value equal to k,

n = Σn_k (from k=0 to 255) is the total number of pixels in an image.

The histogram plot reveals easily which gray level occurs frequently and which occurs rarely.

Two kind of histogram can be calculated: linear and cumulative histograms. In both cases the horizontal axis represents the gray-level value that ranges from 0 to 255.

For linear histogram, vertical axis represents the number of pixels n_k set to the value k. In this case, the density function is simply given by (10).

The probability density function is (Gonzalez & Woods, 1992):

$$P_{Linear}(k) = n_k / n \tag{11}$$

where $P_{Linear}(k)$ is the probability that a pixel is equal to k.

For cumulative histogram instead, the distribution function is (NI Vision Concept Manual, 2005):

$$H_{Cumulative}(k) = \sum_{i=0}^{k} n_i \tag{12}$$

where $H_{Cumulative}(k)$ is the number of pixels that are less than or equal to k. The cumulated probability function is (NI Vision Concept Manual, 2005):

$$P_{Cumulative}(k) = \sum_{i=0}^{k} \frac{n_i}{n} \tag{13}$$

where $P_{Cumulative}(k)$ is the probability that a pixel is less than or equal to k. For the application illustrated below, only linear histograms will be used.

7.1 Colour intensity determination

First application has the aim to determine colour intensity of aesthetic goods. The problem is due to the production process, in fact for this kind of products, only a mould is used and different colours of the same particular are obtained mixing masterbatch colour to the virgin polymer and also recycled material (like sprue and cold runners). If the mixing process is not constant and regular, different percentage of colour can be mixed and the particular can be darker or lighter of the master. Figure 20.a show the PP cap here examined as example.

To determine the intensity of the colour the software acquires an image, from which extracts a portion; figure 20.b shows the acquired image and the extracted area in the yellow rectangle. On the extracted image, it calculates the histogram and the mean value of gray-level intensity of the pixels.

From figure 21 it is evident that the mean value of the histogram of the dark particular (grey-level=41,54) is lower than mean value of histogram of bright one (grey-level=64,21). So, it is possible to set a range of median value in which the extracted image have to be comprised. In this example, six ranges have been determined (figure 22 from a to f) to separate different intensity colours. For each group the median grey-level intensity is reported below. Consider that six groups represent a resolution approximately three times bigger than human control (where only two groups are usually determined) and with much more objectivity.

a) The particular examined

b) The area extracted

Fig. 20. The particular examined

Finally, it must be noted that it is possible to deal with coloured parts by means of a monochromatic camera only if one is interested in measuring the colour intensity; obviously no considerations can be done on colour tonality. So the system is able to detect problems of incorrect mixing (as showed in next section) or insufficient/excessive masterbatch in the mix, but it cannot guarantee that the masterbatch used is correct.

a) A light particular

b) A dark particular

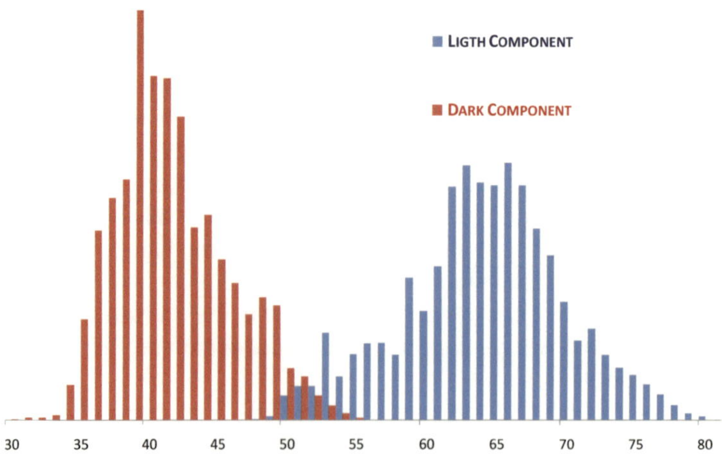

c) Histograms on comparision

Fig. 21. Analysis of a light (a) and a dark (b) component by histogram (c)

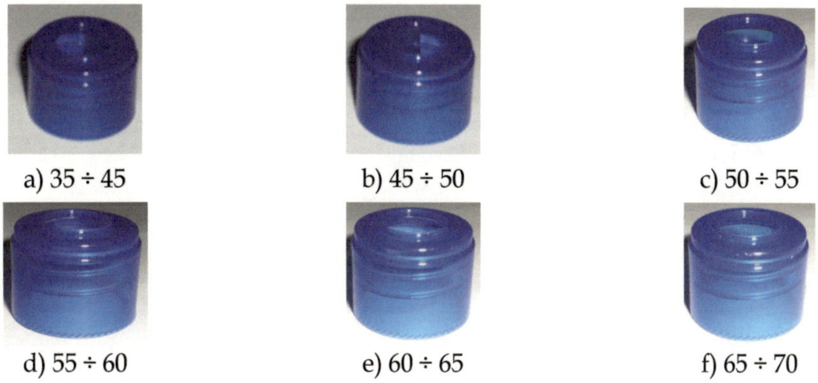

Fig. 22. Six groups of different intensity colours

7.2 Colour uniformity determination

The second application developed regards the mixing problems between virgin polymer and masterbatch colour. In this case, components appears with non uniform colour as showed in figure 23. Figure 23.a shows a particular in which white material hasn't been correctly mixed, while the component of Figure 23.b was obtained with a correct mixing, with uniform rose. Figures 23.c and 23.d show the acquired images of the same particulars with the system. Extracting the same area from the images, as illustrated in fig. 20b, the histograms of figure 23e are obtained; it's evident that histogram of first particular has a standard deviation bigger than the second. So, it is possible to establish a tolerance range in which the variation can be accepted while outside the parts have to be discharged.

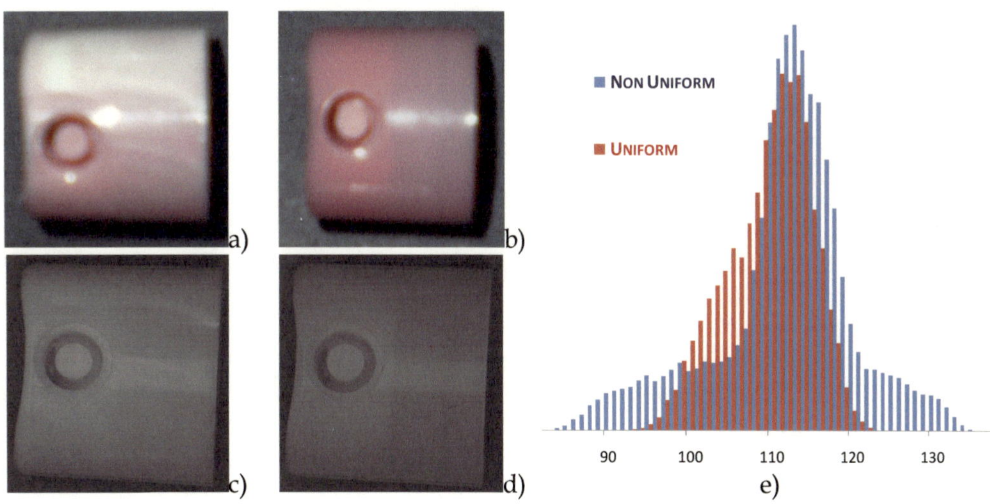

Fig. 23. Example of non uniform colour

7.3 Detection of small black dots

This problem is generally due to small impurities included in the material, which appear as small dots in the final product. Often, this kind of control is still demanded to visual inspection of expert operators, who look at the particulars from a distance of about 40 cm,

and if they can see the black dots the pieces are rejected. In this way, the control totally depends on the operator judgment and is totally arbitrary.
The optical system here presented uses an algorithm based on particle analysis: firstly it performs a binarization, finding all pixels with an intensity value lower than a fixed threshold; then, it considers all the particles (agglomerates of pixels) and counts the number of pixels each particle is made of. So, by knowledge of the pixel size in real dimensions, it is yet possible to establish maximum dimension of acceptable black dots, and the control will not depend from the operators estimation.
In fig. 24 a cap with black dots of different dimensions is reported. In this case the software looks for object with dimension bigger than 4 pixels. Dots with dimension bigger than the threshold value are highlighted in red circles, while black dots smaller than four pixels are highlighted in green circles. If only green circles are present, the piece can be accepted, while if red circles are present, the piece have to be rejected.

a)

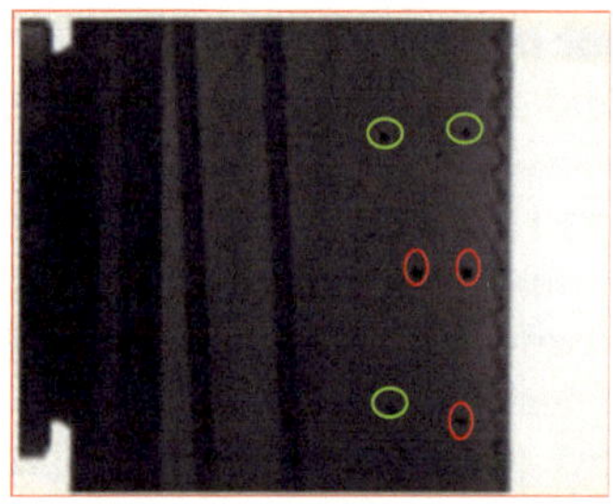
b)

Fig. 24. An example of black dots on a particular

8. References

Feigenbaum A.V. (1986), *Total quality Control*, McGraw-Hill, 0-07-020353-9, Singapore.

Gonzales R. C.; R. E. Woods (1992), *Digital image processing*, Addison-Wesley, 0-201-50803-6, United States of America.

Juran J.M.; F.M.Gryna (1988), *Juran's quality control handbook fourth edition*, McGraw-Hill, 0-07-033176-6, United States of America.

Montgomery D. C. (1985), *Introduction to statistical quality control*, j. Wiley & sons, 0-471-80870-9, New York.

National Instruments (2005), *Labview user manual*

National Instruments (2005), *NI Vision concept manual*

Russ J.C. (1994); *The image processing handbook second edition*, CRC Press, 0-8493-2516-1, United States of America.

Power Quality Control in Wind/Fuel Cell/Battery/Hydrogen Electrolyzer Hybrid Micro-grid Power System

Xiangjun Li, Dong Hui, Xiaokang Lai and Tao Yan
Electrical Engineering and New Material Department
China Electric Power Research Institute
15 Xiaoying East Road, Qinghe. Beijing 100192
P.R.China

1. Introduction

Generally, such as bio-fuels, wind, solar, small hydropower, marine and geothermal energy, etc. have been defined as new renewable energy. Photovoltaic and wind power generation have been in rapid growth, but they have characteristics of randomness, intermittent, uncontrollable. Although the idea of effective renewable energy use as a means of coping with environmental and resource problems, especially of reducing CO_2 emissions, is globally attractive, inappropriate application of distributed renewable power generation systems can be a cause of insecure power supply. That is, the widespread use of various kinds of distributed renewable power sources which would impact the quality of the power supply within a micro-grid power system, causing many control problems (Li et al., 2008).

Micro-grid is one such system, consisting of distributed generators, loads, power storage devices and heat recovery equipment, among other components. The main advantages of the micro-grid system are that (1) it can be operated independently from conventional utility grids, (2) it can make use of power and heat sources collectively, (3) it can be interconnected to the utility grids at one point (Lasseter et al., 2002; Lasseter & Paigi., 2004; Li et al., 2008).

A micro-grid system (see Fig. 1) comprised of a control and monitoring system, a housing load, a 1MW load-controllable electrolyzer system to manufacture hydrogen, a hydrogen tank and renewable-energy-utilizing generators such as 3MW wind power generation system (WPGS), 500kW proton exchange membrane fuel cell (FC), 1MWh battery energy storage system (BESS), and others, is considered in this chapter. The power supply-demand balance control of the micro-grid system is performed by the control and monitoring system through a Control Area Network (CAN) composed communication network. Moreover, the electricity of electrolyzer system is supplied mainly by the wind power energy source and the hydrogen produced by the electrolyzer system is stored in the hydrogen tank to be converted back to electricity in the proton exchange membrane fuel cells. The wind power is considered as primary source in this chapter. However, when the lack of power supply from renewable power sources is occured, the missing power will be provided via utility grid (Li et al., 2007; Li et al., 2008).

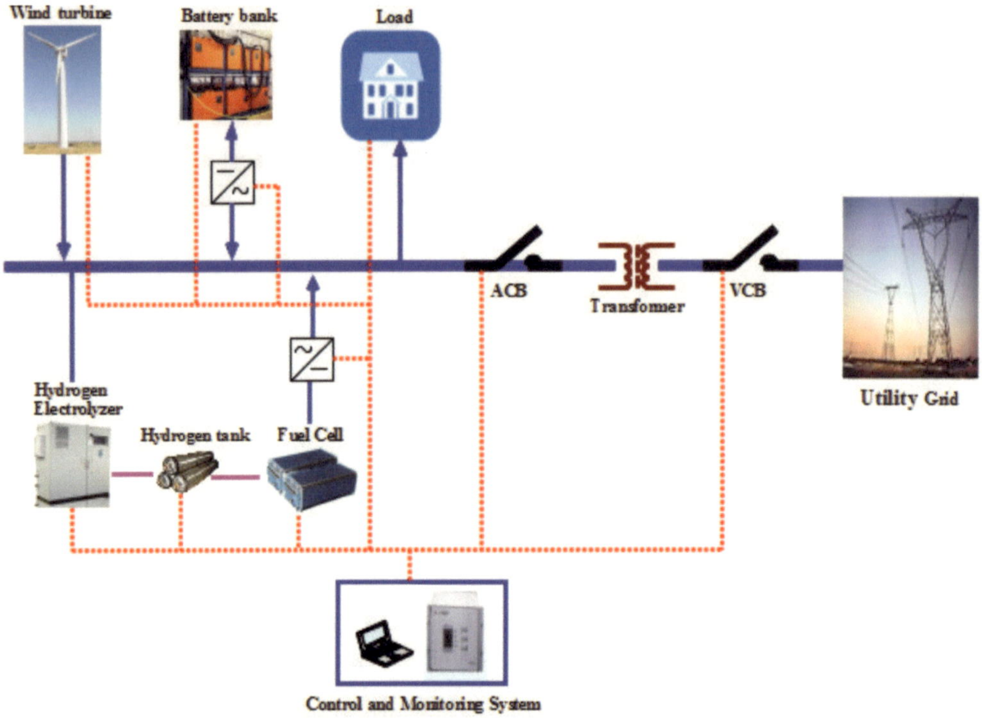

Fig. 1. System configuration of micro-grid network

Wind power and photovoltaic generators have the disadvantage of an unstable power output. Therefore, in these kinds of hybrid power systems, a sudden real power imbalance or a large frequency fluctuation can easily occur. Moreover, when the type of power line designed to be interconnected to a utility grid is 380V three-phase AC line, applying a DC or AC source to the AC grid would lead to harmonic distortion of voltages and currents. The harmonics resulting mainly from the operation of power electronic systems (e.g. converters) have ignored in this study (Hingorani & Gyugyi, 2000; Li et al., 2008). Here, we have studied the frequency fluctuations resulting mainly from real power imbalances and applied the BESS and electrolyzer system to smooth the power fluctuation of micro-grid power system.

The so-called hydrogen economy is a long-term project that can be defined as an effort to change the current energy system to one which attempts to combine the cleanliness of hydrogen as an energy carrier with the efficiency of fuel cells as devices to transform energy into electricity and heat. As an energy carrier, hydrogen must be obtained from other energy sources, in processes that, at least in the long term, avoid or minimize CO_2 emissions (Gregorio & Teresa, 2007; Li et al., 2008). Electrolyzer system fed by renewable energies (such as photovoltaic solar panels or windmills) or biomass reformers is the distributed resource of interest to generate hydrogen. Moreover, the electrolyzer system can offer flexible controllability functions to compensate for system's real power imbalances (Li et al., 2007; Li et al., 2008). In (Li et al., 2008), with regard to such a capability, a combination of a micro turbine and the fuel cell and electrolyzer hybrid system to deal with real-time frequency fluctuations and sudden real power imbalances has been proposed. In this chapter, a fuel cell and electrolyzer hybrid system is also considered to deal with real-time real power imbalances.

In addition, energy storage technology would resolve above-mentioned random fluctuation issues of the new energy power generation. It can achieve a smooth power output of renewable power sources. Therefore, a flexible energy storage system consisting of the battery, rectifier and inverter devices is getting more and more attention. For example, an integration of battery energy storage system with wind power generation system can provide flexible energy management solution for improving power quality of the renewable energy hybrid power generation system (Li et al., 2010b, 2010c & 2010d). Moreover, with the rapid development of batteries, as one possible means, BESS will be utilized to smooth the output fluctuation of renewable energy such as PV, WP and FC (Yoshimoto et al., 2006; Rajendra Prasad & Natarajan, 2006; Senjyu et al., 2008; Muyeen et al., 2009; Lalouni et al., 2009; Teleke et al., 2009; Li et al., 2009; Khalid & Savkin, 2010; Li et al., 2010b-d). However, if there is no control method to regulate the outputting-level of wind power and the charging-level of BESS effectively, a large energy storage capacity is necessary. Therefore, this chapter will introduce a fuzzy-logic based power conditioning strategy of the load-controllable electrolyzer system to deal with real-time fluctuations of micro-grid's real power imbalance and system frequency. Meanwhile, the power of fuel cell system and that of BESS will be adaptively changed according to the feedback values of battery state of charge (SOC) and micro-grid's real power imbalance.

This chapter is organized as follows. Section 2 presents modeling of each power sources in micro-grid power system. That is, (1) modeling of fuel cell and electrolyzer system; (2) modeling of battery energy storage system; (3) modeling of wind power generation system; (4) modeling of system load, etc. will be included, respectively. Section 3 presents an objective function for frequency control in the micro-grid system. Section 4 presents a fuzzy control strategy of hydrogen electrolyzer system based on battery SOC and real power imbalance to expand the micro-grid system's ability to solve power quality issues. Section 5 discusses simulation results based on the considered power system model. Section 6 is the conclusions.

2. Modeling of each power sources

2.1 Modeling of fuel cell and electrolyzer system

Here, the power output characteristics of fuel cell and electrolyzer systems are linearized by the following first-order transfer functions based on different time constants, T_{FC} and T_{ES} in reference to the operation of a battery-energy-storage facility, as shown in (Kottick et al., 1993, Li et al., 2008). As a result, the dynamic behaviors of fuel cell and electrolyzer systems can be formulated as follows.

$$P_{FC} = \frac{1}{T_{FC}s+1} P_{FC}^{CMD} \tag{1}$$

$$P_{ES} = \frac{1}{T_{ES}s+1} P_{ES}^{CMD} \tag{2}$$

2.2 Modeling of battery energy storage system

An 100KW lithium-ion BESS has been modeled in reference to the R_{int} model presented in (Li et al., 2009). In general we know that I_{bat} and V_{bat} can be expressed as Eqs. (3) and (4), respectively, when the battery power consumption is P W. Moreover, as shown in Eqs.(5) and (6), V_{ocv} and R_{bat}^{int} are determined by using look-up tables based on experimental data presented in Figs. 2 and 3. The battery SOC is calculated by using Eqs.(7) and (8), and the η is calculated depend on battery charge/discharge status. The model of BESS based on Matlab/Simulink is shwon in Fig.4 (Li et al., 2009 & 2010a). In this chapter, 1MWh BESS is designed by integrating ten 100kWh BESSs in parallel to minimize real power imbalance in micro-grid power system.

$$I_{bat} = \frac{V_{ocv} - \sqrt{V_{ocv}^2 - 4R_{bat}^{int}P}}{2R_{bat}^{int}} \tag{3}$$

$$V_{bat} = V_{ocv} - R_{bat}^{int} I_{bat} \tag{4}$$

Where

$$V_{ocv} = f_1(SOC) \tag{5}$$

$$R_{bat}^{int} = \begin{cases} R_{ch} = f_2(SOC) & charging \\ R_{dis} = f_3(SOC) & discharging \end{cases} \tag{6}$$

$$SOC = SOC_{ini} - \int \frac{\eta I_{bat}}{Q} dt \tag{7}$$

$$\eta = \begin{cases} \eta_{ch} = \dfrac{V_{ocv}}{V_{ocv} - I_{bat}R_{ch}} & charging \\ \eta_{dis} = \dfrac{V_{ocv} - I_{bat}R_{dis}}{V_{ocv}} & discharging \end{cases} \tag{8}$$

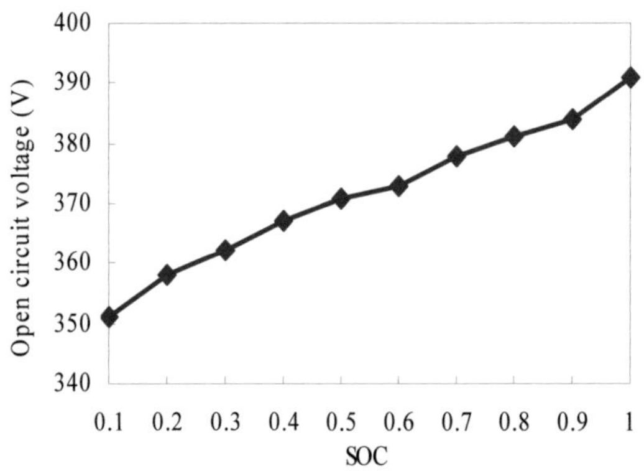

Fig. 2. Characteristic of open circuit voltage via battery

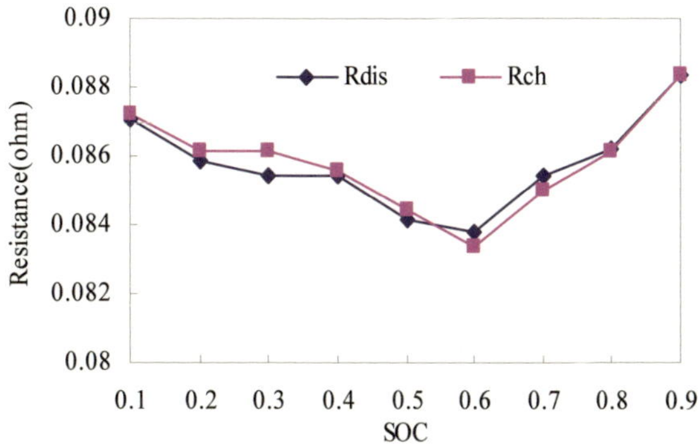

Fig. 3. Interior resistance via battery SOC

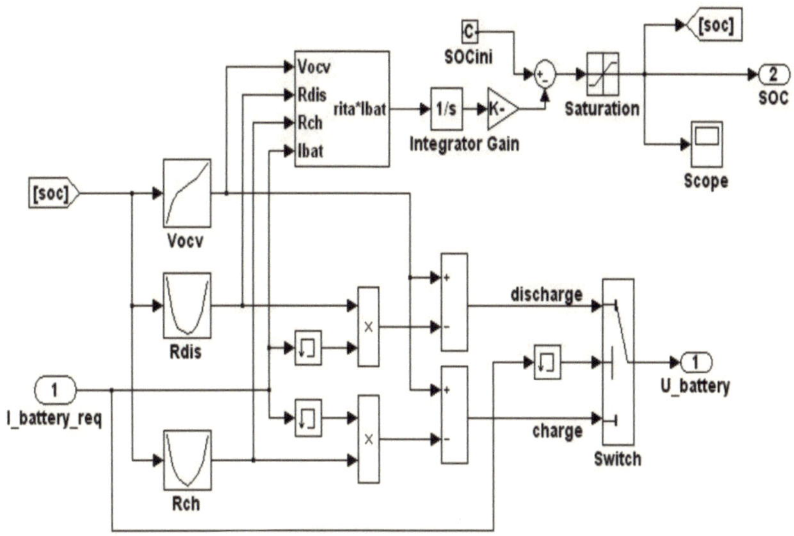

Fig. 4. Model of BESS

2.3 Modeling of wind power generation system

The extracted power from the wind can be expressed as follows:

$$P_w = 0.5\rho\pi R^2 V_w^3 C_p(\lambda,\beta) \tag{9}$$

Where, P_w is the power from wind, ρ is air density [kg/m³], R is blade radius[m], V_w is the wind speed[m/s] and C_p is the power coefficient which is a function of both tip speed ratio λ and blade pitch angle β[deg].

$$C_p(\lambda,\beta) = 0.22\left(\frac{16}{\lambda_i} - 0.4\beta - 5\right)e^{\frac{-12.5}{\lambda_i}} \tag{10}$$

where λ_i is given by

$$\frac{1}{\lambda_i} = \frac{1}{\lambda + 0.08\beta} - \frac{0.035}{\beta^3 + 1} \tag{11}$$

The WPGS modeled by MATLAB/SIMULINK is shown in Fig. 5. In this chapter, the wind speed is modeled by multiplying a random speed fluctuation derived from the white noise block in MATLAB/SIMULINK as shown in Fig. 6. As a result, wind power output profile used in this chapter can be calculated as shown in Fig. 7 (Li et al., 2010b & 2010c).

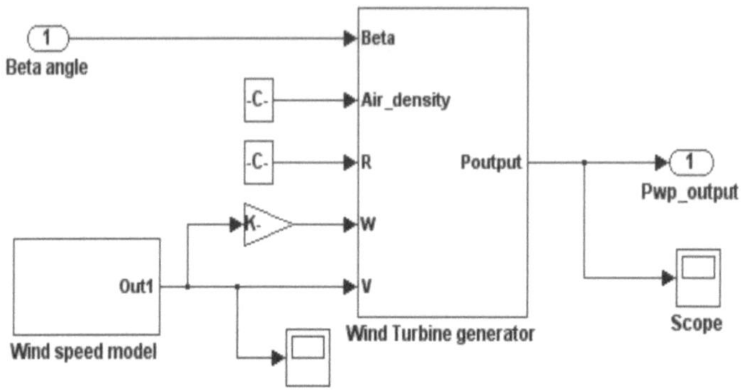

Fig. 5. Model of WPGS using MATLAB/SIMULINK

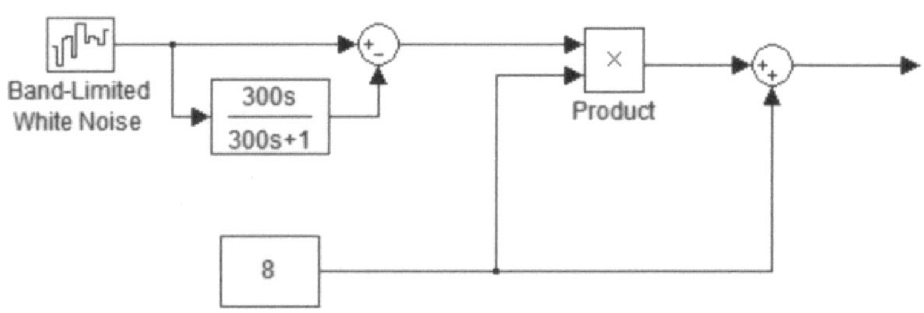

Fig. 6. Wind speed model using MATLAB/SIMULINK

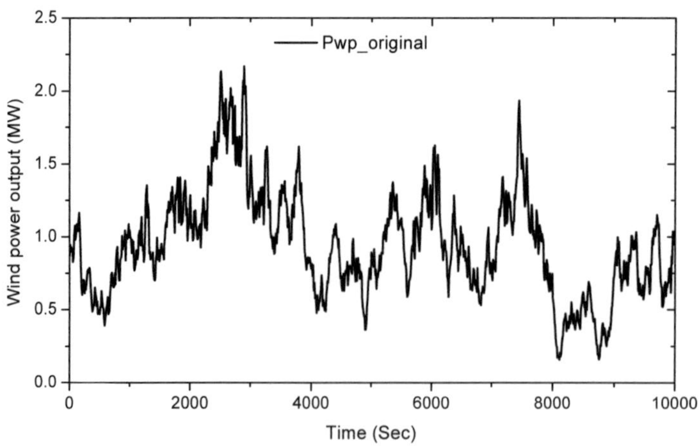

Fig. 7. Wind power output of micro-grid power system

2.4 Modeling of system load

The input power variability of micro-grid system load is determined by considering the deviations from the initial value. And the following standard deviation, dP_{Load} is multiplied by the random output fluctuation derived from the white noise block in MATLAB/SIMULINK in order to simulate the real-time random power fluctuation on the load side. The deviation for the system load is simulated close to an actual change wave by the following functions as presented in (Matsubara et al., 2005; Shinji et al., 2006; Li et al., 2007; Li et al., 2008).

$$dP_{Load} = 0.6\sqrt{P_{Load}} \tag{12}$$

That, the system load power is defined as follows.

$$P_{Load} = P_{Load}^{ini} + dP_{Load} \tag{13}$$

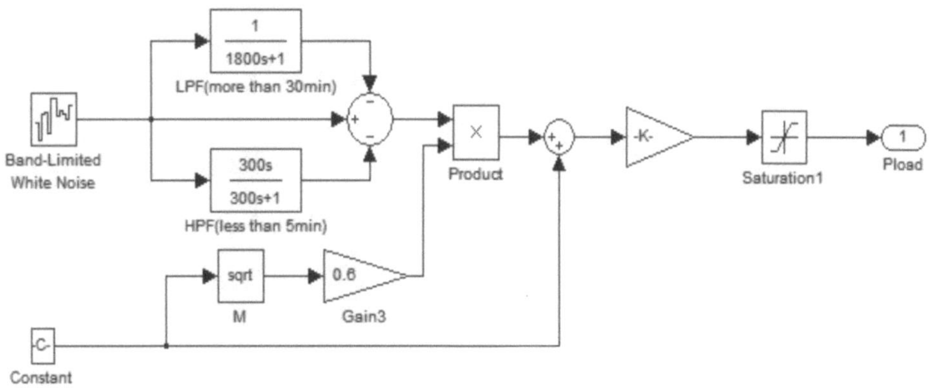

Fig. 8. Load model using MATLAB/SIMULINK

3. Problem formulation

Since frequency fluctuation, Δf, is effected mainly by the fluctuation of real power, power quality problems threatening frequency stability can be solved or avoided by satisfying the real power supply-demand balance constraint in the micro-grid power system (Li et al., 2007; Li et al., 2008). Therefore, an objective function for frequency control in the micro-grid system is formulized as follows.

$$\Delta P = P_{FC} + P_{WP} + P_{BESS} - P_{Load} - P_{ES} \to 0 \tag{14}$$

subject to

$$P_{WP}^{\min} \le P_{WP} \le P_{WP}^{\max} \tag{15}$$

$$P_{FC}^{\min} \le P_{FC} \le P_{FC}^{\max} \tag{16}$$

$$P_{ES}^{\min} \le P_{ES} \le P_{ES}^{\max} \tag{17}$$

$$P_{BESS}^{\min} \le P_{BESS} \le P_{BESS}^{\max} \tag{18}$$

4. Fuzzy control strategy of micro-grid power system

The battery energy storage system is the current attractive means of smoothing intermittent wind or solar power generation. Such battery, wind and/or solar power hybrid micro-grid power systems require a suitable control strategy that can effectively regulate fluctuant power output levels and battery state of charge (SOC); otherwise, an inconveniently and problematically large energy storage capacity is necessary.

Moreover, as above-mentioned, the power consumption of hydrogen electrolyzer system can be controlled in realtime. Therefore, to make full use of such response controllability, the power of electrolyzer system need to be increased in such a case that there is enough WP generation compared with system load in the micro-grid system. Otherwise, the power of electrolyzer system need to be decreased when there is less affluent WP generation compared with system load. These cases can be judged by calculating the difference between P_{WP} and P_{Load}.

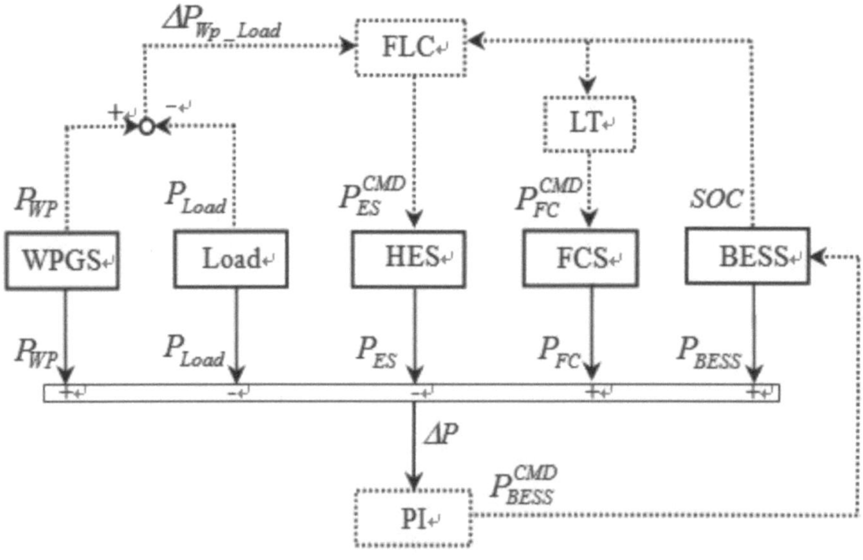

Fig. 9. Schematic diagram of proposed control strategy

For these considerations, in this chapter, a fuzzy logic controller is applied to adaptively regulate the power consumption of electrolyzer system and a PI controller is also introduced to minimize the real power imbalance ΔP in micro-grid power system. That is, the Mamdani-type fuzzy logic controller, FLC, shown in Fig. 9, for which the two inputs are battery SOC and ΔP_{WP_Load} and the one output is $P_{ES}{}^{FUZZY}$, is proposed. And, the PI controller, for which input is ΔP and output is $P_{BESS}{}^{CMD}$, is applied. Moreover, an one dimensional look up table (LT) block, shown in Fig.9, for which input is battery SOC and output is $P_{FC}{}^{CMD}$ is also used to modify output power of FC timely based on battery charging-level. So, the proposed method can enhance the minimizations of tie-line power and imbalance power levels for interconnected micro-grid power system and supervise the battery SOC to secure the charging-level of the BESS. The target power of BESS and that of ES can be determined by the following Eqs. (19)-(22).

$$P_{BESS}^{CMD} = \Delta P\left(K_p + K_i \int dt\right) \tag{19}$$

$$P_{ES}^{CMD} = P_{ES}^{FUZZY} = f\left(SOC, \Delta P_{WP\_Load}\right) \tag{20}$$

$$P_{FC}^{CMD} = f_{LT}\left(SOC\right) P_{FC}^{Initial} \tag{21}$$

Where,

$$SOC^{\min} \le SOC \le SOC^{\max} \tag{22}$$

Fuzzy membership function used in work is shown in Fig. 10. Fuzzy rule table for $P_{ES}{}^{FUZZY}$ is shown in Table 1. For the PI controller as shown in Fig. 9, the gain values of K_P and K_I, to reduce ΔP, are 1 and 10, respectively.

No	Logic of judgment rules
1	If *SOC* is *Big* & ΔP_{WP_Load} is *Big* then $P_{ES}{}^{FUZZY}$ is *Big*
2	If *SOC* is *Medium* & ΔP_{WP_Load} is *Big* then $P_{ES}{}^{FUZZY}$ is *Big*
3	If *SOC* is *Small* & ΔP_{WP_Load} is *Big* then $P_{ES}{}^{FUZZY}$ is *Small*
4	If *SOC* is *Big* & ΔP_{WP_Load} is *Medium* then $P_{ES}{}^{FUZZY}$ is *Medium*
5	If *SOC* is *Medium* & ΔP_{WP_Load} is *Medium* then $P_{ES}{}^{FUZZY}$ is *Medium*
6	If *SOC* is *Small* & ΔP_{WP_Load} is *Medium* then $P_{ES}{}^{FUZZY}$ is *Small*
7	If *SOC* is *Big* & ΔP_{WP_Load} is *Small* then $P_{ES}{}^{FUZZY}$ is *Small*
8	If *SOC* is *Medium* & ΔP_{WP_Load} is *Small* then $P_{ES}{}^{FUZZY}$ is *Small*
9	If *SOC* is *Small* & ΔP_{WP_Load} is *Small* then $P_{ES}{}^{FUZZY}$ is *Small*

Table 1. Fuzzy rule table for $P_{ES}{}^{FUZZY}$

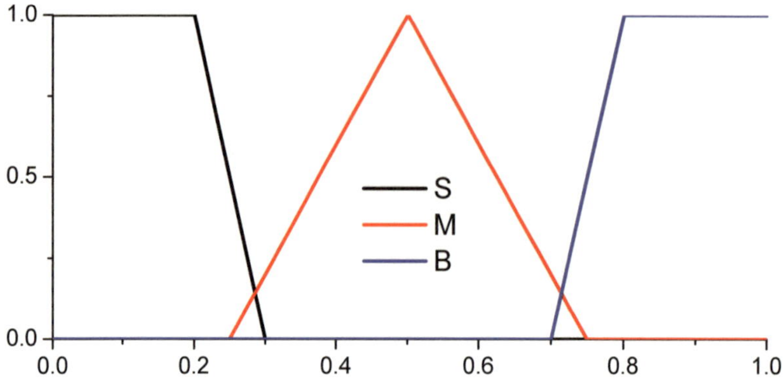

(a) SOC

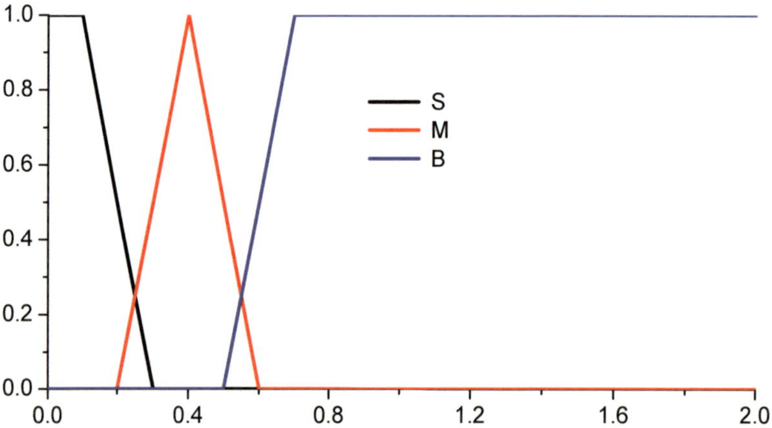

(b) ΔP_{WP_Load}

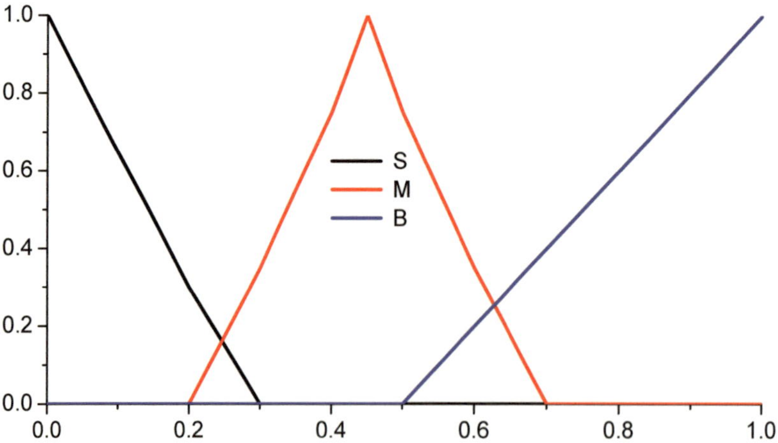

(c) P_{ES}^{FUZZY}

Fig. 10. Fuzzy membership function

5. Simulation results and discussion

The initial real power balance in the micro-grid is set to zero at the simulation starting point and the feasibility of the electrolyzer system control is examined over the course of 2000 seconds. The simulation results for islanding operation is presented as follows. The reference parameters of tie-line power and real power supply-demand imbalance are all set to zero. That is, in the case of interconnected operation with utility grid, the tie-line power flow target is set to zero. In the following, two cases for without/with fuzzy controller are discussed to illustrate the proposed fuzzy-logic-based control strategy. It is noted that the initial SOC of BESS is all set to 80% for the two cases, and the target power of electrolyzer system is set to a fixed value of 0.3MW in the case that fuzzy controller is ignored.

Figs. 11 and 12 show the power profiles of system load and WPGS. Figs. 13 and 14 show the power and SOC profiles of BESS, respectively. It is obviously apparent that the battery SOC is effectively controlled by introducing SOC based fuzzy controller. Otherwise, added capacity of BESS or additional operation is required. Figs. 15 and 16 show power profiles of electrolyzer and fuel cell systems with/without fuzzy controller. Real power imbalance profile in micro-grid system is shown in Fig. 17. Comparing Figs. 15, 16 and 17 shows that the proposed control method controls the power of the electrolyzer system and that of the fuel cell system to relax the wind power fluctuations and match the real power balance. Moreover, as shown in these figures, the result is that the real power imbalance is suitably controlled by the adaptive control of the electrolyzer system, and the power quality of frequency fluctuation is accordingly improved by applying the proposed fuzzy controller.

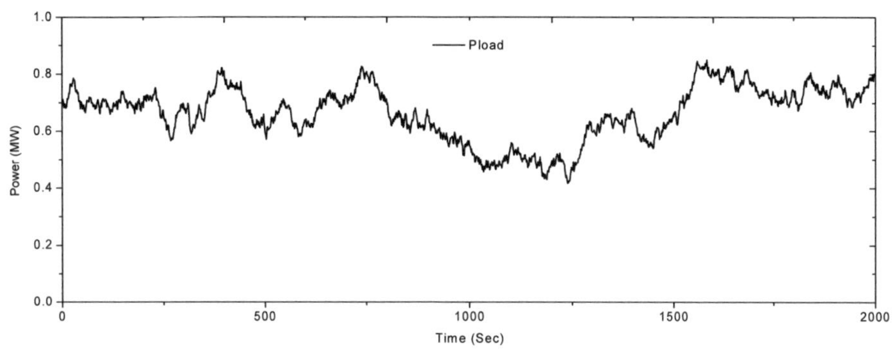

Fig. 11. Power profile of system load

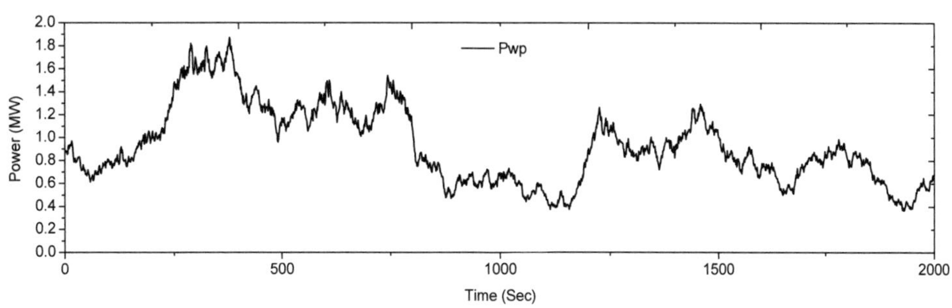

Fig. 12. Power profile of WPGS

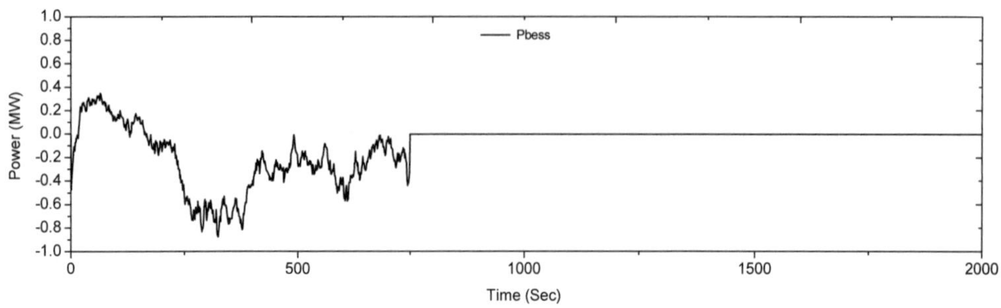

(13-1) without fuzzy controller

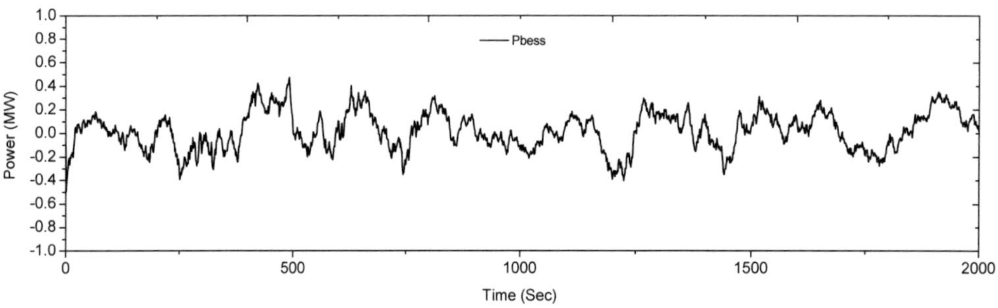

(13-2) with fuzzy controller

Fig. 13. Power profiles of battery energy storage system with/wothout fuzzy controller

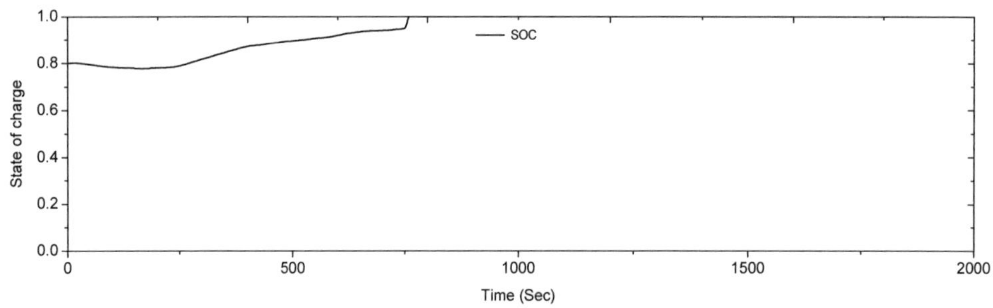

(14-1) without fuzzy controller

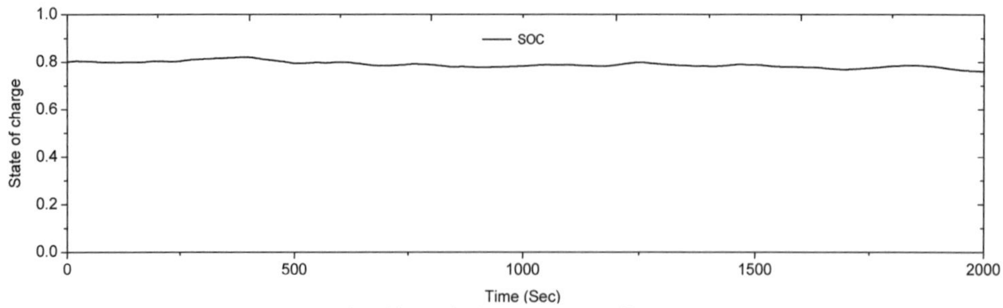

(14-2) with fuzzy controller

Fig. 14. SOC profile of battery energy storage system with/without fuzzy control

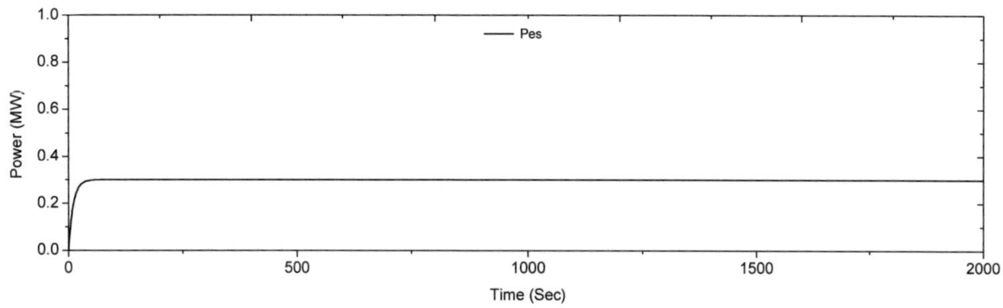

(15-1) without fuzzy controller

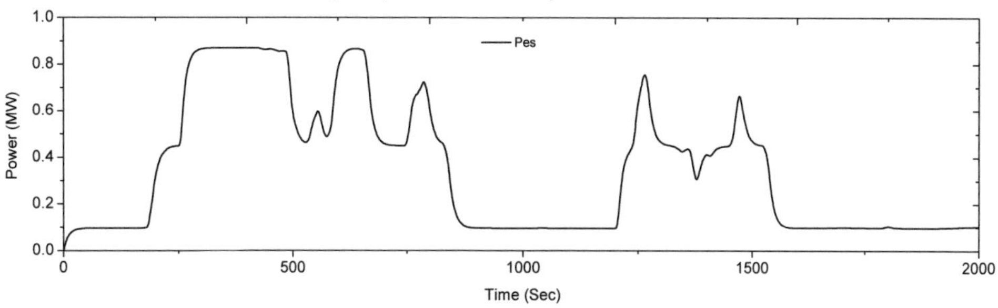

(15-2) with fuzzy controller

Fig. 15. Power profile of electrolyzer system with/wothout fuzzy controller

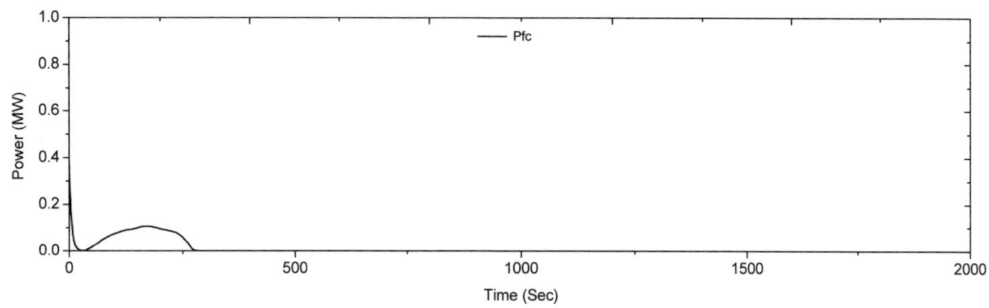

(16-1) without fuzzy controller

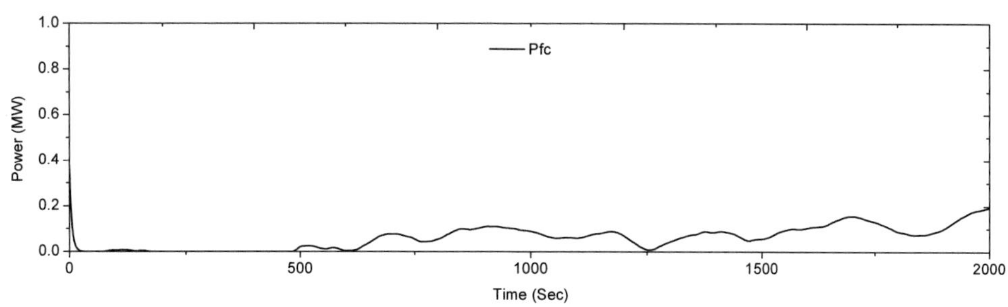

(16-2) with fuzzy controller

Fig. 16. Power profile of fuel cell system with/wothout fuzzy controller

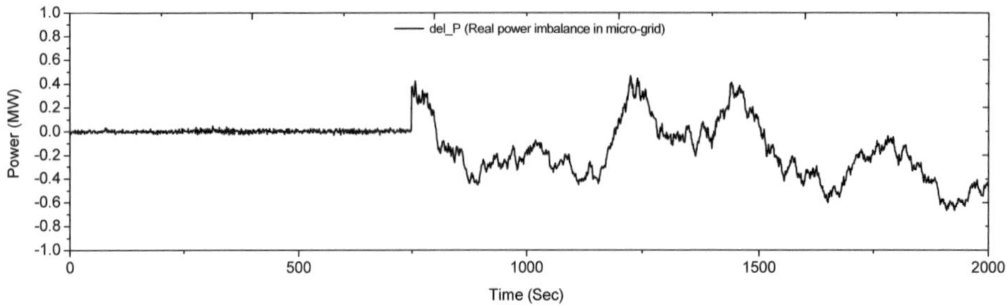

(17-1) without fuzzy controller

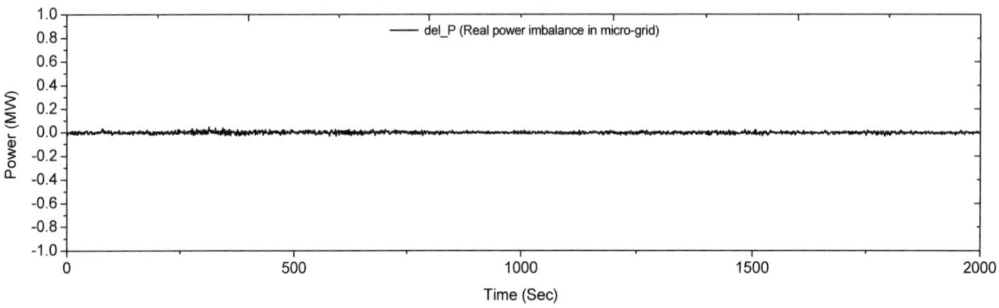

(17-2) with fuzzy controller

Fig. 17. Real power imbalance profile in micro-grid system with/without fuzzy controller

7. Conclusion

In this chapter, an electrolyzer system's dynamic control method, which secures a real power balance and enhances the operational capability to handle frequency fluctuation in multiple renewable energy hybrid micro-grid power system, is proposed. The PI and fuzzy controllers are introduced to compensate the power fluctuations in micro-grid power system. The effects of their usages are also explored through simulation experiment. Moreover, the fluctuation of real power imbalance under the islanding operational mode is discussed in demonstrating the effectiveness of the proposed control strategy. It is concretely shown that a power quality improvement for frequency can be achieved by actively utilizing battery energy storage system and load-controllable electrolyzer system. Moreover, the battery SOC can be managed within a specified target region. Therefore, the proposed control scheme helps to solve power quality issue resulting from real power imbalance fluctuations.

8. Acknowledgements

This work is supported by the National Program on Key Basic Research Project (973 Program) (Grant No. 2010CB227206) and by the Scientific Research Foundation for the Returned Overseas Chinese Scholars, State Education Ministry (Grant No. 20091001).

9. References

N. G. Hingorani & L. Gyugyi, (2000) Understanding FACTS-Concepts and Technology of Flexible AC Transmission Systems, New York, IEEE Press.

R. H. Lasseter, A. Akhil, C. Marnay, J. Stephens, J. Dagle, R. Guttromson, A. S. Meliopoilous, R. Yinger & J. Eto, (2002) Integration of Distributed Energy Resources–The CERTS MicroGrid Concept, *CERTS white paper*, Apr. 2002

R. H. Lasseter & P. Paigi, (2004) Microgrid: A Conceptual Solution, *in Proc. 35th Annu. Conf. IEEE PESC'04*, pp.4285-4290, 2004

Gregorio Marban. & Teresa Valdes-Solis, (2007). Towards the hydrogen economy? *International Journal of Hydrogen Energy* 32, pp.1625-1637, 2007

M. Matsubara, G. Fujita, T. Shinji, T. Sekine, A. Akisawa, T. Kashiwagi & R. Yokoyama, (2005) Supply and Demand Control of Dispersed Type Power Sources in Micro Grid, *in Proc. 13th Int. Conf. ISAP'05*, pp.67-72, 2005

T. Shinji, T. Sekine, A. Akisawa, T. Kashiwagi, G. Fujita, M. Matsubara, (2006) Discussion on Reduction of Power Fluctuation by Distributed Generation in Micro Grid, *IEEJ Trans. PE.* Vol.126. No.1, 2006. (In Japanese)

K. Yoshimoto, T. Nanahara, G. Koshimizu, (2006) New Control Method for Regulating State-of-Charge of a Battery in Hybrid Wind Power/Battery Energy Storage System, *IEEE Power Systems Conference and Exposition*, pp. 1244-1251, 2006.

A. Rajendra Prasad & E. Natarajan (2006) Optimization of integrated photovoltaic-wind power generation systems with battery storage, *Energy*, Vol. 31, Issue 12, pp.1943-1954, Sep. 2006

T.Senjyu, Y. Kikunaga, A.Yona, H. Sekine, A.Y. Saber & T. Funabashi, (2008) Coordinate control of wind turbine and battery in wind power generator system, *IEEE Power and Energy Society General Meeting-Conversion and Delivery of Electrical Energy in the 21st Century*, pp.1-7, Jul. 2008

S.M. Muyeen, Rion Takahashi, Toshiaki Murata, Junji Tamura & Mohd. Hasan Ali, (2009) Application of STATCOM/BESS for wind power smoothening and hydrogen generation, *Electric Power Systems Research*, Vol. 79, Issue 2, pp. 365-373, Feb. 2009

S. Lalouni, D. Rekioua, T. Rekioua, E. Matagne, (2009) Fuzzy logic control of stand-alone photovoltaic system with battery storage, *Journal of Power Sources*, Vol. 193, Issue 2, pp. 899-907, Sep. 2009,

Sercan Teleke, Mesut E. Baran, Alex Q. Huang, Subhashish Bhattacharya & Loren Anderson, (2009) Control Strategies for Battery Energy Storage for Wind Farm Dispatching, *IEEE Transactions on Energy Conversion*, Vol. 24, No. 3, pp. 725-732, Sep. 2009

M. Khalid & A.V. Savkin, (2010) A model predictive control approach to the problem of wind power smoothing with controlled battery storage, *Renewable Energy*, Vol 35, Issue 7, pp. 1520-1526, Jul. 2010

D. Kottick, M. Blau & D. Edelstein (1993) Battery Energy Storage for Frequency Regulation in an Island Power System, *IEEE Transactions on Energy Conversion*. Vol.8, pp.455-459, Sep. 1993

Xiangjun Li, Yu-Jin Song & Soo-Bin Han, (2007). Study on Power Quality Control in Multiple Renewable Energy Hybrid MicroGrid System, *IEEE PowerTech2007*, pp.2000-2005, Jul. 2007

Xiangjun Li, Yu-Jin Song, Soo-Bin Han, (2008) Frequency control in micro-grid power system combined with electrolyzer system and fuzzy PI controller, *Journal of Power Sources*, Vol.180, Issue 1, pp. 468-475, May. 2008

Xiangjun Li, Liangfei Xu, Jianfeng Hua, Xinfan Lin, Jianqiu Li, Minggao Ouyang. (2009) Power management strategy for vehicular-applied hybrid fuel cell/battery power system, *Journal of Power Sources*, Vol. 191, Issue 2, pp. 542-549, Jun. 2009

Xiangjun Li, Jianqiu Li, Liangfei Xu, Minggao Ouyang, Xuebing Han, Languang Lu, Chengtao Lin. (2010a) On-line management of lithium-ion battery based on time-triggered controller area network for fuel cell hybrid vehicle applications, *Journal of Power Sources*, Vol. 195, Issue 10, pp.3338–3343, May 2010

Xiangjun Li, Dong Hui, Xiaokang Lai, Minggao Ouyang, Yong Min. (2010b) Fuzzy logic based coordinated controller for wind/battery/IDSMS hybrid micro-grid power system, *The 2nd International Conference on Computer Engineering and Technology (ICCET2010)*, Vol. 4, pp. 263-267, Chengdu, Apr. 2010.

Xiangjun Li, Dong Hui, Xiaokang Lai, Minggao Ouyang. (2010c) Control Strategy of Wind Power Output by Pitch Angle Control using Fuzzy Logic, *IEEE International Symposium on Industrial Electronics (ISIE2010)*, pp.120-124, Bari, Jul. 2010.

Xiangjun Li, Dong Hui, Li Wu, Xiaokang Lai. (2010d) Control Strategy of Battery State of Charge for Wind/Battery Hybrid Power System, *IEEE International Symposium on Industrial Electronics (ISIE2010)*, pp.2723-2726, Bari, Jul. 2010.

A Discussion on Current Quality-Control Practices in Mineral Exploration

Armando Simón Méndez
AMEC International Ingeniería y Construcciones Limitada
Chile

1. Introduction

Vallée (1998) indicated that few exploration and mining companies have explicit and systematic quality-assurance policies, and identified three main approaches: laissez-fair, catch-as-catch-can, and systematic quality control, the latter being very uncommon. In the author's experience, this situation has not significantly improved in the intervening twelve years. Results of numerous recent audits and due-diligence jobs conducted on exploration and mining projects in South and North America, Asia, Africa, and Europe (many of them managed by North American and Australian companies) indicate that comprehensive geological quality-control programs are still relatively infrequent.

As a result of new regulations in place, junior and major companies are increasingly interested in implementing such programs, particularly when public financing is required. Unfortunately, the initial interest is often followed by shock and, sometimes, even by anger, when project management realizes that implementation of a quality-control program involves certain undesired modifications of the exploration budget.

Reluctance to implement comprehensive quality-control programs does not arise only from management or from budgetary constraints. Implementing such programs demands improved organization in the sampling process, database preparation, and data processing, and a dynamic and cooperative relationship with the laboratories. Due to the lack of appropriate training, project geologists do not always understand the need for such additional efforts, and often complain about these supposedly redundant control measures. In addition, laboratory personnel commonly identify geological quality control with lack of confidence in their results.

The purpose of this paper is to compare best quality-control practices with some current practices observed on international exploration and mining projects, and to discuss their direct implications on resource estimation and classification.

2. QA/QC definitions and principles

Quality assurance and quality control are the two major components of any quality management system. ISO (1994) defines quality assurance as "*the assembly of all planned and systematic actions necessary to provide adequate confidence that a product, process, or service will satisfy given quality requirements*", and quality control as "*the operational techniques and activities that are used to satisfy quality requirements*".

The quality-assurance program usually consists of written protocols that describe at least the sampling, sample preparation and assaying protocols, as well as the quality-control protocol (Shaw, 2009). Quality control includes the quality evaluation, or "*the system of activities to verify if the quality control activities are effective*" (Van Reeuwijk, 1998). While quality assurance aims at preventing problems, the purpose of quality control is detecting them, in the event that they occur, assessing their extent, and taking the appropriate actions to minimize their effects.
Rogers (1998) states that quality assurance policies and quality-control procedures in mineral exploration involve continuous monitoring of work processes and information flows, in order to ensure precise, accurate, representative, and reliable results, and to maximize data completeness (SMST, 2006). In the same line, quoting the Vancouver Stock Exchange Mining Standards Guidelines, Bloom (1999) writes that "*quality assurance programs should be routinely implemented as part of any exploration program that is generating analytical results. Such a program should verify the validity of sample collection, security, preparation, analytical method and accuracy*". Similarly, current international mining standards (JORC, 2004; CIM, 2003a, 2003b, 2005; CSA, 2005a) require that a program of data verification accompany any exploration program to confirm the validity of exploration data.

2.1 Precision

JCGM (2008) defines precision as the "*closeness of agreement between indications or measured quantity values obtained by replicate measurements on the same or similar objects under specified conditions*". Such definition can be in principle extended to repeatability (precision under repeatability conditions) and to reproducibility (precision under reproducibility conditions). Repeatability conditions imply the same measurement procedure, same operators, same measuring system, same operating conditions and same location, and replicate measurements on the same or similar objects over a short period of time; reproducibility conditions include different locations, operators, measuring systems or procedures, and replicate measurements on the same or similar objects. However, precision is usually measured on repeatability conditions.
Precision refers to the random measurement error, the component of the measurement error that in replicate measurements varies in an unpredictable manner (JCGM, 2008). Precision is commonly assessed through a series of repeated measurements on the same sample material, or through successive measurements on different original-duplicate sample pairs. For assessing precision on repeatability conditions, the original and the repeated (or duplicate) measurements should consider similar sampling intervals and similar sampling and sample-preparation procedures, and should be assayed at the same laboratory, with the same analytical techniques, the same equipment, the same reagents and the same personnel. However, such ideal situation can only be attained if the original and the duplicate samples are included in the same submission batch.
Precision should be treated as a qualitative attribute (i.e. low or lower precision, high or higher precision), and is quantitatively measured through parameters expressing imprecision, such as standard deviation, variance, or coefficient of variation in case of a series of repeated measurements (JCGM, 2008). For assessing the random error of a series of original-duplicate pair measurements, the most commonly used parameter is the average of the absolute value of the difference between the original and duplicate values, which is known as ARD, for absolute relative difference, or simply relative difference (Long, 1998, 2000), or relative error (Simón, 2004), as well as a related parameter, the HARD, for half

absolute relative difference (Shaw, 1997). Most recently, Stanley and Lawie (2007) and Abzalov (2008) have suggested the use of the average coefficient of variation CV_{AVR}, which is described as an unbiased estimate of the random measurement error. Whatever the parameter used for assessing the random error, though, there is an inverse relationship between precision and the corresponding quantitative parameter: the higher the random error, the lower the precision, and vice versa.

2.2 Accuracy

Measurement accuracy, or simply accuracy, is defined as the "*closeness of agreement between a measured quantity value and a true quantity value of a measurand*" (JCGM, 2008). All measurements have an uncertainty attached to them. Furthermore, different samples of the lot will have different concentrations of the element being measured, and this contributes additional uncertainty to the hypothetical true quantity value of the material's grade, which makes it, in fact, unknowable.

Accuracy refers to the systematic measurement error, the component of the measurement error that in replicate measurements remains constant or varies in a predictable manner (JCGM, 2008). Accuracy is commonly assessed through standard reference materials (SRMs), which are usually prepared from natural materials, or metallurgical concentrates, or tailings. When establishing the grade of these materials, the true quantity value is never known; however, under very controlled conditions, an accepted grade or "best value" of a particular element in a SRM can be established with a sufficient level of assurance through a round-robin test, consisting of multiple measurements at a series of reputable laboratories. The best value represents, in fact, the consensus value resulting from a round-robin test.

By inserting SRMs in sample batches, it is possible to compare, indirectly, the performance of the particular laboratory where measurements are conducted to the performance of other reference laboratories and, therefore, to assess the possible existence of bias between that particular laboratory and the consensus of other laboratories.

Accuracy should be treated as a qualitative attribute (i.e. low or lower accuracy, high or higher accuracy), and should not be given a quantity value (JCGM, 2008). Accuracy is quantitatively measured through the bias, which is the estimate of a systematic measurement error, or the difference between the expectation of the test results and an accepted reference value (AMC, 2003). The bias is usually calculated as the difference between the average value of a series of measurements of the SRM grade over a certain period of time and its best value, divided to the best value. However, there is an inverse relationship between accuracy and bias: the higher the absolute value of the bias, the lower the accuracy, and vice versa.

Accuracy can also be assessed through check samples, pulp samples initially assayed at a primary laboratory and re-submitted to a secondary laboratory. This way, it is possible to quantify the systematic error existing between the two laboratories. Nevertheless, this method should be complementary to the use of SRMs. Whereas in a sampling campaign few SRMs only characterize some fixed-grade values, the check samples reassayed at a secondary laboratory usually cover a wider value range. The combination of both methods leads to a more representative quantitative appraisal of the accuracy.

It is essential that a highly reputed and reliable laboratory be always chosen as the secondary laboratory. However, in spite of the fact that the secondary laboratory is considered as a reference laboratory, its accuracy should also be established or confirmed through the insertion of SRMs in the check-sample batches.

2.3 Contamination

Contamination consists of the inadvertent transference of material from one sample or the environment to another sample, and may take place during sample preparation and/or assaying, or merely through sample manipulation.

Contamination is assessed through blank samples, which are barren samples on which the presence of the elements undergoing analysis has been confirmed to be below the corresponding detection limit. A significant level of contamination is identified when the blank sample yields values exceeding several times the detection limit of the analysed element.

In order to be effective, blank samples should always be inserted after mineralized samples. Whenever possible, the matrix of the blank samples should be similar to the matrix of the material being routinely analyzed.

3. A comprehensive quality-control program

A comprehensive quality-control program should monitor various essential elements of the sampling-preparation-assaying sequence, in an effort to control or minimize the total possible measurement error:

- Sample collection and splitting (sampling precision)
- Sample preparation and sub-sampling (sub-sampling precision; contamination during preparation)
- Analytical accuracy, analytical precision, and contamination during assaying.

Monitoring these aspects is achieved through the random insertion (or submittal) of various control samples, each of them having a particular purpose in the quality-control protocol. The control samples will also be useful to alert about possible mix-ups or mislabelling produced during manipulation. Whenever possible, the identity of the control samples must remain "blind" to the analytical laboratory. In addition to control sample insertions (or submittals), some control operations are often conducted to assess certain aspects of the preparation process.

A quality-control program also examines the reporting (clerical or data transfer) accuracy. Best practice for monitoring reporting accuracy is double data entry of the manually-entered data, which consists of using two independent teams to enter the most sensible data into two independent databases, and subsequently cross-checking both data sets. A simpler alternative is regular checking, with the same method, of a representative proportion (at least 5%) of the entered data.

3.1 Controlling sampling quality

Sampling precision, the main indicator of sampling quality, is monitored through coarse-grained, uncrushed control samples that are inserted in the submission batches during the sampling operation. Such samples usually include twin samples and field duplicates.

- Twin samples:
  In the case of half-core sampling, a twin sample would ideally be the second half of the core, usually kept as backup. However, companies and auditors are reluctant to leave portions of the hole with no material geological record. Alternatively, twin samples are collected as a quarter-core, resulting from the double-split of the original core sample. In this case, the sample is initially cut in two halves, and then each half is again cut in two quarters, one quarter representing the original sample, and the adjacent quarter

(preferably from the opposed half) representing the twin sample; the remaining two quarters are usually stored as backup.
An alternative solution, which can be applied if the core is sufficiently compact and wide (HQ or larger diameter), is cutting a fillet or slice, which is kept as backup, and cutting the remaining core piece in half, one portion representing the original sample, and the remaining portion representing the twin sample (Glacken, 2006).
In the case of channel samples, the twin sample should be taken as a channel adjacent to the original channel, using the same interval and the same sampling procedure. Blast-hole twin samples are taken using a spear or auger device, a sampling tray or a slotted hole on the blast-hole cone.
Twin samples are mainly indicated to assess sampling precision and, indirectly, mineralization homogeneity. In order to ensure repeatability conditions, both the original and the twin samples should be taken by the same crew, be submitted to the same laboratory (the primary laboratory), in the same sample batch and under a different sample number, so that preparation and assaying follow similar procedures. The term "duplicate" is herein avoided, since the original and the twin sample do not occupy, formally, the same spatial position (Long, 2000).

- Field Duplicates:
These are samples taken from the first split of the original bulk reverse-circulation samples, without any previous crushing. The field duplicates are mainly used to assess the reverse-circulation sampling precision. In order to ensure repeatability conditions, both the original and the field duplicate samples should be taken from the same splitting step, be submitted to the same laboratory (the primary laboratory), in the same sample batch, and under a different sample number, so that preparation and assaying follow similar procedures.

3.2 Controlling preparation quality

Sub-sampling precision and contamination during preparation, reflecting preparation quality, are monitored through coarse-grained control samples that are inserted in the sample batch prior to or during preparation. Such samples usually include coarse blanks and coarse duplicates.

- Coarse Blanks:
These are coarse samples of barren material, emulating the granulometry of the ordinary samples (with fragments over 1" diameter for diamond drilling or channel samples, or over ¼" diameter for samples from reverse-circulation drilling). These samples are used to assess contamination during preparation, and should be inserted into the submission batch prior to dispatching the samples for preparation. In order to be most effective, the coarse blanks should be prepared immediately after highly mineralized samples. Blanks sometimes inserted in the first position of the batch are not considered part of the quality-control program, but just a quality-assurance provision, since they are actually used to clean the preparation equipment.
- Coarse Duplicates:
Coarse duplicates (also called preparation or coarse reject duplicates) are duplicate samples taken immediately after the first crushing and splitting step. The coarse duplicates will inform about the sub-sampling precision. In order to ensure repeatability conditions, both the original and the coarse duplicate samples should be submitted to the

same laboratory (the primary laboratory), in the same sample batch, and under a different sample number, so that pulverization and assaying follow similar procedures.

3.3 Controlling assaying quality

Assaying quality (analytical accuracy, analytical precision, and contamination during assaying) is assessed through fine-grained, previously pulverized control samples that are inserted in the sample batch after preparation and prior to assaying. Such samples usually include pulp duplicates, SRMs and fine blanks.

- Pulp Duplicates:
  These duplicates consist of second splits of finally prepared samples, analyzed by the same laboratory as the original samples under different sample numbers. The pulp duplicates are indicators of the analytical precision, which may be also affected by the quality of pulverization and homogenization. In order to ensure repeatability conditions, both the original and the pulp duplicate samples should be submitted to the same laboratory, in the same sample batch, and under a different sample number, so that assaying follows a similar procedure.
- Standard Reference Materials (SRMs)
  SRMs are samples with well established grades, prepared under specially controlled conditions. These samples should be included in regular submissions to the primary laboratory, as well as in check-sample submissions to the secondary laboratory, and will be used to assess the analytical accuracy.
  It is considered best practice to use at least three different SRMs for the most economically important elements and significant contaminants, covering the expected range of relevant concentrations. Minimum requirements are a low-grade SRM, with a grade close to the deposit cutoff; a medium-grade SRM, with a grade close to the average grade of the deposit; and a high-grade SRM, taking into consideration the grade level that for the particular deposit can be judged as high-grade (i.e., the grade corresponding to the 95th percentile).
  When choosing the SRMs, it is always recommended to minimize the matrix-related analytical effect by using SRMs of a composition as similar as possible to the composition of the routine samples. The ideal situation would be to prepare the SRMs from the same type of material that will be evaluated. If possible, the SRMs should not be used to evaluate the accuracy of the same laboratory where they were prepared.
- Fine Blanks:
  The fine blanks are pulverized samples of barren material. These samples will assess the eventual contamination during assaying. In order to be most effective, the fine blanks should be assayed immediately after highly mineralized samples.
  When coarse and fine blanks are inserted, the following order is recommended: after a highly mineralized sample, the first one should be a fine blank, and the second one should be a coarse blank. Hence, the fine blank will be assayed immediately after the high-grade sample, whereas the coarse blank will be prepared immediately after the high-grade sample.

3.4 Check samples

Check samples consist of second splits of finally prepared samples, routinely analyzed by the primary laboratory, and resubmitted to a secondary laboratory, under a different sample number. These samples are used to assess the assay accuracy of the primary laboratory relative to the secondary laboratory.

3.5 Control operations

In order to monitor the quality of the preparation process, it is highly recommended that particle-sizing tests be conducted at each stage of the comminution process (Shaw, 2009). In particular, regular sieve tests of the crushed and pulverized material are required, so that the proportions of material passing the stipulated sieve sizes are adequately known. Such tests are usually conducted by the primary laboratory after each crushing and grinding step. Other recommended control operations are weight checks before and after crushing, splitting and pulverization, to determine if significant losses of weight are produced.

When submitting check samples to a secondary laboratory, sieve checks should be requested to a certain proportion of the check samples, in order to obtain an independent assessment of the grinding quality achieved at the primary laboratory.

4. Control-sample insertion frequency

4.1 Current practice

During an examination of industry quality-control practices, the author reviewed current trends in control-sample insertion, using four main sources: well known international QA/QC consultants, SEDAR[1]-filed technical reports, documents from regulatory bodies, and information published in the Internet by exploration and mining companies in web sites and press releases (Simón, 2007).

- International QA/QC Consultants:
  A general agreement seems to exist between international QA/QC consultants about recommending an overall insertion rate of control samples close to 20% (Table 1).
- SEDAR-filed Technical Reports:
  The author reviewed a random selection of published NI 43-101 technical reports (16) resulting from placing Google® queries[2] for "technical report", "insertion rate", "qa/qc" and "43-101", with no preference for region, size of the company or type of mineral. Not all of the consulted reports had definite figures to describe the quality-control protocols, but those with detailed data are listed below:
  - Porcupine Project, Canada (GoldCorp): Coarse rejects: 5%; coarse blanks: 5%: pulp duplicates: 5%; SRMs, 5%; check samples: 5%. Total: 25%. Source: AMEC (2006)
  - Modder East Load Project, South Africa (sxr Uranium One Inc. and Aflease Gold Ltd.): Coarse blanks: 2%; SRMs, 9%; pulp duplicates: 11%; check samples: 2%. Total: 23%. Source: SRK (2007).
  - Perama Hill Project, Greece (Frontier Pacific Mining Corporation): Duplicates, 10%; other control samples: 9%. Total, approx. 19%. Source: RPA (2004).
  - Nuestra Señora, Mexico (Scorpio Mining Corporation): Coarse duplicates, 2.5%; SRMs+blanks, 2.5%; pulp duplicates, 5%; pulp check samples, 5%; coarse reject check samples, 2.5%. Total: 17.5%. Source: CAM (2006).

---

[1] SEDAR: System for Electronic Document Analysis and Retrieval, Canadian Securities Administration. www.sedar.com.

[2] Date of queries: 27 October, 2006.

Source	Details	Suggested Proportion of Control Samples
Rogers (1998)	Duplicates, SRMs, blanks: one in twenty; external checks: 5%	Approx. 20%
Vallée (1998)	10% duplicate plus SRMs, a 'somewhat lower figure' for rock sampling (?)	Approx. 15% (?)
Neuss (1998)	2%-5% field duplicates, 2%-5% coarse duplicates, 5%-10% internal pulp duplicates, 5%-10% external pulp duplicates, plus one SRM and one blank in every submission	Approx. 19% to 25%
Long (1998, 2000)	5% coarse reject duplicates, 5% pulp duplicates, 5% SRMs, one blank per batch (approx. 3%), check assays, a portion of the pulp duplicates (3%)	Approx. 21%
Sketchley (1999)	In a twenty-sample batch: one blank, one SRM, one duplicate; in addition, all pulp duplicates should be re-assayed at check lab	Approx. 20%
Bloom (1999)	In a twenty-sample batch: one blank, one SRM; in addition, sending one in ten sample pulps to an umpire lab	Approx. 20%
Lomas (2004)	In a twenty-sample batch: one blank, one SRM, one coarse duplicate and one pulp duplicate; in addition, 5% of the pulps should be re-assayed at check lab (including SRMs)	Approx. 25%

Table 1. Quality-Control Programs: Suggested Insertion Rates by Various Authors (After Simón, 2007)

- Twangiza Project, Congo (Banro Corporation): 2% coarse blanks, 8% SRMs; in addition, check assays (proportion not specified). Total: 15% (?). Source: Skead (2006).
- Mirador Project, Ecuador (Corriente Resources): Coarse duplicates: 5%; pulp duplicates: 5%; SRMs: 5%. Total: 15%. Source: MDA (2006).
- HC Property, Nevada (J-Pacific Gold): 5% twin samples, 5%; coarse duplicates, 5%; pulp duplicates, 5%. Total: 15%. Source: Durgin (2005).
- Pueblo Viejo, Dominican Republic (Placer Dome): 10% of SRMs and blanks. Source: AMEC (2005).

A general trend for using a 4% to 5% insertion rate for each type of control samples (duplicates, SRMs, blanks, check assays) could be observed, although in some cases particular sample subtypes are ignored. The insertion rate is less than 17% only when check assays are not included. An acceptable average is approximately 18%, with minor differences in some particular types of samples. The lack of duplicates in the Pueblo Viejo program invalidates it as an element for comparison.

In many of the studied examples, only one SRM was included in the quality-control program. When various SRMs were considered, sometimes there was no correlation between the grade levels of the SRMs and the actual sample grades. Not infrequently, the author has reviewed projects where SRMs have below cut-off values, or even close-to-detection-limit levels.

- Information from Exploration and Mining Companies:
  This information has been obtained mainly from press releases published in the Internet by a random selection of exploration and/or mining companies. The selection resulted from placing Google® queries[3] using "exploration", "mining", "qa/qc" and "insertion rate" as key words. Unfortunately, most companies do not offer details of their quality-control protocols in the press releases, but the author could find some examples:
  - Carpathian Gold (Colnic, Romania): Coarse blanks, 5%; CRMs, 4%; Check samples, 20% (before AMEC's Technical Report in 2006). Total: 29%. Source: www.carpathiangold.com/site06/images/CheckCode.pdf.
  - African Copper (Dukwe Project, Botswana): approx. 20% control samples. Source: www.mineweb.net/co_releases/302480.htm.
  - Aurelian Resources, FDN epithermal Au-Ag: CRMs, duplicates and blanks, 15%; in addition, samples from significant drill intercepts are sent to two reference laboratories. Total: 18% (?). Source: www.aurelian.ca/dynamic/press/pr-2006-08-21.pdf.
  - GlobeStar Mining. Regular practice: Duplicates: 4%; CRMs, 4%; blanks, 4%; check samples, 4%. Total: 16%. Source: www.globestarmining.com/content/standards.php.
  - Cambridge Mineral Resources: blanks, 5%; duplicates, 5%; CRMs; 5%. Total: 15%. Source: www.cambmin.co.uk/?page=press_releases&num=61.
  - Scorpio Mining Corporation (Nuestra Señora, Mexico): Coarse duplicates, 2.5%; CRMs, 2.5%; check assays: 5% pulps, 2.5% coarse rejects. Total: 12.5%. Source: www.scorpiomining.com/i/pdf/QAQC-NS.pdf.
  - Belvedere Resources: 12% control samples (only CRMs, blanks and duplicates). Source: www.belvedere-resources.com/rss/

  The same general trend for using a 4% to 5% insertion rate for each type of control samples (blanks, duplicates, SRMs, check assays) is observed. With the exception of the Nuestra Señora Project, the insertion rate is less than 16% only when check assays are not included. An acceptable average is approximately 20%, with minor differences in some particular types of samples.
- Regulatory Bodies:
  In 1999, the Toronto Stock Exchange and the Ontario Securities Commission prepared a document which later became the basis of NI 43-101 (CSA, 2005a). The document, named Setting New Standards (TSE-OSC, 1999), recommended that in a sample batch of 20 samples there should be a duplicate sample, a coarse blank, and an SRM. The document also recommended that previously assayed pulps be re-submitted to the same laboratory (rate not stated) and to another laboratory as check assays (rate not stated). The first three control samples would represent an overall 15% insertion rate, and the additional pulp re-assays (internal and external to the primary laboratory) would probably take the total to a figure close to 20%.

In conclusion, a general agreement appears to exist between the consulted sources: a 20% control-sample insertion rate has practically become an industry standard. However, most sources did not distinguish between duplicates or blank subtypes (twin samples, field duplicates, coarse and pulp duplicates, coarse and fine blanks), all of them with different functions in a comprehensive and properly conducted quality-control program.

[3] Date of the queries: 27 October, 2006.

In addition, the insertion of reference materials or the submission of samples for external control is often erratic and inconsistent. In many of the studied examples, only one SRM was included in the quality-control program. When various SRMs were considered, sometimes there was no correlation between the grade levels of the SRMs and the actual sample grades.

4.2 Recommended insertion rates

A comprehensive quality-control program should include all types and subtypes of control samples, so that precision, accuracy and possible contamination at the various points in the sampling-preparation-assaying sequence are properly assessed.
The quality-control programs should be tailored to the specific needs of each project and the average size of the analytical batches. Whereas an overall insertion rate of 20% is in principle recommended, the individual proportions of the various types of control samples should reflect those problems with higher probability of occurrence. With the advance of the project and the identification and correction of those problems, the absolute and relative proportions of control samples can be adjusted accordingly. A suggested, general-purpose insertion rate table is presented in Table 2.
It should be emphasized that the external check batches should also include pulp duplicates, SRMs and pulp blanks in appropriate proportions, so that precision, accuracy and possible contamination at the secondary laboratory could be independently assessed.

Sample Type	Sample Sub-Type	Suggested Insertion Rate	
Duplicates	Twin Samples	2%	6%
	Coarse Duplicates	2%	
	Pulp Duplicates	2%	
SRMs	SRMs	6%	6%
Blanks	Coarse Blanks	2%	4%
	Pulp Blanks	2%	
Check Samples	Check Samples	4%	4%

Table 2. Core-Drilling Quality-Control Program: Suggested Insertion Rates

5. Geological quality-control programs in the real world

As mentioned above, in spite of the fact that new regulations in place impose more strict requirements regarding the implementation of proper QA/QC programs (CSA, 2005a, 2005b; JORC, 2004), the author's experience on numerous audits and due-diligence jobs conducted in recent years on exploration and mining projects in South and North America, Asia, Africa, and Europe clearly indicates that exploration and mining companies should give additional attention to geological quality control.
The author evaluated the performance of the geological quality-control programs on 46 projects from South and North America, Africa, Asia and Europe audited or reviewed between 2007 and 2010. Each project received a qualitative score, based on how the author evaluated the actual degree of assessment of precision, accuracy and contamination by project geologists. The scores were assigned based on the following considerations:

- A (excellent): when precision, accuracy and contamination were fully monitored on real time, using a comprehensive control-sample selection and suitable procedures, and taking appropriate correction actions when required
- B (acceptable): when precision, accuracy and contamination were partially monitored on real time with an incomplete, but still reasonable control-sample selection, using appropriate procedures, and taking some corrective actions when required
- C (inadequate): when precision, accuracy and/or contamination were partially monitored with a very limited control-sample selection, unsuitable procedures, and not taking correction actions when required
- D (nonexistent): when a regular quality-control program was lacking.

In total, 14 projects (31%) were qualified as A or B, only three of them (7%) being ranked as A, versus 32 projects (69%) with qualifications of C or D, 13 of them (28%) being ranked as D (Table 3). This outcome represents an improvement when compared to a similar evaluation conducted three years ago on projects reviewed between 2003 and 2006 (Simón, 2008), when the proportions of projects ranked A-B and C-D were 16% and 84%, respectively (Table 3). Nevertheless, the current situation should still be considered unsatisfactory. Furthermore, a review of industry quality-control practices showed that, in spite of the fact that 43-101-compliant technical reports should describe in detail the QA/QC program and the nature and limitation of the verification, and should explain any problems encountered during data verification, no relevant details on the quality-control programs in place could be found in half of the consulted SEDAR-filed technical reports (Simón, 2007).

	2003-2006		2007-2010		Total	
Rank	**Number**	**Proportion**	**Number**	**Proportion**	**Number**	**Proportion**
D	11	42%	13	28%	24	33%
C	11	42%	19	41%	30	42%
B	3	12%	11	24%	14	19%
A	1	4%	3	7%	4	6%
Totals	26	100%	46	100%	72	100%

Table 3. Performance of Quality-Control Programs in Reviewed Projects (2003 to 2010)

6. Examples of poor quality-control practices

During the review of quality-control programs implemented in numerous exploration and mining projects elsewhere, the author has identified certain common practices that could lead to the inadequate assessment of precision, accuracy and contamination.

6.1 Practices leading to an inadequate assessment of precision

An inadequate assessment of precision commonly results from not respecting the repeatability conditions, but other factors are equally present in many reviewed quality-control programs. Examples of frequently observed poor practices are listed below.

- Making useless the assessment of precision by ignoring the results of the quality-control program:
    - Storing the duplicate data into the database and not processing them on a timely fashion, or not processing them at all.

- Providing an inadequate assessment of precision by putting in place an incomplete quality-control program:
  - Inserting only one or two duplicate types in the submission batches
  - Not maintaining a regular duplicate insertion frequency
  - Not conducting sieve tests to monitor the crushing and pulverization granulometry.
- Precluding the proper assessment of precision by not observing the repeatability conditions:
  - Using different crews to collect the original and the duplicate samples
  - Using different sampling methods to collect the original and the duplicate samples
  - Using different sampling intervals when collecting the original and the duplicate samples
  - Collecting the original and the duplicate samples at very different times
  - Collecting the field or the coarse duplicates from the last split of the bulk initial sample, instead of collecting them from the first split
  - Not inserting the duplicate samples in the same batches as the original samples
  - Using different analytical methods or conditions (aliquot size, detection limit) to assay the original and the duplicate samples
  - Using core, reverse-circulation material or coarse rejects to prepare new pulp samples and using them as pulp duplicates in order to monitor analytical precision, instead of using a split of the original pulp duplicate
  - Submitting the original samples to one laboratory and the duplicate samples to another laboratory.
- Affecting the assessment of precision due to poor handling practices that may produce sample mix-ups, wrong labelling and/or contamination.
- Preventing the assessment of precision by:
  - Not inserting duplicates in the regular submission batches
  - Not inserting duplicates in the check sample batches.

6.2 Practices leading to an inadequate assessment of accuracy

As explained above, accuracy can be assessed at the same laboratory, through the insertion SRMs in the batches, or by submitting check samples to a secondary laboratory. Examples of frequently observed poor practices are listed below.

- Making useless the assessment of accuracy by ignoring the results of the quality-control program:
  - Storing the SRM and check sample data into the database and not processing them on a timely fashion, or not processing them at all.
- Providing an inadequate assessment of accuracy by putting in place an incomplete quality-control program:
  - Inserting an insufficient number of SRMs in the submission batches
  - Not maintaining a regular SRM insertion frequency
  - Not submitting check samples to a secondary laboratory
  - Submitting twin samples or coarse or reject duplicates to a secondary laboratory instead of submitting pulverized check samples
  - Not inserting SRMs in the check-sample batches.

- Precluding the proper assessment of accuracy by not using suitable SRMs:
  - Inserting inadequately prepared or documented SRMs
  - Inserting SRMs not corresponding to the grades and/or composition of the mineralization of the regular samples.
- Jeopardizing the assessment of accuracy by not maintaining the anonymity of the SRMs:
  - Allowing the laboratory to identify the identity of the SRMs
  - Communicating the laboratory the best values of the SRMs.
- Affecting the assessment of accuracy due to poor handling practices that may produce sample mix-ups, wrong labelling and/or contamination.
- Preventing the assessment of accuracy by:
  - Not inserting SRMs in the regular submission batches
  - Not inserting SRMs in the check sample batches
  - Inserting supposedly valid SRMs, but not previously documented on a round-robin test.

A common problem in the assessment of accuracy is originated as a result of the construction and interpretation of the control charts used to process the SRM data. Control charts are helpful in assessing the stability of the analytical process. With that purpose, the center line of a control chart is defined as the average of a series of periodic measurements, around which the values vary at random, and process-dependant control limits are established on each side of the center line, usually at plus/minus three standard deviations (Mullins, 2003). The notion of control, therefore, has to do with precision, not with accuracy. For assessing the accuracy with SRMs, the first step should be identifying the in-control values, those values of the SRM that lie within the control limits, which implies that they are sufficiently precise to be used to estimate the bias during that particular period.

Many geologists prepare the control charts using the best value of the SRM as the center line, and define the control limits as a function of the standard deviation resulting from the round robin. However, the control limits for one process (the assays of a SRM at one laboratory) should not be established on the basis of another process (the assay of the SRM at various laboratories during the round-robin test). Furthermore, very often when supposedly out-of-control samples are identified, the laboratory is requested to repeat the entire batch.

Out-of-control values can be produced as a result of random errors, affecting only the SRM, or of systematic errors, affecting the entire batch. It is recommended that only the outliers (out-of-control samples) be repeated, together with some neighboring samples, to determine if the odd SRM value resulted from a random or a systematic error. Only in the latter case the repetition of the entire batch should be requested.

6.3 Practices leading to an inadequate assessment of contamination

As explained, above, contamination may be produced during preparation and during assaying. Examples of frequently observed poor practices in the assessment of contamination are listed below.

- Making useless the assessment of contamination by ignoring the results of the quality-control program:
  - Storing the blank sample data into the database and not processing them on a timely fashion, or not processing them at all.
- Providing an inadequate assessment of contamination by putting in place an incomplete quality-control program:

- Only inserting one type of blanks in the submission batches
- Not maintaining a regular blank insertion frequency.
- Precluding the proper assessment of contamination by not using suitable blanks:
  - Inserting blanks that were not previously analyzed to ensure that the grades of the studied elements are sufficiently low to allow the detection of contamination.
  - Inserting blank samples very different in composition from the regular samples
  - Using alluvial boulders and/or sand as the source of blank samples.
- Jeopardizing the assessment of contamination by not maintaining the anonymity of the blanks:
  - Allowing the laboratory to identify the identity of the blanks.
- Affecting the assessment of contamination due to poor handling practices that may produce sample mix-ups, wrong labelling and/or contamination.
- Preventing the assessment of contamination by:
  - Inserting blanks at the beginning of the batch, or following barren or poorly mineralized samples
  - Not inserting blanks in the regular submission batches
  - Not inserting blanks in the check sample batches.

7. Implication of poor quality-control programs for resource estimation

The importance of data quality for resource estimation can hardly be overemphasized. The use of the increasingly sophisticated mathematical processing methods on resource estimation can only be justified if the data on which they rely are sufficiently precise and accurate. On the other hand, resource classification is based on confidence of local and global estimations.

The first step in resource estimation is assessing the data reliability. Lacking proper quality-control data often results in having to exclude entire drilling campaigns from the estimation database, so affecting the resource tonnage, grade and classification. Poor precision due to the presence of coarse gold could lead to serious grade overestimation if this problem is not timely detected and addressed. Similarly, poor assay accuracy often require painful corrections, sometimes affecting the project feasibility. There is only one way to avoid unpleasant surprises when the critical moment arises: implementing from day one a comprehensive quality-control program, and following it strictly during the entire project life.

8. References

Abzalov, M. (2008). Quality Control of Assay Data: A Review of Procedures for Measuring and Monitoring Precision and Accuracy. *Exploration and Mining Geology*, Vol. 17; No. 3-4, July-October 2008, 131-144, ISSN 0964-1823.

AMC (2003). Terminology - the key to understanding analytical science. Part 1: Accuracy, precision and uncertainty. *AMC Technical Brief*, Nr. 13, September 2003. Available on line at http://www.rsc.org/Membership/Networking/InterestGroups/Analytical/AMCTechnicalBriefs.asp.

AMEC (2005): *Pueblo Viejo Project, Province of Sánchez Ramírez, Dominican Republic. NI 43-101 Technical Report*. Report prepared by AMEC Americas Limited for Goldcorp Inc. Project # 147107, October 16, 2005. Available on line at www.sedar.com.

AMEC (2006): *Review of Porcupine Joint Venture Operation, Ontario, Canada. NI 43-101 Technical Report*. Report prepared by AMEC Americas Limited for Goldcorp Corporation. Project # 152236, August 28, 2006. Available on line at www.sedar.com.

Bloom, L. (1999). *The role of economic geologists in evaluating assay data quality*. Unpublished manuscript.

CAM (2006): *Nuestra Señora Project, Sinaloa, Mexico. NI 43-101 Technical Report*. Report prepared by Chlumsky, Armbrust and Meyer, LLC. for Scorpio Mining Corporation. Project # 047251, April 17, 2006. Available on line at www.sedar.com.

CIM (2003a): *Exploration Best Practices Guidelines*. Adopted by CIM Council, August 20, 2000. Canadian Institute of Mining, Metallurgy and Petroleum. Available on line at http://www.cim.org/standards/MenuPage.cfm?sections=177&menu=178.

CIM (2003b): *Estimation of Mineral Resources and Mineral Reserves. Best Practice Guidelines*. Adopted by CIM Council, November 23, 2003. Canadian Institute of Mining, Metallurgy and Petroleum. Available online at http://www.cim.org/standards/MenuPage.cfm?sections=177&menu=178.

CIM (2005): *CIM Definition Standards for Mineral Resources and Mineral Reserves*. Prepared by the CIM Standing Committee on Reserve Definitions; adopted by CIM Council, December 11, 2005. The Canadian Institute of Mining, Metallurgy and Petroleum; 10 pp. Available online at http://www.cim.org/standards/MenuPage.cfm?sections=177&menu=178.

CSA (2005a). *National Instrument 43-101, Standards of Disclosure for Mineral Projects, Canadian Securities Administrators*. Available on line at http://www.cim.org/standards/MenuPage.cfm?sections=177,181&menu=229.

CSA (2005b). Companion Polity 43-101CP to National Instrument 43-101, Standards of Disclosure for Mineral Projects, Canadian Securities Administrators. Available on line at http://www.cim.org/standards/MenuPage.cfm?sections=177,181& menu=229.

Durgin, D. (2005): *2004 Year-End Technical Report Hc Property Eureka County, Nevada*. Report prepared for J-Pacific Gold, Inc. March 17, 2005. Available on line at www.sedar.com.

Glacken, I. (2006). Getting to the core of field duplicates. *Sampling & QAQC E-News*, May 2006, Snowden.

ISO (1994). *ISO 8402. Quality Management and Quality Assurance – Vocabulary*. International Organization for Standardization, Geneva, Switzerland.

JCGM (2008). *JCGM 200 :2008. International vocabulary of metrology – Basic and general concepts and associated terms (VIM)*. Joint Committee for Guides in Metrology. Available on line at http://www.iso.org/sites/JCGM/VIM/JCGM_200e.html.

JORC (2004). *Australasian Code for Reporting of Exploration Results, Mineral Resources and Ore Reserves* (The JORC Code, 2004 Edition). The Joint Ore Reserves Committee of the Australasian Institute of Mining and Metallurgy, Australian Institute of Geoscientists and Mineral Council of Australia.

Lomas, S. (2004). *QAQC Program*. General Discussion. AMEC Internal document.

Long, S. (1998). Practical quality control procedures in mineral inventory estimation. *Exploration and Mining Geology*, Vol. 7, No. 1-2, April 1998, 117-127, ISSN 0964-1823.

Long, S. (2000). *Assay Quality Assurance-Quality Control Program for Drilling Projects at the Pre-feasibility to Feasibility Report Level*. Mineral Resource Development Inc., Internal Report.

MDA (2006): *Technical Report Update on the Copper, Gold, and Silver Resources and Pit Optimizations, Mirador Project, Ecuador*. Report prepared by Mine Development Associates for Corriente Resources Inc. Project # 775-856-5700, May 18, 2006. Available on line at www.sedar.com.

Mullins, E. (2003). *Statistics for the Quality Control Chemistry Laboratory*. The Royal society of Chemistry, ISBN 0-85404-671-2, Cambridge.

Neuss, I. (1998). Sampling Legends - What Can We Learn From Busang? In : *More Meaningful Sampling in the Mining Industry*. Brian Davis and Susan E. Ho (eds.). Australian Institute of Geoscientists, Bulletin 22, 109-117, ISSN 0812-6089.

Rogers, R.S. (1998). Forensic Geology and Mineral Exploration Projects. *Exploration and Mining Geology*, Vol. 7, No. 1-2, April 1998, 25-27, ISSN 0964-1823.

RPA (2004): *Report on Perama Hill Gold Deposit Mineral Resource Estimate, Greece. NI 43-101 Technical Report*. Report prepared by Roscoe Postle Associates Inc. for Frontier Pacific Mining Corporation, May 13, 2004. Available on line at www.sedar.com.

Shaw, W.J. (1997). Validation of sampling and assaying quality for bankable feasibility studies. *Proceedings of The Resource Database Towards 2000*, 69-79, ISBN 1-875776-54-0, Wollongong, May 1997, Australasian Institute of Mining and Metallurgy.

Shaw, W.J. (2009). *Sampling QAQC*. Golder Associates, April 2009. Available on line at http://minewiki.org/index.php/Sampling_QAQC.

Simón, A. (2004). *Evaluation of Twin and Duplicate Samples : The Hyperbolic Method*. AMEC Perú Ltda. Internal document.

Simón, A. (2007). Control Sample Insertion Rate: Is There an Industry Standard? *Program and Abstracts of the 23rd International Applied Geochemistry Symposium (IAGS)*, ISBN 978-84-690-6461-0, Oviedo, Spain, 14-19 June 2007, The Association of Applied Geochemistry, Oviedo.

Simón, A. (2008). Geological QA/QC in the Real World. *PDAC 2008 Geochemistry Workshop*, Toronto, 1 March 2008.

Skead, M.B. (2006). *NI 43-101 Technical Report, Twangiza Project, South Kivu Province, Democratic Republic of the Congo*. Report prepared by Banro Corporation, March 30, 2006. Available on line at www.sedar.com.

Sketchley, D. (1999). Case history guidelines for establishing sampling protocols and monitoring quality control. *Proceedings of CIMM Annual General Meeting Symposium on Quality Control of Resource Estimations: an ISO Perspective*.

SMST (2006). *Idaho National Laboratory, Site Environmental Report, Calendar Year 2005. Environmental Surveillance, Education and Research Program*. Report prepared for the U.S. Department of Energy Idaho Operations Office under Contract DE-AC07-06ID14680 by the S. M. Stoller Corporation.

SRK (2007): *Dominion Uranium Project North-West Province, Republic of South Africa. NI 43-101 Technical Report*. Report prepared by SRK Consulting for sxr Uranium One Inc. Project # 370896, March 2, 2007. Available on line at www.sedar.com.

Stanley, Cliff R. and Lewie, David (2007). Average Relative Error in Geochemical Determinations: Clarification, Calculation, and a Plea for Consistency. *Exploration and Mining Geology*, Vol. 16, No. 3-4, July-October 2007, 267-275, ISSN 0964-1823.

TSE-OSC (1999). *Setting New Standards: Recommendations for Public Mineral Exploration and Mining Companies*. Toronto Stock Exchange-Ontario Securities Commission Mining Standards Task Force.

Vallée, M. (1998). Sampling Quality Control. In: *Quality Assurance, Continuous Quality Improvement and Standards in Mineral Resource Estimation*. Exploration and Mining Geology, Vol. 7, Nos. 1 and 2, January 1998, 107-116. ISSN 0964-1823.

Van Reeuwijk, L.P. (1998). Guidelines for Quality Management in Soil and Plant Laboratories. *FAO Soils Bulletin*, 74, Food and Agriculture Organization, Rome, ISSN 0253-2050.

31

Sampling Errors and Control of Assay Data Quality in Exploration and Mining Geology

Marat Abzalov
37 Belmont Avenue, Belmont, WA6104
Australia

1. Introduction

The fundamental cause of the errors of samples of rocks and minerals collected by geologists for evaluation of mining projects is heterogeneity of the sampled materials (Gy, 1982; Francois-Bongarcon, 1993; Pitard, 1993). Constitution heterogeneity and distribution heterogeneity (Pitard, 1993) both are important and cause geological sampling errors. The more heterogeneous the sampled material the more difficult it is to obtain a representative sample and infer characteristics of the geological object from samples. The current chapter overviews sampling theory explaining sampling error types and their likely causes, and also describes the practical approaches used in the mining industry for estimating sampling errors and monitoring them at an acceptably low level. It is based on numerous case studies by the author (Abzalov & Both, 1997; Abzalov, 1999, 2007, 2008; Abzalov & Humphreys, 2002; Abzalov & Mazzoni, 2004; Abzalov & Pickers, 2005; Abzalov et al., 2007; Abzalov & Bower, 2009) and also reviews of the recently published QAQC procedures used in the mining industry (Taylor, 1987; Vallee et al., 1992; Leaver et al., 1997; Long, 1998; Sketchley, 1998).

2. Types of sampling errors

Sampling errors are traditionally determined in terms of precision and accuracy of the data (Fig. 1). Precision, or repeatability, is a measure of how close sample values are to one another (Fig. 1) and accuracy is a measure of how close the sample value to the true grade (Fig. 1). Both of these parameters have to be estimated and strictly monitored during evaluation and the eventual exploitation of mineral deposits.
These errors can be generated at any stage of the samples extraction, preparation and the eventual analytical assaying. Depending on the factors causing sample errors they are grouped into three types (Eq. 1):

$$\text{TOTAL ERROR} = \text{Err. }1^{st}\text{ Group} + \text{Err. }2^{nd}\text{ Group} + \text{Err.}3^{rd}\text{ Group} \qquad (1)$$

Where:
Err.1st Group – are sampling errors related to a chosen sample extraction and preparation procedure, referred as sampling protocol. An example is poor repeatability of assays when sample sizes are disproportionately small in comparison with the degree of heterogeneity of

material. The main error of this type is known as Fundamental Sampling Error (Gy, 1982). It is always present and can not be fully eliminated as it is related to intrinsic characteristics of the sampled material, such as mineralogy and texture of mineralisation. The Fundamental Sampling Error (FSE) can be minimised through optimisation of the sampling protocols, which will be discussed in the next section. The first group also includes Grouping-Segregation error which is a consequence of the distribution heterogeneity of the sampled material (Pitard, 1993) and therefore this error also relates to the intrinsic characteristics of the sampled material.

Err.2nd Group - is the group of errors related to sampling practise, in other words the errors which depend on how rigorously the sampling protocol was developed, implemented and followed. The group includes delimitation, extraction, preparation and weighing errors. These errors are caused by incorrect extraction of the samples from a lot, their suboptimal preparation procedures, contamination and incorrect measurements. Human errors, such as mixed sample numbers, can also be included in this group. These types of errors can be minimised by upgrading practices of the samples extraction and preparation, which usually needs an improvement of the quality control procedures and often requires equipment upgrading.

Err.3rd Group - analytical and instrumental errors occurred during the analytical operations (Gy, 1982). The group includes assaying, moisture analysis, weighing of the aliquots, density analysis, precision errors and bias caused by suboptimal performance of analytical instruments. These errors are considered in the current study separately from the two first groups because of the different factors causing them.

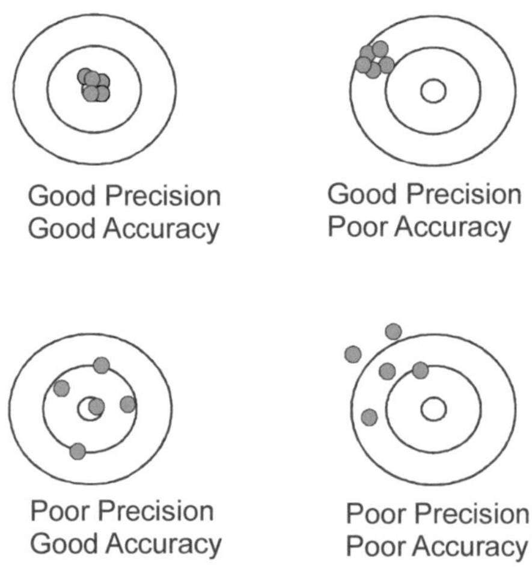

Fig. 1. Sketch explaining precision and accuracy of the data

2.1 Fundamental Sampling Error

Fundamental Sampling Error (FSE) is error related to constitution heterogeneity of the sampled material. It depends on the shape and size of the particles which constitute the sampled material, the size at which the critical components are liberated and also on mineralogy and density of gangue and valuable components. It is the only error which can be theoretically determined as it directly related to the constitutional characteristics of the sampled materials.

a. Theoretical background.

The theoretical approach for estimating the FSE was proposed by P.Gy (1982) and further developed by F.Pitard (1993) and D.Francois-Bongarcon (1993, 1998, 2005). The theory states that FSE, representing precision of the samples expressed as their relative variance, can be estimated as follows (Eq. 2):

$$\sigma^2_{FSE} = fgcld^3_N \left(\frac{1}{M_S} - \frac{1}{M_L}\right) \tag{2}$$

σ^2_{FSE} - Fundamental Sampling Error, representing relative variance of the precision error;
f - shape factor. This parameter represents geometry of the particulate materials. It is a dimensionless factor varying from zero, when particles are ideal cubes, to one, when they represented by ideal spheres. Most types of mineralisation have shape factor varying in a narrow range from 0.2 (gold or mica flakes) to 0.5 (isometric grains).

g - granulometric factor, which is also called a particle size distribution coefficient or size range factor. This factor is dimensionless and taking into account the fact that fragments do not have the same size (d). If all fragments have had exactly the same size the factor (g) would be equal to 1. This theoretically is possible only in an ideal case when studied material is perfectly sorted. In practice it never happens, therefore the (g) factor is less than one and can be as small as 0.1 when particles show a wide range of distribution. Default values of (g) factor are summarised in Table 1. In the mining industry the value of 0.25 is usually used as default value as it suits for most types of mineralisation and corresponds to a case when 95% of particles pass the nominal mesh size.

Type	**Explanation**	**Default (g) value**
Non sorted material	Output of jaw crusher	0.25
Sorted material	Material between two consecutive screen openings	0.55
Naturally sorted material	Grains, e.g. rice	0.75

Table 1. Default values of the granulometric factor (Pitard, 1993)

d_N - a nominal particle size in centimetres. This is size (diameter) of a mesh retaining the upper 5% of particles.

M_S - mass of sample in grams.

M_L - mass of lot in grams.

c - mineralogical composition factor (g/cm³) calculated using (Eq. 3)

$$c = \left(\frac{1-t_L}{t_L}\right) \times (\rho_M (1-t_L) + \rho_G t_L) \tag{3}$$

where:

t_L - absolute grade of a lot expressed as decimal proportions of ore mineral, it changes from 0 to 1 (e.g. 1g/t = 0.000001), ρ_M - specific gravity of ore minerals, ρ_G - specific gravity of gangue.

The formula (Eq. 3) can be simplified (Francois-Bongarcon, 1998) and represented by it's concise version (Eq. 4).

$$c = (\frac{1\text{-}t_L}{t_L}) \times (\frac{\rho_M \times \rho_G}{\rho}) \qquad (4)$$

In the equation (Eq. 4) (ρ) denotes the average specific gravity of mineralisation at a given grade (t_L), other variables are the same as in Eq. 3.
For low-grade ores, a mineralogical factor (c) can be further simplified and approximated as ratio of the density of the mineral of interest by the average grade of the studied material (Eq. 5):

$$c = \frac{\rho_M}{t_L} \qquad (5)$$

The mineralogical factor (c) relates the sampling variance given by formula (Eq. 2) to the grade of mineralisation (lot) being sampled. D.Francois-Bongarson and P.Gy (Francois-Bongarson & Gy, 2001) have noted that 'any use of the formula, or any sampling nomogram derived from it, only makes sense when the grade level at which it is established is duly stated'.
l - liberation factor, estimated as ratio of liberation size to a nominal particle size (Eq. 6).

$$l = (\frac{d_L}{d_N})^A \qquad (6)$$

where: d_N - a nominal particle size in centimetres, d_L - liberation size in centimetres, representing a liberation diameter of a mineral of interest, A - exponent.
Substituting liberation factor (Eq. 6) to equality defining FSE it becomes (Eq. 7):

$$\sigma^2_{FSE} = fgc\,(\frac{d_L}{d_N})^A d_N^3\,(\frac{1}{M_S} - \frac{1}{M_L}) \qquad (7)$$

If the exponent (A) is expressed as ($3-\alpha$) and after reduction of (d_N^3) the FSE formula becomes (Eq. 8)

$$\sigma^2_{FSE} = f\,g\,c\,d_L^{3-\alpha}\,d_N^{\alpha}\,(\frac{1}{M_S} - \frac{1}{M_L}) \qquad (8)$$

Product of ($f\,g\,c\,d_L^{3-\alpha}$) is known as the sampling constant (Francois-Bongarson, 1993; De Castilho et al., 2005) and usually is denoted as (K).

$$K = f\,g\,c\,d_L^{3-\alpha} \qquad (9)$$

Substituting sampling constant (K) to equality (Eq. 2) leads to formula of FSE (Eq. 10) which is most commonly used in practice.

$$\sigma^2_{FSE} = Kd_N^{\alpha}\,(\frac{1}{M_S} - \frac{1}{M_L}) \qquad (10)$$

The value of exponent (α) changes depending on (d_N). When (d_N) is smaller than liberation size (d_L) the exponent (α) is equal 3. Above the liberation size (d_L) the exponent (α) can be smaller, within the range of 1 to 3.
Equality (Eq. 10) can be further simplified by removing the ratio ($\frac{1}{M_L}$) which becomes negligibly small when the mass of a lot (M_L) is significantly larger than sample mass (M_S), which leads to concise version of the FSE formula (Eq. 11)

$$\sigma^2_{FSE} = \frac{K\, d_N^{\alpha}}{M_S} \tag{11}$$

Equality (Eq. 10) and its concise version (Eq. 11) are practically the most convenient tools for experimental definition of the FSE because parameters (K) and (α) can be calibrated experimentally (Francois-Bongarson, 1993, 2005). Methods of calibration are discussed in the next section of the book. When calibrated parameters are not available D.Francois-Bongarcon (1993) has suggested default (K) and (α) values for low-grade mineralisation, such as gold veins, which are K = 470 and α =1.5. However, great care should be taken as actual values of the sampling constant (K) can significantly differ from the default value (Sketchley, 1998).
b. Experimental calibration of sampling constants.
Several techniques have been proposed (Gy, 1982; Pitard, 1993; Francois-Bongarson, 1993, 2005; Bartlett & Viljoen, 2002; Minkkinen & Paakkunainen, 2005; De Castilho et al., 2005; Minnitt et al., 2007) for experimental determination of sampling constants. The most common approach is the technique developed by Francois-Bongarson (2005), representing a modified version of the 'sampling tree experiment' (Francois-Bongarson, 1993), and 'heterogeneity test' of Pitard (1993). '30-Pieces Experiment' developed by D.Francois-Bongarson (1993) has many similarities to above mentioned 'heterogeneity test' (Pitard, 1993) representing a simplified version of it.
'Sampling Tree Experiment' was first proposed by D. Francois-Bongarcon in 1993 and then modified in 2005 (Francois-Bongarson, 2005). The modified version represents analysis of the series of the duplicate samples (Fig. 2) cut from a lot at various comminution degrees (Table 3) allowing to experimentally obtain the (K) and (α) parameters of the Fundamental Sampling Error (Eq. 11).

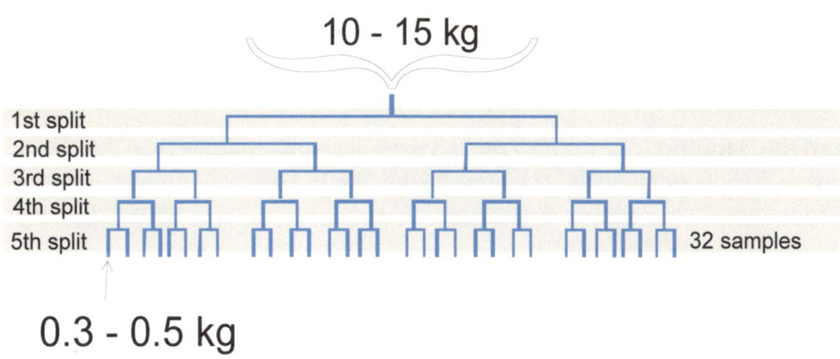

Fig. 2. Flow sheet of the 'Modified Sampling Tree Experiment' (MSTE). The shown binary sampling tree is applied to each of the four nominal size fractions (Table 2)

Theoretical background of this method is as follows. Firstly the formula (Eq. 11) can be logarithmically transformed to the equality Eq. 12:

$$\mathrm{Ln}\,(M_S\sigma^2_{FSE}) = \alpha\ \mathrm{Ln}(d_N)\ + \mathrm{Ln}(K) \qquad (12)$$

According to this expression the values of $[\,\mathrm{Ln}\,(M_S\sigma^2_{FSE})\,]$ are plotted against the particle sizes $[\,\mathrm{Ln}\,(d_N)\,]$ as a straight line because the equality (Eq. 12) represents equation of the line ($Y = AX + B$). Tangent of the angle between this line and the abscissa axis (A) is equal to exponent (α) in the equality Eq. 12 and constant (B) is equal to $[\,\mathrm{Ln}(K)\,]$. The objective of the 'MSTE' is to deduce parameters (A) and (B) of a linear function describing relationships between ($M_S\sigma^2_{FSE}$) values and the particle sizes (d_N). In practise, to infer parameters of a linear function $[\,\mathrm{Ln}\,(M_S\sigma^2_{FSE}) = \alpha\ \mathrm{Ln}(d_N)\ + \mathrm{Ln}(K)\,]$ it is sufficient to experimentally obtain several points which are plotted onto the diagram $\mathrm{Ln}\,(M_S\sigma^2_{FSE})$ vs. $\mathrm{Ln}(d_N)$ and then a linear function is inferred by a suitable best fit algorithm.
'MSTE' method is based on collecting a representative sample of 40-60kg which is then dried, successively crushed and split following the flow sheet shown on the Fig. 2. The nominal particle sizes for the four groups of subsamples depend on mineralogy and texture of the mineralisation. Examples of the particle sizes that have been used at the 'MSTE' are shown in Table 2 which can be used as a reference when 'MSTE' is planned; however, best practise is to determine experimentally the sample weight and the nominal particle size of each sampling series.

		Sampling Series				Elements of Interest	Reference
		First	Second	Third	Forth		
Deposit type	**Orogenic Gold**	2.5	0.3	0.1	0.05	Au, As	a, b
	Ni-S: Komatiitic-type	3	1	0.5	0.1	Ni, Cu, As	b
	Cu-Au-U: IOCG-type	2.5	0.5	0.1	0.05	Cu, U, Au, S	b
	U: unconformity-type	2.5	1	0.1	0.01	U	b
	Bauxite	3	1	0.5	0.1	Al_2O_3, SiO_2, Fe, LOI	b
	Iron ore: BIF-derived	3	1	0.5	0.1	Al_2O_3, SiO_2, Fe, LOI, P	b
	Cu-Au: porphyry-type	2.5	1	0.1	0.05	Cu, Mo, Au, As	b

LOI – Loss on Ignition
Data used: a – (Minnitt et al., 2007); b – Abzalov, M. (unpublished data).

Table 2. Examples of the nominal particle sizes (cm) of the samples at the "Modified Sampling Tree Experiment"

Procedure of the 'Modified Sampling Tree Experiment' is as follows (Fig. 2):
Representative sample of 40-60kg is collected and dried;

- The whole sample is crushed at jaw crusher to a nominal size of 95% passing the mesh size chosen for Series 1 (Table 2);
- One-quarter of the sample (lot) is split out and forms the first subsampling series;
- Remaining material is crushed to a nominal size of 95% passing the mesh size chosen for Series 2 (Table 2);
- One-third of these secondary crushed material is split out and forms the second subsampling series;

- Remaining two fractions are recombined and crushed to a nominal size of 95% passing the mesh size chosen for Series 3 (Table 2);
- The crushed material is split by riffle splitter onto two equal subsamples, one of them split out and forms the third subsampling series;
- The remaining material is crushed to a nominal size of 95% passing the mesh size chosen for Series 4;
- Using a riffle splitter each of these portions is now split into 32 samples (Figure 2). Each of the produced samples is weighed, pulverised and assayed. Minnitt et al. (2007) recommends to use 2 samples for granulometric analysis. These samples are randomly chosen from the group of 32 and remaining 30 samples are assayed and used for statistical inference of the (K) and (α) parameters (Fig. 3).

This approach produces 4 groups of 32 samples. Each group includes samples of the same nominal size of the particles and approximately of equal weight.

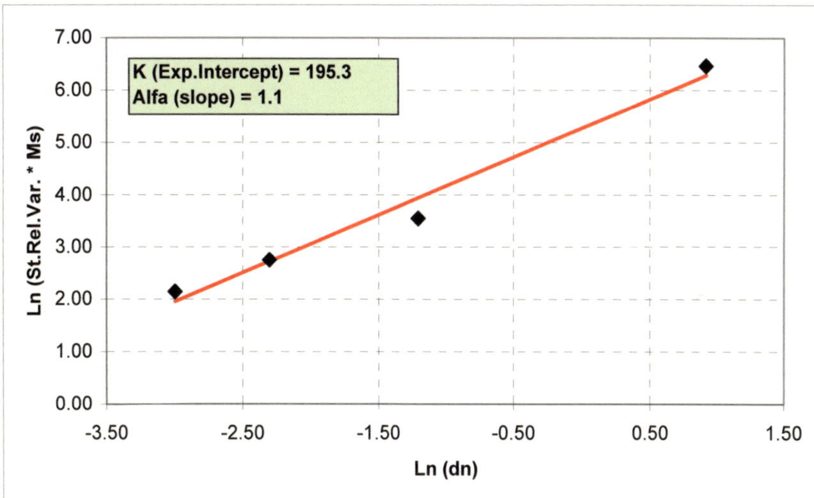

Fig. 3. Experimental calibration of the (K) and (α) parameters of the Eq. 12

c. Sampling nomogram.

Variance of the fundamental sampling error (FSE) can be graphically expressed as a function of sample weight (M_S) and the nominal particle size (d_N). Diagram representing relationships between these parameters is called nomogram (Fig. 4). This is achieved by plotting the FSE vs. the given sample mass (Pitard, 1993; Francois-Bongarcon, 1993). For practical reasons all values are plotted on the nomogram in the logarithmic coordinates (Fig. 4).

On the sampling nomogram (Fig. 4) the crushing and grinding stages, which do not contribute to sampling variance, are represented by vertical lines. The sample reduction stages, when a smaller sample is extracted from a larger sample, in other words a sample mass reduction at constant rock particle size, are represented on the diagram as a path along the straight lines of a slope -1. The actual position of the line depends on particle size (d_N) and also sampling constants (α) and (K) therefore only one line can be constructed for each sub sampling stage at the given sample particle size (d_N). Every stage when particle size is reduced (i.e. comminution) is lowering the FSE and therefore it is represented on the nomogram by vertical line which extended until it intersects a new straight line of slope -1 corresponding to a new particle size. Combination of the vertical and diagonal segments corresponding to all stages of the sampling protocol graphically visualises the entire sample preparation protocol. Each stage contributing to the overall precision error is clearly represented on the nomogram and this allows to use this diagram as an effective tool for assessment, control and improvement of the sampling procedures.

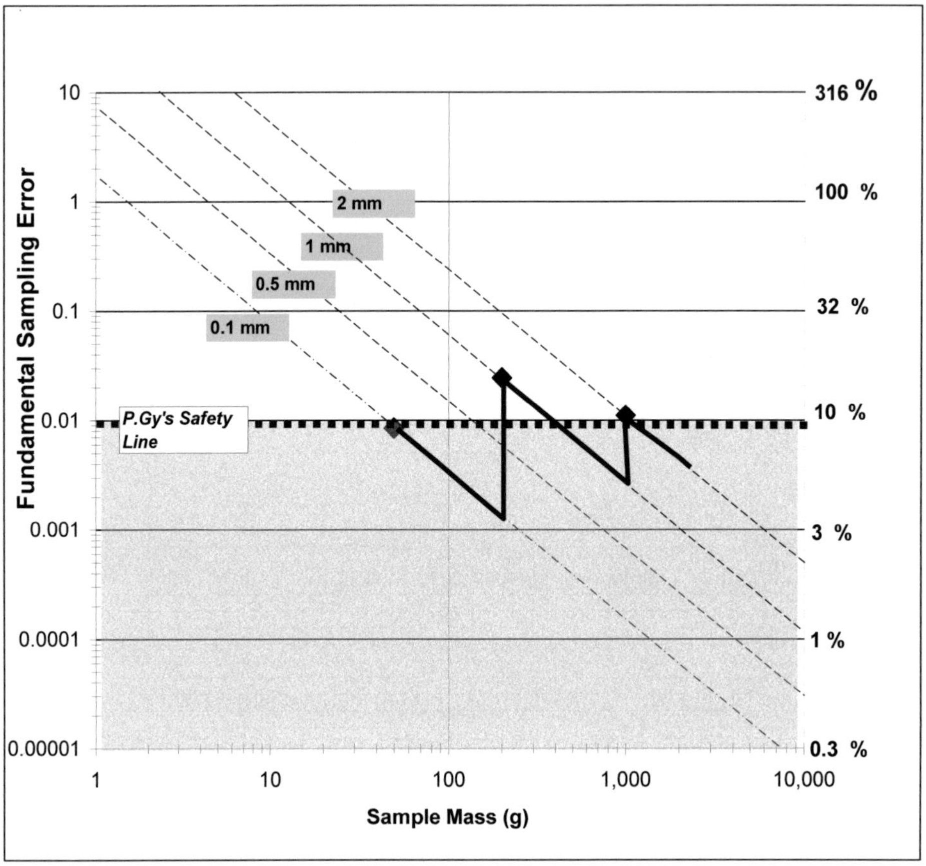

Fig. 4. Sampling nomogram, gold deposit, Canada

To facilitate interpretation of this diagram certain values of acceptable precision errors can be shown. A common practice is to draw a line corresponding to 10% relative error, or in other words the relative variance equal to 0.01 (Fig. 4). This threshold is known as P.Gy's safety line (Gy, 1982). All suboptimal sample preparation stages which FSE exceeds the chosen threshold are easily diagnosed on the nomograms. The example presented in Fig. 3 shows that the largest FSE is introduced at the second subsampling stage when 200g of sample is collected from a pulp ground to -0.5mm. This stage introduces approximately 20% precision error whereas first sub-sampling, when 1 kg was split from the material crushed to

-1mm causes a smaller error, less than 10%. Final stage when 50g aliquot was collected from a 200g pulp pulverised to -0.1mm is also characterised by less than 10% of FSE. Based on this nomogram it is obvious that improvement of the sampling protocol should be focused on optimisation of the stage 2, which can be easily achieved by collecting larger sub-sample, approximately 500g.

2.2 Grouping and segregation error

Grouping-Segregation error is generated by small scale distribution heterogeneity and reflects the differences in the content of a metal of interest between groups of fragments (increments) collected at very small intervals (Gy, 1982; Pitard, 1993). Examples of Grouping - Segregation error includes separation of fine particles from larger fragments when they are discharged from the sampling devices or conveyor belts (Fig. 5A). Another example of a possible error caused by segregation factor is accumulation of the heavier and smaller particles at the bottom of blast hole cones (Fig. 5B). This can be gold grains or sulphide minerals liberated from a gangue. Segregation can also be caused by different physical or chemical properties of the sampled materials, including magnetic or electrostatic properties, moisture content, adhesiveness (Pitard, 1993).

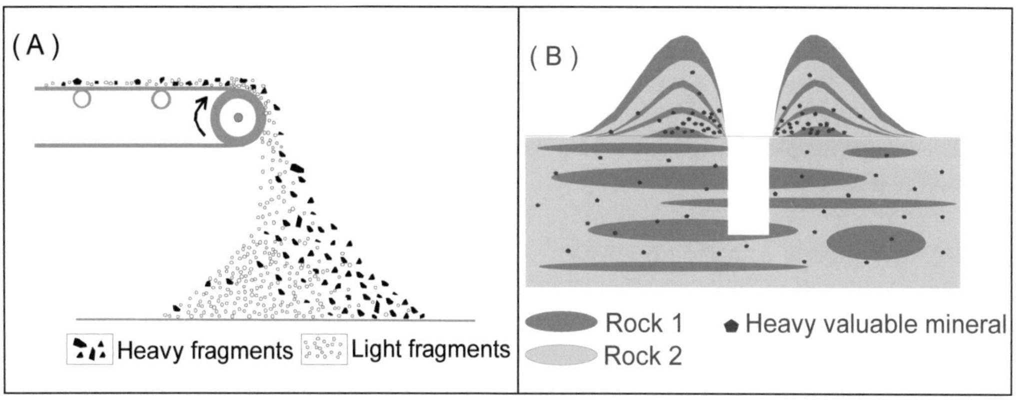

Fig. 5. Examples of segregation of the rock fragments: (A) segregation of the heavy and lighter fragments at the discharge from conveyor belt, modified after Pitard (1993); (B) uneven distribution of fragments in a blast hole cone caused by segregation of the smaller and heavier fragments (e.g. liberated gold particles, or sulphides grains) and their accumulation at the bottom of the cone

The Grouping-Segregation error represents a complementary part of the Fundamental Sampling Error (Eq. 13) and therefore has been assigned here to the first group of errors (Eq. 1) because of its direct relationships with fundamental error. However, in practice this error as much depends on sampling protocol, being controlled by the value of fundamental error (σ^2_{FSE}), as it is a consequence of the practical implementation of the sampling protocol. For example Grouping-Segregation error can be generated by insufficient homogenisation of the material prior to sampling.

The Grouping-Segregation error by definition is a function of two factors, grouping factor (f_G) and segregation factor (f_S), both being a consequence of the small scale distribution heterogeneity of the material (Eq. 13).

$$\sigma_{GS}^2 = f_G f_S \sigma_{FSE}^2 \tag{13}$$

Grouping factor in practise characterises the size of the increments (i.e. fragments) making up the sample. A segregation factor characterises the number of increments. Both of these factors are dimensionless and can not be dissociated from each other (Pitard, 1993).

Grouping-Segregation error can not be theoretically estimated, however, in practice, it can be minimised if factors causing this error are well understood. F.Pitard (1993) has pointed out that grouping factor can be minimised by taking as many and as small increments as practically possible, assuming that all aspects of extraction and preparation of these increments is carried out correctly. Practical implication of this finding is that when selecting a riffle splitter preference should be given to that with a more riffles. When collecting samples by randomly chosen fragments, as a rule of thumb, the sample should be made of at least 30 increments (Pitard, 1993). The same rational applies to the choice of rotary splitter. To minimise the grouping-segregation error it is necessary to assure that sample was collected by at least 30 cuts from the lot.

The minimisation of errors caused by segregation factor is more difficult than for errors related to the grouping factor (Pitard, 1993). The only practical approach is to homogenise the material prior to sampling, which is not always technically possible. It is important to understand the segregation mechanisms in the proposed sampling operation and adjust the procedures so that a samples extraction will be less affected by inhomogeneous (segregated) distribution of the sampled particles. For example, to minimise the errors caused by segregation factor the blast hole in Figure 5B should be sampled by using a sampling device which is radial in plan view and taking several increments positioning them at random around the blast hole pile (Pitard, 2005).

2.3 Errors related to sampling practices

This group includes delimitation, extraction, preparation and weighing errors, which occur as a result of incorrect extraction of samples from a lot and their suboptimal preparation procedures. In other words the errors included in this group are not directly related to the sampling protocol but in fact are associated with the practical implementation of the protocol. Simply this means, that having the best possible protocol does not guarantee that your samples will be accurate and repeatable as this requires that protocol should be implemented in a practical way that eliminates, or at least minimises, the errors related to sampling practices which are described here as Group 2 errors. The most common example of this type of error is contamination of the laboratory samples during their preparation. Contamination of the samples caused by incorrect preparation procedures can completely destroy the integrity of the samples and all efforts to obtain the samples with the lowest possible fundamental error will be wasted.

Delimitation error occurs when not all material in the sampled lot has equal probability to be selected to the sample (Gy, 1982; Pitard, 1993). The most common situation leading to delimitation errors is sampling of broken ore by collecting material from the top of the pile (Fig. 6A). The bottom of the pile is not represented in the sample causing biased results due to delimitation error. Another common example of delimitation error is shown in Figure 6B. When the sample is taken by a rounded scoop the material at the bottom of the pile has less likelihood to be part of sample then fragments at the top of the pile.

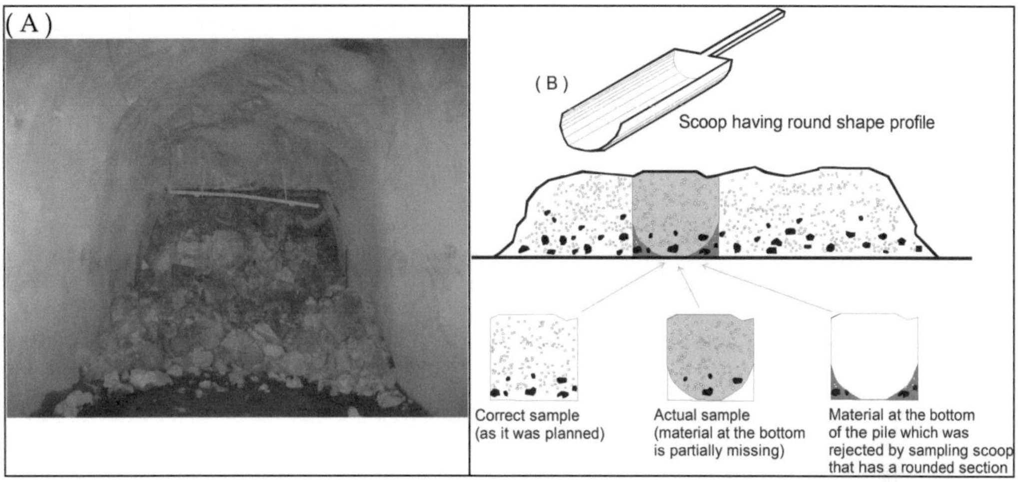

Fig. 6. Examples of delimitation errors: (A) sampling broken ore at the draw point of underground stope represents an unresolvable issue of the delimitation error.; (B) sampling of crushed material (e.g. blast hole cone) by a scoop having round shape profile

Extraction error is the result of a sampling tool which is selectively taking fragments, therefore these errors also known as sample recovery errors (Pitard, 1993) because they are caused by selective sampling systems. This type of error can be frequently observed in geological exploration and mining geology applications (Fig. 7). One of the most common examples of extraction error is a preferential caving of soft material, such as clay pods, when drilling rocks of variable strength. Sampling of blast hole cones using incorrectly designed auger drill, which rejects large fragments is another type of extraction error often occurring in geological applications. In all these cases the extraction error can cause significant biases of the analytical results. It is important to note that having the correct equipment does not guarantee high quality results because inappropriate use of equipment can also lead to significant extraction error. The example shown in Figure 7B demonstrates an extraction error caused by the incorrect use of a riffle splitter which is quickly fed on one side. Such an approach leads to disproportional distribution of the fragments and segregation of heavier particles on one side.

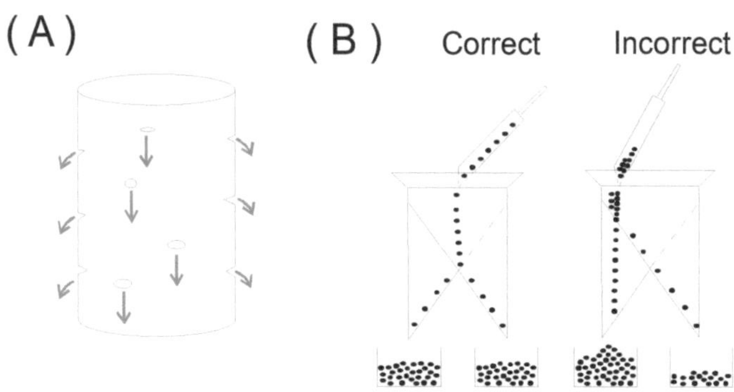

Fig. 7. Examples of Extraction errors: (A) extraction error caused by selective plucking of mineral grains (e.g. gold grains) from the drill core surface; (B) extraction error caused by incorrectly used riffle splitter, modified after Pitard (1993)

Preparation errors are the changes of chemical or physical characteristics of the material caused by its crushing, grinding, pulverising, homogenising, screening, filtering, drying, packaging and transportation. These errors take place during processing of the samples, and include contamination of the samples, preferential losses and alteration of sampled material. For example, in the blast holes some portion of the drill cuttings fall back into the hole which can be a source of sample biases due to preparation error.

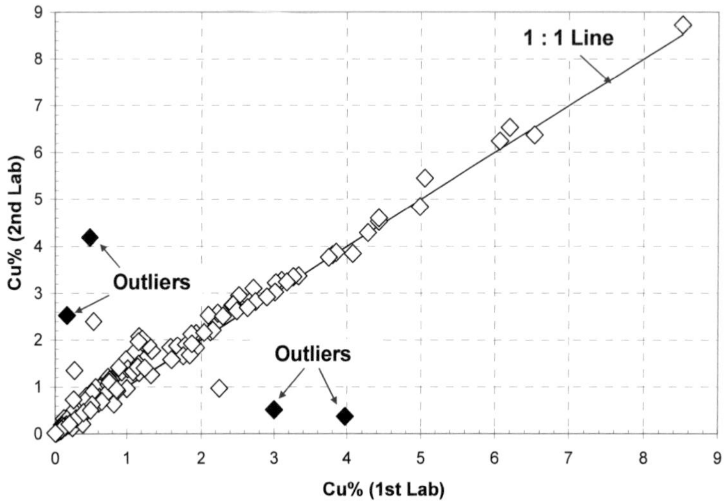

Fig. 8. Scatter-diagram of the Cu grades determined in the drill core samples and their duplicates. Cu-Au bearing skarn, Russia. All data, except several outliers, are compactly distributed along the 1:1 line indicating a good repeatability of the sample grades. Several erratic results (outliers) have been caused by mixing of sample numbers in the lab

Weighing errors are the errors which are introduced by weightometers and scales. The author has observed a case where a laboratory had been equipped by the state of art analytical equipment including robotic XRF instruments whereas the sample preparation stage was still very old, out of date and poorly calibrated scales were used, entirely eliminating all benefits of having high precision analytical instrument.
The second group also includes different types of human errors, such as mixing sample numbers, transcript errors, and incorrect instrument readings. These errors, except in the cases of deliberate fraud, are accidental by their nature and can create extremely erratic values, abnormally high or low in comparison with the true sample grades. When such extreme values are present, these accidental-type of human errors can be easily recognised by presence of outliers on the scatter-diagrams where sample duplicates are plotted against the original samples (Fig. 8).

2.4 Instrumental errors

This group includes errors related to various analytical and instrumental measurements during weighing of the final aliquots and their assaying. A typical example of these types of error is instrumental drift causing biased assays. It also can be incorrect calibration of the instruments. The instrumental errors also include those which are caused by use of the out-of-date equipment, in particular if it is characterised by poor detection limits.

A special case of instrumental errors is represented by incorrectly chosen analytical techniques, which is suboptimal for the given type of mineralisation and/or grade ranges. For example, fire assay with atomic absorption finish is used for low-grade gold mineralisation, whereas high grade gold is better assayed by fire assay with gravimetric finish. Application of the atomic absorption finish to high grade gold samples can lead to their incorrectly determined grades because of instrumental error.

3. Control of analytical data quality (QAQC)

Control of analytical data quality is usually referred to in the mining industry as Quality Assurance and Quality Control (QAQC) and this involves quantification and systematic monitoring of the samples accuracy and precision (i.e. repeatability), and timely diagnostics of sample errors and identification of error sources.

3.1 Accuracy of the analytical data

Accuracy of samples is usually monitored by including samples with known grade to the assayed samples batch (Kane, 1992; ISO Guide, 1989; Leaver et al., 1997; CANMET, 1998; Sketchley, 1998). The samples with *a priori* known grade are called standard samples or simply standards. They can be obtained from commercial laboratories where standards have been prepared and assayed rigorously following the appropriate procedures (Kane, 1992) and results statistically certified. The best practice is to prepare standards from material mineralogically similar to that of the studied mineralisation; these are called the matrix matched standards.

Possible contamination is controlled by inserting blank samples. Blanks are samples where the grade of a metal of interest is negligibly low, usually below detection at the given laboratory. It is a common practise to make the blank samples from barren quartz.

Standards and blanks are called reference materials and used for detection a possible sample biases and quantification the accuracy errors. Leaver et al. (1997) have suggested a definition of a reference material as 'a substance for which one ore more properties are established sufficiently well to calibrate a chemical analyser or to validate a measurement process.' A reference material is classified as a certified reference if it was issued and certified by agencies, such as Canadian Centre for Mineral and Energy Technology (CANMET), the National Institute of Standards and Technology and other governmental agencies, whose technical competency is internationally recognised (Leaver et al., 1997).

Accuracy of laboratory analyses is assessed by statistical tests which are largely based on a comparison of the arithmetic mean of the replicate analyses of a certified standard against its certified mean. The statistical tests of the standard samples repeatability can be based on replicate assays of a certified standard in one laboratory or, conversely, inter laboratory (Round Robin) analyses. Different tests are available (Kane, 1992; ISO Guide, 1989; CANMET, 1998; Sketchley, 1998) and most common are summarised in the current section of this book. It is noted that a good practice is to support testing of the assay accuracy by estimation of standard deviation of the replicate analyses of the standard samples, as poor precision of the standard sample assays can prohibit reliable detecting of analytical biases.

a. Statistical tests for assessing performance of standard samples: single laboratory case

The most common situation is when one or several certified standard samples have been included in a batch of samples and analysed together in the same laboratory. In such case when several repeat assays of a given standard sample are available the analytical accuracy is assessed using a statistical test (Eq. 14). If this condition (Eq. 14) is satisfied the analytical results are considered acceptable with regard to accuracy.

$$|m - \mu| \leq 2\sqrt{\sigma_L^2 + \frac{S_W^2}{n}} \tag{14}$$

where:
μ - certified mean of a given standard sample;
σ_L - certified between-laboratory standard deviation of a given standard sample;
m - arithmetic mean of the replicate analyses of this certified standard sample in the assay batch;
S_W - estimated within-laboratory standard deviation of the replicate analysis of the standard sample included in assay batch;
n - number of replicate assays of a given certified standard in analytical batch.
The base formula (Eq. 14) can be simplified as the usual empirical relationships is $\sigma_L \approx 2\,S_W$ (CANMET, 1998). Consequently, for large (n > 10) number of replications the above condition (Eq. 20) can be simplified as follows (Eq. 15):

$$|m - \mu| \leq 2\,\sigma_L \tag{15}$$

Certified value of the between-laboratory standard deviation (σ_L) of the reference materials is not always available. If this is the case, the condition of the accuracy acceptance (Eq. 15) can be further simplified (Eq. 16):

$$|m - \mu| \leq 4\,S_W \tag{16}$$

When multiple analyses of a certified standard have been made in the same laboratory the results can be used for estimation of analytical precision of laboratory. Analytical precision is acceptable if results of the replicate analysis of the certified standards assayed in this laboratory satisfy the statistical test (Eq. 17) (ISO Guide 33):

$$(S_W / \sigma_C)^2 \leq \frac{\chi^2_{(n-1);\,0.95}}{n\text{-}1} \tag{17}$$

where:
S_W - standard deviation of the replicate analysis of the standard
σ_C - the certified value of the within-laboratory standard deviation of the standard
$\chi^2_{(n-1);\,0.95}$ - critical value of 0.95 quartile ($\alpha = 0.05$) of the (χ^2) distribution at (n-1) degrees of freedom, where (n) is the number of replicate analysis of the standard. In practise, at least 3 repeat analysis of the certified standards should be available for this test (CANMET, 1998).

b. Statistical tests for assessing performance of the standard samples: different laboratories case

When exploration samples and standards inserted to the samples batch have been analysed in (*p*) several different laboratories, the overall accuracy of the analytical results can be tested (Kane, 1992; ISO Guide, 1989; CANMET, 1998) using the following statistical condition (Eq. 18):

$$|m-\mu| \leq 2\sqrt{\frac{S_{LM}^2 + \frac{S_W^2}{k}}{p}} \tag{18}$$

where:

μ - certified mean of a given standard sample;

m - arithmetic mean of the replicate analyses of this certified standard sample in the assay batch;

S_W - estimated within-laboratory standard deviation of the replicate analyses of the standard samples;

S_{LM} - estimated between-laboratory standard deviation of the replicate analyses of the standard samples;

$k = \frac{n}{p}$, is a ratio of total numbers of replicate analysis (n) of certified standards to a number of laboratories (p) participating in Round Robin test.

The assays of the certified standards obtained from different laboratories can be used for estimation analytical precision (Eq. 19).

$$(S_W/\sigma_C)^2 \leq \frac{\chi^2_{p(k-1);\ 0.95}}{p(k\text{-}1)} \tag{19}$$

where:

S_W - estimated standard deviation of the replicate analysis of the standard;

σ_C - the certified value of the within-laboratory standard deviation of the standard;

$k = \frac{n}{p}$, is a ratio of total numbers of replicate analysis (n) of certified standards to a number of laboratories participating in Round Robin test (p);

$\chi^2_{p(k-1);\ 0.95}$ - critical value of 0.95 quartile ($\alpha = 0.05$) of the (χ^2) distribution at $p(k\text{-}1)$ degrees of freedom.

Inter-laboratory precision can be assessed indirectly using the equality Eq. 20. Precision is considered satisfactory if this equality is true.

$$\frac{S_W^2 + kS_{LM}^2}{\sigma_C^2 + \sigma_L^2} \leq \frac{\chi^2_{(p-1);\ 0.95}}{(p\text{-}1)} \tag{20}$$

where:

σ_C - certified value of within-laboratory standard deviation of reference standard;

σ_L - certified value of between-laboratory standard deviation of reference standard;

S_W - estimated standard deviation of the replicate analyses of the standard samples;

S_{LM} - estimated between-laboratory standard deviation of the replicate analyses of the standard samples;

$k = \frac{n}{p}$, is a ratio of total numbers of replicate analysis (n) of certified standards to a number of laboratories participating in Round Robin test (p);

$\chi^2_{(p-1);\ 0.95}$ - critical value of 0.95 quartile (α = 0.05) of the (χ^2) distribution at (p-1) degrees of freedom.

c. Statistical tests for assessing performance of the standard samples: single assay case

Repeat assays of certified standards may not be available in a single analytical batch, which often can contain only one standard sample. In such case, when repeat assays of certified standards are not available a decision on the analytical accuracy should be made using a single value (X) of the certified standard included into the analysed batch the above mentioned statistical tests are inapplicable. Decisions on the accuracy of the method need to be based on the different statistical test (Eq. 21) which uses a single assayed value of the certified standard (Kane, 1992; CANMET, 1998):

$$|X - \mu| \leq 2\,\sigma_C \tag{21}$$

where:

X - assayed value of the certified standard;

σ_C - certified within-laboratory standard deviation of the standard;

μ - certified mean of a given standard sample.

d. Assay biases diagnostic diagrams: pattern recognition method

The pattern recognition technique is applied to certified standards (Leaver et al., 1997; Sketchley, 1998; Abzalov, 2008).This method is based on the fact that specific types of analytical problems have recognisable patterns on the special types of the diagrams. The different distribution patterns of analytical results are indicative of error sources and types.

The assayed values of certified standards are plotted onto the diagram on a batch/time basis (Fig. 9). Good quality analysis will be characterised by random distributions of the data points around the certified mean value on this diagram (Fig. 9A) and 95% of the data points will lie within two standard deviations of the mean and only 5% of assays should lie outside the interval of two standard deviations from the mean. It is essential that the same number of samples should occur above and below mean. When distribution of the standard sample assays differ from the pattern shown on the Figure 9A this likely is indicate a possible analytical errors. For example, the presence of 'outliers' (Fig. 9B) is most likely an indication of data transcription errors. This feature does not imply data bias but nevertheless indicates a poor data management system suggesting possible random errors in the database. Consistent shift of a standard sample assay (Fig. 9C) is indicative of analytical bias which could be caused by equipment calibration issues or changed analytical procedures in the lab.

Less common distribution patterns occur when dispersion of the standard sample grades rapidly decrease (Fig. 9D). Such a decrease in the standards variability is commonly interpreted (Sketchley, 1998) as indicative of data tampering, or suggests that standard samples have been recognised by laboratory personnel and treated more carefully than other samples (Abzalov, 2008). Such assay values of standard samples cannot be used as confirmation that the assayed samples lack bias.

Accurate analyses are also characterised by lack of systematic data trends on grade versus order of analysis diagrams. Trends can be recognised by a systematic increase or decrease of the assayed values of the standards (Fig. 9E). Another commonly used criteria for identification of possible trends are distributions where two successive points lie outside of the two standard deviation range or four successive points lie outside the one standard deviation range(Leaver et al., 1997).

Systematic drifts (Fig. 9E) of assayed standard values usually indicate a possible instrumental drift. Alternatively this can be also caused by degradation of the standard samples. The author of the current paper is familiar with cases when characteristics of the standard samples have degraded in comparison with their certified values because of inappropriate storing conditions of the standards when they were kept in the large jars and have not been protected against vibrating due to operating equipment.

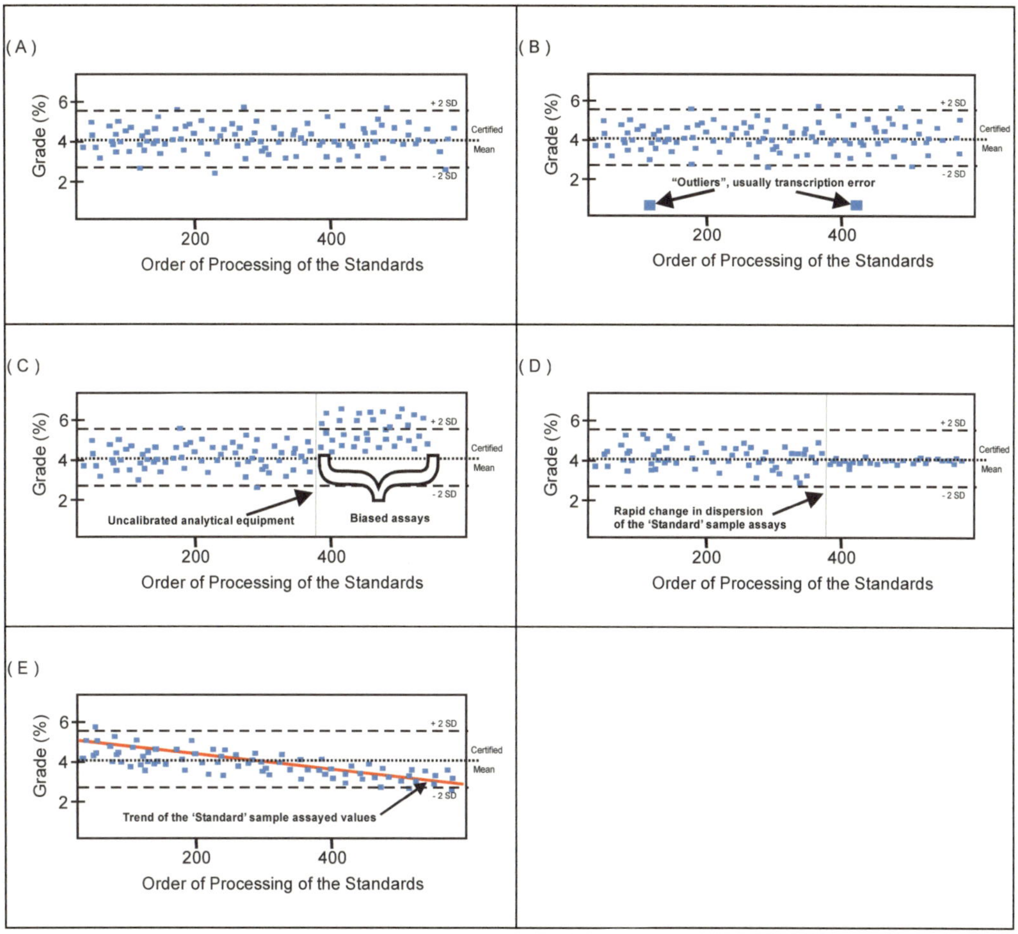

Fig. 9. Schematic diagrams showing quality control pattern recognition method (Abzalov, 2008): (A) Accurate data, statistically valid distribution of the standard values; (B) Presence of 'outliers' suggesting transcription errors; (C) Biased assays; (D) Rapid decrease in data variability indicating for a possible data tampering; (E) Drift of the assayed standard values

Blank samples (i.e. material that have very low grade of a metal of interest) are usually inserted after high grade mineralisation samples. The main purpose of using blanks is to monitor laboratory for a possible contamination of samples which is mainly caused by poor house keeping and insufficiently thorough cleaning of equipment. Blank assays are also presented on the grade versus order of analysis diagram (Fig. 10) and if equipment have not been properly cleaned the blank samples will be contaminated which is reflected on the diagram as increased value of a metal of the interest. The case presented on the Figure 10 shows that laboratory procedures have degraded during the course of the project because the blank samples after approximately blank with sequential number 150 have started to systematically indicate contamination by base metals derived from processed Ni-Cu sulphide samples.

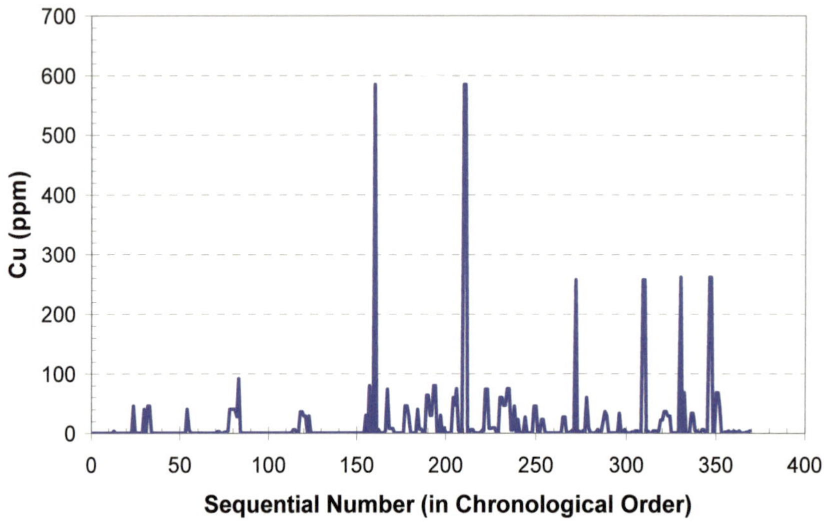

Fig. 10. Cu grade of blank samples plotted versus order of their analysis, Ni-Cu project, Australia

3.2 Precision of the analytical data

Precision error of assayed samples is estimated and monitored by using matching pairs of samples (Abzalov, 2008). Sample pairs (duplicates) represent the most common type of matching pairs of data. The first sample is usually referred to as original sample, and the second is called duplicate. The duplicate sample is merely another sample collected from the same place and following the same rules as used for collecting the initial (original) sample. In practise, the duplicate sample can be a second blast hole sample collected from the same blast hole cone or another half of the drill core. It also can be a duplicate sample collected at a particular stage of the sampling protocol, such as coarse rejects after crushed material has been split using an appropriate sample reducing devise or second aliquot collected from the same pulverised pulp. When sampling protocol includes several stages of comminution and subsampling it is a good practise to take duplicate samples at each subsampling stage. Another type of matching data pairs are twin drill holes (Abzalov, 2009a).

a. Statistical analysis of sample pairs

The most common approach in assessing paired data is to plot the duplicate sample results vs. original assays and the precision error can be quantified from the paired data through assessing the scatter of the data points with corrections for their deviation from y=x line. The differences between an assayed sample and it's duplicate is caused by errors associated with sample preparations and assaying. Precision error is mathematically deduced from differences between matching pairs of data, and is usually represented as variance of the assayed values normalised to the means of the corresponding pairs of the data.

The different statistical methods are available for estimation precision error from paired data (Garrett, 1969; Thompson & Howarth, 1978; Bumstead, 1984; Shaw, 1997; Francois-Bongarson, 1998; Sinclair & Bentzen, 1998; Sinclair & Blackwell, 2002). All these methods have been reviewed by the author (Abzalov 2008) and compared by applying the same sets of duplicated pairs of samples collected in operating mines and mining projects. The study (Abzalov, 2008) concurs with the suggestion of Stanley and Lawie (2007) that the use of average coefficient of variation (Eq. 22 and 23) as the universal measure of relative precision error in geological applications is appropriate.

$$CV\% = 100\% \text{ x } \frac{\text{Standard Deviation}}{\text{Mean}} = 100\% \text{ x } \frac{|a_i - b_i| / \sqrt{2}}{\frac{a_i + b_i}{2}} \qquad (22)^1$$

Average CV% can be calculated from the (N) pairs of the duplicated samples using equality (Eq. 23).

$$CV\% = 100\% \text{ x } \sqrt{\frac{1}{N}\sum_{i=1}^{N}\frac{\sigma_i^2}{m_i^2}} = 100\% \text{ x } \sqrt{\frac{2}{N}\sum_{i=1}^{N}\left(\frac{(a_i - b_i)^2}{(a_i + b_i)^2}\right)} \qquad (23)$$

Stanley (2006) has shown that absolute difference between paired data is directly proportional to standard deviation ($|a_i - b_i| = \sqrt{2}$ x Standard Deviation)[2] and consequently the various measurements representing a ratio of the absolute differences between paired data to the means of the corresponding pairs (Table 3) are directly proportional to coefficient of variation.

Hence, such statistics as AMPD and HARD (Table 3), commonly used by geoscientists (Bumstead, 1984; Shaw, 1997; Roden & Smith, 2001), represent nothing else but products of CV% by constants ($\sqrt{2}$) and ($\frac{\sqrt{2}}{2}$) respectively (Table 3) and therefore offer no more information than the CV% itself.

---

[1] For consistency with other measurements coefficient of variation (CV) (one standard deviation divided by mean) is expressed as percentage (CV%)

[2] Variance (σ^2) is determined by the following equality: $\sigma^2 = \frac{\sum_i (x_i - \frac{\sum x_i}{N})^2}{n-1}$. In case when only a single pair of data is available the formula can be transformed as follows:

$$\sigma^2 = \frac{(a_1 - \frac{a_1 + b_1}{2})^2 + (b_1 - \frac{a_1 + b_1}{2})^2}{2-1} = \frac{(\frac{2a_1 - a_1 - b_1}{2})^2 + (\frac{2b_1 - a_1 - b_1}{2})^2}{1} = \frac{(a_1 - b_1)^2 + (b_1 - a_1)^2}{4}$$

$$= \frac{(a_1 - b_1)^2}{2};$$

Taking square root of the variance we arriving to a final formula of standard deviation of the paired data: $\text{Standard Deviation} = \sqrt{\sigma^2} = \frac{|a_1 - b_1|}{\sqrt{2}}$

Conventional Name of the Error Estimator	Single Duplicate Pair Formula and relationship to CV%	Reference	Comments
AMPD	$\text{AMPD} = 100\% \text{ x } \frac{\lvert a_i - b_i \rvert}{(a_i + b_i)}$ $\text{AMPD} = \sqrt{2} \text{ x CV\%}$	Bumstead, 1984	This estimator is also known as MPD (Roden and Smith, 2001) or ARD (Stanley & Lawie, 2007)
HARD	$\text{HARD} = 100\% \text{ x } \frac{\lvert a_i - b_i \rvert}{(a_i + b_i)}$ $\text{HARD} = \frac{\sqrt{2}}{2} \text{ x CV\%}$	Shaw, 1997	HARD is simply the half of the corresponding AMPD
where (a_i) is original sample and (b_i) is its duplicate			

Table 3. Measures of Relative Error Based on Absolute Difference Between Duplicated Data Pairs

b. Reduced Major Axis (RMA)

This is a linear regression technique which takes into account errors in two variables, original samples and their duplicates (Sinclair & Bentzen, 1998; Sinclair & Blackwell, 2002). This technique minimises the product of the deviations in both the X- and Y- directions. This, in effect, minimises the sum of the areas of the triangles formed by the observations and fitted linear function (Fig. 11). Calculation of the RMA parameters is explained in detail in (Abzalov, 2008) and only briefly summarised below.

The general form of the reduced major axis (RMA) is as follows (Eq. 24):

$$b_i = W_0 + W_1 a_i \pm e \tag{24}$$

where (a_i and b_i) are matching pairs of the data, (a_i) denotes a primary samples plotted along X axis and (b_i) is duplicate data, plotted along the Y axis. (W_0) is Y-axis intercept by the RMA linear model, (W_1) is the slope of the model to X-axis, (e) standard deviation of the data points around the RMA line.

The parameters (W_0, W_1 and e) estimated from the set of the matching pairs of data (a_i and b_i), plotted along X-axis and Y-axis, respectively. The slope of RMA line (W_1) is estimated as ratio of standard deviations of the values (a_i and b_i) (Eq. 25).

$$W_1 = \frac{\text{St.Dev}(b_i)}{\text{St.Dev}(a_i)} \tag{25}$$

Intercept of the RMA model with Y axis is estimated from (Eq. 32)

$$W_0 = Mean(b_i) - W_1 Mean(a_i) \tag{26}$$

RMA model allows quantifying the errors between matching data pairs (Eq. 27 - Eq.30). Dispersion of the data points about the RMA line (S_{RMA}) is estimated using Eq. 27.

$$S_{RMA} = \sqrt{[2(1-r)(Var(a_i) + Var(b_i)]} \tag{27}$$

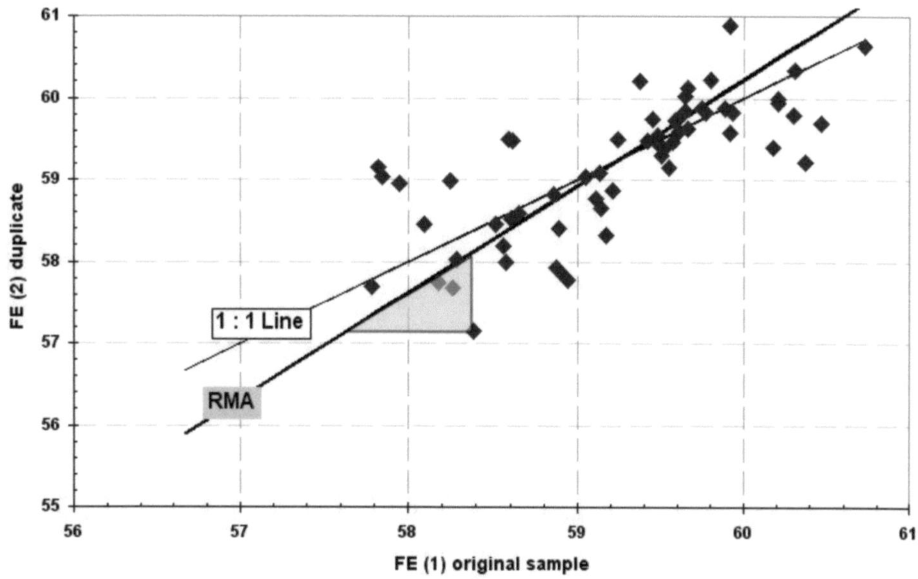

Fig. 11. Scatter- diagram and the RMA model (thick line) fitted to paired Fe grades of the blast hole samples (Abzalov, 2008). 1:1 line (fine) is shown for reference. Grey triangle shows an area formed by projection of the data points and RMA line. RMA technique minimises the sum of the areas of the all triangles

Error on Y-axis intercept (S_0) is estimated using Eq. 28. and the error on the slope (S_{SLOPE}) is estimated using Eq. 29.

$$S_0 = St.Dev(b_i)\sqrt{\left[\frac{(1-r)}{\mathrm{N}}\right]\left[2+\left(\frac{Mean(a_i)}{St.Dev(a_i)}\right)^2(1+r)\right]} \tag{28}$$

$$S_{SLOPE} = \frac{\mathrm{St.Dev}(b_i)}{\mathrm{St.Dev}(a_i)}\sqrt{\frac{(1-r^2)}{\mathrm{N}}} \tag{29}$$

where (r) is correlation coefficient between (a_i and b_i) values, (N) is the number of the data pairs and (Var) are variances of the (a_i and b_i) values.
The relative precision error ($P_{RMA}(\%)$)[3] can be estimated from the RMA model:

$$P_{RMA}(\%) = 100 \text{ x } \frac{\sqrt{\frac{S^2_{RMA}}{2}}}{\frac{(\sum_i^N a_i) + (\sum_i^N b_i)}{2N}} \tag{30}$$

[3] For consistency with CV% and other estimators discussed in this section the ($P_{RMA}(\%)$) value is estimated at 1 standard deviation and reported as per cents (i.e., multiplied by 100 in Eq. 30).

c. Relative Difference Plot (RDP)

Relative difference plot (RDP) has been suggested by F.Pitard (1998) as graphic tool for diagnostics of the factors controlling precision error (Fig. 12). Relative difference is estimated as the differences between matching pairs of data which are normalised to the means of the corresponding pairs of the data (Eq. 31). This relative difference is expressed as per cents.

$$\text{RD (Relative Difference) \%} = \frac{1}{\text{N}}\sum_{i}^{N}[100 \text{ x } \frac{(a_i - b_i)}{(a_i + b_i)/2}] \quad (31)$$

where (a_i) and (b_i) are the matching pairs of the data and N is the number of pairs.
Calculated RD(%) values are arranged in increasing order of the average grades of the data pairs and then RD(%) values are plotted against sequential numbers of the pairs whereas the calculated average grades of the data pairs are shown on the secondary y-axis (Fig. 12). Link between average grades of the data pairs and their sequential numbers are established by adding a 'Calibration curve' to the RDP diagram (Fig. 12). RD(%) values usually exhibit a large range of variations; therefore, interpretation of this diagram can be facilitated by applying the moving window technique to smooth the initially calculated RD(%) of the data pairs (Fig. 12).

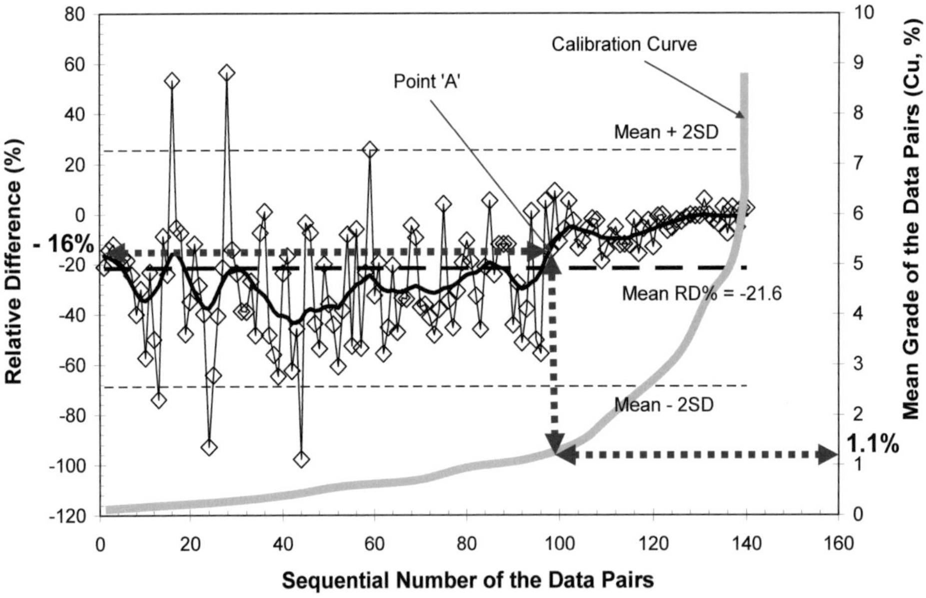

Fig. 12. Relative Difference Plot (RDP) showing Cu(%) grades of duplicated drill core samples assayed in different labs (Abzalov, 2008). Open symbols (diamond) connected by fine tie-lines are RD(%) values calculated from matching pairs of data (i.e., original sample and duplicate). Average RD(%) value (thick dashed line) and ± 2SD values (fine dashed lines) are shown for reference. The solid line is a smoothed line of the RD(%) values calculated using a moving windows approach. The 'Calibration Curve' sets the relationship between RD% values on the primary y-axis, the sequential number of the data pairs plotted along the x-axis and the average grades of the data pairs plotted on the secondary y-axis. For example point 'A' (sequential number of the data pairs is 100) is characterised by RD% value equal to -16% and average grade of that pair of data is 1.1% Cu

The example in Figure 12 is based on data collected from a massive Cu-sulphide project in Russia. Approximately 140 core duplicate samples have been analysed in an external reputable laboratory as part of project due diligence. The results, when plotted on an RDP diagram (Fig. 12), show that copper assays of low grade samples (Cu < 1.1%) are biased. Assayed values have significantly underestimated the true copper grade of those samples. Another feature revealed by this diagram is that low grade samples (Cu < 1.1%) exhibit excessive precision error which is not shown by higher grade samples (Cu >1.1%). These findings have triggered a special investigation of the laboratory procedures, with a particular emphasis on differences between high-grade and low-grade samples.
The RDP diagram can be used for testing the impact onto data precision the different factors, for example sample size, core recovery, depth of the samples collection. In that case the RD% values are arranged according to a factor under investigation. In Figure 12 the RD% values are plotted against the sequential number of the samples arranged by their sizes (i.e. length of the sampling interval). The diagram (Fig. 13) shows that small samples, less than 0.5m in length, are characterised by approximately twice as larger precision error than samples of 1 m in length.

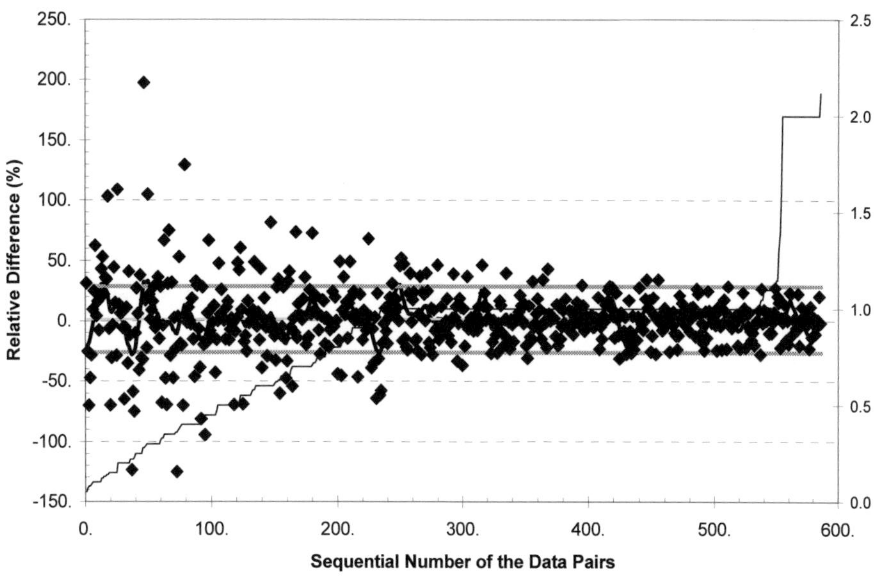

Fig. 13. Cu% grades of the duplicated drill hole samples plotted on the RDP diagram. RD% values are arranged by length of the samples

4. Twin holes

Drilling of twinned holes (i.e., drilling of a new hole, or "twin", next to an earlier drill hole) is a traditional technique used in exploration and mining geology for verification of mineralization grades. The importance of this technique is emphasized in the JORC Code, which specifically queries if twinned holes have been used for 'verification of sampling and assaying' (JORC, 2004, p.15). A well known example of a (negatively) successful twinned holes program is the confirmatory drilling at the Busang project of Bre-X Minerals Ltd.

Fraud was revealed by seven twinned holes (Lawrence, 1997). Each of the seven new holes was drilled only 1.5 m away from supposedly extensively gold mineralized holes, and all failed to substantiate the originally stated gold values (Lawrence, 1997). Application of the twinned holes technique is not limited to verification of the high-grade intersections and also includes verification historic data (Abzalov, 2009a). Such a need usually occurs where results from earlier drill holes are causing concerns either by being noticeably different to other drilling campaigns in the same area, or where poor quality of drilling has been revealed from previously documented parameters, such as core recovery. In some cases, historic data might be suspect (e.g., biased) because of suboptimal drilling technology or inappropriate drilling parameters. For example, air-core drilling of a mineral sands deposit might be biased, and the issue could be examined by twinning the historic air-core holes with sonic drill holes.

Verification of earlier drilled holes by twinning them with new holes is a common technique used by geological due diligence teams when reviewing third party projects (Gilfillan, 1998; Abzalov, 2009a). Drilling twinned holes becomes particularly important where the reviewed project is based on historical data that were collected without rigorously applied sampling quality control procedures, or where the relevant quality assurance/quality control (QAQC) documentation is lacking. Where bias of historic data has been proven, the data from twinned holes can be used for quantification of this bias, and integrated into the resource model using multivariate and/or non-stationary geostatistical methods for correcting the resource estimates (Abzalov & Pickers, 2005; Abzalov, 2006).

The number and locations of twinned holes, drilling methods, and the tested variables are selected to accomplish the objectives of the proposed study. For example, if the purpose of drilling twinned holes is to verify high grade intersections, new holes should be drilled next to previous holes that reportedly intersected high grade mineralization. Conversely, if the objective is to test and statistically quantify a possible bias in historic data, the twinned holes should be drilled in such a manner that assures that a wide range of grades is tested, including both low- and high- grade mineralization. In this latter case, twinned holes should be distributed throughout the entire deposit, and the number of twinned holes should be sufficient for statistical analysis of the data.

It should be remembered that twinned holes should be located as close as possible to the original holes to minimize the effects of short range variability in the studied variables. A study undertaken by Abzalov et al. (2007) at the Yandi iron-ore open pit has shown that CV% of Al_2O_3 grades estimated from matching pairs of blast holes and reverse circulation (RC) holes increase from 23.8% to 35.7% when the distance between twinned holes increases from 1 m to 10 m. These results suggest that many unsuccessful twinned holes studies could have failed because twinned holes have been drilled too far apart. Personal experience of the author indicates that good practice is not to drill matching pairs of holes more than 5 m apart.

When planning twinned drill holes it is necessary to remember that important decisions may have to be made based on a limited number of verification holes. Therefore, it is necessary to ensure that their quantity and quality are sufficient to make a conclusive decision. It is necessary to ensure that verification holes are of the best practically achievable quality. This often requires the use of a different drilling technique, commonly a more expensive one to that used for the previous drilling. For example, air core holes in mineral sands projects are preferably twinned by sonic drilling, auger holes in bauxite deposits by diamond core holes, and RC holes in coal deposits by PQ diamond core holes. It is important to realize that verification holes, even if they are drilled using the most up to date technology, can still produce biased and/or excessively variable results. For example, the quality of confirmatory diamond drilling can be marred by poor core recovery, or sonic drilling can be adversely affected by self-injection of sands into core barrels caused by lithostatic pressure.

The number of twinned holes depends on the aim of the twinning program and the range of variation in the studied variables. Where studied variables are characterized by large short range variability, the number of twinned holes should be sufficient for conclusive statistical and geostatistical analysis. In the author's experience this may often be achieved by drilling a relatively small number of holes. However, special studies that require 20–30 twinned holes are not uncommon (Abzalov, 2009a).

Prior to comparing assayed grades in the twinned holes it is necessary to compare geological logs, because the geological features are usually more continuous than grades (Sinclair and Blackwell, 2002). Twinned holes are initially compared by mineralized intersections. Comparisons should not be limited to average grades of intersections, but should also include thicknesses and positioning of the contacts. The latter is particularly important when down hole contamination of the samples is suspected. More detailed studies include comparison of twinned holes by matching intervals. Practically achievable outcomes are often obtained through grouping samples by geological units (Abzalov, 2009a). This approach is particularly useful when twinned holes are drilled through stratified mineralization or regularly intercalated low- and high-grade zones.

In particular cases, for example when twinned holes are drilled through mineralization exhibiting gradational zoning, comparison can be made directly between samples where they are of an equal length; alternatively comparisons may be made between equal length composites (Abzalov, 2009a). Compositing of samples of small size into composites of a larger size smoothes away the impact of outliers and converts samples into common length data, which are necessary for geostatistical resource estimation. In twinned holes analysis, when comparing variables with high statistical variation, grouping samples to larger composites is commonly necessary to minimize the noise exhibited by individual samples (Abzalov, 2009a). It is important to ensure that samples are composited to geological boundaries.

5. Database

Modern mining operations use a vast amount of different sampling data, coming from drill hole, trench, surface, grade control and mine workings samples. The total number of collected samples can be in the hundreds of thousands which are supported by hundreds of QAQC samples and they all need to be stored together in a relational database which is used for storing assayed sample results and for data grouping and analysis using different selection criteria (Lewis, 2001; Abzalov, 2009b,c). Therefore, the quality control of the analytical data in the mining industry should include procedures of relational databases management.

Good relational databases should not be limited to storing data which are required for geological modelling and resource estimation. They should also include metadata or "data about the data". A vast amount of auxiliary information is usually received from the drilling contractors, field geologists and analytical laboratories, all which need to be systematically stored in the relational database. S.Long (1998) has suggested a practical way to store all additional information without making the database tables overly cumbersome. Most of the important auxiliary information are recorded in the headers of the data sheets, received from the personnel and laboratories undertaking drilling, documentation, sample collection and assaying. For example almost all auxiliary information related to analytical techniques is usually reported at the header of assay certificate received from the laboratory. Entering this information into a separate table within the same relational database enables storage of all key facts related to analytical laboratory, technique, personnel and dates of analyses.

6. Discussion and conclusions

6.1 Planning and implementation of sampling protocols

The sampling protocol should be optimised and rigorous quality control systems implemented at the early stage of project evaluation just after it has been realised that a discovered mineral occurrence warrants a detailed delineation and technical and economic evaluation. At this stage, usually referenced as Order of Magnitude or Scoping study, the given deposit has to be systematically drilled and sampled, based on which the geological characteristics of the deposit will be quantified and the resources estimated. The biased or otherwise suboptimal assay results can lead to incorrectly assessed mining projects with costly consequences. Use of suboptimal sampling protocols or a lack of rigorous QAQC procedures can cause delays in the mining project development because of additional work requested by auditors to verify the earlier drilling results.

The first step is to design sample preparation procedures assuring that they are optimally suited for the studied mineralisation. The chosen protocol should allow the cost effective collection of representative samples of a quality sufficient for the correct evaluation of the deposit. A good starting point is to estimate Fundamental Sampling Error and plot the proposed protocol on a sampling nomogram (Fig. 4). This approach allows the optimisation of such parameters as weight of the initial sample, particle sizes after each comminution stage and sizes of reduced samples. Based on this study all stages of the sample preparation can be clearly determined and parameters of the process are quantified. These parameters need to be considered when the project team chooses equipment for sample preparation. In particular it is necessary to assure that capacity of equipment matches the proposed protocol.

Particular attention should be given to the minimum size of the samples which can be taken without dangerously increasing the samples precision error. The latter is unfortunately often forgotten when geologists are sampling drill core. It is very important to record the length of the sampled intervals and mass of the collected samples. Good practice is when both weights of the samples wet and dry, are measured and documented.

Established sampling protocols need to be documented and represented graphically, as the sample preparation flow chart (Fig. 14) and be made easily accessible to the geological team. The next step after the sampling protocol has been optimised is to add quality control procedures. At this stage it is necessary to decide the stages within the sampling protocol where duplicates are collected and the procedures and frequency of collecting of duplicates. It is also necessary to decide how many reference materials shell be inserted with each sample batch and develop procedures allowing the disguise of standards and blanks. The project management at this stage are making the decision to develop matrix matched standard, specially prepared for the studied mineralisation, or use commercially available certified standards.

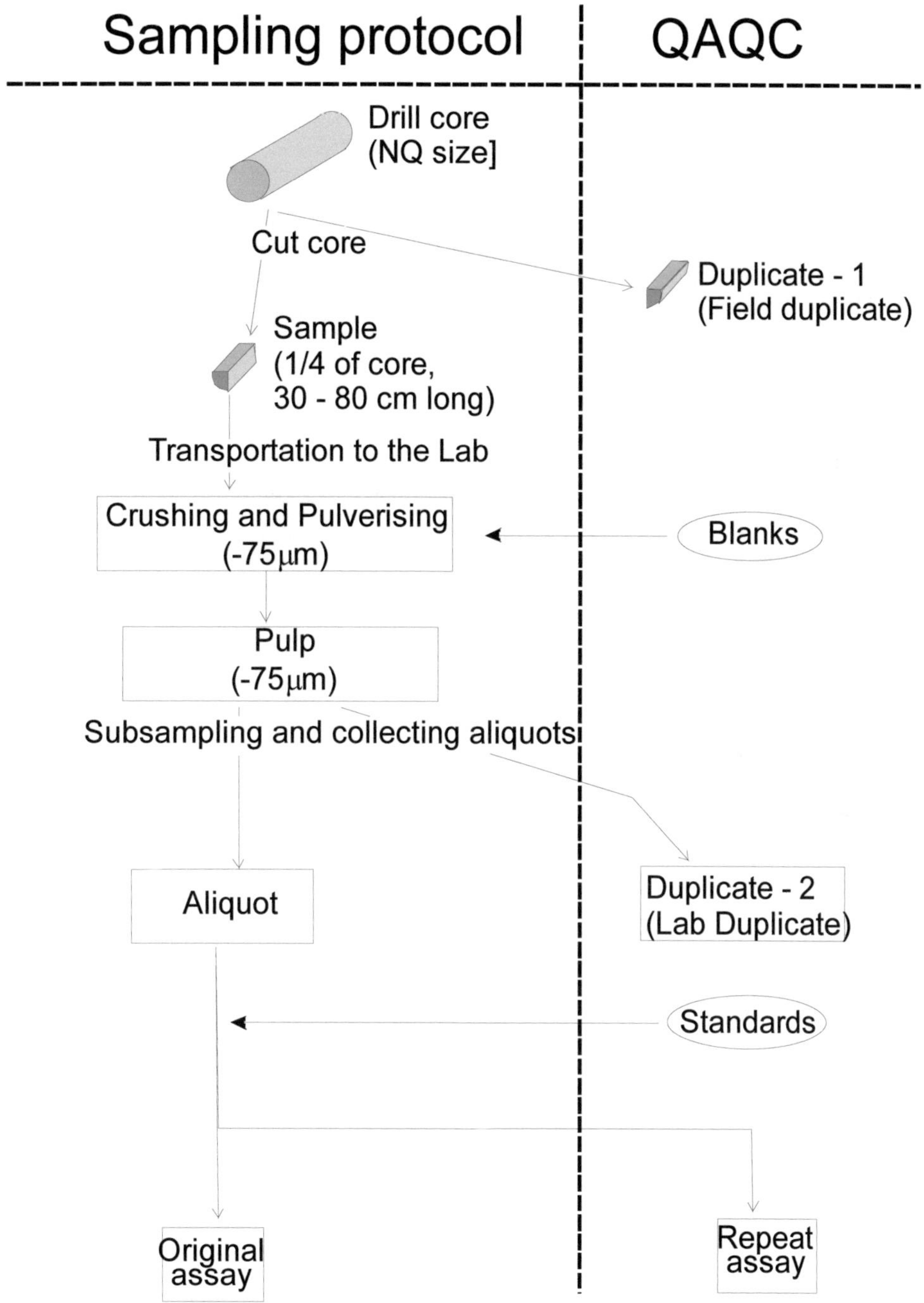

Fig. 14. Sample preparation flow chart combined with the samples quality control map, West Musgrave Ni-sulphide project, Australia

Quality control procedures should be shown on the sample preparation flow-chart (Fig. 14). Such combined diagrams, where sampling protocol is joined with the samples quality control map, are useful practical tools helping to implement and administrate the QAQC procedures assuring that all sampling and preparation stages are properly controlled. It is also important to note establishing the references between the quality control map and sample preparation flow chart help to better understand stages monitored by different quality control actions facilitating interpretation of QAQC results. When the sampling protocol is changed it should be documented and the associated QAQC procedures updated.

Finally, all procedures should be documented and personnel responsible for their implementation and control determined and instructed. It is necessary to assure that geological team, working at the mine or developing project, regularly reviews the QAQC results. Alternatively, all good efforts can be wasted if sampling errors have not been timely diagnosed. The author after reviewing many different mines found the most effective and practically convenient periodicity for the data quality review is when the QAQC results were checked by mine geologist with every analytical batch and summary QAQC reports have been prepared for approval by the chief geologist on a monthly basis. The monthly reports should contain several diagnostic diagrams showing performance of the reference materials and duplicates. The reports also should present the calculated precision variances which shell be compared with their acceptable levels. The levels of acceptable precision errors and deviation of the standards from their certified values should be clearly determined and documented as part of the QAQC procedures.

6.2 QAQC practice

a. Frequency of inserting QAQC materials into the sample batches

The reliable control of sample precision requires approximately 5% to 10% of field duplicates, and 3% to 5% of pulp duplicates. Duplicate samples should be prepared and analysed in the primary laboratory. The bias in analytical results is detected by including 3% to 5% of standard reference materials with every sample batch. Anonymity of the standards is a serious issue, because standards can be easily recognised in the sample batches and be treated more thoroughly by laboratory personnel. To overcome this problem some duplicate samples should be analysed in an external, reputable laboratory as part of accuracy control. It is suggested that at least 5% of the total analysed duplicates, including pulp duplicates and coarse rejects, should be analysed in a reputable external laboratory.

b. Distribution of reference materials

Standards should be inserted with such frequency that allows the constant monitoring of possible instrumental drifts and biases. Distribution of reference material within the batch should allow the detection of possible biases of the results and at the same time these reference samples should remain anonymous. In general, the best practice is to insert standards and blanks to every sample batch. A good practice is to use more than one standard, so that their values span the practical range of grades in the actual samples. However, standard samples alone cannot identify biases introduced at different stages of sample preparation.

c. Distribution of duplicate samples

Duplicate samples should be chosen in a manner whereby all stages of data preparation are properly addressed and the precision errors associated with all sub-sampling stages are adequately estimated by the sample duplicates. This means that sample duplicates should include field duplicates, coarse reject duplicates and the same-pulp duplicates (Long, 1998). Special attention should be given to the field duplicates as they are mostly informative for estimation of the overall precision of samples. When rotary drilling is used, the field duplicates, which in this case are called "rig" duplicates, are collected from the sample splitting devices built-in to the drill rigs. These can be rotary, cone or riffle slitters. In case of diamond core drilling the field duplicates are represented by another portion of core. Field duplicates of the blast hole samples should be another sample, taken from the same blast hole cone as the original sample and following exactly the same procedures.

Coarse reject duplicates consists of material representing output from crushers. There often can be more than one type of coarse reject when a samples preparation requires several stages of crushing and splitting. In this case, coarse reject duplicates should be collected for each stage of crushing and/or grinding followed by sample reduction. Operations, using large pulverisers, usually don't have coarse rejects as all collected sample is pulverised to a fine pulp. In this case, it is extremely important to collect and analyse field duplicates to understand the overall repeatability of the assayed results.

Same-pulp duplicates should be analysed in the same laboratory with the original samples and externally, sending a certain portion of them to an independent and reputable laboratory.

When collecting sample duplicates it is necessary to remember that they should be identical to the sample which they are duplicating. Unfortunately this is not always possible as the chemical or physical characteristics of the remaining material can be altered after sample has been collected or its amount is simply insufficient to make a representative duplicate. For example, if drill core is sampled by cutting a half core it is unwise to collect quarter core as a duplicate as this will produce a duplicate samples of a twice smaller weight than the original samples. Such duplicates are likely to produce a larger precision error than that of the original samples. Using all remaining second half of core as duplicate is also suboptimal practice as it breaches auditability of the data. The problem can be partially resolved if to take duplicate using the approach shown on the Figure 15.

In case when all drill samples are assayed, the latter is a common case at the bauxite mines, the only possibility to assess precision of the drill samples is to use twin holes (Abzalov, 2009a).

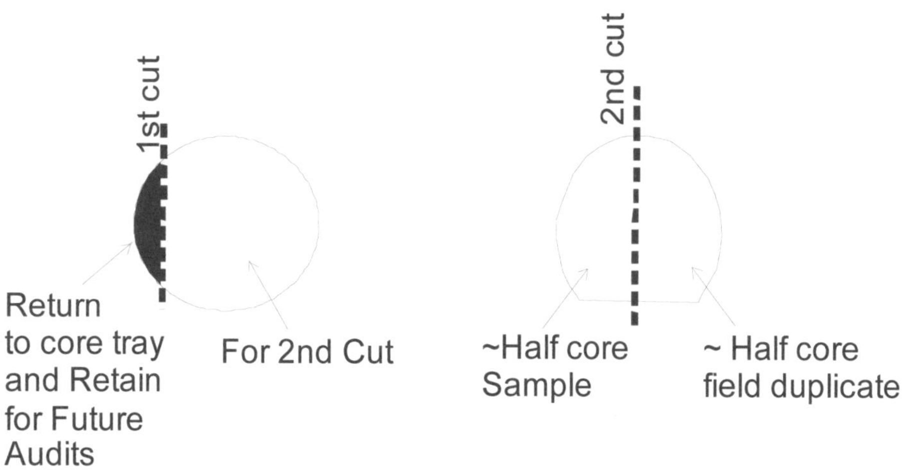

Fig. 15. Sketch explaining collection the duplicate drill core samples in case if routine samples represent the half of the core cut by diamond saw

It is important to assure that duplicate samples should be representative for the given deposit covering the entire range of the grade values, mineralisation types, and different geological units and have a good spatial coverage of the deposit. It is difficult to achieve the representativeness of the duplicates when they are collected at random from the sample batches. The disadvantage of randomly selected duplicates is that most of them will represent the most abundant type of rocks, usually barren or low-grade mineralisation. The ore grade and, in particular, high-grade intervals are often poorly represented in randomly chosen duplicates. They can also be non-representative regarding mineralisation types, their spatial distribution or grade classes. To meet these two conditions, a dual approach is recommended: approximately half of the duplicates should be collected by a project geologist with instructions to ensure that they properly represent all mineralisation styles and grade ranges, and that they spatially cover the entire deposit. The other half of the duplicates should be selected and placed at random in each assay batch.

The position of sample duplicates in a given analytical batch should not be immediately after their original samples and also should not be systematically related to it. It is important that the original samples and their duplicates are included in the same analytical batch which will allow using them for within-batch precision studies.

Numbering of the duplicate samples should disguise them in the analytical batch. The duplicate samples should be taken through the same analytical procedure as the original sample.

6.3 Precision and accuracy estimates

Repeatability of the samples assay is best represented by coefficient of variations of the sample assays estimated from differences between original samples and duplicates (Eqs.22 - 23). Based on numerous case studies, the appropriate levels of sample precisions are proposed (Abzalov, 2008) for different types of deposits which are summarised in Table 4. It is suggested that these levels be used as approximate guidelines for assessing analytical quality. When the estimated CV% exceeds the allowed tolerance level (Table 4), an RDP diagram (Figs. 12 and 13) can be used for a more detailed analysis and diagnosis of errors.

It is important to remember that the values in Table 4, although based on case studies of the mining projects, may not be always appropriate because mineral deposits can significantly differ by the grade ranges, statistical distribution of the studied values, mineralogy, textures and grain sizes.

The Reduced Major Axis (RMA) model (Fig. 11) is another QAQC technique that is becoming increasingly popular among geoscientists (Sinclair and Bentzen, 1998; Sinclair and Blackwell, 2002). This method consists of the calculation of the RMA line and related parameters (Eqs.24 - 30), and is particularly useful for identifying bias between paired data. This method can be used to calculate the precision error estimate from dispersion of the data points about the RMA function. However, it was shown (Abzalov, 2008) that the errors estimated from the RMA model can significantly differ from CV% values. These differences are likely due to sensitivities of the RMA model to the presence of outliers, which makes this technique inappropriate for strongly skewed distributions.

Accuracy of the data should be controlled using standard reference materials and also by assaying a certain portion of the pulp duplicates in a reputable external lab. Standard reference materials alone don't provide full coverage of all possible causes of the sample biases and therefore don't guarantee their timely detections. For example, even matrix matched standards can create a false impression of security when they not fully reflect the level of problems and complexity presented in the actual samples.

Analysis of the standards should be routinely carried out by the mine geologists after results are received from a laboratory. Analysis of standard performance should commence with graphic presentation of the standard assay results (Fig. 9) and a review their distribution patterns. A pattern recognition method is applied for qualitative diagnostics for possible accuracy issues. When distribution of assayed values of reference standards don't exhibits any systematic drifts or other types of specific patterns indicating possible accuracy errors the statistical techniques are applied for quantitative estimation of accuracy (Eqs. 20 - 27). Obtained statistical estimates should be compared with the certified means and the standard deviations. Only when the assayed values of the standards match their recommended values the analytical batch can be considered sufficiently accurate.

All obtained data should be stored together in a relational database which is regularly backed up assuring transparency of the data handling procedures and their auditability. It should be remembered that database is a final storage for large volume of data and therefore the quality control of the analytical data in the mining industry impossible without accurate management of the data flow and database administration procedures.

Mineralisation Type / Deposit	Metal	Best Practise	Acceptable Practise	Sample type
Gold, very coarse grained and nuggety	Au (g/t)	20 (?)	40	Coarse rejects
Gold, coarse to medium grained	Au (g/t)	20	30	Coarse rejects
	Au (g/t)	10	20	Pulp duplicate
Cu-Mo-Au porphyry	Cu (%)	5	10	Coarse rejects
	Mo (%)	10	15	
	Au (g/t)	10	15	
	Cu (%)	3	10	Pulp duplicate
	Mo (%)	5	10	
	Au (g/t)	5	10	
Iron Ore, CID type	Fe (%)	1	3	Field duplicate
	Al2O3 (%)	10	15	
	SIO2 (%)	5	10	
	LOI (%)	3	5	
Cu-Au-Fe scarn and Iron oxide associate Cu-Au	Cu (%)	7.5	15	Coarse rejects
	Au (g/t)	15	25	
	Cu (%)	5	10	Pulp duplicate
	Au (g/t)	7.5	15	
Ni-Cu-PGE - sulphides	Ni (%)	10	15	Coarse rejects
	Cu (%)	10	15	
	PGE	15	30	
	Ni (%)	5	10	Pulp duplicate
	Cu (%)	5	10	
	PGE (g/t)	10	20	
Detrital Ilmenite sands	Total Heavy Minerals (%)	5	10	Field duplicate

Table 4. Best and acceptable levels of the precision errors (CV%) at the mining projects (Abzalov, 2008)

7. Acknowledgements

Financial support for, and permission to publish, the paper by Rio Tinto Exploration is gratefully acknowledged. Author also would like to thank A. Faragher and C.Welton for critical review of the paper and useful comments.

8. References

Abzalov, M.Z. (2009a). Use of twinned drill - holes in mineral resource estimation. *Exploration and Mining Geology,* Vol.18, No.1-4, p.13-23, ISSN 0964-1823

Abzalov, M.Z. (2009b). Principals and Practice of Use of the Relational Databases in the Mining Industry, Part 1. *The AusIMM Bulletin,* No.5, (October 2009), p.39-41, ISSN 1034-6775

Abzalov, M.Z. (2009c). Principals and Practice of Use of the Relational Databases in the Mining Industry, Part 2. *The AusIMM Bulletin,* No.6, (December 2009), p.57-60, ISSN 1034-6775

Abzalov, M.Z. (2008). Quality control of assay data: a review of procedures for measuring and monitoring precision and accuracy. *Exploration and Mining Geology,* Vol.17, No 3-4, p.131-144, ISSN 0964-1823

Abzalov, M.Z. (2007). Granitoid hosted Zarmitan gold deposit, Tian Shan belt, Uzbekistan. *Economic Geology,* Vol.102, No.3, p.519-532, ISSN 0361-0128

Abzalov, M.Z. (2006). Localised Uniform Conditioning (LUC): a new approach for direct modelling of small blocks. *Mathematical Geology,* Vol.38, No.4, p.393-411, ISSN 0882-8121

Abzalov, M.Z. (1999). Gold deposits of the Russian North East (the Northern Circum Pacific): metallogenic overview, *Proceedings of the PACRIM '99 symposium,* pp.701-714, ISBN 9781875776719, Bali, Indonesia, 10-13 October, 1999, AusIMM, Melbourne

Abzalov, M.Z. & Bower, J. (2009). Optimisation of the drill grid at the Weipa bauxite deposit using conditional simulation, 7th *International Mining Geology Conference,* pp.247 - 251, ISBN 978-1-921522-05-5, Perth, Australia, 17-19 August, 2009, AusIMM, Melbourne

Abzalov, M.Z. & Both, R.A. (1997). The Pechenga Ni-Cu deposits, Russia: Data on PGE and Au distribution and sulphur isotope compositions. *Mineralogy and Petrology,* Vol.61, No. 1-4, p.119-143, ISSN 0930-0708

Abzalov, M.Z. & Humphreys, M. (2002). Resource estimation of structurally complex and discontinuous mineralisation using non-linear geostatistics: case study of a mesothermal gold deposit in northern Canada. *Exploration and Mining Geology,* Vol.11, No.1-4, p.19-29, ISSN 0964-1823

Abzalov, M.Z. & Mazzoni, P. (2004). The use of conditional simulation to assess process risk associated with grade variability at the Corridor Sands detrital ilmenite deposit, *Ore body modelling and strategic mine planning: uncertainty and risk management,* pp.93-101, ISBN 1-920806-22-9, Perth, Australia, 22-24 November, 2004, AusIMM, Melbourne

Abzalov, M.Z. & Pickers, N. (2005). Integrating different generations of assays using multivariate geostatistics: a case study. *Transactions of Institute of Mining and Metallurgy. (Section B: Applied Earth Science),* Vol.114, No.1, p.B23-B32, ISSN 0371-7453

Abzalov, M.Z.; Menzel, B.; Wlasenko, M. & Phillips, J. (2007). Grade control at Yandi iron ore mine, Pilbara region, Western Australia – comparative study of the blastholes and Reverse Circulation holes sampling, *Proceedings of the iron ore conference,* pp. 37 – 43, ISBN 978-1-920808-67-5, Perth, Australia, 20-22 August, 2007, AusIMM, Melbourne

Bartlett, H.E. & Viljoen, R. (2002). Variance relationships between masses, grades, and particle sizes for gold ores from Witwatersrand. *South African Institute of Mining and Metallurgy Journal*, Vol.102, No. 8, p.491-500, ISSN 0038-223X

Bumstead, E.D. (1984). Some comments on the precision and accuracy of gold analysis in exploration. *Proceedings AusIMM*, No. 289, p.71-78, ISSN 1034-6783

CANMET. (1998). *Assessment of laboratory performance with certified reference materials*, CANMET Canadian Certified Reference Materials Project Bulletin, 5p

De Castilho, M.V.; Mazzoni, P.K.M. & Francois-Bongarcon, D. (2005). Calibration of parameters for estimating sampling variance, *Proceedings Second World Conference on Sampling and Blending*, pp.3-8, ISBN 1-92086-29-6, Sunshine Coast, Queensland, Australia, 10-12 May, 2005, AusIMM, Melbourne

Francois-Bongarcon, D. (1993). The practise of the sampling theory of broken ore. *CIM Bulletin*, Vol.86, No.970, p.75-81, ISSN 0317-0926

Francois-Bongarcon, D. (1998). Error variance information from paired data: applications to sampling theory. *Exploration and Mining Geology*, Vol.7, No.1-2, p.161-165, ISSN 0964-1823

Francois-Bongarcon, D. (2005). Modelling of the liberation factor and its calibration, *Proceedings Second World Conference on Sampling and Blending*, pp.11-13, ISBN 1-92086-29-6, Sunshine Coast, Queensland, Australia, 10-12 May, 2005, AusIMM, Melbourne

Francois-Bongarcon, D. & Gy, P. (2001). The most common error in applying "Gy's formula" in the theory of mineral sampling, and the history of the liberation factor, In: *Mineral resource and ore reserve estimation – the AusIMM guide to good practice*, A. C. Edwards (Ed.), pp. 67-72, AusIMM, ISBN 1-875776-80-X, Melbourne

Garrett, R.G. (1969). The determination of sampling and analytical errors in exploration geochemistry. *Economic Geology*, Vol.64, No.5, p.568-574, ISSN 0361-0128

Gilfillan, J.F. (1998). Testing the data—The role of technical due diligence, *Ore reserves and finance seminar*, pp. 33–42, Sydney, Australia, 15 June, 1998, AusIMM, Melbourne

Gy, P. (1982). *Sampling of particulate materials, theory and practice,2dn edition*, Developments in Geomathematics 4, Elsevier, ISBN 0-444-42079-7, Amsterdam, 431p

ISO Guide 33. (1989) *Uses of certified reference materials*. Standards Council of Canada, 12p

JORC Code (2004). *Australaisian Code for Reporting of Exploration Results, Mineral Resources and Ore Reserves*, AusIMM, Melbourne, 20p

Kane, J.S. (1992). Reference samples for use in analytical geochemistry: their availability preparation and appropriate use. *Journal of Geochemical Exploration*, Vol.44, No.1-3, p.37-63, ISSN 0375-6742

Lawrence, M.J. (1997). Behind Busang: the Bre-X scandal: Could it happen in Australia? *Australian Journal of Mining*, Vol.12, No.134, p.33–50, ISSN 0375-6742

Leaver, M.E.; Sketchley, D.A. & Bowman, W.S. (1997). The benefits of the use of CCRMP's custom reference materials. Canadian Certified Reference Materials Project MSL No 637, *21st International precious metals conference*, pp., ISBN 1-881825-18-3, San Francisco, 15-18 June, 1997, International Precious Metals Institute, Florida

Lewis, R. W. (2001) Resource database: now and in the future, In: *Mineral resource and ore reserve estimation – the AusIMM guide to good practice*, A. C. Edwards (Ed.), pp. 43-48, AusIMM, ISBN 1-875776-80-X, Melbourne

Long, S. (1998) Practical quality control procedures in mineral inventory estimation. *Exploration and Mining Geology*, Vol.7, No.1-2, p.117-127, ISSN 0964-1823

Minkkinen, P. & Paakkunainen, M. (2005). Direct estimation of sampling variance from time series measurements - comparison to variographic analysis, *Proceedings Second World Conference on Sampling and Blending*, pp.39-44, ISBN 1-92086-29-6, Sunshine Coast, Queensland, Australia, 10-12 May, 2005, AusIMM, Melbourne

Minnitt, R.C.A.; Rice, P. M. & Spangenberg, C. (2007). Part 2: Experimental calibration of sampling parameters K and alfa for Gy's formula by the sampling tree method. *The Journal of South African Mining and Metallurgy*, Vol.107, No.8, p.513-518, ISSN 0038-223X

Pitard, F.F. (1993). *Pierre Gy's sampling theory and sampling practise, 2nd edition*, CRC Press, ISBN 0-8493-8917-8, New York, 488p

Pitard, F.F. (1998). A strategy to minimise ore grade reconciliation problems between the mine and the mill, *Mine to Mill Conference*, pp.77-82, ISBN 1875-7766-3X, Brisbane, Australia, 11-14 October, 1998, AusIMM, Melbourne

Pitard, F.F. (2005). Sampling correctness - a comprehensive guidelines, *Proceedings Second World Conference on Sampling and Blending*, pp.55-66, ISBN 1-92086-29-6, Sunshine Coast, Queensland, Australia, 10-12 May, 2005, AusIMM, Melbourne

Roden, S. & Smith, T. (2001). Sampling and analysis protocols and their role in mineral exploration and new resource development, In: *Mineral Resources and Ore Reserve Estimation - The AusIMM Guide to Good Practise*, A.Edwards (Ed.), pp.73-78, AusIMM, ISBN 1-875776-80-X, Melbourne

Shaw, W.J. (1997). Validation of sampling and assaying quality for bankable feasibility studies, *The resource database towards 2000*, pp.69-79, AusIMM Illawara branch, Wollongong, Australia

Sinclair, A.J. & Bentzen, A. (1998). Evaluation of errors in paired analytical data by a linear model. *Exploration and Mining Geology*, Vol.7, No.1-2, p.167-173, ISSN 0964-1823

Sinclair, A.J. & Blackwell, G.H. (2002). *Applied mineral inventory estimation*, Cambridge University Press, ISBN 0-521-79103-0, Cambridge, 381p

Sketchley, D.A. (1998). Gold deposits: establishing sampling protocols and monitoring quality control. *Exploration and Mining Geology*, Vol.7, No.1-2, p.129-138, ISSN 0964-1823

Stanley, C.R. (2006). On the special application of Thompson-Howarth error analysis to geochemical variables exhibiting a nugget effect. *Geochemistry: Exploration, Environment, Analysis*, Vol.6, No.4, p.357-368, ISSN 1467-7873

Stanley, C.R. & Lawie, D. (2007). Average relative error in geochemical determinations: clarification, calculation and a plea for consistency. *Exploration and Mining Geology*, Vol.16, No.3-4, p.265-274, ISSN 0964-1823

Taylor, J.K. (1987). *Quality assurance of chemical measurements*, Lewis Publishers, ISBN 0873710975, Michigan, 135p

Thompson, M. & Howarth, R. (1978). A new approach to the estimation of analytical precision. *Journal of Geochemical Exploration*, Vol.9, No.1, p.23-30, ISSN 0375-6742

Vallee, M.; David, M.; Dagbert, M. & Desrochers, C. (1992). Guide to the evaluation of gold deposits, *Geological Society of CIM, Special Volume 45*, ISBN 0-919986-31-4, 299p

32

Adaptive Calibration and Quality Control of Smart Sensors

Matej Možek, Danilo Vrtačnik, Drago Resnik,
Borut Pečar and Slavko Amon
Laboratory of Microsensor Structures and Electronics (LMSE)
Faculty of Electrical Engineering, University of Ljubljana
Tržaška 25, Ljubljana,
Slovenia

1. Introduction

Smart sensors represent an attractive approach in sensor applications due to their adaptability, achieved by means of digital signal processing. Sensor adaptability can be further turned into a major advantage by introduction of smart calibration systems.
Smart sensors are generally integrated with signal conditioning circuits. Signal conditioning circuits are needed to adjust the offset voltage and span, for compensation of temperature effects of both offset voltage and span, as well as to provide an appropriately amplified signal. The proposed approach is based on a special case of smart pressure sensors, but the developed calibration system is generally applicable for any kind of smart sensor.
In manufacturing of modern electronic devices achieving and maintaining high yield level is a challenging task, depending primarily on the capability of identifying and correcting repetitive failure mechanisms. Yield enhancement is defined as the process of improving the baseline yield for a given technology generation from R&D yield level to mature yield. Yield enhancement is one of the strategic topics of ITRS (International Technology Roadmap for Semiconductors, Test And Test Equipment, 2006). This iterative improvement of yield is based on yield learning process, which is a collection and application of knowledge of manufacturing process in order to improve device yield through the identification and resolution of systematic and random manufacturing events (International Technology Roadmap for Semiconductors, Yield Enhancement, 2006). Yield improvement process will consequentially increase the number of test parameters and hence the calibration system complexity. One of advantages of increasing system complexity is the ability to integrate the input testing processes and output final testing processes into the calibration process itself, thus shortening the total time for calibration.
Several types of smart sensors with integrated signal conditioning have been presented over the past few years (Takashima et al., 1997) & (IEEE Std. 1451.2 D3.05, 1997). The calibration processes and temperature compensating methods for these sensors are based either on analog, digital or mixed approaches. Analog approach usually comprises an amplifier with laser trimmable thin film resistors (Chau et al., 1997) & (Wang et al., 2005) or off-chip trimmable potentiometers (Schnatz et al., 1992) & (Lee et al., 1999), to calibrate the sensor span and offset voltage and to compensate for their temperature drift. Analog compensation

techniques are relatively slow, inflexible and cost-ineffective. In digital approach, sampling for raw digital pressure and temperature values is first performed, followed by an evaluation of the output digital values via polynomials for describing sensor characteristic, and finally converting the computed pressure values to according analog voltages (ZMD31020 description, 2002) & (ZMD31050 description, 2005). Mixed approach retains strictly the analog signal conversion path, while smart sensor offset and span are adjusted by setting of operational amplifiers by digital means (MLX90269 datasheet, 2006).

This paper will focus on the problem of adaptive calibration any quality control of smart sensors with digital temperature compensation, which is one of the most time consuming steps in sensor production. In order to advance calibration system performance, smart calibration system is conceived as a digitally controlled closed loop system capable of adaptive learning. Presented concept of calibration system is directly implemented in the iterative yield enhancement process in the production of piezoresistive pressure sensors for automotive applications. The calibration system operation and quality control is illustrated on the case of Manifold Absolute Pressure (MAP) sensors. The emphasis will be on MAP sensors, although the proposed approach can be implemented in other fields of application.

2. Calibration procedure

Main calibration procedure starts with measurement of sensor coarse gain and offset and optimization of sensor parameters to the sensor signal conditioner front end stage. After initial optimization procedure the calibration conditions are set according to calibration scenario. Raw sensor readouts of supplied reference quantities are acquired at each calibration point. After acquisition, digital description of sensor characteristic is evaluated and the results are stored back to sensor. A detailed description of calibration procedure is given in (Možek et al., 2008). Calibration scenario defines the sequence of reference quantities, which are applied to sensors under calibration. In case of temperature compensation of pressure sensor, the reference quantities are pressure and temperature. Minimal number of calibration points is 4. This is defined by using the lowest (i.e. linear) degree of polynomial for sensor characteristic description (ZMD31020 description, 2002) & (ZMD31050 description, 2005) in the temperature and pressure direction.

The calculation function used in the ZMD31020 signal conditioner is given in equation (1). It covers variations in sensor offset, sensitivity and first order nonlinearity.

$$p = 2^{12}\frac{2^2 \cdot p_{oc} + 2^{-24} \cdot a_0 \cdot p^2_{oc} + a_1 + 2^{-9} \cdot a_2 \cdot (T_{oc} - 550) + 2^{-18} \cdot a_3 \cdot (T_{oc} - 550)^2}{a_4 + 2^{-9} \cdot a_5 \cdot (T_{oc} - 550) + 2^{-18} \cdot a_6 \cdot (T_{oc} - 550)^2} \qquad (1)$$

where p is the resulting corrected sensor signal, p_{oc} is an uncorrected raw analog to digital converter readout from pressure sensor, T_{oc} is a offset corrected raw analog to digital converter readout from temperature sensor and a_0 through a_6 are calibration coefficients of pressure sensor.

Maximal number of calibration points is primarily limited by total calibration time. In case of pressure sensors, both calibration axes consist of three calibration points, thus enabling compensation of second order nonlinearity in both directions, as depicted in Fig. 1.

Maximal number of calibration points for pressure sensor can cover nonlinearities up to third order in pressure direction. Actual number of calibration points is a compromise between calibration precision and total calibration time. To shorten total calibration time, the slower settling axis should be used for definition of the calibration points order. In case of MAP sensor, the temperature axis defines the calibration scenario.

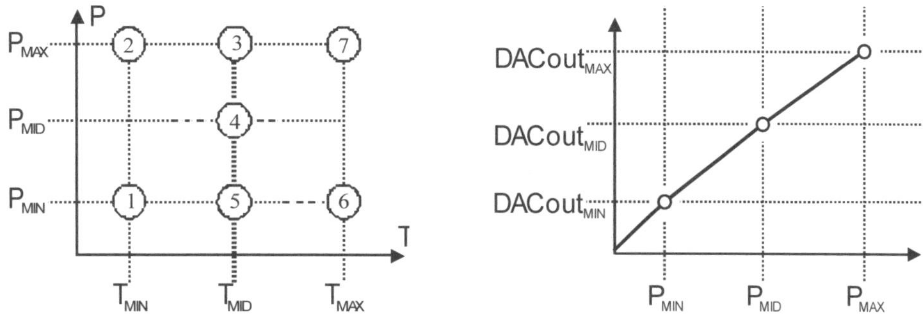

Fig. 1. Calibration scenario

2.1 Sensor range optimization

Optimization algorithms are varying the signal conditioner front-end parameters in order to fit sensor response variations into desired analog to digital converter measurement range. The ZMD31020 preamplifier features three settings of preamplifier gain (15.66, 24 or 42). The preamplifier gain settings cover a wide sensor sensitivity range. The corresponding value of gain is found by altering the gain setting at full scale pressure applied in calibration point 2 of the calibration scenario. Note, that gain optimization is performed at minimal temperature, due to negative temperature coefficient of pressure sensor sensitivity. The signal conditioner readout is evaluated at each setting. If the signal conditioner readout exceeds the limit of maximal readout available (4095), the value of signal conditioner gain must be lowered. Maximal gain, which complies with gain optimization criterion, is selected. For relative pressure sensors, both negative and positive full-scale signal conditioner responses must be evaluated for each gain setting. Maximal gain, which complies with signal conditioner limits for both pressure values, is selected. Output reference bias (zero signal response of pressure analog to digital converter) can be offset to either 1/16, 1/8, 1/4 or 1/2 of analog to digital converter measurement range (0...4095). In case of relative pressure sensors, the value of digital range shift (CRROB) is usually fixed at 1/2, since the measurement range should be centered. In case of absolute pressure sensor, the range shifting is performed at calibration point 1. Sensor zero scale response is read and the values are evaluated. The values are compared to zero response. If the output is clamped to zero, higher range shift option is selected. The selected option represents the minimal value of range shift. Measurement and acquisition of raw data is normally a two stage process described in equation (2) for temperature measurement.

$$T_1 = 2^{10}\left[\frac{V_{offsetT}[V]}{0.98V}+\frac{1}{16}\right] \quad T_2 = 2^{10}\left[\frac{V_{temp}[V]+V_{offsetT}[V]}{0.98V}+\frac{1}{16}\right]$$
$$T_{OC} = T_2 - T_1 = 2^{10}\frac{V_{temp}[V]}{0.98V} \tag{2}$$

where $V_{offsetT}$ is the offset voltage of the temperature sensor measurement path, V_{temp} is the signal from temperature sensor and T_{OC} is the offset corrected value of temperature sensor readout. Addition factor of 1/16 represents the value of corresponding bias voltage, used to adapt the temperature analog to digital converter conversion range. The voltage 0.98 V is the differential temperature analog to digital converter reference voltage. Similarly, the pressure measurement process is a two stage process described by equation (3).

$$P_1 = 2^{12}\left[\frac{A\cdot V_{offsetP}}{0.98\cdot V_{REF}} + CRROB\right] \quad P_2 = 2^{12}\left[\frac{A\cdot\left(V_{pressure} + V_{offsetP}\right)}{0.98\cdot V_{REF}} + CRROB\right]$$
$$P_{OC} = P_2 - P_1 = 2^{12}\frac{A\cdot V_{pressure}}{0.98\cdot V_{REF}} \tag{3}$$

where $V_{offsetP}$ is the offset voltage of the pressure sensor measurement path, $V_{pressure}$ is the signal from pressure sensor and P_{OC} is the offset corrected value of pressure sensor readout. Addition factor of CRROB represents the value of conversion range referenced output bias, used to adapt the pressure analog to digital converter conversion range to different sensor applications (absolute, relative). The voltage $0.98V_{REF}$ is the differential ratiometric analog to digital converter reference voltage. The conversion stage can be supplied by arbitrary V_{REF} voltage within $[V_{SS}...V_{DD}]$ limits. Normally, the value equals V_{DD}. Voltage V_{REF} can be derived from supply voltage, thus making the conversion full ratiometric.

2.2 Acquisition of raw pressure and temperature readout

The readouts from pressure and temperature acquisition are gathered by the calibration station. The calibration station also controls the stability of calibration point conditions. The major advantage of digital calibration system is that stability of a reference pressure and temperature can be controlled by using the sensors being calibrated. The latter becomes effective in case of temperature stabilization, where temperature stability of ΔT=0.2 °C must be assured within calibration chamber. Such temperature stability would normally require an expensive temperature chamber, which still would not solve the problem of temperature stability on the sensors themselves. The same applies for reference pressure value. In order to achieve the most stable reference conditions on the sensors, raw pressure and temperature sensor response have to be filtered. This is performed by means of digital implementation of moving average low-pass filter as depicted in Fig. 2..

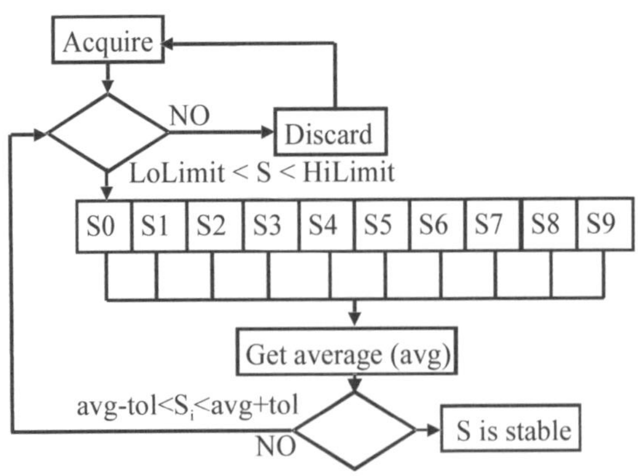

Fig. 2. Moving average filter algorithm for pressure and temperature acquisition

The input sample is first checked for valid limits. If it complies with preset limits, it enters the moving average filter loop. The sampling continues and the complying measured samples are stored in the filter array. At every measurement, filter average is calculated, and the resulting filter average is compared to every sample (S_i) residing in the filter array. The tolerance from current filter average (noted as *tol* in Fig. 2) is used to determine the level of residual measurement noise. This parameter is set in the calibration system setup according to acquisition noise level during calibration. The value of filter tolerance (ε) is calculated in percent of full scale response.

$$\varepsilon = \frac{tol}{2^{RADC}} \cdot 100\% \tag{4}$$

For typical application, where R_{ADC}=13 bits, filter tolerance is 0.06 %FS. The amount of acquisition noise, filtered by such moving average filter can be minimized by either lowering the tolerance parameter or by increasing the filter length.
Filter lengths are different for temperature and pressure acquisition. For pressure sampling point acquisition, ten filter elements have proven to be enough for system noise minimization according to described system setup. On the other hand, for temperature setpoint, filter length is 60 elements, and additional sampling delay of 1 s was introduced. Increasing the filter length also increases the filter stabilization time - i.e. filter step response, hence a compromise between noise level filtering and step response time must be achieved. Both finite and infinite impulse response filter implementations (FIR, IIR) were tested. Equation (5) represents the IIR realization of moving average low-pass filter, while the FIR representation is given in equation (6).

$$y_n = \frac{x_n - x_{n-N}}{N} + y_{n-1} \tag{5}$$

$$y_n = \frac{x_n + x_{n-1} + x_{n-2} + \ldots x_{n-(N-1)}}{N} \tag{6}$$

IIR filter realizations are, although simpler to evaluate, inappropriate due to their instability and non-linear phase response. Stability and linear filter phase response are prerequisite for such application as sensor noise filtering, because they offer instant information about stabilization of measurement quantity.

2.3 Calibration of signal-conditioner analog output stage

This step was introduced after the initial version of ZMD31020 calibration system was completed. During the initial calibrations of sensors with analog output, a large output voltage error of calibrated sensors was noticed. This error was attributed to temperature dependency and chip-to-chip differences of output digital to analog converter characteristic. After the raw pressure and temperature sensor output have been measured at a given calibration point, the sensor analog output digital to analog converter has to be set to match the desired voltage response at a given calibration point. Signal conditioner 11 bit digital to analog converter response exhibits large chip-to-chip differences as shown in Fig. 3. In Fig. 3 the digital to analog converter digital setting histogram at calibration point 5 is shown at a calibration point of 17 kPa and temperature 35 °C. Calibration results based on 36049 sensors are depicted in the Fig. 3.

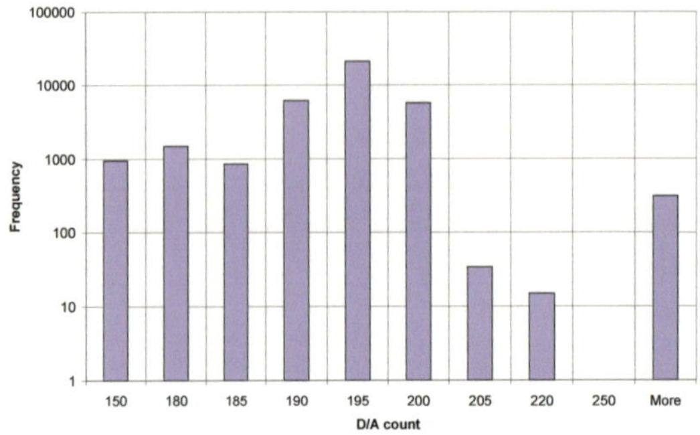

Fig. 3. Digital to analog converter readout at calibration point 5 of calibration scenario

At a given point, the desired sensor output should be at 0.25 V at a power supply voltage of 5 V, so the corresponding ideal linear digital to analog converter setting would be at 205 counts.

$$DAC = \frac{0.25V}{5V} \cdot (2^{12} - 1) = 205 \quad (7)$$

However, if this calculated value is fed into calculation of calibration coefficients, the sensor output actually varies from ideal value. The results show that most signal conditioners don't comply with the calculated ideal value of 205 counts and furthermore there may be large variations (up to 50 digital to analog converter levels) from this ideal value.

One reason for variations is the output stage itself: The signal conditioner digital to analog converter characteristic is described with no load on the sensor output, however this is not the case in real-world sensor application. The signal conditioner namely "features" a rather poor output current driving capability, which is deteriorated by introducing additional pull-up resistance on the sensor output. Second reason lies in the technological chip-to-chip variations of digital to analog converter.

In order to produce an accurate analog output signal, these variations must be compensated. The adjustment of sensor analog output is performed by iterative algorithms, based on successive approximation (SA) principle or P regulator principle. The sensor analog output calibration mechanism provides means for minimization of system specific errors (e.g. calibration point to calibration point errors) as well as sensor specific errors (e.g. signal conditioner chip to chip errors). The digital to analog converter calibration mechanism is based on the successive approximation method, where every sensor digital to analog converter output is measured by a 40 channel data acquisition (DAQ) system. Sensors are also connected to a host computer via digital interface. This arrangement of instruments forms a closed regulation loop which adjusts the sensor digital to analog converter output according to successive approximation algorithm in such a manner that the desired output response (V_{SET}) is obtained at every temperature of calibration. The register of the sensor output digital to analog converter (DAC_{OUT}) can be set according to successive approximation method. The digital to analog converter response voltage (V_{DAC}) is measured and compared to desired output response (V_{SET}). If the V_{DAC} value exceeds the V_{SET} voltage, the preset DAC_{OUT} bit has to be set to zero, or it remains set to one. Algorithm starts with MSB (most significant bit) and has to be iterated 11 times, which equals the number of digital to analog converter bits. The process is time-consuming, especially if output voltage scanning is performed upon 128 or more sensors. Instead of calibrating the digital to analog converter via successive approximation principle, a novel P regulator based calibration mechanism was introduced. This principle shortens the number of iterations from initial 11 to typically 2 or 3 iterations. Initially, the digital to analog converter register value DAC_{OUT} is calculated according to "ideal" digital to analog converter transfer characteristics:

$$DAC_{OUT} = \frac{V_{supply}}{2^{11}} \cdot V_{SET} \tag{8}$$

The initial value covers two variables during digital to analog converter calibration. First one is the supply voltage variation, which is "ideally" considered to be 5 V, and the second one are chip-to-chip variations of the digital value of digital to analog converter. The supply voltage has to be measured before any corrections are made to the digital to analog converter output. The digital to analog converter register value is transmitted to the sensor and the corresponding output voltage V_{OUT} is set. The V_{OUT} voltage is then measured and error is calculated as a difference between calibration point digital to analog converter set value V_{SET} and the measured voltage V_{OUT}. This voltage error is converted to digital value by multiplying with 2^{11} and the resulting digital error value is added to digital representation of digital to analog converter voltage. The loop is iterating until the error becomes less than 0.5 LSB (least significant bit). The exit criterion was chosen to be <0.5 LSB, so that LSB is also set for minimal V_{OUT} error.

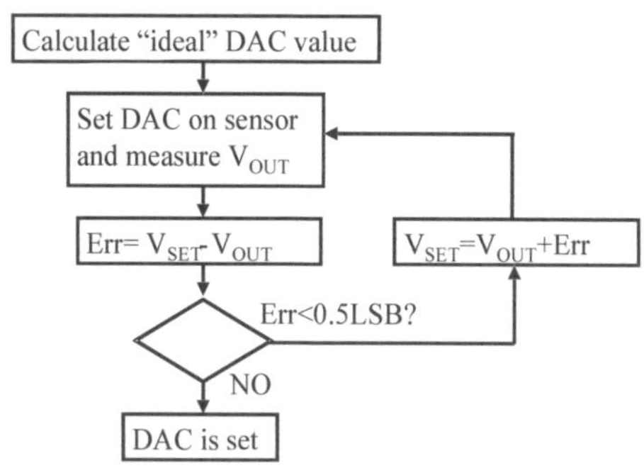

Fig. 4. P regulator based digital to analog converter calibration flow diagram

This algorithm can be executed either on sensors with ratiometric voltage output or fixed voltage output.

2.4 Calibration coefficients calculation

A set of seven coefficients (a_0 ... a_6) must be evaluated in order to take full advantage of calculation precision of sensor characteristic description, expressed in equation (1). Maximal number of acquired data points is limited to seven. Equation (1), which defines the ZMD31020 sensor characteristic, contains only non-involuted coefficients, enabling simple computation of unknown sensor coefficients. If the equation (1) is arranged into form (9) we obtain

$$\begin{aligned} &-2^{-12} \cdot P_{OC}^2 \cdot a_0 - 2^{12} \cdot a_1 - 2^3 \cdot T_{OC} \cdot a_2 - 2^{-6} \cdot {T_{OC}}^2 \cdot a_3 + p \cdot a_4 \dots \\ &\dots - 2^{-9} \cdot p \cdot T_{OC} \cdot a_5 - 2^{-18} \cdot p \cdot T_{OC}^2 \cdot a_6 = 2^{14} \cdot P_{OC} \end{aligned} \tag{9}$$

If seven acquired calibration points are numbered $CP_1..CP_7$, each CP_X having its corresponding raw pressure and temperature output and desired output (P_{OCX}, T_{OCX}, and p_X respectively), the coefficients can be determined by resolving a linear system of seven equations:

$$[C]\cdot[A]=[P]$$

$$C=\begin{bmatrix} -2^{-12}\cdot P_{OC1}^2 & -2^{12} & -2^3\cdot T_{OC1} & -2^{-6}\cdot T_{OC1}{}^2 & p_1 & -2^{-9}\cdot p_1\cdot T_{OC1} & -2^{-18}\cdot p_1\cdot T_{OC1}^2 \\ -2^{-12}\cdot P_{OC2}^2 & -2^{12} & -2^3\cdot T_{OC2} & -2^{-6}\cdot T_{OC2}{}^2 & p_2 & -2^{-9}\cdot p_2\cdot T_{OC2} & -2^{-18}\cdot p_2\cdot T_{OC2}^2 \\ -2^{-12}\cdot P_{OC3}^2 & -2^{12} & -2^3\cdot T_{OC3} & -2^{-6}\cdot T_{OC3}{}^2 & p_3 & -2^{-9}\cdot p_3\cdot T_{OC3} & -2^{-18}\cdot p_3\cdot T_{OC3}^2 \\ -2^{-12}\cdot P_{OC4}^2 & -2^{12} & -2^3\cdot T_{OC4} & -2^{-6}\cdot T_{OC4}{}^2 & p_4 & -2^{-9}\cdot p_4\cdot T_{OC4} & -2^{-18}\cdot p_4\cdot T_{OC4}^2 \\ -2^{-12}\cdot P_{OC5}^2 & -2^{12} & -2^3\cdot T_{OC5} & -2^{-6}\cdot T_{OC5}{}^2 & p_5 & -2^{-9}\cdot p_5\cdot T_{OC5} & -2^{-18}\cdot p_5\cdot T_{OC5}^2 \\ -2^{-12}\cdot P_{OC6}^2 & -2^{12} & -2^3\cdot T_{OC6} & -2^{-6}\cdot T_{OC6}{}^2 & p_6 & -2^{-9}\cdot p_6\cdot T_{OC6} & -2^{-18}\cdot p_6\cdot T_{OC6}^2 \\ -2^{-12}\cdot P_{OC7}^2 & -2^{12} & -2^3\cdot T_{OC7} & -2^{-6}\cdot T_{OC7}{}^2 & p_7 & -2^{-9}\cdot p_7\cdot T_{OC7} & -2^{-18}\cdot p_7\cdot T_{OC7}^2 \end{bmatrix} \quad (10)$$

$$A=\begin{bmatrix} a_0 & a_1 & a_2 & a_3 & a_4 & a_5 & a_6 \end{bmatrix}^T$$

$$P=\begin{bmatrix} 2^{14}\cdot P_{OC1} & 2^{14}\cdot P_{OC2} & 2^{14}\cdot P_{OC3} & 2^{14}\cdot P_{OC4} & 2^{14}\cdot P_{OC5} & 2^{14}\cdot P_{OC6} & 2^{14}\cdot P_{OC7} \end{bmatrix}^T$$

where [C] represents a calibration point matrix, [A] an unknown coefficient vector and P represents pressure values from calibration points. The most prevalent Gaussian elimination was implemented into the calibration system. However, as the calculation of calibration coefficients is well known procedure, care must be taken when deciding upon the final set of coefficients. The computation of linear system of equations is performed using floating point calculations. Sensor correction microcontroller on the other hand is evaluating equations (2) and (3) using 16 bit integer arithmetic. In order to minimize the sensor calculation error, all coefficient rounding combinations have to be evaluated versus ideal (desired - p_x) sensor response at every calibration point. A dedicated algorithm for simulating actual sensor digital response by forcing 16-bit arithmetic was also implemented. Combinations of rounded coefficients (a_0 ... a_6) are subsequently inserted into implementation and normalized difference at ε_{CPi} between desired output $p_{DESIRED}$ and computer simulted output $p_{GetResult}$ evaluated at every calibration point.

$$\varepsilon_{CPi}=\left|\frac{p_{DESIRED}-p_{\mathrm{GetResult}}}{p_{DESIRED}}\right|\cdot 100\% \quad \rightarrow \quad \varepsilon=\frac{1}{N}\sqrt{\sum_{i=1}^{N}\varepsilon_{CPi}^2} \quad (11)$$

where N denotes number of calibration points and ε_{CPi} denotes the calibration error at a given calibration point and ε denotes total calibration error. The normalization with square root of sum of squares was used to amplify large errors and make better distinction of faulty sensors.

2.5 Evaluation of parameters at calibration input

Calibration scenario enables the assessment of essential input parameters to calibration procedure, which enables early fault detection on sensors before they enter actual calibration process. Input parameters comprise the properties, such as offset, sensitivity and nonlinearity of uncompensated sensing element (e.g. pressure sensor). Evaluation of such properties is essential for determination of decision criteria for adaptive concept of calibration system.

Sensor sensitivity can be evaluated at three temperatures. At each temperature, sensitivity is obtained as a difference of pressure sensor voltage response, normalized to corresponding pressure change.

Temperature coefficient of pressure sensor sensitivity is evaluated as a difference between sensor sensitivities at two temperature endpoints (T_{MIN} and T_{MAX} in Fig. 1). Resulting difference is normalized to temperature corresponding temperature change. Sensor offset at room temperature can be evaluated at calibration point 5 as T_{MID} in Fig. 1 is normally set at room temperature. Digital sensor offset readout is transformed into voltage according to analog to digital ASIC stage parameters using equations (2) and (3).

Temperature coefficient of sensor offset is estimated from endpoint calibration points offset values normalized to corresponding temperature difference. In presented calibration scenario the calibration endpoints for estimation of temperature coefficient are marked 1 and 6. Obtained result is recalculated to temperature response at 0°C.

Nonlinearity is calculated by using calibration points 3, 4 and 5 in Fig. 1. Nonlinearity is evaluated as a difference of midpoint pressure response at calibration point 4 from ideal linear sensor response, formed by calibration points 3 and 5. Resulting difference is normalized to calibration span, defined by calibration points 3 and 5. For practical purposes, evaluation of sensor nonlinearity is performed only at room temperature.

Temperature sensor sensitivity in (mV/°C) is calculated by evaluation of raw temperature sensor response difference according to (2) in calibration scenario (Fig. 1) between calibration points 1 and 6. Similar response difference can be obtained from calibration points 2 and 7. Evaluated sensitivities are stored in the calibration database as parameter TCSx, where x denotes the number of calibration point.

Temperature sensor nonlinearity is calculated by evaluation of raw temperature response difference according to equation (2) in calibration scenario (Fig. 1) between calibration points 1, 5 and 6. Linear temperature sensor response is calculated between endpoints 1 and 6. The difference between calculated value at temperature T_{MID} (see Fig. 1) and actual readout from temperature sensor are normalized to previously calculated temperature sensor sensitivity on a corresponding calibration point. Resulting temperature nonlinearity stored in the calibration database as parameter TNLx, where x denotes the number of calibration point. Note, that temperature sensor nonlinearity cannot be evaluated at calibration point 4 according to calibration scenario in Fig. 1. Calculated nonlinearity has the same value for calibration points 1, 5 and 6 and calibration points 2, 3 and 7.

Temperature coefficient of offset is calculated after evaluation of sensor offset values at three temperatures in calibration scenario in Fig. 1. The temperature coefficient of offset is defined as a difference of previously obtained sensor offset values over a temperature interval, formed by temperature T_{MIN} and T_{MAX}. Resulting values from calculation are stored as parameter TCOFx, where x denotes the number of calibration point. Units are [μV/°C]. For applied calibration scenario, the TCOFx parameter value is the same for every calibration point. Temperature coefficient of sensitivity is calculated after evaluation of sensor sensitivities at a given temperature in calibration scenario. The temperature coefficient of sensitivity is defined as a difference of sensor sensitivity over a given temperature interval. The temperature coefficient of sensitivity is defined as a difference of previously obtained sensor sensitivity values over a temperature interval, formed by temperature T_{MIN} and T_{MAX}. Resulting values from calculation are stored as parameter TCSx, where x denotes the number of calibration point. Units are [(mV/V/bar)/°C]. For applied calibration scenario, the TCSx parameter value is the same for every calibration point.

2.6 Evaluation of parameters at calibration output

Calibration output parameters are directly related to compensation of unwanted dependencies. In case of presented MAP pressure sensor this is the temperature compensation. Temperature error is evaluated at every calibration point immediately after evaluation of calibration coefficients. It is calculated by calibration computer as a difference between output of ideal characteristic of MAP sensor and the ASIC simulation of sensor characteristic. Total temperature error comprises RSS (root square of sum of squares) of temperature errors, calculated at every calibration point. Total calibration error is comprised of RSS sum of total temperature error and the combined standard uncertainty for output analog stage, if the sensor features analog output. The ASIC features 16 bit integer arithmetic, therefore a rounding error, which occurs during coefficients calculation, is further minimized by evaluation of total temperature error on all rounding combinations. Rounding combination of calibration coefficients, that yields minimal temperature error at each calibration point is written to ASIC.

2.7 Sensor failure analysis

The analysis of acquired parameters can unveil several causes of sensor failure. Not all causes are universally implied by the large calibration error, therefore failure analysis must be performed on each calibrated sensor. The causes of failure can be either sensor related or system related. Furthermore, the sensor related failures can be divided into signal conditioner failures and sensing element (either pressure or temperature sensor) failure. Detailed detection of system related causes makes the calibration system itself smart.

Non-zero sensor error can be caused by inadequate raw pressure or temperature sensor response: The temperature is measured by an ASIC by measuring a voltage drop on internal diode or an external temperature sensing element (resistor, diode). The available voltage response on internal diode is predefined within interval [294 ... 805] counts for ZMD31020. First, the sensor temperature response is checked against these limits and the response is evaluated. If the sensor response doesn't comply with these limits, an error is raised and sensor is excluded from further evaluation. This error is related to ASIC failure or ASIC bonding failure. Furthermore, the temperature sensor response is also checked at different calibration temperatures - if the response stays within predefined limit, the temperature sensing element is clearly faulty.

2.8 ASIC response related causes of failure

Inadequate raw pressure sensor response is detected by comparing the sensor sensitivity response at a given temperature after successful range optimization was performed. Range optimization technique adapts the ASIC amplifier parameters so that sensor sensitivity should be adapted to analog to digital converter measurement range. Normal raw pressure sensor response difference between minimal and maximal pressure is therefore in the range of 2^R counts, where the R represents the analog to digital converter resolution. If the sensor sensitivity stays well below 2^R, then the sensor is considered faulty due to inadequate pressure response. The "well below 2^R" limit is set at a 3/4 of analog to digital converter measurement range. Calibration errors from ASIC can be related to faulty temperature sensor. The calibration setup was based on a temperature sensing element, located on the ASIC. Furthermore, electrical connection failure can be detected and calculation related errors, which are caused by 16-bit arithmetic used in the ASIC. Electrical connections can fail on the ASIC side, which is immediately detected by the calibration system, because of digital interface error. If any of the sensor bonds fail, this error may be considered same as inadequate sensor response, because the sensor readout will be constant for different excitations. The main ASIC related cause of non-zero calibration error ε_{cpi} is related to calculation error. The arithmetic in ASIC is namely 16-bit, hence calibration coefficients must not exceed the integer value interval [-32678...32767]. Therefore a solution to the system of equations may exist, but the coefficients are clamped to either limit of validity interval, leading to erroneous digital output values.

2.9 Calibration system related causes of failure

Inadequate raw pressure sensor response can be also related to pressure controller failure. To distinguish between pressure leakage and inadequate raw pressure sensor response, a separate reference pressure sensor is connected to a pressure line and it constantly monitors the pressure. Moreover, if there is any leakage in the pressure connection, the controller will not stabilize and the calibration procedure stays put. The calibration system warns the operator for leakage, but finding an actual spot of leakage is still a human operator related issue.

Second major system failure can be related to temperature instability. As the temperature stabilization is achieved by sensors themselves, the temperature stability is checked versus all orthogonal calibration points in calibration scenario. If the temperature varies more than 10 counts over all orthogonal temperature points, a warning is raised due to unstable temperature conditions. Another cause of system failure is that the system fails to stabilize after predefined maximum temperature stabilization time. If temperature stabilization timeout is detected, the system analyzes the temperature stabilization filter contents for each temperature sensor. Sensors that exhibit large fluctuations in temperature response are marked bad and disregarded in further calibration. This type of error can be detected at room temperature and faulty sensors can be replaced before entering further calibration process.

2.10 Calibration error estimation – quality control

Total calibration error represents the main instrument for separation of faulty sensors. At the same time the calibration error relates directly to sensor quality and can provide means for sensor classification. The sensor signal conditioner uses digital representation of sensor characteristic described by equation (3). This description enables immediate evaluation of sensor properties, by running simulation algorithm of sensor response upon acquired calibration data. Total calibration error is calculated by simulating actual ASIC 16-bit arithmetic calculation, analyzed in equation (11).

At a first glance, the calibration error at every calibration point ε_{cpi} of the equation (11) should always yield 0. The calibration coefficients are obtained by resolving a system of equations. Calibration error is caused form coefficient rounding to integer value and by 16-bit integer arithmetic in signal conditioner. Calibration error value may differ ±1 LSB at every calibration point. Cumulative (RSS) error of the sensor characteristic approximation error should be as close as possible to zero. Large calibration error ε_{cpi} evaluation of sensor solution error may have several different causes. The causes for large calibration errors are analyzed by the sensor failure analysis software.

3. Results

Presented results are based on 34422 calibrated manifold absolute pressure sensors. Sensor properties investigation is presented on ZMD31020 signal.

In order to evaluate the input properties of uncalibrated sensors a histogram was plotted. The input temperature coefficient of pressure sensitivity at calibration point 3 in Fig. 1 is in the range of [-8%...-0.2%], which represents a insurmountable span of temperature coefficients, if analog calibration was to be made upon such sensors.

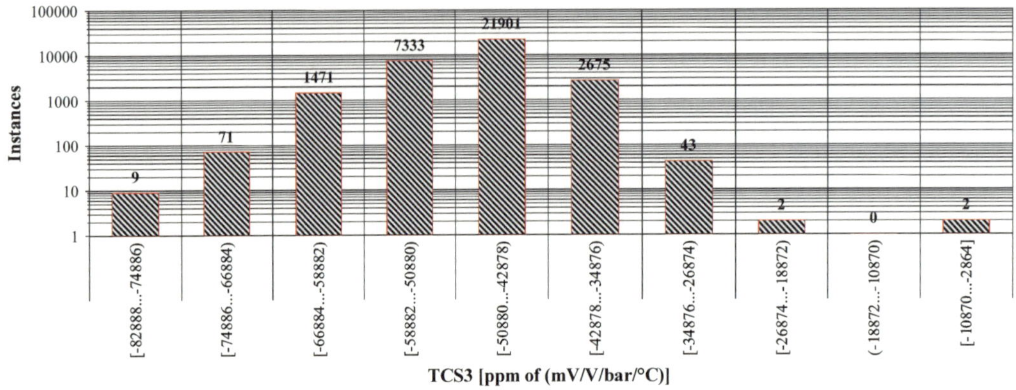

Fig. 5. Input temperature coefficient of sensitivity

Average value of input temperature coefficient of pressure sensitivity in the histogram, depicted in the Fig. 5 is -4.9% (mV/V/bar). Standard deviation from this value is 0.51% of (mV/V/bar). Sensors, based on analog signal conditioners with operational amplifiers (Schnatz et al., 1992), can compensate temperature coefficient of sensitivity up to 0.2%/°C. The latter demonstrates the advantage of the digital temperature compensation based signal conditioners. Input temperature coefficient of offset voltage is depicted in Fig. 6. Again, the plotted histogram depicts large variations for temperature coefficient of offset voltage. Analog calibration system could not calibrate the sensor with temperature coefficient of offset voltage in the range of 1mV/°C.

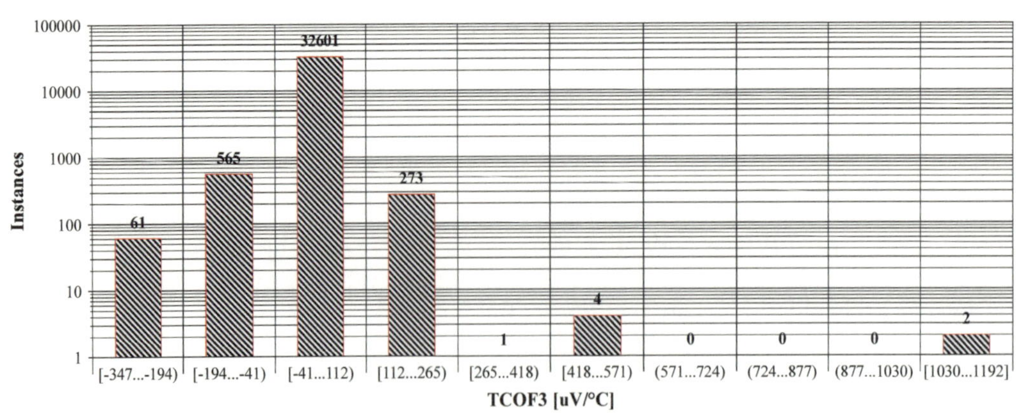

Fig. 6. Input temperature coefficient of offset voltage

In case of calibration of ZMD31020 based MAP sensors, further 11 test points were introduced to calibration scenario. Output temperature error histograms were evaluated at 5%, 10%, 20%, 30%, 40%, 50%, 60%, 70%, 80% and 90% of power supply voltage pressure response at 85°C and 20°C upon a set of 5828 sensors. From initial 5828 sensors, 366 were evaluated as bad. Among them were 182 sensors, lacking the results from testing at 20°C. Calculated histograms are a clear demonstration of effectiveness of digital temperature compensation. The histogram in Fig. 7 depicts the magnitude of temperature error in test point 1 (T=85°C, P=17kPa, V_{OUT}=5%V_{CC}). Presented result was subtracted with an ideal value and the resulting error was normalized in ppm. The data in the Fig. 7 shows temperature error in the range of [-0.2% ... 0.38%] for 5075 sensors out of 5462 total, whereas the admissible range of temperature errors lies within ±1.7%.

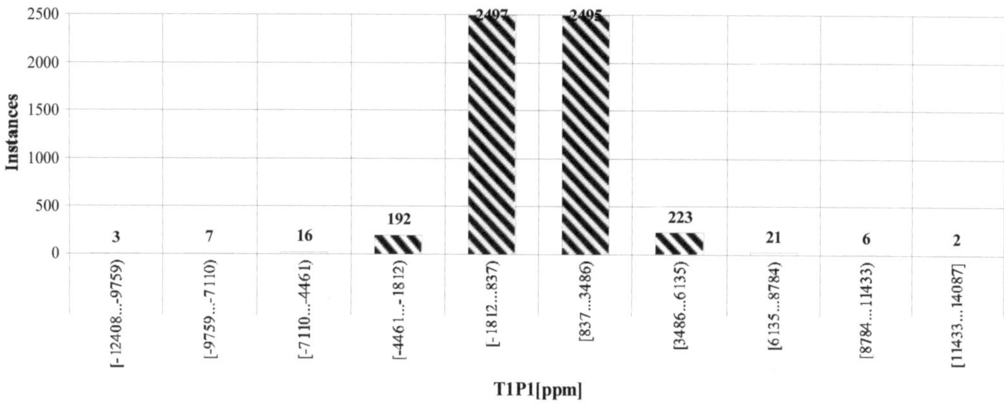

Fig. 7. Temperature error at test point 1

Mean histogram value, representing a typical calibration temperature error is 0.086%. The standard deviation from this value is 0.16%. Similar histogram was evaluated at test point 11 (T=20°C, P=105kPa, V_{OUT}=95%V_{CC}) and the resulting temperature error is depicted in Fig. 8. Mean histogram value is now 0.15%, while the standard deviance is 0.19%.

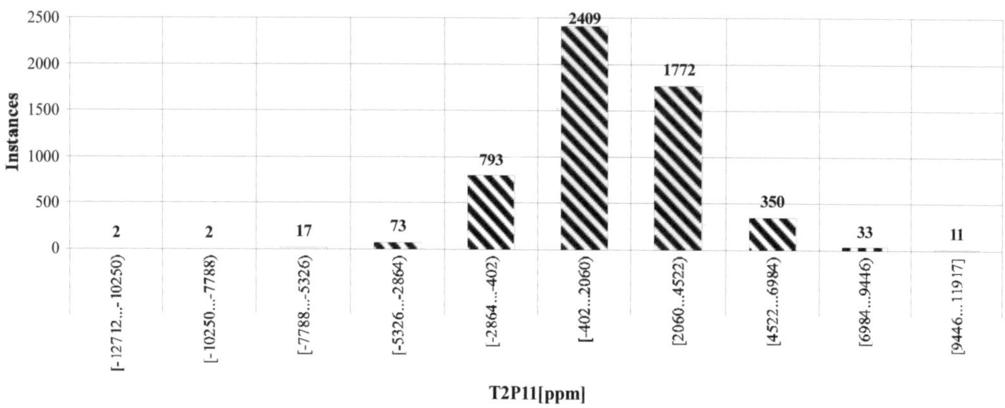

Fig. 8. Temperature error at test point 11

If remaining 184 sensors (366-182) bad sensors are further analyzed, the output stage failure is noted on 82 sensors, which can be attributed to faulty connection of the sensor output, because the sensor output stays the same on every test point. The same cause of error can be attributed to faulty output stage – fault in signal conditioner. The actual cause can be determined with combined insight into calibration database. Remaining 102 sensors were calibrated with temperature error out of MAP sensor specification. One of them (ID=58800) was rejected by calibration process due to inadequate pressure response. Upon analysis of calibration database upon these 101 sensors, it becomes apparent that most of the tested sensors passed the calibration, but failed the test. The analysis of maximal sensor temperature error was performed on all tested temperatures. Resulting maximal errors were divided into 10 bins and the result was evaluated in the histogram. Resulting histogram is summarized in Table 1. The histogram depicts sensor classification in ten classes. The majority (5189 of 5462) of sensors are well within 0.7% limit of temperature error. Important is, that classification can be performed on each and every calibrated sensor. Because of complete sensor traceability, we are able to identify the class and quality for each calibrated sensor. The calibration yield upon 5462 sensors is 98.1%.

Class	Limit (ppm)		Sensors
	upper	lower	
1	767	2239	992
2	2239	3711	2190
3	3711	5183	1477
4	5183	6655	530
5	6655	8127	182
6	8127	9599	44
7	9599	11071	24
8	11071	12543	16
9	12543	14015	4
10	14015	15490	3

Table 1. Classification of calibrated sensors

The cause of failed sensors is attributed to change of sensor properties after calibration during packaging process. This was counter measured by performing the packaging process prior to calibration and performing the calibration as a last step of production process.

The adaptivity of the calibration system is based upon determining the limits of all system parameters, which define the criteria for quality of calibrated sensors. The result from criteria adaptation is the calibration interval for a given sensor property, based on sensors which comply with predefined output response. The limit optimization process is performed upon every sensor that enters the calibration process. Primary acquired sensor parameters are obtained directly from acquisition – raw pressure and temperature sensor readouts. The raw values are recalculated to analog measured quantity according to preamplifier settings, including sensor offset compensation and preamplifier gain.

An illustrative case of sensor limit adaptation is presented when a new sensor enter the calibration process. After initial acquisition of raw values the new sensor response is evaluated and its response is inserted in the histogram, which depicts the raw sensor readouts at the first calibration point (17kPa, -20°C). Entering sensor was assigned identification number (ID=31326).

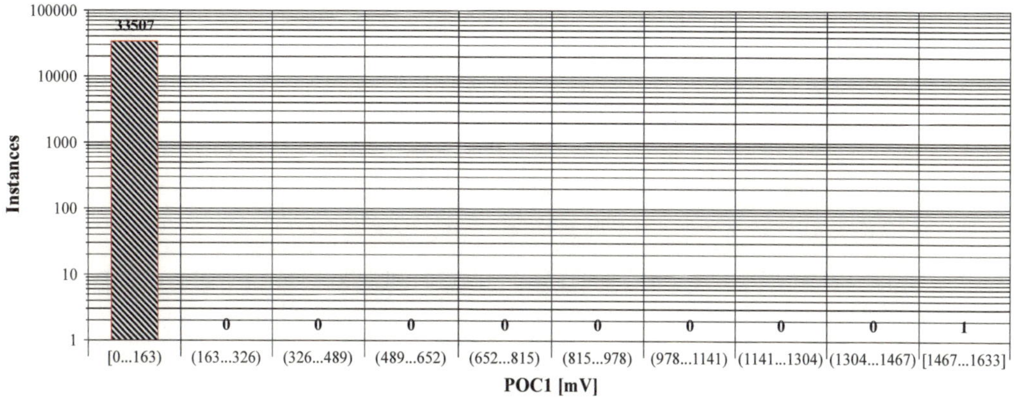

Fig. 9. Raw pressure sensor response at calibration point 1

From the histogram on Fig. 9, it is obvious, that the tested sensor extremely deviates in raw response from all other sensors. However, an automated analysis must establish other sensor properties in order to determine whether a given sensor will enter a full calibration process or not. Sole evaluation of the magnitude of raw pressure response is not sufficient for final estimation, because the sensors at the input can be e.g. from different manufacturers and their responses may vary. The calibration system is designed to adapt also to new type of sensor with different input properties. If several pressure points are scanned, the sensor properties can be evaluated (sensitivity, offset and nonlinearity). First, the sensor sensitivity is calculated as a difference of two pressure responses. If the sensor readout is approximately ten times larger than normal, then the sensitivity should be in proportion with raw readouts. Otherwise, the sensor response can be considered inadequate - this indicates failure in offset or gain optimization process. The system calculates the sensor sensitivity and depicts the result in the histogram for comparison with other sensors. The resulting histogram is depicted in the Fig. 10. The sensitivity was evaluated in the range between 300 and 338, which is in proportion with sensor readouts.

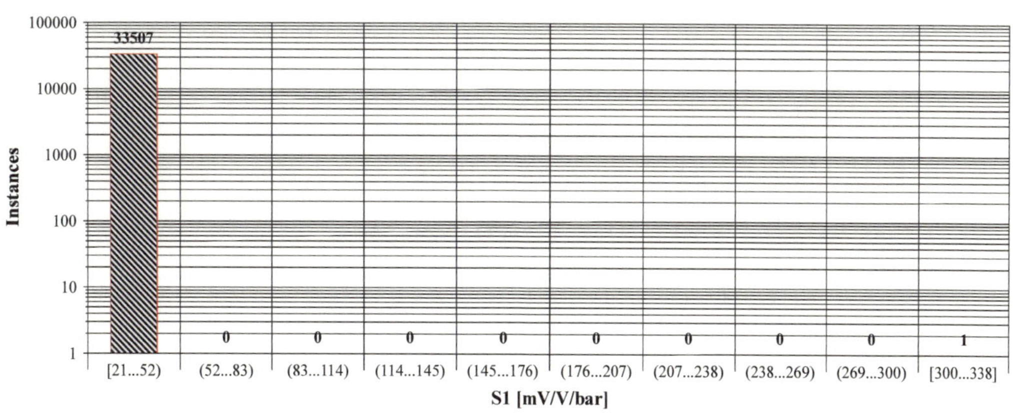

Fig. 10. Sensitivity at calibration point 1

Therefore, further analysis is performed and sensor nonlinearity is evaluated and the results are depicted in the Fig. 11. When the sensor nonlinearity is compared to other sensors in histogram, it becomes obvious, that the sensor is highly nonlinear (55.8%). Therefore, the sensor is discarded from further calibration process.

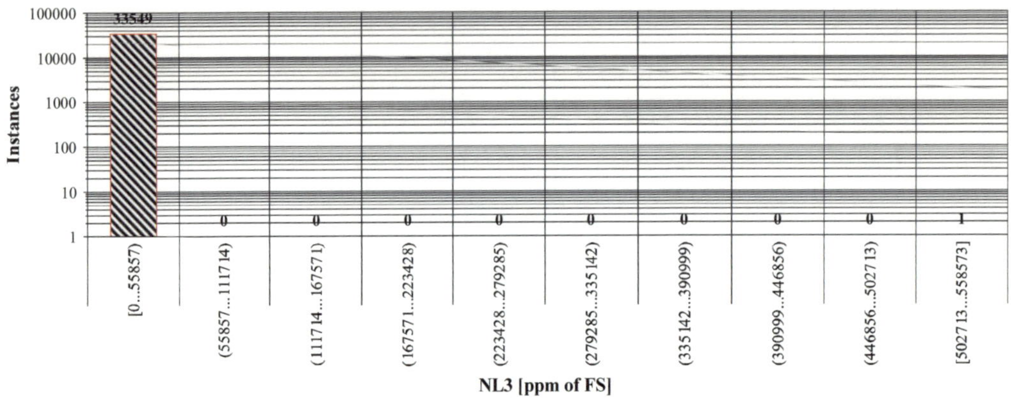

Fig. 11. Nonlinearity at calibration point 1

The sequence of high sensitivity and excess nonlinearity failures implies that a pressure sensor was not designed for calibration on a high pressure range: A low pressure sensor was exhibited to calibration on a high pressure range. Such a low pressure sensor exhibits larger sensitivity but also nonlinear response, when exposed to overpressure. Sensors such with nonlinearity can be calibrated, but not with the seven point calibration scenario, which was used during calibration process of manifold absolute pressure sensor.

Maximal nonlinearity of uncalibrated pressure sensors was limited to 2%. Sensor is discarded from further calibration and resulting histograms of raw pressure readout are evaluated again. Resulting histograms after discarding are depicted in Fig. 12 and Fig. 13.

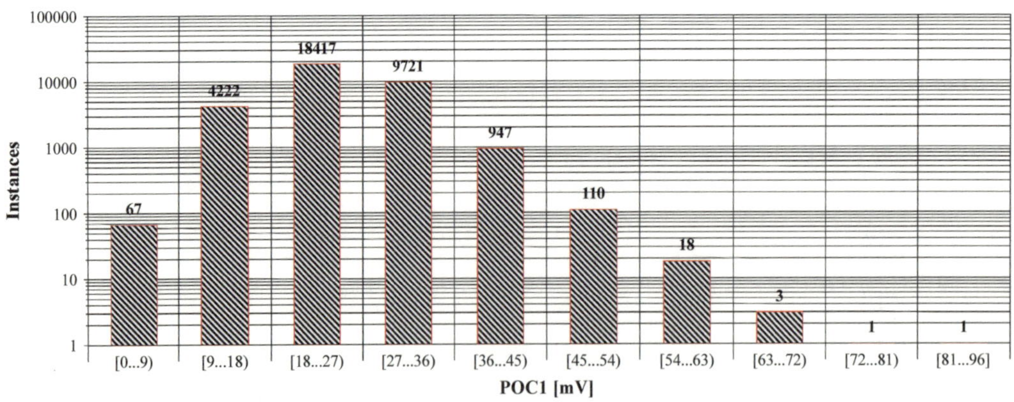

Fig. 12. Corrected raw pressure sensor response at calibration point 1

The resulting limits for raw pressure response stay between 0 and 96mV as can be seen in the Fig. 13, and for the pressure sensitivity in interval [21…41mV/V/bar].

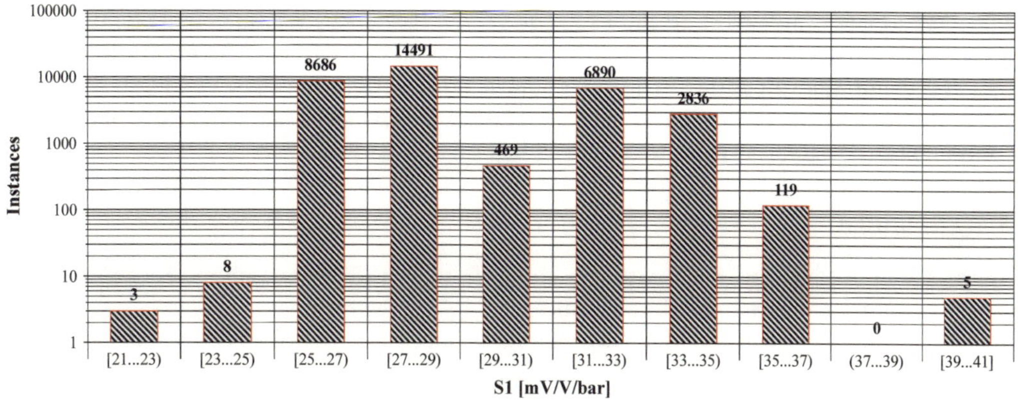

Fig. 13. Corrected pressure sensor sensitivity at calibration point 1

Failure analysis was performed upon a set of calibrated sensors. A detailed insight of failure analysis results is summarized in Table 2. Calibration yield, which would be calculated disregarding failure analysis, would yield 93.7%, since there are 2289 failed sensor out of 36711 calibrated. However, the calibration database stores everything including failed attempts related to system causes, which are not caused by failed sensors. Most of system failures are attributed to improper sensor connection (operator error). Therefore the system related causes must be removed from analysis to obtain actual yield of calibration. After this, the calibration yield improves to 96.8%, since there are only 1127 failed sensors of 35549.

Cause of failure	Origin of failure	Nr. of sensors
Inadequate response of pressure sensor	Sensor	220
Inadequate response of temperature sensor	Conditioner	81
Calibration coefficients clamped	Calibration	373
Communication failure	System	1108
Inadequate temperature stabilization	System	34
Excess nonlinearity	Sensor	384
Tolerance error during coefficient calculation	Calibration	36
Calibration of sensor output stage failure	System	20
Output stage clamped to maximum level	Conditioner	33
Total		2289

Table 2. Failure analysis of calibrated sensors

4. Conclusions

Adaptive calibration and quality control were presented. During initial calibration stage early detection of faulty sensors has proven essential for calibration system yield improvement. Yield enhancement is achieved by thorough specification of sensor related failure causes. Further refinement of calibration failure causes gives a detailed insight into sensor related failures by thorough evaluation of essential sensor properties such as nonlinearity, offset and sensitivity. Described quality control mechanisms enable automatic sensor classification. Proposed calibration system shortens the total time for calibration of smart sensors by implementing the input testing of the sensor parameters as well as final testing of the calibrated sensors. Final testing was achieved by inserting excess test points into the calibration scenario. In its maximal extension, which offers calibration of 2048 sensors at a time, the calibration time was reduced to 3 seconds per sensor - in its current configuration, the total calibration time is 42 seconds per sensor.

5. References

International technology roadmap for semiconductors. 2006 update - Test And Test Equipment Electronic file available at: http://www.itrs.net/Links/2006Update/FinalToPost/03_Test2006Update.pdf, 2008

International technology roadmap for semiconductors. 2006 update - Yield Enhancement Electronic file available at: http://www.itrs.net/Links/2006Update/FinalToPost/13_Yield2006Update.pdf, 2008

Takashima, Y.; Adachi, T. & Yoshino, T. & Yamada, T. (1997). Temperature compensation method for piezoresistive sensors. *JSAE Review*, Volume 18, Issue 3 (1997), pp. 317 319

IEEE Std. 1451.2 D3.05-Aug1997. IEEE standard for a smart transducer interface for sensors and actuators - Transducer to microprocessor communication protocols and transducer electronic data sheet (TEDS) formats. Institute of Electrical and Electronics Engineers, September 1997

Chau, M.; Dominguez, D. & Bonvalot, B. & Suski, J. (1997). CMOS fully digital integrated pressure sensors. Sensors and Actuators A, Volume 60, Issues 1-3 (1997), pp. 86 89

Wang, Q.; Ding, J. & Wang, W. (2005). Fabrication and temperature coefficient compensation technology of low cost high temperature pressure sensor. *Sensors and Actuators A*, Volume 120, Issue 2 (2005) 468 473

Schnatz, F. V. et al. (1992). Smart CMOS capacitive pressure transducer with on-chip calibration capability. *Sensors and Actuators A*, Volume 34, Issue 1 (1992), pp. 77 83

Lee, Bo-Na et al. (1999). Calibration and temperature compensation of silicon pressure sensors using ion-implanted trimming resistors. *Sensors and Actuators A*, Volume 72 (1999), pp. 148 152

ZMD31020 Advanced Differential Sensor Signal Conditioner Functional Description Rev. 0.75, (2002) ZMD AG

ZMD31050 Advanced Differential Sensor Signal Conditioner Functional Description Rev. 0.75, (2005), ZMD AG

MLX90269 Absolute Integrated Pressure Sensor datasheet, June 2006, *Melexis- Microelectronic Integrated Systems Inc., rev 2*

Možek, M.; Vrtačnik, D. & Resnik, D. & Aljančič, U. & Penič, S. & Amon, S. (2008). Digital self-learning calibration system for smart sensors. *Sensors and Actuators A*. 141 (2008), 141, pp. 101 108

33

Hyperspectral Imaging for Raw Material Sorting and Processed Product Quality Control

Pilar-Beatriz García-Allende, Olga M. Conde
and José Miguel López-Higuera
University of Cantabria
Spain

1. Introduction

Agricultural and industrial sectors are more competitive and quality conscious than ever before. To be profitable, industry requires equipment that would ensure pure, high-quality production, and efficient work and cost. This fact is even more important in developed countries to aid companies to defend their position and competitiveness against others where labour cost does not account for such a significant part of the overall manufacturing cost. Raw material sorters, conveyors and processing systems easy to install, energy efficient, reliable, low-maintenance and simple to adjust and control are sought. Even the inclusion of network connectivity capabilities for remote monitoring and communication is also desirable.

Optical Spectroscopy (OS) becomes highly appropriate for this kind of applications because it covers all types of qualitative and quantitative analytical methods based on the interaction of light with living and non-living matter (Schmidt, 2005). For more than 200 years it has been utilized in various fields of science, industry and medicine, particularly in (bio-) chemistry, biology, physics and astronomy. OS is highly specific since each substance is discernible from all others by its spectral properties. In addition, the requirements of the samples are not particularly restrictive. Measurements of different optical parameters as a function of wavelength/energy ("spectrum") provide valuable insights that are not, or not readily, attainable by other analytical methods. Traditional OS techniques generate information on the bulk properties of a sample or a portion taken from it (Millar, 2008). However, there are many aspects of sample properties which result from heterogeneity in composition and the monitoring of this aspect of quality is considerably more challenging. In the last few years *Hyperspectral Imaging Spectroscopy* (HIS) that integrates conventional imaging and moderate resolution spectroscopy, which was primarily developed for applications in remote sensing and astronomy, has been developed for laboratory use and it has even slowly transitioned into other areas as life sciences or industrial production. The key difference is that, in this technique, entire spectra are collected for each pixel within the image so its advantages in agricultural and industrial procedures such as material discrimination are obvious, i.e. reduced measurement time due to the simultaneous acquisition without the need for scanning mechanics. By assessing specific spectral features at each pixel corresponding to a material point, or by using calibration to quantify individual components, an *optical parameter map* of the material may be generated. There exists a wide variety of systems to obtain this both spatial and spectral information from an object, the so-called imaging spectrometers that differ in their acquisition methods (Aikio, 2001). A review of the existing methods and technologies is performed and compared in terms of their suitability for on-line quality applications in agricultural and industrial environments.

In quality assurance systems the final output should be delivered in a straightforward way facilitating its application, i.e. they have to provide a clear added value to the user (*usability* concept). For instance, in raw material discrimination, only the presence and position of spurious materials are of interest. No further information is required to determine which jets of pressurized air need to be activated to blow the unwanted material into a container, while the rest of the material continues to another conveyor. Imaging spectrometers collect, however, underpinning information about material optical properties. Therefore, to obtain efficient quality indicators, considerable data management and analysis is required to convert this huge information into the desired operative indicators. The investigation into novel and time efficient compression and interpretation techniques, suitable for laboratory use as well as for real time production lines in industrial sectors, is addressed in this chapter. To this end, Section 2 describes the principles and different approaches for HIS operation. The implementation of the sensor system is provided in Section 3, whereas Section 4 is focused in how the HIS processing unit is structured. Section 5 discusses different classifiers for automatic material sorting. Section 6 presents the results obtained with each proposed alternative comparing their performances in terms of classification accuracy as well as time and space complexity. Finally, some conclusions and future lines of work are outlined.

2. HIS applied to industry processes

Raw material on-line characterization and discrimination processes are a key issue in several agricultural (Park, 2004; Xin, 2005) and industrial sectors (Reich, 2008; Herrala, 1997; Garcia-Allende, 2008c). Technologies for quality evaluation detect physical properties which correlate well with certain quality factors of the products such as density, firmness, vibrational characteristics, X-ray and gamma ray transmission, optical reflectance and transmission, electrical properties and nuclear magnetic resonance. Each technique has advantages and limitations, such as image resolution, imaging duration, safe handling, sample-specific requirements and availability of details such as surface color, texture, and internal details.

- X-rays, due to their high energy, can penetrate many objects (Mesina, 2007; Kotwaliwale, 2007). However, there is variability in penetration through different materials depending on their properties. Although X-rays great potential, the associated high equipment cost and safety issues make them less practical for on-line applications.
- Colorimetric analysis based on cameras and machine vision are another option (Abbasgholipour, 2010) but they are constrained to characterize the outer layer of the material with high probability of confusing different materials presenting similar colours or due to problems derived from variations on the illumination.
- Optical spectroscopy, based on diffuse reflectance measurements, is another promising method able to extract the chemical composition from the spectrum of the material, which can be employed for sorting.

- Thermal imaging (Varith, 2003) measures infrared energy emitted from the material. The detected energy is converted into a thermal map called a *thermogram*. Quality indicators are derived from temperature differences on material points. In some cases, thermal imaging does not require an illumination source.
- LIBS (Laser Induced Breakdown Spectroscopy) determines the chemical composition of laser evaporated material on surfaces (Wiggenhauser, 1998). When material is vaporized by a high energy laser pulse, the composition elements emit their specific fluorescence. This radiation is used to identify them in the evaporated material by spectroscopic methods (Solo-Gabriele, 2004).

Among all the previous techniques, the *non-destructive* and *non-contact* characteristics of optical techniques, along with the low equipment cost and absence of material pre-processing, make them a perfect candidate for material characterization and sorting. Imaging spectroscopy measures the optical spectrum of interactions of light and matter (absorption, scattering, reflectance, etc.) and the spatial location of an object in a surface (Lu, 1999). This ability offers a great potential for identifying raw materials while they are transported by the conveyor belt in production plants. Analysis systems can be easily installed over the conveyor belt with minimum interference in the industrial set-up. As a result, work in this area has gained prominence in the last few years and it is being adopted for product development and control of a wide variety of industrial processes (Bonifazi & Serranti, 2006; Bonifazi, 2008; Gomez-Sanchis, 2008; Leitner, 2009).

Usually, raw materials need to be classified on different quality levels. Present experience has been acquired working for a tobacco industry located in the region of Cantabria (Spain). Apart from the company interest in tobacco sorting procedures, few studies were found about the spectral characterization of tobacco chemical components (nicotine, polyphenols, etc.) (McClure, 1997). Even less work is reported in open literature regarding tobacco quality control in industry which is, at present, mainly performed by time-consuming machine vision techniques. On account of this, it was definitely a real problem-oriented niche for both knowledge and technology production. Moreover, the developed strategies could undeniably be extended to other industrial fields and materials.

The problem to be solved in the tobacco industry consists in the classification of the raw material at the entrance of the production plant. Tobacco leaves are intermingled with a great variety of materials such as plastics, cords, cardboard, papers, etc., mainly due to the manual harvest procedure. These spurious elements or "defects" have to be rejected before starting the process of making cigars. At first, the leaves undergo a mechanical classification procedure, where rejection due to differences of weight, size and density is performed. However, some of the undesired materials go through this coarse classification and additional discrimination systems are required. The design of a completely non-intrusive, non-contact and real-time system capable of performing this identification task, employing an imaging spectroscopic technique, is the aim of this research.

2.1 HIS basics

Diffuse reflectance is the most appropriate light-matter interaction for material sorting. It is produced when an incident radiation penetrates the material sample. Once this radiation travels through the sample, it becomes partially absorbed and partially scattered. At the end, some radiation gets out of the material on arbitrary directions. The spectrum of this reflected radiation has now imprinted the effects of the chemical composition of the sample.

The HIS concept is applied over this parameter. The diffuse reflectance measurement generates a three-dimensional data set, *datacube*, defined by spatial (x and y) and wavelength (λ) coordinates (Willoughby et al., 1996; Green et al., 1998). Figure 1 depicts the datacube consisting of the light reflected by the material shown at the foremost part of the cube. Hyperspectral images and the associated reflectance spectra of the vertical lines located at x_1 and x_2 positions are shown at the lower part of Fig. 1. The discontinuous lines highlight the correlation between the real and spectral images of each segment. A colour gradation scale is employed to represent the spectra of each spatial position y_i of the vertical lines. When the colour of a pixel is close to red, it means that the intensity of the reflected radiation at that particular wavelength is high. On the other hand, if the pixel approaches blue, the material shows low reflectance.

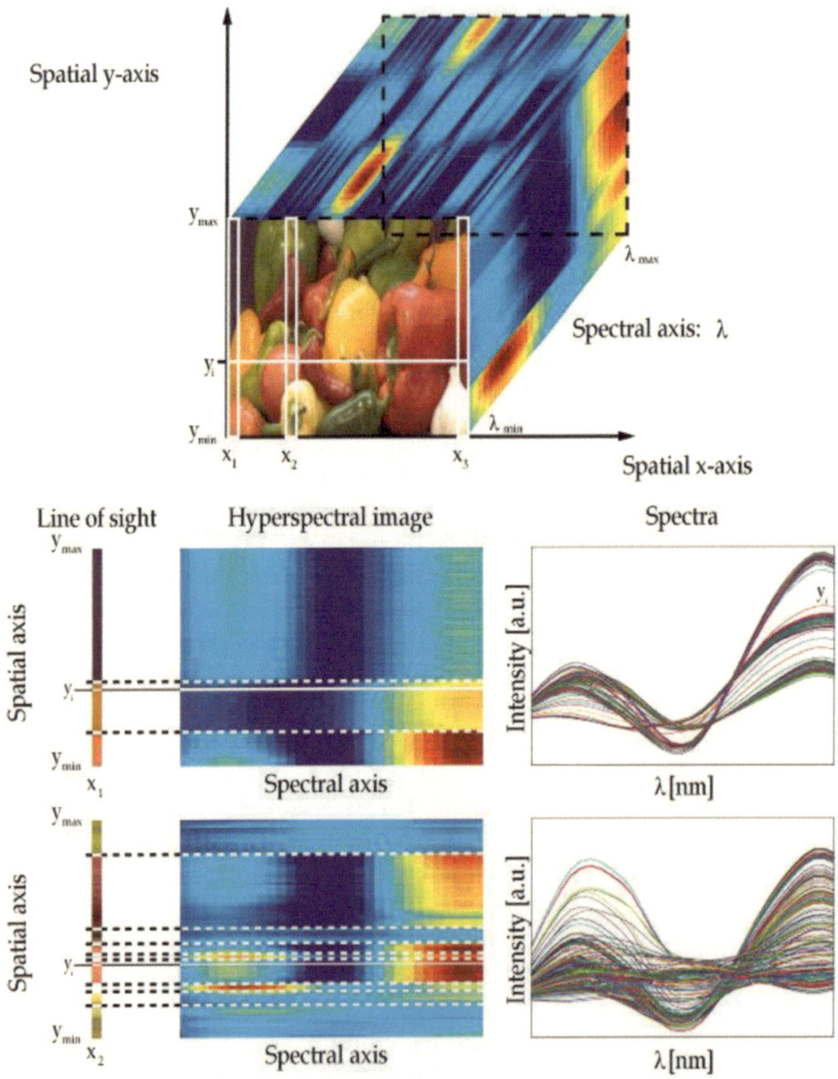

Fig. 1. HIS concept showing how the spectrum of a spatial point is recovered from an image

2.2 HIS alternatives

The datacube provided by imaging spectrometers can be classified depending on the data-acquisition mode. The most conventional types are shown in Figure 2:

- Non-imaging spectrometers are called *whiskbroom scanners* because they provide the spectrum of a single point within the image, i.e. a column within the image cube. The recording of the whole image implies a double spatial scanning step, and, either scanning mechanics or several single point instruments are, therefore, required. These approaches are respectively time-consuming and expensive.
- *Staring imagers*, as filtered cameras or tuned wavelength illumination source, electronically scan the spectral domain. The image is collected one spectral plane at a time. Although easy to implement by means of tuneable filters or LED's activation, this configuration has a major drawback. The spectra for each pixel are not taken simultaneously, and, consequently, pixel-to-pixel spectral mixing could occur if the scene is moving while the spectra are being scanned (Willoughby, 1996).
- *Pushbroom scanner* employs an array of detectors to simultaneously scan the spectral dimension and one spatial dimension. This acquisition mode allows data collection in hundreds or thousands of spectral bands (hyperspectral) to produce an image. This makes a difference respect to colour cameras that collect information in the three red, green, and blue bands or multispectral imaging achieved by tuneable filters. In this way, not only spectral mixing is prevented, but also sensitivity to minor components is improved. There still remains the scanning of the other spatial dimension, but this can be easily achieved by the relative motion between the scene and the imaging spectrometer field of view. As a result, this configuration is particularly suitable for industrial applications, which exhibit a predominance of conveyor belt systems.

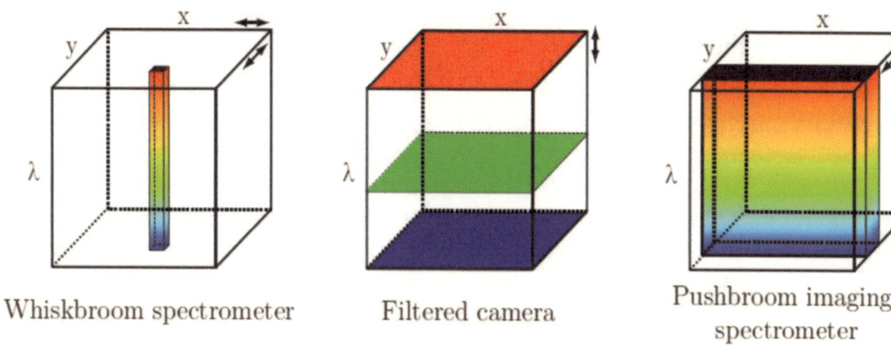

Fig. 2. Spectral imaging sensors depending of their data acquisition modes

Different imaging spectrometers designs can employed depending on the selected dispersive configuration (Aikio,2001). Among the transmissive configurations, imaging spectrometers could be based on PGP (prism-grating-prism) devices, HoloSpec design or Fourier-Transform configurations. On the reflective side, Offner configurations, Echelle gratings, Czerny-Turner layouts and Hadamard transform approaches can be employed.

3. HIS acquisition and system approach

A non-destructive, non-contact and real-time optoelectronic system based on a PGP (prism-grating-prism) imaging spectrograph was designed. Figure 3 shows its block diagram. HIS images are first recorded by a camera through a PGP device in the "hyperspectral image acquisition" unit, where spectra of the diffuse reflectance of materials are obtained.

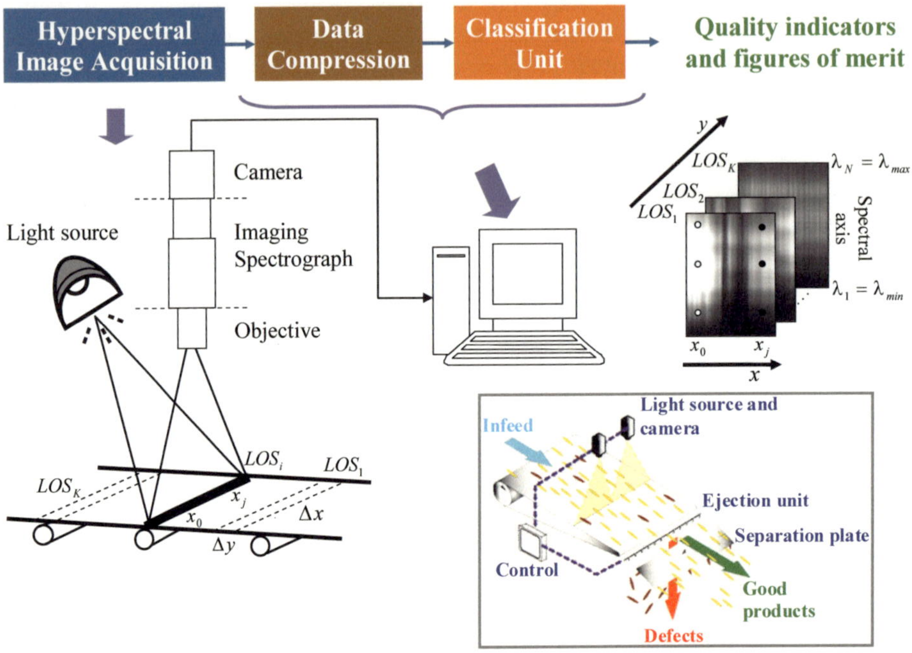

Fig. 3. HIS system for material sorting in industrial processes. Inset: schematic implementation for on-line sorting

In the final set-up, the commercial imaging spectrograph ImSpector (Specim Ltd.) was used. It is a pushbroom scanner that images the spectral dimension and one of the spatial dimensions. The temporal displacement of the sample provides the imaging of the second spatial dimension. The light source consists of two halogen floodlights with a power of 500 W each. The diffuse reflectance from the material surface is collected by a Navitar Zoom 7000 lens that delivers it into the entrance slit of the imaging spectrograph. This slit fixes the instantaneous "field of view" of the system in the spatial direction to a length Δx and a width Δy. This is the so-called observation line (LOS, Line Of Sight). Each point x_i of the observation line has its specific image on the entrance slit. The radiation at the slit is then collimated by a lens and then dispersed whose output direction depends on the wavelength of the radiation. The dispersive element of ImSpector is a volume type holographic transmission grating. This grating is used in a PGP-structure, providing high-diffraction efficiency and good spectral linearity (Vaarala, 1997). Material analysis was performed in the Vis-NIR (Visible-Near Infrared) and NIR (Near Infrared) ranges. Imspector devices V10E (400-1000 nm) and N24E (1000-2400 nm) were employed. Dispersed radiation is then focused into the detector plane of a CCD or a CMOS sensor. Every point x_i is represented by a line of monochromatic points that represents a continuous spectrum in the direction of the spectral axis. In the Vis-NIR range, a high-performance monochrome digital camera, Pixelink PL-A741 (1280×1204 pixels) was used. It has an IEEE1394 communication interface and its spectral range goes from 350 up to 1000 nm. In the NIR range, the sensor module is a

monochrome camera XEVA-FPA-2.5-320 (320×256 pixels) by Xenics with Camera Link interface and a spectral range that covers from 850 to 2500 nm.
Once the system set-up was arranged, spectral and spatial calibration processes were performed before the acquisition of HIS images. These procedures are particularly relevant in the NIR range, because, few imaging spectroscopic systems based on a PGP device working in the NIR range were previously reported. The *wavelength* or *spectral calibration* defines the spectral axis. In the Vis-NIR range (García-Allende, 2007), it was performed using two different light sources, whose emission wavelengths were previously known: a laser at 670 nm and a Hg-Ar lamp with multiple emission lines. To deal with the unavailability of such a multi-line light source in the NIR spectral range, a custom-developed set-up consisting of a tuneable laser source in the L-band and two laser sources were employed (Garcia-Allende, 2008d). The axis alignment was performed by rotating the camera with respect to the spectrograph. The system is properly calibrated when a particular wavelength corresponds to the same pixel number at both sides of the spatial axis. The *spatial calibration* calculates the dimensions, length and width, of the observation line. Apart from the entrance slit of the ImSpector, it depends on the lens focal length and on the distance between the material and the lens. A pattern of uniformly distributed black and white lines were used to obtain the field of view of the system. Finally, and in a last step, the *responsivity calibration* is performed. Due to the non-linear spectral responses of the cameras, flat standard reflectance measurements do not result in flat curves. Thus, a reflectance calibration is required within the working spectral range. This is achieved through dark and white reference images. The dark image is obtained by blocking the light into the system, while the white reference comes from the reflectance measurement of Spectralon, a thermoplastic resin with a spectrally flat behaviour. Once the image sample is captured, it becomes compensated generating the calibrated one $R(\lambda_i, x_i)$:

$$R(\lambda_i, x_i) = \frac{S(\lambda_i, x_i) - D(\lambda_i, x_i)}{W(\lambda_i, x_i) - D(\lambda_i, x_i)} \tag{1}$$

where $S(\lambda_i, x_i)$ is the sample value at wavelength λ_i and spatial position x_i, $D(\lambda_i, x_i)$ is the value for the dark reference and $W(\lambda_i, x_i)$ is the equivalent for the white reference. This calibration compensates offsets due to the CCD dark current, the light source colour temperature drift, and the lighting spatial non-uniformity across the scene line. In this way, the sample reflectance is extracted from the system response.

4. From HIS measurements to significant information

Once the diffuse reflectance is registered in the datacube, valuable automatic quality indicators must be generated. Accurate and real-time operation can only be accomplished if appropriate and intelligent criteria are implemented extracting relevant information from the data cube. The upper part of Figure 3 shows the two additional stages needed to get the desired quality indicators: data compression and data classification.
The "data compression" carries out the reduction of data volume. Too much not-relevant information could distract the classifier and decrease the identification accuracy. On the other hand, the loss of information if compression is excessive becomes another risk. A wide variety of methods (Workman & Springsteen, 1998) are available. Basically, they are classified between (Figure 4):

- *"feature extraction"* where the spectral data is projected into to a lower dimension space where only some features, or projections, are kept to represent the general information. PCA (Principal Component Analysis) and its variations are examples of this category (Moeslund, 2001-2002; Mignani et al., 2003; Koren et al., 2004; O'Farrell et al., 2005; Park & Choi, 2004).
- *"feature selection"* methods that identify and ignore those variables that do not contribute to the classification. They work directly with the spectral data space providing advantages such as the interpretability of results. Sequential Floating Feature Selection (SFFS) algorithm (Ferri et al., 1994; Gomez-Chova, 2003) belongs to this category.

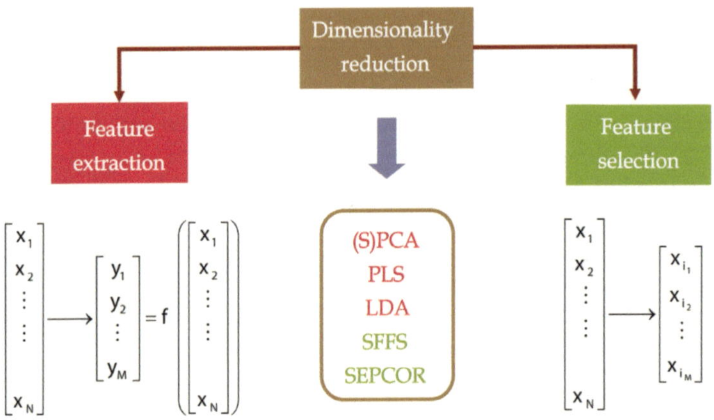

Fig. 4. Data compression alternatives: "feature extraction" and "feature selection"

4.1 Feature extraction

PCA is a feature extraction compression method that obtains the vectors or projection directions by maximizing the projected variance of a given data set. For dimensionality reduction purposes, the number of projection directions should be truncated trying to lose as less information as possible. PCA traditionally selects and filters vectors as a function of their corresponding eigenvalues. This approach is the so-called "m-method" (Moeslund, 2001) and, since it does not take into account any kind of class or category pertinence, it has several drawbacks. "J-measure" and "SEPCOR" (Moeslund, 2001) are also unsupervised methods like the "m-method". However, the first alternative selects those features that best separate the mean of the individual data categories and SEPCOR (SEPCOR, SEParability and CORrelation) takes into account both mean and variance of the classes, since the components are selected in decreasing order with respect to a variability measure:

$$V(\vec{b}_i) = \frac{\textit{Deviation of the class means}}{\textit{Mean of the deviation of each class}} = \frac{\sum_{j=1}^{K}\left(\mu_{ji} - \mu_i\right)^2}{\sum_{g \in clase\, j}\left(y_{gi} - \mu_{ji}\right)^2} \qquad (2)$$

where $V(\vec{b}_i)$ is the variability of the i'th component, $\vec{b}_i$ is the i'th eigenvector, K stands for the number of classes, μ_{ji} is the mean of class j in the input data at the i'th component and y_{gi}

denotes a sample from the g class in the i'th component. The numerator measures the distance between the class means and it would be desired to be large. On the other hand, the smaller the denominator, the better since it represents the compactness or clustering degree of the classes. Consequently, the larger the variability $V(\vec{b}_i)$ is, the better the component $\vec{b}_i$ behaves to discriminate classes. Section 6 will also compare, both qualitatively and quantitatively, the traditional approach ("m-method") with a supervised PCA alternative (SPCA, Supervised Principal Component Analysis). SPCA is not a different method for selecting eigenvectors. It is based on the modification of the original data set to include information related to the data class. This class information is considered afterwards for the projection (Chen, 2005).

Figure 5 represents the clustering degree obtained by the different PCA alternatives when spectra of tobacco leaves ("target", green dots) are compared with spectra of spurious material ("non-target", red dots). "Non-target" spectra appear to be well dispersed due to the different chemical properties of this set of materials (wood, cardboard, leather, cellophane, other vegetable leaves, etc.). Figure 6 shows tobacco spectra only compared with diffuse reflectance spectra from other vegetable leaves in different maturity levels ("closest-non-target"). Although both groups remain still separated due to the specific chemical components of tobacco rich in alkaloids (nicotine, nornicotine, anabasine, etc.), sometimes they become mixed. This could induce identification errors in the classification stage. However, if the extracted PCA features become ordered following the variability criteria (2) (Conde, 2007) the enhancement in cluster separation is evident as shown in Figure 7, where SPCA exhibits a slightly better separation degree.

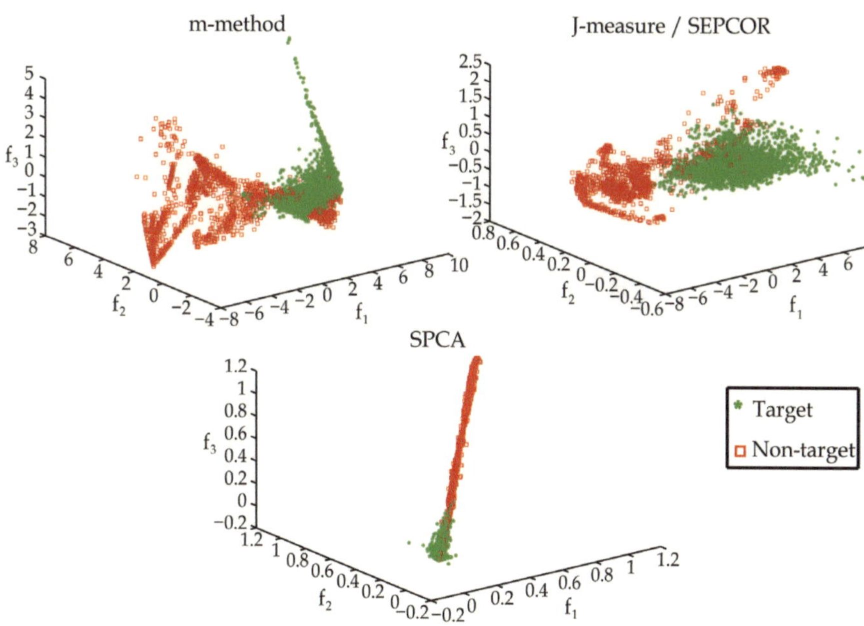

Fig. 5. Clustering of PCA alternatives for "target" material (tobacco leaves) and "non-target" material (all the spurious matter)

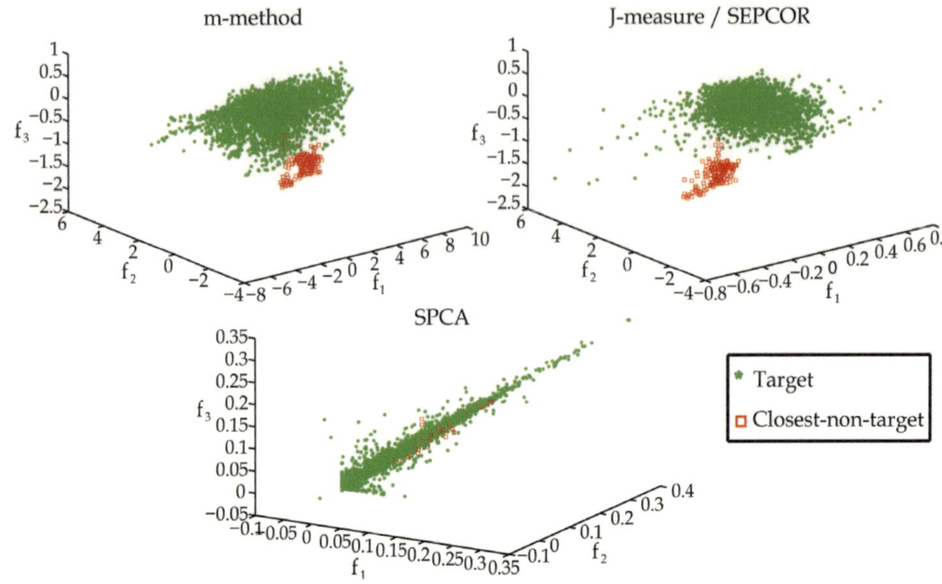

Fig. 6. Clustering of PCA alternatives for "target" material (tobacco leaves) and "closest-non-target" material (other vegetal spurious leaves)

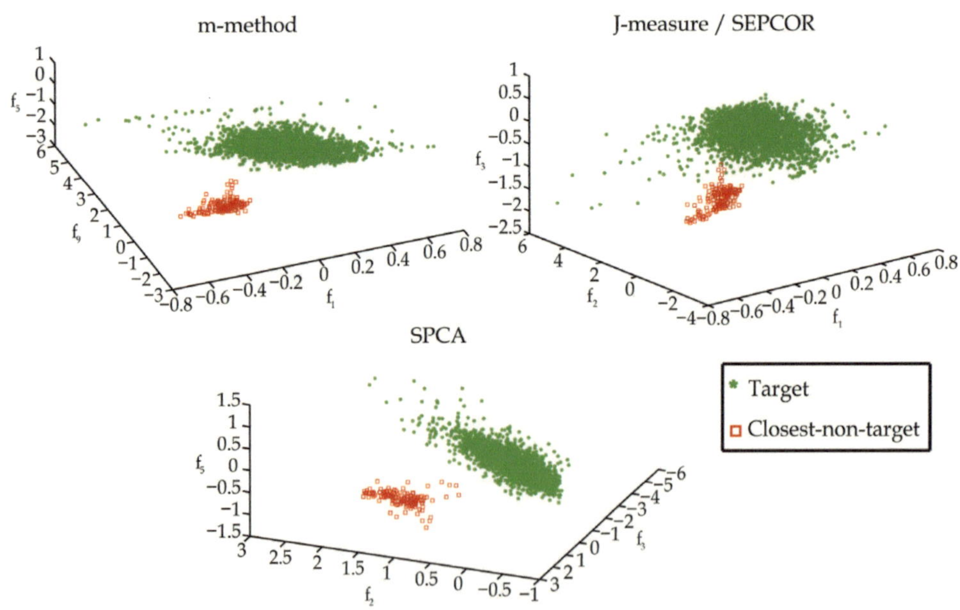

Fig. 7. Effect of variability ordering in PCA clustering for "target" material (tobacco leafes) and "closest-non-target" material (other vegetal spurious leaves)

4.2 Feature selection

Feature selection methods come to solve some drawbacks of feature extraction techniques:

- When analyzing an unknown spectrum it is absolutely necessary to measure the intensity at all the spectral bands to be able to perform its projection.
- It is not possible to extrapolate the results to other spectrometers with different number of spectral bands.
- Result interpretation becomes a complex task as the extracted features cannot be associated with spectral bands with physical meaning.

Feature selection algorithms are first structured in terms of their *selection criterion*. "Class separability-based" algorithms assume that there are several classes, or statistical distributions, to be discriminated and choose those spectral bands that maximize the distance among classes in the lower dimensional space. On the other hand, those algorithms based on "classification performance" select the features that provide higher accuracy in subsequent classification stages. The latter is not so interesting because they have two major drawbacks. First, they have high computational loads, since it is necessary to train and test the classifier for each selected feature subset. Secondly, the obtained results cannot be extended to other classifiers.

The feature selection criterion is usually based on the calculation of a probabilistic distance. Chernoff, Bhattacharyya or Patrick-Fisher measurements are typically used. The Bhattacharya distance measures the class or category separability between the spectra of distinct materials, it is given by:

$$J_B = \frac{1}{4}\left(\mu_2 - \mu_1\right)^T \left[\Sigma_1 + \Sigma_2\right]^{-1}\left(\mu_2 - \mu_1\right) + \frac{1}{2}\ln\left(\frac{\left|\Sigma_1 + \Sigma_2\right|}{2\left(\left|\Sigma_1\right| \cdot \left|\Sigma_2\right|\right)^{\frac{1}{2}}}\right) \quad (3)$$

where μ_i and Σ_i are the mean and the covariance matrix for class i. Gaussian distribution of the classes is assumed. In a multiclass discrimination problem, the separability measurement, J, would require to calculate the distance between every two classes:

$$J = \sum_{i=1}^{C} \sum_{j=i+1}^{C} P_i P_j J_{ij} \quad (4)$$

where C is the number of classes, P_i is each class probability and J_{ij} is the distance between classes i and j.

Apart from the selection criterion, an *optimal search strategy* of the feature bands must be followed because, even if the number of final selected features is known, the evaluation of the performance of each selected subset is unapproachable due to time consumption. A SFFS approach (Gomez-Chova, 2003) is interesting because it eliminates the redundant features after the addition of more useful ones and it also reconsiders the utility of some features that were possibly discarded before. Data decorrelation is compulsory prior to the application of SFFS. Blocks of correlated spectral bands are identified applying a threshold for the correlation coefficient, thus obtaining the correlation. Afterwards, only one wavelength of each block is selected. SFFS is then applied over this reduced set of wavelengths (Garcia-Allende, 2006; Garcia-Allende, 2009).

5. HIS classification alternatives for quality control

After the discussion about how to obtain representative features of the material, "data classification" follows as depicted in Figure 3. Although data compression is essential, it is also compulsory the employment of an algorithm who identifies pixels whose spectrum exhibits a high degree of correlation ("matches"') to the expected material signature. Otherwise, the system will be useless in field applications. Several classification units can be implemented: Artificial Neural Networks (ANN), Fisher Linear Discriminant Analysis (FLD), Spectral Angle Mapper (SAM) and K-Nearest Neighbors (KNN). ANNs were initially employed because of their ability to handle non-linearity, their parallel processing of information and their quick adaptability to system dynamics (Nogueira, 2005). ANNs turned out to be very useful in this foreign object detection application and high classification accuracies (Garcia-Allende, 2007) were achieved, suggesting that maybe simpler algorithms could be employed instead. Consequently, classification units based on FLD and SAM were considered for spurious element detection because, due to their linearity, they become time efficient and then more appropriate for their employment in production lines of industrial environments. But with all of them, the addition of new products requires training again the system. To solve this, K-NN could be used instead (Fukunaga, 1990), given that no training is required for the classification. A brief introduction to these methods is presented in this section.

5.1 SAM, Spectral Angle Mapper

Spectral Angle Mapper is a simple algorithm based on the measurement of the spectral similarity between two spectra. This spectral similarity, α, is obtained considering each spectrum as a vector in a n-dimensional space (Girouard et al., 2004):

$$\alpha = \cos^{-1}\left(\frac{\sum_{i=1}^{n} s_1 s_2}{\left(\sum_{i=1}^{n} s_1^2\right)^{1/2} \cdot \left(\sum_{i=1}^{n} s_2^2\right)^{1/2}} \right) \tag{5}$$

where n is the number of spectral bands, and s_1 and s_2 are the two compared spectra. The angle between two spectra is therefore proportional to their similarity.

To implement a classification method based on SAM, two stages should be followed (Garcia-Allende, 2008f): training and test. In the *training stage*, reference spectra of the materials should be collected. For the wanted material a reference spectrum, s_w, is calculated by averaging *homogeneous* (the same kind of material in all the points of the line of vision) images. The unwanted material reference spectrum, s_{uw}, is determined by averaging images of the typical spurious materials in the selected application (foil, leather, plastics, etc.). In the *test stage*, an unknown image is classified. The spectral similarities α_{t-w} (between the spectrum of each spatial point, s_t, of the test image and s_w) and α_{t-uw} (between the test spectrum and s_{uw}) are calculated. If $\alpha_{t-w} < \alpha_{t-uw}$, s_t is closer to s_w and, therefore, that position contains tobacco. On the other hand, $\alpha_{t-uw} < \alpha_{t-w}$ indicates that s_t corresponds to unwanted material. A schematic of the classification procedure is depicted in Figure 8, using a dimensionality of n=3 for visualization purposes.

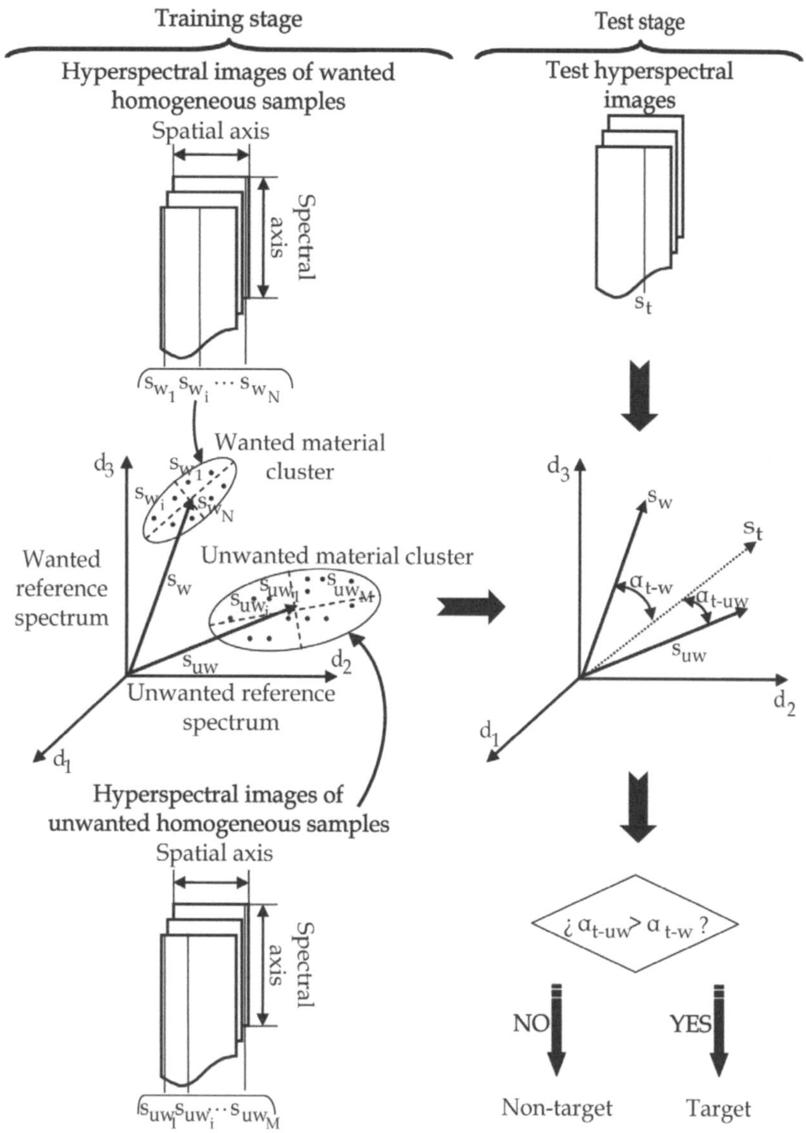

Fig. 8. SAM schematic description. "Training stage": calculation of the reference spectrum of each class (s_{uw} and s_w). "Test stage" an unknown spectrum s_t is compared with these references for classification

5.2 FLDA, Fisher Linear Discriminant Analysis

This method projects high-dimensional data onto a lower-dimensional space in a way that best separates the classes in a least-squares sense. In the particular case of two material categories, the discriminant is one-dimensional and the direction that best discriminates the data is given by:

$$w = S_W^{-1}(m_1 - m_2) \tag{6}$$

where S_W is the within-class scatter matrix, which is proportional to the sample covariance matrix, and m_1 and m_2 are the mean for each class. Then, each spectrum or data point, r_i, has a projection y_i on w given by

$$y_i = w^T r_i \tag{7}$$

The projections on w of the two classes, "target" (wanted) and "non-target" (unwanted) are shown in Figure 9. For visualization purposes, each spatial point of the image is considered as a vector in a 3-dimensional space, where d_i stands for each dimension.

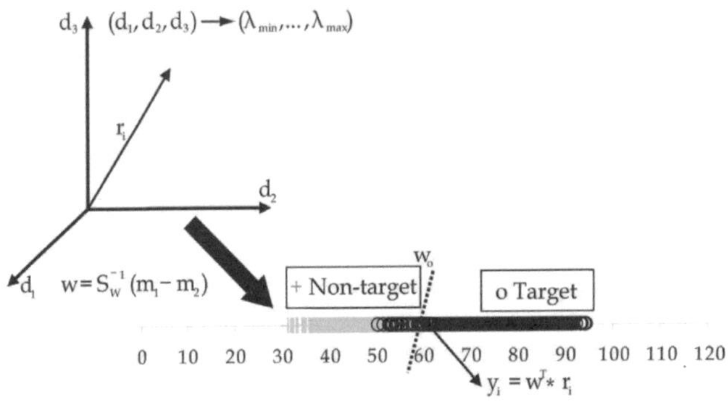

Fig. 9. Projections of the training set on the Fisher's linear discriminant

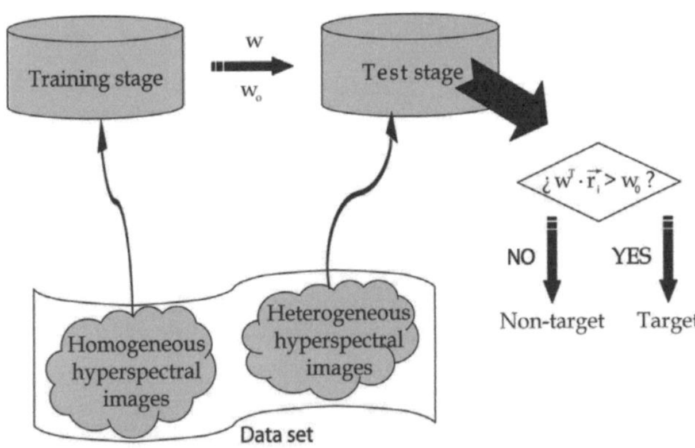

Fig. 10. Block diagram of image interpretation based on FLD

Since the discriminant is one dimensional, the projections of the material spectra on w are numerical values. As a consequence, a threshold w_o has to be fixed (Garcia-Allende, 2008e). When a material needs to be identified, its spectrum is projected onto w. Depending on its projection lies to the right or left of the decision threshold, w_o, the classification of the material will vary. The choice of the appropriate threshold is very important as it directly affects the classification error. Different strategies can be followed. The first one fixes the threshold, w_{str1}, as the middle point of the means of each projected class. But, data variance could also taken into account, and the threshold w_{str2} could be determined as the middle point of the minimum of the "target" spectra projections and the maximum of the projections of "non-target" spectra. Figure 10 depicts a schematic description of the whole data processing system based on FLD.

$$w_{str_1} = \frac{m_2 + m_1}{2} \qquad w_{str_2} = \frac{(m_2 - \Delta_2) + (m_1 + \Delta_1)}{2} \tag{8}$$

5.2 ANN, Artificial Neural Networks

ANN have become really popular for their flexibility to different problems: chemical analysis (Eytan, 2000), arc welding (Garcia-Allende, 2008a), information technology (Messom, 2005), etc.

Neural networks emulate the human brain because they consist of layers of interconnected nodes. Each node produces a non-linear function of its input. The input to a node may come from other nodes or directly from the input data. Finally, some nodes are identified with the output of the network. The complete network, therefore, represents a very complex set of interdependencies which may incorporate any degree of nonlinearity, allowing very general functions to be modeled. There are multiple types of neural networks depending on the number of layers, the number of nodes per layer, transfer functions that are the transformation that each node produces of its input, training algorithm, etc. (Chen, 1996).

5.3 K-NN, K-Nearest Neighbors

The most significant advantage of a KNN classifier is that no training is required. When a new and unknown spectrum needs to be classified a comparison with the set of previously known spectra is performed. Consequently, KNN provides an enormous improvoment in foreign object detection applications in industrial scenarios because, since no training is required, new products to be discriminated can be added at any time.

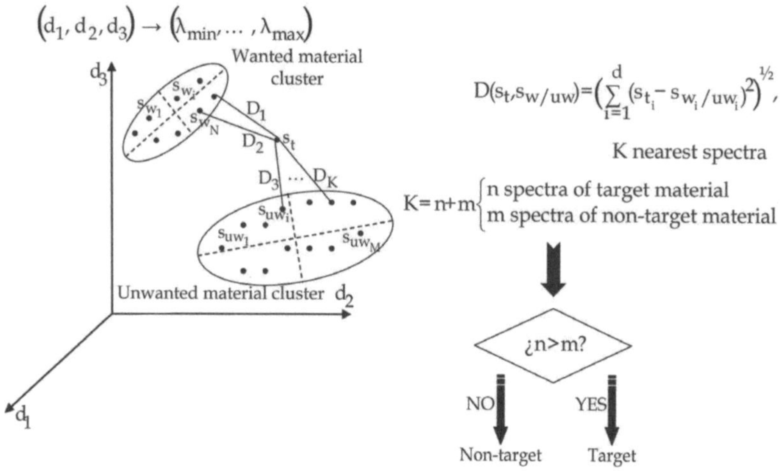

Fig. 11. KNN classifier where each feature is considered as a coordinate axis and the distance between points in this space determines how similar or different they are

A schematic of the KNN process for classifying the hyperspectral images is depicted in Figure 11 (Garcia-Allende, 2008g). The map is populated by points, each of which represents a spatial position of a particular image and as in previous cases. Since KNN is based on considering that similar data values should belong to the same material, then *K* pixels with the most similar spectral properties to the unclassified spectrum are initially determined. This similarity is estimated in terms of the Euclidean distance. The unclassified pixel will be assigned to the most numerous class ("target" or "non-target") among the *K* closest neighbors. Therefore, this classifier has only one independent parameter, *K* or the number of neighbors, to consider.

6. HIS performance evaluation for raw material sorting

In essence, all considered methods are *supervised* algorithms where a training set of spectra is available with their corresponding materials labels. In the training dataset the expected classification outputs are one ("target") or zero ("non-target"'). For the test phase, "target" and "non-target" material are combined in the same line to confirm that the system is able to spatially discriminate them. A thorough and comprehensive work programme that included a wide variety of test samples was performed in this last stage. Figure 12 illustrates a couple of samples and their corresponding results. Its horizontal lines contain the following information:

- A photograph of the material sample under analysis. The white lines delimit the observation line where the horizontal axis corresponds to the spatial axis of the HIS image.
- A zoom of the observation line.
- The spectral images. The spatial axis is the horizontal one with the origin on the left as it appears in the real photograph. The spectral axis is the vertical one with the smallest wavelength of the spectral range at the bottom of the image.
- The results of the classification provided by the neural network when data dimensionality reduction is based on PCA (m-method).
- Idem but data compression is performed in this case by means of SFFS.
- Classification provided by the linear discriminant. The projection on *w* of the reflectance of each spatial pixel of the observation line is also plotted in black. The decision threshold is plotted in green, showing those pixels of the observation line where the projection is higher than the threshold. These points are classified as "target" material.
- The last line shows SAM outputs.

Figure 13 shows a comparison between discrimination methodologies from a qualitative point of view. In order to accurately estimate the performance of the classification strategies, distinct cross-validation procedures should be applied. This way, the dependency of the results on the training or test sets is removed. Table 1 quantitatively compares different PCA algorithms in terms of the classification error attained by ANNs. A 3-fold cross-validation procedure (Zhu, 1996; Goutte-1997) was followed employing 11520 spectra from 120 images, 80 images of "target" material and 40 of "non-target" material. The ANN training-and-test procedure was repeated 3 times and the average values of success probability are displayed in Table 1. Although classification percentages are similar for all PCA methods, SPCA is confirmed as the best alternative. It is not possible to include also SFFS in this quantitative comparison in terms of the ANN classification indicators, because the network topology was not the same. Multilayer feed-forward networks with back-propagation learning algorithms were employed in both cases. The goal error considered in the training phase was 0.001%. When data compression was performed by PCA methods, a network topology consisting of an input layer with 10 neurons and an output layer with one neuron was sufficient to achieve the training goal error (Garcia-Allende, 2007). When SFFS was implemented, the ANN only converged to the specified training error if the complexity of the topology was increased (a hidden layer with 5 neurons was included and the number of neurons in the input layer was reduced also to 5). A maximum classification accuracy of 94.62% was obtained by selecting either 10 or 12 spectral bands. Since compression rate is higher, 10 spectral bands would be selected in a practical implementation.

Selection algorithm	Number of eigenvectors	Processing time [mseg]	Success probability [%]
m-method	19	12	99.557
J-measure	19	11	99.601
SEPCOR	19	11	99.531
SPCA	16	12	99.818

Table 1. Quantitative comparison of dimensionality reduction for material sorting

Table 2 compares the classification capabilities of different strategies when the whole gallery of material samples was considered (again through a 3-fold cross validation procedure to attain generality on the training and test sets). Data compression was performed employing PCA ("m-method" discarding those eigenvectors whose associated eigenvalues are smaller than a thousandth of the maximum) in the case of classification units based on ANN and SAM, respectively. There is no need to use a pre-processing stage in the case of FLD because dimensionality reduction is directly obtained by the spectral projection onto the discriminant.

Classification algorithm	Success Probability [%]
ANN	99.557
FLD	92.870
SAM	90.540

Table 2. Quantitative comparison of classification strategies

Test samples shown in Figure 12 and the success probability figures displayed in Tables 1 and 2 suggest that ANNs provide best classification performance than FLD and SAM. However, this does not imply that ANNs are better in a general sense. They only offer higher classification performance. FLD and SAM have, due to their linearity and simplicity, better time performance so, if identification errors are acceptable, they would be preferable. This situation resembles the actual one also in remote sensing. It has not been encountered yet, as pointed by Manolakis et al. (Manolakis et al., 2009), the best hyperspectral detection algorithm and it depends on the particular requirement applications. In the context of applications and experiences of quality control both, time consumption and difficulty for designers of hyperspectral imaging systems, should be taken into account to look through the existing literature. Consequently, the estimation of the expected level of performance of the variety of schemes presented throughout is a great asset for different future applications of imaging spectroscopy systems within the field. To aid in this selection, all the developed processing schemes are quantitatively compared in terms of the attained accuracy and their space complexity. The latter is the number of elementary objects that need to be stored during execution, i.e. the complexity of the hardware implementation compulsory to their employment in an embedded system in production lines. This comparison is presented in

Figure 13 using a pseudo-color scale, where "green" corresponds to the best estimated performance. In a practical application, a trade-off has to be reached between accuracy and space complexity.

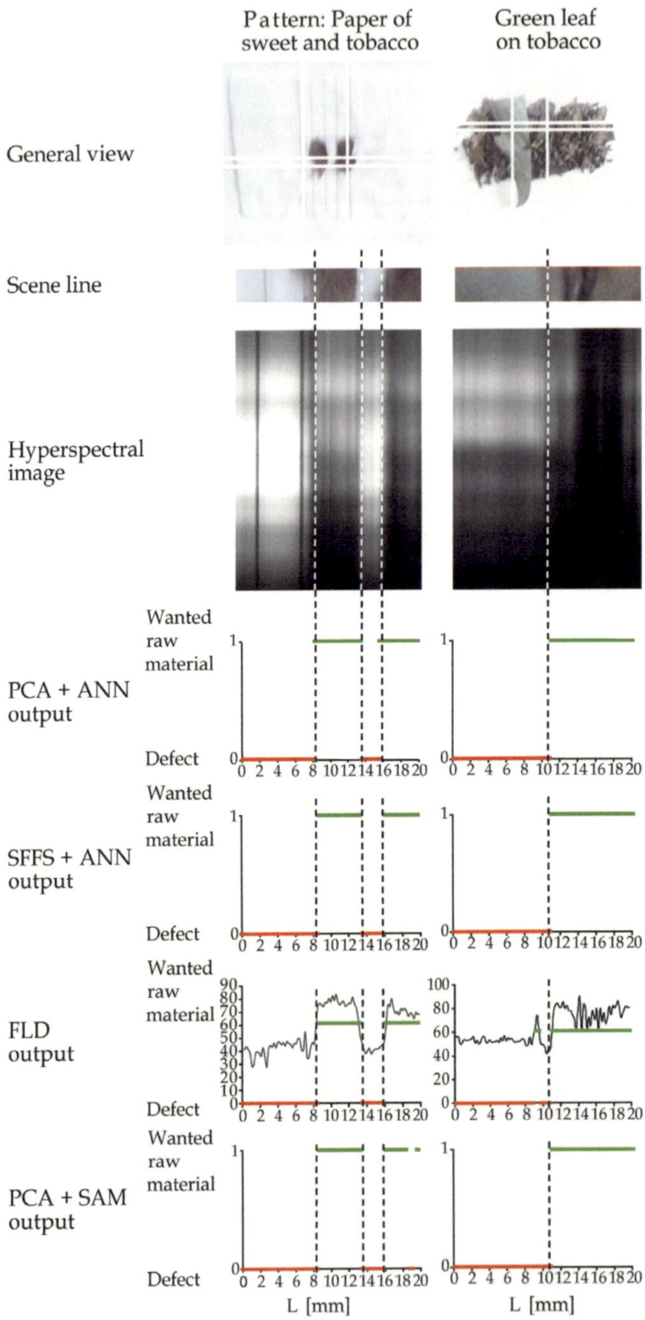

Fig. 12. Material classification results. Vertical dotted lines help to correlate discrimination outputs with the hyperspectral (third row) and visible (second row) images of material

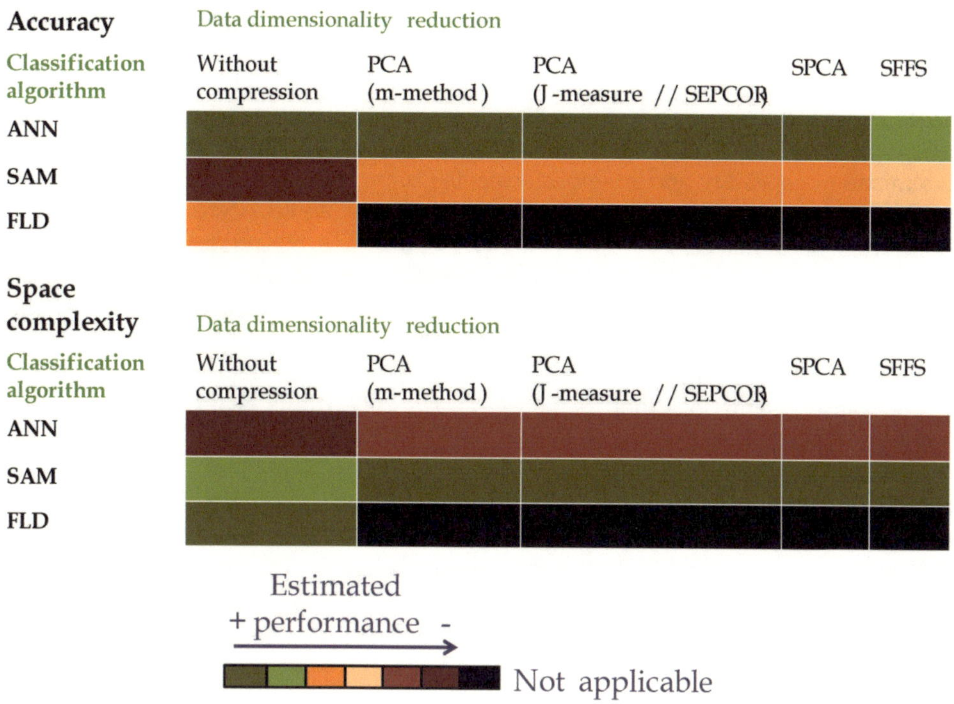

Fig. 13. Qualitative comparison of developed image processing strategies

7. Conclusions

In short, the main motivation of this chapter was to address two technological challenges that remained in HIS, i.e. the improvement of the actual acquisition systems and of the spectral interpretation methodologies. This improvements help to solve some issues in the niches focused on industrial applications, where high speed, in-line inspection and enhanced quality control are of primary concern. HIS systems on Vis-NIR and NIR spectral ranges have been designed, developed and validated. Special attention has been paid to calibration processes, wavelength and spatial calibration, since these procedures are particularly relevant in the NIR range. As shown in literature, HIS has gained prominence in recent years for solid material sorting.

The final output from the systems should be delivered in a straightforward way that facilitates its use. In raw material sorting, system usability, suggests that only the presence and position of spurious materials are of interest. Alternatives have been presented throughout this work related to the spectral interpretation of the HIS images to enhance the industrial usability of the technique. This includes proposed methodologies to achieve both dimensionality reduction and accurate determination of quality indicators. A variety of image analysis schemes were successfully validated in the tobacco industry. Different versions of PCA, both supervised (SPCA) and unsupervised ("m-method", "J-measure" and SEPCOR) were implemented to attain data compression. Their performances were qualitatively (clustering degree) and quantitatively (discrimination capability) compared. A feature selection method (SFFS) was also considered to reduce the data volume collected by

the HIS system, and its advantages and drawbacks respect to feature extraction alternatives were specified and discussed. Finally, the sorting stage was addressed through different classification algorithms where their capabilities were also compared. A non-linear solution based on ANN was initially explored and very accurate spurious material detection was achieved with the appropriate training. This suggested that simpler algorithms could also be employed. Therefore classification units based on FLD and SAM were also proposed. Discrimination capability and time performances were compared but it was not possible, however, to conclude the superiority of any of them. While ANNs provided the best classification performance, FLD and SAM had better time behaviour. In some industrial contexts, probably FLD or SAM will be preferable if the provided sorting error is assumable.
Some open research lines have been identified in the course of this work. A key challenge that remains in system design is that the scanning speed is tightly subject to the exposure time (several tenths of a second). Brighter sources would be required to perform more efficient measurements. A strong reduction in acquisition time could be achieved by the employment of high brightness supercontinuum sources (Alfano, 2006; Kaminski, 2008; Mignani, 2009). Additionally, studies on the possibility of using several light sources with narrower spectral ranges are ongoing. Specifically, an adaptive illumination source consisting of high-bright Light Emitting Diodes (LED) has been already envisaged (Conde, 2008) and applied with success to material sorting. Selected LED central wavelengths were those selected by the SFFS algorithm as the most discriminant of the spectral datacube.
Finally, a variety of interpretation/segmentation schemes of the acquired HIS images were proposed and all of them can be successfully extended to other sorting problems related to different materials.
Authors want to thank to M. Amado, F. Anabitarte, P. Anuarbe, A. Cobo for their contributions to this work.

8. References

Abbasgholipour, M.; Omid, M.; Keyhani, A. & Mohtasebi, S.S. (2010). Color image segmentation with genetic algorithm in a raisin sorting system based on machine vision in variable conditions. *Expert Systems with Applications, in press.* 0957-4174.

Aikio, M. (2001). *Hyperspectral prism-grating-prism imaging spectrograph,* PhD Thesis, VTT Electronics.

Alfano, A.R. (2006). *The Supercontinuum Laser Source,* 2nd Ed, Springer, Berlin Heidelberg New York. 03-87245049.

Bonifazi, G. & Serranti, S. (2006). Hyperspectral imaging based techniques in fluff characterization. *Proc. of SPIE: Advanced environmental, chemical and biological sensing technologies IV*, 6377, p. 63770O. 0277-786X.

Bonifazi, G.; Serranti, S. & D'Aniello, L. (2008). Compost quality control by hyperspectral imaging. *Proc. of SPIE*, 7003, p. 700321. 0277-786X.

Chen, W.H.; Hsu, S.H. & Shen, H.P. (2005). Application of SVM and ANN for intrusion detection. *Comput. Oper. Res.*, 32, 10, pp. 2617-2634. 0305-0548.

Chen. C.H. (1996). *Fuzzy logic and neural network handbook.* McGraw Hill. 978-0070111899.

Conde, O.M.; Amado, M.; Garcia-Allende, P.B.; Cobo, A.; Lopez- Higuera, J.M. (2007) Evaluation of PCA dimensionality reduction techniques in imaging spectroscopy for foreign object detection. *Proc. of SPIE*, 6565, p. 65650M. 0277-786X.

Conde, O.M.; Cobo, A.; Cantero, P.; Conde, D.; Mirapeix, J., Cubillas, A.M. & Lopez-Higuera, J.M. (2008). Adaptive illumination source for multispectral vision system applied to material discrimination. *Proc. of SPIE*, 7000, p. 70000Z. 0277-786X.

Eytan, O.; Sela, B.A. & Katzir, A. (2000). Fiber-optic evanescent-wave spectroscopy and neural networks: application to chemical blood analysis. *App. Opt.*, 39(19) pp. 3357-3360. 1559-128X.

Ferri, F.; Pudil, P.; Hatef, M. & Kittler, J. (1995). *Comparative study for techniques for large-scale feature selection in Pattern Recognition in Practice IV: Multiple Paradigms, Comparative Studies, and Hybrid Systems*. E.S. Gelsema and L.N. Kanal

Fukunaga, K. (1990). *Introduction to statistical pattern recognition*. Academic Press, San Diego, 2nd Ed. 0-12-2698517.

Garcia-Allende, P.B.; Conde, O.M.; Mirapeix, J.; Cubillas, A.M. & Lopez-Higuera, J.M. (2006). Relevant Information Extraction from Hyperspectral Images for Quality Control Industrial Processes. *ODIMAP V (5th Topical Meeting on Optoelectronic Distance / Displacement, Measurements and Applications)*, pp. 271-276. 8-469009-38-9.

Garcia-Allende, P.B; Conde, O.M.; Cubillas, A.M.; Jauregui, C. & Lopez-Higuera, J.M. (2007). New raw material discrimination system based on a spatial optical spectroscopy technique; *Sensors and Actuators A: Physical*, 135(2), pp. 605-612. 0924-4247.

García-Allende, P.B.; Mirapeix, J.; Conde, O.; Cobo, A. & López-Higuera, J.M. (2008a). Arc-welding quality assurance by means of embedded fiber sensor and spectral processing combining feature selection and neural networks, *Sensors*, 8(10), pp. 6496-6506, 1424-8220.

Garcia-Allende, P.B.; Anabitarte, F.; Conde, O.M.; Madruga, F.J.; Lomer, M. & Lopez-Higuera, J.M. (2008b). Infrared imaging spectroscopic system based on a PGP spectrograph and a monochrome infrared camera. *Proc. of SPIE*, 6941, p. 694118. 0277-786X.

Garcia-Allende, P.B.; Mirapeix, J.; Cobo, A.; Conde, O.M. & Lopez-Higuera, J.M. (2008c). Arc welding quality monitoring by means of near infrared imaging spectroscopy. *Proc. of SPIE*. 6939, p. 69390Q. 0277-786X.

Garcia-Allende, P.B.; Conde, O.M.; Madruga, F.J., Cubillas, A.M. & Lopez- Higuera, J.M. (2008d). Industrial defect discrimination applying infrared imaging spectroscopy and Artificial Neural Networks. *Proc. of SPIE*, 6939, p. 69390H. 0277-786X.

Garcia-Allende, P.B.; Conde, O.M.; Mirapeix, J.; Cobo, A. & Lopez-Higuera, J.M. (2008e). Quality control of industrial processes by combining a hyperspectral sensor and Fisher's linear discriminant analysis. *Sensors and Actuators B: Chemical*, 129(2), pp. 977-984. 0925-4005.

Garcia-Allende, P.B.; Conde, O.M.; Mirapeix, J.; Cubillas, A.M. & Lopez-Higuera, J.M. (2008f). Data Processing Method Applying Principal Component Analysis and Spectral Angle Mapper for Imaging Spectroscopic Sensors. *IEEE Sensors Journal*, 8(7), pp. 1310-1316. 1530-437X.

Garcia-Allende, P.B.; Conde, O.M.; Amado, M.; Quintela, A. & Lopez-Higuera, J.M. (2008g). Hyperspectral data processing algorithm combining Principal Component Analysis and K Nearest Neighbours. *Proc. of SPIE*, 6966, p. 69660H. 0277-786X.

García-Allende, P.B.; Mirapeix, J.; Conde, O.; Cobo, A. & López-Higuera, J.M. (2009). Defect detection in arc-welding processes by means of the line-to-continuum method and feature selection, *Sensors*, 9(10), pp. 7753-7770, 1424-8220.

Girouard, G.; Bannari, A.; El Harti, A. & Desrochers, A. (2004) Validated spectral angle mapper algorithm for geological mapping: comparative study between Quickbird and Landsat-TM. *Geo-imagery bridging continents*, Istanbul, July 12–23, pp. 599–604

Gomez-Chova, L.; Calpe, L.; Camps-Valls, G.; Martin, J.D.; Soria, E.; Vila, J.; Alonso-Corda, L. & Moreno, J. (2003). Feature selection of hyperspectral data through local correlation and SFFS for crop classification. *IEEE Int. Geosci. Remote Sens. Symp. Proc.*, 1, 555 - 557.

Gomez-Sanchis, J.; Gomez-Chova, L.; Aleixos, N.; Camps-Valls, G.; Montesinos-Herrero, C.; Molto, E. & Blasco, J. (2008). Hyperspectral system for early detection of rottenness caused by Penicillium digitatum in mandarins. *J. Food Eng.*, 89(1), pp.80 - 86. 0260-8774.

Goutte C. (1997) Note on free lunches and cross validation. *Neural Comput.*, 9(6), pp. 1245-1249. 0899-7667.

Green, R.O.; Eastwood, M.L.; Sarture, C.M. ; Chrien, T.G.; Aronsson, M.; Chippendale, B.J.; Faust, J.A. ; Pavri, B.E. ; Chovit, C.J. ; Solis, M.; Olah, M.R. & Williams O. (1998). Imaging spectroscopy and Airborne Visible/Infrared Imaging Spectrometer (AVIRIS). *Remote Sens. Environ.*, 65(3), pp. 227-248. 0034-4257.

Herrala, E.; Hyvarinen, T.; Voutilainen, O. & Lammasniemi, J. (1997). An optoelectronic sensor system for industrial multipoint and imaging spectrometry. *Sensor. Actuat. A-Phys.*, 61, 1-3, pp. 335 - 338. 0924-4247.

Kaminski, C.F.; Watt, R.S.; Elder, A.D.; Frank, J.H. & Hult, J. (2008). Supercontinuum radiation for applications in chemical sensing and microscopy. *Applied Physics B*, 92, pp. 367-378. 0946-2171.

Koren, Y.; Tian, Q.; Zhou, X.S. & Huang, T.S. (2004). Robust linear dimensionality reduction. *IEEE T. Vis. Comput. Gr.*, 10, 4, pp. 459 - 470

Kotwaliwale, N.; Weckler, P.R.; Brusewitz, G.H.; Kranzler, G.A. & Maness, N.O. (2007). Non-destructive quality determination of pecans using soft X-rays. *Postharvest Biology and Technology*, 45(3), pp. 372-380. 0925-5214.

Leitner, R.; McGunnigle, G.; Kraft, M.; de Biasio, M.; Rehrmann, V. & Balthasar, D. (2009). Real-time detection of flame-retardant additives in polymers and polymer blends with NIR imaging spectroscopy. *Proc. of SPIE*, 7312, p. 73120M. 0277-786X.

Lu, R.; Chen, Y. R.; Park, B. (1999). Hyperspectral imaging for detecting bruises in apples. *ASAE Annual International Meeting*, Paper No. 99–3120.

Manolakis, D.; Lockwood, R.; Cooley, T. & Jacobson, J. (2009). Is there a best hyperspectral detection algorithm? *Proc. of SPIE*, 7334, p. 733402. 0277-786X.

McClure, W.F.; Horris, K.H. & Weeks, W.W. (1997). Rapid spectrophotometric analysis of the chemical composition of tobacco, Part 1: Total reducing sugars. *BeitrÄage zur Tabakforschung International*, 9(1), pp. 13 - 18.

Mesina, M.B.; de Jong, T.P.R. & Dalmijn, W.L. (2007). Automatic sorting of scrap metals with a combined electromagnetic and dual energy X-ray transmission sensor. *International Journal of Mineral Processing*, 82(4), pp. 222 - 232. 0301-7516.

Messom, C.H.; Sarrafzadeh, A.; Johnson, M.J. & Chao, M.J. (2005). *Affective state estimation from facial images using neural networks and fuzzy logic in* Neural *networks: applications in information technology and web engineering*. Borneo Publishing.

Mignani, A.G.; Smith, P.R.; Ciaccheri, L.; Cimato, A. & Graziano, S. (2003) Spectral nephelometry for making extravirgin olive oil fingerprints. *Sensor. Actuat. B- Chem.*, 90(157), pp. 57 - 62. 0925-4005.

Mignani, A.G.; Ottevaere, H.; Ciaccheri, L.; Thienpont, H.; Cacciari, I.; Parriaux, O. & Johnson, M. (2009). Innovative spectroscopy of liquids: a fiber optic supercontinuum source and an integrating cavity for scattering-free absorption measurements. *Proc. of SPIE,* 7503, p. 750377. 0277-786X.

Millar, S.J.; Whitworth, M.B.; Chau, A. & Gilchrist, J.R. (2008). Mapping food composition using NIR hyperspectral imaging, *New Food,* 3, pp. 34-39.

Moeslund, T.B. (2001-2002) Principal component analysis. An introduction. *Technical report,* Aalborg University CVMT 01-02

Nogueira, A.; de Oliveira, M.R.; Salvador, P.; Valadas, R. & Pacheco, A. (2005). Classification of internet users using discriminant analysis and neural networks. In *NGI 2005: Next Generation Internet Networks: Traffic Engineering,* pp. 341 - 348.

O'Farrell, M.; Lewis, E.; Lyons, W.B. & Jackman, N. (2005). Combining principal component analysis with an arti¯cial neural network to perform online quality assessment of food as it cooks in a large-scale industrial oven. *Sensor. Actuat. B-Chem.*, 107, pp. 104 - 112. 0925-4005.

Park M.S. & Choi, J.Y. (2004). Theoretical analysis on feature extraction capability of class-augmented PCA. *Pattern Recogn.*, 42, pp. 2353-2362. 0031-3203.

Park, B.; Windham, W.R.; Lawrence, K.C. & Smith D.P. (2004). Classification of hyperspectral imagery for identifying fecal and ingesta contaminants. *Proc. of SPIE* 5271, 1, pp. 118-127. 0277-766X.

Reich, G. (2005). Near-infrared spectroscopy and imaging: basic principles and pharmaceutical applications. *Adv. Drug Deliver. Rev.*, 57(8), pp. 1109 - 1143. 0169-409X.

Schmidt, W. (2005). *Optical spectroscopy in chemistry and life sciences,* Wiley-VCH, Weinheim. 978-3527299119.

Solo-Gabriele, H.M.; Townsend, T.G.; Hahn, D.W.; Moskal, T.M.; Hosein, N.; Jambeck, J. & Jacobi, G. (2004). Evaluation of XRF and LIBS technologies for on-line sorting of CCA-treated wood waste. *Waste Management,* 24(4), pp. 413-424. 0956-053X.

Vaarala, T.; Aikio, A. & Keraenen, H. (1997). An advanced prism-grating-prism imaging spectrograph in on-line industrial applications. *Proc. of SPIE,* 3101, pp. 322 - 330. 0277-766X.

Varith, J.; Hyde, G.M.; Baritelle, A.L.; Fellman, J.K. & Sattabongkot, T. (2003). Non-contact bruise detection in apples by thermal imaging. *Innovative Food Science & Emerging Technologies,* 4(2), pp. 211-218. 1466-8564.

Wiggenhauser, H.; Schaurich, D. & Wilsch, G. (1998). LIBS for non-destructive testing of element distributions on surfaces. *NDT & E International,* 31 (4), pp. 307-313. 0963-8695.

Willoughby, C.T.; Folkman, M.A. & Figueroa M.A. (1996). Application of hyperspectral imaging spectrometer system to industrial inspection. *Proc. of SPIE,* 2599, pp. 264 - 72. 0277-766X.

Workman Jr., J.; Springsteen, A.W. (1998) *Applied Spectroscopy: A compact reference for Practitioners,* London Academia Press Limited, 0127640703, London.

Xing, J.; Bravo, C.; Jancsók, P.T.; Ramon, H. & Baerdemaeker J. (2005). Detecting bruises on *Golden Delicious* apples using hyperspectral imaging with multiple wavebands. *Biosystems Engineering*, 90(1), pp. 27 - 36. 1537-5110.

Zhu, H. & Rohwer, R. (1996). No free lunch for cross-validation. *Neural Comput.*, 8(7), pp. 1421-1426. 0899-7667.

34

Quality Control of the Microlenses Array

Rafał Kasztelanic
University of Warsaw
Poland

1. Introduction

Microlenses and microlens arrays, whose use in contemporary science and industry constantly increases, are objects which modify the phase of the light passing through those objects. Thus, in order to assess their characteristics and their quality one can measure either their shape, or the phase change they introduce.

Visualization and measurement of phase is an important element of various contemporary domains of science and industry. The first domain to use setups for measuring phase objects was biology. The possibility to observe almost transparent microorganisms or objects without the prior introduction of artificial coloring, has added to the development of biology and medicine. Another domain which has benefitted from measuring phase objects is astronomy. The needs of biologists on the one hand, and astronomers on the other, has forced the development of new technologies, methods of fabrication and measurement.

Yet another domain which benefits greatly from setups for measuring phase objects is the industry, including optical industry producing glasses and contact lenses. Changes in the setups for measuring phase objects have thus been induced by mass production for industrial purposes, miniaturization and by the need to add new characteristics and parameters. The methods which worked well for macroscopic objects did not bring satisfying results for microscopic ones. In the case of mass production dynamic quality control of phase objects and measuring the quality of large numbers of similar objects has become a must.

This chapter first focuses on describing various groups of methods used for measuring phase objects. Further on it presents the results of computer simulations and experimental realizations of an optical setup for automatic quality control of microlens arrays.

2. Methods used for measuring the phase

There are numerous methods which allow for the examination of phase objects. Generally such methods can be divided into four main groups: interferometric methods, methods based on focusing light, methods based on analyzing the direction of the light ray and methods based on modifying the spatial frequency of the phase object examined. The choice of the method depends on the precision of the measurement required, the specificity of the object to be measured and their cost-effectiveness.

2.1 Interferometric methods

Interferometric methods, used for transparent objects and objects reflecting light, do not directly measure the shape of the object or the phase change introduced by this object. The measurements are indirect: the methods allow for deducing the shape of the object on the basis of the distribution of the interferometric fringes and the distribution results from the phase change. A single light beam passes through or is reflected by the object examined, changing the phase of the light wave. This beam interferes with a reference beam to create a fringe pattern, where the fringes represent lines of constant phase. The location of the fringes bears information about the shape of the object examined.

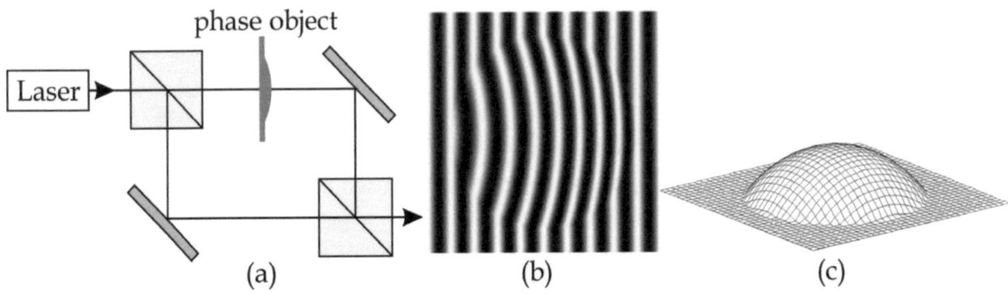

Fig. 1. An interferometric method. A scheme of the set-up a), an example of the interferogram b), shape reconstruction c)

The shape reconstruction of the object is carried out from the set of fringes with the use of relevant mathematical procedures, such as the Fourier transform, the windowed Fourier transform, the wavelet transform, the phase shifting or the phase stepping method, as well as frequency tracking methods etc.

For example, the phase stepping method uses three, four or five phase shifted fringe pattern images. In the case of four images it is possible to use a symmetrical phase shift by $\pi/2$:

$$\varphi=\left\{-\tfrac{3}{4}\pi,-\tfrac{1}{4}\pi,\tfrac{1}{4}+\pi,+\tfrac{3}{4}\pi\right\} \tag{1}$$

In a two-beam interferometer, the corresponding phase-shifted intensity signals for a single image point in the fringe pattern are:

$$g_{\varphi}(\theta,v)=Q\left[1+V\cos\left(\theta+v\phi_{\varphi}\right)\right] \tag{2}$$

where θ is the phase that we are interested in for surface profiling, V is the fringe visibility, $v\approx 1$ is the phase-shift frequency, with respect to the nominal or expected phase-shift sequence. The overall coefficient Q is the dc level of the signal. The algorithm which extracts the phase uses the following formula:

$$\theta=\arctan\left[\frac{-g_0-g_1+g_2+g_3}{-g_0+g_1-g_2+g_3}\right] \tag{3}$$

The main advantages of interferometric methods are their precision in the range of 0.1 nm, and the possibility of simultaneously measuring numerous parameters of the microlens, both optical and geometrical. The main disadvantage, however, is the size and the cost of

the setup. Apart from that, the precision of measuring small microlenses may often be problematic, as well as taking dynamic measurements in real time. Yet another problem when using the methods may be the presence of artifacts resulting from Fresnel diffraction, on the lens apertures, for instance.
Apart from all the above-mentioned problems, interferometric methods are most widely used because they are the most comprehensive.

2.2 Methods based on focusing light

Methods based on focusing light are direct methods which consist in measuring the shape of the surface. If the refraction index of the element examined is constant, it is possible to obtain information about the phase change.
The basis for the method is scanning the surface in order to measure the distance between the objective and the surface of the object. In each point of the object examined the setup is focused to obtain a sharp image, only one for each point. In this way information is gathered about each point of the object. As a result, information about its shape is reconstructed.
There are many types of focusing setups. One of the most widely used is the setup based on the confocal microscope.

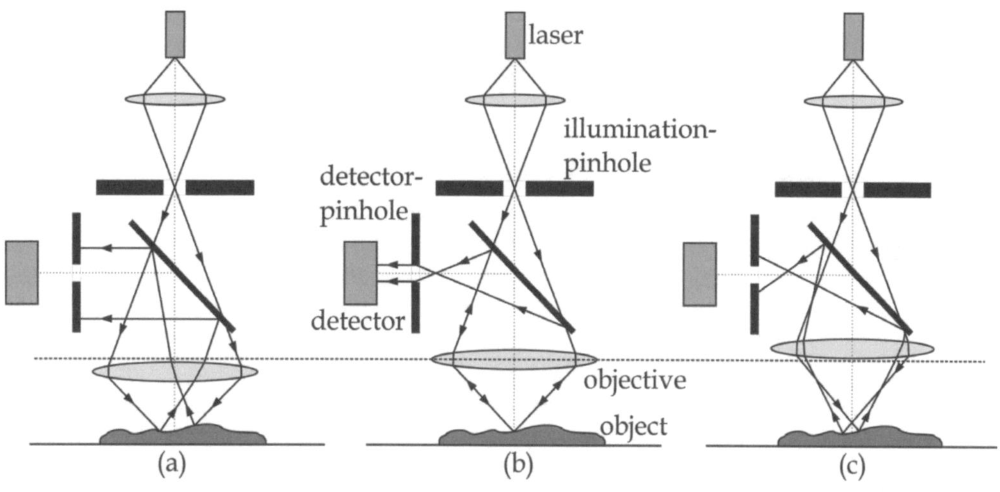

Fig. 2. Shape measuring with the use of confocal microscopy. Error in measuring the distance a) and c), the right measuring of the distance b)

The advantage of such setups is high precision of their vertical measurements, reaching 0.1 nm. Their horizontal resolution is lower, around 1 μm. The disadvantage is the time required for the measurement to take place, resulting from scanning the surface point by point, and the limitation in the size of the elements measured, also resulting from the scanning process. A strong point of the method is the possibility of using the stroboscopic light, which allows for measuring moving objects.

2.3 Methods based on analyzing the direction of the light ray

These indirect methods involve the analysis of the angles of the light rays leaving the phase object examined. The best known of these is the Shack-Hartmann filter based on microlens arrays. When illuminated by a plane wave, each of the microlenses focuses the light in its own focal point. The non-uniformity in the location of the focal points reveals the distortions of the light wave. When the light beam is reflected, or passes through the object examined, its wave front changes. Its shape is now coded in the distribution of the bright points, the analysis of which allows for the reconstruction of the shape of the object examined.

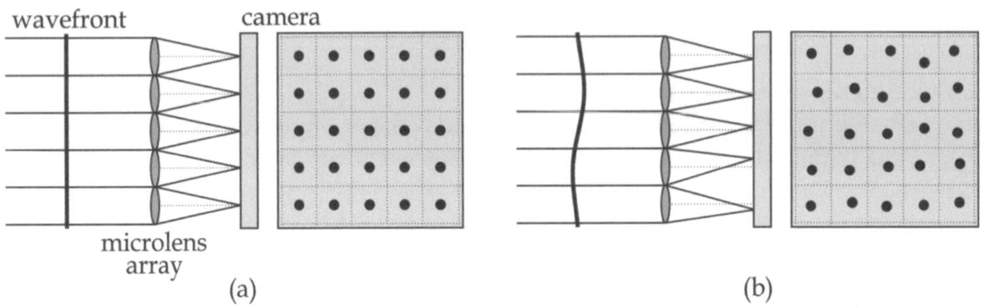

Fig. 3. Shack-Hartmann detector. Plane wavefront a), distorted wavefront b)

Such methods are predominantly used for examining the shape of the wavefront and find their uses, e.g. in astronomy and adaptive optical setups. Their main advantage is the short time of measurement. The main drawback is their relatively low horizontal resolution within the 30-50 μm range, as well as the fact that the results for large phase changes are difficult to interpret. That is why these methods are not used for measuring microlenses.

2.4 Methods based on modifying the spatial frequency

The last group comprises direct methods based on the modification of the light wave in the spatial frequency plane. Here, information about the phase of the object examined is based on the analysis of the intensity of light reflected or passing through the object, which is dependent on the real phase change. The method, where the light wave interferes with the light whose phase is shifted by $\pm \pi/2$, comprises phase-contrast Zernike filters (Fig.4), as well as the schlieren, or knife edge method (also called a binary amplitude Foucault filter), and the Hilbert filter when realized as cutting or shifting a part of the spectrum of the object visualized. There is also the three steps Hoffman filter, where various fragments of the spectrum are blocked, or their intensity is diminished. The results obtained with these methods are primarily of qualitative character. The complex relation between the phase change of the input object and the light intensity on the output obtained in the measurement does not allow for the quantitative analysis, or limits such an analysis in a considerable way. That is why the use of these methods is highly limited. A possible solution is the use of spatial frequency filters where the intensity signal at the output of the setup carries information about the phase of the object examined, and is easy to interpret. The group includes the linearly graded filter (Settles, 2001), and the semiderivative filter (the square root filter) (Kasztelanic & Sagan, 2009).

3. Amplitude real filter

The use of the amplitude real filter to measure pure-phase objects is rooted in the methods for modifying spatial frequency. Its basis is a 4f correlator setup with a coherent light source. Its scheme is presented in Fig.5.

The phase object to be examined e.g. a microlens array is placed in the input plane of the correlator. The distorted wavefront is directed to the first lens (L1 - Fig.5). As a result, a distribution of the spatial frequencies of the examined field is obtained. This spectrum is filtered by the amplitude filter and the result of the transformation through the second lens (L2 - Fig.5) is an image in the output plane. The change of the light intensity in this image codes the information about the phase of the object examined. The exact form of the changes depends on the amplitude transmittance of the amplitude filters.

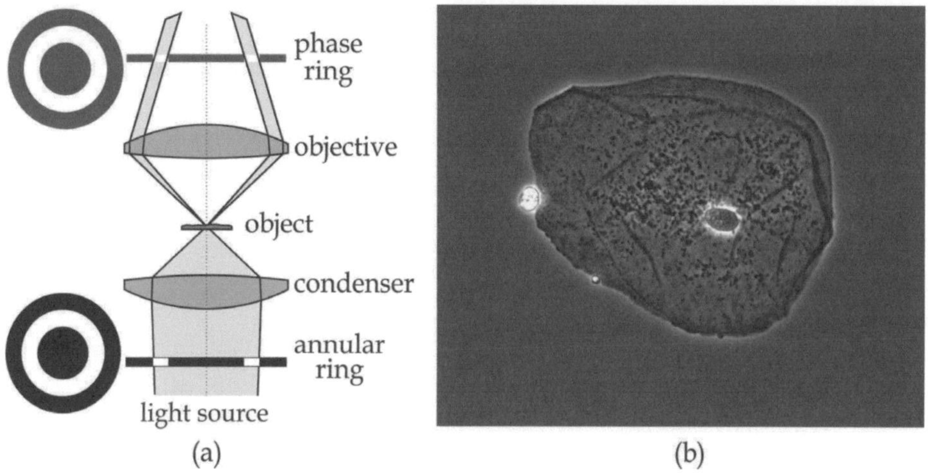

Fig. 4. Phase measuring with the use of phase-contrast Zernike filters a), sample result b)

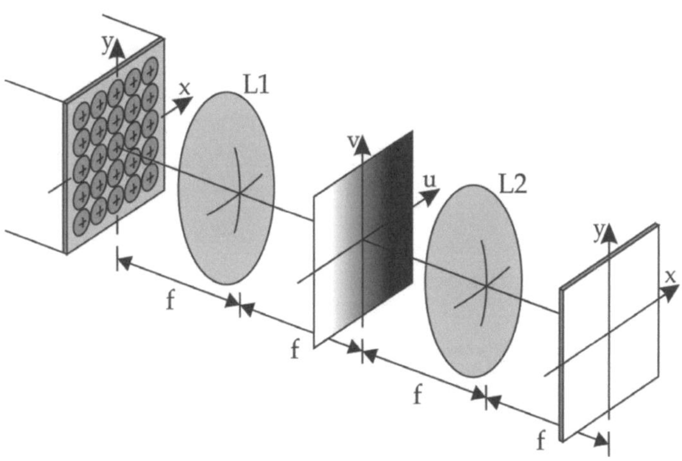

Fig. 5. The scheme of the setup for measurement of the quality of the microlens arrays

3.1 Linearly graded filter

Let us assume that the element examined is a pure-phase object and the object phase is described by the function $\theta(x,y)$. The transmittance of this object is equal to:

$$t(x,y) = \exp\left[i\theta(x,y)\right] \tag{4}$$

The amplitude transmittance of the linearly graded filter is defined as:

$$t(u,v)=\begin{cases}0 & u<-w/2\\ a+u/w & -w/2\le u\le w/2\\ 1 & u>w/2\end{cases} \tag{5}$$

where a is the constant bias transmittance, typically equal to 0.5, and w is the spatial extent of the filter. Examples of amplitude transmittance profiles of this filter are presented in Fig.6.a). For a pure-phase object, it gives an output intensity distribution of:

$$I(x,y)=A_0^2\left[a+\frac{\lambda f}{2\pi w}\frac{\partial\theta(x,y)}{\partial x}\right]^2 \tag{6}$$

where A_0^2 is the intensity of the illuminating beam, λ is wavelength of light and f is a focal length of lenses in the correlator.

3.2 Square root filter

The amplitude transmittance of the semiderivative real filter is defined as:

$$t(u,v)=\begin{cases}0 & u<-w/2\\ \sqrt{a+u/w} & -w/2\le u\le w/2\\ 1 & u>w/2\end{cases} \tag{7}$$

where a, like in the previous equations, is the constant bias transmittance, typically equal to 0.5, and w is the spatial extent of the filter. Examples of amplitude transmittance profiles of this filter are presented in Fig.6.b). For a pure-phase object, it gives the output intensity distribution of:

$$I(x,y)=A_0^2\left[a+\frac{\lambda f}{2\pi w}\frac{\partial\theta(x,y)}{\partial x}\right] \tag{8}$$

It is worth mentioning here that the result is not exact, while it is obtained through the rejection of the higher number of terms. However, as further shown, it does not influence the quality of the results in any considerable way.

The main advantage of using these filters lies in the fact that in the output the image of the light intensity is proportional to the first derivative of the gradient of the phase change in the object. The intensity is registered on the CCD camera and its distribution informs us about the shape of the object examined. The size of the observation area depends on the scale of the setup we use. The main disadvantage of the filter is the fact that the information about the gradient is one-directional only. In order to reconstruct the shape of the element examined, information about the gradient in two perpendicular directions is necessary. That is why all the measurements are carried out twice for each of the filters: once with the filter set in the x direction and once in the direction perpendicular to the first one.

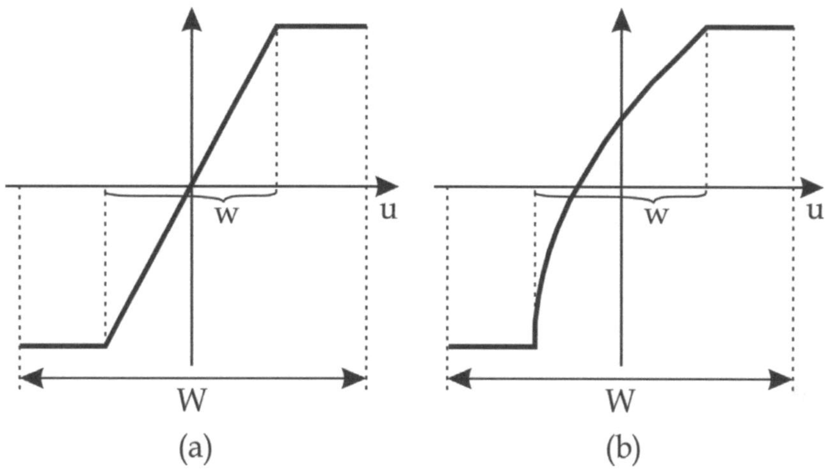

Fig. 6. The cross-section of the amplitude real filter: linearly graded filter a), square root filter (semiderivative) b). (W is the total width of the filter.)

4. Shape reconstruction with the use of the amplitude real filter

As a result of the work of the setup described above, an intensity signal is obtained. It codes the information about the phase gradient of the object examined. In order to reconstruct the real shape of the object on the basis of the data various techniques are used. Generally, the techniques use algorithms based on zonal estimation, modal estimation or a combination of both in a least-squares sense. These methods are used to analyze the wave front in the Shack-Hartmann detector. Of these, two methods will be presented at length. The Zernike polynomial expansion method gives more precise results but assesses the shape of the individual microlenses one by one when used for the analysis. The Fourier-based modal method allows for a simultaneous measurement of the shape of numerous elements. It is less precise but suitable for a fast analysis of the microelements' quality.

4.1 Zernike polynomials expansion method

In the case where the wave front distortion is introduced by such objects as a lens or a microlens arrays, in order to reconstruct the distortion, it is natural to use a method which gives information not only about the shape of the elements but also about the aberrations the shape introduces. One of such methods is the modal estimate method, which usually uses Zernike or Legendre polynomials, or complex exponentials as expansion functions. In the case of Zernike polynomials expansion method (Wyant & Creath, 1992), the shape of the wavefront W is given by the equation:

$$W(\rho,\theta) = \sum_m \sum_n C_n^m Z_n^m(\rho,\theta) \tag{9}$$

where C_n^m are the expansion coefficients, and Z_n^m are Zernike polynomials:

$$\begin{Bmatrix} Z_n^m(\rho,\theta) \\ Z_n^{-m}(\rho,\theta) \end{Bmatrix} = R_n^m(\rho) \begin{Bmatrix} \sin(m\theta) \\ \cos(m\theta) \end{Bmatrix} \tag{10}$$

and where R_n^m is the radial coefficient equal:

$$R_n^m(\rho) = \sum_{k=0}^{(n-m)/2} \frac{(-1)^k (n-k)!}{k!\left[\frac{n+m}{2}-k\right]!\left[\frac{n-m}{2}-k\right]!} \rho^{n-2k} \quad n = 0,1,2,\dots \ (n-m)\text{even} \tag{11}$$

Such a description opens up the possibility of an easy comparison of the shape of the lens examined with the ideal lens. An additional advantage here is the fact that some polynomials of Zernike series have recognizable optical properties. For instance, the polynomial Z_2^0 describes the optical property of 'defocus', the polynomial Z_4^0 describes the 'spherical aberration' and the polynomials Z_2^{-2} and Z_2^2 describe the 'astigmatism'. In the case of a microlens arrays polynomial expansion opens an additional possibility of comparisons between the particular lenses.

Further on, single indexation of Zernike polynomials with the j parameter is used, due to the fact that it is easier to use and present the results. The j parameter is calculated on the basis of m and n parameters as $j=[n(n+2)+m]/2$.

The reconstruction of the shape of the optical element examined is carried out on the basis of the information about the phase gradient change measured in two perpendicular directions and is possible after differentiation of Eq. 9:

$$\frac{\partial\theta}{\partial x} = \sum_{k=0}^{M} a_k \frac{\partial Z_k(x,y)}{\partial x}, \quad \frac{\partial\theta}{\partial y} = \sum_{k=0}^{M} a_k \frac{\partial Z_k(x,y)}{\partial y} \tag{12}$$

where $Z_k(x,y)$ are the kth Zernike polynomials and M is the number expansion of coefficients a_k. In the matrix form it could be written as:

$$S = Aa \tag{13}$$

where A is a gradient rectangular matrix and M the number of columns. The number of rows is equal to the sum of pixels in the two images of the gradient along the x and y axes. The solution of Eq. 13 can be found on the basis of equation:

$$a = \left(A^T A\right)^{-1} A^T S. \tag{14}$$

4.2 Fourier-based modal method

In the case when the wavefront distortion is introduced by numerous objects such as microlenses, it is more convenient to reconstruct the shape of the distortion for all the objects simultaneously. An example of such an approach is the Fourier-based modal method (Guang-ming Dai, 2008).

Assuming that A is the shape of the wavefront, and a is the spectrum of this front, it is easy to note a relation between the two:

$$A(x,y)=F^{-1}\{a(u,v)\}=\iint a(u,v)\exp[2\pi i(xu+yv)]du\,dv \tag{15}$$

$$a(u,v)=F^{-1}\{A(x,y)\}=\iint A(x,y)\exp[-2\pi i(xu+yv)]du\,dv \tag{16}$$

where F{ }, F^{-1}{ } mean the Fourier transform and the inverse Fourier transform, respectively. In the correlator setup with a linear or square root filter, the output image is proportional to the gradient of the phase change. Using such two setups with mutually perpendicular filters we obtain two signals:

$$\begin{cases} \dfrac{\partial A(x,y)}{\partial x}=2\pi i\iint u\, a(u,v)\exp[2\pi i(xu+yv)]du\,dv \\ \dfrac{\partial A(x,y)}{\partial y}=2\pi i\iint v\, a(u,v)\exp[2\pi i(xu+yv)]du\,dv \end{cases} \tag{17}$$

Introducing the notifications:

$$b_x(u,v)=2\pi i u\, a(u,v) \tag{18}$$

$$b_y(u,v)=2\pi i v\, a(u,v) \tag{19}$$

and consequently multiplying by u on both sides of Eq.18 and v on both sides of Eq.19 combining them yields:

$$u b_x(u,v)+v b_y(u,v)=2\pi i(u^2+v^2)\,a(u,v) \tag{20}$$

to finally obtain the spectrum of the wevefront examined:

$$a(u,v)=-i\frac{u b_x(u,v)+v b_y(u,v)}{2\pi(u^2+v^2)} \tag{21}$$

The final shape of the wavefront distortion is obtained after the inverse Fourier transform:

$$A(x,y)=F^{-1}\left\{-i\frac{u b_x(u,v)+v b_y(u,v)}{2\pi(u^2+v^2)}\right\}=F^{-1}\left\{-i\frac{uF\left\{\frac{\partial A(x,y)}{\partial x}\right\}+vF\left\{\frac{\partial A(x,y)}{\partial y}\right\}}{2\pi(u^2+v^2)}\right\}. \tag{22}$$

5. Computer simulation

Computer simulations were carried out in order to examine the work of the setup presented in Fig.5 and to establish how the kind of the amplitude filter and its parameters influence the results obtained.

As test objects various lenses were used, whose shape can be given by the equation:

$$h(r) = \frac{h_0}{R} \frac{r^2}{1+\sqrt{1-(K+1)r^2/R^2}} \tag{23}$$

where h(r) is the height of the lens as a function of the distance r to the optical axis, R is the radius of curvature at the vertex, h_0 is the maximum height and K is the aspherical constant. The lens profile h(r) might be spherical (K=0), elliptic (1<K<0 or K>0), parabolic (K=-1) and hyperbolic (K<-1). In the simulations it was assumed that K=-1.5 for hyperbolical lenses and K=-0.5 for elliptical lenses.

The parameters of the optical setup used for the simulations were the following: the wavelength of λ=632.8 nm, the focus L1 and L2 of the setup from Fig.5 f=100 mm, the lens diameter D≈8.2 mm, the size of a single pixel 1μm. The propagation method used was Rayleigh-Sommerfeld propagation.

In the first steps of the simulation the pure phase element to be examined was a single microlens, 100 μm in diameter, placed in the optical axis of the setup. The first simulations checked the influence of the spatial extent of the filter w on the quality of the lens reconstruction. In order not to block high spatial frequency the spatial extent should be as large as possible. However, in accordance with Eq.6 and Eq.8 it cannot be too large because this parameter influences the range of modulation of the light intensity in the output of the setup. In the simulation it was assumed that the maximum size of the filter W is calculated on the basis of the maximal gradient of the examined lenses. The change of the light phase of the λ wavelength, in the lens with refraction index n, and shape given by Eq.23 can be defined as:

$$\theta(r) = h(r)\frac{2\pi(n-1)}{\lambda} \tag{24}$$

The maximal gradient can be found on the edge of the lens, for r=R/2, and thus we obtain:

$$\left.\frac{\partial\theta(r)}{\partial r}\right|_{\substack{\max \\ r=R/2}} = \frac{h_0 2\pi(n-1)}{\lambda\sqrt{3-k}} \tag{25}$$

Calculating further the spatial frequency and passing on to the object domain after the transform by the lens of focal length f we obtain the maximum size of the filter:

$$W = 2f h_0(n-1)/\sqrt{3-k} \tag{26}$$

As the measurement of the error made when reconstructing the element examined it was assumed that:

$$E = \sum_S (H_S - h_S)^2 \tag{27}$$

where the sum comprises the whole the whole area of the lens S, while H_S and h_S are the height of the original and the measured lens, respectively.

As the measurement of the error made when reconstructing the element examined with Zernike expansion it was assumed that:

$$\sigma^2 = \sum_{k=0}^{M}\left(a_k - c_k\right)^2 \tag{28}$$

where c_k are Zernike coefficients of the lens examined and a_k are Zernike coefficients calculated on the basis of the images obtained for linear graded and semiderivative filters of various width.

Fig. 7 presents the simulation of the change in the quality of reconstruction with the use of Fourier modal method. The reconstruction is carried out for the test lenses and for two kinds of filters: the semiderivative and linearly graded filter, with regards to the filter width w and the camera of 8 Bit dynamic range.

The results show that there is an optimal width of the filter w, for which minimal errors are obtained. For both the filters the optimum width is within the range of 0.1 and 0.2 of the maximum filter with, given by Eq.26, independently from the shape of the lens examined. If the filter is too wide, it gives bigger errors due to the low modulation of the signal obtained. Changing the dynamic range of the camera over 10 Bit improves the results if the filter size is not precisely adjusted, however, it does not influence the minimal reconstruction error in a considerable way (Fig.8.a). A very narrow filter generates big errors because then it acts similarly to the knife edge filter. As a result of the work of such a filter, the intensity obtained in the output, which carries information about the phase change, has a more complicated form than the simple proportion to the gradient of the phase changes.

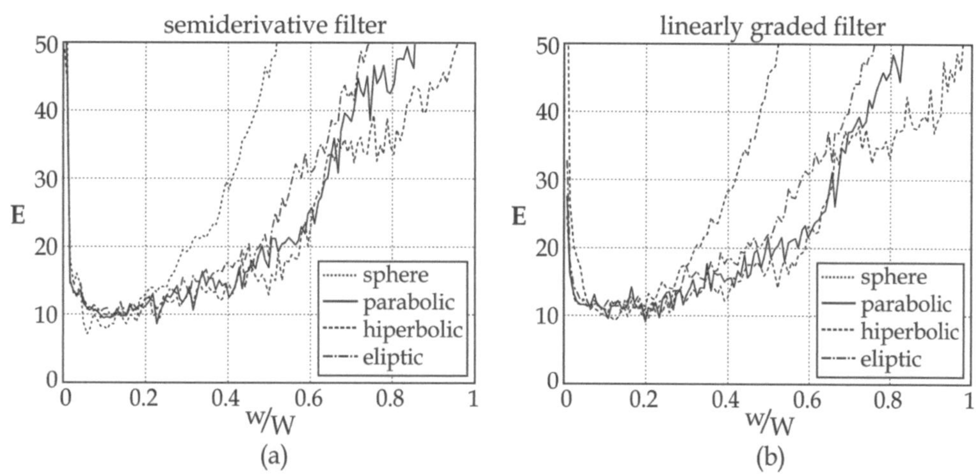

Fig. 7. E error in the reconstruction of the shape of the lenses carried out with the use of the Fourier modal method, depending on the kind of filter and its width: semiderivative filter a), linearly graded filter b)

The occurring reconstruction errors (Fig.8b) reach 1% of the total thickness of the microlens (100% error is obtained for flat surfaces). However, the size of the error depends only on the slight degree on the number of the microlenses simultaneously undergoing reconstruction (Fig.9). Thus, the Fourier-based modal method will not be successful where it is important to precisely establish the shape of the lens. On the other hand, the reconstruction errors are small enough to be disregarded when the method is used for a fast quality control. Thus, the method allows for a fast assessment of the lenses' shape, which in turn, allows for depicting errors in microlens arrays at the fabrication stage.

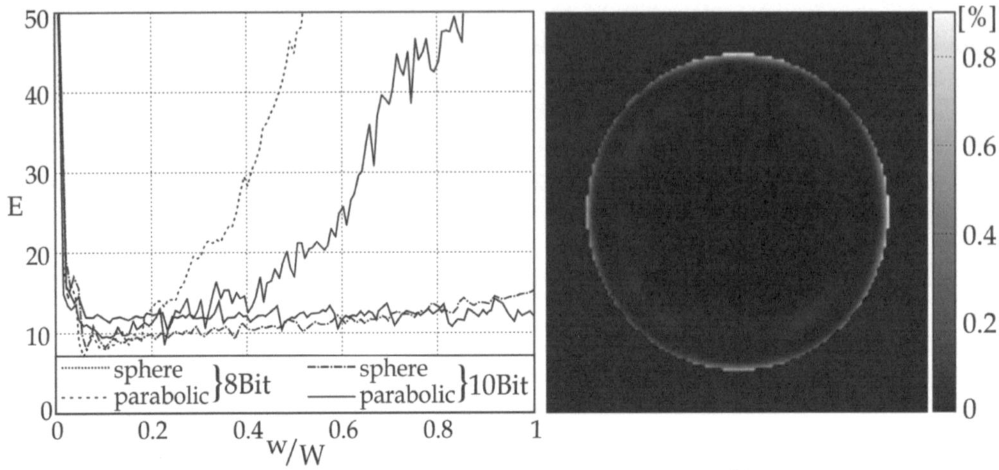

Fig. 8. Error E in the reconstruction of the lens shape with the use of the Fourier modal method for the camera with 8 and 10 Bit dynamic range a) and the image of the reconstruction error of the spherical lens reconstructed with the use of the semiderivative filter for a camera with an 8 Bit dynamic range b)

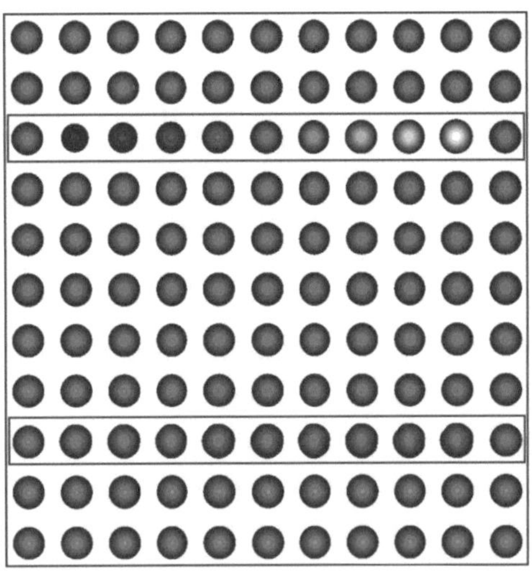

	sag of lenses / sag of the ideal lens											
	1	0.5	0.63	0.75	0.88	1	1.13	1.25	1.38	1.5	1	
E=	0	16.7	9.4	4.2	1	0	1	4.2	9.4	16.7	0	[%]

	distortion [%]											
	0	6.3	7.5	8.8	10	11.3	12.5	13.8	15	16.3	0	
E=	0	0.27	0.38	0.51	0.64	0.79	0.95	1.11	1.28	1.45	0	[%]

Fig. 9. The results of the reconstruction of the shape of a microlenses array with the use of Fourier modal method a) error E of the shape reconstruction for two modified rows of microlenses b)

Fig.9 presents the reconstruction of an 11x11 array of microlenses, 75 µm in diameter and lying within the distance of 100 µm. Apart from the two marked rows, all the remaining lenses have the same shape. The lenses marked in the upper row have a different height, and vary from 0.5 to 1.5 as compared to the height of the original lens. The lenses in the lower row have shapes different from the original. Their shape has been distorted by adding 36 consecutive Zernike polynomials, which has resulted in the shape changes between 6.25% and 16.25% of the original lens shape.

The results obtained show that all the lenses with the regular shape have been correctly reconstructed, and the reconstruction error is close to zero. On the other hand, while reconstructed, the distorted lenses differ from the original ones. The reconstruction error, as compared to the original shape, allows for the assessment of the size of the distortion.

In the case of the reconstruction using the Zernike polynomials expansion method, the reconstruction errors are smaller. Fig.10 presents the simulation of the change in the quality of the reconstruction of the test parabolic lens with an added distortion, with regards to the filter width w and for the camera of 8 and 10 Bit dynamic range. The results show that there is an optimal width of the filter w, for which minimal errors are obtained, both for the shape reconstruction and Zernike polynomial expansion. The width depends on the type of the amplitude filter used.

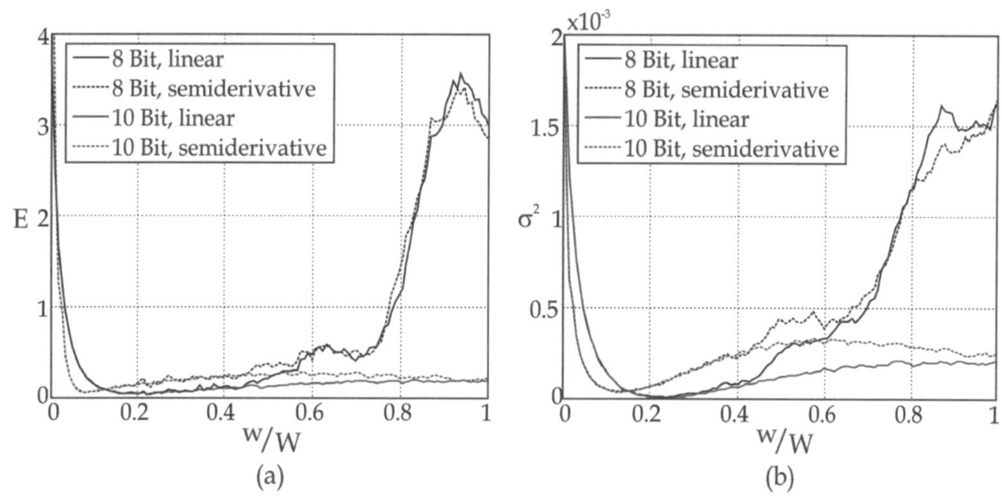

Fig. 10. E error in the reconstruction of the shape of the parabolic lens with the use of the Zernike polynomials expansion method a) and σ^2 error of Zernike polynomial expansion b) with regard to the kind of filter for the camera with 8 and 10 Bit dynamic range

Another series of simulations helped to estimate how the basic shape of the microlens examined influences the quality of the reconstruction and whether it is possible to recognize such a shape. In the simulations expansion for 36 Zernike polynomial and 10 Bit dynamic range of the camera have been assumed. Fig.11.a) presents the results for a linear filter. The results for the semiderivative filter are similar, but the error is bigger. In the case of spherical lenses the error is increased by 1.5 factor, and the results for the remaining lenses decrease by about an order of magnitude. From the simulation it is also possible to infer the shape of the lens examined, by choosing the characteristic elements of the 0, 4th and 12th terms. Especially, it is the 12th element of Zernike expansion which allows for establishing this shape. On the other hand, having the data about the original shape of the lens, it is possible, to obtain information only about the deviations of the real shape from the ideal one.

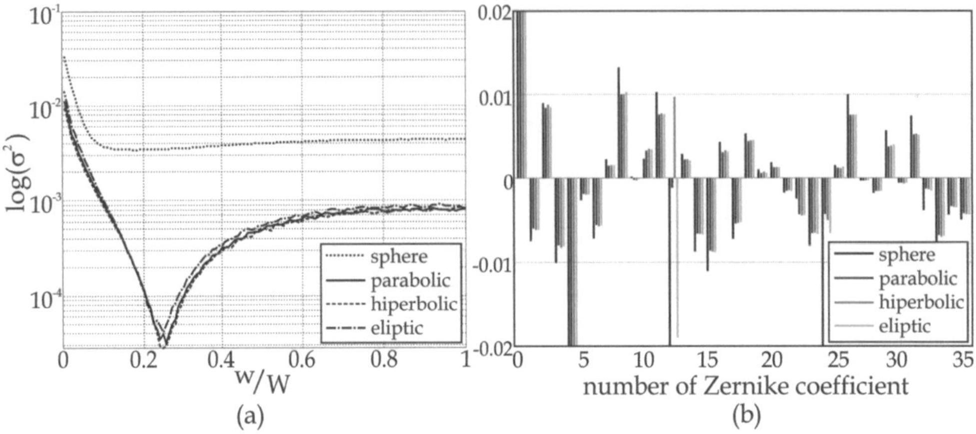

Fig. 11. The results of the reconstruction of variously shaped lenses by the linear filter with the use of the Zernike polynomials expansion method. The error made during the reconstruction a), values of Zernike expansion coefficients b)

It has also been checked how the preciseness of reconstruction changes depending on the location of the lens, with regards to the optical axis and on the size of the microlens arrays. The simulations carried out for the 15x15 array of lenslet pitch equal 250 μm show that there are slight differences in the reconstruction depending on the place of the location. In the case of σ^2, the spread of the results is within 1% for all the lens types and for both filters used.

6. Experimental realization of the 4f correlator

The experimental realization of the 4f correlator based on lenses with the focus of 100 mm and the diameter of 51.2 mm, as well as a CCD camera with 14 Bit dynamic range. A low-cost method based on commercial slide imagers was used to fabricate grayscale amplitude filters. The smallest feature of the filter obtained in this way was about ~4.5 μm (Fig.12).

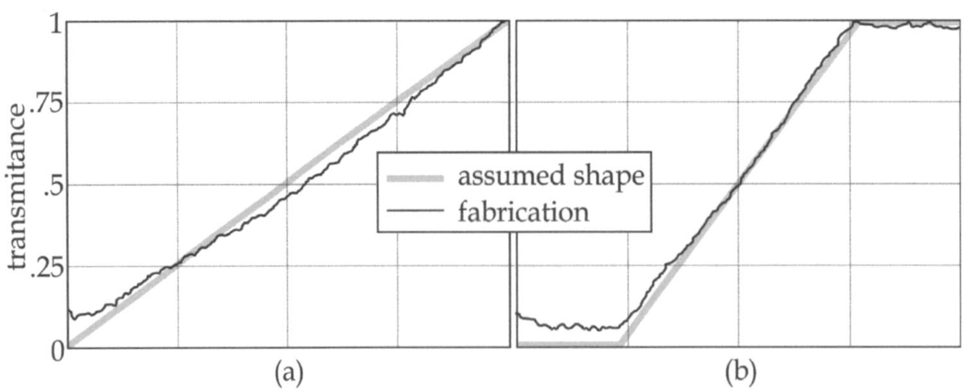

Fig. 12. Intensity transmittance of the fabricated square root filter before optimization a) and after optimization b)

The first experiments examined the quality of reconstruction of the microlens arrays. The experiment used a commercially available array of plano-convex microlenses arranged in a square grid of 146 µm in diameter, 150 µm lenslet pitch and 6.7 mm focal length produced by ThorLabs. Two images of microlenses were registered with the two mutually perpendicular locations of the amplitude filter (Fig.13).

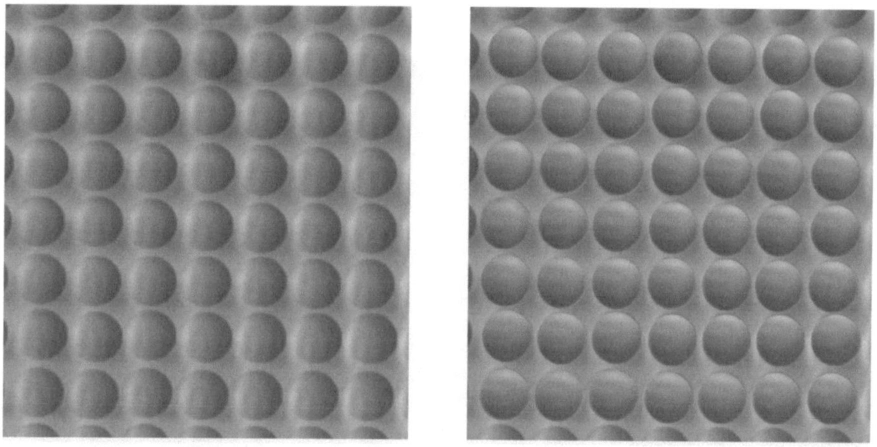

(a) (b)

Fig. 13. Image registered on CCD camera for horizontal a) and vertical b) location of the linear filter

In the case of the reconstruction carried out with the use of the Fourier modal method, the results obtained (Fig.14) show the lens whose shape is different from the remaining ones (the central column, bottom row). The reconstruction error E of the lens shape against average for the remaining ones is above 12%. The differences in the shape of the remaining lenses fall within 2%, similarly to the differences in the maximum height of the individual lenses, while the error between the reconstructed shape of the lenses and the theoretical shape of the lenses in the array is within 3%. Such results can be obtained for the 75% of the central surface of the lenses, but on the edges the errors reach 10%. The main difference between the computer simulation and the experimental realization can be noticed in the area between the lenses, which originally is flat. The differences result from the light reflections on the elements of the experimental optical setup, adjustment errors and the non-homogeneity of the amplitude filter used.

In the case of the reconstruction carried out with the use of the Zernike polynomials expansion method, the analysis consisted in the automatic location of the centers of all the lenses, choosing the image of the single lens and applying Zernike expansion method. The operation was carried out for each individual lens. The exemplary results are presented in Fig.15. They show that for the theoretically identical lenses the results obtained are very similar. The maximal difference in the reconstruction between the presented lenses calculated on the basis of Eq.28 is less than 0.5%.

In the next experiment a matrix of 15x5 microlenses was examined. They were all 125 µm in diameter and their shape had been earlier examined with the Wyko profilometer and with the use of Twyman-Green interferometer. The comparison of the results obtained for the linear filter with 36 Zernike expansion coefficients with the results from interferometry is presented in Fig.16. The horizontal and the vertical cross-sections are very similar, which proves the efficiency of the method used. However, the reconstruction of the full shape of the lenses with the use of the amplitude filter differs from the interferometric results. The differences may result from the errors in adjusting the setup or the tiny scratches on the surface of the test plate around the microlens arrays.

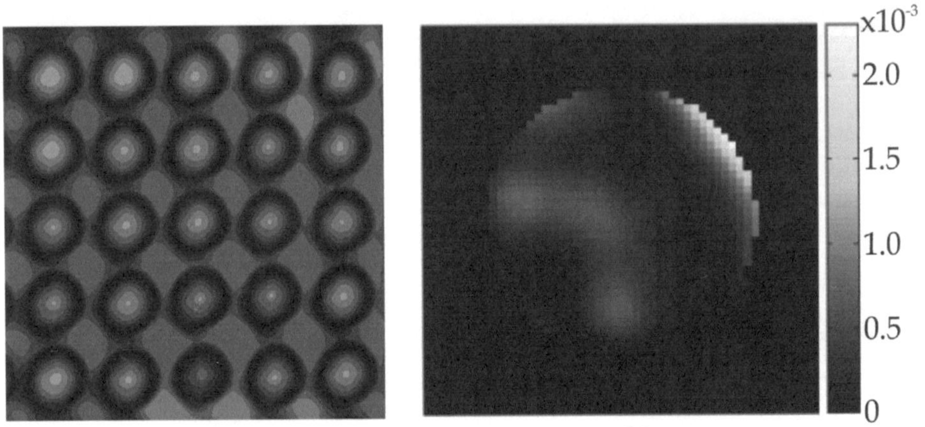

(a) (b)

Fig. 14. Fourier-based model reconstruction of the microlenses array a), error E as compared with the ideal lens b) (the lens in the upper left corner)

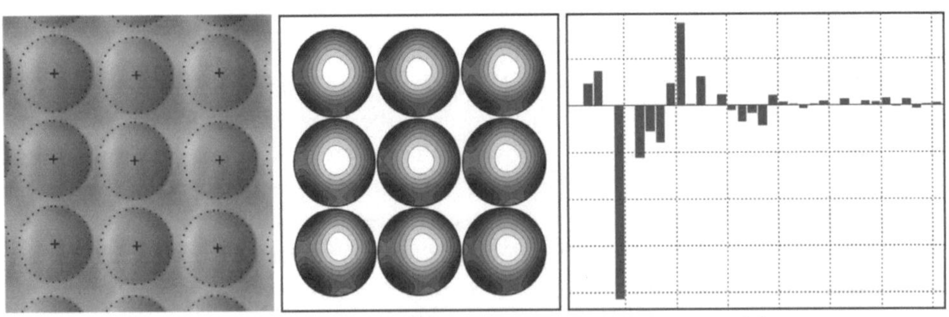

(a) (b) (c)

Fig. 15. Zernike reconstruction of the shape of the microlens arrays. The automatically located centers and areas of the individual lenses a), an example of Zernike reconstruction of nine microlenses b) exemplary Zernike expansion coefficients c)

6. Conclusion

The chapter has presented various methods for measuring phase objects and its potential uses. Then it has focused on the use of the amplitude real filter for quality control of microlenses. It presented computer simulations and an experimental realization of an optical setup for automatic quality control of microlens arrays. It has shown that with the use of the simple 4f correlator setup and a single measurement (in the case of two parallel paths with two perpendicular amplitude filters), as well as a non-complicated computer analysis it is possible to obtain results very close to much more complex, time-consuming and expensive methods. The methods of analysis described here give results slightly less precise than interferometry, especially with the use of the Fourier modal method. On the other hand, they are effective enough for most of the commercial applications in which quality assessment is important.

The main disadvantage of the proposed solution is the necessity to precisely adjust the setup. Also the dependency of the optimal filter width on the parameters of the object examined requires checking several variants of the filters to be used. However, the main advantage of the method, apart from its simplicity, is the possibility of using the whole resolution of the camera, and not its fraction as in the case of the Shack-Hartman method. Using the Zernike polynomials method make the results resistant to the noises present in all optical setups. Moreover, for the Fourier modal method the fact that both the measurement and the analysis are carried out simultaneously for all the objects makes the method much more effective than the ones used so far when fabricating large numbers of microoptical elements.

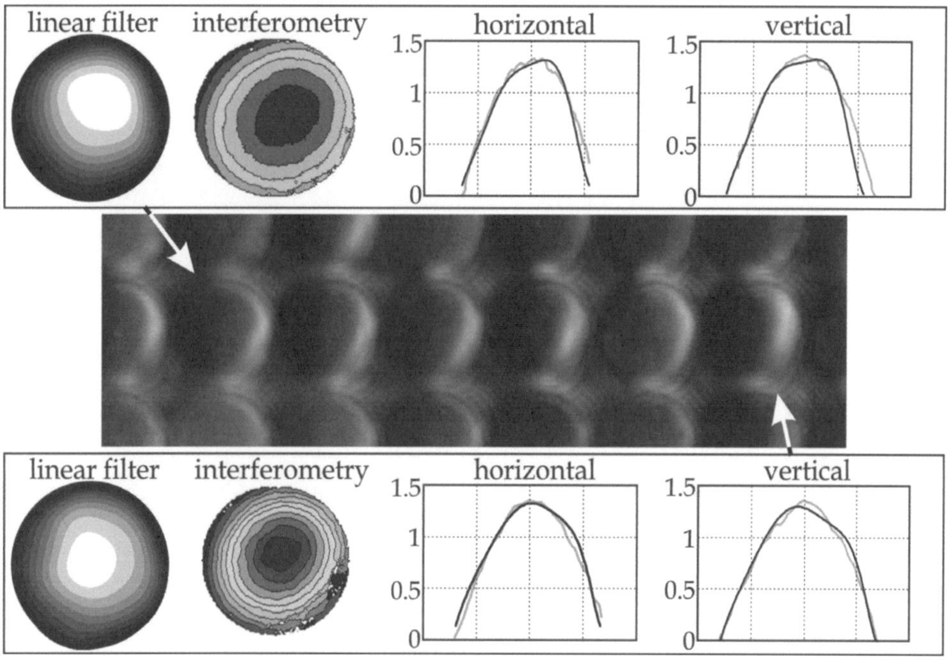

Fig. 16. Zernike reconstruction of the shape of the microlenses (all dimensions given in μm)

7. References

Guang-Ming Dai (2008). *Wavefront Optics for Vision Correction*, SPIE Press Book, ISBN 978-0-8194-6966-3, USA.

Kasztelanic, R. & Sagan, A. (2009). Semiderivative real filter for microoptical elements quality control, *Optical Review*, Vol. 16, No. 3, pp. 252–256, ISSN 1340-6000, Japan.

Settles, G. (2001). *Schlieren and Shadowgraph Techniques*, Springer-Verlag, ISBN 3-540-66155-7, Berlin, Germany.

Wyant, J. & Creath, K. (1992). *Basic wavefront aberration theory for optical metrology*, Applied Optics and Optics Engeenering, Vol. XI. Academic Press, ISBN 0-12-408611-X, New York, USA.